D0169322

TERRESTRIAL ECOSYSTEMS

Second Edition

TERRESTRIAL ECOSYSTEMS

Second Edition

John D. Aber

University of New Hampshire

Jerry M. Melillo

Marine Biological Laboratory
Woods Hole, Massachusetts

A Harcourt Science and Technology Company

San Diego San Francisco New York Boston
London Toronto Sydney Tokyo

Sponsoring Editor: Jeremy Hayhurst
Production Editor: Ellen Sklar
Managing Editor: Frank Messina
Editorial Assistant: Nora Donaghy
Production Manager: Alicia Jackson
Marketing Manager: Marianne Rutter
Cover Design: Jacqueline LeFranc
Composition: Modern Graphics, Inc.
Printer: RR Donnelley

This book is printed on acid-free paper. ♾

Academic Press
A Harcourt and Science Technology Company
525 B Street, Suite 1900, San Diego, CA 92101-4495, USA
http://www.academicpress.com

Academic Press
Harcourt Place, 32 Jamestown Road, London, NW1 7BY, UK
http://www.academicpress.com

Harcourt/Academic Press
A Harcourt and Science Technology Company
200 Wheeler Road, Burlington, MA 01803, USA
http://www.harcourt-ap.com

Library of Congress Catalog Card Number: 00-111102

International Standard Book Number: 0-12-041755-3

Printed in the United States of America
01 02 03 04 05 RRD 9 8 7 6 5 4 3 2 1

Dedication

We dedicate this book to Lynn, Lalise, Ted, Patrick, Colleen, and Caitlin, and also with gratitude and respect to F. Herbert Bormann.

Preface

Preface to the First Edition

From the clearing of tropical forests to the deposition of air pollutants, humans are altering the way the Earth looks and functions. The cumulative effects of our actions are clearly visible in the changing chemistry of the global atmosphere and alterations in forest health and agricultural production.

Managing this increasing human presence on the Earth requires an understanding of how terrestrial ecosystems, the basic units of the landscape, function. It is from these systems that we obtain food and fiber, clean air and water, and the esthetic and social benefits of contact with the natural world. Climate change, acid rain, increasing concentrations of carbon dioxide in the atmosphere, all of these affect, and are affected by, the metabolism of the biotic communities that cover the land surface.

Our purpose in writing this book has been to integrate information from several different traditional disciplines in order to present a holistic view of ecosystem function. Topics as diverse as photosynthesis, soil chemistry, and population dynamics of herbivores are discussed as they relate to the movement of energy, water, and elements through units of the landscape. We believe that this interdisciplinary approach is required for the successful application of ecosystem concepts to the pressing problems resulting from human use of the biosphere.

The book has been written for advanced undergraduates and beginning graduate students in a wide range of disciplines. To achieve this, we assume only an introductory-level knowledge of the workings of soils, plants, and animal populations and have made every effort to minimize the use of specialized terms (scientific "jargon"). Concepts and processes are presented in ways that should be accessible to a wide audience. Figures, photographs, and diagrams are used liberally to illustrate the major points in each chapter, and references are provided for those who wish to pursue particular topics in greater detail.

The structure of the book reflects our goal of providing an integrative, interdisciplinary discussion of the function of terrestrial ecosystems. Part I introduces basic concepts of ecosystem studies, including a focus on the movement of energy and matter through the landscape, and some first steps in measuring function at the ecosystem level. A brief introduction to systems analysis reemphasizes the system-level viewpoint of the text.

Part II deals with the key processes that control ecosystem function. The transfers of carbon and nutrients are used to structure the presentation of factors controlling net photosynthesis and water use at the ecosystem level, nutrient uptake and release through decomposition, the role of soil chemistry in controlling nutrient availability, and the importance of herbivory and fire as alternate "sinks" for fixed carbon and other elements.

Part III returns to the ecosystem level, presenting an integrated view of how the processes described in Part II interact to produce the patterns of change and distribution visible in three very different landscapes. The role of human use of each of these three systems, and the interaction of management with natural processes, is an important part of this discussion.

In Part IV, the growing dominance of human activity in controlling ecosystem distribution and function is dealt with in detail. Forest management, agricultural practices, and the indirect effects of pollution loading, which represent three of the spatially dominant uses of the landscape, are presented. The final chapter deals with terrestrial ecosystems as components of the global system and describes methods for measuring and predicting the net effect of terrestrial ecosystem function on the regional, continental, and global environment.

Preface

Preface to the Second Edition

The study of ecosystems remains at the heart of the effort to understand and predict the environmental future of the Earth. The basic concepts of ecosystem studies have not changed significantly in the 10 years since the first edition of this text appeared (and so the purpose for writing this book remains the same as described in the preface to the first edition). However, the knowledge base and the sophistication with which problems such as production management, climate change, and acid deposition are treated have increased tremendously. The interactions between ecosystem function and biodiversity have also received increased attention.

This second edition reflects these realities. The basic four-part structure of the text remains the same, but certain chapters have been added or combined. Within this structure, every chapter has been revisited; every topic has been reviewed; and, in each case, the recent literature has been consulted in order to present the most complete and accurate picture of the state of ecosystem studies in the new millenium.
The most important changes include:

1. The transfer of the discussion of modeling from the forest management chapter to the concepts section in Chapter 5. Discussions of several other important tools, including the use of isotopes and mass balances, have been added to this chapter. Eddy covariance appears in Chapter 3, and in several other locations as well.

2. The discussion of ecosystem theory, which was scattered throughout Part 2, has been moved and expanded in a new Chapter 18.

3. The geological time scale studies of the forests of the Hawaiian islands represents one of the major, new, large-scale, integrated ecosystem studies of the past decade. A chapter on this system (Chapter 22) has been added.

4. The chapters on forest and agricultural management have been combined (Chapter 23).

5. A new chapter on the interactions between biodiversity and ecosystem function has been added (Chapter 24).

Many, many colleagues have added invaluable information, graphics, ideas and discussions that have been included from the first edition or added in the second. We would like to thank the following people in particular for their help: Dan Binkley, Rich Birdsey, Herb Bormann, Alice Cialella, Wally Covington, Eric Davidson, Lisa Dilling, Charley Driscoll, David Foster, Jim Fownes, Rita Freuder, Andy Friedland, Steve Frolking, Solomon Gbondo-Tugbawa, Henry Gholz, Christine Goodale, Jim Gosz, Joe Hackler, Steve Hagen, Warren Heilmann, David Hollinger, John Hom, Richard Houghton, Feng Sheng Hu, Clive Jones, Lou Iverson, Julian Jenkins, Jed Kaplan, Tom

Lee, Robert Mahoney, Mary Martin, Kimberly McCracken, Steve McNulty, Knute Nadelhoffer, Ron Neilson, Julie Newman, Scott Ollinger, John Pastor, Donna Poulin, Anantha Prasad, Kurt Pregitzer, Colin Prentice, Francis Putz, Peter Reich, Barry Rock, Mike Routhier, Steve Running, Dan Schiller, Dave Schimel, William Schlesinger, Daniel Schoen, Nellie Stark, Harald Sverdrup, Lloyd Swift, Jim Tucker, Keith Van Cleve, James Vilkitis, Peter Vitousek, Jim Vogelmann, Dick Waring, Carol Wessmann, Neil West, Diane Wickland, Keith Winterhalder, Steve Wofsy, Don Zak.

We also greatly appreciate the efforts of all of the people at Harcourt/ Academic Press who have brought this book into existence, especially: Jeremy Hayhurst, Stephanie Stevens, Nora Donaghy, Ellen Sklar, who orchestrated the editing and layout, Jacqueline Lefranc, and Alicia Jackson.

We would also like to thank the hundreds of students who have taken the Terrestrial Ecosystems course over the past 20 years—either at the University of Wisconsin or the University of New Hampshire—for their constant feedback, corrections, ideas, inspiration; and most of all, enthusiasm for the study of ecosystems.

Content Overview

PART *3* Synthesis: Dynamics of Selected Ecosystems *347*

PART *4* Application: Human Impacts on Local, Regional, and Global Ecosystems *425*

Table of Contents

Part 1

INTRODUCTION

The Nature of Ecosystem Science

The study of ecosystems has developed rapidly over the last four decades, yet the field is still young as scientific disciplines go. The ecosystem approach has been used with increasing frequency in the analysis of large-scale environmental problems because it deals explicitly with large units of the landscape; the scale at which forest and range managers and agencies concerned with environmental protection need to operate. Increasingly, the ecosystem is the basic landscape unit of study for regional and global environmental problems.

For all of this, it is often difficult to visualize the scope and purpose of ecosystem studies. Just what are ecosystems? What are their important characteristics? What distinguishes one system from another? Perhaps one of the most troubling questions is: How do you measure the function of such large landscape units?

The purpose of Part 1 is to present an introduction to the study of ecosystems. These five chapters present a very brief historical view of the development of the ecosystem concept and some basic definitions (Chapter 1); a discussion of how ecosystems vary in structure and function in different climates (Chapter 2); an introduction to the measurement of carbon, water, and nutrient balances over terrestrial ecosystems (Chapters 3 and 4); and a brief introduction to some general methods frequently used in the analysis of ecosystem function (Chapter 5).

In establishing the history and scientific foundation of ecosystem research, Part 1 purposely refers to older, classic studies in the field, with a few newer "classics" added in. More current literature is presented in Parts 2 to 5.

1

DEVELOPMENT OF CONCEPTS IN ECOSYSTEM SCIENCE

WHY STUDY ECOSYSTEMS?

The term "ecosystem" is a familiar word to most readers of this book, although the actual meaning of the term might not be clear. The term is used with regularity in media descriptions of environmental problems. We hear of air pollution damaging forest ecosystems and development or dumping of chemicals altering wetland ecosystems. Increasingly, we encounter descriptions of changes in the global ecosystem, such as increasing concentrations of carbon dioxide in the atmosphere or the conversion of tropical forests to pastures. These presentations generally convey a sense that ecosystems are something more than the plants and animals residing in a certain place. They also generally suggest that something will go wrong with the system, that some important function will be damaged or altered.

These general conceptions actually capture two important aspects of ecosystem studies: (1) a focus on the interactions between organisms and the environment within a specific area and (2) a focus on processes that cause plants to grow and decay, soils or sediments to form, and the chemistry of water and air to change. The study of terrestrial ecosystems can be thought of as the study of the metabolism or physiology of units of the landscape, concentrating on energy, chemical, and water balances. Indeed, the subject matter of ecosystem studies is defined as the movement of energy and materials, including water, chemicals, nutrients, and pollutants, into, out of, and within ecosystems.

This sounds pretty dry. Yet the study of ecosystems has grown rapidly over the past 40 years as we have become aware of the extent to which human society depends on the continuing health of ecosystems. We rely on ecosystem function to clean polluted water and air. The food we eat and the fiber and energy we use often are the products of agricultural or managed forest ecosystems. One of the strengths of ecosystem studies is that the approaches and concepts discussed in this book apply just as well to intensively managed systems, such as cornfields and forest plantations, as they do to wild, unmanaged systems, such as grasslands and deserts. Scholarly studies have even been done on suburban lawns as ecosystems! In fact, one of the values of ecosystem studies is in determining just how much human use of the landscape changes the energy, water, and chemical balances of ecosystems. In this way, we monitor the "health" of managed systems, or their effect on adjacent urban or wild ecosystems.

One of the great realizations of the past decade is that many of our most pressing environmental problems are not local or even regional in scope, but rather global. For example, in examining the complexity of the global cycling of carbon, ecosystems have become the smallest and most basic unit of the Earth's surface that can be identified and measured. Ecosystems are the basic building blocks we use to understand the function of the Earth as a single, interconnected system.

From a purely intellectual point of view, the scientific study of ecosystems has offered a new and unique challenge to natural scientists in the broadest sense. It is most fundamentally an interdisciplinary or holistic science, involving experts from several traditional fields. For example, an atom of nitrogen may, in a single growing season, move from an organic compound (such as protein) in the soil to an inorganic one (such as ammonium); into the soil solution; and then into the plant root, stem, and leaf. From the leaf it may be washed by rainfall back into the soil, down past the roots, and into a stream. So far we have mentioned processes that traditionally lie in the provinces of soil science, microbiology, plant physiology, water chemistry, micrometeorology, and hydrology. Others could easily be included. Integration of these fields has been a continuing challenge to ecosystem scientists (and has led to the invention of new interdisciplinary "disciplines," such as biogeochemistry).

Another characteristic of ecosystem studies is the effort to determine which among the vast array of complex connections among biotic (living) and abiotic (nonliving) components that make up an ecosystem are most significant. Identifying critical interactions is a central theme throughout this book. Everything that happens in nature happens within ecosystems. This does not mean that all processes are equally important in terms of controlling ecosystem function. For example, plants take both nitrogen and sodium out of soils and incorporate them into plant tissues. However, nitrogen is a required element, whereas sodium generally is not. A low availability of nitrogen frequently limits overall growth of plants in terrestrial ecosystems by limiting the total amount of leaf biomass or chlorophyll. Thus, the movement of nitrogen receives more attention than does the movement of sodium. This book emphasizes the most important interactions—those that control the overall rates of energy, water, and nutrient flux.

DEVELOPMENT OF ECOSYSTEM CONCEPTS

The concept of an ecosystem as a natural entity involving the interrelationships of biotic and abiotic factors in nature is an ancient one. However, the analysis of entire natural systems as a rigorous science in terms of factors controlling structure and function is a newer phenomenon derived from several disciplines. Involving both a description of the biological community and the interaction with nonliving parts of a landscape (soils, water, and energy), the study of ecosystems has been built from the traditions of several very different fields. English and American ecologists exploring the nature of large-scale natural communities and patterns of change over time brought the notion of communities as something more than an assemblage of species. Russian soil scientists and geochemists contributed a systematic view of the chemical development of soils across landscapes. European and American limnologists (limnology is the study of lakes) applied ecosystem concepts before the word was coined to the study of their contained and easily identified units of study.

A detailed examination of both the intellectual roots of ecosystem science and the influences of technology and social history on the ecosystem concept can be found in Golley's *A History of the Ecosystem Concept in Ecology* (1993). One distinction Golley makes is between a top-down approach to ecosystem studies, which emphasizes ecosystems as units of study subject to direct experimentation, and a bottom-up approach, which emphasizes initial understanding of many of the component parts. This book takes the first approach, beginning with the measured responses of units of the landscape to disturbance or climate, and then explaining the interactions within the system that control that overall response. Two outcomes of this approach to ecosystem studies are (1) the use of large areas as the unit of study (such as whole lakes or large blocks of forest or prairie) and (2) an attempt to study each of the system's components (soils, sediments, plants, animals, microbes) with an intensity that reflects their overall importance in the system's energy and material movement.

A. G. Tansley, an English plant ecologist, first coined the term "ecosystem" in 1935, partially in an attempt to heal a growing rift between ecologists who emphasized static classification of plant communities along the lines used for species and those who emphasized the dynamic but predictable change in communities over time. He also linked ecology to current advances in the physical sciences and mathematics by inclusion of the term "system." Tansley clearly envisioned ecosystem studies as an integration of ecology with other disciplines and tried to encourage the study of interactions between biotic and abiotic components:

> The more fundamental conception is . . . the whole system (in the sense of physics), including not only the organism-complex, but also the whole complex of physical factors forming what we call the environment . . . the habitat factors in the widest sense. . . . Our natural human prejudices force us to consider the organisms . . . as the most important parts of these systems, but certainly the inorganic "factors" are also parts, . . . and there is constant interchange of the most various kinds within each system, not only between the organisms but between the organic and inorganic. These ecosystems, as we may call them, are of the most various kinds and sizes. (Tansley 1935)

While several other terms conveying similar meanings have surfaced at different times throughout the world (Table 1.1), "ecosystem" has become most widely accepted. This may be at least partly due to the rapid growth of "systems analysis" as a method for expressing interactions between components of any system, biological, electrical, or mechanical, that has been applied frequently to ecological data (see Chapter 5).

Acceptance of the ecosystem as a new unit of study creates some serious problems. If all parts of the system are to be treated in a similar way, what is the common factor that expresses their interdependence? A classic paper by a young American ecologist named Lindeman (1942) suggested one factor that is common to all life processes: energy. He used the storage and movement of energy to express connectedness between organisms within a system,

TABLE 1.1 In the late 1800s and early 1900s, several different terms were proposed to capture the idea of the interactions between organisms and the environment within units of the landscape.

The following is a partial list drawn from Golley (1993), and the listing of other terms for units of the landscape which have been used to convey the same notion as "ecosystem" (Golley 1995).

Biochora	Elementary Landscape
Bioecos	Epifacies
Biogeocenosis	Epimorph
Biosystem	Facies
Diatope	Geocenosis
Ecotope	Holocoen

in this case, a lake, and also used it to compare rates of important processes between very different types of ecosystems:

> . . . a lake, for example, might be considered by a botanist as containing several distinct plant aggregations. . . . The associated animals would be "biotic factors" of the plant environment, tending to limit or modify the development of the aquatic plant communities. To a strict zoologist, on the other hand, a lake would seem to contain animal communities, although the "associated vegetation" would be considered merely as a part of the environment of the animal community. . . . The trophic dynamic viewpoint, as adopted in this paper, emphasizes the relationship of trophic or "energy-availing" relationships within the community unit. From this viewpoint . . . a lake is considered as a primary ecological unit in its own right, since all the lesser 'communities' . . . are dependent upon other components of the lacustrine food cycle for their very existence. (Lindeman 1942)

Lindeman emphasized that the complex set of interactions between living and dead organic material that existed in any ecosystem made separation of spatially distinct "communities," such as the "plankton community," meaningless. The key to the function of the system was the transfer of energy from green plants (producers) to animals (consumers) and microbes (decomposers). This occurred throughout the lake ecosystem. Thus the energy transfers within the lake ecosystem became the focus of his study, and different spatial or organismal components were treated as important in relation to their contribution to this total energy flow. Lindeman also produced one of the first energy flow diagrams, again for a lake ecosystem (Figure 1.1). Such diagrams have become the most common method for expressing interrelationships between components of ecosystems. Many are presented in this text. Energy has become one of the "universal currencies" of ecosystems and ecosystem studies.

Another point mentioned by Lindeman but not fully developed in his paper is the inseparable relationship between the flow of energy and the flow of nutrient elements (see earlier quote). Plants require several chemical ele-

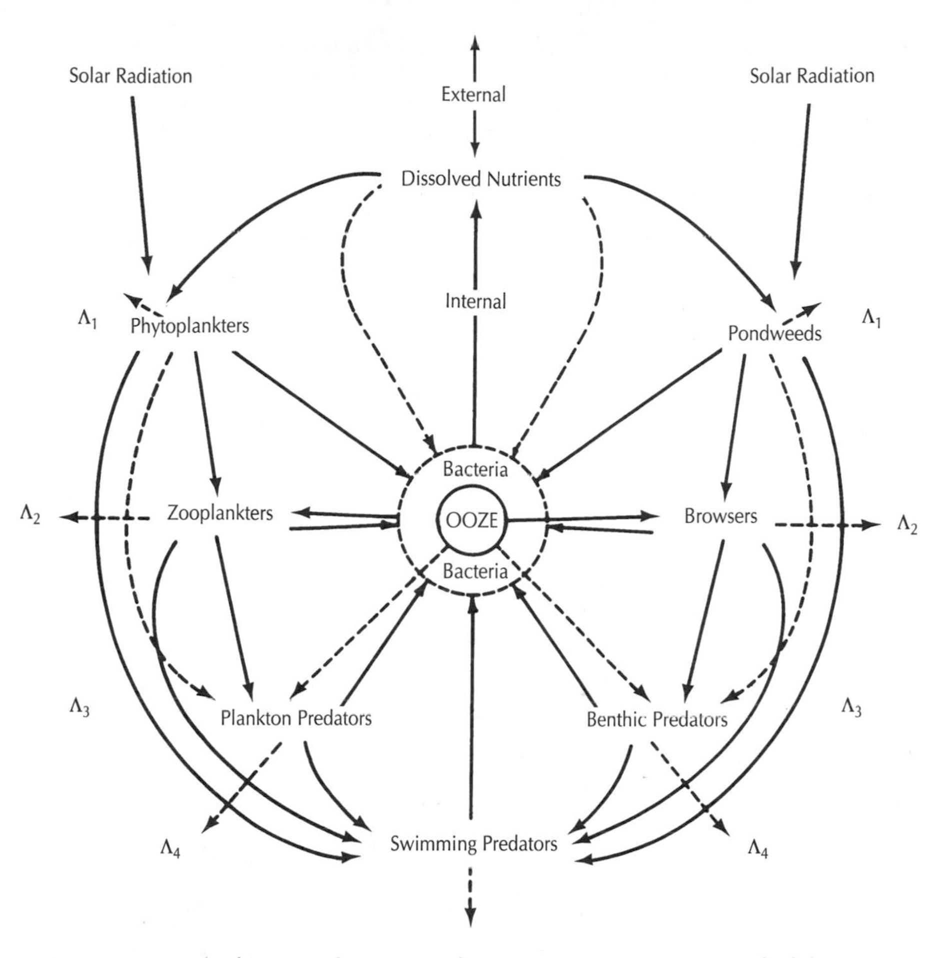

FIGURE 1.1 Early diagram of energy exchanges among components of a lake ecosystem. (Lindeman 1942)

ments in order to grow, photosynthesize, and reproduce. Pioneering work and early synthesis in this area was carried out by J. D. Ovington, an English forester and silviculturalist. In another classic paper, he set forth the basic outline of terrestrial ecosystem studies, which holds largely unaltered to this day.

As a forester, Ovington was concerned with the quantity of water and nutrients required to grow a given amount of tree biomass. He pulled together all existing data on forest production, water use, and nutrient content and tried to express nutrient and energy movements as yearly rates. Transfers within the system were examined, such as those detailed by Lindeman for lakes. However, Ovington realized that in an era of increasing demands for wood products, the continued productivity of forests might depend on their input–output balances. Removing wood from the forest also removes the nutrients contained in that wood. These must eventually be balanced by inputs in rainfall or by other processes. Ovington thus began to consider all the possible inputs and outputs of nutrients in forests. His concept of the movement of energy, water, and nutrients within and through ecosystems was therefore more complete (Figure 1.2). His statement on the need for more basic understanding of ecosystem function (1962) was one of the first to envision the importance of ecosystem analysis in the wise management of natural resources:

> Of all terrestrial communities, woodlands are probably the most complex and massive, and not unexpectedly, woodland ecologists have attempted to restrict their research to whatever features they felt were of greatest importance or could be recorded most readily. . . . The resultant lack of integration has caused an oversimplified approach to woodland ecology which has

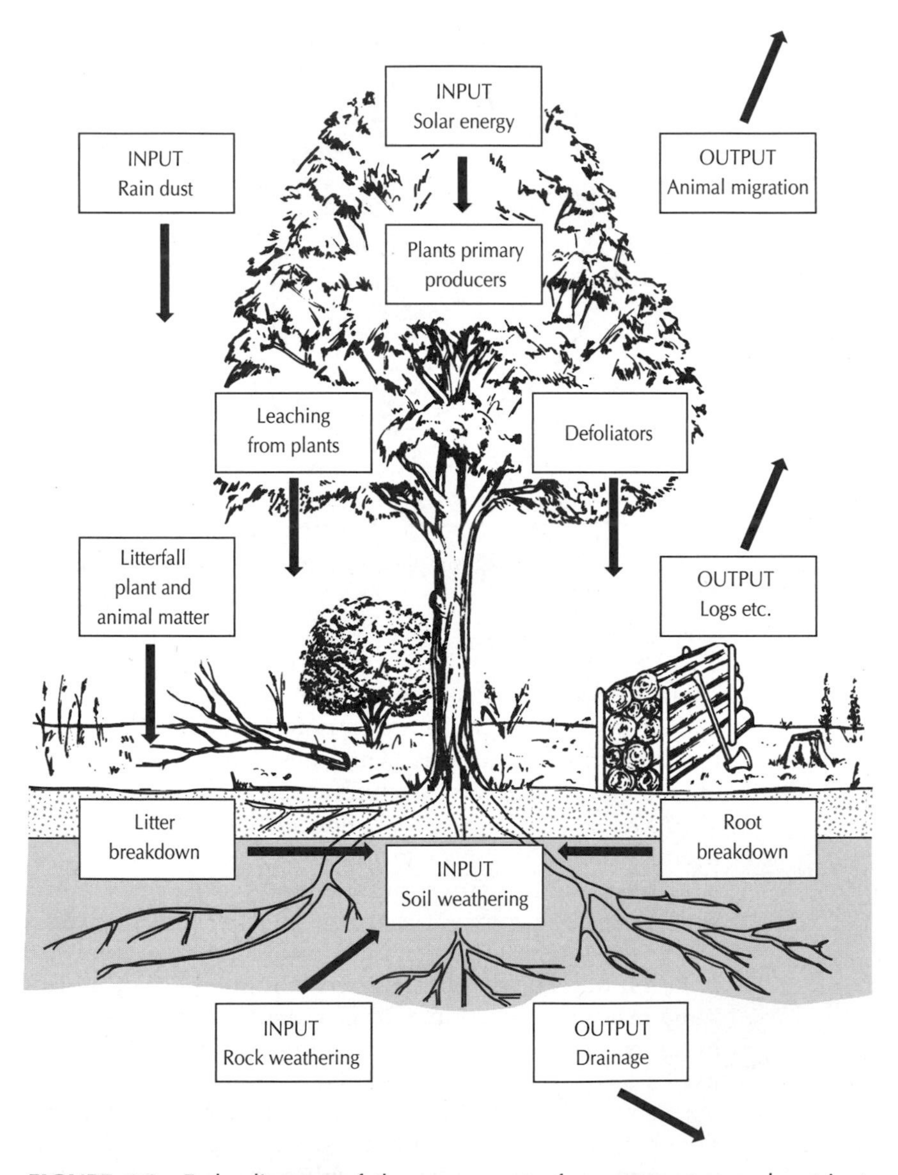

FIGURE 1.2 Early diagram of the movement of energy, water, and nutrients through forest ecosystems. (Ovington 1962)

> further hindered the effective co-ordination of results from different disciplines and areas. Multiple and more intensive use of forest land is inevitable as the world population multiplies, and the more thorough and comprehensive our knowledge of woodland ecology, the better will be the prospect for wise use and long-term conservation of the woodland resource. Some unifying concept embracing all aspects of woodland ecology is needed to bring forth a deeper understanding of woodland systems and to serve as the basis for their more rational utilization. It seems that the concept of ecosystem may provide the universal backcloth against which to show woodlands in all their patterned complexity. (Ovington 1962)

Thus Ovington largely defined the focus of terrestrial ecosystem studies as the movement of energy, water, and nutrients within and through ecosystems, with particular emphasis on factors controlling those rates of flow. He admitted that the necessary data were not available at the time of his paper to answer the crucial questions on controlling factors or for predicting effects of management on ecosystems.

In the 40 years since Ovington's paper, there has been an explosion in the amount of information available on the structure and function of terrestrial ecosystems. With this has come a significant increase in our ability to predict the short-term futures of specific systems under specific conditions. However, this same period has also seen a rapid acceleration in the rate of change in demands placed on ecosystems both to produce goods and services and to resist pollutants and environmental change. Our ability to predict the future in this more complex world is still limited. We will return to this topic in the last part of this book.

This same 40-year period has seen ecosystem studies develop from a raw area of research laden with theory and short on observations to a mature, data-rich science in which the fundamental interactions and processes have been described for many areas of the Earth. There is still much to be learned, especially in relation to the changing global environment, but considerable progress has been made. That progress is the subject of this book.

DELIMITING THE ECOSYSTEM

Thus far, ecosystems have been discussed in the abstract as combinations of biotic and abiotic components, using examples such as lakes and forests. In order to develop a rigorous set of methods for measuring processes, particularly inputs and outputs of energy, water, and nutrients, a definite boundary must be placed around the system. How are the boundaries of an ecosystem determined?

To begin with a general and not very satisfying answer, the boundaries of the system are determined by the purposes of the study or the questions posed. All of the ecosystems of the world, however defined, are linked by their inputs and outputs to all other ecosystems, so the largest definition is the Earth itself. This was demonstrated most dramatically by studies on the global distribution of pollutants, such as DDT or radioisotopes from nuclear bomb tests.

For example, large-scale application of the chlorinated hydrocarbon DDT began in the 1940s. DDT was one of several potent, synthetic pesticides that were to rid the world of insect-borne diseases and revolutionize agriculture. However, it was quickly learned (see popular account in Carson's *Silent*

Spring [1962]) that severe side effects resulted from its use. Among these was the disruption of reproduction in predatory birds, including some of the rarest and most valued, in whose tissues the chemical became concentrated. Surveys of world animals revealed measurable concentrations of DDT even in Antarctic penguins far removed from areas of direct application. Thus, the Earth's intertwined food chains had carried this new chemical to the farthest corners, indicating that all of the distinct ecosystems are indeed linked.

At the other extreme, researchers have found that the ecosystem concept is applicable to areas as small as a single leaf. Enclosing these in small chambers and measuring gas balances (oxygen, carbon dioxide) allows calculations of photosynthesis and respiration, describing the carbon balance over a single organ (see Chapter 3).

In terrestrial ecosystem studies, two definitions are most common: (1) the watershed and (2) the stand. A watershed is a topographically defined area such that all of the precipitation falling into the area leaves in a single stream. Figure 1.3 shows the outline of a small watershed used for hydrologic research in a forested region. Watersheds range in size from a few to hundreds of thousands of hectares. The watershed of the Mississippi River, for example, includes roughly one third of the 48 contiguous United States.

The watershed concept has been used extensively because of the importance of water balances in the study of ecosystems. The water balance itself is

FIGURE 1.3 A watershed-ecosystem at the Coweeta Hydrologic Laboratory in North Carolina. Note the use of the complete hydrologic basin (watershed) for an experiment on the effects of disturbing the plant community. (Courtesy of the Coweeta Hydrologic Laboratory, US Forest Service)

of interest because of the role of water availability in limiting plant growth and because of the importance of water yield from ecosystems for urban and agricultural uses. As both rainfall and streamflow contain measurable concentrations of chemical elements, precipitation inputs and stream water outputs are also important in nutrient balances over ecosystems. Watersheds allow for accurate measurements of nutrient losses because the stream acts as a sampling device, giving an average concentration of nutrients in water leaving the watershed-ecosystem. Most watersheds used in ecosystem studies are relatively small (10–50 hectares) and occur in areas of extreme topography (Figure 1.3; additional discussion of the use of watersheds is presented in Chapter 4).

The second important spatial definition of an ecosystem is the stand, a term borrowed from forestry or agriculture but applicable to any form of terrestrial vegetation. A stand is an area of sufficient homogeneity with regard to vegetation, soils, topography, microclimate, and a disturbance history to be treated as a single unit. This is clearly less precise than that of the watershed as variation over small distances is the rule in natural systems and a strict definition of "sufficient homogeneity" is lacking. However, stands can be much smaller than watersheds (usually an area less than 5 hectares is actually sampled). This reduces the effort required to measure processes within the system. In areas of little or no topographic variation where distinct watersheds cannot be identified, it is the only practical definition. In arid areas, such as deserts or short-grass prairies, where soil water percolation and streamflow are minimal, the watershed loses its value in sampling exports from the system. Thus, although less satisfying than the definition of the watershed, the definition of the stand is the only practical approach in many situations.

COMPONENTS OF TERRESTRIAL ECOSYSTEMS

What do grasslands, deserts, forests, shrub lands, tundra, and all other terrestrial ecosystems have in common? If we want to devise a general structure for terrestrial ecosystems, what components do all of these systems share? They are very different in terms of plant species, soil structure, and other visible characteristics (see Chapter 2). Building on the early efforts of Lindeman and Ovington, can we discern a common set of functional components?

For the flow of energy, associated in terrestrial systems with carbon-based compounds, components can be separated according to their source of energy (Figure 1.4*a*). Green plants are called autotrophs (*auto* = self, *troph* = energy), or primary producers. Animals that graze on plants, ingesting living tissues, are called heterotrophs (*hetero* = other), secondary producers, or consumers. Consumption of live plants is generally less important quantitatively than the pathway through saprotrophs (*sapro* = dead), which feed on and decompose dead organic matter shed by plants, animals, and other decomposers. More simply stated, more plant production dies and is shed as litter (such as the falling of leaves in the autumn) than is consumed live. Notable exceptions to this generalization are discussed in Chapters 15 and 16.

The flow of energy and carbon compounds through plants, consumers, and decomposers is complicated by the wide array of complex biochemical compounds formed by plants from the products of photosynthesis (Chapter 12). This is further complicated by an even more complex set of organic com-

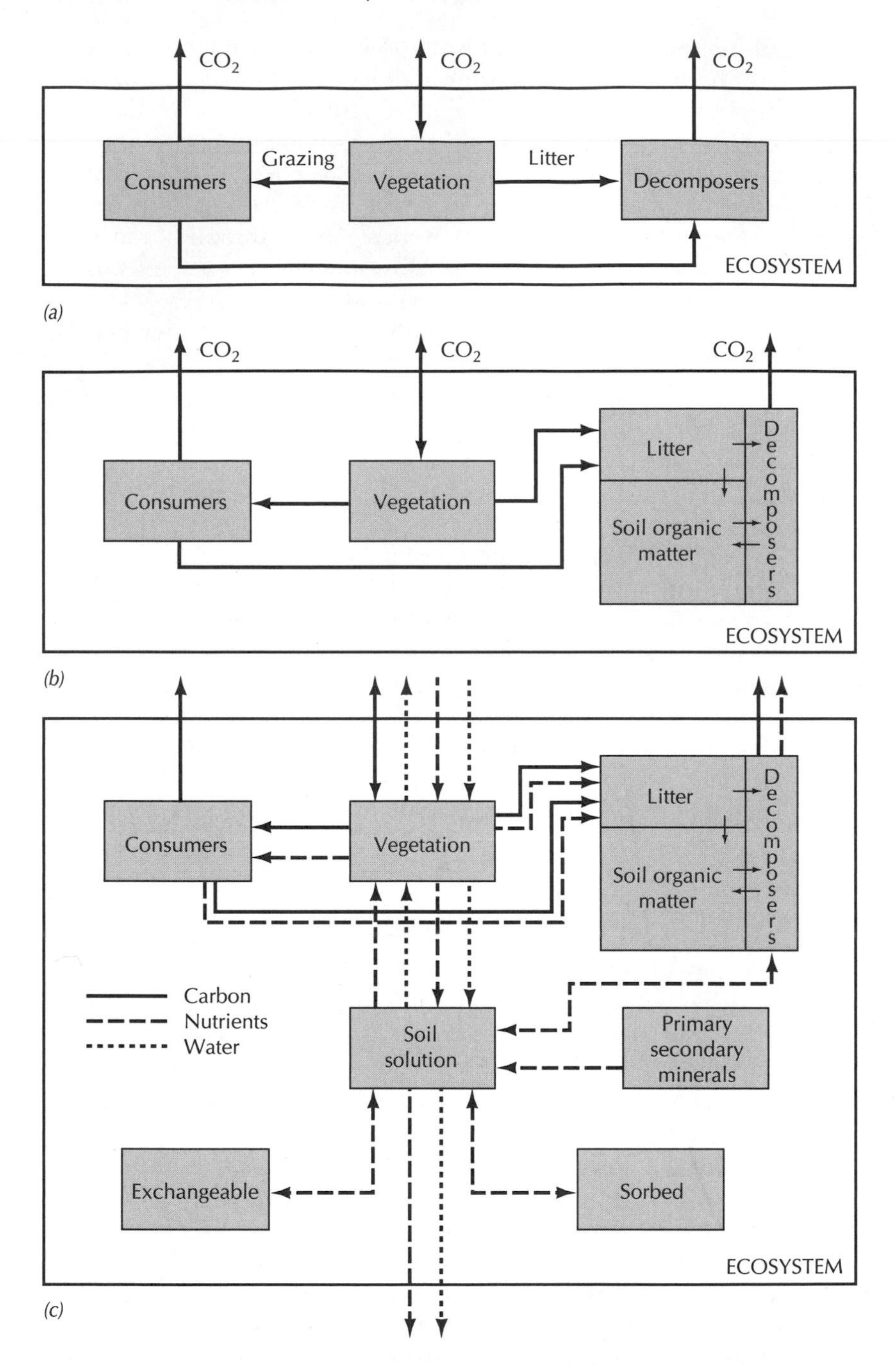

FIGURE 1.4 Components common to all terrestrial ecosystems. (*a, b*) Components important in the carbon cycle. (*c*) Enlarged model containing components important in nutrient and water dynamics.

pounds that are produced by decomposer organisms and that accumulate in soils in terrestrial ecosystems. Separating relatively fresh litter (such as this year's leaves or roots) from that which has been decomposed to some extent results in the situation illustrated in Figure 1.4*b*, in which energy and carbon in the ecosystem reside in either plants, consumers, decomposers, or fresh or partially decomposed organic matter. The last three are spatially combined within the soil subsystem.

Each of these categories could be divided further. Vegetation consists of different species, perhaps of different growth forms (such as trees, shrubs, and grasses) and of different tissue types (such as leaves, stems, and roots). Soil organic matter is really a composite of the simple and complex carbon substances produced by plants, decomposers, and strictly chemical reactions. The decomposer component can be a vast and diverse assemblage of microorganisms and soil animals operating as an integrated community or can be dominated by a few major organisms. The consumer community includes several trophic layers, including herbivores and predators. The number of components recognized within an ecosystem is much like the boundaries of the system in that they can be selected to reflect the goals of the study. At different points in the following chapters, it will be necessary to dissect the basic components presented here to achieve greater understanding and accuracy. For now, we will use this simpler model.

Production of organic matter cannot occur without nutrients essential to plant function. Thus the flow of energy and carbon through an ecosystem depends on a parallel flow of essential nutrients. Unlike carbon, hydrogen, and oxygen, which make up as much as 98% of plant tissue and are derived from atmospheric gases and water, mineral nutrients, such as nitrogen, phosphorous, and calcium, reside largely within the system and are recycled. Inputs and outputs are relatively small compared to the total amount cycled annually.

In addition, most nutrients can occur in numerous forms within the system. They can be part of the vegetation, litter, soil organic matter, animals, and microbes—components shown in Figure 1.4*b*. However, plants take up nutrients mainly in simple organic and inorganic forms, not directly from the large, complex molecules that constitute the bulk of soil organic matter. Microbes have to "digest" organic material and release the simpler forms. These simpler forms can be found in the soil water solution (see Figure 1.4*c*).

Mineral nutrients in solution can have several fates. They can be taken up and used either by plants or by microbes. They can also, depending on the element, participate in several strictly chemical interactions. They can be held on the surfaces of soil particle surfaces either through simple electrostatic attraction (exchangeable) or by stronger surface complexation reactions (sorbed). Both of these transfers are reversible, although exchangeable elements are more readily available to plants. Elements previously held in completely unavailable forms in rocks can be released to the soil solution as simple inorganic forms through weathering. Finally, rain water washing down through the system can both bring nutrients in and leach nutrients below the rooting zone, out of the biologically active area, and into streams or groundwater. Under certain conditions, selected elements can be lost from the system as gases. Thus our list of components and transfers is much larger than for carbon alone (Figure 1.4*c*).

One of the important advances in the past decade of ecosystem studies has been the growing awareness of the importance of small molecules of organic matter that can become dissolved in the soil solution and move through soils to streams. These dissolved organic forms are not a quantitatively important part of the total carbon cycle, but the chemical activity of these forms, and the potential for plant uptake, gives them an important role in soil and plant chemistry.

The movement of water is less complex conceptually (Figure 1.4*c*), although no easier to measure. Precipitation can first contact plant surfaces, such as leaves or stems, or fall directly on the soil. In the former case, it can either evaporate or fall to the soil surface. Soil water becomes part of the soil solution and either evaporates, is taken up by plants and evaporated from leaf surfaces, or percolates below the rooting zone. Water carries nutrients throughout these pathways and exchanges them with plant and soil surfaces contacted.

What emerges in Figure 1.4*c* is a fairly complete but not very detailed diagram or model of the pathways by which energy, water, and nutrient elements are transported into, out of, and within terrestrial ecosystems. We return frequently to this diagram in later chapters.

Contained in this diagram is a conception of the function of ecological systems that was unknown 40 years ago. As we will see, rates of important ecosystem processes, such as plant production and decomposition, are largely determined by the relative availability of energy, water, and nutrients and the efficiency with which they are used. Thus, plants and decomposers have evolved to optimize rates of function under the constraints of the availability of these resources. As environments vary widely over the face of the Earth, organisms and communities that carry out the same basic functions have developed vastly different structures and appearances. These differences, which produce the great diversity of terrestrial communities on Earth, are the subject of the next chapter.

REFERENCES CITED

Golley, F. B. 1993. *A History of the Ecosystem Concept in Ecology.* Yale University Press, New Haven.

Lindeman, R. L. 1942. The trophic-dynamic aspects of ecology. *Ecology* 23:399–418.

Ovington, J. D. 1962. Quantitative ecology and the woodland ecosystem concept. *Advances in Ecological Research* 1:103–192.

Tansley, A. G. 1935. The use and abuse of vegetational concepts and terms. *Ecology* 16:284–307.

Additional References

Bormann, F. H., and G. E. Likens. 1967. Nutrient cycling. *Science* 155:424–429.

Carson, R. 1962. *Silent Spring.* Houghton Mifflin, Boston.

Kormondy, E. J. 1969. The nature of ecosystems. In *Concepts of Ecology.* Prentice-Hall, New York.

Odum, E. P. 1971. Principles and concepts pertaining to the ecosystem. In *Fundamentals of Ecology.* W. B. Saunders, Philadelphia.

Odum, E. P. 1969. The strategy of ecosystem development. *Science* 164:262–270.

Swift, M. J., O. W. Heal, and M. Anderson. 1979. Decomposition processes in terrestrial ecosystems. In *Decomposition in Terrestrial Ecosystems.* University of California Press, Berkeley, California.

Woodwell, G. M. 1967. Toxic substances and ecological cycles. *Scientific American* 216:24–31.

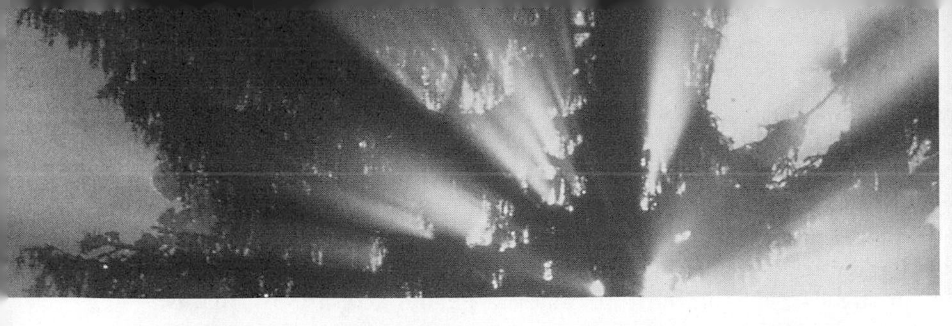

2

STRUCTURE OF TERRESTRIAL ECOSYSTEMS

INTRODUCTION

In Chapter 1, a single conceptual model that emphasized major processes common to all terrestrial ecosystems was developed. Thinking about different systems in this uniform way is useful for drawing comparisons between them, and we use this basic structure throughout the book. However, this model also masks the tremendous variation that exists in the physical structure of ecosystems, which both results from and contributes to the rates at which important processes occur in different systems.

The purpose of this chapter is to provide a brief, descriptive tour of some of the major terrestrial ecosystem types. Differences in the structure of both vegetation and soil are presented, along with a brief description of the environmental factors driving those differences. Variation in key processes, such as plant growth (primary production), water use (transpiration), and soil formation, are also described. The mechanisms controlling vegetation structure and soil development are numerous and complex. The simplified discussion presented here is intended only to describe trends in vegetation and soil that result from the range of climatic conditions found over the Earth. Greater detail is provided in following chapters.

We also take this opportunity to introduce a newer perspective on the distribution of ecosystem types and ecosystem function. There is growing urgency in assessing the current state of terrestrial ecosystems worldwide and changes in condition over time. This new global perspective has led to the combination of traditional field measurements, with new information obtained from remote sensing satellites as synthesized by computer models. The integration of large-scale, detailed images of the Earth's surface, with simple models for converting those signals into estimates of ecosystem function, is a powerful approach for assessing the state of the biosphere. Both remote sensing and modeling are critically discussed in Chapter 5, and examples are given throughout the book. In this chapter, we present images capturing the distribution of ecosystem type and function made possible by this combination of technologies.

DISTRIBUTION AND CHARACTERISTICS OF MAJOR ECOSYSTEM TYPES

On a global scale, climate plays the largest role in determining the structure of both vegetation and soils in natural ecosystems. Temperature and the balance between precipitation and evaporation (including evaporation through plant surfaces, known as transpiration), are particularly important because these largely determine the rate at which biological and chemical reactions occur. Both the production of new organic matter by plants and decomposition of dead organic matter by microbes are temperature- and moisture-dependent processes. The amount of water available for use by plants, relative to the potential for that water to evaporate, determines the amount of water stress encountered by plants and often limits the length of the growing season and total plant growth. The amount of water percolating down through the soil profile and the chemistry of that water play a large role in determining soil structure and nutrient content.

VEGETATION TYPE, PLANT STRUCTURE, AND MAJOR PROCESSES

Traditionally, the global distribution of ecosystem types has been mapped through correlations between aerial extent and mean climatic conditions (Figure 2.1*a*). The shaded area in the figure represents the range of mean precipitation and mean temperature values on the surface of the earth. Figure 2.1*b* maps the geographical distribution of these biomes.

Figure 2.2 summarizes differences in structure and function between the major ecosystem types identified in Figure 2.1. Three characteristics are used to describe the structure of the vegetation: (1) total height, (2) the distribution of the fine roots that are most active in nutrient and water uptake, and (3) total above-ground biomass (B) or the dry weight of all live plants.

Three important processes by which ecosystems are compared are also included. The first is annual net primary production above ground (P). This is the total dry weight of all the new plant growth in a year, excluding roots and other below-ground parts. Second is annual nitrogen uptake or use by above-ground vegetation (N). This is determined by measuring the concentration of nitrogen in the different kinds of tissues produced (leaves, wood, and fruits), and multiplying by the annual production of those tissues. The third process is actual evapotranspiration (AET). This is the sum of water lost to the atmosphere by evaporation from the soil surface (evaporation) and through plant tissues (transpiration) over the course of a year. It is often compared with potential evapotranspiration (PET), or the amount of evapotranspiration that would occur with unlimited supplies of water. In hot, dry climates, AET is limited by the amount of water available to be evaporated and may be much lower than PET. Regular droughts occur in such areas and little water moves down through the soil. In cool, rainy climates, less solar energy and more humid air reduce evaporative demand, and PET is lower. AET may then be equivalent to PET, and drought and water stress on plants is lower. When AET is limited by evaporative demand (PET), more water is available for movement down through the soil.

Simple models of the effects of mean climate on physiological processes, such as photosynthesis and transpiration, have been invoked to attempt a more mechanistic prediction of the distribution of both ecosystem type and water and carbon balances. One such model uses first a simple set of rules to

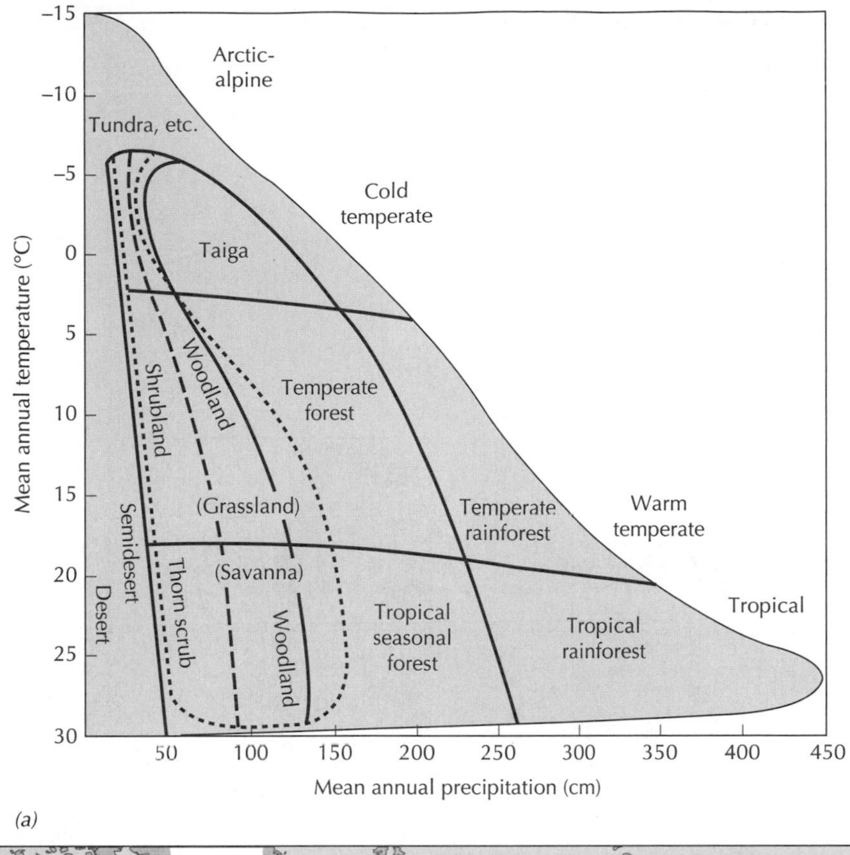

(a)

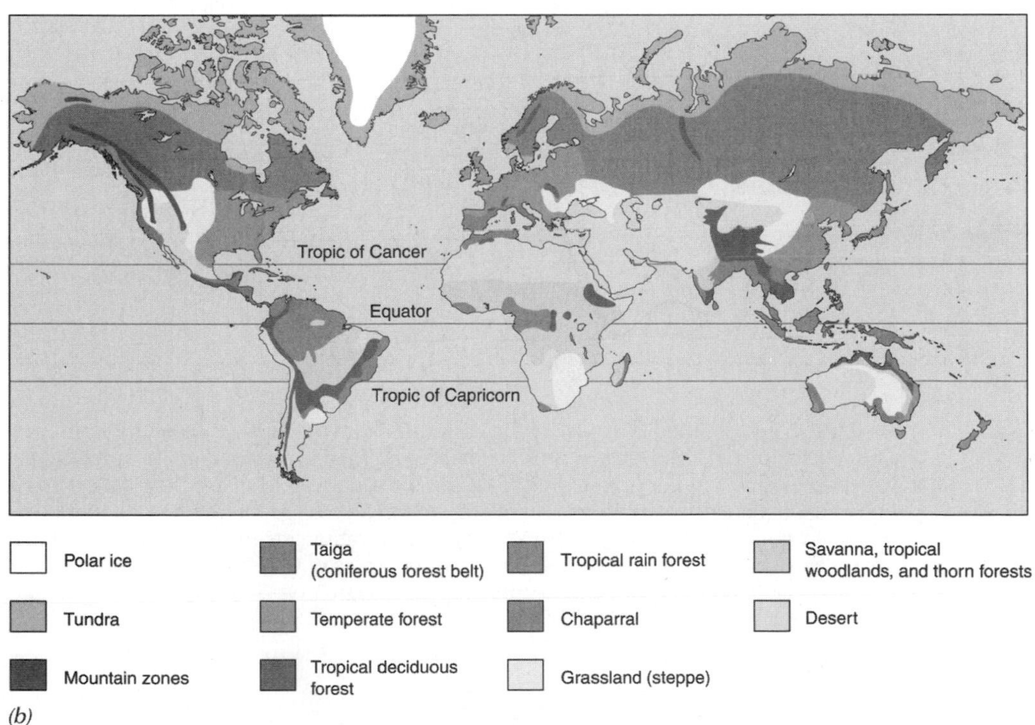

(b)

FIGURE 2.1 *(a)* Distribution of major vegetation types or biomes in relation to mean annual precipitation and temperature (Whittaker 1975). (b) Geographic distribution of major vegetation types of the world. (Solomon, World of Biology, 5th ed, Saunders College Publishing, 1995) **See plate in color section.**

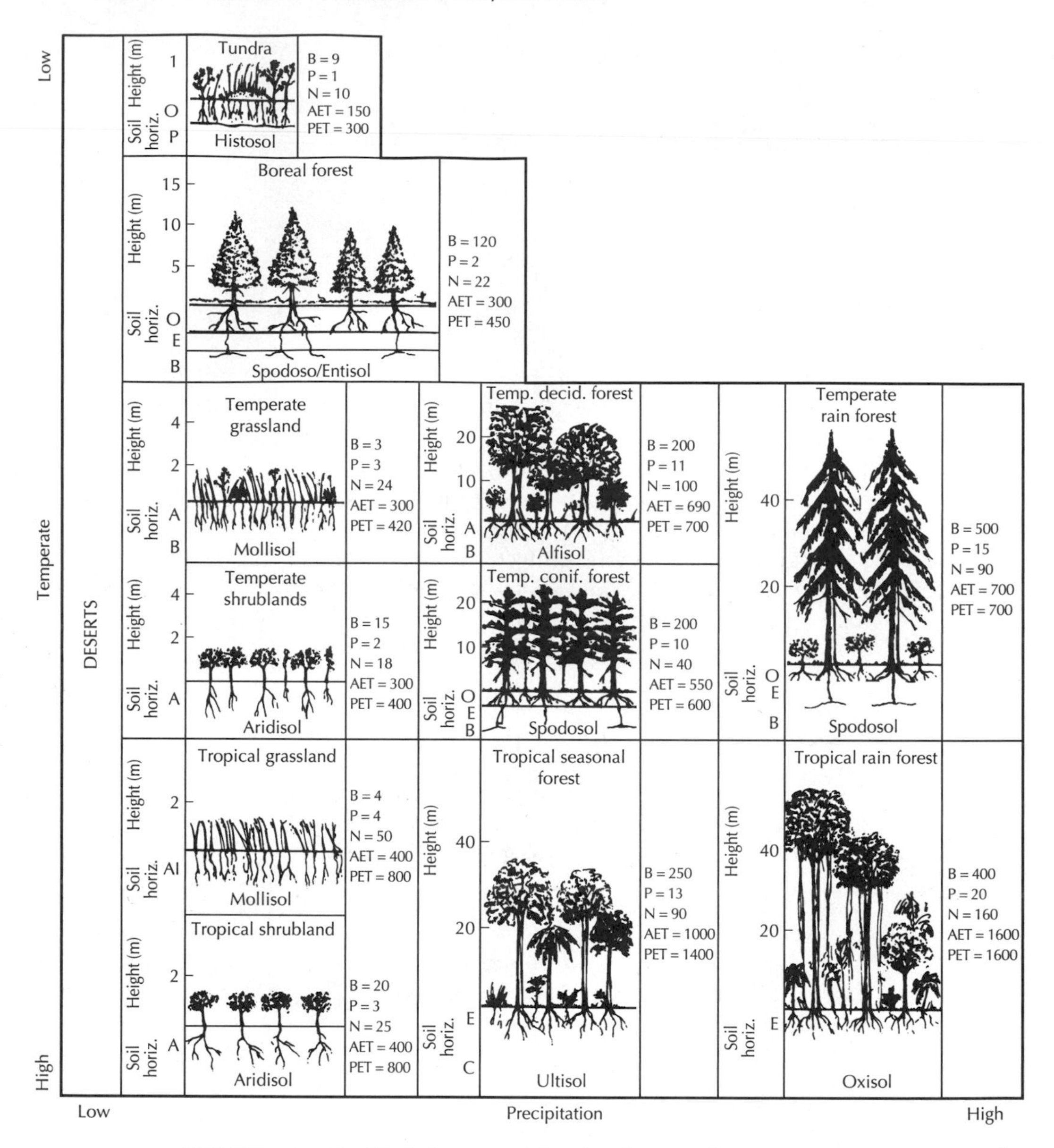

FIGURE 2.2 Modified diagram of the distribution of major ecosystem types in relation to precipitation and temperature. Values for elements of structure and function that are used to characterize ecosystems are given for each type. B = total plant biomass in grams per square meter, P = total plant production above ground in nitrogen uptake by plants in $g \cdot m^{-2} \cdot yr^{-1}$, AET= actual evapotranspiration in millimeters of water per year, PET = potential evapotranspiration in millimeters of water per year. (Bliss et al. 1973, Bokhari and Singh 1975, Gray and Schlesinger 1981, Murphy and Lugo 1986, Nadelhoffer et al. 1985, Pastor et al. 1984, Sims et al. 1978, Sinclair and Norton-Griffiths 1979, Van Cleve et al. 1983, Vorosmarty et al. 1989, and Vitousek and Sanford 1986)

determine which vegetation types could grow in a given location (e.g., evergreen forests of the far north can withstand minimum temperatures well below 0°C, while tropical species cannot). Climate data and physiological descriptions of different growth forms are then combined to predict the amount of foliage and the total production of biomass, which can be sustained at each point. Predicted production is generally constrained either by a cold-shortened growing season or by drought. With certain restrictions, the vegetation type that can achieve the highest leaf mass and production is predicted to dominate at that location. Coarse-scale predictions by this model match well with maps produced from a summary of field observations worldwide (Figure 2.3).

If both the observations and model agree, and we already know from observations what the global distribution of vegetation types is, then what is the advantage of the model approach? There are two: (1) when the model predicts well, then we find that our understanding of the factors that control ecosystem distribution does not contradict the observed pattern and (2) in a changing world, we may want to predict the *future* distribution of these same communities. A model driven by current climate patterns can also make predictions using future climate patterns (the value and process of modeling are discussed in more detail in Chapter 5).

SOIL PROCESSES AND DISTRIBUTION OF SOIL TYPES

The important characteristics of soils include depth of the different soil horizons or layers and the chemical characteristics of each that result from the interaction between climate and geology. There are five major soil horizons (Figure 2.4).

1. The O, or organic, horizon is a surface accumulation of partially decomposed litter and organic matter that has not been mixed into the mineral soil.

2. The A horizon is the upper layer of mineral soil altered mainly by the mixing of organic matter from the O horizon and the deposition of organic matter by root production and death. Chemical weathering also occurs in this horizon, but the results of weathering are masked by organic matter content.

3. The E horizon is defined as that mineral soil horizon the chemistry of which has been altered mainly by the weathering and leaching of minerals. Many soils show *either* an A or an E horizon, depending on the relative importance of chemical weathering versus biological incorporation of organic matter in determining the characteristics of the top layer of mineral soil.

4. The B horizon is that part of the soil profile that has been altered by the chemical deposition or precipitation of material leached from the overlying horizons.

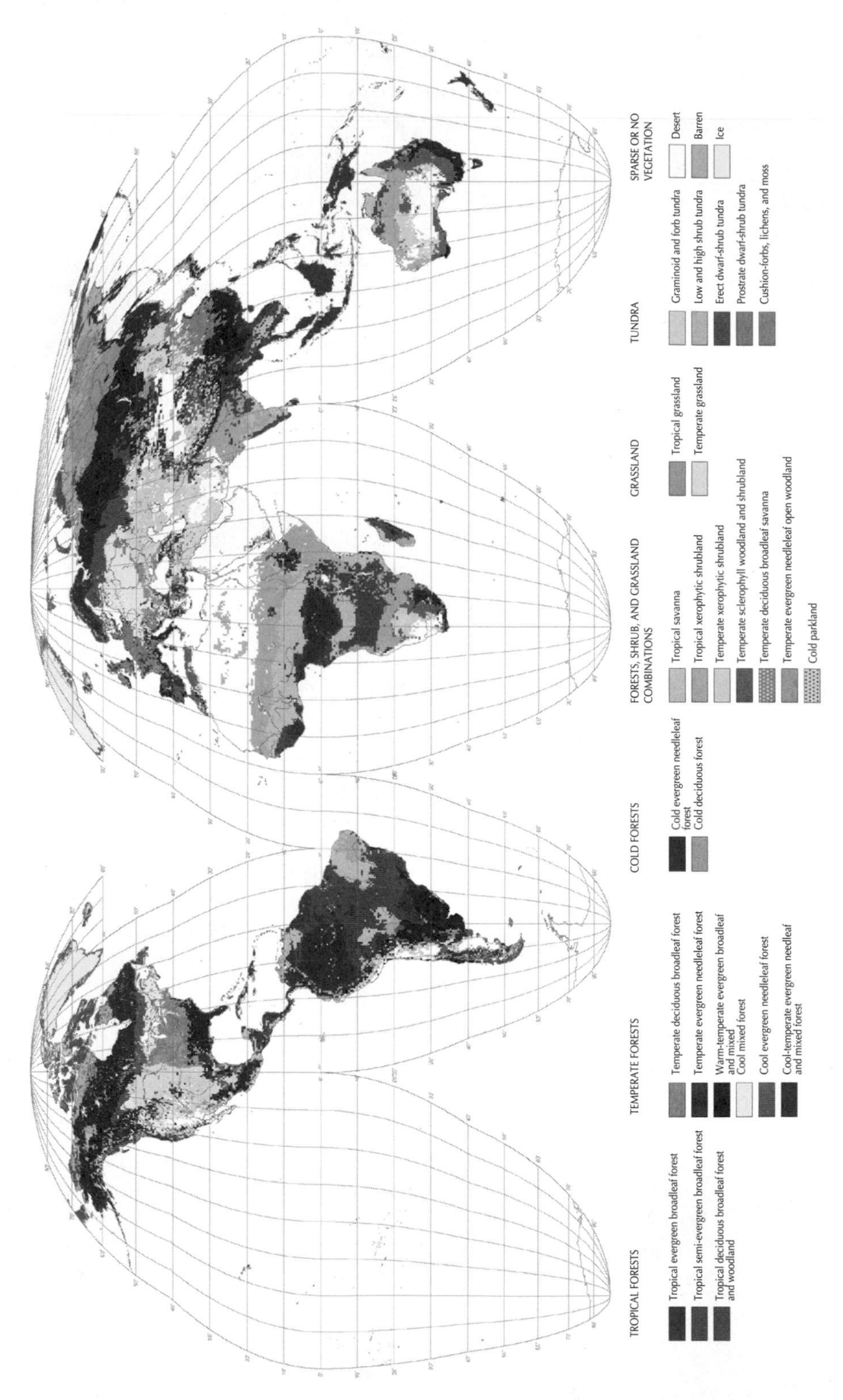

FIGURE 2.3 Comparison of current distribution of major biomes of the world as summarized from field observations (*a*) and as predicted by the BIOME3 model. (Haxeltine and Prentice 1996) **See plate in color section.**

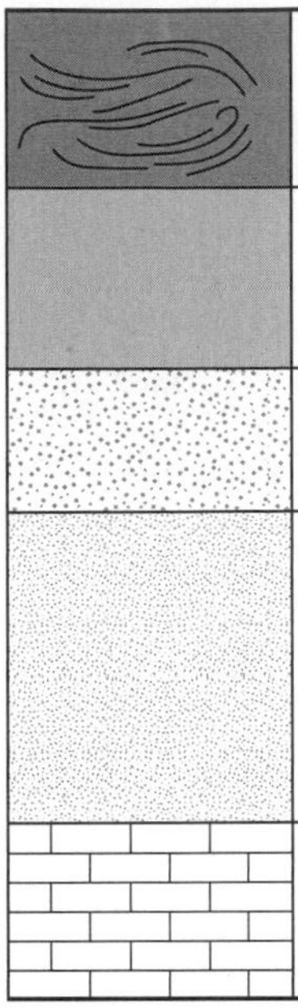

Horizon	Characteristics
O	Surface accumulation of partially decayed organic matter
A	Mineral soil in which mixing of surface organic matter and root growth masks effects of chemical leaching
E	Mineral soil in which weathering and leaching dominate horizon characteristics
B	Mineral soil zone affected by chemical deposition of material leached from A horizon
C	Parent material or mineral horizon largely unaffected by soil development

FIGURE 2.4 Major soil horizons and their characteristics.

5. The C horizon is mineral material that has not been affected by soil development. In some cases, it represents the original material from which the A, E, and B horizons have developed.

Which substances are leached from the E horizon is determined largely by climate but also partly by changes in water chemistry caused by the presence of different types of vegetation and their decay products. Four basic processes can be defined.

1. **Podzolization** is the leaching of iron, aluminum, and organic matter along with very little formation of clays, producing an E horizon with a light gray color that is enriched in silicon (sand) and of very low fertility. Podzolization is most effective under cool, moist, and acidic conditions. A sizable O horizon is commonly found over podzol soils, and organic acids leaching from the O horizon can increase the rate of podzolization. The B horizon can be enriched in iron and aluminum oxides (rust colored) leached from the E horizon.
2. **Laterization** is the converse of podzolization. Under hot, moist conditions, silicon is weathered and leached more readily than are iron and aluminum oxides or clays, and the E horizon become enriched in the latter. This process creates the red clay soils of the southeastern United States and the tropics. The leached silicon is not redeposited lower in the soil profile but is leached away to groundwater and streams, and there is no visible B horizon.

3. **Melanization/Lessivage** occurs in temperate areas with fine, textured soils and less acidic conditions. The formation and leaching of clays (lessivage) is important in determining the structure of the B horizon, while the A horizon tends to be well mixed by the activity of soil animals and is enriched and darkened by the addition of organic matter (melanization). The boundary between the A and B horizons can be very difficult to distinguish, and there may be no visible E horizon.

4. **Calcification/Salinization** occurs in arid regions where percolation of water through the soil profile is minimal. Even very soluble compounds, such as calcium carbonate (lime) and sodium chloride and other salts, remain within the soil column, although they may be moved down from the surface to an extent dependent on precipitation.

In many soils, some combination of these processes is evident. In all but the driest conditions, the E horizon increases in depth through time. Its depth at any one time reflects the intensity of soil-forming processes and the length of time that they have been at work.

MAJOR VEGETATION AND SOIL TYPES OF THE EARTH

Tropical rain forests (Figure 2.2) occur where temperatures are warm and constant and where rainfall is high and occurs throughout the year. The growing season lasts all year, and plants grow quickly and continuously, leading to high levels of production and biomass accumulation within a very tall and vertically stratified canopy. A tremendous diversity of growth forms is present, including vines and epiphytes, as well as the greatest diversity of species. High production levels require the uptake and cycling of large amounts of nitrogen. Although evapotranspiration is high, rainfall still exceeds evaporative demand, and leaching of water down through the soil column is intense. Many soils under this vegetation type are also very old, so the E horizon can be exceptionally deep. Laterization is the dominant soil-forming process, and the E horizon is high in iron and aluminum oxide and clay content. Conditions are also optimal for decomposition, so little or no surface litter accumulates, but there is an abundance of organic matter in the mineral soil. The occurrence of large stores of soil organic matter and nitrogen in almost all terrestrial ecosystems is a fundamental and perplexing property that receives more attention in later chapters.

Such fully weathered soils with large accumulations of clays and iron and aluminum oxides are called Oxisols. They are very nutrient-poor soils. A paradox of the tropical rain forest is that it supports the richest and most luxuriant vegetation on the poorest soils. Much of the nutrient capital is present in the vegetation at any one time and is quickly recycled to the vegetation after litter fall occurs. This is reflected in the concentration of fine roots at the soil surface, which take up nutrients as they are made available by decomposers from litter.

Tropical areas in which regular dry seasons occur support tropical seasonal forests. Periods of drought reduce plant growth, so total plant production and associated litter fall and nitrogen cycling over a full year are lower than in the rain forest, as is evapotranspiration. Soils are less severely leached but otherwise similar to the Oxisols and are called Ultisols. The major soil-forming process is still laterization, which results in a red clay E horizon rich in iron and aluminum. This soil type may also be found in subtropical forests, such as the pine and hardwood stands in the southeastern United States.

In still drier tropical areas, forest gives way to grassland/savanna or shrub "thorn" forest. Both are shown in the same climatic square because both can occur under similar mean annual climatic conditions. As discussed in more detail for temperate grasslands and shrublands later, disturbance, particularly fire and grazing, might play a role in determining which growth form predominates. The severely limiting moisture conditions lead to much lower rates of evapotranspiration, although evaporative demand (PET) in this hot, dry climate is very high. Total production and nitrogen cycling are also reduced. Growth form reflects these limitations; there is a shift from trees to shorter vegetation and lower total biomass. Roots, especially under shrub vegetation, may penetrate very deeply into the soil to reach ground water stored well below the soil surface.

The potential for evaporation in these hot and relatively dry areas reduces leaching, thus soil-forming factors are less active. Under shrub vegetation on moderate to steep slopes, occasional heavy rainstorms may cause extensive erosion, which removes material from the top, most developed horizon. Root masses are lower than for grasslands, and soils are relatively low in organic matter. In grasslands, fine root masses are larger, creating a deep A horizon. This increases water percolation into the soil, decreases runoff, and reduces erosion. The profile shown for the shrubland in Figure 2.2 is called an Aridisol, while the grassland soil is called a Mollisol. Calcification and salinization are dominant in the aridisol. Melanization tends to mask these processes in the Mollisol.

A similar sequence of vegetation types may be described for temperate regions. Temperate rain forests occupy a much smaller area than their tropical counterparts, usually occurring on the windward side of coastal mountains or in other coastal areas. The vegetation is totally different from the tropical systems in species composition, but, in general, structure and function are quite similar (Figure 2.2). In North America, the temperate rain forests are dominated by needle-leaved evergreen trees rather than the broad-leaved trees of the tropics, but vines and epiphytes and a deep, stratified canopy are all similar to tropical conditions. Production is somewhat lower than in the tropics because temperatures are less favorable, but AET is close to PET due to abundant rainfall. Litter production is somewhat lower than the tropical forests, while nitrogen cycling is considerably lower because the dominant needle-leaved evergreens have lower concentrations of nitrogen in the biomass produced, including leaf litter. This low nitrogen content, along with lower temperatures, reduces decomposition and soil animal activity, creating large accumulations of surface organic matter (O horizon). With the cooler climate and more acidic litter, podzolization is the predominant soil forming process. There is often a distinct E horizon. Accumulations of iron; aluminum; and, to some extent, organic matter occur in the B horizon. The soil type is a Spodosol. Root distribution is similar to that of organic matter. Roots are concentrated in the O horizon, very low in the E horizon, and somewhat higher in the organically enriched top of the B horizon.

Temperate, seasonal forests may be either deciduous or evergreen. The two types occupy climatic regions that may appear very similar in terms of annual mean temperature and precipitation (Figure 2.2). However, the timing of precipitation and annual temperature extremes are important in determining the occurrence of one type over the other. Deciduous forests occupy areas where summer precipitation is high and winter temperatures are low. Needle-leaved evergreens are dominant in areas where winters are warm and summers are dry because the evergreen habit allows for significant photosynthesis during periods when deciduous trees would be leafless. Evergreens also occur on poor sites within the areas normally dominated by deciduous species largely because of lower demands for nutrients and water.

Needle-leaved evergreen and broad-leaved deciduous species can also have large differences in the chemical constituents from which they are made. When living tissues are shed as litter (e.g., the loss of leaves and needles in autumn), these chemical differences can affect soil formation and fertility. In general, deciduous litter decomposes more quickly than do evergreen needles and tends to support higher levels of soil animal activity, leading to the creation of an A horizon, with moderately dense rooting throughout. Leaching is primarily of clays, so the B horizon is enriched in these and may contain some fine roots. The soil is called an Alfisol. Production is lower than in other forests described previously but higher than in tropical grasslands and shrublands, as is biomass. AET is close to PET and drought stress is relatively slight. Nitrogen cycling can be higher than in temperate rain forest because of higher nitrogen concentrations in the biomass produced by deciduous trees.

Productivity in temperate evergreen forests may be similar to deciduous forests, but less of it may take the form of leaf litter because needles are replaced only every 2 to 5 years. Longer needle retention may result in higher total foliar biomass in evergreen stands, even when foliage production is lower than in deciduous stands. Nitrogen cycling is significantly less in most cases, again because less production goes into relatively nitrogen-rich needles and more into nitrogen-poor wood. Needle litter also has less nitrogen than broadleaf litter, which reduces decomposition and mixing. A substantial O horizon can develop. The acidifying effect of conifer litter fosters podzolization, resulting in the formation of an E horizon. Rooting is concentrated in the O and upper B horizons.

In drier temperate regions, grasslands and shrublands occur that are somewhat similar to their tropical counterparts. Again, the reason for the predominance of one growth form over the other is not always clear, but the timing of precipitation and the occurrence of disturbance seem important. For example, the shrub-dominated chaparral vegetation type occupies coastal areas of the southwestern United States, the Mediterranean region, and western South America, where winters are warm and wet. In central North America, where winters are cold and most of the precipitation occurs in the summer, grasslands were predominant in presettlement times. However, fire exclusion and excessive grazing have encouraged a transition to forest at the humid end of the prairie region and to semiarid shrubland at the dry end. In either case, AET is far below PET, and the length and timing of the growing season is determined mainly by the timing and quantity of precipitation.

Differences in growth form produce large differences in soil properties under shrubs and grasses. Although production may be similar for the two, grasses form much more extensive fine root systems that incorporate much more organic matter into lower soil levels. This creates a very deep A horizon with no clear line between the A and B. Soil animal activity is high and surface organic accumulations are minimal, due perhaps as much to fire as to rapid decomposition, and the soil type is called a Mollisol.

Under shrub communities, as discussed for tropical shrublands, the soil-forming factors of calcification and salinization are more active, causing the accumulation of carbonates and salts.

To the north of the temperate forests and grasslands in the northern hemisphere is a circumpolar region of coniferous forest known as the boreal forest, or taiga. Even though precipitation can be much lower here than in the tropical or temperate forest zones, cold temperatures reduce evaporation and transpiration so that water limitations are episodic rather than chronic. Production is lower than for other forest types because of low temperatures, and nitrogen cycling is further reduced by the low N content of needles, which in turn creates very poor litter quality. This, along with cold temperatures, reduces decomposition rates. Soil animal activity is minimal, and podzolization is the dominant process. However, soil development has generally not proceeded very far due to the cold temperatures and because the soils have been in place for only a few thousand years since the last glaciation. Such soils have large O horizons but less pronounced E horizons and are called Entisols or Inceptisols (actually, these two classes of soils can be found in any region where the soil-forming processes have been interrupted or restarted by such factors as glaciation, flooding, volcanic flow, erosion, etc.). Mature or old growth forests, or those over shallow water tables, are susceptible to invasion by the ground moss *Sphagnum,* which can lead to creation of Histosols, discussed later. Forests similar to the taiga in structure and growth form also occur in the temperate zone at high elevation. These are called subalpine forests.

The lowland tundra is a mixture of grasslike sedges, low shrubs, lichens, and mosses growing at the northern limit of vegetation in the northern hemisphere or above timberline in mountains elsewhere (alpine tundra). Production, decomposition, and evaporation are severely limited by low temperatures. Soils can be waterlogged during the short growing season, further reducing decomposition and plant root activity. Large amounts of organic matter accumulate over a layer of rock or frozen water ("permafrost"). This soil type is called a Histosol (again, this soil class can be found in any region in wetlands where decomposition is restricted and large amounts of organic matter accumulate). Rooting occurs throughout the thawed organic horizon but is concentrated toward the top.

Under the driest conditions at all mean temperatures occur the sparsely vegetated "deserts." These can range from the moving sand dunes of the Sahara region to the uniquely vegetated Sonoran region in North America, to the moss and lichen systems of the high arctic and the antarctic. In temperate and tropical regions, desert soils are dominated by the processes of salinization and calcification.

The boundaries drawn on the maps in Figures 2.1*b* and 2.3 should not be considered either precise or permanent. Vegetation types grade into each other through areas of mixed vegetation called ecotones as environmental factors change continuously rather than suddenly across the landscape. Before the last glacial retreat, only 15,000 years ago, all of the northern temperate and boreal systems would have occurred hundreds of miles farther south than they do today, and many of there current locations would have been under thick sheets of ice.

Natural disturbance factors can also be important in determining the boundary between one vegetation type and another. The predominance of tall grass prairie systems in the upper Midwest in the United States, where climatic conditions would support the growth of forests or shrublands, has long been attributed to the frequent prairie fires that swept through this region, burning back trees and shrubs and favoring the growth of grasses. Another model of biome distribution has addressed this question by including fire as an important mechanism and predicting both the frequency of fire in this region and its effects on vegetation. The results (Figure 2.5) support the idea that fire has played an important role, with greatly increased westward ex-

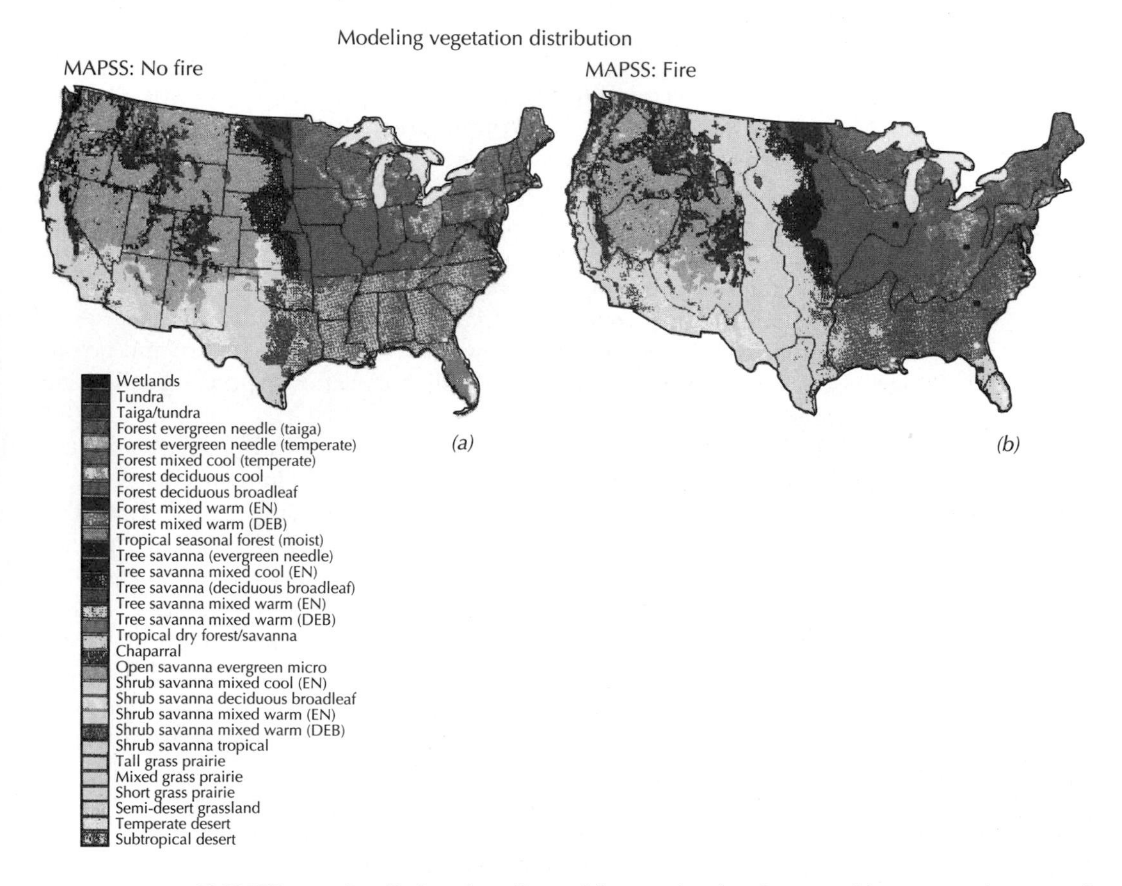

FIGURE 2.5 Predicting the effect of fire on the distribution of biomes in the United States with and without fire using the MAPPS model. (from R. P. Neilson, "A model for predicting continental-scale vegetation distribution and water balance" from *Ecological Applications* 5 (1995). Copyright © 1995 by the Ecological Society of America. Reprinted with the permission of the Ecological Society of America) **See plate in color section.**

pansion of grasslands in the presence of fire into areas predicted to contain savanna/shrublands in the absence of fire.

Human activities can also alter these boundaries either directly or indirectly. The term desertification is applied to the creation of deserts out of dry grasslands and shrublands through human disturbance. The climatic extremes in arid and semiarid systems allow plants only a tenuous existence. Ecosystems develop very slowly and are easily disrupted. Once disturbed, recovery may be slow or may not occur at all. Overgrazing and fuel wood harvesting are two of the main causes of desertification. Alteration of the Earth's atmosphere through increases in carbon dioxide concentration may increase the average temperature at the Earth's surface, driving gradual shifts northward in many vegetation types (Chapter 26), and direct conversion of natural ecosystems to agriculture has been an important form of ecosystem modification for centuries (or millennia; see Chapter 23).

CORRELATIONS BETWEEN CLIMATE AND ECOSYSTEM FUNCTION

Some fairly clear and not-too-surprising trends emerge from this brief survey of terrestrial vegetation and soils. The first is that total plant growth, expressed as primary production, is strongly related to temperature and moisture. These two variables are combined in AET. Figure 2.6 shows a relationship between AET and total above-ground production for several sites around the world. Total above-ground biomass in mature or relatively undis-

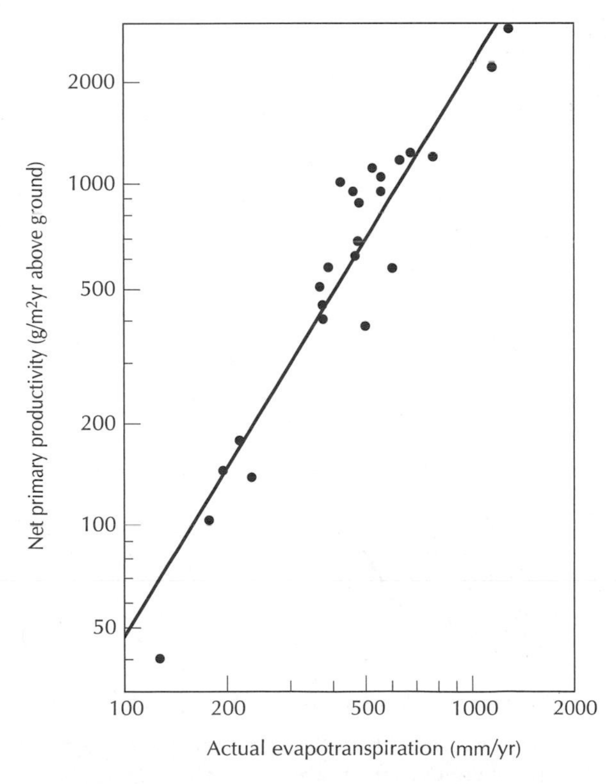

FIGURE 2.6 Net primary productivity of terrestrial ecosystems in relation to actual evapotranspiration. (Rosenzweig 1968, cited in Whittaker 1975)

turbed terrestrial ecosystems is also closely related to both AET and production (Figure 2.7). Any disturbed or young systems will have less biomass than their mature counterparts.

Simple physiological models, like the one used to produce Figure 2.3, also do well in predicting global patterns of ecosystem production using more complex relationships between climate and vegetation response (Figure 2.8*a*). An even simpler class of models defines a single parameter that expresses the efficiency of conversion of sunlight into biomass. This efficiency factor changes between vegetation types depending on climate and plant factors but can be defined either from field observations or from more complex models. One such model uses satellite remote sensing data on the fraction of incoming solar radiation absorbed by ecosystems on the ground to predict that efficiency. This approach assumes that the more leaf mass or the more chlorophyll present in a canopy (chlorophyll is the pigment that converts sunlight to plant-available energy), the more efficiently sunlight is utilized. This model has been used to derive a map of predicted total biomass production for the world (Figure 2.8*b*). We say more about remote sensing in Chapter 5.

Soil profiles show the effects of the relative intensity of the four major soil-forming processes. Laterization becomes less important with cooler temperatures. Podzolization becomes less important either with warmer temperatures or with decreasing precipitation. Lessivage/melanization increases in importance under near-neutral soil pH and where soil and vegetation

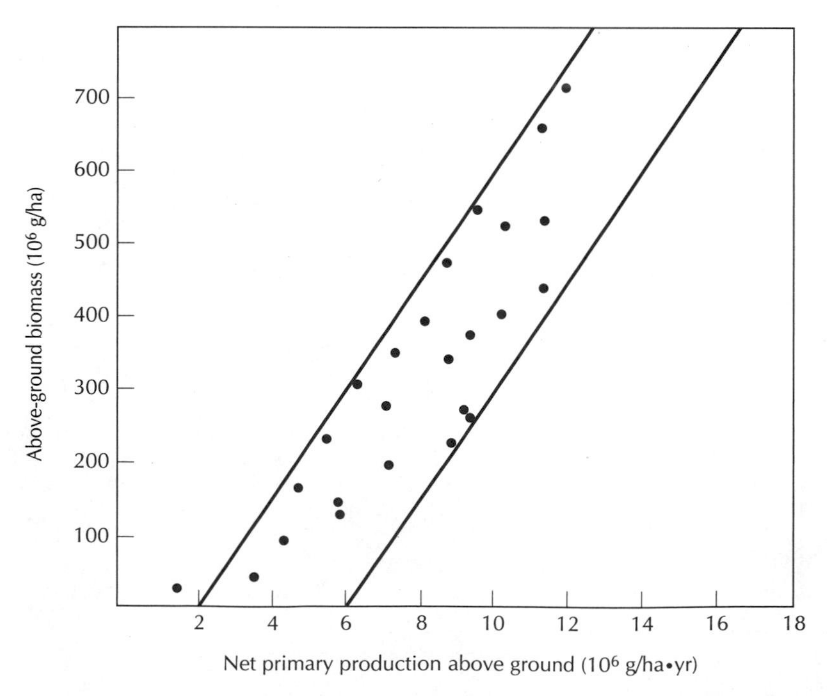

FIGURE 2.7 Above-ground biomass of terrestrial ecosystems as a function of above-ground net primary production. (After Whittaker et al. 1974)

allow for large and active populations of soil-mixing animals. Calcification/salinization is important where the potential for evaporation greatly exceeds precipitation and soil-water leaching is minimal. Surface organic accumulations tend to increase with decreasing temperature, being a minor component of total soil organic matter in the tropical rain forest and representing much of the active soil profile in the tundra Histosol.

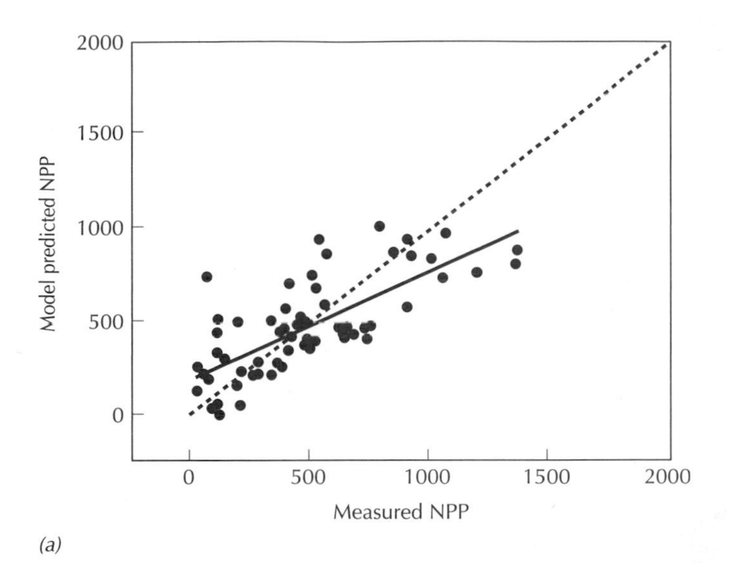

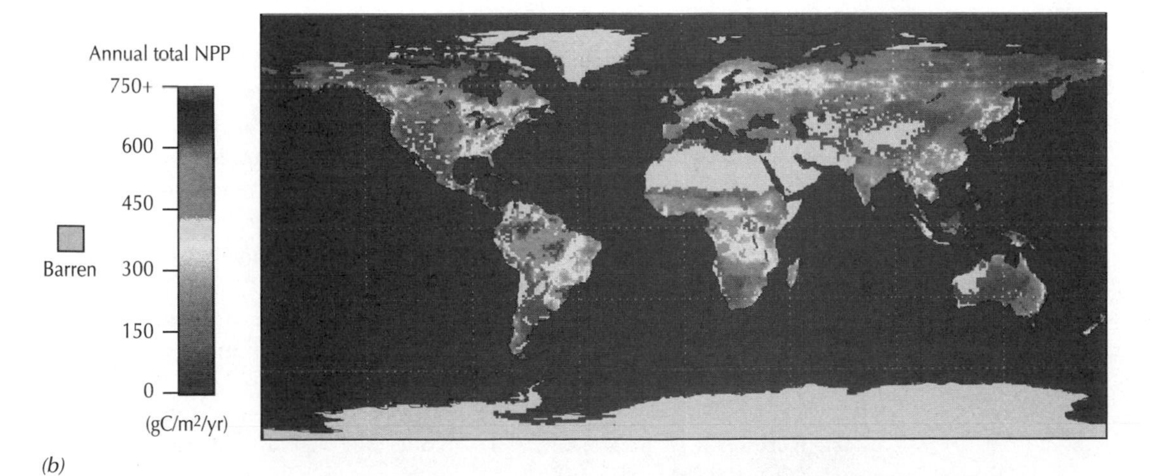

FIGURE 2.8 Predictions of biomass production for ecosystems of the world. (*a*) A comparison of observed production values with predictions from the BIOME3 model. (Haxeltine and Prentice 1996) (*b*) A global map of net primary production derived using a value for the efficiency of conversion of sunlight to chemical energy by plants based on remote sensing data. (Waring and Running 1998) **See plate in color section.**

VARIATION WITHIN LARGE CLIMATIC REGIONS

As we stated at the beginning of this chapter, this is a very oversimplified view of the interactions between climate, plants, and soils. Exceptions to every generalization presented can be found. Spodosols can be found in the tropics. Shrublands can be found in the heart of the deciduous forest. Sites with very low productivity occur in tropical rain forest regions. A closer examination of Figures 2.2 and 2.6 shows significant variation in both soil structure and production for a given climatic condition. For example, production ranges from 300 to 1000 $g \cdot m^{-2} \cdot year^{-1}$ at 500 mm AET in Figure 2.6. Temperate forest soils show very different structures, ranging from Spodosols to Alfisols, depending on soil texture, bedrock geology, species present, and other factors. In Parts II and III, we discuss the extent to which the interaction of soil, plants, consumers, decomposers, and disturbance can alter ecosystem structure and function within the limits established by climate.

REFERENCES CITED

Bliss, L. C., G. M. Courtin, D. L. Pattie, R. R. Riewe, D. W. A. Whitfield, and P. Widden. 1973. Arctic tundra ecosystems. *Annual Review of Ecology and Systematics* 4:359–399.

Bokhari, U. G., and J. S. Singh. 1975. Standing state and cycling of nitrogen in soil-vegetation components of prairie ecosystems. *Annals of Botany* 39:27–285.

Emanuel, W. R., H. H. Shugart, and M. P. Stevenson. 1985. Climatic change and the broad-scale distribution of terrestrial ecosystem complexes. *Climate Change* 7:29–43.

Gray, J. T., and W. H. Schlesinger. 1981. Biomass, production and litterfall in the coastal sage scrub of Southern California. *American Journal of Botany* 68:24–33.

Haxeltine, A., and I. C. Prentice. 1996. BIOME3: An equilibrium terrestrial biosphere model based on ecophysiological constraints, resource availability, and competition among plant functional types. *Global Biogeochemical Cycles* 10:693–709.

Murphy, P. G., and A. E. Lugo. 1986. Ecology of dry tropical forests. *Annual Review of Ecology and Systematics* 17:67–88.

Nadelhoffer, K. J., J. D. Aber, and J. M. Melillo. 1985. Fine root production in relation to total net primary production along a nitrogen availability gradient in temperate forests: A new hypothesis. *Ecology* 66:1377–1390.

Neilson, R. P. 1995. A model for predicting continental-scale vegetation distribution and water balance. *Ecological Applications* 5:362–385.

Pastor, J., J. D. Aber, C. A. McClaugherty, and J. M. Melillo. 1984. Above-ground production and N and P cycling along a nitrogen mineralization gradient on Blackhawk Island, Wisconsin. *Ecology* 65:256–268.

Sims, P. L., J. S. Singh, and W. K. Lauenroth. 1978. The structure and function of ten western North American grasslands. I. Abiotic and vegetational characteristics. *Journal of Ecology* 66:251–285.

Sinclair, A. R. E., and M. Norton-Grifiths (eds.). 1979. *Serengeti: Dynamics of an Ecosystem.* The University of Chicago Press, Chicago and London.

Van Cleve, K., L. Oliver, and R. Schlentner. 1983. Productivity and nutrient cycling in taiga forest ecosystems. *Canadian Journal of Forest Research* 13:747–766.

Vorosmarty, C. J., B. Moore III, A. L. Grace, M. P. Gildea, J. M. Melillo, B. J. Peterson, E. B. Rastetter, and P. A. Steudler. 1989. Continental scale models of water balance and fluvial transport: An application to South America. *Global Biogeochemical Cycles* 3:241–265.

Vitousek, P. M., and R. L. Sanford, Jr. 1986. Nutrient cycling in moist tropical forests. *Annual Review of Ecology and Systematics* 17:137–168.

Waring, R. H., and S. W. Running. 1998. *Forest Ecosystems: Analysis at Multiple Scales.* Academic Press, San Diego.

Whittaker, R. H. 1975. Production. In *Communities and Ecosystems.* MacMillan, New York.

Whittaker, R. H., F. H. Bormann, G. E. Likens, and T. G. Siccama. 1974. The Hubbard Brook Ecosystem Study: Forest biomass and production. *Ecological Monographs* 44:233–254.

Additional References

Buol, S. W., F. D. Hole, and R. J. McCracken. 1980. *Soil Genesis and Classification.* Iowa State University Press, Ames, IA.

Jenny, H. 1980. *The Soil Resource.* Springer-Verlag, New York.

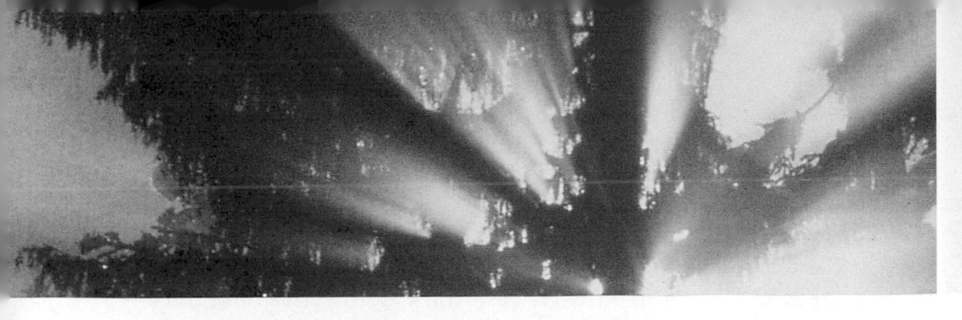

3

MEASUREMENT OF ECOSYSTEM FUNCTION I

THE CARBON BALANCE

INTRODUCTION

We have now described ecosystems in terms of their common functional components (Chapter 1) and the tremendous diversity of structure and function that they can exhibit (Chapter 2). It may be difficult to imagine just how the kind of information presented in Figure 2.2 is obtained. Just how do you go about measuring biomass or production or nitrogen cycling in something as large as an ecosystem? How might inputs and outputs of elements or water be measured in something as difficult to enclose or define as a forest stand? For the ecosystem concept to be useful, methods must be developed for measuring the structure and function of these systems. In this chapter and the next, we present several classic studies demonstrating different approaches to the measurement of whole-ecosystem function. These two chapters should help to convey something about the nature of ecosystem studies and to visualize ecosystems the way a research scientist sees them.

In this chapter, several methods are introduced for measuring the overall carbon balance between terrestrial ecosystems and the atmosphere. Each method works from a slightly different set of assumptions and from one of two conceptual models of the system. Both model and method are presented simultaneously.

THE CARBON BALANCE OF TERRESTRIAL ECOSYSTEMS

The exchange of carbon as gaseous carbon dioxide (CO_2) between ecosystems and the atmosphere reflects the balance between photosynthesis by plants and respiration by plants, animals, and microbes. This balance, called net ecosystem production (NEP) or net ecosystem exchange (NEE), has become a critical characteristic of terrestrial ecosystems because of the increase in CO_2 concentration in the Earth's atmosphere resulting from the combustion of fossil fuels and the clearing of tropical forests (Chapter 26). Some controversial analyses predict that temperate forests in particular are important sites for storage of added CO_2, reducing the impact of human activities on the atmosphere. Verifying these predictions by direct measurement and understanding how, and for how long, this increased storage might continue is a crucial scientific issue with important implications for environmental policy.

Two basic approaches to calculating the total carbon balance over an ecosystem are presented. Each begins with a different conceptual model of the system. The method employed in each case must define the boundary separating the terrestrial ecosystem from the atmosphere and capture whatever internal parts of the system are to be measured separately. Ideally, the method would allow for accurate measurements while minimizing the effect of the method on the system. In other words, the method should not alter the function of the system significantly.

Approach 1. Whole-System Balances

If a method is used that measures only the net CO_2 exchange between the ecosystem and the atmosphere, then the conceptual model is extremely simple because we do not need to know the fate of carbon within the system. Using Figure 1.4*b*, reproduced here as Figure 3.1*a*, we need only know the net flux of CO_2 across the boundary separating the ecosystem and the atmosphere. We need not know the total flow in either direction, only the balance between them. Thus the conceptual model of system function is reduced to Figure 3.1*b*.

The Giant Cylinder – A method that matches this conceptual diagram (Figure 3.1*b*) most closely is simply enclosing the system in question, pumping air

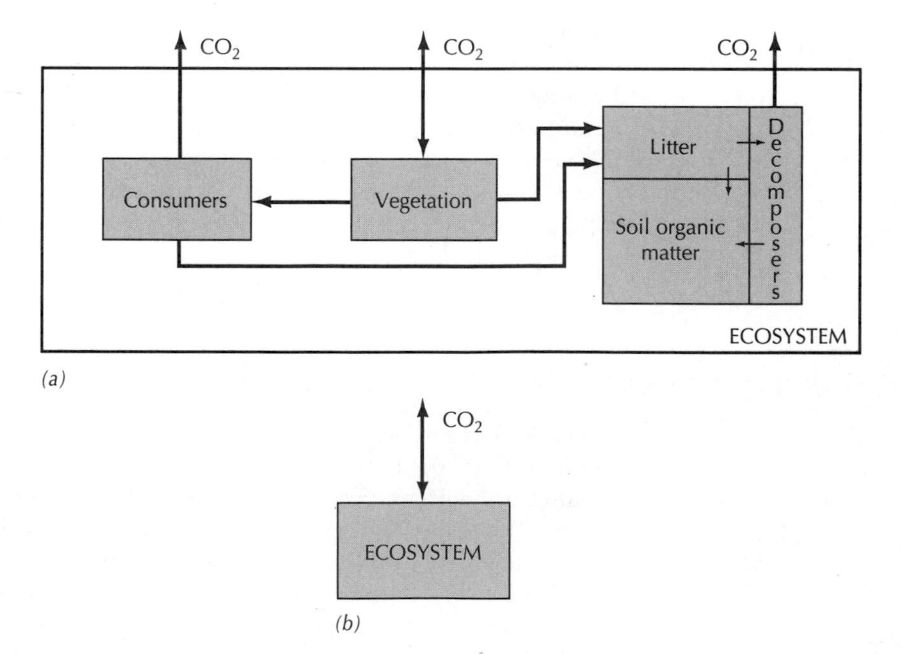

FIGURE 3.1 Conceptual models for the study of carbon balances in terrestrial ecosystems (*a*) as developed in Chapter 1, and (*b*) as required for the measurement of total carbon balances only.

through it, and measuring the changes in CO_2 concentration between input and output air. The most spectacular example of this was also one of the earliest attempts to measure whole-ecosystem carbon balances in a tropical rain forest and was carried out in the El Verde Forest of Puerto Rico (Figure 3.2). An entire stand (213 m^2) was enclosed in a giant cylinder of plastic over 20 m tall. A fan was installed at the base of the cylinder to draw air down through it. A tall (21.5 m) instrument tower was constructed in the center to carry devices for measuring CO_2, moisture, and other parameters of incoming air. The same measurements were made on the air leaving the fan, and differences were ascribed to photosynthesis, respiration, and transpiration by the enclosed vegetation and soil. Although the cylinder did not extend down into the soil, CO_2 generated within the soil would diffuse into the atmosphere within the cylinder and would be measured at the fan.

The CO_2 balance over the course of a single day is shown in Figure 3.3. Note that a net carbon gain by the system within the cylinder is recorded as a negative difference between the intake air and the air at the fan. When the air passing through the cylinder loses CO_2, the ecosystem gains carbon. The mean change in concentration between intake and outlet air for this day is +3.33 parts per million CO_2. Concentration values can be multiplied by the rate of air movement through the fan to calculate the total CO_2 exchange for the day. Divided by the land area enclosed (213 m^2), this gives the rate of CO_2 taken up or given off per square meter of ground area per day. For the day

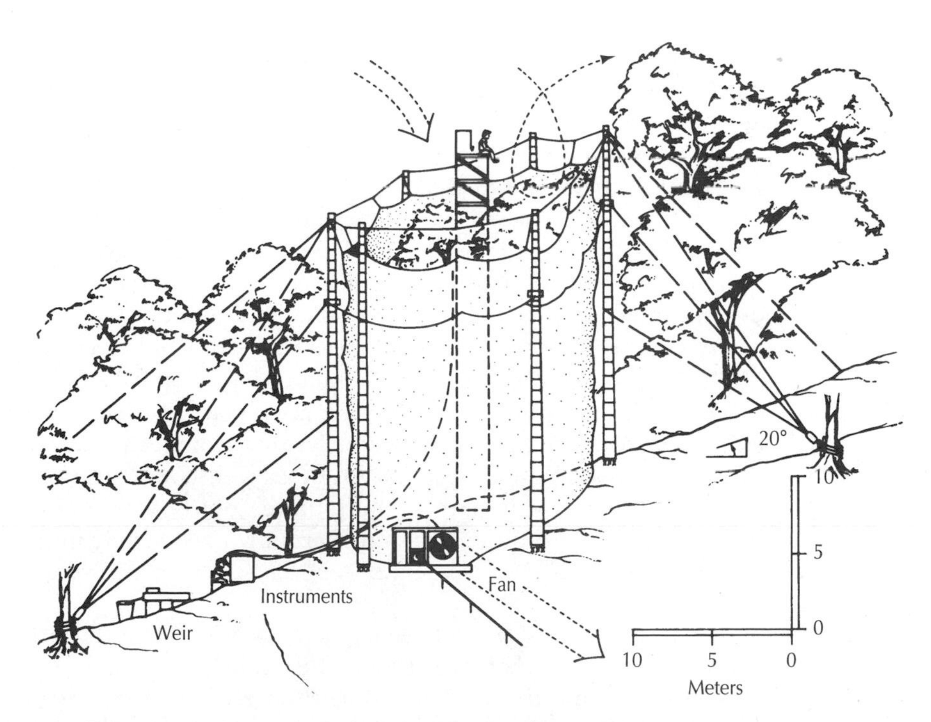

FIGURE 3.2 The "giant cylinder" experiment with air exchange and measurement system for determining total carbon balances. (Odum and Jordan 1972)

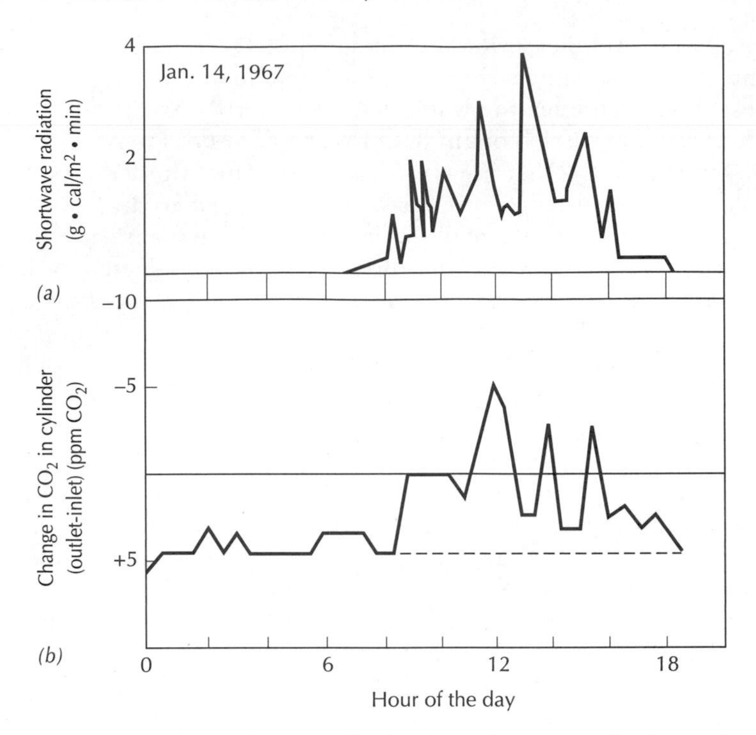

FIGURE 3.3 Data for a single day from the giant cylinder study. (*a*) Net radiation. (*b*) Carbon balance. (Odum and Jordan 1972)

recorded in Figure 3.3, the net carbon loss from the system was 10.2 g · m^{-2} CO_2. To obtain a balance for a full year, measurements can be taken daily for the full period, or a relationship can be obtained between the CO_2 balance and such factors as temperature, soil moisture, and time of year, and a mathematical model developed to predict rates of CO_2 exchange for times when measurements are not made.

Two characteristics of ecosystem studies are apparent here. The first is the conversion of data from whatever original form (in this case, CO_2 concentration and air flow rates) to rates expressed in units of mass per unit land area per unit time. The second is that a tremendous amount of time and effort and a very large number of individual measurements can be required to obtain a single number, in this case, the CO_2 balance of a tropical rain forest.

It has been argued that enclosing all or part of the ecosystem in plastic containers alters the system enough to cause changes in function. Temperature, wind speed, humidity, and CO_2 or oxygen concentrations could be directly affected. While researchers have put considerable effort into minimizing these differences, some certainly exist. Their importance is still subject to debate.

Aerodynamic Analysis of the Boundary Layer – The second method for measuring total net CO_2 balances avoids the problems of altered conditions. Instead of enclosing the system, this method requires frequent measurements of the change in CO_2 concentration with height above the canopy and combines these with an estimate of the rate of air movement into or out of the canopy. Multiplying these two together gives an estimate of CO_2 flux. A study of this type has been carried out over a shortgrass prairie area in Saskatchewan. The instrumentation for the carbon balance measurement is a sampling mast 6.4 m tall, with several CO_2 analyzers at various heights, combined with a second tower for measuring gradients in wind speed, humidity, and temperature (Figure 3.4). From these micrometeorological data, the average rate of turbulent air flow up from the canopy and away into the atmosphere can be estimated if certain crucial assumptions are made. Multiplying this by the gradient in CO_2 concentrations allows for the calculation of total CO_2 flux. By this method, seasonal changes in carbon gain during the daytime and loss at night were estimated for this site (Figure 3.5).

The Eddy Covariance Method – A newer method, called eddy covariance, has revolutionized the measurement of whole-ecosystem carbon balances. The wide application of this technique promises to increase the information available on NEE significantly.

Conceptually, the method is very simple (Figure 3.6). Movement of air through a canopy occurs as a rapidly shifting set of eddies, small packets of

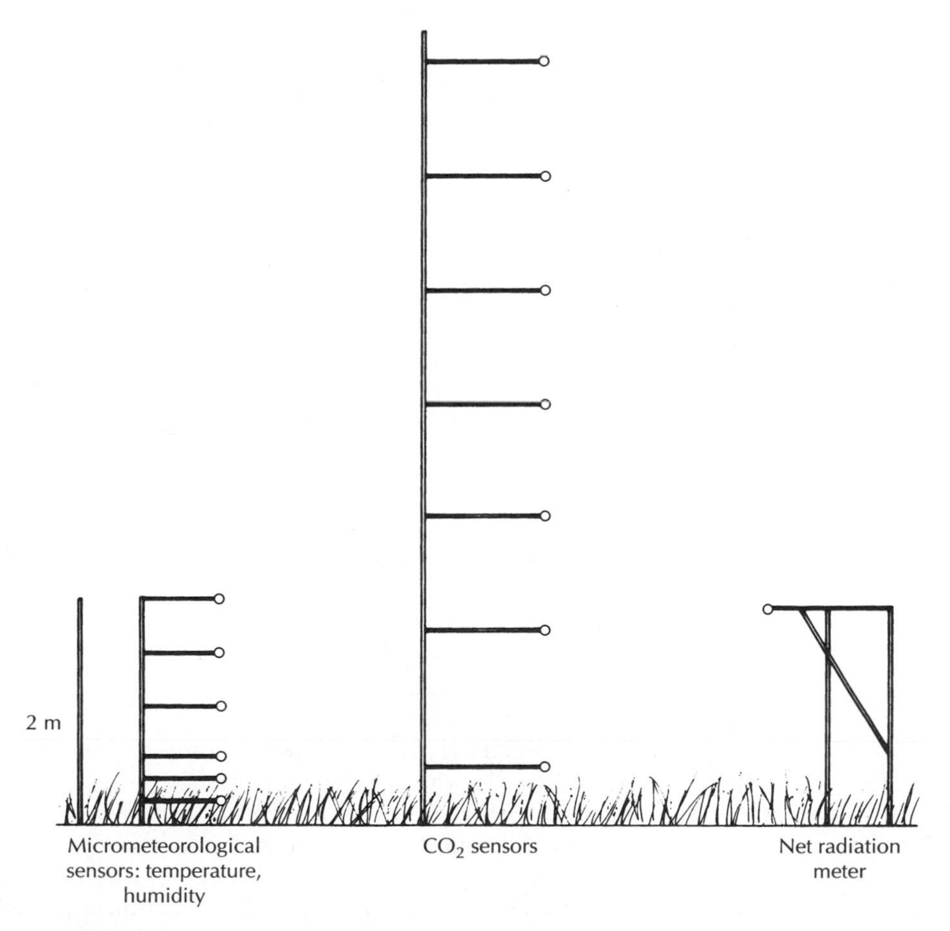

FIGURE 3.4 Sampling towers used for the aerodynamic determination of net carbon balance over a prairie ecosystem. (Ripley and Saugier 1974)

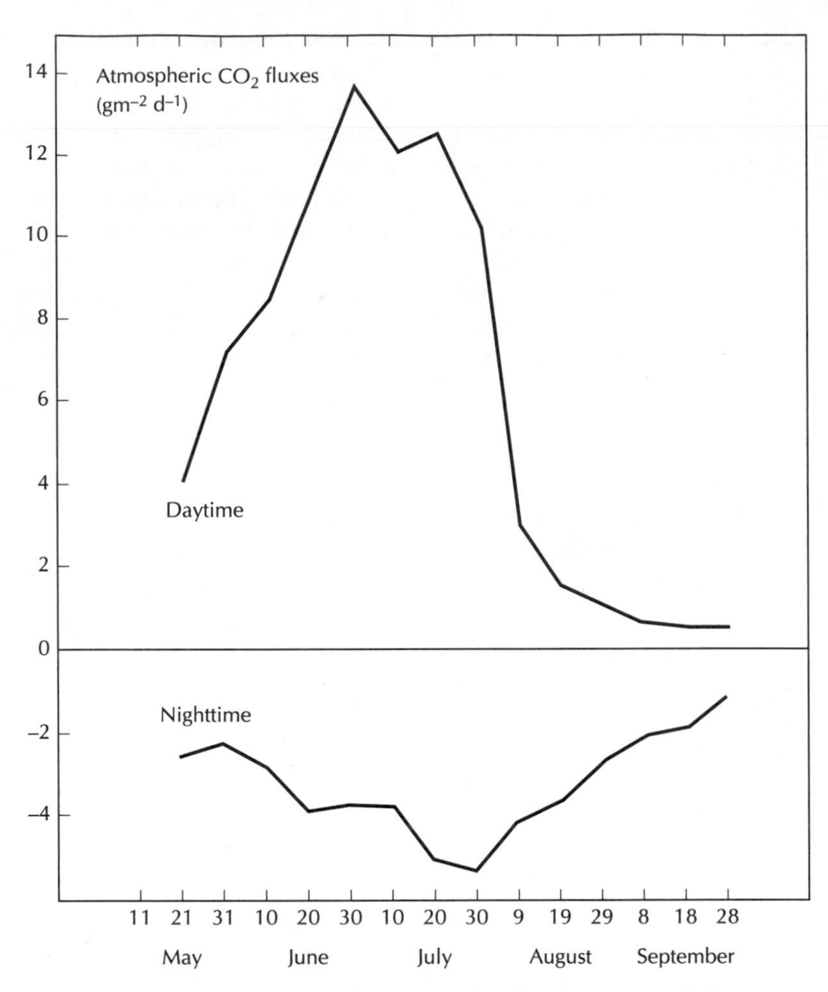

FIGURE 3.5 Seasonal variation in daytime and nighttime CO_2 balances over a prairie ecosystem. (Ripley and Saugier 1974)

turbulent air moving in response to the interaction of wind currents above the canopy with the rough structure presented by foliage, branches, etc. The eddy covariance method attempts to capture this rapid, fine-scale movement and measure the CO_2 concentration of the air moving in different directions. If the air moving up and out of the canopy has a lower CO_2 concentration than the air moving down and in, then a net carbon accumulation within the system is occurring. Effectively, this method attempts a direct measurement of the very rapid exchange of air between canopy and atmosphere that is modeled or predicted by the aerodynamic method.

The measurements required to realize this simple concept are complex and technologically demanding. One of the key components of the system is a sonic anemometer, which measures wind speed in three dimensions simultaneously. This is placed above the canopy. A plastic tube is placed at the same height to capture samples of air that are then drawn down to ground level for the analysis of CO_2 concentration. These measurements are repeated as often

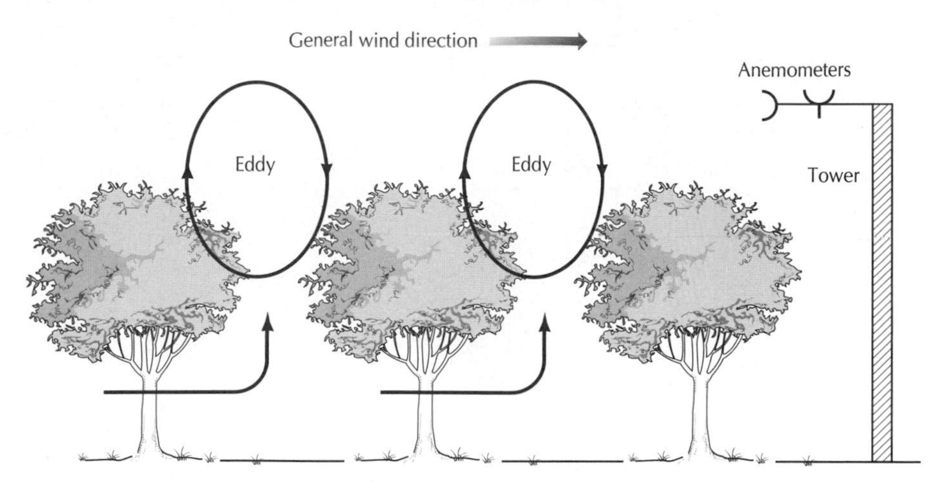

FIGURE 3.6 Eddy covariance approach to the measurement of CO_2 balances over unenclosed ecosystems. Friction between directional air movement in the atmosphere and foliage, branches, and stems in the canopy cause irregular, turbulent flow (eddies) of air through the canopy. These eddies capture air from below canopy level as well and transfer CO_2 contained there to the atmosphere, moving it past the tower. The sonic anemometer and inlets for gas measurements sit atop a tower that extends several meters above the general canopy level. Measurements of wind speed in three directions and simultaneous measurements of CO_2 concentration are made six times per second. (After Munger et al. 2000)

as six times per second. Matching a CO_2 measurement with the right anemometer reading, and so linking the right concentration with right direction and speed of air movement, is the major challenge. In addition, measurements of CO_2 storage within the canopy are made using a series of intake tubes distributed down through the canopy. Changes in storage are added to flux across the top of the canopy to determine instantaneous CO_2 production or consumption.

When the technical challenges are met, this method produces large amounts of data and allows for a detailed examination of NEE. Long-term measurements using this method have been obtained in only a few locations, but international efforts are under way to expand the number of sites significantly and to standardize methods (see Chapter 26).

In one comparison of measurements, a temperate deciduous forest was found to have much higher rates of gross carbon exchange than a boreal pine forest, while both responded to changes in the amount of incoming sunlight (Figure 3.7). As we will see in Chapter 6, this is the same type of measurement that physiologists have made on individual leaves for decades. Now they can be made for whole ecosystems.

Continuous operation of eddy covariance towers provide very detailed estimates of total carbon exchange over the course of a year. In an old-growth spruce forest at the Quebec site, 4 years of data have demonstrated how variation in weather patterns from one year to the next alter NEE (Figure 3.8). In 1996 and 1997, total NEE was near zero. 1995 saw a large accumulation of carbon, while 1998 produced almost an equivalent loss. Over the entire 4 year

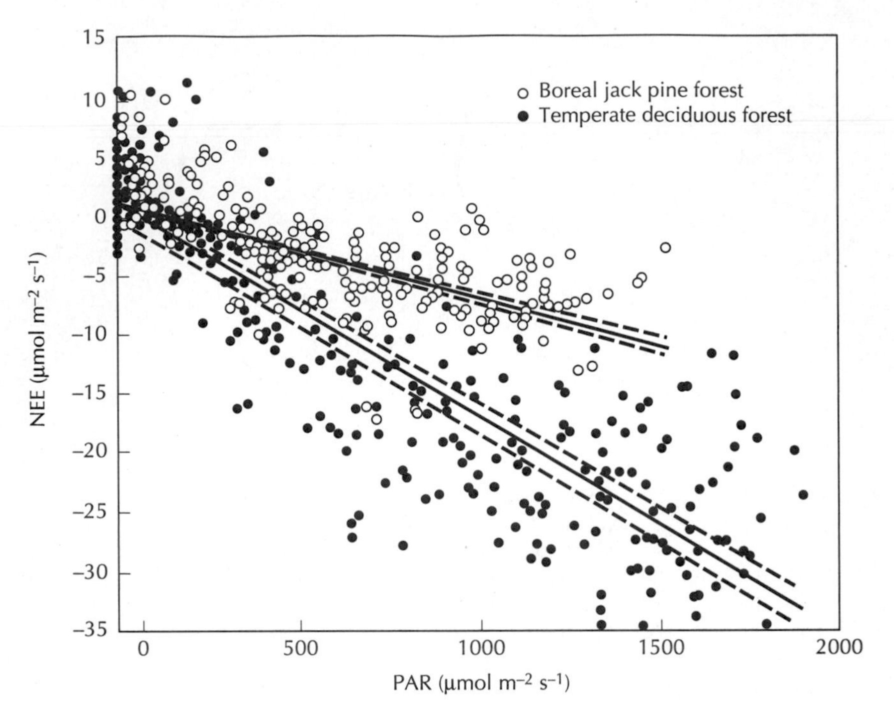

FIGURE 3.7 Comparison of relationship between net ecosystem exchange (measured in micromoles of CO_2 per square meter per second) and radiation intensity (measured as photosynthetically active radiation in micromoles of photons per square meter per second) for a temperate deciduous forest and a boreal pine forest. Data collected by eddy covariance. (Baldocchi and Vogel 1996)

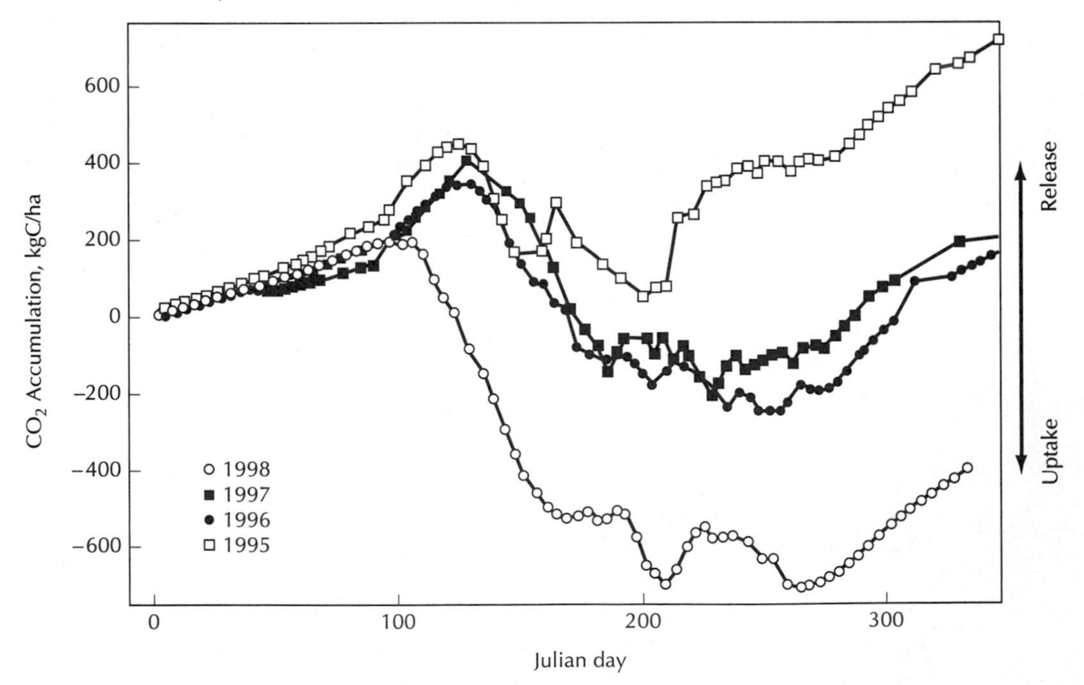

FIGURE 3.8 Eddy covariance data for 4 years of net ecosystem exchange data for a mature black spruce stand in Ontario, Canada. (Fan et al. 1995)

period, the entire stand around the tower gained only about 30 gC · m^{-2} · $year^{-1}$.

Estimating Gross Photosynthesis – A slight modification of the conceptual model used with whole-stand methods, and a separation of nighttime and daytime measurements, can yield another important ecosystem parameter, gross ecosystem exchange (GEE). Plant physiologists would call this gross photosynthesis. Respiration, or the production of CO_2 as a by-product of metabolism, occurs in plants, animals, and soils. These different sources cannot be separated by these whole-system methods. However, photosynthesis occurs only in plants. We can redefine the conceptual model in Figure 3.1*b* to separate these two functions (Figure 3.9; note that the respiration function overlaps the vegetation compartment). If we could add the sum of all of the respiration fluxes, then we could estimate the flux of CO_2 into the system, gross photosynthesis by plants, by using the equation:

$$\text{GEE} = \text{NEE} + \text{Respiration}$$

where respiration is the total for plants, soils, animals, etc.

Net ecosystem flux measurements made at night, in the absence of any sunlight, do not include photosynthesis, only respiration. By making many measurements at night, a relationship can be developed between this flux and soil temperature, the primary variable controlling respiration (Figure 3.10). This relationship is then used with measurements of soil temperature during the daytime to estimate total respiration for the system. Adding this to the measured net ecosystem exchange (NEE) during the day yields an estimate of GEE.

Measurements of GEE can be divided into periods with different conditions of sunlight, temperature, and humidity, and the effect of each of these can be determined. For the Quebec site, there was a very strong relationship

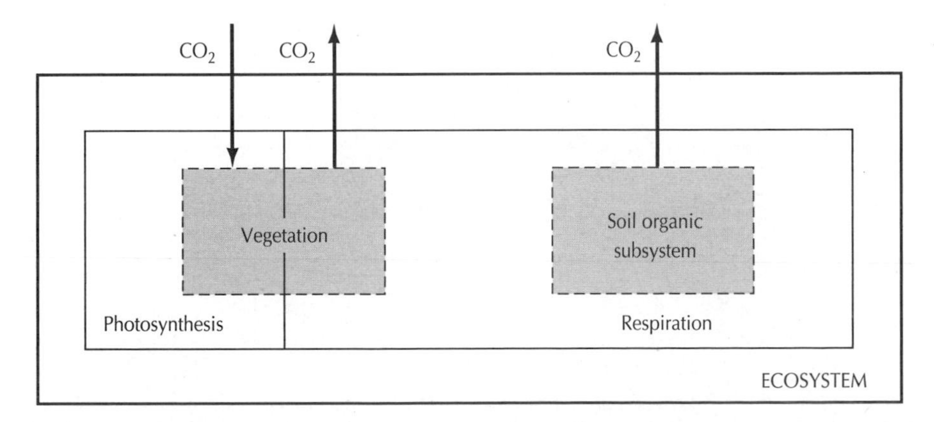

FIGURE 3.9 Conceptual model for the calculation of gross photosynthesis by plants and total respiration by plants, animals, and microbes using data from the giant cylinder.

with sunlight and a measurable response to humidity but, surprisingly, no response at all to temperature (Figure 3.11).

Approach 2. Small-Chamber Enclosures

The second approach to estimating NEE is to measure the most important fluxes within the system individually, and then sum them for the entire stand. The same approach as in the tropical rain forest example can be used, that of enclosing the system and measuring inputs and outputs, but the conceptual model is different, and the cylinders will be much smaller and more numerous. The model separates those parts of the system that are most important for total carbon flux. For a terrestrial ecosystem, this means measuring leaves, the non-leaf parts of the plants above ground, and soils (Figure 3.12).

You might ask why such a detailed approach would be taken to measuring just the single NEE value. It's a good question. In general, there is little value in adding complexity to a study if a simpler design will provide the same answer. In this case, there are two reasons. One is to act as a check on other methods. The small-chamber method was developed before eddy covariance and when questions still existed about the aerodynamic methods. Attempting measurements of the same flux with two very different methods is one way of testing the accuracy (or at least the consistency) of the different methods.

The second reason for the more complex design was to understand the contributions made to total ecosystem carbon balance by several different processes. NEE is the difference between gross photosynthesis (discussed earlier) and respiration by several different components. The small-chamber ap-

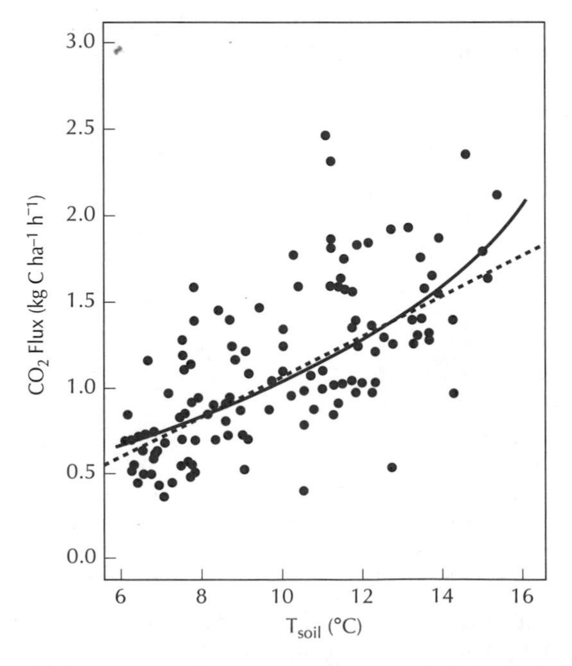

FIGURE 3.10 Relationship between soil temperature and total ecosystem respiration rate for a mature black spruce forest in Ontario, Canada. (Fan et al. 1995)

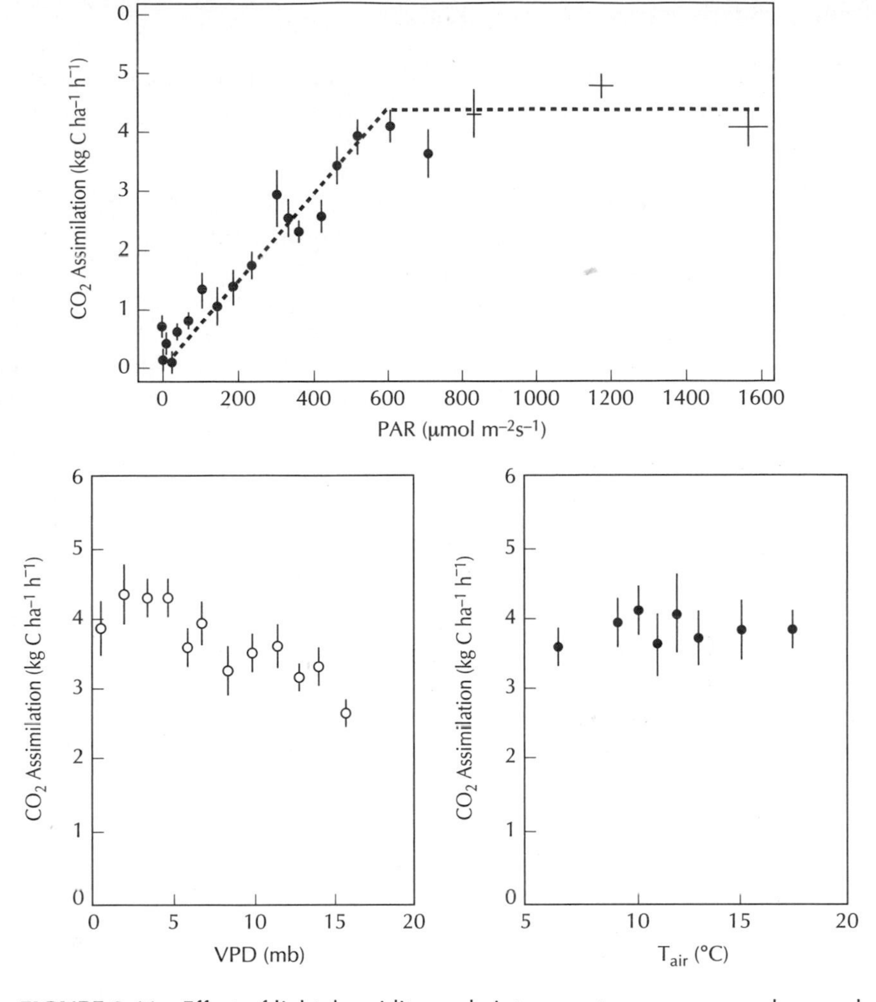

FIGURE 3.11 Effect of light, humidity, and air temperature on gross carbon exchange in a mature black spruce forest in Ontario, Canada (Fan et al. 1995). Light is measured as photosynthetic photon flux density (PPFD), and humidity as vapor pressure deficit (VPD; larger numbers indicate drier air).

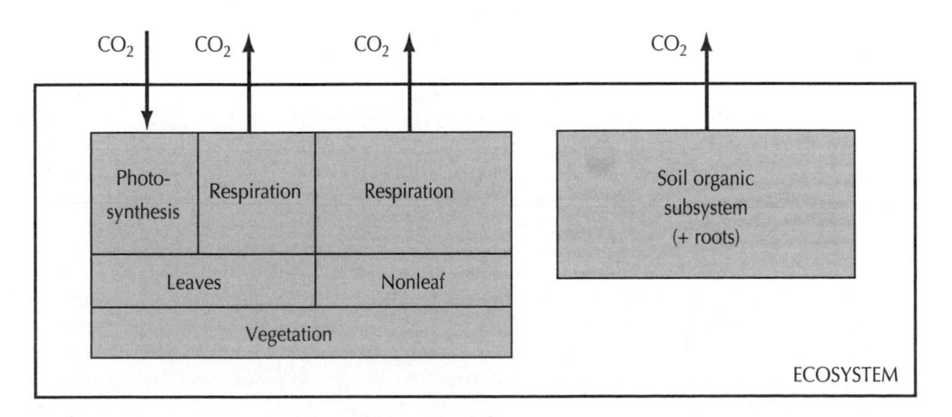

FIGURE 3.12 Conceptual model for measuring ecosystem carbon balances using the small-chamber-enclosure technique.

proach allows some apportioning of respiration among different components of the ecosystem. A standard hierarchy of terminology as applied to the fixation, transformation, and respiration of carbon is given in Table 3.1. These terms are used throughout this book.

One of the first large-scale attempts to apply the small-chamber method was carried out in an oak–pine forest on sandy soils at the Brookhaven National Laboratory on Long Island, New York (Figure 3.13). A problem encountered here that does not occur in the whole-system methods is that of selecting the portion of the system to be measured or sampled. If only a small portion of all the leaves in the stand are to be enclosed, care must be taken to measure a representative sample from each species, as well as the right proportion in full sun, partial shade, and deep shade. Similar problems occur for enclosing stems.

In soils, an additional problem is that plant roots and microbes cannot be physically separated without severely disturbing the soil system. Chambers placed over the soil to measure efflux of CO_2 actually include decomposer respiration plus plant root respiration. Thus, roots and the soil organic subsystem are measured together (Figure 3.13; Table 3.1). The difficulty of separating the functions of plants and microbes within soils is a recurring problem in ecosystem studies.

The Brookhaven carbon balance study involved a total of 75 cylinders, each connected by a maze of plastic tubing to a central, computer-controlled sample laboratory that obtained CO_2 exchange data from each cylinder once

TABLE 3.1 Process Values from Brookhaven Forest (Carbon, g/m²/yr)*

Botkin et al. 1970, Woodwell and Botkin 1970.

Property	Process	Value
	Gross photosynthesis	5199 (sum of next two)
Minus	Leaf respiration	1524 (measured)
Equals	Net photosynthesis	3675 (measured)
Minus	Root and stem (nonleaf) respiration	2049 (stem only)
Equals	Net primary production	(1586)
Minus	Litterfall, herbivory, mortality	
Equals	Net biomass accumulation	
Minus	Heterotrophic respiration	(1724) (root and soils)
Equals	Net ecosystem production or net ecosystem exchange	−138

*Photosynthesis (daytime measures of carbon balance on chambers holding leaves) is added to leaf respiration (leaf chambers in the dark) to estimate gross photosynthesis. Only the stem contribution to nonleaf respiration can be measured directly because root and soil respiration cannot be separated, thus the estimate of net primary production is actually too high. This is offset by the inclusion of root respiration with soil values, such that net ecosystem production can be estimated accurately. Separating root and soil function is a constant challenge in ecosystem studies and greatly complicates estimates of basic system parameters such as net primary production (we will return to this in Chapters 10 and 11).

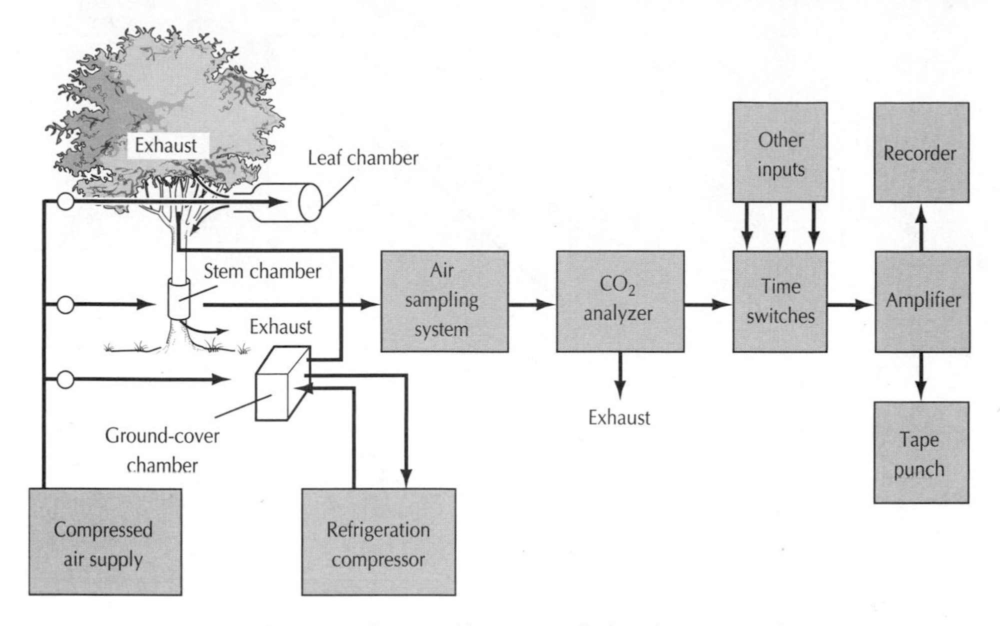

FIGURE 3.13 The Brookhaven small-chamber system for monitoring CO_2 exchange rates. (Woodwell and Botkin 1970)

every 25 minutes. Computations are identical to those for the large cylinder but are carried out separately for each chamber and expressed in per gram leaf weight, or stem or soil surface area. Mean values for each cylinder type (e.g., oak leaves or pine stem) are then obtained and multiplied by the total weight of leaves or surface area of stem in the stand.

Total gross photosynthesis in this forest, assuming daytime respiration rates equal to dark respiration, was 5199 g · m^{-2} · $year^{-1}$. Chambers placed over both stem and soil components always show increases in CO_2 concentration as respiration greatly predominates over photosynthesis. The annual pattern of respiration per square meter of bark surface reflects changes in temperature (Figure 3.14). Soil respiration shows similar trends.

Comparisons Between Ecosystems

Although different methods were used in each of the four systems described in this chapter, comparable data were obtained for both gross photosynthesis and total ecosystem CO_2 balance (Table 3.2). In accordance with our idealized discussion in Chapter 2, the tropical rain forest shows the largest gross CO_2 fixation as a result of both faster rates of fixation and a longer growing season. The temperate forest at Brookhaven is second in both daily and total gross photosynthesis, followed by the grassland, and boreal forest. It is interesting that the boreal forest has a much lower total value for gross photosynthesis than the grassland, even though the trees are active for a longer period of time because of the much lower inherent rate of photosynthesis in spruce

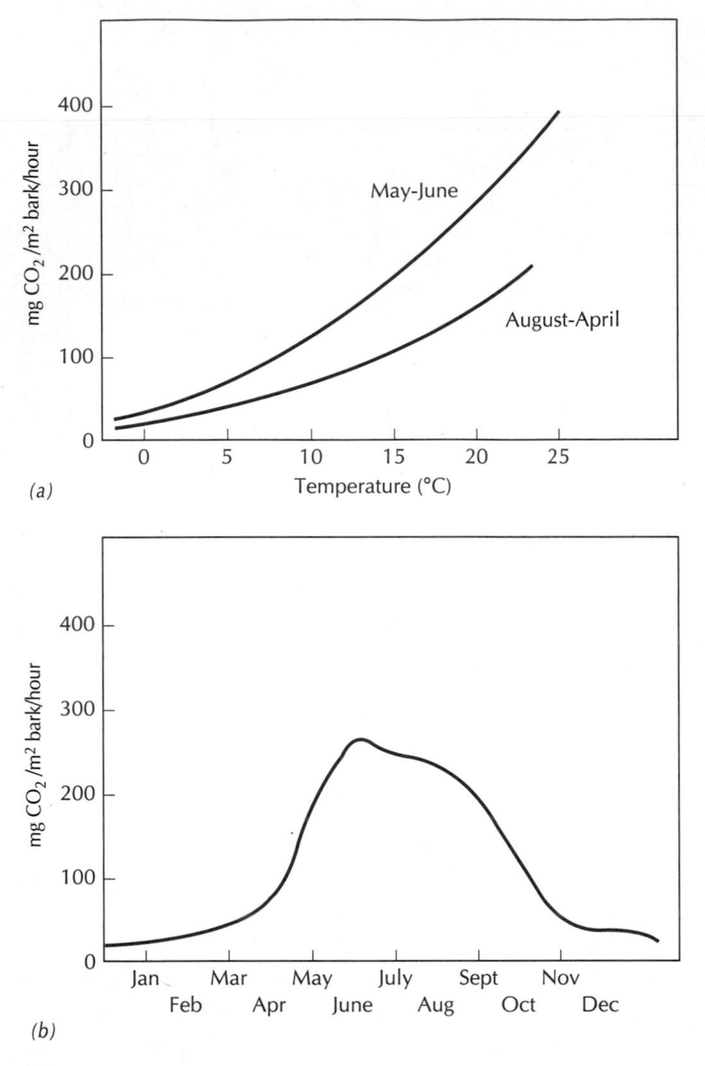

FIGURE 3.14 Respiration rates by tree stems in the Brookhaven study as a function of (*a*) air temperature and (*b*) seasonal changes. (Woodwell and Botkin 1970)

trees. It is also interesting that the evergreen trees in the boreal forest can extend the growing season so that it is longer than either the deciduous forest or the grassland despite much colder temperatures. This is because the evergreen needles are in place and ready to function as soon as daytime temperatures exceed 0°C, while the deciduous forest and grassland must create the entire leaf canopy from scratch each year.

Total carbon balances are quite different from gross photosynthesis. Respiration rates are also higher at higher temperatures and can balance high rates of CO_2 fixation in the tropics. All three of the forest sites show near zero net accumulation of carbon (actually, a zero accumulation was assumed at El

TABLE 3.2 Gross Photosynthesis and Total Carbon Balance for Three Terrestrial Ecosystems

System	Gross Photosynthesis: Mean Rate for Growing Season ($g \cdot m^{-2} \cdot day^{-1}$)	Length of Season ($days \cdot year^{-1}$)	Total ($g \cdot m^{-2} \cdot year^{-1}$)	Annual Carbon Balance ($g \cdot m^{-2} \cdot year^{-1}$)
Tropical rain forest	53	365	19,345	0
Temperate forest	29	180	5,199	138
Grassland	14	120	1,680	720*
Boreal forest	4.5	194	776	20

*8 mo only.

Verde in calculating gross photosynthesis). The grassland site shows a positive carbon balance, but data were collected only for the growing season (May–September). Respiration during the remainder of the year may have brought this balance much closer to zero.

It may at first appear contradictory that ecosystems with such different rates of CO_2 fixation should all show similar and near zero total CO_2 balances. However, this is a basic characteristic of mature (not recently disturbed) ecosystems. A constant, positive carbon balance would mean a continuous increase in total carbon within the system, approaching an infinite amount stored. This is impossible. Terrestrial ecosystems all tend toward a balance between CO_2 uptake and CO_2 evolution (such need not be the case for aquatic or wetland systems, where sediments or peat can accumulate for very long periods of time). All mature and undisturbed ecosystems should have net carbon balances near zero in the long term, although individual years can show gains or losses depending on differences in weather patterns. In contrast, disturbances, such as fire and clearing for agriculture, can release large amounts of carbon previously stored in plant biomass and soil organic matter.

An Experiment at the Ecosystem Level

The studies discussed so far in this chapter have made use of existing stands and the natural variation in climatic patterns to learn about the factors controlling gross and net carbon exchange. Not all questions can be answered by observation alone, however. For example, the rapidly changing concentration of CO_2 in the atmosphere leads us to wonder what effect this will have on terrestrial ecosystems. Natural variation in CO_2 over time is not enough to allow for measurements that would be relevant to an atmosphere with twice the current CO_2 (as is predicted to occur by the end of the current century). Approximate answers can be obtained by small-scale laboratory or microcosm experiments using seedlings in controlled conditions, but serious questions arise as to how these results relate to mature trees in the field encountering other forms of stress. On the other hand, running an experiment in the field in which CO_2 concentrations are increased without enclosing the system presents a formidable challenge.

Just such an experiment has been devised in the Duke Forest in North Carolina. A complex set of pumps and tubes, with constant feedback control providing instantaneous measurements of CO_2 concentration within the canopy, has been established in a large and relatively homogeneous stand of loblolly pine trees (Figure 3.15). Unfortunately, the small size of the areas involved rules out the use of eddy covariance or other techniques for measuring whole-stand carbon balances. Instead, a large number of studies are under way to examine the effects on plant and soil processes within the CO_2-enriched rings. At the time of this writing, the first results are beginning to appear, including a significant increase in tree diameter growth in years 2 and 3 of the experiment (Figure 3.16). Similar experiments are under way in other vegetation types as well.

FIGURE 3.15 Aerial view of the Free Air CO_2 Enrichment experiment in the Duke Forest, North Carolina. Rings are created by a series of large pipes through which CO_2 is pumped into the atmosphere within the ring. High-frequency sampling of wind speed and direction are used to determine which pipes should receive the CO_2 to maximize exposure within the rings and minimize the use of CO_2.

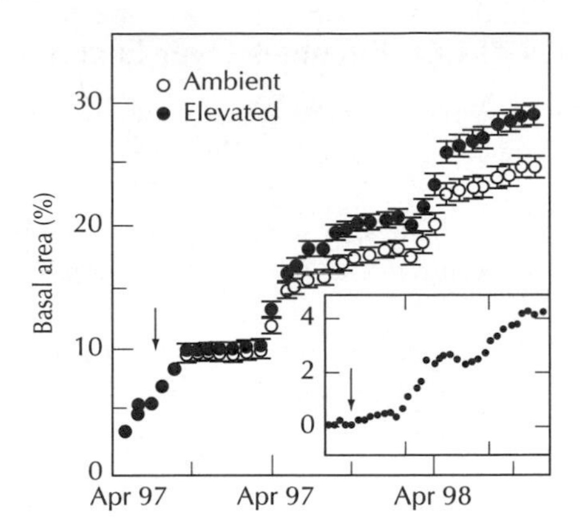

FIGURE 3.16 Response of loblolly pine forest in the Free Air CO_2 Enrichment experiment to increased CO_2 concentration, expressed as a percentage increase in basal area relative to initial volume. (The arrow marks the beginning of the treatment.) Note that the increase in tree growth only began to occur in the second year of treatment. The inset shows the absolute difference in basal area between treatments (DeLucia et al. 1999).

REFERENCES CITED

Baldocchi, D. D., and C. A. Vogel. 1996. Energy and CO_2 flux densities above and below a temperate broad-leaved forest and a boreal pine forest. *Tree Physiology* 16:5–16.

Botkin, D. B., G. M. Woodwell, and N. Tempel. 1970. Forest productivity estimated from carbon dioxide uptake. *Ecology* 51:1057–1060.

DeLucia, E. H., J. G. Hamilton, S. L. Naidu, R. B. Thomas, J. A. Andrews, A. Finzi, M. Lavine, R. Matamala, J. E. Mohan, G. R. Hendry, and W. H. Schlesinger. 1999. Net primary production of a forest ecosystem with experimental CO_2 enrichment. *Science* 284:1177–1179.

Fan, S. M., M. L. Goulden, J. W. Munger, B. C. Daube, P. S. Bakwin, S. C. Wofsy, J. S. Amthor, D. R. Fitzjarrald, K. E. Moore, and T. R. Moore. 1995. Environmental controls on the photosynthesis and respiration of a boreal lichen woodland: a growing season of whole-ecosystem exchange measurements by eddy correlation. *Oecologia* 102:443–452.

Munger, W., C. Barford, and S. Wofsy. 2000. Atmospheric exchanges. In Foster, D., and J. Aber (eds.), *Forest Landscape Dynamics in New England* (in preparation).

Odum, H. T., and C. Jordan. 1972. Carbon balance of the large cylinder. In Odum, H. T., and R. F. Pigeon (eds.), *A Tropical Rain Forest.* Atomic Energy Commission, Washington, DC.

Ripley, B. and E. A. Saugier. 1974. Microclimate and production of a native grassland: A micrometeorological study. *Oecologia Plantarum* 9:333–363.

Woodwell, G. M., and D. B. Botkin. 1970. Metabolism of terrestrial ecosystems by gas exchange techniques: The Brookhaven approach. In Reichle, D. (ed.), *Studies in Ecology.* Springer-Verlag, New York.

Additional References: Other Examples of CO_2 Balances Over Enclosed Systems

Curtis, P. S., B. G. Drake, and D. F. Whigham. 1989. Nitrogen and carbon dynamics in C3 and C4 estuarine marsh plants grown under elevated CO_2 in situ. *Oecologia* 78:297–301.

Hillbert, D. W., and W. C. Oechel. 1987. Response of tussock tundra to elevated carbon dioxide regimes: Analysis of ecosystem CO_2 flux through non-linear modeling. *Oecologia* 72:466–472.

Tissue, D. T., and W. C. Oechel. 1987. Response of *Eriophorum vaginatum* to elevated CO_2 and temperature in the Alaskan tussock tundra. *Ecology* 68:401–410.

4

MEASUREMENT OF ECOSYSTEM FUNCTION II

NUTRIENT AND WATER BALANCES

INTRODUCTION

The carbon balance discussed in the previous chapter is only one part of the "metabolism" of ecosystems. Plants, animals, and microbes also require nutrients and water in order to function. The input and output balances for water and nutrients in terrestrial ecosystems are linked to crucial global cycles that affect climate and environmental quality; they are equally as important as the carbon cycle. Unlike carbon, inputs and outputs of mineral nutrients and other chemicals do not occur primarily in gaseous form. The movement of water through ecosystems provides a second important pathway, as both precipitation and streamflow can carry compounds either crucial or detrimental to life. Measuring the flow of water through ecosystems, and its nutrient or chemical content, has become a major activity in ecosystem studies.

The water balance of a terrestrial ecosystem is important in its own right. The amount of water available for transpiration plays a large role in determining the productivity of the plants in the system. The water that is not used by plants and that finds its way to groundwater or to streams becomes the source for human use. Some of the earliest ecosystem-level studies and experiments were carried out by hydrologists attempting to divert water from transpiration to streamflow. These experiments were conducted on whole watersheds so that the streams draining these watersheds could be used to measure total liquid water output.

The purpose of this chapter is to present an introduction to the watershed-ecosystem concept, to discuss its advantages and limitations, and to present important findings obtained by this and subsequent nonwatershed nutrient-balance methods. The correlation between conceptual model and methods again is presented, as in Chapter 3. Differences between different ecosystems also are presented, although, for watershed studies, the spectrum is limited to areas with appreciable streamflow (mainly forested regions). One of the most powerful applications of the watershed technique is in assessing the integrated response of all of the components of the ecosystem to human use, such as forest harvesting. This response can be "read out" as the change in the quantity and chemical quality of the water draining from the system. Results of this kind of research are directly applicable to questions of both forest productivity and water quality management.

NUTRIENT AND WATER BALANCES

An ecosystem's CO_2 balance reflects the difference between fixation through photosynthesis and respiration by plants, animals, and decomposers. The only major storage for carbon in the ecosystem is in some form of organic matter (actually, a great deal of carbon can be found in limestone-type rocks within a system, but this carbon is part of a very different cycle; see Chapter 9). In contrast, required nutrients are found in a wide array of organic and inorganic forms in soils. Inorganic forms vary widely in availability to plants, while concentrations in organic forms vary widely with the type of material. The increased complexity of transformations within the system is reflected in the differences between Figure 1.4*b* and *c* (Figure 1.4*c* is repeated here as Figure 4.1). Nutrient balances over ecosystems need not parallel those for carbon.

Water balances can also be very different for different ecosystems. The importance of precipitation and evaporative demand (or potential evapotranspiration) in determining the amount of actual evapotranspiration was mentioned in Chapter 2. Of particular importance for watershed-ecosystem studies is the division of the remaining water between surface flow (streams), and deep seepage (to groundwater). Storage within the system can also occur at different depths and changes with time.

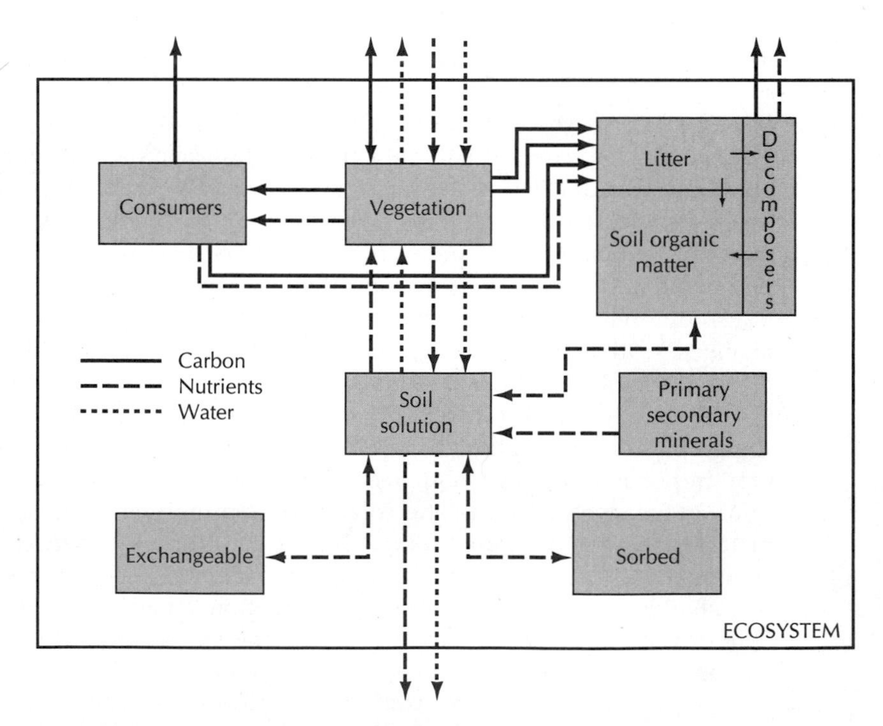

FIGURE 4.1 Conceptual model of nutrient dynamics in terrestrial ecosystems.

These complexities are treated at length in Part II. They are mentioned here to introduce the idea that different ecosystems may show very different input–output balances and responses to disturbance even if total primary production or carbon balances are similar. At the watershed-ecosystem level, the first goal is to measure balances, so this internal complexity is initially ignored, and the conceptual model is greatly reduced to that in Figure 4.2.

METHODS IN WATERSHED-ECOSYSTEM STUDIES

The two important criteria for a watershed study site are that (1) it be of reasonably small size so that conditions of vegetation, soils, geology, and microclimate are sufficiently uniform, and (2) all liquid water leaving the system be measurable as streamflow. These two criteria together generally limit watershed-ecosystem studies to regions of extreme topography and relatively young, unweathered soils over hard rock formations such as granites. Extreme topography, as in mountainous areas, creates many small, distinct watersheds in a limited area. Shallow soils over hard bedrock keep water that leaches below the rooting zone near the surface so it will appear in surface streamflow rather than being lost through deep seepage to groundwater.

Water and nutrient inputs to a watershed are measured with precipitation gauges for both rain and snow. From each collection, a subsample is taken to the laboratory and analyzed for nutrient concentration. Water quantity and concentration are multiplied together to give total nutrient inputs, usually expressed in grams per square meter per year ($g \cdot m^{-2} \cdot year^{-1}$). Collectors are placed in small clearings at several locations within each watershed to account for variation in precipitation amounts and chemistry within the ecosystem, particularly with elevation.

Gauging water and nutrient losses from the system requires a weir of some type. The most common is the V-notch weir embedded in a concrete slab which is set directly upon the bedrock at the base of the watershed. This forces all stream and subsurface water to flow up and over the weir (Figure 4.3). The height of the water column flowing through the weir is directly

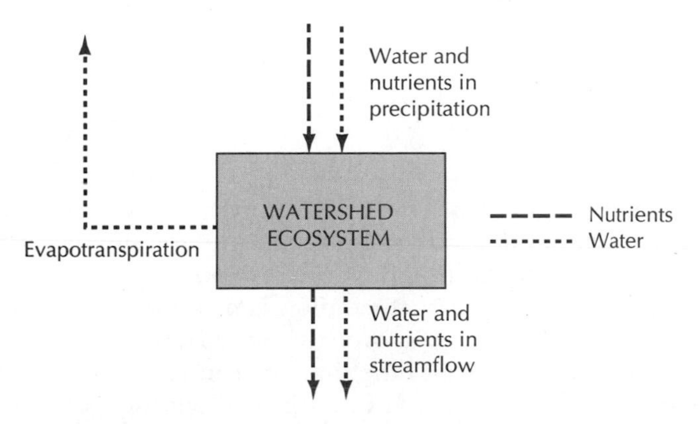

FIGURE 4.2 Simplified conceptual model for measuring water and nutrient balances over watershed ecosystems.

(a)

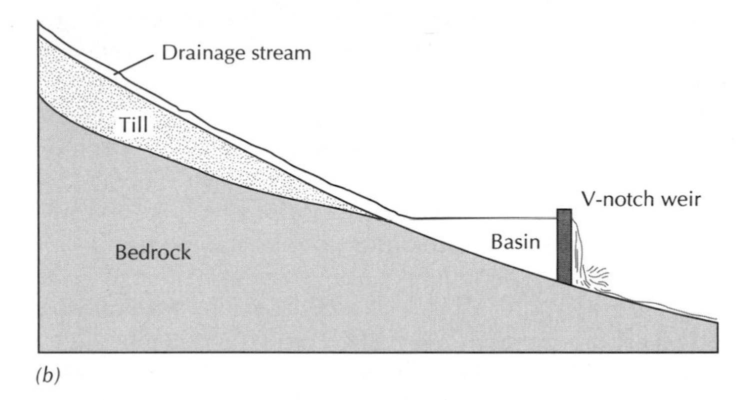

(b)

FIGURE 4.3 A stream gauging station for measuring losses of water and nutrients from a forested ecosystem. (*a*) The V-notch weir placed at the base of the watershed. (From G. H. Likens, F. H. Bormann, R. S. Pierce, J. S. and N. M. Johnson, *Biogeochemistry of a Forested Ecosystem*. Copyright © 1977 by Springer-Verlag. Reprinted with the permission of the publishers) (*b*) Schematic view of how the weir, when placed on water-tight bedrock, forces all liquid water losses from the ecosystem to pass through the notch. (Likens et al. 1977)

related to rate of streamflow in volume per unit time. The small structure next to the weir (Figure 4.3*a*) houses a recorder that continuously monitors water height. A mean flow rate can be converted to rates of water loss in the same units as precipitation (depth per unit time).

Nutrient concentration in streamflow is measured by collecting a small amount of the water as it spills over the weir. However, this concentration varies with the rate of flow and the time of year, so concentrations are measured under a wide range of flow rates throughout the year. Concentrations are again matched with the amount of water lost at different flow rates to yield a weighted mean concentration in stream water. Multiplying this by the total streamflow gives total nutrient outputs. Again, as with the measurement of annual carbon balances, the tremendous number of carefully collected and analyzed samples required to calculate this single number is apparent.

SOME RESULTS FROM WATERSHED-ECOSYSTEM STUDIES

An Early Water–Nitrogen Experiment

One of the first watershed-ecosystem studies in the United States was begun in the southern Appalachian Mountains in the 1950s. The work was not conceived as an ecosystem study, as that term was not widely known or used. It was, rather, one of several very applied studies in forest hydrology.

The Coweeta Experimental Forest (see Figure 1.3) had been established in the Great Smoky Mountains of North Carolina as a site for hydrologic research. It was known that forests used a considerable amount of water through transpiration and that increases in streamflow could be realized by cutting trees and removing the forest canopy. Unfortunately, at least for water yield, cutover forests regrow quickly in humid climates and increases in water yield were short-lived. One idea for providing permanently higher yields was to convert the forest to grass, which might not transpire as much. A watershed-level experiment was established where the existing forest was cut and removed, and the area was replanted to grass. With the addition of a fertilizer treatment, this experiment actually became one of the first, at this scale, on the interactions between primary productivity and water and nutrient use in ecosystems.

Figure 4.4 shows the results of this experiment in terms of differences between expected and measured streamflow. Those bars above 0.0 indicate increases in water yield due to the conversion to grass. In the first year (1960–1961), total water yield was roughly the same under grass as under forest, but more of it occurred in the fall and winter, with less in the spring (Figure 4.4*a*). In the 4 succeeding years, water yield rose. This coincided with a decline in the productivity of the grass cover (Figure 4.4*b*). To test the idea that this reduction in grass growth was due to reduced nutrient availability, a fertilizer treatment was carried out in 1965. Grass production returned to near first-year levels following the fertilization, and water yield declined markedly. These results showed that water use by grass could be just as high as by an intact forest when nutrient availability was sufficient to meet the plants' growth requirements. As nutrient availability declined, so did plant growth. This resulted in a decrease in transpiration and an increase in streamflow.

What happened on that converted watershed to cause nutrient losses and reduced grass growth? Increased water yield occurs because of decreased transpiration, which in turn results from reduced plant production. Is this reduced production somehow linked to the development of nutrient shortages? What role does productivity play in the retention of nutrients in an ecosystem? These kinds of questions on the interactions between important processes, such as productivity, water use, and nutrient retention have been examined in detail at several other watershed-ecosystem sites, including a second research site initially established by the US Forest Service.

Nutrient Balances in Watershed Ecosystems

The Hubbard Brook Experimental Forest in New Hampshire was also established as a outdoor hydrologic laboratory and was also used initially for forest

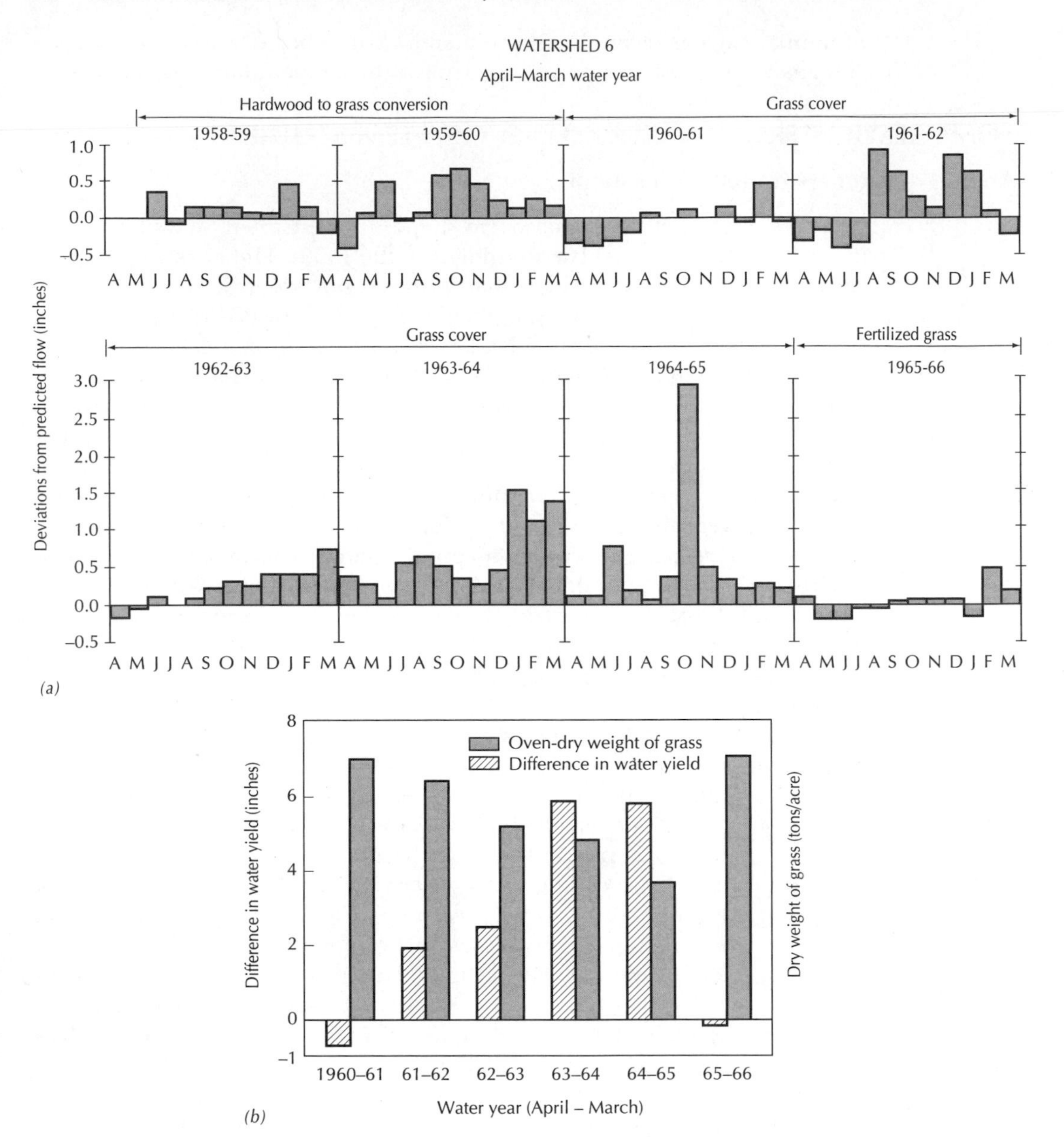

FIGURE 4.4 Results of forest-to-grass conversion at the Coweeta Hydrologic Laboratory. (*a*) Changes in seasonal streamflow in response to conversion and the additions of fertilizer. (*b*) Relationship between increases in streamflow and the net primary production of grass. (Hibbert 1969)

hydrology research. However, since the early 1960s, Hubbard Brook has been the site of one of the longest running studies on the water and nutrient dynamics of forest ecosystems. Initially, nutrient concentration measurements were added to the regular monitoring of precipitation inputs and stream water outputs. This allowed calculations of total nutrient balances over the watershed ecosystems much like the carbon balances discussed in Chapter 3.

One of the major findings of the Hubbard Brook study was that undisturbed forests exhibit regularity and predictability in their input–output balances. This was particularly true for water. Figure 4.5 shows stream water output at Hubbard Brook as a function of total precipitation. The difference between these two is evapotranspiration, which is nearly constant over this wide range of precipitation values. Most of this is transpiration by trees, indicating that water use by vegetation varies little from year to year in this forest. Water that is not evaporated or used by plants appears in the stream. Thus, an initial result of this study was an equation that predicts annual water yield from forests of this type as a function of precipitation. Such predictability can be important in the management of surface water resources.

The predictability at Hubbard Brook extends to the losses of certain chemical elements as well. Figure 4.6*a* shows losses for four major cations as a function of annual streamflow. Nitrogen shows a more complex but still regular pattern of stream concentrations (Figure 4.6*b*). The seasonal changes in

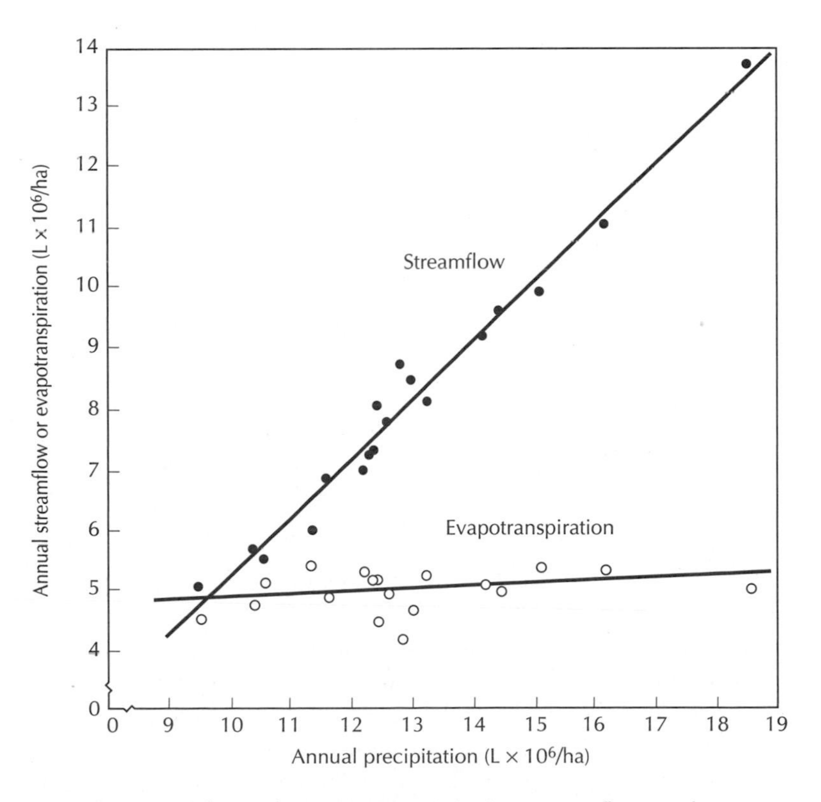

FIGURE 4.5 Relationship among precipitation, streamflow, and evapotranspiration for the control watershed at Hubbard Brook. (Likens et al. 1977)

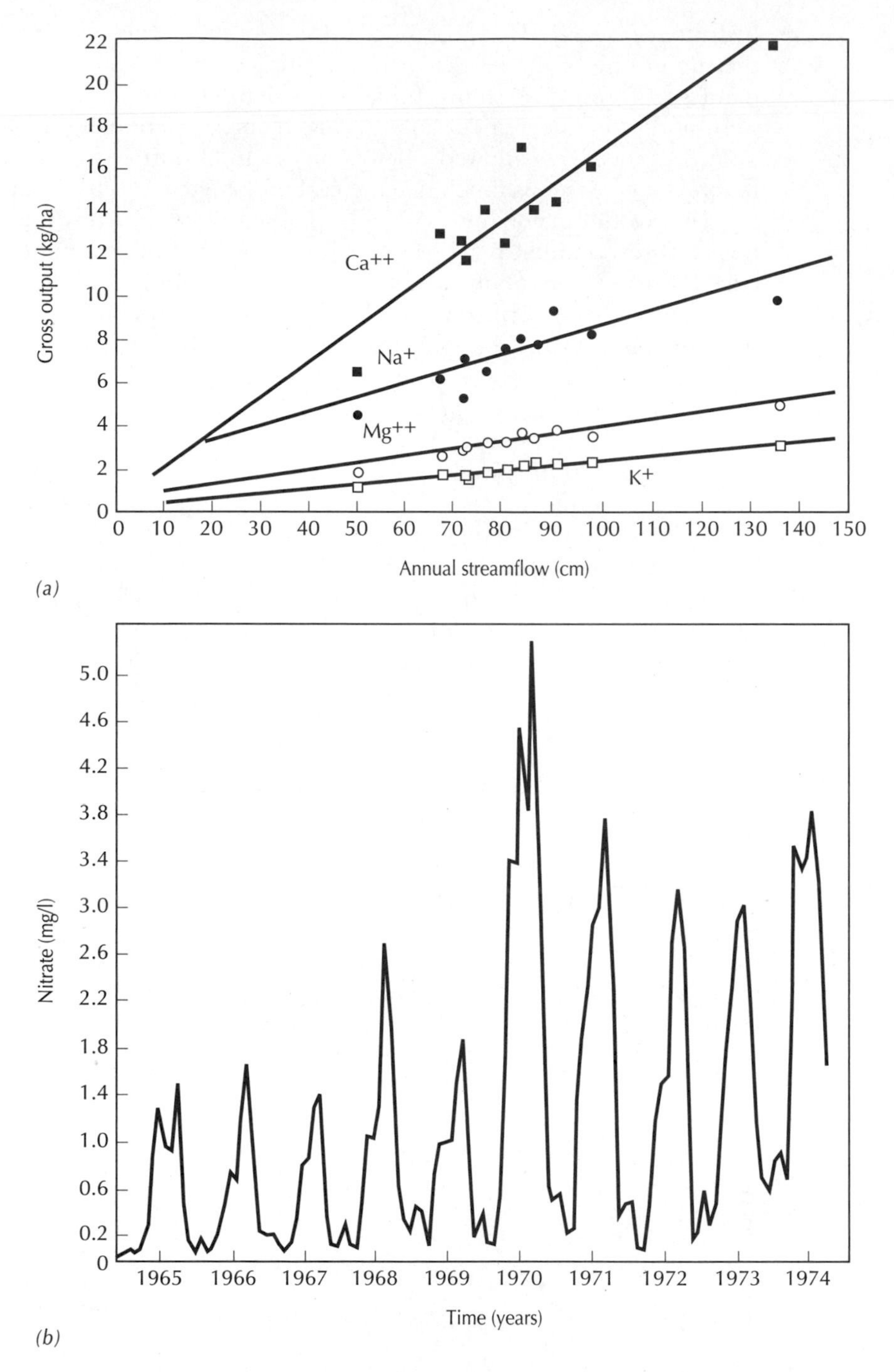

FIGURE 4.6 Nutrient outputs from the control watershed at Hubbard Brook. (*a*) Losses of calcium, sodium, magnesium, and potassium in relation to total annual streamflow. (*b*) Seasonal pattern of nitrate losses. (Likens et al. 1977)

nitrogen losses has been linked to biological demand for nitrogen by plants and microbes, which keeps losses near zero during the growing season. Losses are higher in the dormant season when biological activity is greatly reduced.

An early assumption of watershed studies, derived mainly from geologic methods, was that the storage of nutrients within the watershed ecosystem was not changing. This is called the steady state assumption. If this were strictly true and precipitation and streamflow were the only significant inputs and outputs, then, using the conceptual model in Figure 4.2, these inputs and outputs should be equal. They are not for most systems (Table 4.1).

For several important ions (calcium, magnesium, sodium, and potassium), outputs are almost always larger than inputs. Geologists have long used this information with a slightly different conceptual model of the system to estimate rates of rock weathering (the physical and chemical breakdown of rocks, with the release of chemicals they contain) by difference. Large amounts of these and other ions are present in most systems as minor components of primary minerals or rocks. In limestone formations, calcium and occasionally magnesium are major constituents. Several reactions release these chemicals in simple ionic forms at a slow but continuous rate (Chapter 9). By removing primary minerals from the ecosystem box as in Figure 4.7, measuring precipitation inputs and stream water outputs, and assuming steady state for the remaining parts of the system within the box (soils and biological components), the weathering rate can be calculated as outputs minus inputs.

In general, results from this method coincide with expectations based on bedrock geology (Table 4.1). Streams draining the Hubbard Brook and Coweeta sites, both having granitic bedrock, which is low in calcium and magnesium and slow to weather, have low concentrations of these elements. The Walker Branch watershed in Tennessee, in contrast, is on dolomitic limestone, which weathers more rapidly and is rich in both calcium and magnesium. The andesite (a rock of volcanic origin) under the H. J. Andrews forest in Oregon is intermediate in content of calcium and magnesium and in rate of weathering.

These results can be compared with nutrient balances estimated by nonwatershed techniques (see subsequent discussion) for a tropical rain forest

TABLE 4.1 Element Input–Output Balances for Five Forest Ecosystems (kg/ha/yr)
Likens et al. 1977, Henderson et al. 1978, Vitousek and Sanford 1986.

Watershed Ecosystem	Calcium		Magnesium		Potassium		Sodium		Nitrogen	
	In	Out	In	Out	In	Out	In	Out	In	Out
Hubbard Brook	2.2	13.7	0.6	3.1	0.9	1.9	1.6	7.2	6.5	3.9
Coweeta	4.5	7.5	1.0	3.8	2.5	5.5	4.5	11.0	4.9	0.2
Walker Branch	12.0	148	2.0	77	3.0	7.0	4.0	4.5	5.8	0.4
Andrews	4.5	55.0	0.5	12.0	0.3	2.0	2.0	29.0	0.7	0.2
Tropical	28.0	3.9	3.0	0.7	24.0	4.6				

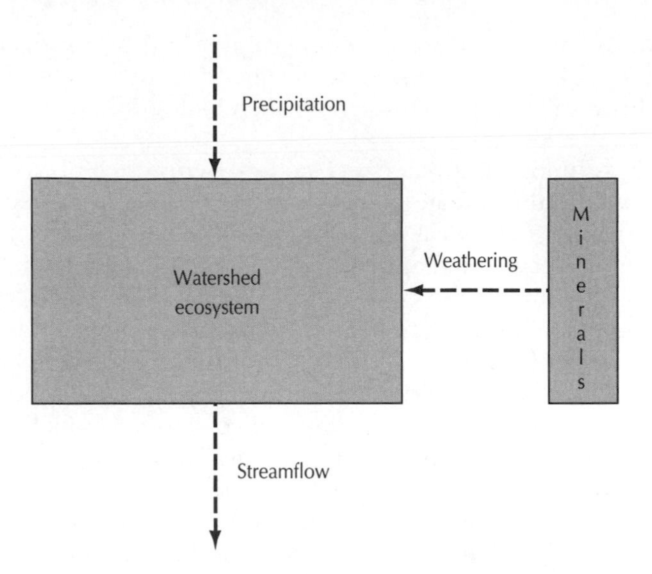

FIGURE 4.7 Modified conceptual model of nutrient balances over ecosystems with primary minerals and weathering rates considered external to the ecosystem.

on a highly weathered soil (Oxisol, Table 4.1). Outputs below the rooting zone are actually less than inputs. The vegetation is taking up and storing nutrients and, in this case, the old, fully weathered soil does not provide significant new cations as weathering products.

Nitrogen balances in the temperate forest ecosystems are quite different from balances of other ions. All four ecosystems in Table 4.1 show net decreases in this nutrient between precipitation and streamflow, similar to those for other ions in the tropical system. The reasons are the same. All four are increasing in total nitrogen content, and weathering rates for nitrogen are negligible (rocks with significant nitrogen content are very rare). From this limited analysis, which does not include gaseous exchanges, all five ecosystems are accumulating nitrogen.

STUDIES ON RESPONSES TO DISTURBANCE

A Devegetation Experiment

One of the major advances made possible by the watershed-ecosystem approach is the experimental manipulation of entire landscape units and the measurement of the integrated response of all components of the system as judged by water and nutrient balances. This approach was first undertaken at Hubbard Brook in the mid-1960s, at a time when much controversy surrounded the use of clear-cutting (the harvesting of all trees on a site) in the national forests. The results were relevant to that controversy because of the similarity between the size of the watersheds used for the experiment and the size of management units in the national forests and because the watershed method allowed for the straightforward measurement of the effects of different practices on water quality.

However, the experiment carried out at Hubbard Brook was not designed to directly test the effects of clear-cutting but rather to assess the effect of higher plant processes on ecosystem function. In particular, the question was asked, To what extent were the input–output balances in the undisturbed watershed (Table 4.1) and their predictability (Figures 4.5 and 4.6) controlled by plant function? Thus, the experiment was not a commercial clear-cutting but rather a complete devegetation. One of the watersheds (watershed 2 or W2), which was determined to be nearly identical to a control, undisturbed watershed (W6), was selected for the devegetation. All trees on the watershed were cut, and the area was sprayed with herbicide to kill all plant regrowth. No logging roads were constructed into the watershed, and all downed trees were left on site.

Figure 4.8 shows that removal of vegetation had a marked effect on water and nutrient balances. Summer streamflow during the devegetation experiment, which lasted 3 years, was nearly fourfold higher than in the control watershed because evapotranspiration was now limited only to evaporation from the soil surface. (Note that all results in Figure 4.8 compare W2 with W6, which was left undisturbed. By this comparison, differences in hydrology due to annual differences in weather patterns are accounted for. The differences between the two lines in each panel expresses the effect of the experiment, not year-to-year variability.) While this increase in streamflow would, by itself, be expected to result in increases in total nutrient losses (see Figure 4.6*a*), increases in the concentration of nutrients within the stream also occurred, especially for nitrate (NO_3^-). These combined to yield increases in loss rates in grams per square meter per year ($g \cdot m^{-2} \cdot year^{-1}$) of 2.5 to 44.0 fold that of undisturbed conditions. Nutrients such as nitrogen and potassium, which are used in large quantities by plants, showed the greatest increases in loss rates.

Nitrate concentrations in the stream water from the devegetated watershed exceeded US Public Health Service standards for drinking water. This created concern over the effects of commercial cuttings as well. However, because this experiment was not an actual clear-cut, and especially because all vegetation was killed for 3 years, it was argued that commercial cuttings would have a lesser effect. On the other hand, some argued that the removal of wood and the construction of logging roads into a real clear-cut would make conditions worse.

To answer these arguments, concentrations of nitrate were measured in stream water draining several commercially clear-cut watersheds. Results showed that concentrations were indeed lower in these streams than on watershed 2, most likely due to nitrogen uptake by the regrowing vegetation. Still, nitrate concentrations in some cases were above Health Service standards.

Comparative Ecosystem Response to Disturbance

A question of both basic and applied interest immediately arose. Were high nitrate losses following cutting a general response of forest ecosystems? Would other systems show the same high concentrations of nitrate? Clear-cutting experiments in other regions and forest types were soon conducted, with varying results. A Douglas-fir forest in the Pacific Northwest showed negligible increases in nitrate loss. A hardwood forest in West Virginia showed slight but measurable increases. An alder forest in Alaska showed very high loss rates.

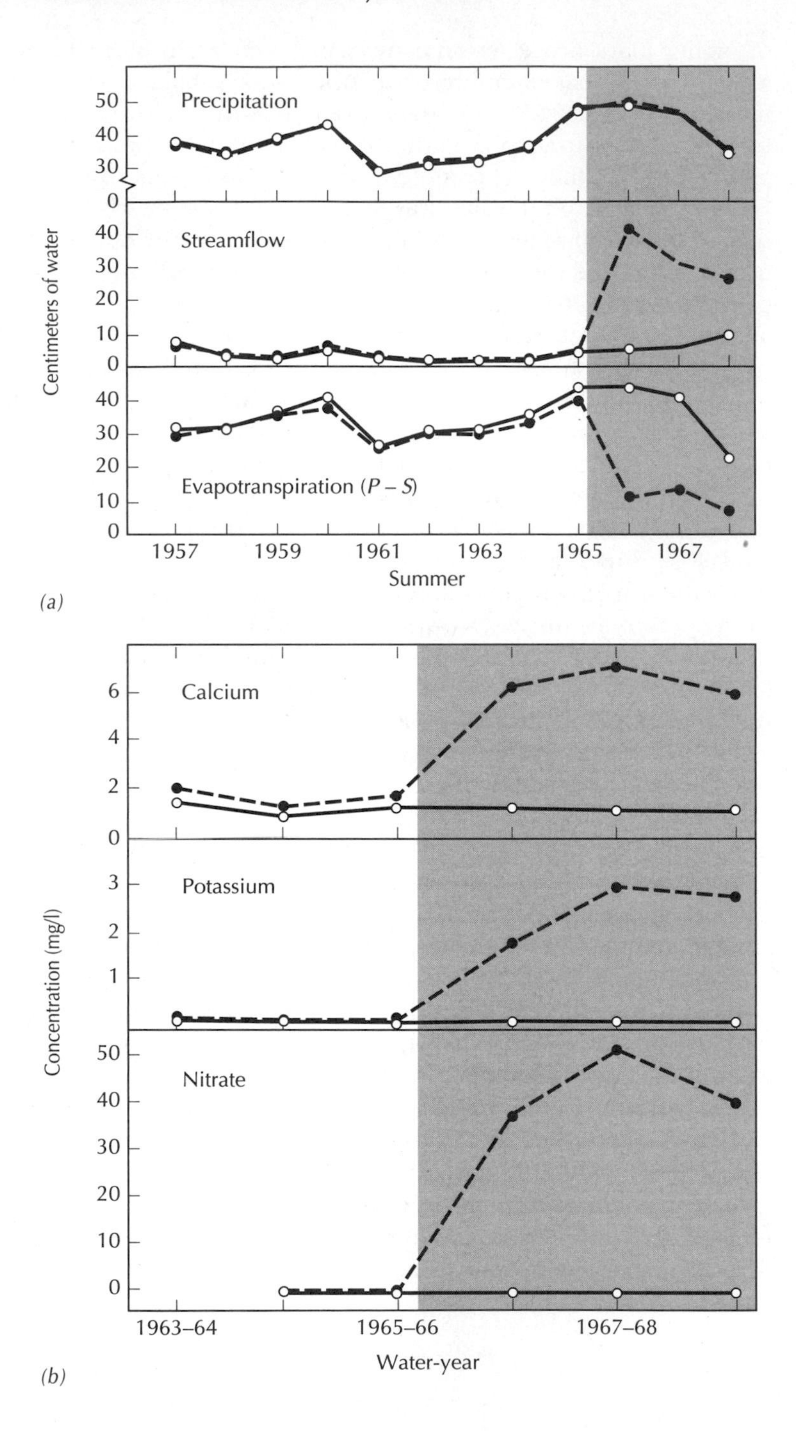

FIGURE 4.8 Effects of devegetation of watershed 2 at Hubbard Brook. (*a*) Changes in summer water balance. (*b*) Changes in stream water chemistry. (Bormann and Likens 1979)

The low nitrogen loss response is the most difficult to explain. In Figure 4.1, nitrogen is cycling within the forest from litter and soil to the available pool due to decomposition and is then taken up by the vegetation. With uptake cut off, one of the other pathways within the system would have to become more important. The leaching of nitrogen to streams is the most obvious, and yet, in many cases, this did not happen. (Some possible mechanisms for this response are discussed in Parts II and III.)

This fundamental difference in the response of forest ecosystems to similar types of disturbance led to a systematic, comparative study of a wide range of forest types. The goal of this study was to measure nitrate leaching loss rates from as wide a range of forest and soil conditions as could be found in the United States, using identical methods of disturbance and measurement in each stand. However, it was not possible to establish watershed study areas for all of these sites because of the requirements of topography, size, and bedrock conditions listed earlier, as well as the cost. Instead, 10 small study plots were established in each stand by digging a trench around small (1–6 m^2) blocks of soil to a depth of 1 m, or below the rooting zone. Thus a good-sized block of soil was severed from the direct influence of living plants.

Without the benefit of whole watersheds with streams to be used for sampling, some other method for collecting water was required. In this case, a tube lysimeter, which is a small plastic tube with a porous ceramic cup at the bottom and a rubber stopper on top, was installed in each soil block. The tube reaches below the rooting zone, and suction is created in the tube to draw water in through the porous cup. While measurements of water volume lost cannot be made by this method, relative concentrations of nutrients in this water can be obtained. The differences in methods of disturbance and measurement between the watershed approach and the trenched plot lysimeter approach indicate that methods of very different scale and requiring very different amounts of effort can be used with the same conceptual model.

Results from this study confirmed a wide range of nitrogen loss rates, even for the same forest type within the same region (see Table 4.2, especially the two Douglas-fir sites in Washington). The general conclusion drawn from this study was that stands on richer or better sites, those in which nitrogen was presumably cycling at a faster rate, lost the most nitrogen following disturbance. However, it became clear from this type of work that more detailed information about the internal mechanisms of nitrogen cycling are required in order to understand, and possibly predict, the nitrogen output response for any particular forest ecosystem.

A Multiple-Site Ecosystem Recovery Experiment

While research on nutrient balances in forest ecosystems in the 1960s and 1970s was focused on harvesting as a major type of disturbance, the 1980s and 1990s saw a shift to concerns over air pollution. Of particular concern were the inputs of nitrogen and sulfur that are the major components of acid rain (see Chapter 25) and are especially high in the heavily industrialized areas of the United States and central Europe. In these areas, the chemistry of both rain and soil water can be dominated by the effects of these added pollutants. A relevant question is, How would the chemistry of soils and water change if

TABLE 4.2 Concentrations of Nitrate in Water Collected Below the Rooting Zone in Control and Trenched Soils from Several Forest Ecosystems
Vitousek et al. 1979.

Site	Lysimeter Nitrate Concentrations (μEq/liter)	
	Control	Trenched
Indiana		
Maple, beech	15	2150
Oak, hickory	12	1510
Shortleaf pine	20	175
Massachusetts		
Oak, pine	0	932
Red pine	0	263
Oak, red maple	1	140
New Hampshire		
Maple, beech	105	1055
Balsam fir	45	570
New Mexico		
Ponderosa pine	1	60
Mixed conifer	0	784
Aspen	0	645
Spruce, subalpine fir	1	24
North Carolina		
Mixed oak	0.5	434
White pine	1.9	610
Oregon		
Western hemlock	25	730
Washington		
Alder	371	1571
Douglas fir (low site quality)	1.4	114
Douglas fir (high site quality)	6.1	779
Pacific silver fir	6.2	5.6

these pollutants were removed and precipitation chemistry were returned to preindustrial values?

Just such an experiment has been carried out as part of the European NITREX (NITRogen EXperiment) project. Due to both fossil fuel combustion and, in some places, intensive livestock management, the deposition of nitrogen is nearly 30-fold higher than background or unpolluted levels (5.9 versus 0.3 g $\cdot$ m^{-2} $\cdot$ $year^{-1}$ nitrogen). In three of the experimental locations used (one in Germany and two in the Netherlands), increased nitrogen deposition has led to very high nitrate losses to groundwater.

The experimental manipulation in this case was to build a roof under the upper canopy in each of these three forest sites (Figure 4.9*a*) and to collect, treat, and reapply the water with a chemical composition similar to unpol-

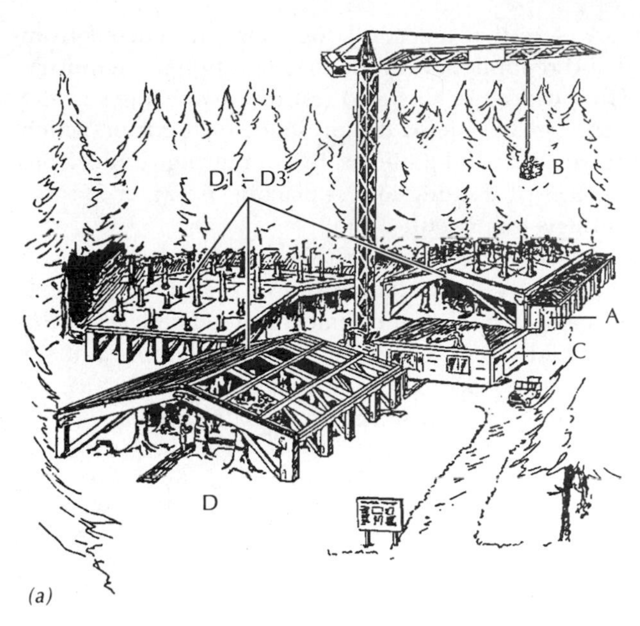

(a)

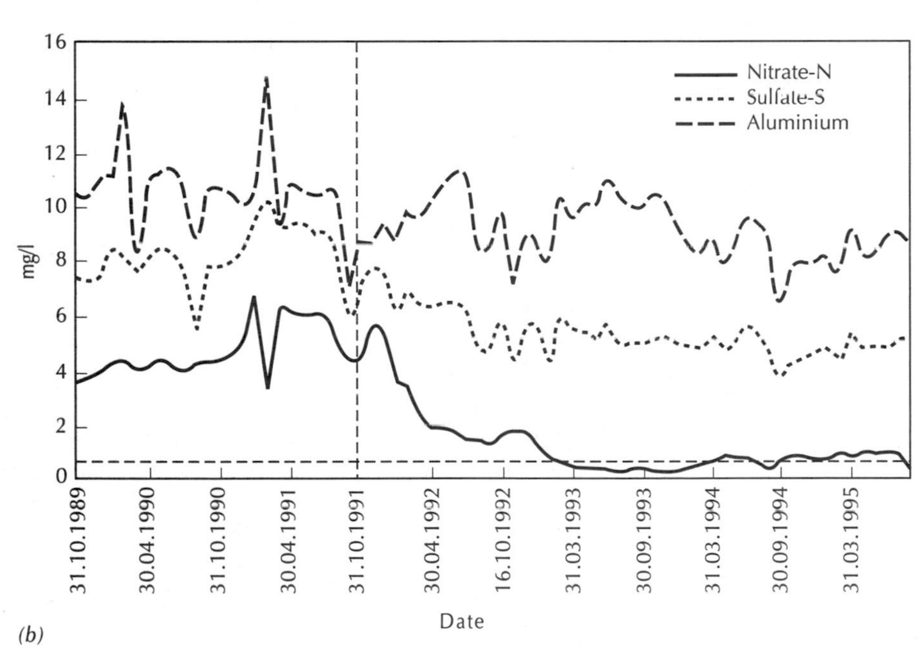

(b)

FIGURE 4.9 Experimental manipulation of precipitation inputs to forests. (*a*) Schematic view of the roof systems used to collect precipitation under the overstory trees. Water was collected and much of the nitrogen and sulfur was removed and then reapplied under the roof to maintain the hydrologic balance (Bredemeier et al. 1995). (*b*) Effects of excluding nitrogen from precipitation. An immediate and dramatic reduction in nitrate leaching occurred. (Bredemeier et al. 1998)

luted rain. Notice in Figure 4.9*a* that a second roof was constructed from which water was collected and reapplied, without any alteration in chemistry, to the soil underneath. The purpose of this is to test for any changes in the function of the forest that might be due to the presence of the roof, which might alter light, temperature, and wind patterns. By comparing results from the two roof systems, the effect of treating and replacing the water is separated from whatever roof effects may occur.

Results from this experiment were dramatic (Figure 4.9*b*). Inputs of nitrogen were reduced by 67% to 93%. The chemistry of the water leaching below the rooting zone in each case changed almost immediately, showing greatly reduced nitrate concentrations. This result has been important for making environmental policy in Europe, showing that efforts to reduce air pollution could result in immediate improvements in water quality.

REFERENCES CITED

Bormann, F. H., and G. E. Likens. 1979. *Pattern and Process in a Forested Ecosystem.* Springer-Verlag, New York.

Bredemeier, M., K. Blanck, N. Lamersdorf, and G. A. Wiedy. 1995. Response of soil water chemistry to experimental "clean rain" in the NITREX roof experiment at Solling, Germany. *Forest Ecology and Management* 71:31–44.

Bredemeier, M., K. Blanck, Y. J. Xu, A. Tietema, A. W. Boxman, B. Emmet, F. Moldan, P. Gundersen, P. Schleppi, and R. F. Wright. 1998. Input–output budgets at the NITREX sites. *Forest Ecology and Management* 101:57–64.

Henderson, G. S., W. T. Swank, J. B. Waide, and C. C. Grier. 1978. Nutrient budgets of Appalachian and Cascade region watersheds: A comparison. *Forest Science* 24:385–397.

Hibbert, A. R. 1969. Water yield changes after converting a forested catchment to grass. *Water Resources Research* 5:634–640.

Likens, G. E., F. H. Bormann, R. S. Pierce, J. S. Eaton, and N. M. Johnson. 1977. *Biogeochemistry of a Forested Ecosystem.* Springer-Verlag, New York.

Vitousek, P. M., and R. L. Sanford, Jr. 1986. Nutrient cycling in moist tropical forests. *Annual Reviews of Ecology and Systematics* 17:137–167.

Vitousek, P. M., J. R. Gosz, C. C. Grier, J. M. Melillo, W. A. Reiners, and R. L. Todd. 1979. Nitrate losses from disturbed ecosystems. *Science* 204:469–474.

Additional References

Norton, S. A., and I. J. Fernandez. 1999. *The Bear Brook Watersheds in Maine.* Kluwer Academic Press, Dordrecht, The Netherlands.

Swank, W. T., and D. A. Crossley, Jr. 1988. *Forest Hydrology and Ecology at Coweeta.* Springer-Verlag, New York.

Vitousek, P. M., and W. A. Reiners. 1975. Ecosystem succession and nutrient retention: A hypothesis. *BioScience* 25:376–381.

Wright, R. F., and L. Rasmussen. 1998. Introduction to the NITREX and EXMAN projects. *Forest Ecology and Management* 101:1–8. (This is the lead article in a special edition summarizing several ecosystem manipulations in Europe.)

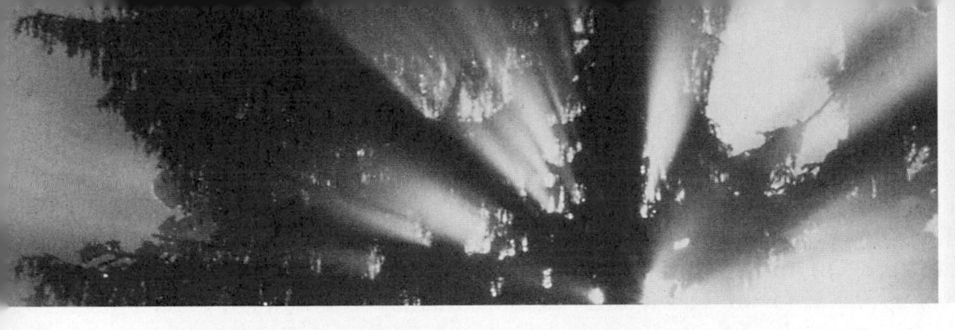

5

ADDITIONAL APPROACHES TO ANALYSIS AND SYNTHESIS IN ECOSYSTEM STUDIES

INTRODUCTION

In the last two chapters, we presented several different methods drawn from several different disciplines that allow for the direct measurement of carbon, water, and nutrients balances at the whole-ecosystem level. We have also seen that measurements at this large scale provide only limited insights into the processes within ecosystems that control those responses. In Part 2, we delve deeply into these basic internal processes. However, before we begin this closer look, there are also some additional generalized analytical and synthetic approaches to understanding ecosystem function that are widely used and recur in the following chapters. These methods are the subject of this chapter.

The methods presented are divided into three categories. Analytical methods include the use of isotopes for tracing elements through systems, the use of mass balance for closing budgets, and the uses and limitations of experimental manipulations involving whole ecosystems. A second category places both ecosystems and important processes in the context of gradients in time and space. The final set are synthetic tools—those derived from the fields of systems analysis and computer modeling that allow for the construction of generalized theories and predictive models.

ANALYTICAL METHODS

Mass Balance, Budgets, and Resource-Use Efficiencies

The concept of the mass balance was used implicitly in Chapters 3 and 4 and relies on the simple physical principle of conservation of mass. For example, the full hydrologic balance at Hubbard Brook (see Figure 4.5) is constructed from measurements of only precipitation inputs and stream water outputs. Conservation of mass simply states that all of the water entering the system must be accounted for through losses to drainage and evapotranspiration, or changes in storage (i.e., the total mass of water must be conserved). By assuming that losses to groundwater and net changes in storage over a whole year are zero, evapotranspiration can be estimated as precipitation minus stream flow. Direct measurement of evapotranspiration at the watershed scale can be difficult. The most accurate estimates of this flux for complex forest ecosystems come from watershed-level measurements combined with the mass balance approach.

The same concept was used implicitly in the discussion of weathering rates in Chapter 4. Where inputs in precipitation and losses in stream water can be measured, then the inputs to the system by weathering can be estimated as the difference between these two (see Figure 4.7). In general, mass balance equations allow for the estimation of one flux that is difficult or impossible to measure directly from the direct measurement of all of the other related fluxes.

An immediate limitation to this approach may come to mind. The assumption that "all other related fluxes" have been measured may not be correct. In the water balance example, a primary assumption is that all of the drainage water leaving the system appears and can be measured in the stream. If a watershed is not underlain by watertight bedrock, drainage directly down into groundwater results in underestimation of drainage by stream measurement and overestimation of transpiration. In the weathering example, if the storage of nutrients in the soils and vegetation within the watershed are increasing, then losses to the stream are reduced and weathering rates are underestimated.

The concept of budgets also extends to processes that occur within ecosystems and may be difficult to measure directly. Fine roots are an excellent example. The production and turnover of these tissues are among the most difficult processes to measure directly in ecosystems (Chapter 11). Because of this, direct measurements that do exist have been compared with those obtained by mass balance of carbon and nitrogen over soils in which all other transfers have been measured. For example, if the fixation of carbon by gross photosynthesis can be measured directly, as well as respiration of carbon in leaves and stems and the allocation to leaf and wood growth, then allocation to roots can be estimated by difference (Figure 5.1). Doing this for nitrogen as well (uptake minus allocation to above-ground parts of the plants; Figure 5.1) provides a second and independent estimate of allocation to roots. If both of these methods agree with direct measurements, then we can have more confidence in our estimates. When they do not, the floor is open for some lively debate! In general, the potential to crosscheck direct measurements against ecosystem-level constraints from mass balance and budgeting methods provides added rigor in constructing an accurate view of ecosystem function.

A final concept related to the simultaneous determination of water, carbon, and nutrient fluxes is resource-use efficiency. This term describes how much of a given resource (energy, water, carbon, nutrients) is required or used to carry out an ecosystem process. Examples include water-use efficiency, which describes the ratio of water lost through transpiration to carbon fixed through photosynthesis, and nutrient-use efficiency, which describes the amount of nutrient required to produce a given amount of plant material (measured as the concentration of that nutrient in biomass). In Figure 2.2, the temperate deciduous forest in the diagram cycles and uses more nitrogen

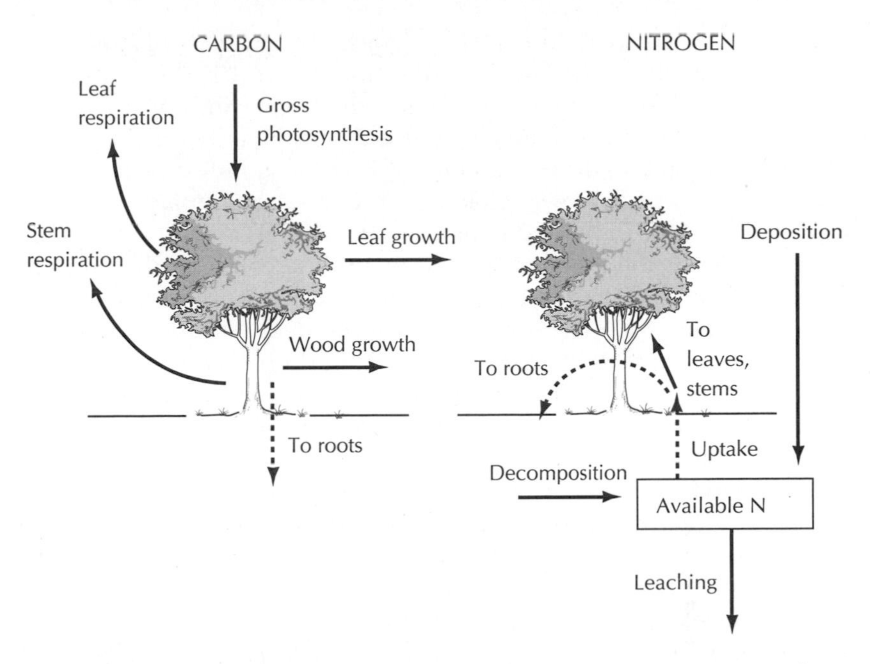

FIGURE 5.1 Examples of estimating root production in ecosystems using the mass balance technique. For carbon balance, direct measurements of gross photosynthesis, leaf and stem respiration, and leaf and stem growth allow for the estimation of allocation of carbon to roots by difference. For nitrogen, uptake is estimated as deposition plus nitrogen released by decomposition of organic matter, minus leaching losses. Root uptake is then estimated as uptake minus allocation of nitrogen to leaves and wood.

per unit of above-ground productivity than does the temperate evergreen forest and so has a lower nitrogen-use efficiency. Resource-use efficiencies are central to the construction of parallel budgets of water, carbon, and nutrients, as they define ratios of transfer through crucial components, such as plants and soils. In Part II, we define several different kinds of resource-use efficiencies and discuss both their importance in ecosystem function and physiological adaptations that allow these efficiencies to be altered.

Isotope Probes

As ecosystem analysts, we want to quantify the movement of nutrients and other elements through forests, prairies, and deserts. One of the most difficult parts of this task is measuring small changes in large pools. For example, suppose that we have hypothesized that increased nitrogen in rainfall is leading to more total nitrogen in a forest and that most of that added nitrogen will be stored in the soil. The problem is that there is a large and variable amount of nitrogen in soils, and so detecting a small increment is very difficult. What if, instead of having to measure changes in total nitrogen content, we could tag individual atoms as they are added to the system and then use that tag to locate those individual atoms later on? This is exactly what the use of isotopes of key elements allows us to do.

What are isotopes? The major building blocks of atoms are protons (positively charged) and neutrons (no charge) in the nucleus, with electrons (negatively charged) found in "shells" or rings around the nucleus. The atomic number of an element is determined by the number of protons and electrons, while the atomic mass is the number of protons plus neutrons (electrons have negligible mass). Carbon has an atomic number of 6; nitrogen, an atomic number of 7. In the most abundant forms of both of these elements, the number of protons and neutrons are the same and the atomic masses are 12 and 14. Isotopes of these elements contain additional neutrons. With an added neutron, the most common form of nitrogen (denoted ^{14}N to show the mass of 14) becomes ^{15}N. Carbon occurs in both ^{13}C and ^{14}C forms.

Isotopes can be further divided into two types: (1) stable and (2) unstable (radioactive). Unstable isotopes "decay" to other forms over time with the emission of high-energy particles that can be biologically destructive (radioactivity). In materials with high concentrations of similarly unstable materials (e.g., processed ores enriched in uranium-238 [^{238}U]), the decay of one atom can trigger the decay of another, initiating a chain reaction resulting in the release of tremendous amounts of radioactivity and energy, such as occurs in nuclear power plants and weapons of mass destruction. Because of this process of "decay" radioactive isotopes have a measurable "half-life," or period of time in which half of the isotope is transformed into a different product.

Stable isotopes such as ^{15}N do not decay and do not emit radioactive particles. As a result, they are much less dangerous to use. However, they are more difficult to measure. The amount of a radioactive isotope in a sample can be determined using simple and inexpensive devices that measure the emission of radioactive particles. Measuring nonradioactive isotopes requires a mass spectrometer, a complex and expensive machine that effectively sorts isotopes by weight (or mass) and determines the number of each type of atom. A summary of isotopes commonly used in ecosystem studies is included in Table 5.1.

There are also two general ways in which isotopes are used in ecosystem studies: (1) natural abundance and (2) experimental enrichment.

Natural Abundance Studies – These rely on the tendency for incomplete chemical reactions to discriminate against heavier isotopes. For example,

TABLE 5.1 Isotopes Commonly Used in Ecosystem Studies

Element	Isotope	Half-Life
Carbon	13	Stable
Carbon	14	5730 years
Nitrogen	15	Stable
Oxygen	18	Stable
Phosphorous	32	14 days
Sulfur	34	Stable
Sulfur	35	87 days

there are two major inorganic forms of nitrogen in soils: (1) ammonium (NH_4^+) and (2) nitrate (NO_3^-). Ammonium is oxidized to nitrate in soils by microbes, but the reaction never goes to completion; there is always some ammonium remaining. In incomplete reactions, the lighter isotope tends to be more reactive than the heavier one, so there is a tendency for the product (nitrate) to be "lighter" or enriched in ^{14}N, while the remaining ammonium is "heavier" (enriched in ^{15}N). As many soil processes involving nitrogen do not use all of the substrate, residual soil organic matter tends to become enriched in ^{15}N over time. The general downward movement of nitrogen through soils over time results in increased ^{15}N concentration with depth (Figure 5.2). As ^{15}N is a stable isotope, the very long-term effects of soil processes on ^{15}N distribution can be observed.

Decaying isotopes can also be used effectively. For example, when plants create new tissues as part of growth, they incorporate carbon with the ^{14}C concentration present in the atmosphere at that time. Knowing this initial concentration and the decay rate, or half-life, of ^{14}C, we can determine how long ago a material was produced by radiocarbon dating, or measuring the current concentration of ^{14}C.

Humans have played a large role in altering the concentration of ^{14}C in the atmosphere. Widespread testing of nuclear weapons in the open atmosphere in the 1950s and 1960s generated huge quantities of ^{14}C, greatly in-

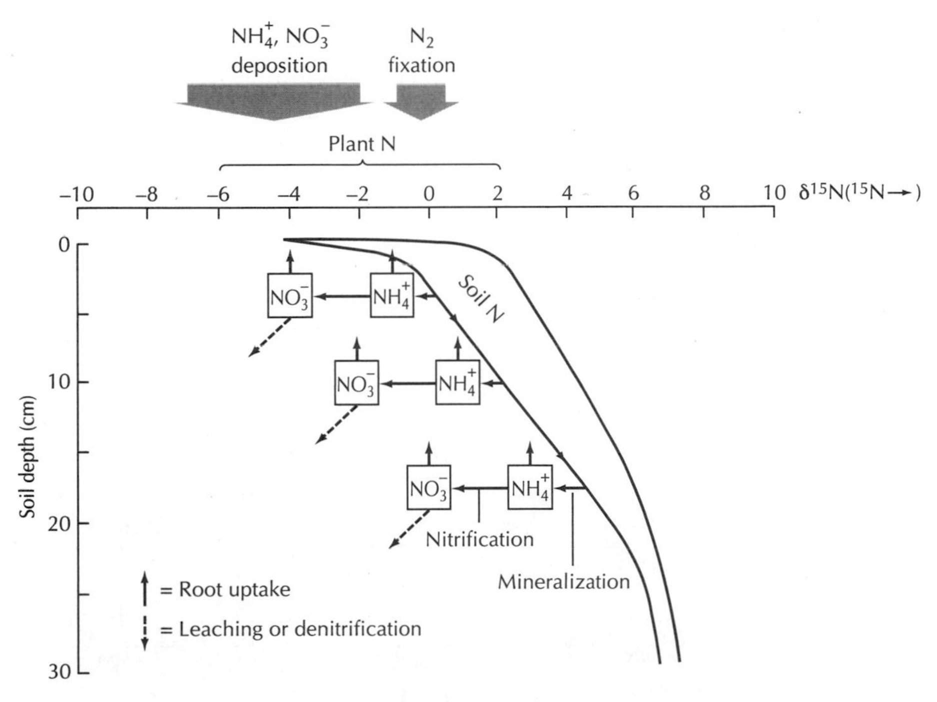

FIGURE 5.2 Example of the distribution of the heavier isotope of nitrogen (^{15}N) through a soil profile (Nadelhoffer and Fry 1994). Processes that convert organic nitrogen to ammonium and nitrate can discriminate against the heavier ^{15}N isotope. Combined with the slow, downward movement of mineral and organic forms of nitrogen, this leads to the accumulation of ^{15}N lower in the soil profile.

creasing the atmospheric concentration (or "natural" abundance) of this isotope (Figure 5.3). As a result, recently produced plant materials (and hence soil and microbial organic matter) carry this enriched signal. This increase in ^{14}C has actually been used to trace the movement of carbon through the Earth's ecosystems over the past several decades.

Enrichment Studies – These bring us back to the beginning of Part I and the desirability of being able to tag individual atoms in experimental additions. Isotopes of carbon and nitrogen (and other elements) can be produced, concentrated, and then transformed chemically or biologically into a tremendous variety of mineral and organic compounds. Adding any of these forms to ecosystems allows us to determine the distribution of specifically those added molecules at any point in the future (subject to loss to decay for radioactive forms).

The time frames over which these remeasurements are performed varies from minutes to years. At the short-term end are studies of the rates at which soil microbes and/or soil chemical processes remove added ammonium and nitrate from soil cores (15 minutes to 24 hours). Midterm studies would, for example, determine the redistribution of added ^{15}N at yearly intervals throughout an entire ecosystem. In effect, studies of the changes in the distribution of "bomb" carbon (^{14}C added to the atmosphere by nuclear bomb testing) throughout the oceans and terrestrial ecosystems of the Earth are a very long-term example of the same kind of study.

Whole-System Manipulations: The Replication Problem

The ability to use mass balances is an important advantage of the ecosystem-level approach to studying the landscape. There are disadvantages as well. An important one is the complexity and expense of carrying out experiments at

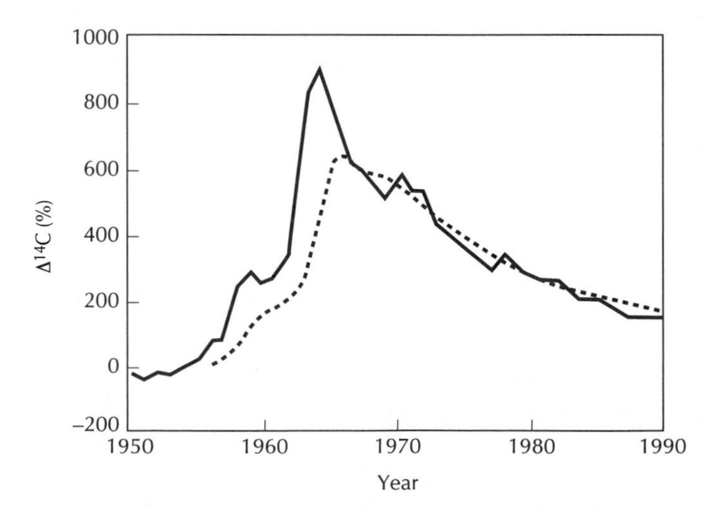

FIGURE 5.3 Variation in ^{14}C concentration in the atmosphere as a result of testing of atomic weapons (After Trumbore 1995). Changes are expressed in parts per thousand (‰) relative to the initial concentration. A +1000 ‰ change is a doubling of ^{14}C concentration in the atmosphere. The solid line represents data from the Northern Hemisphere, the dashed line from the Southern Hemisphere.

this large scale, which often means that ecosystem-scale studies cannot be replicated.

What is replication? If we carry out an experiment once, it is difficult to know the extent to which the unique conditions of that one experiment determined the results. Standard scientific practice is to repeat an experiment under identical conditions (or replicate it) several times. The results of the full set of experiments can then be expressed as the average response, and some expression of the variability of that response can be determined.

Ecosystem studies as a field has always wrestled with the problem of replication. Consider the two forest ecosystem experiments described in Chapter 4 (Hubbard Brook and Coweeta). The expense of carrying out such experiments on one watershed can severely tax the resources of large research organizations such as the US Forest Service and the National Science Foundation. A requirement to carry out such experiments in duplicate or triplicate would make such efforts effectively impossible. As a result, many of our largest and most significant ecosystem experiments are unreplicated. The FACE experiment (see Figure 3.15) is one example of a study of such perceived policy importance that the additional cost of replication was supported.

GRADIENTS IN ECOSYSTEM PROCESSES OVER SPACE AND TIME

What Is a Gradient?

A gradient is any continuous change in an environmental parameter over some distance or over time. There is a gradient of light availability from the top to the bottom of a forest canopy, caused by the interception of light by leaves overhead (Figure 5.4). There is a gradient of temperature from the outside to the inside of the wall of a heated house on a cold day. There is a gradient of CO_2 concentration from the atmosphere to the inside of a leaf caused by the removal of CO_2 from the internal leaf air spaces by photosynthesis. There can also be gradients in the occurrence of species or in the rate of nitrogen cycling over time in the same place.

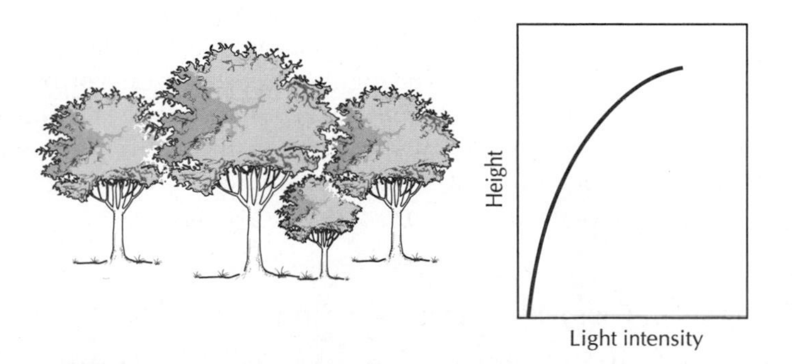

FIGURE 5.4 Example of an environmental gradient. The intensity of light declines down through a forest canopy as light is intercepted by foliage.

The gradient concept is central to ecosystem studies because most changes in ecosystems are continuous and occur along gradients. Gradients in temperature cause heat to move through soils; the greater the difference in temperature, the faster the movement of heat. Water movement in plants is in response to gradients in water potential (a force pulling water from soil to root, leaf, and atmosphere) and again is faster as the gradients get steeper (Chapter 7). While there are some important "switches" in ecosystems that cause rapid changes in function over relatively small changes in environmental conditions, most changes are continuous and in response to gradual changes in conditions.

The Niche: Species Distribution in Response to Resource Gradients

This concept is at the heart of much ecological research and directly relates the idea of gradients to the distribution of species. The niche describes the types of environments for which species are best adapted—those parts of different resource-availability gradients where the species will be found. For example, Figure 5.5*a* shows a two-dimensional field of light and nitrogen availability. Different species might be best adapted to different combinations of availability of these two resources due to specific morphological or physiological characteristics. In this example, species A grows best under high light and high nitrogen conditions, while species B performs best at high light and low nitrogen. A less theoretical example would be the classic distribution of tree species from moist (mesic) to dry (xeric) sites from high-elevation, north-facing slopes to low-elevation, south-facing slopes with a range of mountains (Figure 5.5*b*).

Niches are described in two ways. The fundamental niche is that part of the resource-availability field in which the species can survive in the absence of competition. The realized niche is that portion of the fundamental niche actually occupied in the presence of a particular set of competitors. In Figure 5.5*a*, the realized niches of both species are considerably smaller than their fundamental niches. In Figure 5.5*b*, most of the species described would be capable of growing outside the range at which they are actually found but are constrained by competition. Niches are rarely equivalent to a physical location but relate rather to positions along gradients of resource availability. The moisture gradient in Figure 5.5*b* actually reflects species responses to changes in temperature and precipitation with elevation and topographic position.

Responses to the physical and chemical environment are only one aspect of a species' niche. Much of the rich literature on niche theory relates to species–species interactions involving the processes of herbivory; pollination; competition; and allelopathic, mutualistic, or symbiotic relationships (harmful, beneficial, or physiologically integrated, respectively) between species. Both kinds of interactions (with the environment and with other species) affect ecosystem function.

Succession: Gradients in Ecosystem State over Time

Resource availability within ecosystems can change over time as well. For example, the soil surface in a mature tropical rain forest is heavily shaded, and

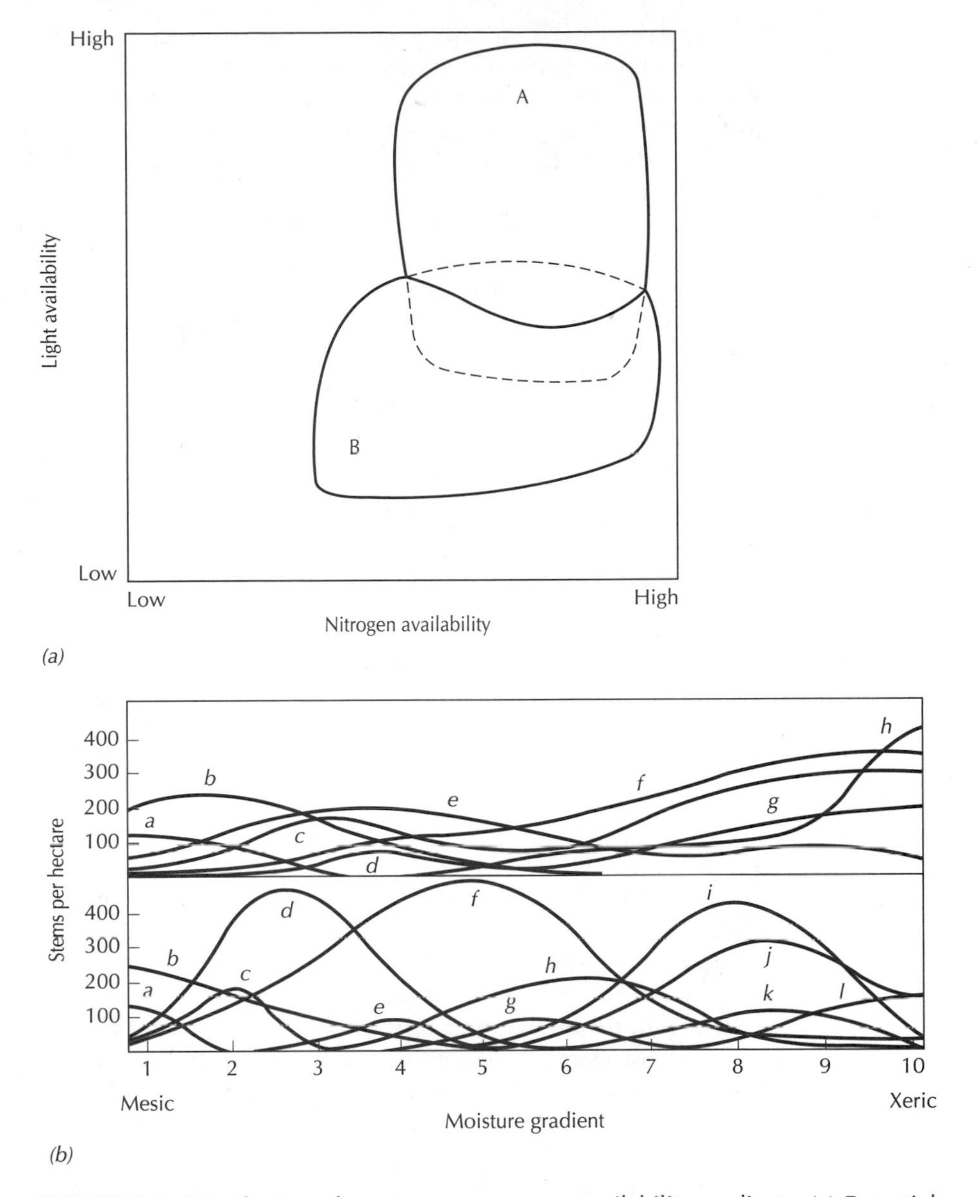

FIGURE 5.5 Distribution of species over resource-availability gradients. (*a*) Potential and realized niches over two gradients for two idealized species. (*b*) Actual distribution of several species over a water-availability gradient in two forests in the western United States. (*Top,* Siskyou Mountains, Oregon; *bottom,* Santa Catalina Mountains, Arizona). The Arizona system shows narrower niche breadth for each species and a greater number of species total across the gradient. (Whittaker 1967)

only plants adapted to this shade are able to survive. Imagine now that a hurricane passes over the forest and brings down all of the overstory trees, exposing the soil to full sun. A different set of species, those whose niche definition includes high light availability, now come to dominate. As these plants grow taller and occupy the site, shade returns to the forest floor and eventually the shade tolerant plants as well. This is a brief description of a successional sequence initiated by the disturbance associated with the hurricane. In general, successional changes are a directional change in species composition or ecosystem state (e.g., light or nutrient availability) initiated by a disturbance and proceeding toward a generally defined "mature" endpoint (the nature and even the existence of such "endpoints" are hotly debated in some quarters).

The term *succession* is generally applied to the changes in species composition and ecosystem state that occur over decades to hundreds of years in response to disturbance. However, very slow but continuous changes also occur naturally in both climate and in soils over periods of thousands to millions of years, ensuring continuous change in ecosystem response and development (see Chapter 22). As a result of this spectrum of frequency of disturbance and speed of environmental change, ecosystems are almost always in the process of recovering from disturbance or adjusting to new conditions.

Remote Sensing: Measuring and Mapping Ecosystems

The largest and most complete ecosystem-level studies can still measure only small pieces of a regional landscape. The Hubbard Brook experiments, for example, were conducted in watersheds covering 10–20 hectares. These were located in a river valley that includes over 3100 hectares of forest and is presented as representative of the White Mountain region in central New Hampshire, which covers more than 300,000 hectares. Given the variety of environmental conditions and biotic communities in any area of this size, how can results from such small areas such as Hubbard Brook be extended to a larger region?

Satellite remote sensing is a rapidly evolving technology that fits rather closely with the needs of large-scale environmental research. A satellite remote sensing system (Figure 5.6) includes the actual sensor for measuring radiation returning to space from the Earth's surface; an optical system for focusing and concentrating this radiation; a computer and data storage system for processing and holding the acquired information until it can be transmitted to receivers on the ground; a solar panel array to provide electricity to run the system; and other physical and electronic systems for temperature control, navigation, and the operation of the sensor. The ground station provides a system for receiving data stored on the platform and for translating it into formats that can be used by the research community.

Important characteristics of remote sensing systems include (1) the portions of the electromagnetic energy spectrum covered, (2) the size of the smallest picture element (or pixel) that can be seen, (3) the area that can be covered in a single scene (the number of pixels), and (4) how often a given portion of the Earth can be resampled.

For example, the LANDSAT satellites, built by NASA, were designed, as the name implies, for the identification and classification of landscape patterns and land uses. The most frequently published LANDSAT images come from the Thematic Mapper sensor, which acquires reflectance data in seven portions of the electromagnetic spectrum, including three bands of visible light; three in the short-wave infrared that respond to vegetation structure, water content, and chemistry; and one in the thermal infrared, recording sur-

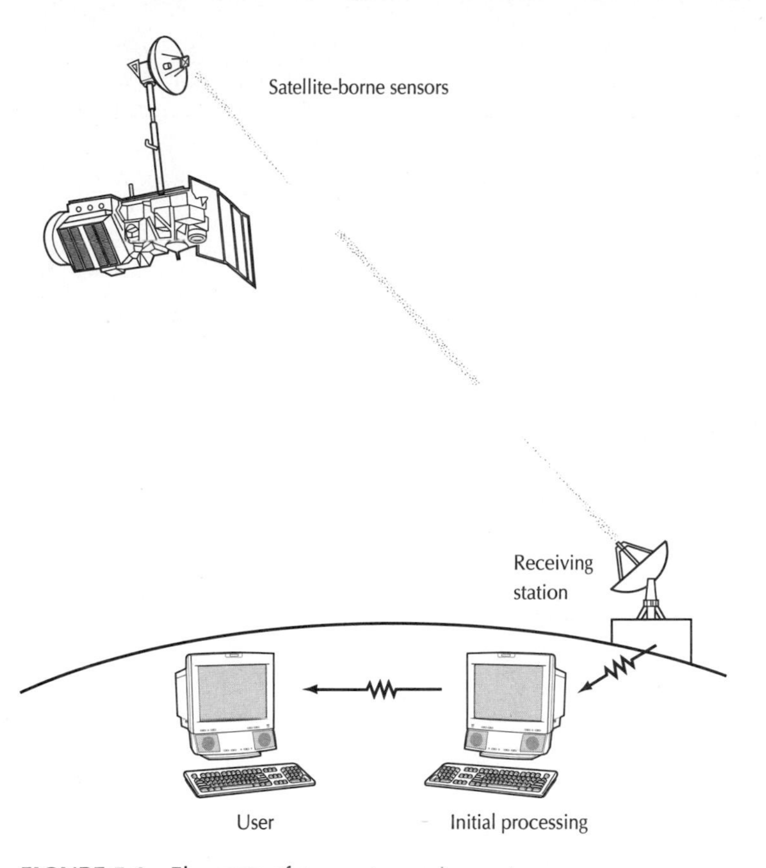

FIGURE 5.6 Elements of a remote sensing system.

face temperature (Figure 5.7). As land-use patterns can change over small distances, LANDSAT has a small pixel size (30 × 30 m), and one full image is just over 6000 pixels wide, allowing for data acquisition over an area 185 km wide. Because of this relatively narrow swath width, LANDSAT returns to a given point over the Earth only once every 16 days.

In contrast, the Advanced Very High Resolution Radiometer (AVHRR), operated by NOAA, was designed for weather observation. Only 5 of its 46 channels produce visual images of the surface, with the remainder used for measurements within the atmosphere. Of the 5 imaging channels, only 1 is in the visible region, with 1 in the near-infrared and 3 more in the long-wave, thermal infrared. The scale of the imaging is also very different from that of LANDSAT. Important climatic features, such as cloud formation along frontal systems and large changes in surface temperatures, occur over much larger areas than do land use changes. As a result, the AVHRR has a pixel size of 500–1000 m, depending on the channel, and covers a swath 2250 km wide, about half the width of the United States. This wider swath allows repeat coverage every day.

The role of spatial resolution or pixel size can be demonstrated with three images of the same location acquired with three, newer instruments with very different spatial resolutions (Figure 5.8). The Moderate Resolution

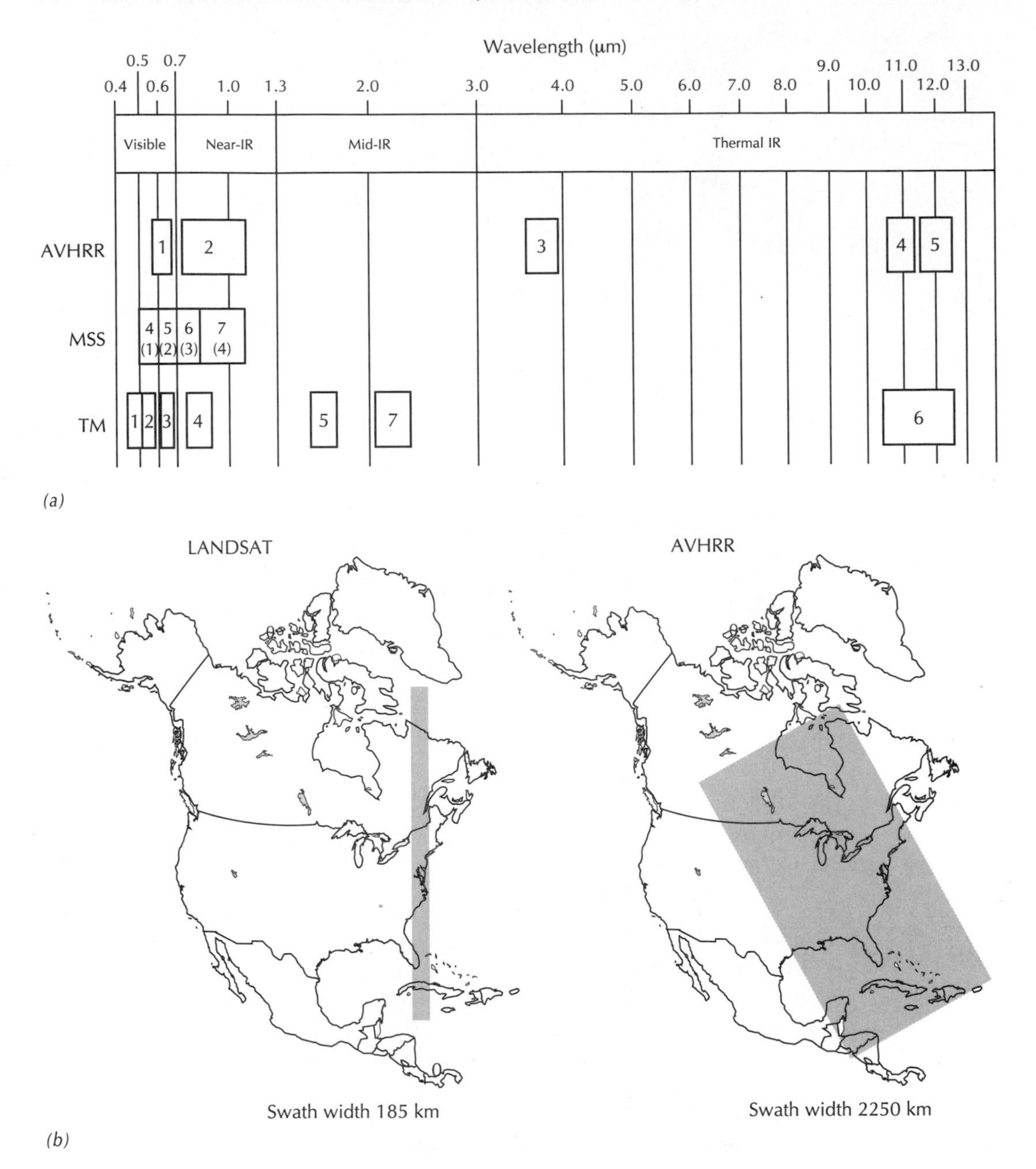

FIGURE 5.7 A comparison of the LANDSAT and AVHRR remote sensing systems. (*a*) Regions of the electromagnetic spectrum sampled by each. (*b*) Sample swath widths showing the portion of North America sampled in one pass.

Image Spectroradiometer (MODIS) is an improved and refined version of the AVHRR intrument with 500 meter resolution in several visible and near-infrared wavelengths. The ETM+ sensor on LANDSAT 8 offers finer spectral resolution than previous Thematic Mapper sensors. IKONOS represents a new class of commercially produced sensors. This instrument—designed and operated by Space Imaging Corp., Tornton, CO—can acquire data with pix-

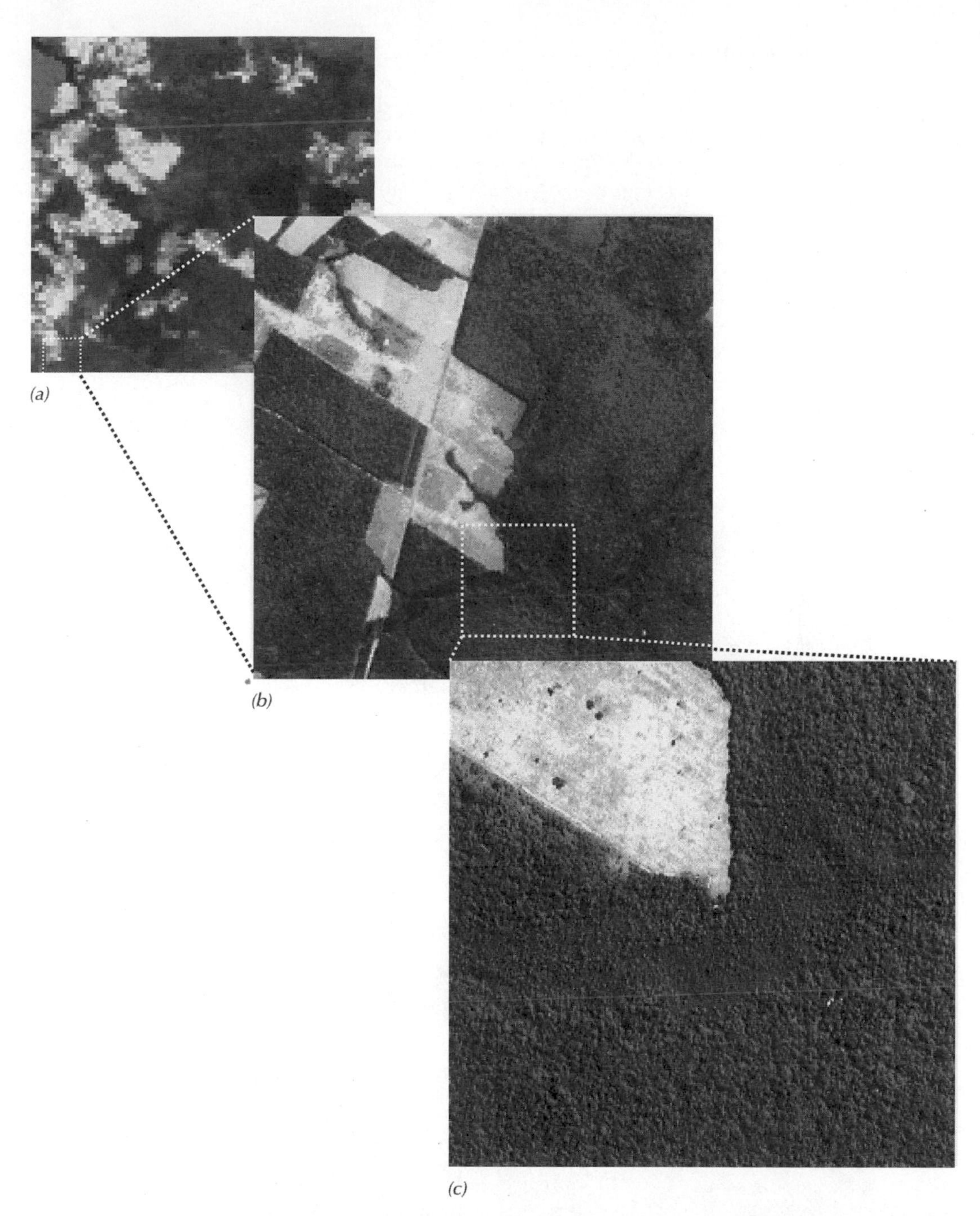

FIGURE 5.8 Comparison of different spatial resolutions (pixel sizes) from different remote sensing systems. All three images are of land use in Brazil. (*a*) MODIS (500 m resolution), (*b*) LANDSAT ETM+ (30 m resolution), and (*c*) IKONOS (4 m resolution). (Steve Hagen, University of New Hampshire) ((c) Reprinted with the permission of Peter and Trish Wolter.) **See plate in color section.**

els as small as one meter. While finer resolution provides information in greater detail, it also requires much more computer power and information storage to describe an area of a given size. The appropriate scale for a remote sensing study depends on the question being asked and the resolution of answer required.

Both of these systems have been used to gather information related to carbon balances and changes in carbon storage in terrestrial ecosystems. For example, LANDSAT imagery has been used to examine the rate of tropical forest clearing. Figure 5.8 is a scene acquired over the state of Rondonia in Brazil that captures, in a single image, a quantitative assessment of the pattern of clearing associated with recent settlement and clearing of forest for agriculture. It would be very difficult and expensive to acquire this kind of in-

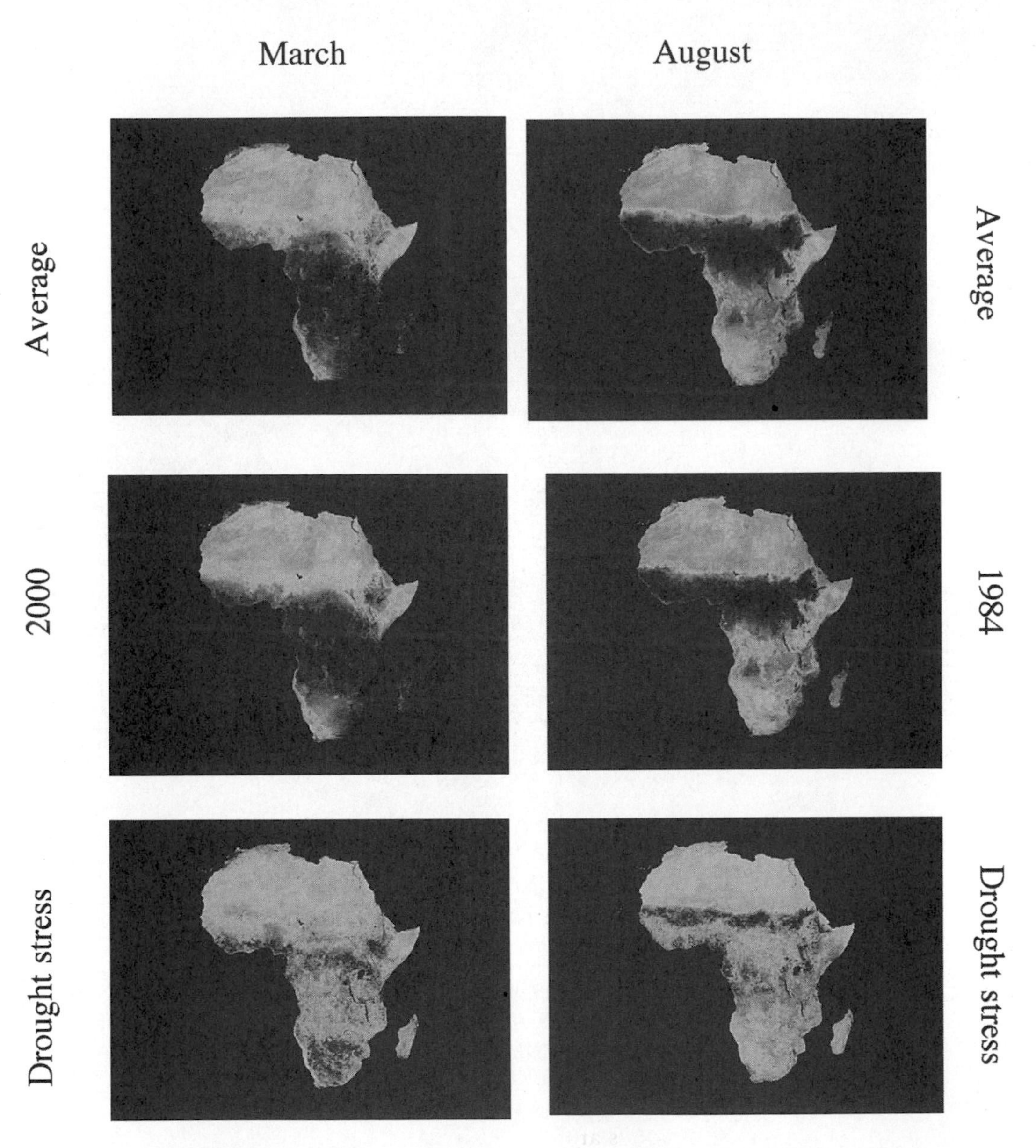

FIGURE 5.9 Mean and anomolous changes in NDVI, or "greenness," of vegetation in Africa. The top two frames show normal seasonal cycles tied to patterns of rainfall and temperature. The center two show maps for the same month, but in years of serious drought. The bottom two are a calculated difference, or anomoly, generated by comparing the two images directly above. (From C.J. Tucker and B.J. Choudhury, Copyright © 1987 by Elsevier Science Limited. Reprinted with the permission of Elsevier Science Limited) **See plate in color section.**

formation, with this degree of accuracy and spatial detail, through ground-based surveys.

One of the most dramatic uses of AVHRR data has been to clock the annual pulses of vegetative growth over the Earth. Figure 5.9 shows the distribution of "greenness" (a value obtained by calculating the ratio of reflectance in the red and near-infrared channels) over Africa at the same season in an average year, and in two years of extreme drought. Particularly striking is the interannual variability in greenness over the Sahel region, the border zone between the Sahara Desert of northern Africa and the tropical forests of central Africa. This is an area of recurring drought and famine. Monitoring greenness over this area by remote sensing offers an early warning system for detecting the vegetative response to drought and offers the hope of speeding international response to relieve famine.

From all of these examples, we obtain a view of important interactions between the human population and the landscape we inhabit. Such clear, quantitative, and persuasive images could not be obtained by land-based methods. As we increasingly recognize and attempt to cope with the truly global-scale human occupation of the Earth, the synoptic views of our habitat provided by remote sensing devices will become a more fundamental method for viewing ourselves and our environment.

SYSTEMS ANALYSIS AND ECOSYSTEM STUDIES

Implicit in the discussion to this point and in most of the following chapters is a "systems analysis" approach to the understanding of ecosystem function. Systems analysis is a method of describing and understanding complex interactions among large numbers of processes or components in a generalized way. It is concerned more with identifying the fundamental units of a system and defining how they interact than with the internal processes of each unit.

Using photosynthesis as an example, textbooks about botany or plant physiology describe this as a biochemical process, including the organelles and metabolic pathways involved, the intermediate products, enzyme systems, and so forth. Relatively less attention is paid to the rate of photosynthesis, how that rate is controlled by leaf and atmospheric conditions, or how photosynthesis interacts with transpiration, leaf energy balance, light absorption by whole canopies, and other factors external to the leaf (for a notable exception, see Lambers et al. 1999).

In contrast, a systems approach stresses factors controlling rates of photosynthesis, including the effects of water stress, nutrient availability, shading within the canopy, climate, and the genetic potential of species. In addition, the further effects of photosynthesis on the movements of carbon, energy, water, and nutrients are of primary importance. The same emphasis and focus holds for other processes as well. Ecosystem analysis is more concerned, for example, with rates of organic matter decomposition as determined by environmental factors and how this process interacts with plant uptake and nutrient cycling than how it is achieved biochemically.

Most people think of scientists as people who do experiments, and in previous chapters we described several such experiments carried out at the ecosystem scale. As we will see in the following chapters, understanding those responses requires a knowledge of many different types of processes that occur within ecosystems—knowledge that is derived from experiments at the leaf, soil, or microbial level. However, it also requires the ability to link these basic processes together by describing the interactions between them—systems analysis.

Systems analysis can be applied to any set of definable components that interact. It was developed first, and has been used most extensively, in electrical engineering, particularly in designing computers, telecommunications systems, and other devices of similar complexity. However, the summary view of ecosystem function presented in Figure 1.4 also fits these criteria. Several separate components common to ecosystems are defined. Those that interact with each other are connected by arrows showing directions of exchange. For the application of systems analysis, the names on the box and arrows are not crucial. They could define traffic patterns in a city, currency flow through an economic system, or material flow through a factory. What matters is how the outputs from any one component vary as a function of inputs and environmental conditions, and how all of the flows in the system affect overall system performance.

This is not to suggest that a meaningful analysis of a system can be carried out without a thorough knowledge of how the components in the system respond. Imagine trying to build a computer without knowing how the individual chips or circuits respond to input signals. The same can be said for ecosystems. An accurate analysis of any ecosystem must be built on a substantial data base of solid information.

A Few Terms from Systems Analysis

Of the full suite of systems analysis terms, a few have particular relevance and are frequently used in the description of ecosystems. These include the following.

Turnover Rate, Residence Time, and Response to Change in Inputs – Figure 1.4 employs a common approach in ecosystem studies, the description of important components of the system (*boxes*), and the flows of energy and materials between them (*arrows*). The relationship between flows in and out of a component and the quantity contained is a fundamental description of the activity of that component and the role it plays in the larger system. These are described by the terms "residence time" and "turnover rate."

Turnover rate is the fraction of material in a component that enters or leaves in a specified time interval. For example, a reservoir that holds 10 million gallons of water and from which 1 million gallons is pumped out every month has a turnover rate of 1/10, or 0.1 (10%), per month. Similarly, if there are 20,000 g of organic matter per square meter (or $g \cdot m^{-2}$) in a prairie soil, and plant death and senescence contribute 400 g per square meter of new organic matter (called *litter*) each year, the turnover rate for all soil organic matter is 400/20,000, or 0.02 (2%), per year (Figure 5.10). Residence time is just the inverse of turnover rate. The residence time for soil organic matter in the prairie example is 1.0/0.02, or 50, years.

Much of the dynamic nature of ecosystems and their response to disturbance or management deals with how quickly and how strongly they respond

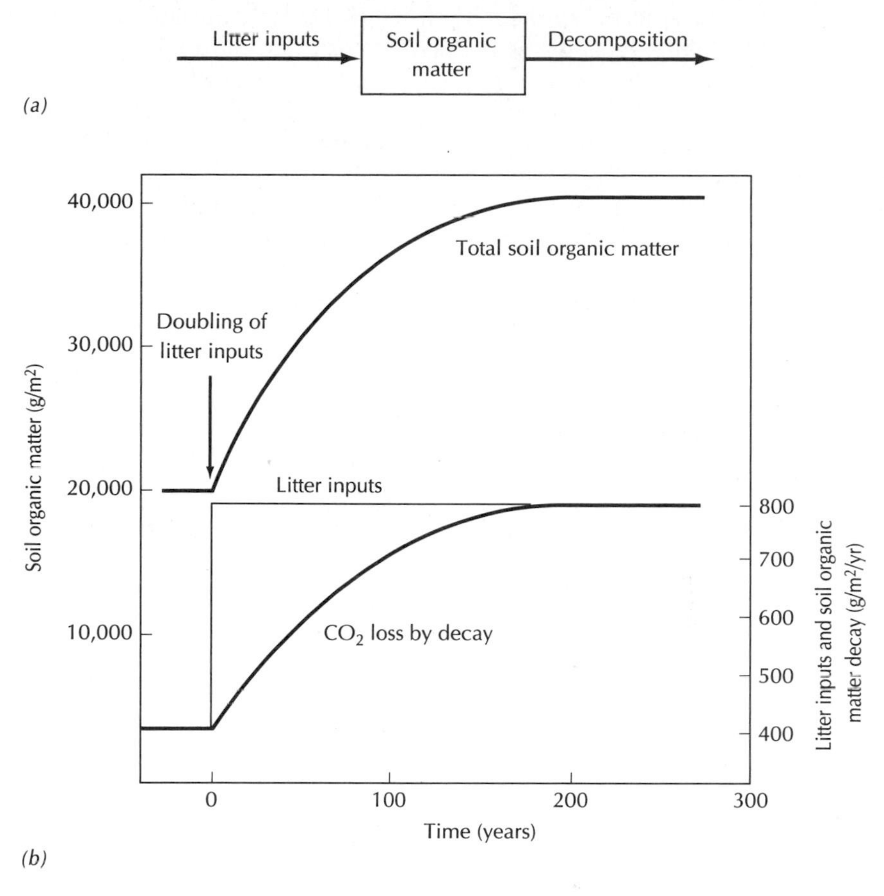

FIGURE 5.10 A simple "box-and-flow" diagram of organic matter dynamics in a prairie ecosystem. (*a*) Structure of the model. (*b*) Predicted effects of a doubling of litter inputs. With inputs and outputs of 400 $g \cdot m^{-2} \cdot year^{-1}$ and a pool size of 20,000 $g \cdot m^{-2} \cdot year^{-1}$, the turnover rate is 0.02 (2% per year), and the residence time is 50 years. Changing litter inputs to 800 $g \cdot m^{-2} \cdot year^{-1}$ and keeping turnover rate at 2% per year causes total soil organic matter to rise slowly to 40,000 $g \cdot m^{-2}$. At that point, outputs ($0.02 \times 40,000$) equal the new input rate of 800.

to changing environmental conditions. Using the prairie soil example mentioned earlier, if litter inputs changed from 400 to 800 $g \cdot m^{-2} \cdot year^{-1}$, but turnover rate remained at 2% per year, it would be quite a long time before loss of organic matter through decomposition increased to be equal with inputs and a new equilibrium value of soil organic matter was reached. At this point, total soil storage of organic matter would be double the initial value of 20,000 $g \cdot m^{-2}$ (Figure 5.10). In general, compartments with large storage and slow turnover rates tend to buffer a system against sudden or large changes in inputs. Indeed, soil organic matter is one of the major stabilizing forces in most terrestrial ecosystems.

Open and Closed Cycles – The concepts of turnover and residence time can be applied to whole ecosystems as well as to individual components. In this case, the content is the total content of the system, and the inputs and outputs are as measured in Chapters 3 and 4. Turnover rates for whole systems are also often used to describe the system as open (inputs and outputs are large relative to internal storage) or closed (inputs and outputs are relatively small). For example, most terrestrial ecosystems in humid regions have a high turnover rate for water, gaining much more from precipitation during the course of a year, and losing much more to stream flow and transpiration than is stored in the system at any one time. In comparison, most terrestrial systems have a very low turnover rate for nitrogen, gaining or losing only a very small fraction of the total content of the system in any 1 year. Such systems can be described as having open cycles for water but closed cycles for nitrogen. The devegetated ecosystem at Hubbard Brook (Chapter 4) went from a relatively closed system of nitrogen recycling to a relatively open one of nitrogen loss.

Resistance/Resilience – The degree to which a system responds to disturbance, for example, whether a forest goes from a closed to an open nitrogen cycle or not, is another crucial characteristic. Systems that show relatively little response to disturbance are said to be resistant. A severe disturbance is required to change the state of the system (Figure 5.11). Using the example of nitrate losses following clear-cutting or trenching presented in Chapter 4 (Table 4.2), those systems that did not show elevated nitrate losses would be termed *resistant.* Often, resistant systems take a relatively long time to return to their initial condition following a disturbance that is strong enough to alter nutrient balances.

Resilience is the opposite of *resistance* (Figure 5.11). Resilient systems can be altered relatively easily but return to the initial state more rapidly. Again,

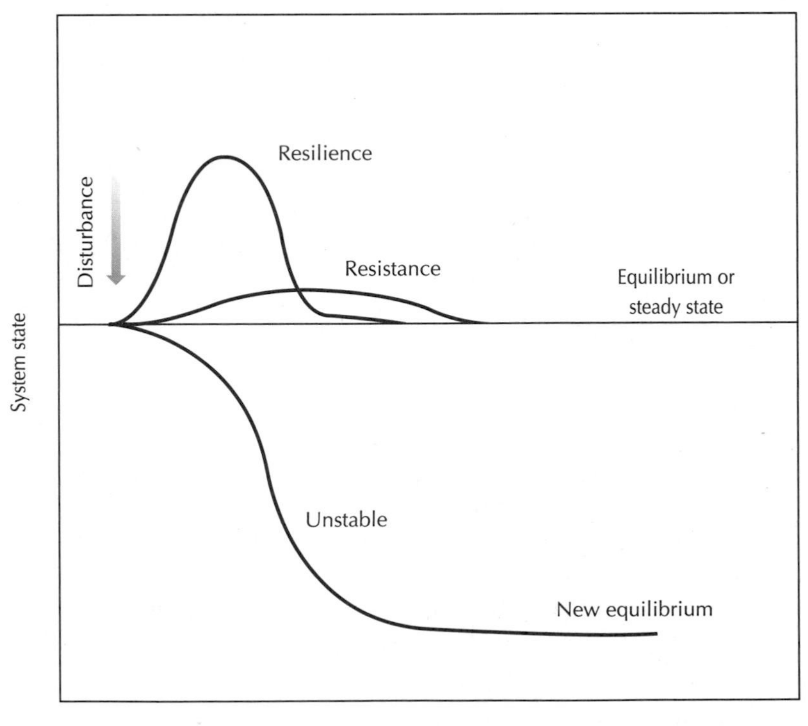

FIGURE 5.11 Comparison of resistant, resilient, and unstable responses to a disturbance event.

using the nitrate loss example, the Hubbard Brook ecosystem would not be called *resistant.* Resilience also requires a rapid return to the initial, low-nitrate-loss condition. At Hubbard Brook, the high nutrient availability evident in high stream water concentrations causes very rapid reestablishment of vegetation. High rates of nutrient uptake by this vegetation then reduces nitrate losses—a resilient response.

Feedback – *Feedback* is a term used to describe a control interaction, rather than a transfer of mass or energy. The classic example of feedback is an office heating/cooling system with a thermostat, an air conditioner, and a furnace. As the temperature falls in an office, a switch in the thermostat causes the furnace to turn on and release heat into the office. As the temperature rises past a set level, the thermostat switches the furnace off. If the temperature gets too high, then the air conditioner will come on. This is negative feedback (Figure 5.12*a*). The cooling of the office causes the furnace to heat the office again, keeping it at a nearly constant temperature. Such responses are often called "homeostatic" mechanisms, meaning that they tend to keep the system in the same state. There are many examples of homeostatic mechanisms in nature. Humans and other mammals are "homeotherms" because several systems in the body work together to maintain a relatively constant temperature. Examples of homeostatic mechanisms at the ecosystem level are presented in detail for two ecosystem types in Part III.

The opposite of *negative feedback* is *positive feedback.* Imagine that the switches on the office thermostat are reversed so that the furnace turns on when the office is too warm, and the air conditioner turns on when it is too cold. If the office is too warm, the furnace comes on and it gets even warmer. This is weak positive feedback. If the furnace burns hotter the further the office temperature gets from the temperature setting on the thermostat, then you have strong positive feedback (Figure 5.12*b*). Positive feedback is very destabilizing and can have destructive effects. There are several examples of positive feedback in ecosystems, particularly in the dynamics of animal populations (Chapter 16) and the effects of nutrient cycling on net primary production and future nutrient cycling (Chapter 18).

Stability – Turnover rate, resistance/resilience, and feedback are all related to the general concept of stability. The Hubbard Brook ecosystem response to devegetation can be described as resilient in that large changes in nitrate losses occurred immediately, but those losses returned to control levels within a few years. The system can also be described as stable relative to devegetation (at least for nitrate losses) as predisturbance conditions were recreated. Imagine if the response at Hubbard Brook had not been either resistant or resilient but had shown continuously high nitrate losses following disturbance with no regrowth of vegetation and eventual deterioration to some severely degraded state. It would then be described as unstable with regard to devegetation. The initial condition is lost, and the system does not have the capacity to return to it (Figure 5.12). Some new condition is reached, and the system continues in that new state. Serious examples of this occur in "desertification," the conversion of semiarid grasslands or shrublands to deserts by overgrazing or by agriculture that disrupts the thin layer of vegetation that holds the soil in place or in extreme fire events that can convert forests to bare soil or rock.

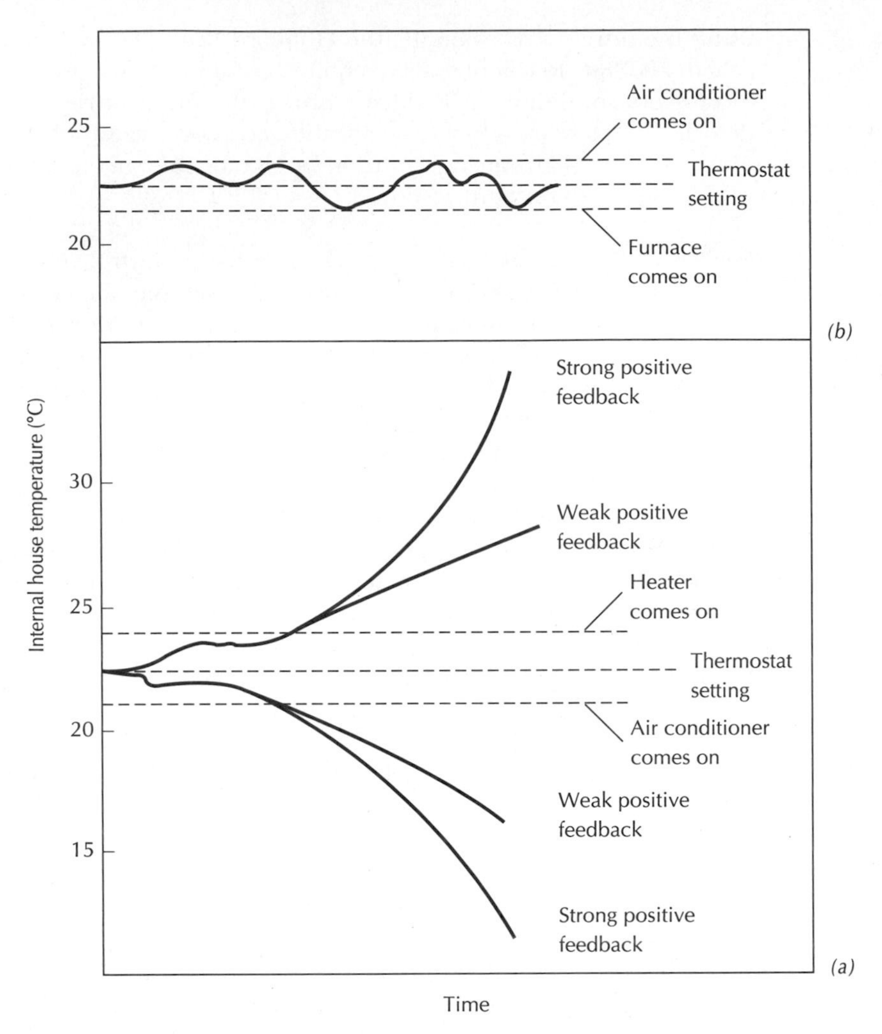

FIGURE 5.12 Effect of (*a*) negative and (*b*) positive feedback on system performance using the thermostat example.

Stability is a very difficult concept to describe quantitatively. The desertification examples are clearly unstable, but is a resilient system more stable because it returns to initial conditions more rapidly, or is a resistant system more stable because it is more difficult to dislodge from the initial state? Volumes have been written on the subject of ecosystem stability, how to define it, and what makes systems stable or unstable. We will use the term only in a relative and intuitive way.

Limitations of the Systems Approach

For all the similarities between the study of natural systems and mechanical or electrical ones, the analogy is not perfect. A major difference is that, in human-made systems, all of the subparts can be described clearly because they have been constructed to particular specifications. This is not true of ecosystems. Often, the responses of individual components are known only dimly for a fairly narrow set of conditions. So systems analysis of ecosystems is not as clean or precise and is always dependent on accurate field measurements by researchers from many disciplines. In that sense, the analysis is never complete, as new information is constantly being made available. Rather, the analysis of ecosystems is more of a continuing set of approximations, hopefully becoming more complete and accurate as the information base accumulates.

COMPUTER MODELING: A SYNTHESIS TOOL FOR ECOSYSTEM STUDIES

While most of the concepts described here have been described in relative or qualitative terms, ecosystem analysis becomes quantitative when the results of specific experiments are expressed in mathematical form, such as an equation resulting from a statistical analysis. To describe the overall function of a complex ecosystem, many experimental studies will be involved, and many separate equations will be developed. While the human mind has a tremendous potential for envisioning the complex interactions that operate in systems (e.g., Figure 1.4), it is not very good at calculating the results of interactions between several (or even a few) equations. Computers provide the perfect complement to the mind's ability to conceptualize, and computer modeling is the primary tool used to synthesize our understanding of ecosystem function.

Computer Models and the Modeling Process

What Are Models, and Why Do We Build Them? – Imagine a central library containing the results from all of the studies relevant to a particular problem expressed in the form of simple mathematical equations. The equations can represent the results of statistical analyses, measurements over time, or whatever form of information is relevant. Now imagine that all of the equations in this library can be linked together, feeding results back and forth and projecting the effects of management, pollution, or climate change into the future. This is basically what a computer model is intended to be.

Similar in structure to many popular game or office software packages, a model is a sequence of instructions and computations that play the "game" of recreating the past or predicting the future of an ecosystem. An operational model generally consists of two parts: (1) the program, which is the series of calculations that the model makes, and (2) one or more sets of data that specify the type of ecosystem and environmental conditions for which the predictions are desired. Equations describing all of the relationships included in the model constitute the commands, or code, which are executed in a model run, using the data provided. In general, the code expresses general relationships that apply across many ecosystem types or environmental conditions, while the data provide values or coefficients that are specific to the ecosystem for which the model is being run.

While the library represented by the computer model is of value, what is lacking in this library is also of interest. The process of constructing a model often reveals gaps in understanding or areas where data are incomplete and so leads directly to additional research. As the library becomes more complete, the model becomes more valuable for predicting responses to specific treatments or changes in environmental conditions.

It is important to realize that no computer model will ever be a *complete* representation of a system, nor should it be. Any number of processes occur in any system that are not important to answering a particular question. For example, including herbivory as a process in systems where it is generally not important only adds to the difficulty of completing the model and the expense of running it. Recurring questions in the modeling process should include: Is this process important enough to be included?, Does it need to be represented in this much detail?, and, Are we looking at processes at the time and space scales appropriate to the purposes of the model?

Steps In Building and Applying a Computer Model – The steps in building a computer model reflect both the distinction between code and data and the three different purposes discussed earlier.

The first step is to put together the structure of the model. The structure is a generalized expression of how the type of ecosystem under consideration operates. For example, Figure 5.13*a*, ignoring for a moment the numbers associated with the boxes and arrows, is a simple flow diagram for the movement of nitrogen through forest ecosystems. This diagram represents the structure of one of the first computer models created for the prediction of forest management effects on nutrient cycling and productivity in forest ecosystems. It is a generalized realization of a forest ecosystem because all forest ecosystems have all of the parts described by the different boxes. This general set of compartments and flows is captured in the code of the computer model.

However, different numbers of components and flows may be described. Figure 5.13*b* shows a second, more complex structure for describing nitrogen cycling that could also be applied to any forest ecosystem. While the more detailed structure may provide a more accurate vision of ecosystem function, it also requires a substantial increase in the amount of information needed to run the model. There is a constant tradeoff in the construction of a model between the desire to increase the complexity of the structure and thereby make the model more complete, and the costs of doing the research to provide the information required to run a more complex model. An additional consideration is the potential for introducing errors into the model predictions by including a poorly described or poorly understood process.

When we put actual numbers on the flows and describe them for a particular system, we parameterize the model for a particular type of forest under a particular set of environmental conditions. For the examples in Figure 5.13, the initial values for nitrogen storage in each of the boxes and the rates at which nitrogen is transferred between boxes are described in the data files that are read by the program and used in running the model. The data files thus contain the *particular* knowledge about the system that has been described in a *general* way in the computer program.

Once the model has been structured and parameterized, two complementary steps are used to test its accuracy and refine the structure. *Validation* is one term for the process of comparing model predictions against measured

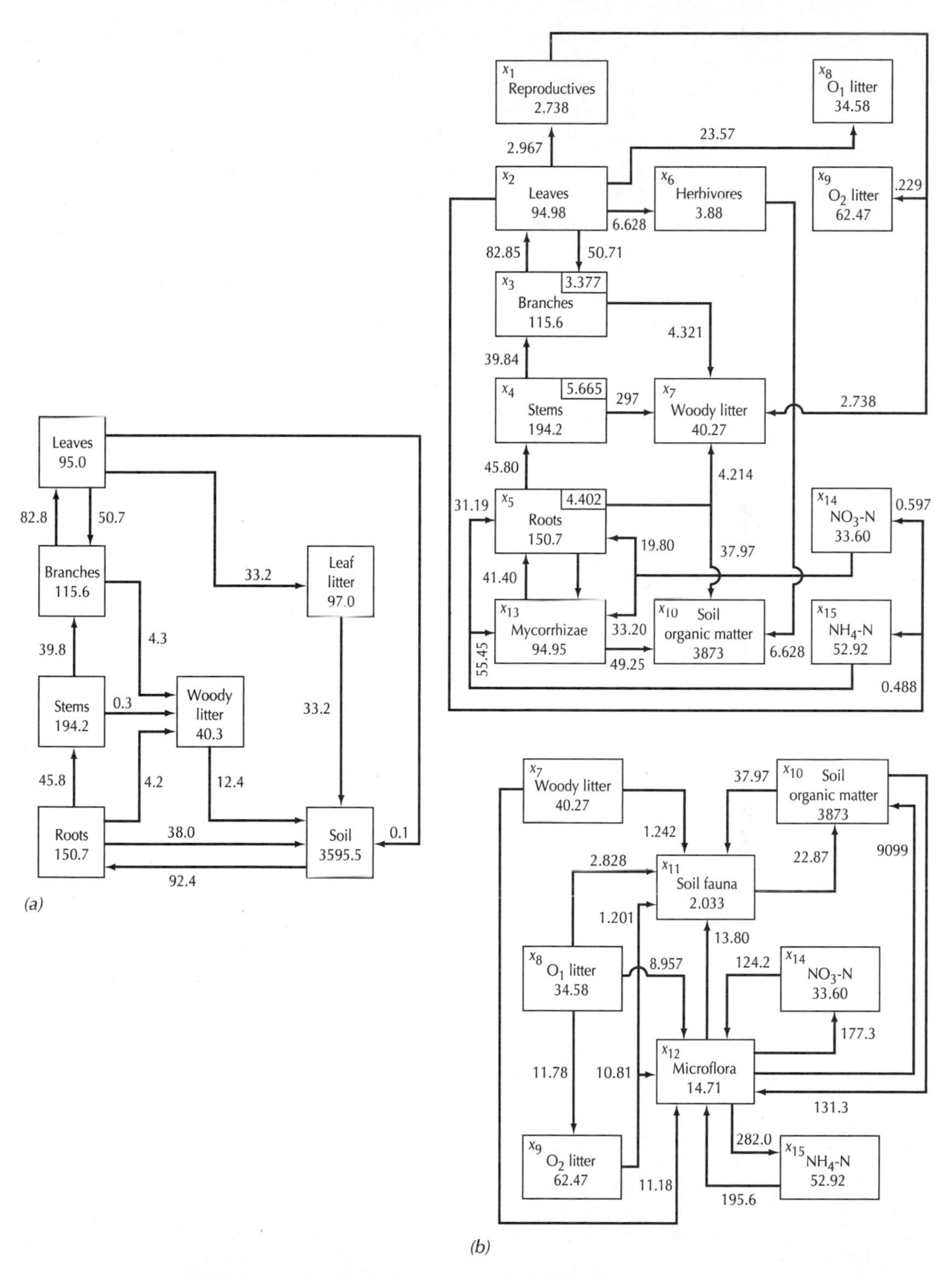

FIGURE 5.13 Two different structures for the definition of a model of nitrogen dynamics in an oak–hickory forest (Waide and Swank 1977). (*a*) A reduced-structure model. (*b*) A model with greater detail and more compartments and transfers (presented in two interconnected parts).

values that *were not used in the construction of the model.* In sensitivity analyses, the response of the model to changes in parameter values is tested. Parameters or processes that do not affect model predictions significantly can be removed from the structure, reducing the complexity of the model and the amount of data required to run it. Both of these steps are repeated to refine the structure and increase the accuracy of the model. The value of the model predictions is determined in part by the accuracy of the validation process.

All of this no doubt sounds pretty tedious and dull! It is an unfortunate fact that there are few things less interesting than describing or discussing computer models. In contrast, actually working with models can be very enlightening and enjoyable. Consider the popularity of video and home-computer games, many of which serve educational purposes. The descriptions of the programs that make those games possible would bore to tears all but the most dedicated hacker.

REFERENCES

Lambers, H., F. S. Chapin, III, and T. L. Pons. 1998. *Plant Physiological Ecology.* Springer, New York.

Nadelhoffer, K. J., and B. Fry. 1994. Nitrogen isotope studies in forest ecosystems. In Lajtha, K., and R. Michener (eds.), *Stable Isotope Studies in Ecology and Environmental Science.* Blackwell Scientific Publications, Oxford.

Trumbore, S. E. 1995. Use of isotopes and tracers in the study of emission and consumption of trace gases in terrestrial environments. In Matson, P. A., and R. C. Harriss (eds.), *Biogenic Trace Gases: Measuring Emissions from Soil and Water.* Blackwell Scientific Publications, Oxford.

Tucker, C. J., and B. J. Choudhury. 1987. Satellite remote sensing of drought conditions. *Remote Sensing of Environment* 23:243–251.

Waide, J. B., and W. T. Swank. 1977. Simulation of potential effects of forest utilization on the nitrogen cycle in different southeastern ecosystems. In Correll, D. L. (ed.), *Watershed Research in North America.* Smithsonian Institution, Washington, DC.

Whittaker, R. H. 1967. Gradient analysis of vegetation. *Biological Review* 42:229.

Part 2

MECHANISMS

Processes Controlling Ecosystem Structure and Function

INTRODUCTION

The transfers of energy, water, and nutrients through ecosystems are driven by the interactions between biotic (plants, animals, microbes) and abiotic (climate, soils) components. The growth of plants (or the production of biomass) is a central process in ecosystem dynamics as it involves the transfer of carbon from CO_2 in the atmosphere into organic forms, and requires both the reciprocal transfer of wat0 soils to the atmosphere and the cycling of nutrients. Once formed, organic materials in plants can become a food source for animals (herbivores) or can be shed as litter (senescent or dead organic matter) into soils for decomposition. In either case, the cycles of carbon and nutrients are completed with CO_2 released into the atmosphere and nutrients released into the soil in plant-available forms. Where conditions are right, the occurrence of fires can also provide a major "shortcut" for these cycles, returning years to decades of accumulated carbon and nutrients to soils and atmosphere in a matter of minutes.

The purpose of Part 2 is to present, in simplified form, a discussion of the major processes that provide for the cycling of carbon, water and nutrients. These are divided into two broad categories: (1) those that control the primary production of organic matter (plant growth) and (2) those that break down or consume organic matter.

The production section begins with a chapter on photosynthesis and transpiration and the factors that determine the rate at which carbon fixation and the evaporation of water occur in individual leaves. These processes are then placed in the canopy context and discussed in terms of total ecosystem photosynthesis on an annual basis. This is followed by a discussion of soil processes that affect nutrient availability and how biological processes alter availability. Having considered the factors that govern biomass production, we then look at allocation of production to leaves, stems, roots, and respiration.

The second half of Part 2 deals with the fate of organic products. We present three main pathways for the breakdown of organic matter: (1) decomposition, (2) herbivory (consumption by plants), and (3) fire. Finally, this section on processes concludes with a presentation of emergent patterns of the variation in canopy structure, production, and nutrient cycling among ecosystem types, as well as emerging theories that begin to explain these large-scale patterns.

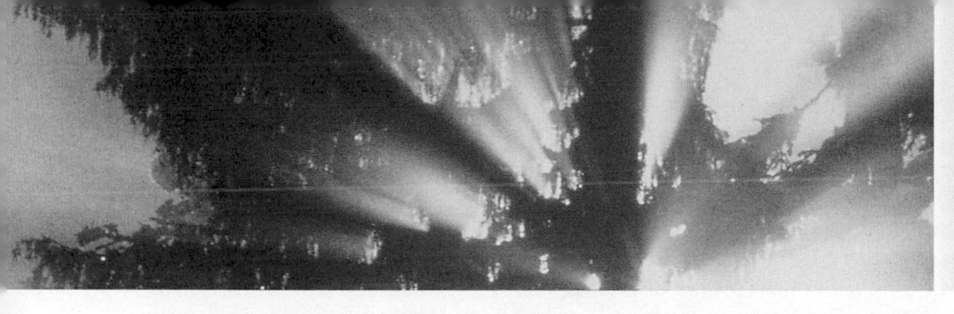

6

ENERGY, WATER, AND CARBON BALANCES OF LEAVES

INTRODUCTION

Leaves are the tissues directly involved in the transfers of energy, carbon, and water between the atmosphere and terrestrial ecosystems (we use the term *leaves* in the generic sense to include needles of coniferous trees, leaf sheaths of grasses, and other photosynthetic tissues). In all but the driest or most degraded systems, leaves intercept most of the sunlight before it reaches the ground. As little as 2% of photosynthetically active solar radiation may reach the soil surface in the most dense (vigorous) forests and grasslands, and the immediate environment around any one leaf varies dramatically from the sunlit top to deeply shaded bottom of a dense canopy.

Radiation absorbed by foliage drives photosynthesis and CO_2 fixation. However, even under ideal conditions, only a very small portion of solar energy is converted to chemical energy through photosynthesis. Over 95% of absorbed light energy is converted to heat that in turn increases leaf temperature. High leaf temperatures can impair leaf function by increasing respiration rates or, at extreme levels, disrupting function. Leaf temperature can be reduced through evaporation of water from cell surfaces, as a large amount of energy is required to convert liquid water to gaseous (vapor) form. Water vapor within the leaf is lost to the atmosphere through pores on the surfaces of leaves called stomata, and the resulting transfer of water to the atmosphere is called transpiration. The availability of water for transpiration is limited by soil water content as well as the ability of plants to extract that water. Under conditions of low availability in the soil and high evaporative demand in the atmosphere, water stress occurs in plants, and stomata are closed to minimize water loss. As CO_2 enters plants through these same pores, stomatal closure also severely reduces photosynthesis. During periods of stomatal closure, heat is lost through conduction and convection to the atmosphere.

Thus, the energy source for photosynthesis, sunlight, drives the fixation of CO_2 required for plant growth but also imposes a heat load on the leaves and a potential for transpirational water loss. The rate at which both photosynthesis and transpiration occur depends on how open the internal spaces of the leaf are to the atmosphere, while changes in the size, shape, and chemical content of leaves affect both how fast these processes occur and how quickly excess heat is lost to the surrounding air.

The morphology of leaves and canopies of individual plants and species are generally viewed as the result of ecological and evolutionary processes in which maximizing total carbon gain to the individual has high adaptive value. All plants need basically the same resources for growth (light, water, certain nutrients), and most use similar physiological mechanisms to obtain those resources. Differences between plant species in morphology, size, and growth form represent subtle shifts in the efficiency with which different resources are acquired and used. Plants as visibly distinct as the giant redwoods of coastal California and tiny shrubs of the arctic tundra are doing basically the same things. The differences in size, shape, longevity, and duration of activity during the year reflect evolutionary adaptations to the physical, chemical, and biological environment in which they grow.

However, there are a few very important differences in biochemical mechanisms of resource acquisition that separate the tremendous number of plant species in the world into a smaller number of distinct categories. The one most relevant to this chapter is the predominant process used in carbon fixation for photosynthesis.

The purpose of this chapter is to discuss the carbon/water/energy interaction at the leaf level. We will see how the basic processes of photosynthesis, transpiration, and energy exchange interact and the effects of differences in leaf morphology, chemistry, and physiology on these processes.

THE ENERGY, CARBON, AND WATER BALANCE OF A LEAF

The key exchanges between leaves and the atmosphere are summarized in Figure 6.1. Direct short-wave radiation (sunlight) is the primary energy input. Long-wave radiation and direct transfers of heat to and from the atmosphere (conduction and convection) can also be important sources or losses of energy for leaves. Transpiration causes the loss of water and can be a very major component of the leaf energy balance due to the energy required to convert water from liquid to vapor form. A small fraction of incoming short-wave radiation is also converted to chemical energy in the simple sugars that are the basic product of photosynthesis. These exchanges can be grouped into net radiation, and the exchange of sensible energy (heat), latent energy (conversion of liquid water to vapor), and chemical energy (photosynthate). Tied to these exchanges are the reciprocal flow of carbon into the plant as CO_2 and water out as water vapor (Figure 6.1).

CHEMICAL AND LATENT ENERGY EXCHANGES

Photosynthesis and Respiration

Nearly all organic compounds found on Earth are the direct or modified products of photosynthesis. Photosynthesis is the energy-driven combination of CO_2 from the atmosphere with liquid water (H_2O) to form simple carbohydrates (sugars). A generalized equation for the process is:

$$CO_2 + H_2O = CH_2O + O_2$$

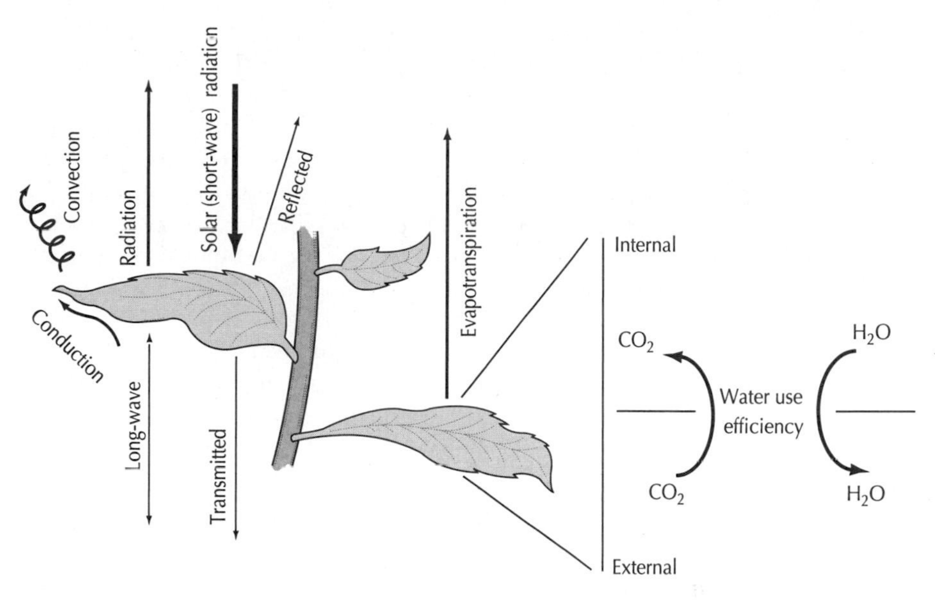

FIGURE 6.1 Energy, water, and CO_2 balances over a leaf. Short-wave solar radiation is transmitted, reflected, or absorbed. The energy in absorbed radiation can be re-emitted as thermal (long-wave radiation), lost through conduction and convection (sensible energy), lost through evaporation of water for transpiration (latent energy), or converted to chemical energy in the products of photosynthesis. Note that evapotranspiration is an important component of both the energy and water balance and is an inevitable consequence of CO_2 uptake. Water use efficiency is the ratio of carbon gained to water lost. Rates of both water and carbon exchange are related to the conductance of the leaf or the degree to which the stomata are open.

Although photosynthesis never accounts for a large proportion of total energy balance of leaves, it provides essentially all of the energy available to plants, animals, and microbes. Respiration can be thought of most simply as the reverse of photosynthesis; the oxidation of organic compounds with a release of energy used by plants, animals, and microbes to drive their own life processes.

As the differences in structure and function of ecosystems over the Earth suggests, there is a good deal of variation in the rate at which carbon is fixed under different conditions. Part of this variation is due to the environmental conditions themselves, and part is due to the variation in physical, chemical, and physiological characteristics of leaves.

Photosynthetic Capacity: The Interaction Between Light and Nitrogen – Photosynthesis is driven by the energy in short-wave radiation and increases with light intensity. However, leaves have a maximum capacity for photosynthesis that is determined by the quantity of light-harvesting chlorophyll and associated enzymes and structures that are required to convert that energy into the chemical energy in the bonds of carbohydrates. There is also a basic energetic cost in maintaining and repairing this system that results in the continuous production of CO_2 through respiration.

The interaction among these processes leads to the characteristic shape of the photosynthetic light-response curve (Figure 6.2), which expresses the net exchange of CO_2 between a leaf and the atmosphere. All light-response curves have a negative value at zero light, which is the dark-respiration rate. As light intensity increases, gross photosynthesis also increases. At the **compensation point,** gross photosynthesis is equal to respiration, and net CO_2 exchange (net photosynthesis) is zero. With further increases in light intensity, the photosynthetic rate approaches an asymptote, reaching the **light-saturated rate,** or maximum rate.

At low light intensities, photosynthesis is limited by energy (light) and increases linearly with increasing light levels. The slope of the line expresses the efficiency with which light is converted to carbohydrates and is called the quantum yield. At high light intensities, the rate is limited by either the rate of diffusion of CO_2 into the leaf or by the rate at which the enzyme systems in the leaf can fix that CO_2. This process of CO_2 fixation is called carboxylation, and, at high light intensities, photosynthesis is carboxylation limited.

There are consistent patterns in the variation of these characteristics across leaf types. For example, leaves of the same species grown at different light levels differ, as shown in Figure 6.2. Those from high-light environments have a higher light-saturated rate and a higher (more negative) dark-respiration rate. They will also reach light saturation at a higher light intensity. As is explained later, these differences in photosynthetic rates in sun and shade are accompanied by differences in leaf morphology.

Differences in maximum photosynthetic rates between species and between leaves grown in full sun and those grown in shade are strongly related to nitrogen concentration within the leaves (Figure 6.3). The generality of this relationship across such a tremendous range of conditions, species, and growth forms suggests that it expresses a fundamental physiological con-

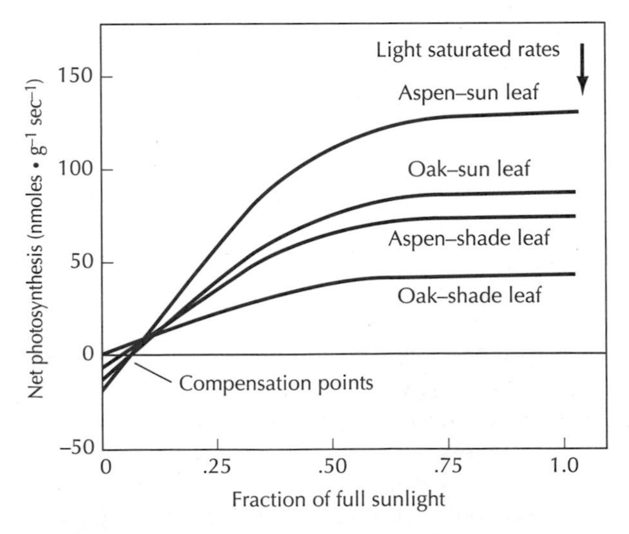

FIGURE 6.2 Photosynthetic response curves for leaves of two different species grown at high and low light intensity. Aspen is a shade-intolerant species. Oak is moderately tolerant of shade. (After Loach 1967)

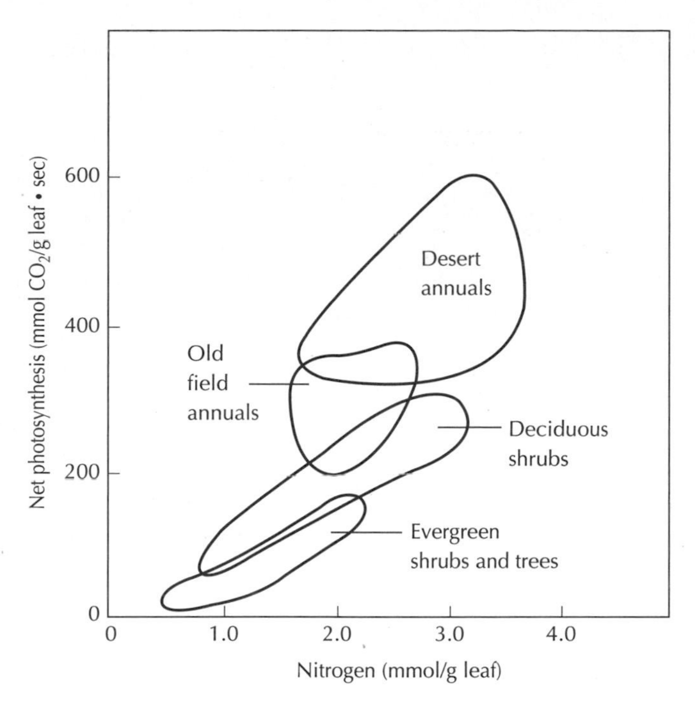

FIGURE 6.3 Maximum photosynthetic rates as a function of leaf nitrogen concentration. (Field and Mooney 1986)

straint on ecosystem function. It reflects the fact that, with the exceptions described later, all plants carry on photosynthesis in the same way using the same system of chlorophyll and associated enzymes. At higher light levels, or on sites with higher nitrogen availability, these light-harvesting systems are "packed" into leaves at higher densities, resulting in both a higher concentration of nitrogen and faster maximum rates of photosynthesis. Similarly, higher nitrogen concentration also results in higher rates of dark respiration. Said in another way, it appears that variations in maximum rate of carbon gain among leaves is determined more by the chemistry of the leaf than by the species that produced it.

An extensive, global scale comparison of leaf function has extended this interaction to two additional leaf characteristics: (1) specific leaf area (related to leaf thickness and measured in square centimeters [cm^2] of leaf area per gram of leaf mass) and (2) leaf life span (the average time a leaf remains on the plant). Figure 6.4 shows that leaves with high nitrogen content are also thinner (higher specific leaf area) and turn over more rapidly (shorter leaf life span). The consistency of these relationships extends the argument for general and fundamental physiological constraints on leaf function from carbon balance to leaf morphology and longevity. All of this is part of a larger theory of ecosystem function that we revisit in Chapter 18.

Temperature – In most plants, there is an optimum temperature for photosynthesis that reflects the combined effects on both gross photosynthesis and respiration (Figure 6.5). In general, photosynthesis peaks at a temperature

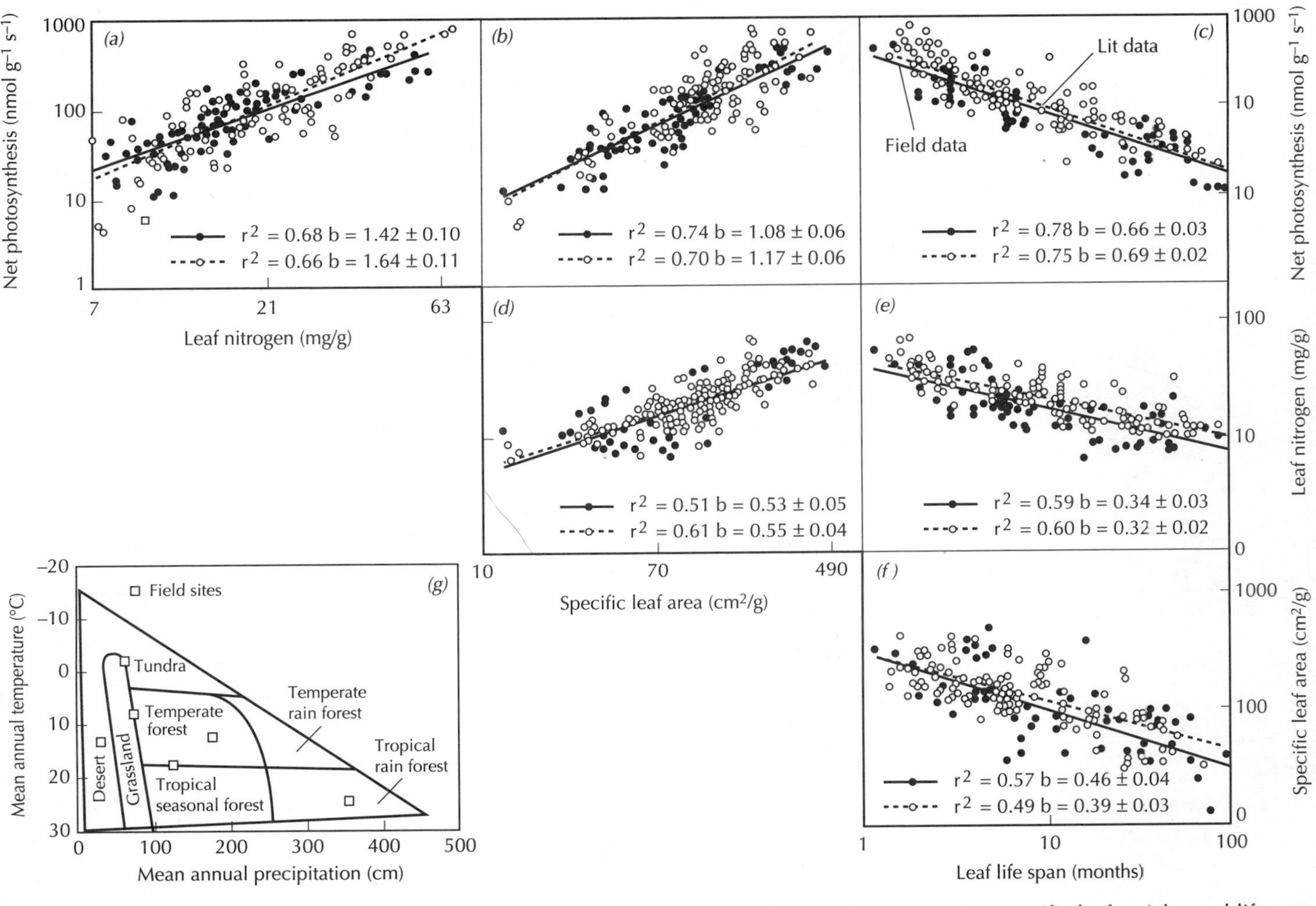

FIGURE 6.4 Relationships among foliar nitrogen concentration, photosynthetic capacity, specific leaf weight, and life span for species from several different biomes around the world. (Reich et al. 1997)

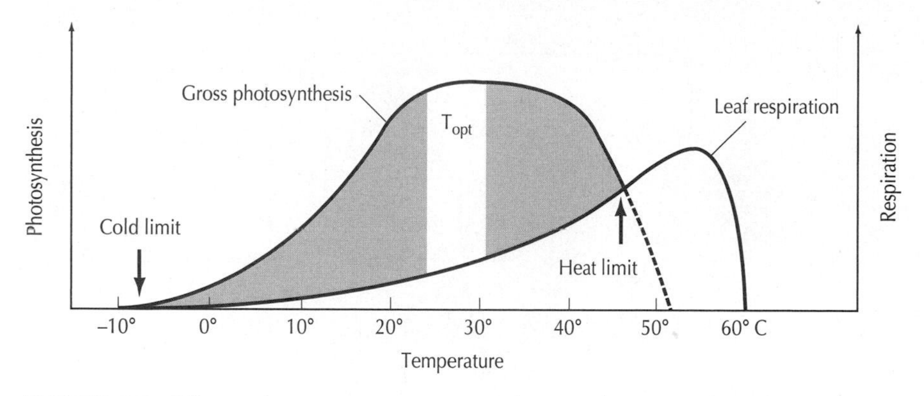

FIGURE 6.5 Effects of temperature on gross photosynthesis (*top line*), leaf respiration (*bottom line*), and net photosynthesis (*shaded area*). (After Larcher 1995)

near the mean daytime temperature of the environment in which the plant is found. This is not strictly a genetic response, as individual plants with the same genetic composition show different optimal temperatures if grown under different conditions. This temperature optimum can also change over the course of the growing season as mean temperature changes.

In contrast, respiration increases exponentially with increasing temperature. The slope of this increase is often summarized in the Q_{10} value, or the proportional increase in rate with a 10°C increase in temperature. An average Q_{10} value is 2.0 (meaning that respiration doubles with each 10°C increase in temperature), but measured rates can vary considerably. As with photosynthesis, respiration rates can acclimate to different temperature regimes. Plants of the same species grown under different temperature regimes show different temperature responses (Figure 6.6).

Vapor Pressure Deficit (Humidity) – A lack of water in the soil can restrict photosynthesis in ways that are presented in Chapters 7 and 8. In addition, many forest species respond directly to relative humidity, or vapor pressure deficit (the difference between the amount of water vapor that the atmosphere can hold and the amount actually present). A linear response has been measured between net photosynthesis and vapor pressure deficit for several evergreen species from seasonally dry regions of the western United States (Figure 6.7). This has been described as a "feed-forward" adaptation system (in contrast to a feedback system), in which plants respond to reduced humidity by restricting gas exchange, thereby reducing transpiration and conserving soil moisture for later in the growing season. This response also restricts stomatal opening to times when humidity is high and evaporation from internal leaf surfaces is low.

Transpiration and Water Use Efficiency

Movement of CO_2 from the atmosphere and into the plant for photosynthesis occurs through openings or pores in the leaves called stomata (see Figure

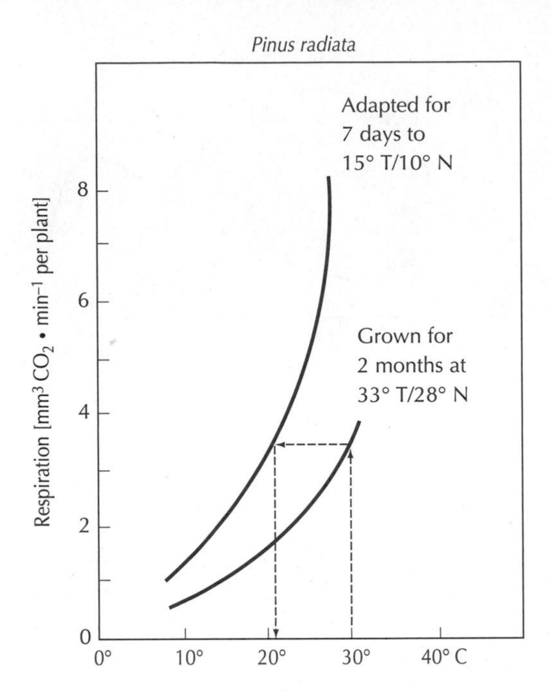

FIGURE 6.6 Respiration rates of pine seedlings grown at two different temperatures. The upper line is for seedlings grown at temperatures of 15°C (day) and 10°C (night). The lower line is for seedlings grown at 33°C and 28°C. Generally plants grown at higher temperatures show lower rates of respiration at any given temperature. (Rook 1969, in Larcher 1995)

6.1). This happens because the fixation of CO_2 at the cell surfaces within the leaf reduces the internal CO_2 concentration, causing more CO_2 to diffuse into the leaf along this concentration gradient. The same cell surfaces are constantly moist, which causes the evaporation of water into the internal leaf (intercellular) spaces. Under conditions of less than 100% humidity and full cloud cover, a gradient of water vapor concentration (relative humidity or vapor pressure deficit) exists in the opposite direction of the CO_2 gradient. Thus, water loss to the atmosphere is an inevitable result of CO_2 uptake. The rate of transpiration is a function of the degree to which the leaf is open to the atmosphere (also called conductance) and vapor pressure deficit. Not surprisingly, conductance is directly related to the rate of photosynthesis (Figure 6.8). This means that, at a given vapor pressure deficit, transpiration is directly proportional to photosynthesis.

Transpiration can be an important component of the leaf energy balance. However, transpiration requires the constant movement of water to the leaf to replace that lost by evaporation. If the plant is unable to take up enough water from the soil, water stress develops, leading eventually to stomatal closure and a great reduction in both water loss and CO_2 uptake for photosynthesis (see Chapter 7).

This tradeoff between water loss and carbon gain, along with the fact that a lack of available water limits plant growth over a large part of the earth, has generated interest in the concept of **water use efficiency (WUE).**

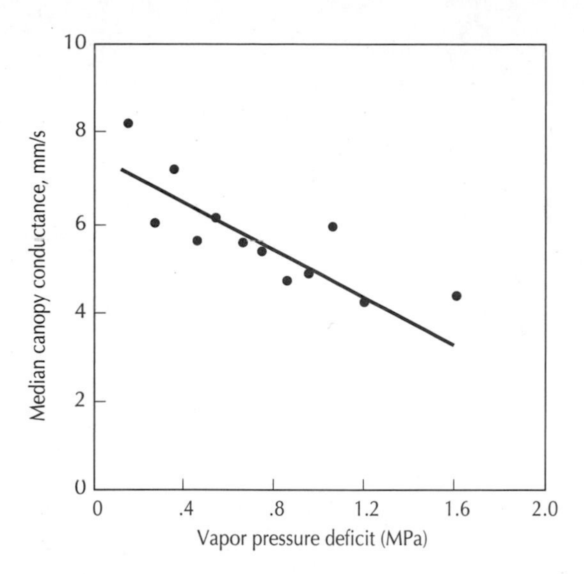

FIGURE 6.7 Reduction in forest canopy conductance (reduction in the opening of stomata) in response to higher vapor pressure deficits (lower relative humidity). (Jarvis 1981, in Waring and Schlessinger 1985)

This is defined as the mass of carbon gained per unit mass of water lost, or:

$$WUE = \frac{C \text{ gain (g)}}{H_2O \text{ lost (g)}}$$

Water use efficiency can be defined at several levels: the leaf, the whole plant, or the whole ecosystem. In general, water use efficiency is high when vapor pressure deficit is low (humidity is high), and low when vapor pressure deficit is high.

The Effect of CO_2 Concentration on Photosynthesis and Transpiration – If CO_2 concentrations in the atmosphere were not changing, the effect of this gas on rates of photosynthesis would be of only academic interest and would not warrant a section in a book on factors affecting ecosystem function. However, human actions have increased atmospheric CO_2 by 25% since the beginning of the industrial revolution (about 1800), and predictions are that this concentration will double again by 2100. This change has major implications for the Earth's ecosystems (discussed in Chapter 26). It also has direct implications for photosynthesis.

The current concentration of CO_2 in the atmosphere is severely suboptimal for photosynthesis. At 370 parts per billion, the diffusion of CO_2 through stomata and into leaves is slower than the rate at which it can be fixed by carboxylation. The response of photosynthesis to increased concentrations of CO_2 in the intercellular spaces within leaves is shown in Figure 6.9. This response is analogous to that described for light in that photosynthesis is limited by the resource (CO_2) at low concentrations and by the capacity to use the resource at high concentrations.

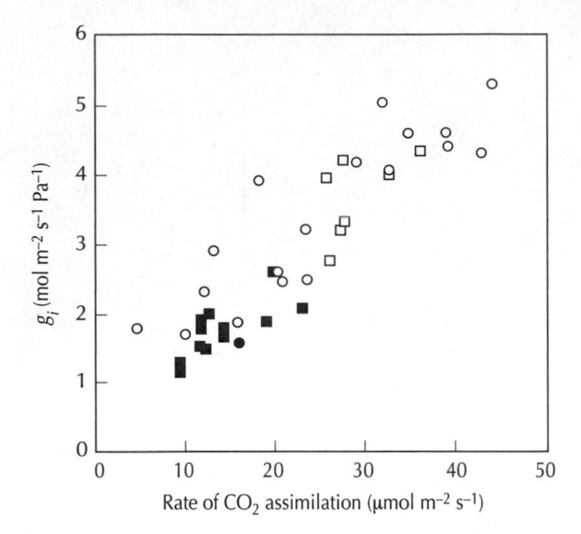

FIGURE 6.8 Relationship between net photosynthesis and conductance for a wide range of species and leaf types. (Evans and Von Caemmerer 1996, in Lambers et al. 1998)

If CO_2 causes no change in conductance and photosynthesis increases, then water use efficiency also increases (carbon gain up, no change in transpiration). Whether and to what degree stomatal conductance changes with increasing CO_2 is still unclear, with different experiments yielding different results. One possibility is that plants are unable to use all of the extra carbon captured by increased photosynthesis because of shortages of nutrients. In this case, photosynthesis could be "down-regulated" through partial stomatal closure, reducing both carbon gain and transpiration. Water use efficiency would still be increased. Again, more on this in Chapter 26.

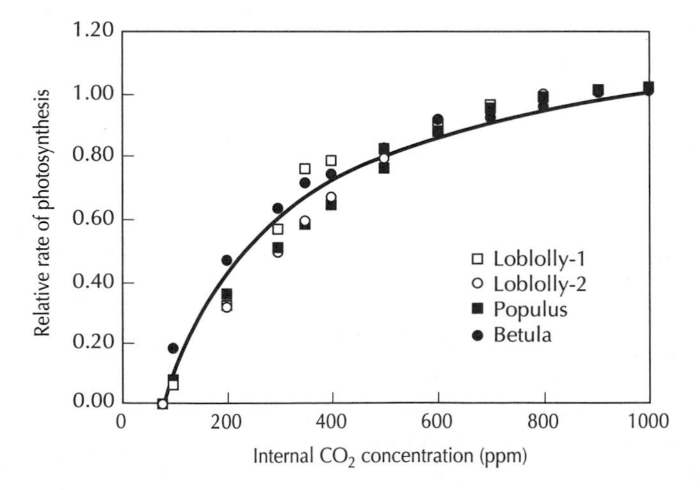

FIGURE 6.9 Relative rates of net photosynthesis as a function of internal leaf CO_2 concentration for four experiments with three different species. (From Ollinger et al. 2000)

NET RADIATION

The sun emits radiant energy over most of the electromagnetic spectrum. After passing through the Earth's atmosphere, some of this energy has been absorbed, particularly in wavelengths associated with water vapor and major atmospheric gases (Figure 6.10). Plant leaves reflect some of this incoming radiation but absorb most. Of the energy absorbed, only a small fraction of that in a very narrow set of wavelengths in the visible part of the spectrum is actually converted by photosynthesis into the chemical energy in the bonds of simple sugars. The rest is converted to heat and increases leaf temperature.

Any object above absolute zero (−273°C, or the temperature at which all molecular motion stops) radiates energy. This is why you feel warmth from glowing coals in a fireplace even when no warmed air reaches you directly. At the temperatures realized by living leaves, this reradiated energy is entirely in the long-wave, or thermal, rather than the short-wave, or visible, part of the electromagnetic spectrum, and increases exponentially with temperature. As the soil and plants surrounding the leaf are all emitting long-wave radiation, this is both an input and an output of energy for leaves.

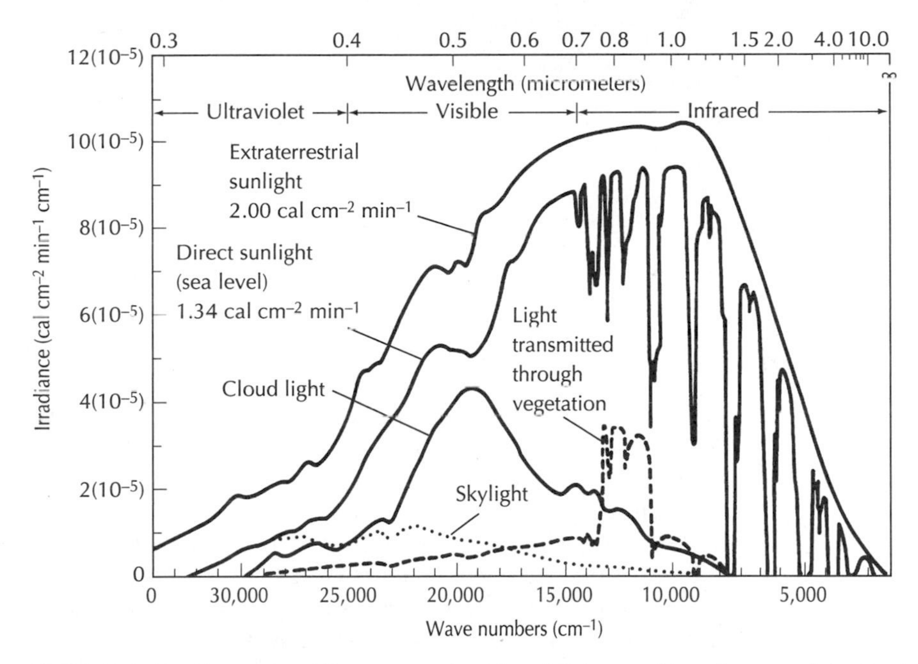

FIGURE 6.10 Energy in different wavelengths of light reaching the top of the Earth's atmosphere and below the atmosphere. (After Gates 1968)

SENSIBLE HEAT EXCHANGE: CONDUCTION AND CONVECTION

Leaves can also transfer heat directly to or from the atmosphere. Two related processes are involved, and leaf morphology can affect the rate of transfer significantly. **Conduction** is the transmission of energy through a material as a function of the gradient in temperature from one side to the other and the

FIGURE 6.11 Turbulent movement of heated air away from the surface of a leaf (convective heat loss). The lighter streaks are currents of warmer air escaping upward from the leaf surface. (Ray 1972, photo courtesy of David M. Gates)

material's ability to conduct heat (e.g., fiberglass insulation in a wall conducts heat more slowly than a glass windowpane). **Convection** is the loss of energy from a heated surface by the warming of adjacent air that then is removed by turbulent air flow (Figure 6.11).

Thicker leaves tend to increase rates of conductance of energy to edges from which it can be removed more readily by turbulent air flow. Smaller or more deeply lobed leaves increase the surface area for rapid heat exchange, thus diminishing the difference between the temperature of the air and the leaf. In contrast, large, round leaves can be at a temperature very different from that of the surrounding air, especially in full sunlight, where the net radiation load is high.

STRUCTURAL AND PHYSIOLOGICAL ADAPTATIONS

Responding to these fundamental controls on photosynthesis and transpiration, plants show both consistent patterns in the structure of leaves under different light environments and some successful physiological adaptations to extreme environments.

Leaf Morphology

Leaves of the same species, and even those found on the same plant, can vary widely in response to differences in the light environment. These differences can be discussed using extreme examples called "sun leaves" and "shade leaves." As shown in Figure 6.2, sun leaves have higher maximum rates of net photosynthesis and dark respiration, as well as higher compensation points. These differences result from a denser "packing" of chlorophyll and associated systems. In broad-leaved plants, sun leaves tend to have two layers of chlorophyll-bearing palisade cells, as opposed to one layer in shade leaves, and so are thicker and have a greater density of chlorophyll (and associated

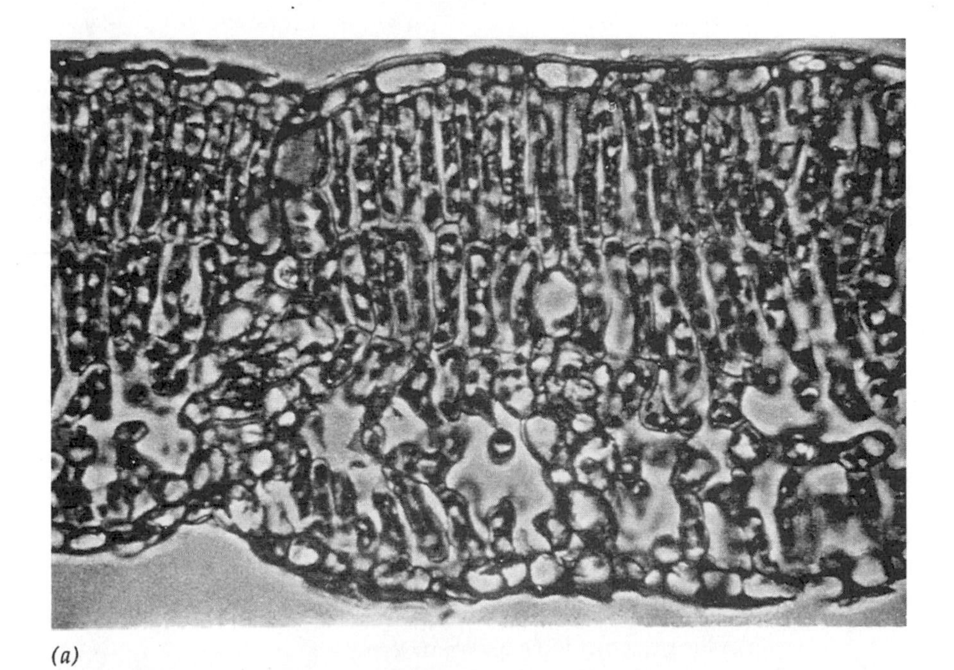

(*a*)

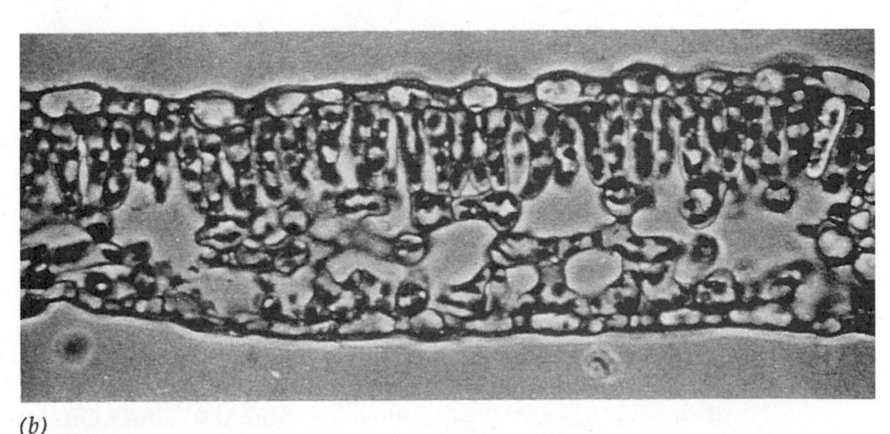

(*b*)

FIGURE 6.12 Electron micrograph showing differences in leaf morphology between (*a*) sun and (*b*) shade leaves of American beech. (Horn 1971)

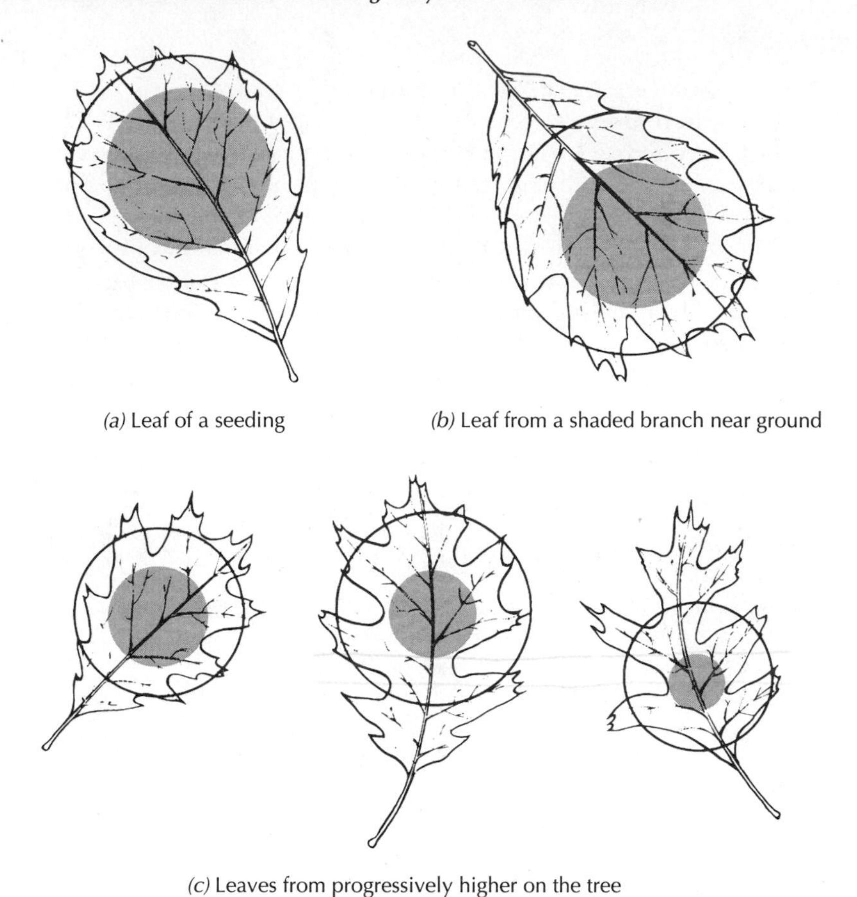

FIGURE 6.13 Changes in leaf size and degree of lobing from shaded seedlings to fully exposed sun leaves of black oak. The shaded area of each leaf is the **critical dimension,** the largest circle that can be contained completely within the leaf outline. The outer circle is equivalent to the total area of the leaf. The ratio between the two is one expression of the degree of lobing. (Horn 1971)

enzyme systems and hence nitrogen) per unit leaf surface area (Figure 6.12). Sun leaves also tend to be smaller and more deeply lobed than shade leaves (Figure 6.13). Both deeper lobing and smaller and thicker leaves enhance rates of heat conduction to leaf edges and reduce the size of the boundary layer around leaves, increasing rates of heat loss and reducing leaf temperature in full sun.

There are also special adaptations for increasing heat loss. For example, leaf hairs, common in plants in arid environments, can give a lighter, more reflective surface to a leaf, reducing radiation loading. One of the most intriguing adaptations in temperate forests is the "tremble" in trembling aspen and other species of the genus *Populus.* In these species, the petiole (or stem) of the leaf is flat and rotated 90° from the plane of the leaf blade (Figure 6.14). A breeze will cause the blade to turn or rotate to present the smallest

FIGURE 6.14 Trembling aspen leaves actually "tremble" in light winds due to the structure of the leaf. The leaf stem or petiole is flattened at a 90° angle to the leaf blade. The causes the entire leaf to be continually rotated back and forth, producing the trembling motion. (Doug Lee/Peter Arnold, Inc.)

area to the wind; when the leaf has turned, the petiole is then struck, causing the leaf to turn back to its original position. This happens very rapidly, and the leaf "trembles," increasing turbulent air flow across the leaf surface, which increases heat loss and rates of water loss and CO_2 gain.

Alternate Pathways for Carbon Fixation: C3 and C4 Plants

As mentioned earlier, photosynthesis is basically the same process in all plants, with differences in rates due largely to differences in structure and the packing of chlorophyll and enzymes as reflected in leaf nitrogen concentration. This is an oversimplification. There is indeed one dominant biochemical mechanism for fixing carbon, but there are also two significantly different adaptations of this process that have evolved in response to extremes of temperature and moisture. The basic and most common process is called C3 photosynthesis, and the two adaptations are called C4 and crassulacean acid metabolism (CAM) photosynthesis.

The C3 pathway was the first to be described, is the most widely occurring, and is the one that was used in describing the general characteristics of photosynthesis earlier. In broad-leaved plants, the leaf structure associated with this process is the one shown in Figure 6.12, with chlorophyll-bearing palisade cells arrayed more or less continuously across the top of the leaf. Carbon fixation occurs within these cells, and the first product of fixation is a 3-carbon organic acid, hence the name.

In C4 plants (so named because the first organic acid formed from fixed CO_2 has 4 carbons rather than 3), both structure and physiology are altered. Palisade cells are absent in C4 plants. Instead, the chloroplasts are located within bundle sheaths concentrated in the center of the leaves (Figure 6.15). These are surrounded by mesophyll cells, in which the initial fixation of CO_2 occurs. Fixed carbon is then transported into the bundle sheath, released, and refixed into a C3 compound for use in photosynthesis. The effect of this is to reduce CO_2 concentrations in the intercellular spaces from about 250 parts per million (ppm) in C3 plants to near 100 ppm. This increases the difference between atmospheric and internal leaf concentrations and causes a faster net flow of CO_2 into the plant. CO_2 concentrations surrounding the chloroplasts are correspondingly higher due to this "CO_2 pump" such that carboxylation occurs more rapidly. As a result, photosynthesis occurs at a faster rate for a given rate of transpiration, so water use efficiency is increased. Respiration is also reduced with this pathway, such that optimum temperatures for photosynthesis are higher than for C3 plants. The cost to the plant for this improved CO_2 fixation mechanism is in the extra energy required to operate it. The reactions require energy and thus are most adaptive in areas where light intensity is high, water is limiting, and/or high temperatures lead to high rates of respiration.

C4 plants are most common in (but not limited to) disturbed areas in the tropics and subtropics and in semiarid areas, where their greater water use efficiency may confer a selective advantage. Some of the most productive crop species (e.g., corn and sugar cane) are C4 grasses. It is interesting that C4 photosynthesis is not limited to one part of the plant kingdom but occurs across widely different families. Within a single genera, some species may be able to carry out C4 photosynthesis, while others are not.

The CAM adaptation provides a means for taking up CO_2 at night and closing stomata during the day, reversing the pattern in C3 or C4 plants (Figure 6.16). At night, CO_2 enters the open stomata; is fixed and converted to a C4 acid; and is stored in the vacuole, or large central cavity of leaf cells. During daylight hours, while the stomata are closed, this acid is removed from the vacuole and transported to the site of photosynthesis, where the CO_2 is re-

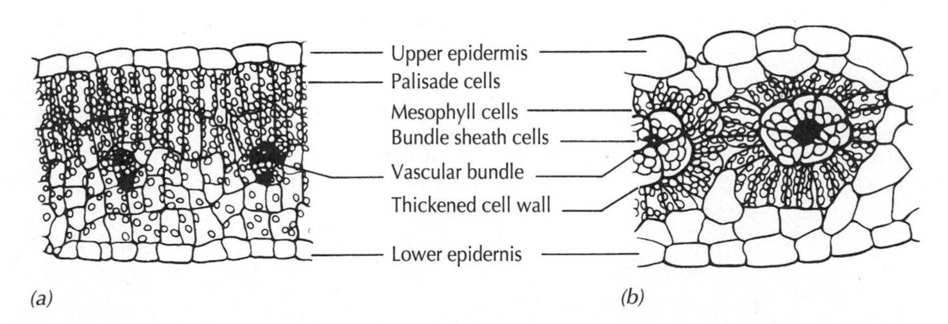

FIGURE 6.15 Comparison of leaf morphology in (*a*) C3 and (*b*) C4 carbon fixation systems in two species of saltbush (*Atriplex*). (Ray 1972)

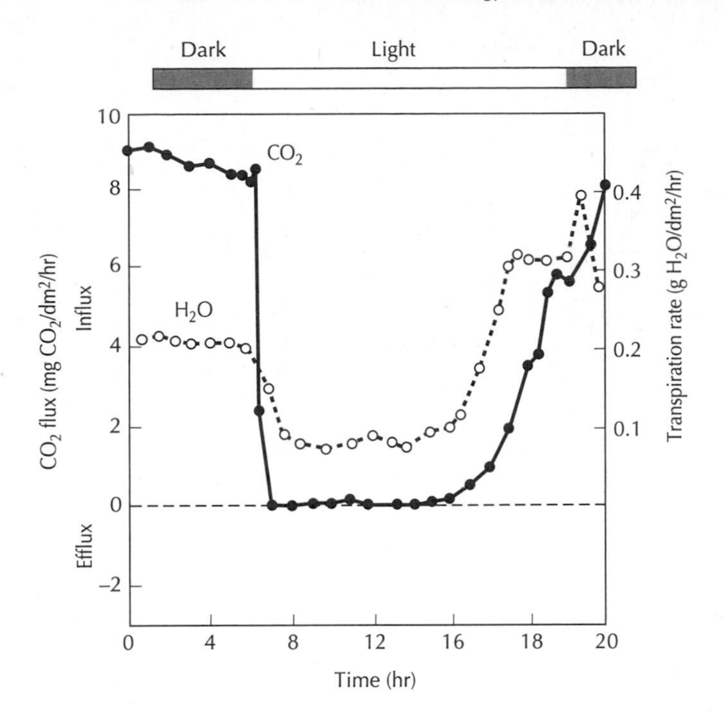

FIGURE 6.16 Pattern of CO_2 uptake and transpiration in a CAM plant. (Salisbury and Ross 1978, after Neales 1975)

leased and refixed by the C3 pathway. As temperatures are lower and humidity higher at night, stomata are open when the atmospheric demand for moisture is lowest. The gradient in concentration of water vapor between leaf and atmosphere is reduced, so water loss is less. Again, water use efficiency is also higher. However, this mechanism is limited both by the amount of CO_2 that can be stored in cells in this way and by the metabolic costs of movement and storage. CAM photosynthesis is most common under extremely dry conditions. Many of the "succulent," or cactuslike desert species are CAM plants.

LEAF STRUCTURE AND FUNCTION IN MAJOR ECOSYSTEM TYPES

Do changes in the size, structure, and physiology of photosynthetic tissues between major vegetation types reflect responses to changing environments? There are trends that fit the patterns predicted by our discussion in this chapter. However, the diversity of leaf types within any region shows that energy balance and water use efficiency are not the only selective forces at work.

Considering the major vegetation types in Figure 2.2, there is a general decrease in leaf size from the tropical rain forest through seasonal forests, grasslands, and shrublands. In the extreme desert environments, leaves are absent entirely from some plants, and instead, the "stems" and "branches" are the photosynthetic tissues (e.g., in cacti). A similar decrease in leaf dimension occurs within the temperate deciduous forest regions of North America. The stands with highest water availability tend to be dominated by species with

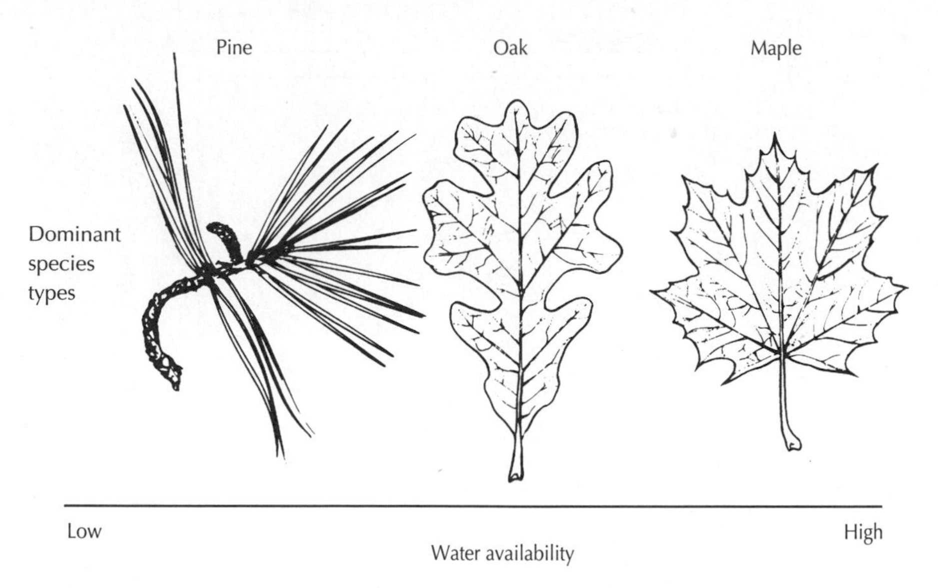

FIGURE 6.17 Changes in leaf shape of dominant species from moist to dry environments within the northern temperate forest region of the United States.

thin, large nonlobed leaves (e.g., maples and basswood); slightly drier sites by species with thicker and deeply lobed leaves (e.g. oak); and the driest sites by needle-leaved species, such as pines. Changes in leaf size and shape ("sun" leaves, in this case) vary as expected (Figure 6.17).

The relative importance of the three different photosynthetic processes also vary as expected. All major species in moist, cool environments are C3 plants. C4 plants are more important in warm and dry to hot and moist regions, such as grasslands, and especially in disturbed environments. CAM plants increase in importance in hot and dry systems, such as deserts.

REFERENCES CITED

Evans, J. R., and S. Von Caemmerer. 1996. Carbon dioxide diffusion inside leaves. *Plant Physiology* 110:339–346.

Field, C., and H. A. Mooney. 1986. The photosynthesis-nitrogen relationship in wild plants. In Givinish, T. J. (ed.), *On the Economy of Plant Form and Function.* Cambridge University Press, Cambridge.

Gates, D. M. 1968. Energy exchange between organism and environment. In Lowery, P. (ed.), *Biometeorology.* Oregon State University Press, Corvallis, OR.

Horn, H. S. 1971. *The Adaptive Geometry of Trees.* Princeton University Press, Princeton, NJ.

Jarvis, P. G. 1981. Stomatal conductance, gaseous exchange and conductance. In Grace, J., E. D. Ford, and P. G. Jarvis (eds.), *Plants and Their Atmospheric Environment.* Blackwell, Oxford.

Lambers, H., F. S. Chapin III, and T. L. Pons. 1998. *Plant Physiological Ecology.* Springer, New York.

Larcher, W. 1995. *Physiological Plant Ecology*, 3d ed. Springer-Verlag, Berlin.

Loach, K., 1967. Shade tolerance in tree seedlings: I. Leaf photosynthesis and respiration in plants under artificial shade. *New Phytologist* 66:607–621.

Neales, T. F. 1975. The gas exchange patterns of CAM plants. In Marcelle, R. (ed.), *Environmental and Biological Control of Photosynthesis.* D. W. Junk, The Hague.

Ollinger. S. V., J. D. Aber, P. B. Reich, and R. J. Freuder. 2000. Tropospheric ozone and land use history affect carbon uptake in response to CO_2 and N deposition. In preparation.

Ray, P. M. 1972. *The Living Plant.* Holt, Rhinehart and Winston, New York.

Reich, P. B., M. B. Walters, and D. S. Ellsworth. 1997. From tropics to tundra: Global convergence in plant functioning. *Proceedings of the National Academy of Sciences* 94:13730–13734.

Rook, D. A. 1969. The influence of growing temperature on photosynthesis and respiration of *Pinus radiata* seedlings. *New Zealand Journal of Botany* 7:43–55.

Salisbury, F. B., and C. W. Ross. 1978. *Plant Physiology.* Wadsworth Publishing, Belmont, CA.

Waring, R. H., and W. H. Schlessinger. 1985. *Forest Ecosystems: Concepts and Management.* Academic Press. Orlando.

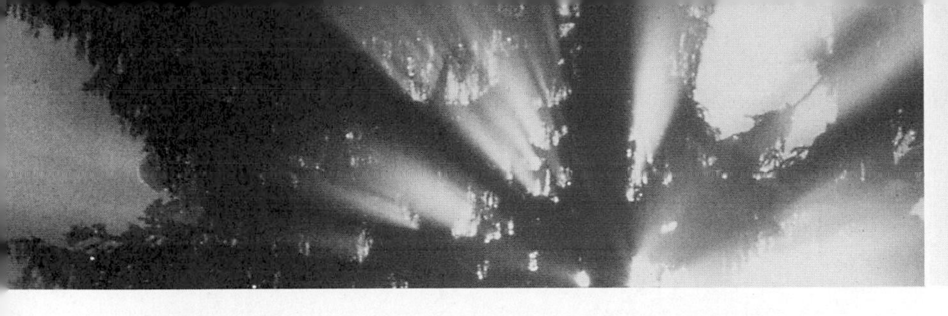

7

WATER USE AND WATER BALANCES IN ECOSYSTEMS

INTRODUCTION

Water loss from leaves through evaporation from internal leaf surfaces (transpiration) is an inevitable consequence of photosynthesis. This water must be replaced by uptake from the soil. In natural ecosystems, the availability of water for transpiration is frequently less than the total needed to maintain photosynthesis at optimum rates (in terms of the presentation in Chapter 2, AET is often less than PET). Under such conditions, water stress develops in plants, and one of several mechanisms comes into play to either reduce the movement of water through the leaf or maintain this flow by physiological modifications to reduce water stress in leaf cells.

Water is required in much greater quantities than nutrients per unit of biomass produced in all terrestrial ecosystems. This is because nutrients allocated to the production of a given tissue tend to remain in that tissue until it is shed as litter. In contrast, water is continuously given up to the environment through transpiration. Nutrients in foliage may even be retranslocated back into perennial parts of long-lived plants for reuse in the following growing season. Up to 40% of annual plant nutrient requirements can be met by this mechanism (Chapter 11). Only very specialized plants, such as desert succulents, can store a significant amount of water relative to daily transpirational demands. In the massive coniferous forests of the Pacific Northwest of North America, perhaps a half-day's worth of water for transpiration can be stored in the woody tissues of tree stems. In most terrestrial ecosystems, the only major storage for water is in the soil. Thus the total amount and seasonal pattern of water availability is determined by two nonbiological (abiotic) factors: (1) precipitation and (2) soil water holding capacity (determined by soil volume and texture).

The purpose of this chapter is to discuss water as a resource for plant growth. Environmental factors determining its availability are presented, along with the process by which water moves from the soil, through the plant, and into the atmosphere. Finally, plant responses to moisture stress and effects of suboptimal moisture availability on photosynthesis are examined.

THE HYDROLOGIC CYCLE OF ECOSYSTEMS

An ecosystem's hydrologic cycle has only a few major components (Figure 7.1). Inputs are mainly as precipitation in the form of rain or snow. At high

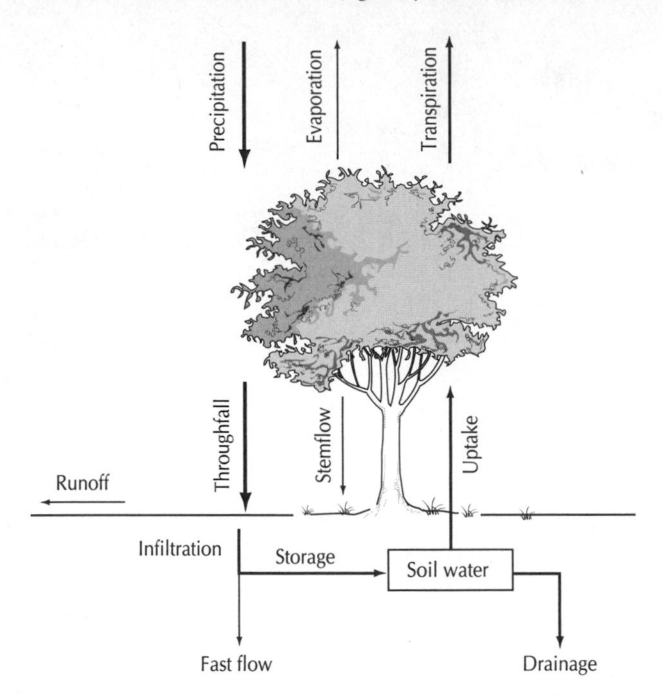

FIGURE 7.1 The hydrological cycle in terrestrial ecosystems. The width of the arrows approximates the relative volume of water following each pathway in a humid forest ecosystem.

elevations or in coastal areas (in any place where clouds meet the ground and become fog), additional inputs of water occur through condensation on leaves and stems. Water intercepted by surfaces can evaporate, drip through to the soil (throughfall), or run down the surface of stems (stemflow). Water that reaches the soil surface in liquid form can run off over the soil surface, a process that is important in soil erosion. Water that infiltrates into the soil can pass quickly below the rooting zone through large spaces called macropores (fast or saturated flow) or can adhere to soil particle surfaces and become part of soil water content (storage). Water on soil particle surfaces can also drain more slowly below the rooting zone in response to gradients of concentration in a process called unsaturated flow (drainage). Depending on a system's position in the landscape, drainage can be impeded, creating elevated water tables that affect many other functions. In addition, if a system is in a downslope position or near a water table, stream, or lake, it may receive water from upslope systems. This movement of water between landscape positions is a major determinant of the distribution of species and ecosystems across a region. Finally, water held on soil particle surfaces can be taken up by plants, transported to leaves, and evaporated back to the atmosphere (transpiration).

THE CONCEPT OF WATER POTENTIAL IN SOILS, PLANTS, AND THE ATMOSPHERE

The concept of gradients in concentrations was presented in Chapter 5 and applied to the movement of CO_2 in photosynthesis in Chapter 6. The movement of water through ecosystems is also driven by gradients, but those gradients are determined by "concentration" expressed in a more complex way, through the concept of "water potential," expressed in units of pressure (megapascals [MPa]).

Technically, water potential is the free energy of water, or its capacity to do work. Pure water in liquid form at 20°C and atmospheric pressure (e.g., distilled water sitting in an unenclosed basin) is defined as having zero water potential. Throughout most of the hydrologic system, water interacts with surfaces or is reduced in concentration such that water potentials are less than that of pure liquid water, and so potentials are generally negative. There are four ways in which water potential is expressed in terrestrial ecosystems: (1) vapor pressure, (2) matric potential, (3) physical or hydrostatic pressure, and (4) osmotic potential.

Water Potential in the Atmosphere

The amount of water vapor that the atmosphere can hold at any time increases with increasing temperature and is called the saturated vapor pressure. The actual amount of water vapor present is the vapor pressure, and the difference between these two is the vapor pressure deficit, or relative humidity. Air of low humidity has a high vapor pressure deficit and a very large capacity to evaporate water. Figure 7.2 summarizes the interactions between temperature, relative humidity, and water potential of the atmosphere. High vapor pressure deficits translate into very negative water potentials. At anything less than 95% relative humidity, the water potential in the atmosphere is much less than can be attained in soils or plants and so creates a very strong demand for evaporation from leaves.

Water Potential in Soils

At the other end of the water transport system, soils retain water against the pull of gravity because of the tendency for liquid water to adhere to particle surfaces. Water in soils that is available to plants resides in thin films on particle surfaces (Figure 7.3). The thicker the film, the less tightly the water at the outer rim is held and the more easily it can be taken up by plants. This attractive force between water and surfaces is called matric potential.

The total amount of water a soil can hold per unit soil volume is a function of the surface area of all the particles within that volume and the amount of air space present between these particles. A soil dominated by clay particles (< 0.005 mm in diameter) would normally hold more water than a soil dominated by silt particles (0.005–0.074 mm) or by sand particles (0.074–2.0 mm) because of the higher ratio of surface area to volume in the smaller particles. These kinds of relationships break down if a soil is severely compacted, as occurs under some type of intensive, mechanized management. In a heavily compacted clay soil, much of the particle surface area would be compressed against other particle surfaces, leaving little space between for the storage of water.

Soil matric potentials less negative than approximately -0.01 MPa are too weak to retain water against the force of gravity. After a heavy rainstorm that can saturate upper soil layers, water drains freely until this matric poten-

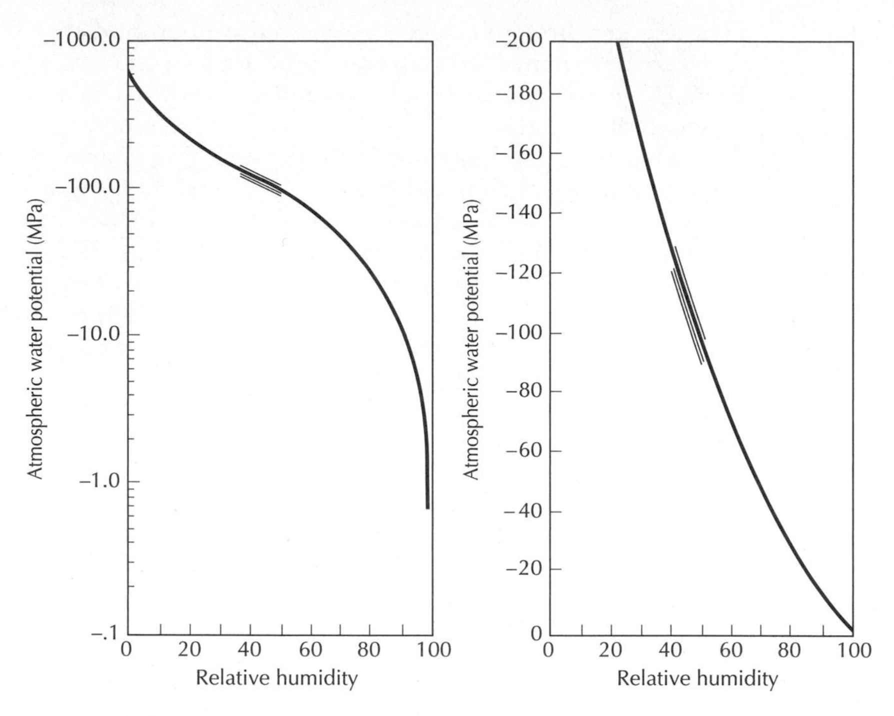

FIGURE 7.2 Atmospheric water potential as a function of relative humidity. The four parallel lines show the relatively small differences caused by changes in temperature over a range from 0 to 30°C. (Salisbury and Ross 1978)

tial is reached. Soil water content at this point is called the field capacity. As soils dry and these films become thinner, the remaining water is held more tightly and matric potential becomes more negative (Figure 7.3). Many plants cannot extract water from soils with matric potentials more negative than approximately −1.5 MPa. This point has been termed the *permanent wilting point,* but that term reflects a humid zone bias. Plants with the adaptations for dry conditions described later can withdraw water from soils with much more negative potentials, so the actual matric potential at which wilting occurs is a function of both soils and the plants growing on them.

Figure 7.4 shows the interaction between water content of soils, water availability, soil matric water potential, and soil texture. Soil water is separated into three categories. The amount of plant-extractable water that a soil can hold (AW) is the difference between the amount present at field capacity and at the permanent wilting point, or between −0.01 and −1.5 MPa (in this example). The loam soil (a loam is a mixture of sand, silt, and clay plus organic matter) has the greatest amount of available water (approximately 22 g · cm^{-3}), followed by the clay soil (approximately 18 g · cm^{-3}) and the sandy soil (approximately 9 g · cm^{-3}). Note that the clay soil contains more water at −1.5 MPa than the sandy soil does at −0.01 MPa (field capacity). Soils with significant measurable water content may be completely "dry" as far as plants are concerned (may have no extractable water)!

This available water storage term is critical for ecosystem function as it represents the total amount of water available for transpiration between pre-

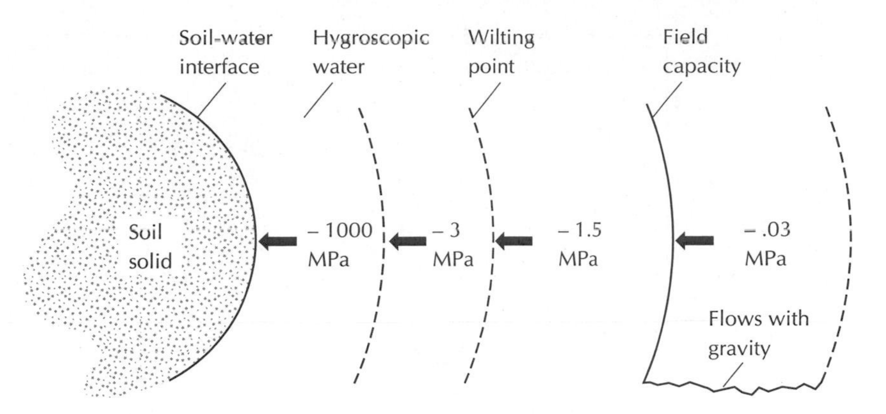

FIGURE 7.3 Relationship between thickness of water films on soil particles and the tension with which water is held. (Buckman and Brady 1969)

cipitation events. In ecosystems with less frequent or less regular precipitation, this "buffer" is important for the continuation of photosynthesis and plant growth. You can determine from Figure 7.4 what the increase in available water storage would be if plants could draw soil matric potential down to −10.0 MPa instead of −1.5.

Water Movement and Water Potential in Plants: The Soil–Plant–Atmosphere Continuum

Water held at moderate matric potentials in soils evaporates into a drier atmosphere. However, evaporation tends to affect only a thin surface layer of soils, as the resistance to movement of both air and liquid water in soils is fairly high. Still, in very hot, dry systems with little precipitation (deserts), evaporation from the soil surface can be a major pathway in the hydrologic cycle.

In systems with a substantial canopy, plants provide an ideal conduit for the movement of water between soils and the atmosphere. Roots grow deep into the soil and tap water reserves far from the surface, providing a pathway between these two reservoirs. Water moves from soil to plant to atmosphere along gradients from high (less negative) to low (more negative) water potential. The rate of water movement increases as the difference between the potentials in different parts of this soil–plant–atmosphere continuum increases and is reduced by the resistance to movement through the plant.

The plant conducts water from the soil to the atmosphere only if the atmosphere has a more negative water potential than the soil. Figure 7.2 shows that this is almost always the case. However, extended periods of high humidity can be important in some ecosystems. The coastal redwood forests of California and Oregon attain high growth rates and tremendous size despite relatively little rainfall. This is thought to be at least partly due to frequent fogs that increase atmospheric water potential (make it less negative) and reduce water loss by plants. This in turn reduces water uptake from soils and slows the rate of soil drying. It also increases water use efficiency.

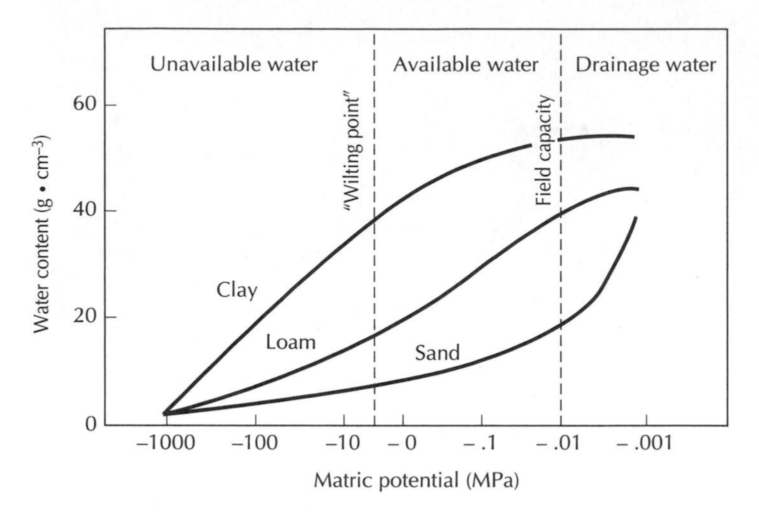

FIGURE 7.4 Water content and soil matric potential for soils with different texture. UW = water unavailable to plants, AW = plant-available water, DW = drainage water. (After Waring and Running 1998, Ulrich et al. 1981)

The movement of water from soil to atmosphere begins when the stomata open and water vapor in the intercellular spaces within the leaf diffuses out into the atmosphere. As the air within the leaf becomes drier, evaporation from the constantly wet cell surfaces increases, and the negative water potential of the atmosphere is transferred to a continuous column of liquid water that begins in the root and ends in the leaf. The water passes through conducting tissues called xylem elements, which provide an uninterrupted pathway for water movement. The negative physical or hydrostatic pressure (also called tension) created at the evaporating surfaces is expressed throughout the xylem elements, such that water moves up the plant toward the leaf, like a continuous rope of water. The tension is conducted out through the roots and, through intimate contact between roots and soil, to the water adhering to soil particles.

This moving water column must be continuous. Any air gaps in the system relieve the tension and stop the movement of water. Transpiration from very dry soils can be limited by the lack of continuous water films across particles so that root surfaces are not in direct contact with these films. Frost damage to trees can also disrupt water flow if the water frozen in xylem elements breaks cell walls and creates air gaps in the water-conducting system.

The actual rate of flow of water up through the plant, and thus from the soil to the atmosphere, is a function of the differences in water potential between these two ends of the gradient (the driving force) and the resistance to flow. Resistance within the plant results mainly from friction between water and the walls of the xylem elements through which it passes. The force of gravity also works against the rate of water movement up the stem.

Physiological Control over Water Loss

So far we have not described any processes that exert biological control over transpiration. We have described, rather, a system by which plants provide "pipes" through which water moves in response to a physical gradient. However, this water pathway passes along the membranes of cells in the leaf, and tension on the water column is transferred to water within those cells (Figure 7.5). Leaf cells need to maintain a positive physical pressure against the cell walls to maintain turgor or avoid wilting. If the tension on the water column adjacent to leaf cells is too great, water is withdrawn from the cells, and they wilt.

How do cells maintain water content against this physical tension expressed at the cell surfaces? To explain this, we need to introduce the final component of water potential: osmotic potential.

Liquid water with any solutes added (e.g., ions, such as the sodium and chloride in dissolved salt, or small organic molecules, such as simple sugars), has a negative water potential caused by the reduction of the free energy of the water present due to interaction with the solutes (a lower effective concentration of water). This is called osmotic potential. The classic example is a beaker of distilled water divided into two sections by a semipermeable membrane that allows water to pass through but not the ions (Figure 7.6). If a solute such as salt (NaCl) is introduced into one side of the beaker, this creates a negative water potential (osmotic potential) on that side. Water then moves through the membrane into the side with the ions and the more negative potential. This continues until the positive physical pressure against the membrane, evident in the expansion of the membrane, counters the osmotic potential.

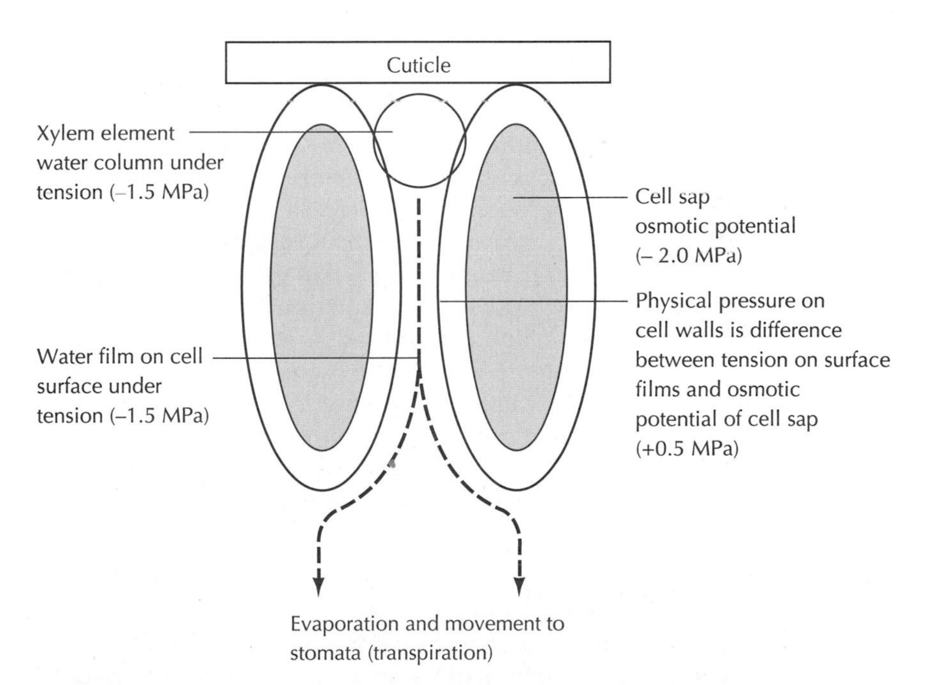

FIGURE 7.5 Expression of different components of water potential on cell walls in a leaf. The tension on the water column passing from the xylem elements along cell surfaces before evaporation is transferred to the cell membranes and exerts a force pulling water out of the cell. This is counterbalanced by an osmotic potential within the cell sap that retains water within the cell. The difference between these two forces (0.5 MPa in this case) is the resulting turgor pressure exerted on the cell walls, which keeps the cells, and the leaf, from wilting.

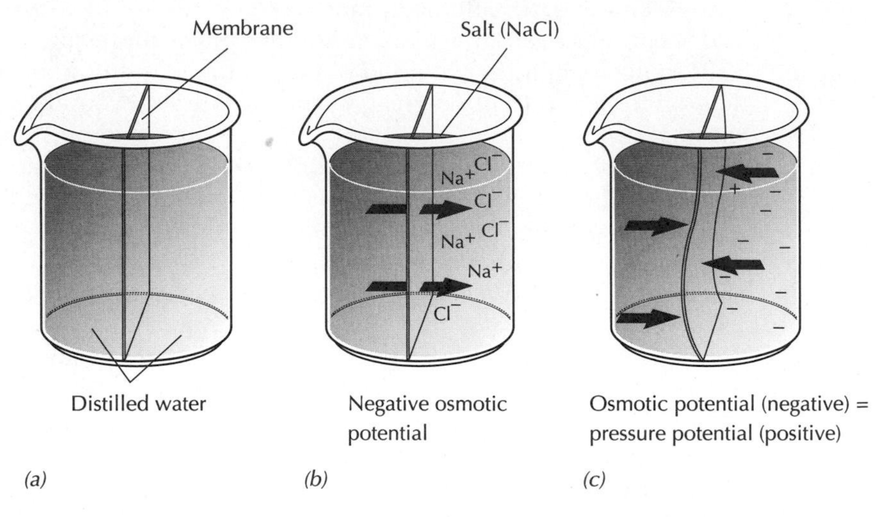

FIGURE 7.6 Example of osmotic potential. (*a*) The beaker contains distilled water and is separated into two compartments by a semipermeable membrane. (*b*) Sodium chloride is added to the right compartment and dissociates into ions that create a negative osmotic potential, pulling water across the membrane. (*c*) Water moves across the membrane until the increase in pressure in the right compartment equals the osmotic potential.

Plants maintain turgor in leaves by creating osmotic potentials in the fluids of cells that are more negative than the physical tension expressed on the liquid water on the cell surfaces.

This is accomplished by the accumulation of low molecular weight ions and simple organic acids within the cells. As is explained later, certain plants can respond to dry conditions by increasing the osmotic potential in leaf cells, allowing them to maintain turgor against more negative water potentials in the transpirational stream flowing across the cell membranes. However, there are limitations to the use of this adaptation. Elevated concentrations of solutes in cell solutions can interfere with cell function. Thus, there are limits beyond which osmotic potential cannot be increased. Because of this, and the very negative water potentials that can occur in the atmosphere and soils, plants often experience water stress—periods when transpiration exceeds the rate of movement of water from soil to leaf. When this occurs, cells can wilt as water is pulled out into the transpirational stream.

Plant Response to Water Stress

Plants have both a short-term and a long-term response to prevent wilting of leaves under conditions of water stress. In the short term, the stomata can close. Stomata are surrounded by guard cells that can be opened and closed rapidly. This is accomplished by rapid movement of solutes into and out of the cells, causing them to wilt and regain turgor. When the stomata are closed, diffusion of water out of the leaf is reduced, the humidity in the air within the leaf increases, and evaporation from cell surfaces declines markedly. At the same time, however, the flow of CO_2 into the plant is also reduced and photosynthesis is suppressed.

A longer-term response is to modify the osmotic potential of the leaf cells. Increasing the concentration of sugars, cations, and other low molecular weight solutes in cells creates a more negative osmotic potential. This increases the length of time that stomata can remain open, increasing CO_2 uptake, but at some cost in reduced metabolic efficiency.

These two mechanisms interact in a plant's response to increasing water stress, as shown by a study of the water relations of two species on north-facing and south-facing mountain slopes in northern Idaho (Figure 7.7). This is a region of winter snows and relatively dry summers, so the soil is moist at the beginning of the growing season and becomes progressively drier from spring to fall.

This study showed first that, during the course of a given day, the water potential of the xylem in the leaf never becomes significantly more negative than the osmotic potential of the leaf cell sap (Figure 7.7*a*). In the early part

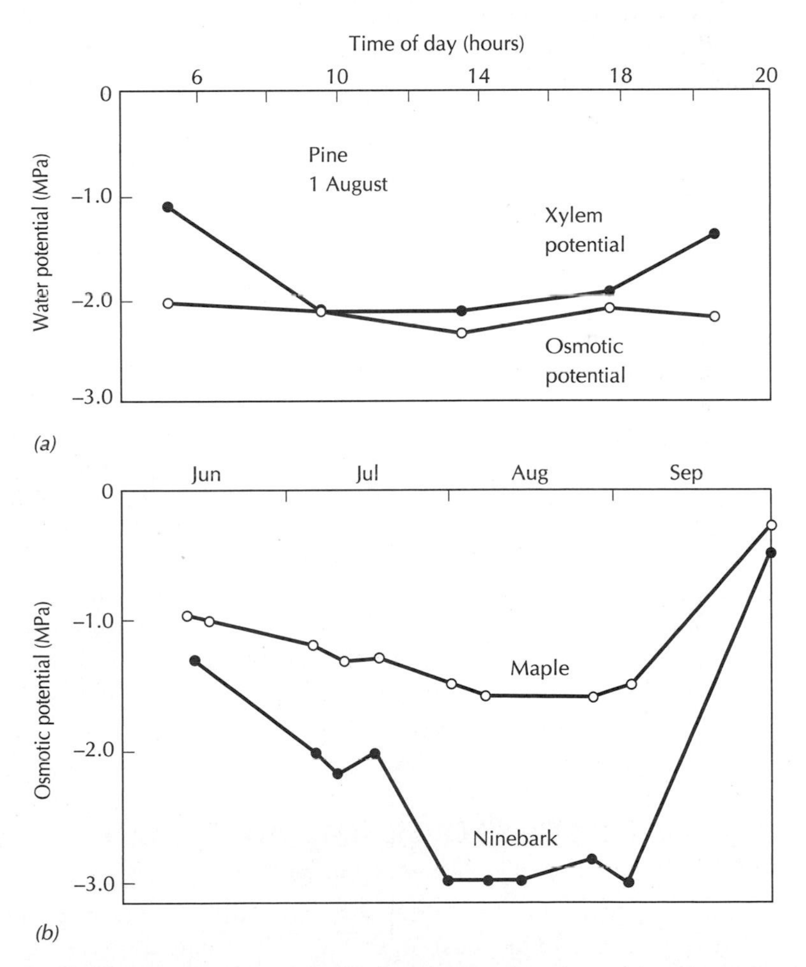

FIGURE 7.7 Changes in leaf xylem potential and osmotic potential. (*a*) Changes in xylem and osmotic potential of pine over the course of 1 day. (*b*) Changes in osmotic potential for two species during the growing season. (Cline and Campbell 1976)

of the day, transpiration decreases the water potential in the xylem. As this tension approaches the osmotic potential, and thus threatens to cause wilting, stomatal closure begins to occur. From 10 a.m. to about 6 p.m., stomata keep opening and closing to keep this xylem tension very near the osmotic potential in the cells, maximizing the time that the stomata are open to allow for CO_2 uptake.

As the summer progresses, the soil becomes drier and the water stress on the plants increases. Stomata would therefore be closed for a greater portion of each day, and photosynthesis further reduced. In response to increased stress, both species reduce (make more negative) the osmotic potential of the cell sap. However, different species show very different osmotic potentials at the same time of year (Figure 7.7*b*). This could result from different degrees of water stress in the different environments in which these two species grew or from different abilities to create negative osmotic potentials and still maintain cell function. The ability to create more negative osmotic potentials in leaves should confer an adaptive advantage and a greater chance of survival on the drier slopes.

If altering osmotic potentials and thereby tolerating more negative xylem potentials (greater tension on the water column and cell surfaces) is an important mechanism for adapting to dry sites, and if midday xylem water potential approaches leaf osmotic potential, then measured midday xylem potentials should be quite different under different levels of water stress. Drier systems should exhibit more negative xylem potentials.

This hypothesis was initially tested using an elegantly simple device known as a Schollander pressure bomb (Figure 7.8). The device measures leaf xylem water potential; the tension being exerted on the water column coming up through the plant. The inventor and several coworkers used this to measure midday xylem potentials across a transect in California running from coastal saline wetlands through dry shrublands (chaparral) up the coastal mountain range, where water stress declines with elevation and over the mountain into the rain-shadow deserts of that region. Measurements from freshwater pond plants were included to yield the widest possible range of water availabilities. Figure 7.9 summarizes their findings. Xylem water potential does indeed become much more negative in the drier habitats.

It is interesting that plants rooted in ocean salt water had intermediate potentials, more negative than freshwater plants. Although these plants are surrounded by water, the salt concentration in the water creates an osmotic potential of approximately -3.0 MPa. This is functionally equivalent to a matric potential of the same magnitude in soils, so this habitat appears fairly "dry" to these rooted, emergent plants.

INTEGRATING WATER STRESS OVER TIME: AN ISOTOPE-BASED METHOD

Stomatal response to water stress is a rapid and short-term process. Over the course of a day, stomata can open and close many times as the water potential on the leaf is maintained close to the osmotic potential of the cells. Because of this, direct measurement of the average conductance of leaves is very

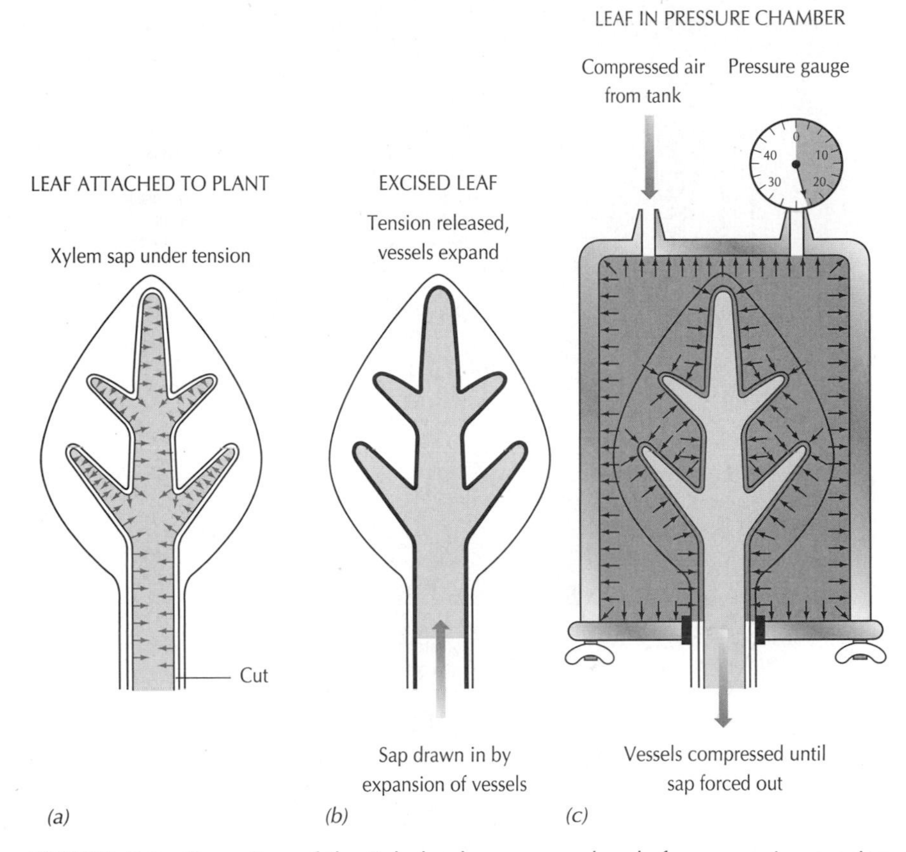

FIGURE 7.8 Operation of the Scholander pressure bomb for measuring tension on the water column in the xylem of plant twigs (xylem water potential). (*a*) Tension on the water column in the xylem results in shrinkage of the xylem elements. (*b*) Cutting the twig releases this tension and the xylem elements expand, drawing water away from the cut edge. (*c*) In the pressure bomb, pressure is exerted on the twig until the water in the xylem returns to the cut edge. It is assumed that this pressure is equal to the tension on the water column before the twig was cut. (Ray 1972)

difficult, requiring constant monitoring over the course of the growing season. Is there a method that can integrate the effects of water stress over time, or estimate the average conductance, or degree of stomatal opening, over longer periods of time?

In Chapter 5, we discussed the use of isotopes in ecosystem research and the tendency for "heavier" isotopes of a given element (e.g., ^{15}N or ^{13}C) to be discriminated against or selectively excluded from biochemical reactions. Discrimination is greater when the substrate for the reaction is present in higher concentrations. Higher stomatal conductance increases the concentration of CO_2 inside the leaf, and hence leads to greater discrimination against the heavier $^{13}CO_2$.

Plant materials reflect the isotopic ratios of the carbon fixed in photosynthesis. Thus, within a species, plants exhibiting higher leaf conductance have higher internal CO_2 concentrations and lower concentrations of ^{13}C in

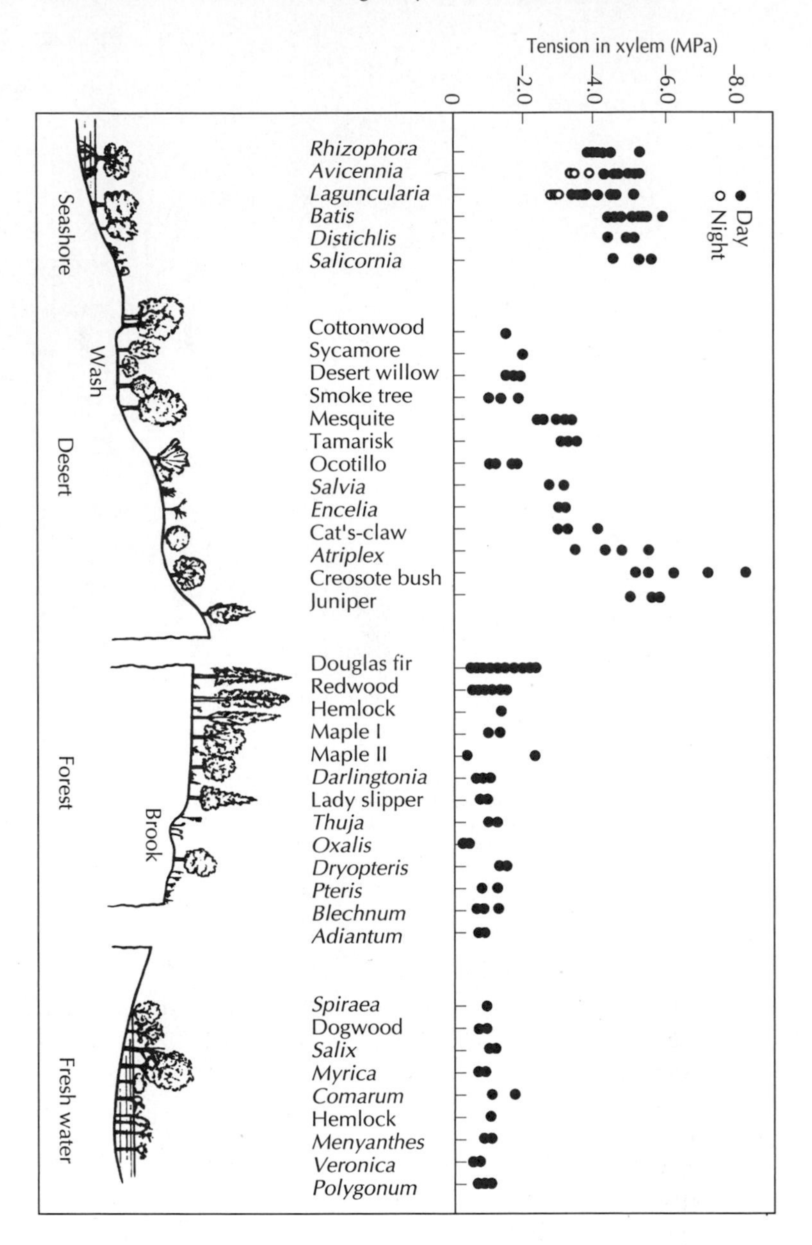

FIGURE 7.9 Range of midday xylem water potentials for plants in several habitats. (Scholander et al. 1965)

tissues (Figure 7.10). In the field, plants experiencing greater water stress have, on average, lower leaf conductance and a higher relative concentration of ^{13}C in tissues.

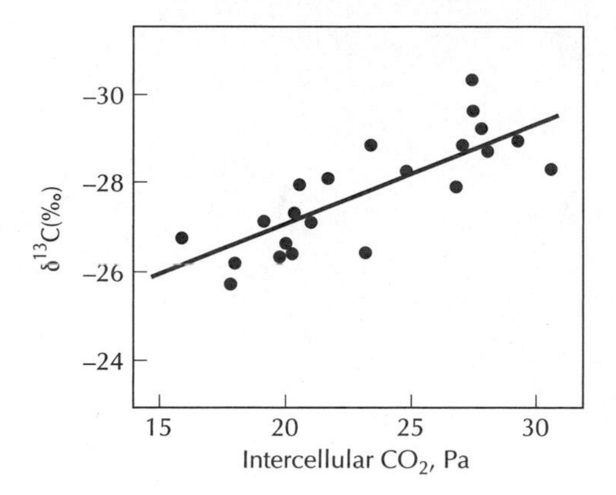

FIGURE 7.10 Effects of leaf conductance and drought on the ^{13}C content of plant tissues. High conductance leads to high internal CO_2 concentrations and increased discrimination against $^{13}CO_2$. Plant tissues reflect this difference. (Data from several Australian species; Ehleringer et al. 1985, in Lambers et al. 1998)

REFERENCES CITED

Buckman, H. O., and N. C. Brady. 1969. *The Nature and Properties of Soils.* MacMillan, New York.

Cline, R. G., and G. S. Campbell. 1976. Seasonal and diurnal water relations of selected forest species. *Ecology* 57:367–373.

Ehleringer, J. R., E. D. Schulze, H. Ziegler, O. L. Lange, G. D. Farquhar, and I. R. Cowan. 1985. Xylem-tapping mistletoes: Water or nutrient parasites? *Science* 227:1479–1481.

Lambers, H., F. S. Chapin III, and T. L. Pons. 1998. *Plant Physiological Ecology.* Springer, New York.

Picon, C., J. M. Guehl, and A. Ferhi. 1996. Leaf gas exchange and carbon isotope composition responses to drought in a drought-avoiding (*Pinus pinaster*) and drought-tolerant (*Quercus petraea*) species under present and elevated atmospheric CO_2 concentrations. *Plant, Cell and Environment* 19:182–190.

Ray, P. M. 1972. *The Living Plant.* Holt, Rhinehart and Winston, New York.

Salisbury, F. B., and C. Ross. 1978. *Plant Physiology,* 2d ed. Wadsworth, Belmont, CA.

Schollander, P. F., H. T. Hammel, E. D. Bradstreet, and E. A. Hemmingsen. 1965. Sap pressure in vascular plants. *Science* 148:339–346.

Ulrich, B., P. Benecke, W. F. Harris, P. K. Khanna, and R. Mayer. 1981. Soil processes. In Reichle, D. E. (ed.), *Dynamic Properties of Forest Ecosystems.* Cambridge University Press, London.

Waring, R. H., and S. W. Running. 1998. *Forest Ecosystems: Analysis at Multiple Scales.* Academic Press, New York.

8

STRUCTURE AND DYNAMICS OF CANOPY SYSTEMS

INTRODUCTION

In the last two chapters, we discussed the energy, carbon, and water balances of individual leaves. In ecosystems, leaves are arranged in structured canopy systems that create unique environments and adaptational challenges. Light, temperature, wind speed, humidity, and CO_2 all change with depth in a closed canopy or even within the crowns of open-grown individuals. Tall, dense canopies of herbaceous species and grasses and the permanently shaded understories beneath evergreen trees are significantly different environments than those beyond a canopy's reach. Across the seasons in a year, canopy structure in turn responds to changes in environmental conditions.

While the canopy environment can be a complex one, there are also emergent properties of canopies that can be compared over large geographic areas. Differences in function between sites can often be predicted or explained by a few generalized characteristics. This has made the large-scale modeling of forest carbon and water balances a fairly rigorous exercise and the depiction of seasonal changes in structure visible at large scales through satellite imagery.

The purpose of this chapter is to describe the canopy environment and how variations in leaf structure and function correspond to position within a canopy. We then describe several consistent trends in canopy display and function across environmental gradients and over time, and conclude with a discussion of how models are being used to predict the carbon and water balances of ecosystems from regional to global scales.

THE CANOPY ENVIRONMENT

Light

The most visible effect that canopies have on their internal environment is the reduction in light. Walking through or under any canopy, you can see that this is a very irregular, patchy effect. Even under the darkest canopies there are some spots, called sun flecks, where the ground is receiving full, direct sunlight. Describing the complete three-dimensional interaction between sunlight and canopies is extremely difficult, involving the plotting of the sun's course through the sky, the precise location of leaves, and how they absorb and reflect light in different wavelengths. However, if abstracted to a larger scale, this problem, like many others, becomes simpler.

In a now classic paper, three Japanese ecologists described a forest canopy as a uniformly dense suspension of light-absorbing particles, much like algal cells suspended in water. The similarity is striking. Both exhibit considerable irregularity in the distribution of individual units. However, the variability in leaf display in forests is on a scale apparent to the human eye, and so some effort has been expended trying to describe it. Few would think of attempting to measure the location of individual algal cells in a water column.

Assuming uniform distribution of leaves, the decrease in light intensity with increasing depth in the canopy can be described by the equation

$$I_L/I_o = e^{-k\,LAI_L}$$

where I_L/I_o is the percentage of incident light at the top of the canopy (I_O) reaching depth L in the canopy, LAI_L is the cumulative leaf area (in square meters of leaf area per square meter of ground area, called leaf area index) from the top of the canopy to depth L, k is a stand or species-specific constant, and e is the base of natural logarithms (2.718). This relationship has the shape shown in Figure 8.1. Examples of forest, grassland, and pasture canopies, their leaf distribution and light distribution are shown in Figure 8.2 and 8.3.

Different types of vegetation can have very different k values, causing very different rates of light extinction for the same amount of leaf area. Said in another way, it is much darker underneath a forest canopy than under a grassland canopy with the same total leaf area. Why is this? There are two main factors affecting k values: the angle at which leaves are displayed and the height and density of the canopy. These affect the two types of visible light that reach the canopy: (1) direct and (2) diffuse.

Direct radiation is that received directly from the disk of the sun. Interception of direct light results in a shadow. When a leaf is displayed at right angles to direct beam radiation, this shadow is larger because more direct

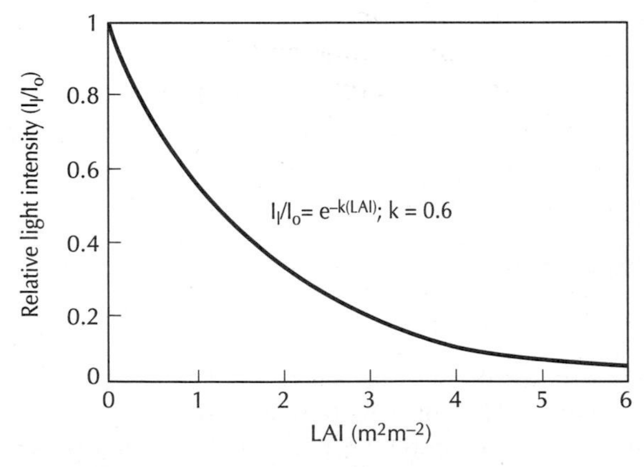

FIGURE 8.1 Exponential decay in light intensity with accumulating leaf area down through a canopy. (After Monsi et al. 1973)

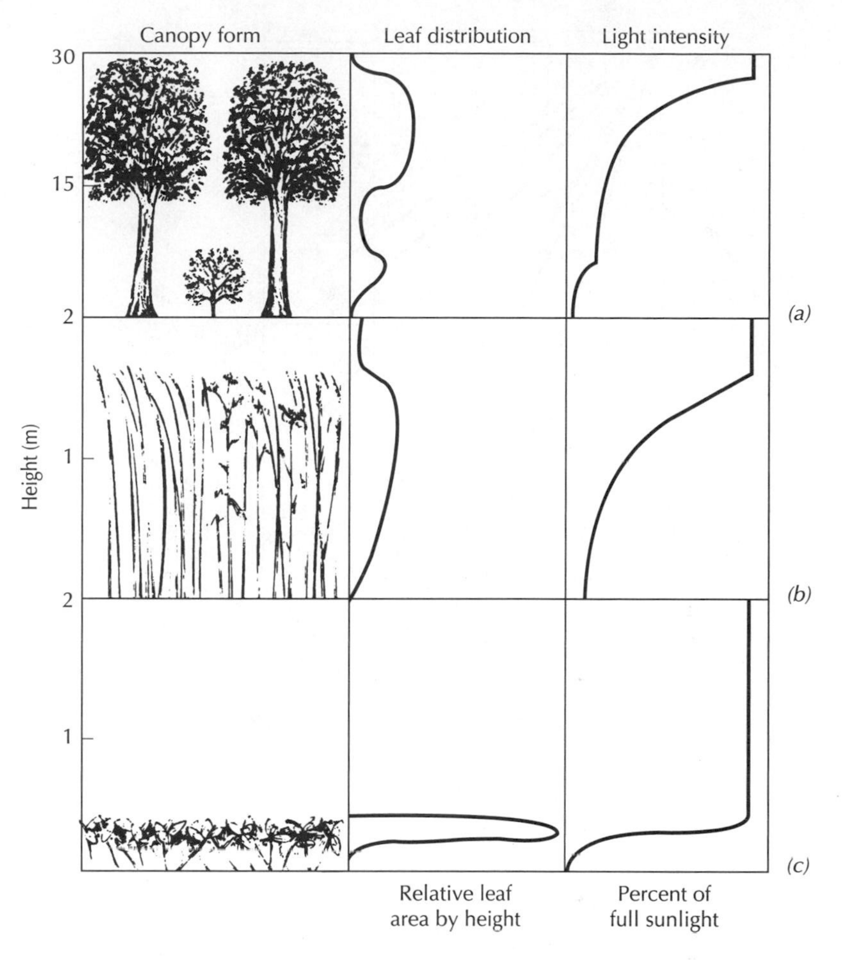

FIGURE 8.2 Examples of leaf distribution and light attenuation in three canopy types: (*a*) broad-leaved forest, (*b*) grassland, and (*c*) pasture dominated by broad-leaved herbs.

radiation is intercepted. As the leaf rotates away from this angle, the shadow shrinks as less direct radiation is intercepted. This allows more radiation to penetrate to lower levels in the canopy, and the *k* value is lower.

Diffuse radiation is that which you would see on a uniformly overcast day. This is energy that has been reflected off clouds, water vapor, and atmospheric gases and particles. It creates no shadow but can be the major form of energy received by canopies on cloudy days. In this case, the taller the canopy or the less densely the leaves are packed into that canopy, the more diffuse radiation can penetrate.

In grasslands, canopies are dominated by erect-leaved grasses and tall, herbaceous plants with photosynthetic stems. Thus, grasslands have the lowest *k* values (Figures 8.2*b* and 8.3). At the other end of the spectrum, short-statured, herbaceous communities, such as pastures that are dominated by species with horizontal leaves, have the highest *k* values, both because of the angle of leaf display and because of the short and densely packed canopy (Figures 8.2*c* and 8.3).

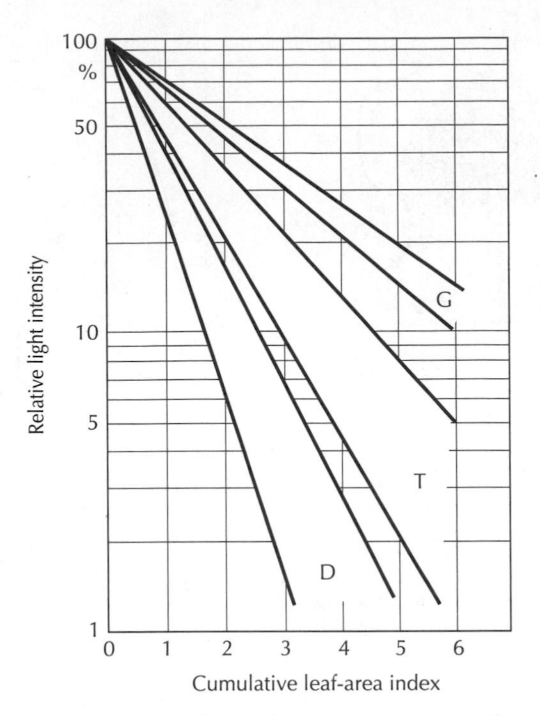

FIGURE 8.3 Relationship between accumulated leaf-area index and light extinction for the three types of plant canopies in Figure 8.2. G = grasses (k less than 0.5); T = forest canopies (k = 0.5–0.65); D = short-stature, broad-leaved canopies ($k > 0.7$). (After Larcher 1995)

Forest canopies are intermediate in both leaf angle and k value, but there is a tremendous amount of diversity both within and between forest tree species. This variability in angle of inclination can help solve the heat load/temperature/respiration problems discussed in Chapter 6 without reducing rates of photosynthesis. For example, if we assume that, on a sunny day, most of the radiation loading to that leaf comes from direct radiation, then changing the angle of display from horizontal to 60° reduces the intensity of radiation at any point on the leaf and the total radiation loading by 50% (Figure 8.4*a*). If that leaf has the photosynthetic response curve shown in Figure 8.4*b*, then this reduction from full sunlight to 50% of full sun causes no reduction in the rate of photosynthesis (Figure 8.4*b*, points 1 and 2). This occurs because the leaves are light saturated at 50% of full sunlight. Inclining the leaf 60° in effect spreads the incident radiation over twice the leaf surface area. The rate of photosynthesis is unaffected, while heat load per unit leaf area is cut in half.

The exact value of this inclination depends on the shape of the photosynthetic response curve and the leaf environment. The same degree of inclination in a leaf receiving only 10% of full sun (Figure 8.4*b*, points 3 and 4) actually reduces photosynthesis more than it reduces heat-loading due to the steep slope in the photosynthetic response curve at this radiation level. Thus, sun leaves might be expected to attain greater angles of inclination than shade leaves. This is generally true (Table 8.1), with species requiring full sun

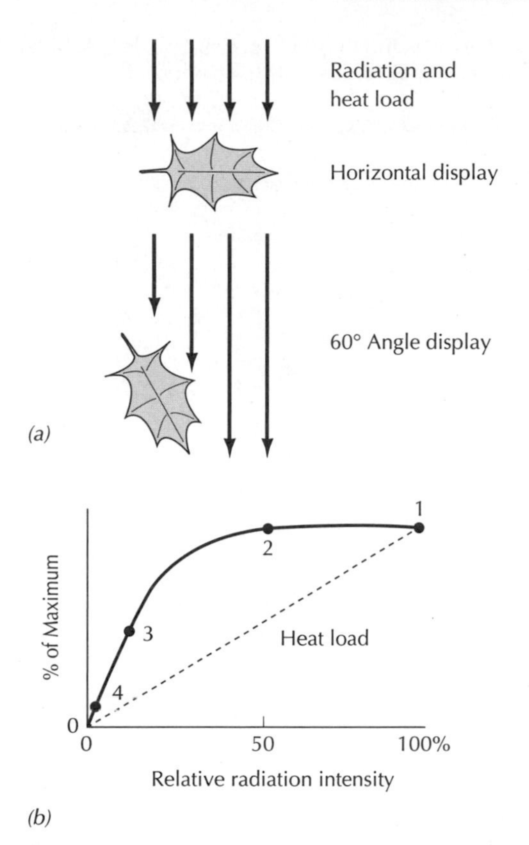

FIGURE 8.4 The effects of angle of leaf inclination on heat loading and photosynthesis. (*a*) A leaf displayed at a 60° angle relative to the sun experiences only half the heat load of a leaf displayed perpendicular to the sun. (*b*) A 50% reduction in radiation intensity due to leaf angle results in no change in carbon gain (see text for meaning of numbers along the curve) but a 50% reduction in heat load.

(e.g., cottonwood) developing greater variations in leaf angle than relatively shade-tolerant species, such as red oak and sugar maple.

Wind Speed, Humidity, Temperature, and CO_2

The physical structure of a canopy also affects the movement of air over the plant surfaces and the distribution of gases. Here again, the different structure of the grassland and forest systems results in very different environments.

A key concept in this case is one borrowed from atmospheric sciences called aerodynamic roughness. This refers to the physical surface presented by the top of the canopy to the atmosphere. Forest canopies are irregular in size and shape, and the height of the tallest trees varies spatially (Figure 8.5*a*). A mapping of the upper surface shows a pattern of peaks and troughs. In addition, the individual elements in the canopy (leaves and branches) are widely and irregularly distributed. Both the irregular surface and the sparse distribution of structures encourage turbulent flow of air across the canopy surface and deep penetration of packets of air (eddies) into the canopy. Such a canopy has a high surface roughness, and air mixes freely between the canopy volume and the air mass above the canopy.

TABLE 8.1 Average Angles of Inclination for Several Species, Arranged Roughly in Order From Shade Intolerant (Top) to Tolerant (Bottom)
(From McMillen and McClendon 1979)

	Degrees from Horizontal		Z-Test (Significance)
Species	**Sun**	**Shade**	
Cottonwood	75.7 ± 7.8 (5,5)	32.3 ± 19.8 (5,5)	0.01
Plum	33.4 ± 15.5 (5,4)	16.4 ± 12.4 (5,3)	0.01
Kentucky coffee tree	64.5 ± 19.2 (5,5)	10.2 ± 7.0 (5,5)	0.01
Catalpa	24.2 ± 15.0 (5,5)	8.2 ± 6.4 (5,5)	0.01
Redbud	35.9 ± 18.8 (4,5)	13.8 ± 14.2 (4,1)	0.01
Green ash	36.8 ± 18.9 (5,4)	14.4 ± 13.8 (5,5)	0.01
Red oak	10.1 ± 10.9 (_,3)	11.5 ± 8.2 (_,4)	NS
Mulberry	34.0 ± 15.4 (5,5)	10.6 ± 8.4 (5,5)	0.01
Silver maple	16.9 ± 15.5 (5,5)	12.7 ± 9.8 (5,5)	NS
Silver maple	18.7 ± 12.5 (_,5)	11.1 ± 8.6 (_,5)	0.01
Sugar maple	14.6 ± 10.2 (_,3)	7.8 ± 5.5 (_,4)	0.01

In contrast, a tallgrass prairie contains plants of more uniform height distribution and much denser concentration of leaves and stems (Figure 8.5*b*). In addition, the individual stems move easily in response to wind, yielding the "rippling" effect that lets you actually visualize the currents of air passing over the surface. All of these characteristics inhibit the mixing of air between the canopy volume and free air above.

As a result, we can see larger differences between the environment within the canopy and above the canopy in grasslands compared with forests (Figure 8.5). In forests, wind speed does diminish with depth, but not enough to cause large gradients in humidity, temperature, or CO_2 concentration. Leaf temperatures in forest canopies generally approximate free air temperatures. In tall-grass prairie systems, all of these characteristics differ by a greater margin over the shorter depth of canopy.

Even in very dry or cold systems where canopy development is minimal, the existence of low-stature and low-density canopies can cause significant changes in the plant environment. One classic example is in desert systems, where canopies intercept and "trap" soil particles carried by the wind (Figure 8.6). This results in an accumulation of fine-textured soil and associated nutrients under the canopies leading to a very uneven distribution of soil characteristics.

STRUCTURED CANOPIES, SUCCESSION, AND LIGHT USE EFFICIENCY

If for a moment we consider light as the only important resource determining whole-system photosynthesis, then we can begin to construct simple models of carbon gain by combining changes in the light environment with

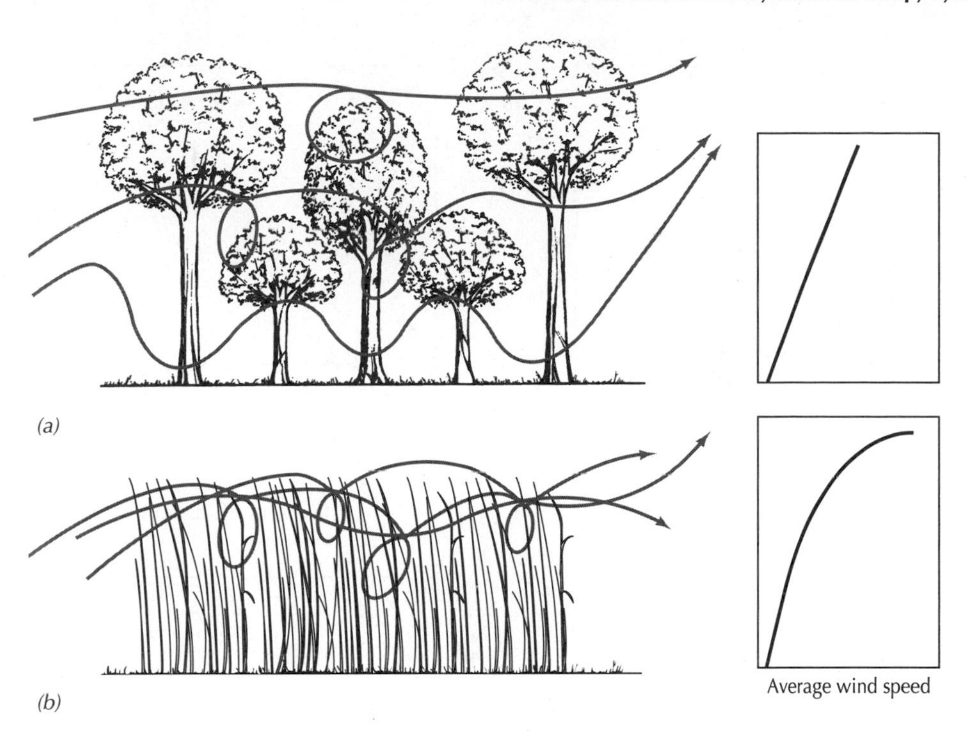

FIGURE 8.5 Differences in canopy structure and gradients in environmental characteristics within a broad-leaved forest and a tallgrass prairie ecosystem. (*a*) A forest presents an open and uneven, or "rough," surface to the atmosphere, leading to rapid mixing of air between that canopy and the atmosphere. (*b*) A tallgrass prairie presents a smoother surface, which impedes exchange such that internal canopy atmosphere is different from the free atmosphere.

changes in the characteristics of foliage with canopy depth. If we treat the canopy as a series of discrete layers, we need estimates of light penetration to that layer, the amount of leaf area by species, and each species rate of photosynthesis at the measured light intensity. The *k* value for the canopy determines the reduction in light as leaf area accumulates with depth. Multiplying the photosynthetic rate by the leaf area for each species in each layer produces the total canopy estimate.

Using these types of calculations, it is an interesting exercise to "design" plant communities for maximum photosynthesis and to compare such optimal canopies with those realized by human management or produced by competition between species in natural communities.

A canopy system "designed" for maximum photosynthesis would simply match photosynthetic response curves with declining light levels down through the canopy. As an example, Figure 8.7 describes hypothetical photosynthetic response curves for three tree species. The response curve for species 3 is typical of a "shade-tolerant" species—one that can survive at low light intensities but cannot grow as fast in full sun. In contrast, species 1 is "shade intolerant"—growing rapidly in full sun but unable to maintain itself in shade. Species 2 would be called "intermediate." As discussed in

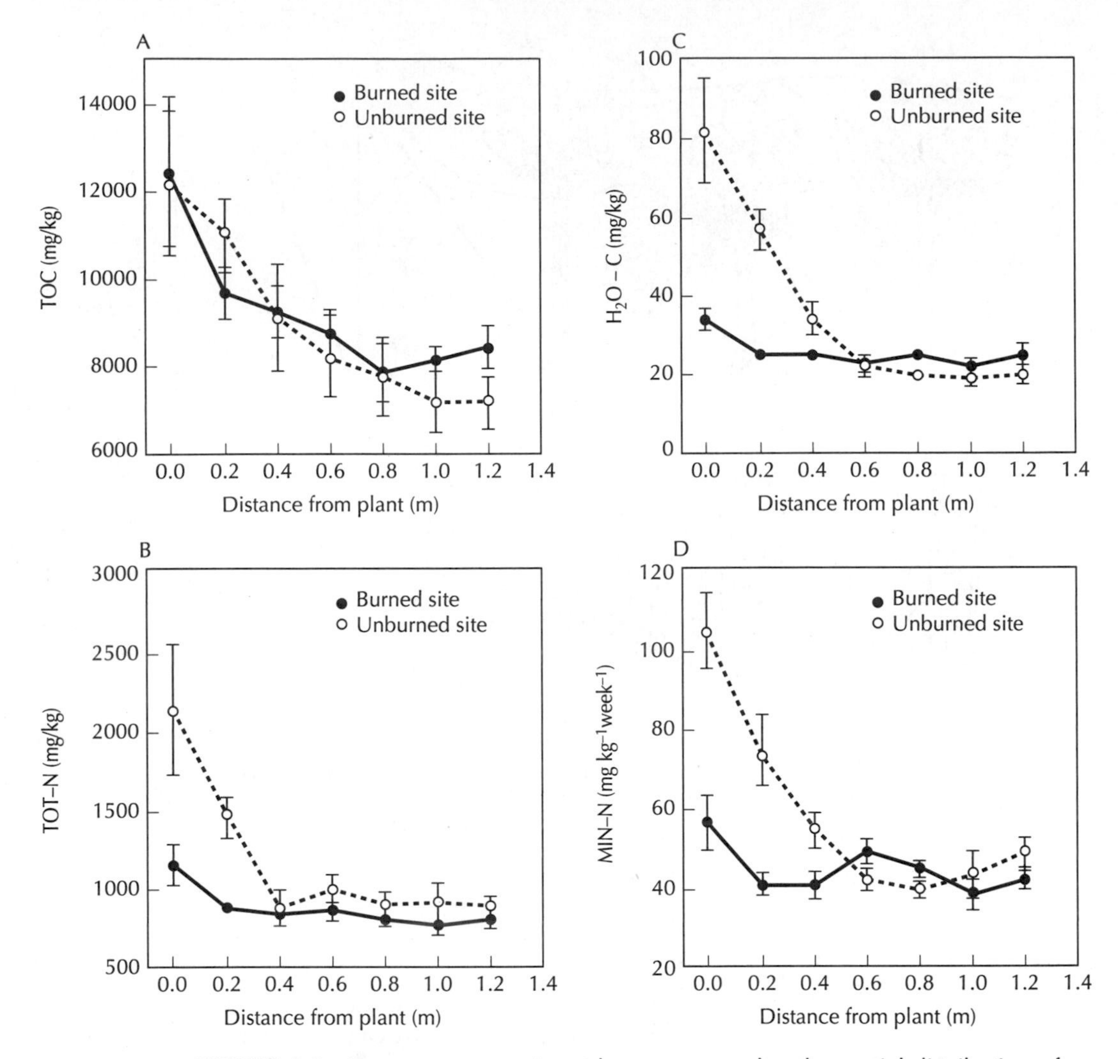

FIGURE 8.6 Canopy structure in arid systems can alter the spatial distribution of resources. In this case, stems of shrubs capture airborne particles that collect around the base, creating enriched microsites. (Halvorson et al. 1997)

Chapter 6, the differences in these response curves would be strongly related to differences in leaf thickness and nitrogen content. We assume that this response includes variation between sun and shade leaves within a species.

A canopy composed solely of species 3 would realize a total photosynthesis equal to the area under curve A (Figure 8.7*b*). Energy at the higher light intensities at the top of the canopy is used inefficiently. A canopy with only species 1 would do better (Figure 8.7*c*) but fails to utilize energy present at low light intensities in the lower canopy. The optimal arrangement would put an upper layer of leaves from species 1, reducing light levels to 60% of full sun, over a middle layer of species 2, reducing light levels to

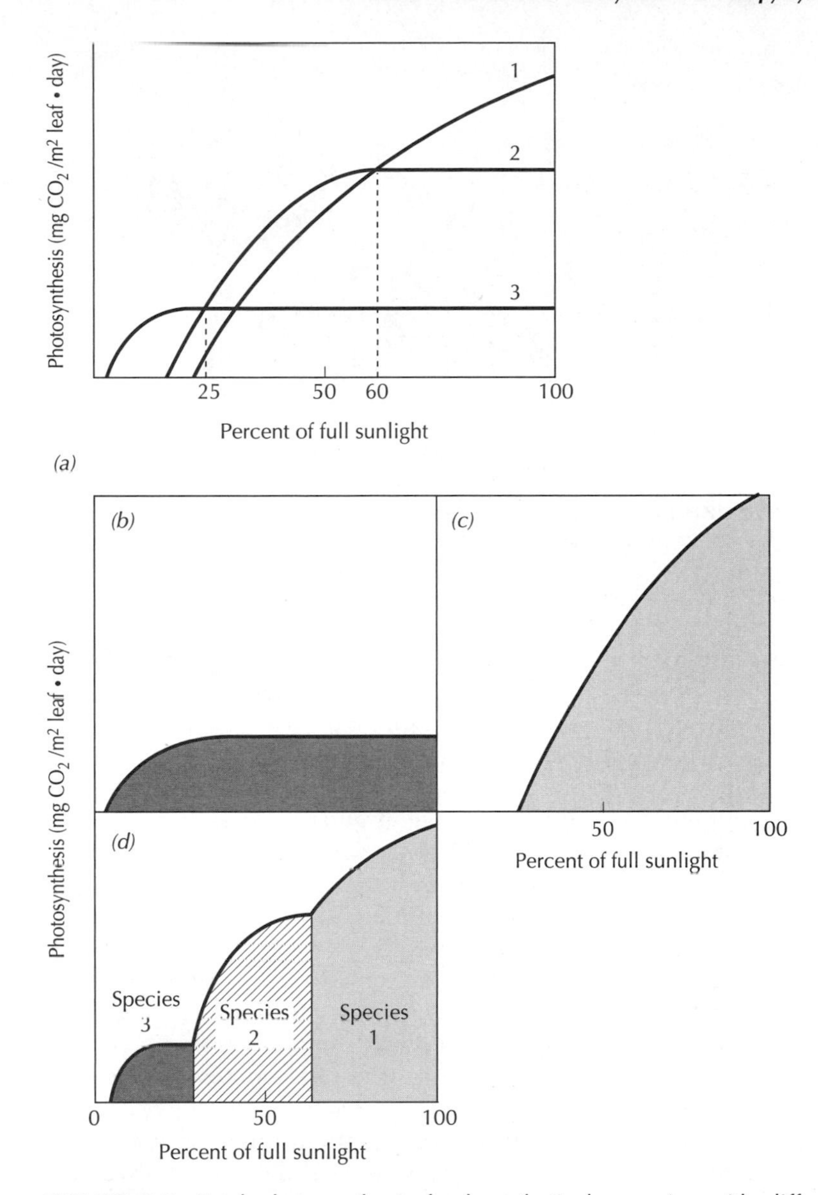

FIGURE 8.7 Total photosynthesis for hypothetical canopies with different species composition. (*a*) The photosynthetic response curves of the three hypothetical species. (*b*) Total photosynthesis for canopy composed only of species 3. (*c*) Total photosynthesis for canopy of species 1. (*d*) Total photosynthesis for "optimal" canopy arrangement.

25% of full sun, over a lower layer of species 3. Photosynthesis carried out by each species is seen in Figure 8.7*d*. Such a canopy might look like that in Figure 8.8.

How do managed ecosystems compare with these idealized canopies? They are totally different. Intensive management of either forests or grasslands usually results in single-species (monospecific) stands (monocultures) of highly productive "intolerant" species, such as species 1 in Figure 8.8. Two

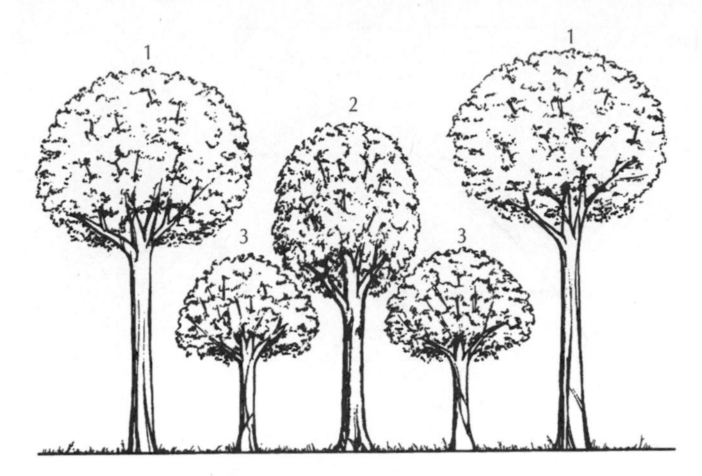

FIGURE 8.8 Diagram of the optimal canopy structure described in Figure 8.7. Numbers refer to species in the previous figure.

of the most widespread examples in temperate North America are loblolly pine forests and cornfields. The structure and light extinction profiles of two such canopies are shown in Figure 8.9.

Why this discrepancy? Three factors are involved. First, it is easier and less expensive to plant, maintain, and harvest uniform stands with mechanized equipment. Second, in managed stands, yield of harvestable material (wood, grain) is the goal, not total photosynthesis. Thus, if a tree produces photosynthate but not wood, it is of little economic value. Slow-growing, tolerant trees, such as species 3 in Figure 8.8, tend to allocate more photosynthate to leaves, twigs, and roots. Restated, total photosynthesis is only one part of yield. The other is how that photosynthate is partitioned or allocated (Chapter 11). Third, breeding and genetic selection of a single species may allow for the development of an optimal range of leaf characteristics within that species so that canopy structure and total photosynthesis may equal those of mixed stands.

There are counter-examples to the two mentioned earlier. The management of tropical agro-forestry systems is moving toward multilayered, multispecies assemblages that both maximize the use of light and provide resistance to pests (see Chapter 23). One of the best established practices is for coffee production, in which tall, nitrogen-fixing (Chapter 10) or commercially valuable species are used as overstory shade trees to create the partially lit conditions in which coffee grows best.

How do canopies in natural plant communities compare with the idealized structure in Figure 8.8? This is a much more complicated question. Natural communities are complex and go through many stages of development. Disturbances such as fire and wind throw are an integral part of community dynamics (see Part III) and alter canopy structure significantly. Should we compare our idealized canopy to old, "undisturbed" stands or younger ones?

An early hypothesis in ecosystem research held that net primary production should be greatest in old-growth, or mature, forests (see Chapter 18). This hypothesis can be tested in part by comparing an optimal canopy structure to different stages of development in natural communities fol-

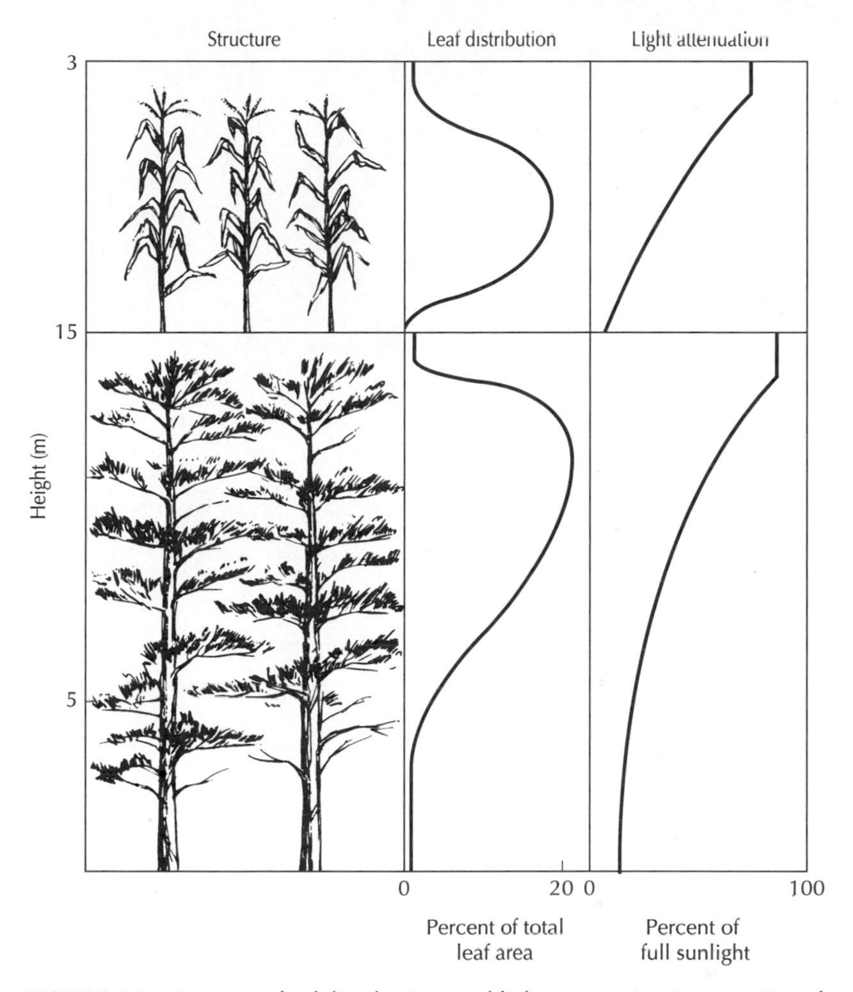

FIGURE 8.9 Structure, leaf distribution, and light attenuation in canopies of commercial corn and forest crops.

lowing disturbance. An example for one type of temperate deciduous forest, the northern hardwood forests of eastern North America, is outlined in Figure 8.10. In this case, the disturbance was a clear-cutting. Progressive, directional change in the structure of a community is called succession, and Figure 8.10 represents a successional sequence (see Chapter 5). A more detailed description of succession in northern hardwood forests is provided in Chapter 21.

Four stages are represented. Within the first stage (age 4 years), total canopy leaf area is already approaching that of the mature forest, even though the height of the canopy is much smaller. Stratification of species has occurred, and the most intolerant fills the top meter, with other, less tolerant species below. By year 30, the canopy is much taller and more evenly distributed. In addition, species are even further stratified by height, with intolerants above species of intermediate tolerance, which are in turn above the tolerant species.

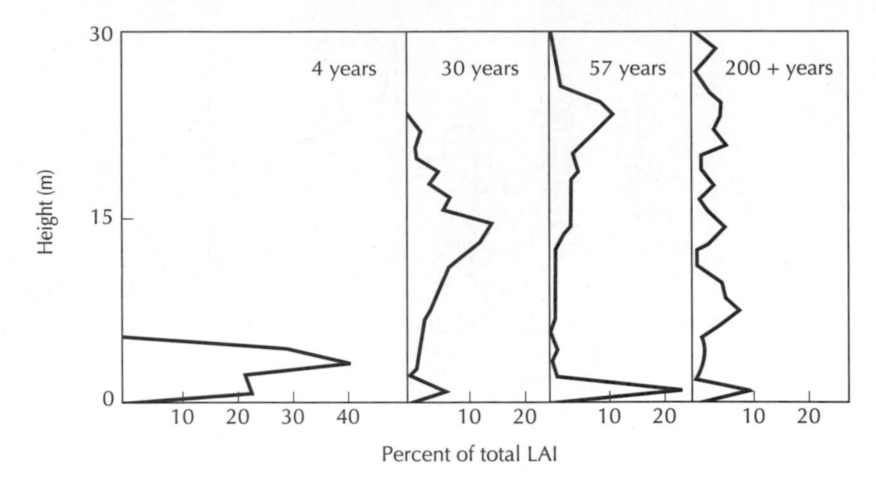

FIGURE 8.10 Four stages of development in canopies of northern hardwood forests. Also shown are the shade tolerance classes of the major species present. (Aber 1979)

By 60 years, intolerant species have died or been shaded out by the now equally tall tolerant trees. Some intermediate trees remain. By year 200+, the oldest stands begin to show a three-layered canopy of tolerant trees over tolerant saplings over extremely tolerant shrubs. Similar three-layered structures have been described for very old, undisturbed tropical rain forests.

Which of these canopies is most similar to the optimal structure? Both ages 30 and 200+ show layering of species, but only the thirty-year-old stand contains a large number of intolerants in the upper canopy. Thus, this relatively early successional stand seems optimal for photosynthesis. This may be a general case for forest ecosystems—that recently disturbed areas with higher numbers of species arranged in height according to tolerance are the most productive, rather than older or mature forests dominated by tolerant species at all levels.

Why are old, mature ("climax") communities not "designed" for optimal photosynthesis? Very simply, intolerant species, with high compensation points, cannot grow in the shade of tolerant ones (see Figure 8.7). Intolerants grow quickly, but, unless they can achieve, permanently, greater height than can tolerants, they eventually are replaced and disappear from the area until a new disturbance reopens the canopy. The presence of species in mature northern hardwood communities is determined more by compensation point than by maximum photosynthetic rate.

PHENOLOGY: SEASONAL VARIATION IN CANOPY STRUCTURE AND FUNCTION

Although light levels may determine rates of total photosynthesis for an existing canopy in the absence of water stress, the actual timing of canopy development and display across the tremendous variety of ecosystems on the Earth is driven more by temperature and the availability of water than by light. These interact to determine the total length of the growing season, that part of the year in which carbon gain through photosynthesis can be maintained. Canopy phenology, or the timing of leaf production and loss, generally tracks this period of potential carbon gain. The length of the growing season and the total amount of foliage displayed determines the total productivity of the system that is also then, in positive feedback, linked to the amount of leaf area that a system will develop during the growing season.

The interactions among climate, phenology, water balance, and carbon gain can be best presented by describing the seasonal patterns of temperature, precipitation, soil water availability, leaf area display, and net photosynthesis for five types of temperate zone ecosystems (Figure 8.11). These represent the four temperate systems in Figure 2.2, and a true desert. See Figure 2.1 for the distribution of these different ecosystem types.

In deserts (Figure 8.11*a*), total production is very low and occurs whenever the rare rainfall events occur. In the example shown here, the annual "blooming" of the desert occurs in midwinter, when rainfall occurs and evaporative demand is relatively low. This allows for a partial recharge of soil water content that causes dormant seeds to germinate and grow, deciduous shrubs to refoliate, and "evergreen" succulents—cacti—to become active. The duration and intensity of this growth flush depends directly on the amount of water available and so is quite variable from one year to the next. Photosynthesis is always less than the potential determined by temperature. Water is always limiting, although somewhat less so during this postwinter period. When the soil water is exhausted, the nonsucculents die or resume dormancy.

As annual precipitation increases, ecosystems come to be dominated by grasses or shrubs. Figure 8.11*b* shows the pattern for a shrub (or chaparral) ecosystem, where precipitation occurs mainly in the winter. Many of the shrub species are evergreens and maintain some leaf area year round, while others are deciduous. The deciduous shrubs, as well as grasses and annual plants, develop leaf area during the wet and cool winter season, when the majority of photosynthesis occurs. As the winter rains end and soil water is depleted, the deciduous shrubs, grasses, and herbaceous plants senesce, leaving only the evergreen shrubs to make use of rare summer rains, deep water tables, or the scarce water reserves in the soil. Photosynthesis is rarely, if at all, temperature limited in these systems.

Grasslands (Figure 8.11*c*) tend to occur in areas of moderate summer rainfall and cold winter temperatures. Here, soil water increases during winter even though precipitation is light because of low temperatures that limit evaporation and disallow plant function. As temperatures increase in the spring, growth and photosynthesis begin and may be more limited by temperature than by water, depending on the amount of winter precipitation. Summer rains tend to recharge the soil but are generally insufficient to eliminate water stress. Depending on the amount of rainfall and the length of the frost-free season, senescence and the end of the growing season may result either from drought or from cold. In the example in Figure 8.11*c*, drought is the cause.

Seasonal evergreen forests are similar in climatic pattern to the chaparral but cooler and with greater precipitation, much of which occurs as snow (Figure 8.11*d*). Cooler temperatures reduce evapotranspiration in winter, and photosynthesis, when it occurs, is mostly temperature limited. Photosynthesis and transpiration increase in early summer with increasing

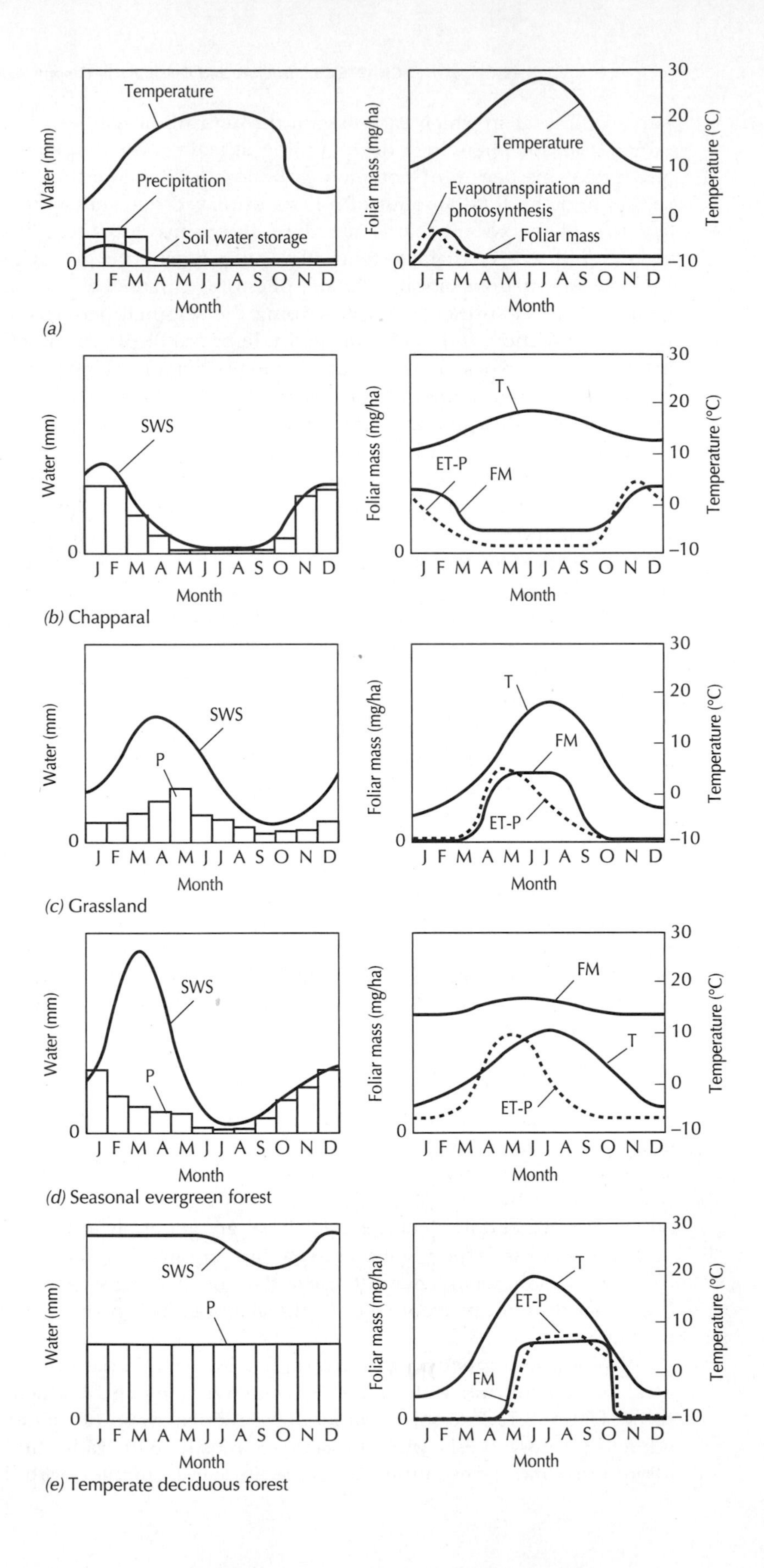

Temperature
Precipitation
Soil water storage
Water (mm)
0
J F M A M J J A S O N D
Month
Foliar mass (mg/ha)
Temperature
Evapotranspiration and photosynthesis
Foliar mass
Temperature (°C)
30
20
10
0
−10
(a)
SWS
T
ET-P
FM
(b) Chapparal
P
(c) Grassland
(d) Seasonal evergreen forest
(e) Temperate deciduous forest

FIGURE 8.11 Opposite page. Seasonal patterns of temperature, rainfall, soil water storage, foliar biomass, and evapotranspiration for five idealized temperate-zone ecosystems. Potential net photosynthesis is proportional to the temperature curve. Realized photosynthesis is proportional to the evapotranspiration curve. (Emmingham and Waring 1977, Walter 1979, Parton et al. 1981, Running and Coughlan 1988, and Vorosmarty et al. 1990)

temperatures until water stress develops in midsummer. Unlike the previous three systems, leaf area of the vegetation does not vary significantly between seasons. Perhaps 20% to 25% of foliage is dropped and replaced each year. Total annual photosynthesis is greater here than in the previous two systems, enough to support respiration by leaves and stems during the dry season, such that standing biomass is also higher.

In the temperate deciduous forest, total and synchronous canopy senescence is one of the most conspicuous processes. This type tends to occur where precipitation exceeds potential evapotranspiration (PET) and extremes of temperature occur (Figure 8.11*e*). Limitations on photosynthesis by water stress are less common and generally less severe than in the seasonal evergreen forest. Low temperatures inhibit leaf-out in the spring and cause senescence in the fall, such that leaves are displayed only during the frost-free season.

Finally, small areas of temperate rain forest occur in which rainfall greatly exceeds PET. Photosynthesis and Actual Evapotranspiration (AET) are driven by temperature, and productivity and standing biomass are very high.

These patterns are general ones for broad classes of ecosystem types. Within types, variations in both climate and soils can also result in significant alterations in canopy structure and function.

A striking example of this can be seen in the temperate evergreen forests along a transect running from coastal Oregon to the deserts in the eastern part of that state. Rainfall is abundant near the coast and at higher elevations in the mountains. Rain shadow effects cause increasingly dry conditions to the east, such that potential evapotranspiration increases from west to east, while precipitation decreases. The intensity of summer drought also increases west to east and causes a very large decline in total leaf area (Figure 8.12).

In the prairie–forest border region of southern Wisconsin, the climate is marginal for forest development and midsummer droughts can be a major factor affecting canopy development. Within this climate regime, soil depth and texture play a large role in determining total soil water storage, which acts as a buffer against drought and extends the growing season. As soils grade from silt clay loams (fine textured) to sandy loams (coarse textured), total soil water storage capacity declines. In response, both mean leaf height and canopy layering change significantly, along with changes in species composition (Figure 8.13)

MODELS OF CANOPY CARBON EXCHANGE

The last three chapters have suggested that canopy photosynthesis is a well-understood process involving interactions of several measurable characteristics, including maximum photosynthetic rates (summarized by nitrogen

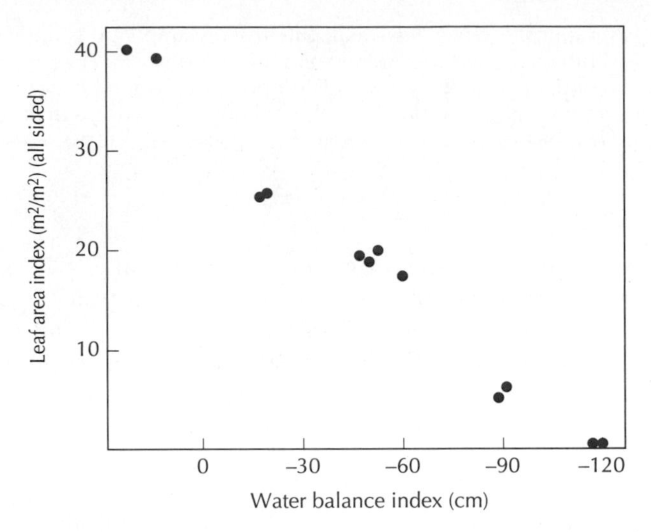

FIGURE 8.12 Changes in forest leaf biomass in relation to an index of annual water stress, along a gradient in water availability in Oregon. (After Grier and Running 1977, Gholz 1982)

concentration); canopy structure and light extinction; and responses to temperature, Vapor Pressure Deficit (VPD), and CO_2. Given this, it should be possible to predict carbon gain with simple models.

This is indeed the case. No important ecosystem process is as well understood as is photosynthesis. As a result, there are a large number of different models that vary widely in complexity and resolution that have been used to predict carbon gain. Three general categories that capture some of the differences between existing models can be described.

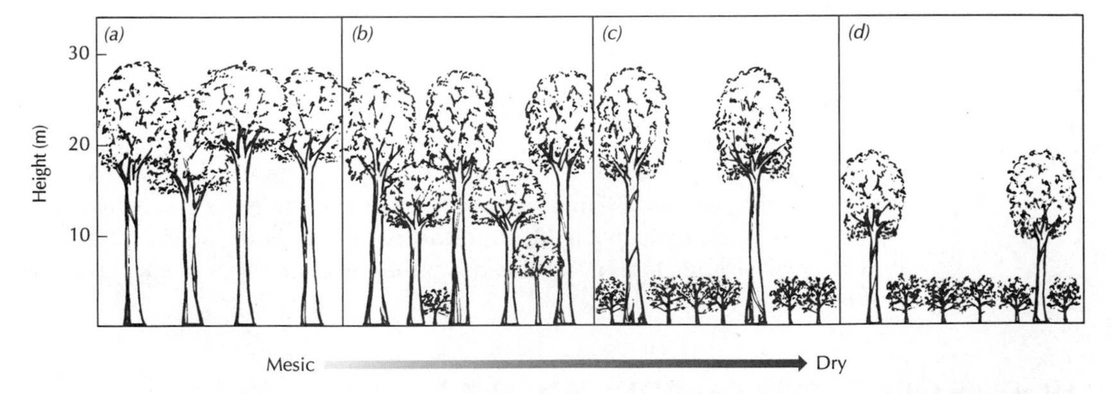

FIGURE 8.13 Changes in forest canopy structure with changes in soil texture in forests of the forest–prairie border region in Wisconsin. (*a*) Fine-texture soils have closed upper canopies and little understory. As soils become progressively drier, canopy structure grades to (*b*) evenly distributed, (*c*) open upper canopy with significant shrub layer, and (*d*) open upper canopy of reduced height, with significant shrub layer. Sites drier than those covered in this study would support prairie systems. (After Aber et al. 1982)

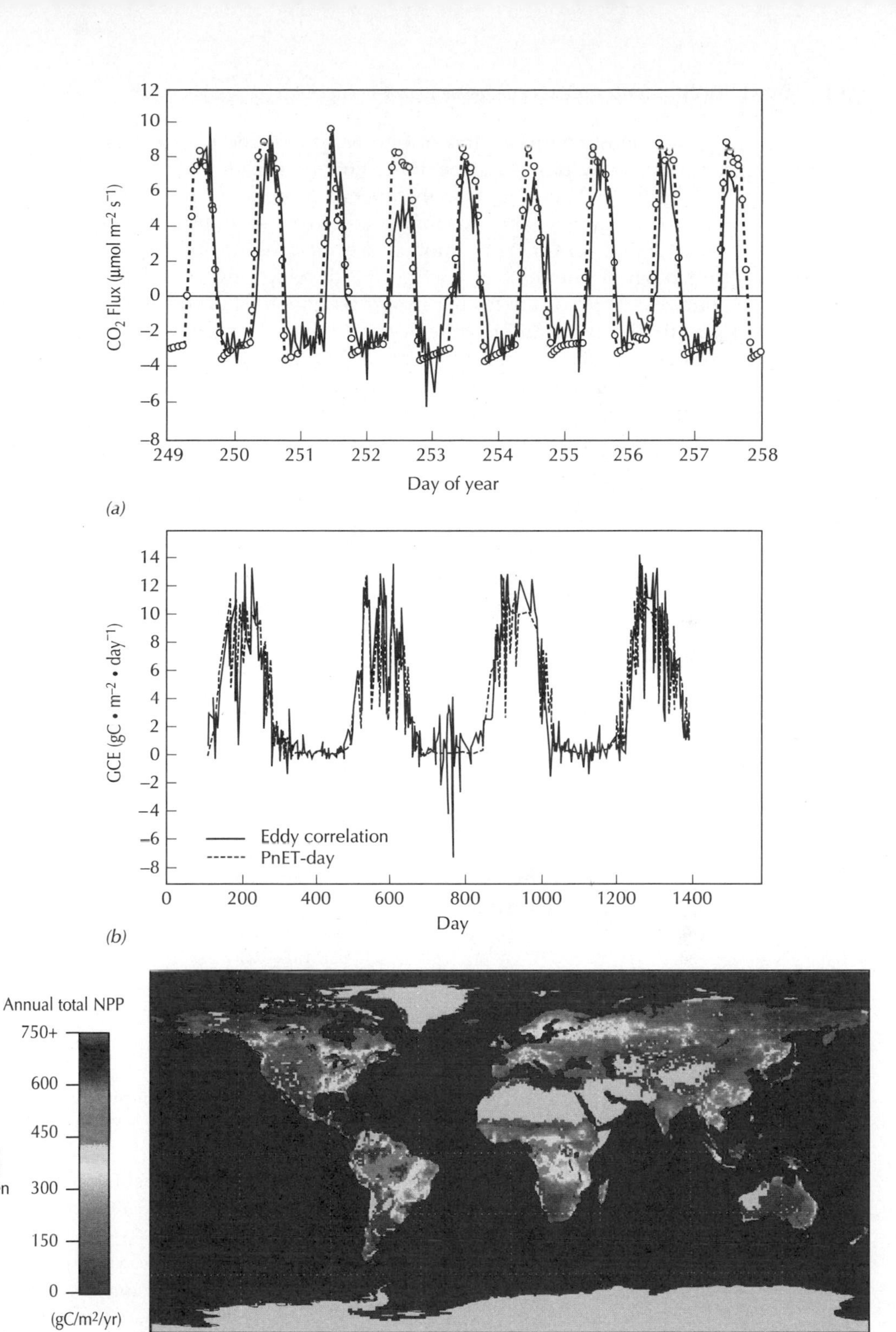

FIGURE 8.14 Examples of outputs from three very different types of models predicting total ecosystem carbon gain. (*a*) Complex models with very short time steps can predict rapid changes in carbon gain with changes in radiation or temperature. (*b*) "Lumped-parameter" models require fewer inputs and are more easily run over longer time period. (*c*) "Efficiency" models are driven primarily by remote sensing data and can be applied at the global scale. (Levy et al. 1997, Aber et al. 1996, Waring and Running 1998)

The most complex class of photosynthesis models describe canopies in great detail, including a three-dimensional realization of leaf distribution and display angle. They also describe the environment over very short time steps (minutes to hours) and can thus predict very short-term changes in carbon gain (Figure 8.14*a*). Such models are most useful for examining complex physiological interactions, such as the development of water stress over the course of a day, or for testing hypotheses about short-term controls on photosynthesis. They can be difficult to run for long periods of time because of the intensity of environmental data required.

A second class of models are relatively simple, or "lumped-parameter," models—those that try to capture the basic dynamics of the physiological processes involved with relatively few equations requiring relatively few measurements as input. For example, the canopy can be described as a simple distribution of leaf area by height, with a single k value to determine light extinction. Mean daily climate data can be used as inputs to equations that are simplified forms of the complex biochemical reaction equations used in the more complex models. Such models can be compared with daily estimates of carbon gain and, because of reduced complexity and a longer time step, are more easily run for several years (Figure 8.14*b*).

The final class of models are simpler still and have been designed to be driven by large-scale remote-sensing data. Such models might have a single parameter that expresses the efficiency of conversion of solar radiation into photosynthate, and this efficiency is determined by the "greenness" of each cell or pixel in a remote-sensing image (Chapter 5). The theory here is that the display of foliage over the year is an integration of all of the environmental site factors, such as drought and temperature. This approach is especially well suited to the prediction of photosynthesis or primary production over large areas (Figure 8.14*c*), although efficiency factors may vary over time and in different ecosystems.

REFERENCES CITED

Aber, J. D. 1979. Foliage height profiles and succession in northern hardwood forests. *Ecology* 60:18–23.

Aber, J. D., J. Pastor, and J. M. Melillo. 1982. Changes in forest canopy structure along a site quality gradient in southern Wisconsin. *American Midland Naturalist* 108:256–265.

Aber, J. D., P. B. Reich, and M. L. Goulden. 1996. Extrapolating leaf CO_2 exchange to the canopy: a generalized model of forest photosynthesis compared with measurements by eddy correlation. *Oecologia* 106:257–265.

Emmingham, W. H., and R. H. Waring. 1977. An index of photosynthesis for comparing forest sites in Western Oregon. *Canadian Journal of Forest Research* 7:165–174.

Gholz, H. L. 1982. Environmental limits on aboveground net primary production, leaf area, and biomass in vegetation zones of the Pacific Northwest. *Ecology* 63:469–481.

Grier, C. C., and S. W. Running. 1977. Leaf area of mature northwestern coniferous forests: Relation to site water balance. *Ecology* 58:893–899.

Halvorson, J. J., H. Bolton, and J. L. Smith. 1997. The pattern of soil variables related to *Artemisia tridentate* in a burned shrub-steppe site. *Soil Science Society of America Journal* 61:287–294.

Lambers, H., F. S. Chapin III, and T. L. Pons. 1998. *Plant Physiological Ecology.* Springer, New York.

Larcher, W. 1995. *Physiological Plant Ecology,* 3d ed. Springer-Verlag, Berlin.

Levy, P. E., J. B. Moncrief, J. M. Massheder, P. G. Jarvis, S. L. Scott, and J. Brouwer. 1997. CO_2 fluxes at leaf and canopy scale in millet, fallow and tiger bush vegetation at the HAPEX-Sahel southern super-site. *Journal of Hydrology* 189:612–632.

Monsi, M., Z. Uchijima, and T. Oikawa. 1973. Structure of foliage canopies and photosynthesis. *Annual Review of Ecology and Systematics* 4:301–327.

Parton, W. J., W. K. Lauenroth, and F. M. Smith. 1981. Water loss from a shortgrass steppe. *Agricultural Meteorology* 24:97–109.

Running, S. W., and J. C. Coughlan. 1988. A general model of forest ecosystem processes for regional applications: I. hydrologic balance, canopy gas exchange and primary production processes. *Ecological Modeling* 42:125–154.

Vörösmarty, C. J., B. Moore, M. P. Gildea, B. Peterson, J. Melillo, D. Kicklighter, J. Raich, E. Rastetter, and P. Steudler. 1989. A continental-scale model of water balance and fluvial transport: Application to South America. *Global Biogeochemical Cycles* 3:241–265.

Walter, H. 1979. *Vegetation of the Earth and Ecological Systems of the Geo-biosphere.* Springer-Verlag, New York.

Waring, R. H., and S. W. Running. 1998. *Forest Ecosystems: Analysis at Multiple Scales.* Academic Press, New York.

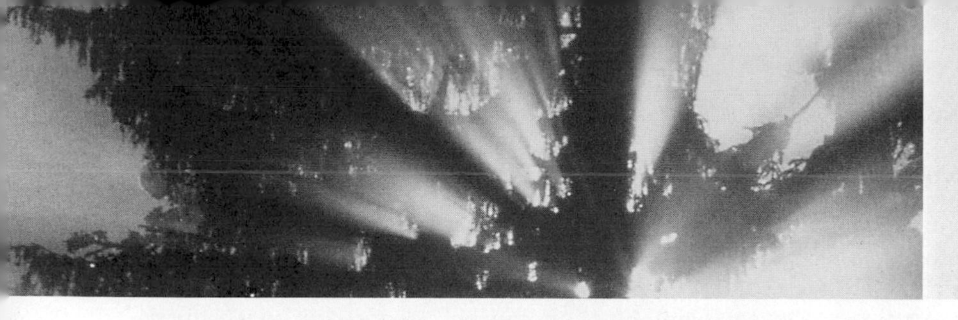

9

SOIL DEVELOPMENT AND THE SOIL ENVIRONMENT

INTRODUCTION

Canopies are visually and structurally complex, yet they are much simpler than soils. Canopies are dominated physically by plant structures and physiologically by the interrelated process of photosynthesis, respiration, and transpiration. The major active structures—leaves—are large, identifiable, spatially separated, and relatively easy to access and to study. These leaves exist in a fluid environment, the atmosphere, which is well mixed and relatively uniform spatially or at least presents identifiable gradients.

In contrast, soils contain a tremendously intricate interweaving of plants and microbes in a complex and heterogeneous solid medium in which chemical and physical conditions vary at the scale of the molecule and the cell. Plants acquire water and nutrients from soils, but root uptake of limiting nutrients often occurs in competition with free-living microbes and must overcome physical and chemical processes that bind those nutrients or inhibit their movement toward the root. At the same time, mutually beneficial symbiotic relationships between roots and microbes, especially fungi, occur in almost all systems. Both biological and nonbiological processes alter basic soil properties, such as acidity and element concentrations. These reactions all occur at such a microscopic scale and are so strongly interrelated that it is nearly impossible to separate them. Because of this, soils are more difficult to study in terms of the rates at which basic processes occur, and our understanding of them is less complete than what has been described in the previous chapters for canopies.

The purpose of Chapters 9 and 10 is to describe the soil system from the perspective of nutrient availability to plants. In this chapter, we focus on chemical reactions that provide and sequester nutrients, setting them in the context of the soil environment and discussing how they affect soil development. We concentrate on reactions involving those chemical elements that most influence ecosystem behavior, those that are required by plants and microbes, and those that dominate the chemical environment. In Chapter 10, we add plant and microbial demand for nutrients to this mix. The complementary process of decomposition, by which nutrients in plant and microbial biomass are returned to soils in available form, is described in Chapters 12 and 13.

THE MAJOR ELEMENTS

The elements that dominate the biogeochemistry of ecosystems are those required for biological processes and those that are major components of the Earth's crust and atmosphere (Table 9.1).

Elements required for plant function can be classified as "macronutrients" and "micronutrients" according to the amounts generally found in plant tissues. Most macronutrients are major components of important structural and metabolic molecules, such as carbohydrates, proteins, chlorophyll, DNA, RNA, sugar phosphates, and phospholipids (Table 9.2). The exception is potassium, which is used in stomatal control, for charge balance during ion movements across membranes, and as a coenzyme in many important biochemical reactions, but does not occur in organically bound forms. The micronutrients are generally used either as structural components of less common molecules or as coenzymes—ions needed to catalyze specific reactions. In terrestrial ecosystems, limitations on important processes by suboptimal availability of micronutrients is the exception rather than the rule,

TABLE 9.1 A Summary of the Relative Concentration (Percentage of Total Mass) of Different Elements in Plants, the Atmosphere, and the Earth's Crust
(Salisbury and Ross 1978, Flint and Skinner 1974, Levine 1989)

Element	Chemical Symbol	Plants		Concentration (%)	
		Leaves	Wood	Atmosphere	Crust
Plant macronutrients					
Carbon	C	45.0	45.0	0.037	—
Hydrogen	H	5.0	5.0	—	—
Oxygen	O	45.0	45.0	21.0	45.2
Nitrogen	N	2.0	0.1	78.0	—
Potassium	K	1.0	0.1	—	1.7
Calcium	Ca	0.8	0.2	—	5.1
Magnesium	Mg	0.2	0.02	—	2.8
Phosphorous	P	0.2	0.01	—	—
Sulfur	S	0.2	0.02	—	—
Additional major crustal components					
Silicon	Si	Not required		—	27.2
Aluminum	Al	Not required/toxic		—	8.0
Iron	Fe	Micronutrient		—	5.8
Additional plant micronutrients					
Chlorine (Cl)					
Boron (B)					
Manganese (Mn)					
Zinc (Zn)					
Copper (Cu)					
Molybdenum (Mo)					

TABLE 9.2 The Macronutrients, Their Biochemical Uses, Form of Uptake by Plants, and Whether They Have Been Shown to be Limiting to Plant Growth and Mobile Within Plants[a]
(After Salisbury and Ross 1978)

Element	Uses	Taken Up As	Mobility	Limiting?
Carbon (C), Hydrogen (H), Oxygen (O)	Carbohydrates and derivatives, basic building blocks for nearly all plant products	CO_2 H_2O	Variable	See Chapters 6–8
Nitrogen (N)	Amino acids, proteins, enzymes, nucleic acids, chlorophyll	NO^-_3, NH^+_4	High	Yes
Phosphorus (P)	Sugar phosphates (ATP, ADP), nucleic acids, phospholipids	$H_2PO^-_4$	High	Yes
Potassium (K)	Not structural, enzyme co-factor catalyzes protein formation; stomata; charge balance across membranes	K^+	Very high	Rarely
Sulfur (S)	Amino acids, proteins, enzymes	SO^-_4	Low	Very rarely
Magnesium (Mg)	Chlorophyll, enzyme cofactor	Mg^{2+}	Very low	No (but see chapter 25)
Calcium (Ca)	Crucial to membrane function; binds wood fibers together; stabilizes waste products in vacuoles	Ca^{2+}	Very low	No (but see chapter 25)

[a]Mobile elements are those that can be retranslocated by plants before leaf senescence. Elements such as sulfur and magnesium are somewhat mobile in plants but are rarely retranslocated due to excess availability in soils.

occurring mainly on unusual or extremely old soils. In keeping with our focus on primary controlling factors, we do not deal extensively with micronutrients in this book.

Elements required by animals are generally the same as for plants, with some notable exceptions. One is sodium. Mammals require relatively large amounts of sodium for proper function of the nervous system. Selective use of sodium-rich plants (for plants do take up elements other than those required for growth) or even the ingestion of sodium-rich soils (saltlicks) are two common animal responses to a lack of this element in plants.

Carbon, hydrogen, and oxygen form the majority of plant biomass—up to 96% for some types of tissues. These three are derived from atmospheric CO_2 and soil water, combined through photosynthesis. All of the remaining macronutrients except for nitrogen are derived ultimately from materials in the Earth's crust through the weathering of rocks (see subsequent discussion), where they are present in low concentrations. Rock types, such as limestone, in which calcium and magnesium are major components, are formed from the products of biological activity.

Although the atmosphere is 78% nitrogen (in the relatively inert form of dinitrogen gas [N_2]), most plants acquire nitrogen through uptake of inorganic forms in soils (Table 9.2). It is an interesting irony of ecosystem studies that plants in many temperate-zone ecosystems experience a shortage of nitrogen, while their leaves are exposed to an atmosphere rich in this element. Certain classes of plants exist in a symbiotic relationship with microbes that allows conversion of atmospheric N_2 into forms usable by plants (Chapter 10).

Silicon, oxygen, aluminum, and iron are major components of the Earth's crust (Table 9.1) and play crucial roles in soil development and chemistry. The preponderance of these four leads to the generic name given to most crustal materials—iron and aluminum silicates. Iron is also a micronutrient in plants and is important in mammals because of its role in the heme structure of red blood cells. Both iron and aluminum play a large role in the availability of phosphorus, and aluminum can accumulate in very acidic soils to levels that inhibit root function. Silicon contributes a major soil component in cold, acidic soils, and while not listed as a required plant nutrient, it is concentrated in certain grasses where it apparently functions as a deterrent to herbivores.

While carbon, hydrogen, and oxygen cycle mainly between the atmosphere and organic matter in plants and soils, the remaining six macronutrients exist in a variety of states in soils, and their cycling and availability to plants are affected by a variety of processes. The summary figure at the end of Chapter 1 (reproduced here as Figure 9.1) lumps these states into four categories: (1) organic, (2) exchangeable, (3) sorbed, and (4) primary and secondary minerals. While these states and classes of reactions represent a grand oversimplification, they provide a basis for discussing the differences in the chemical and biological factors affecting the availability of the soil-derived macronutrients.

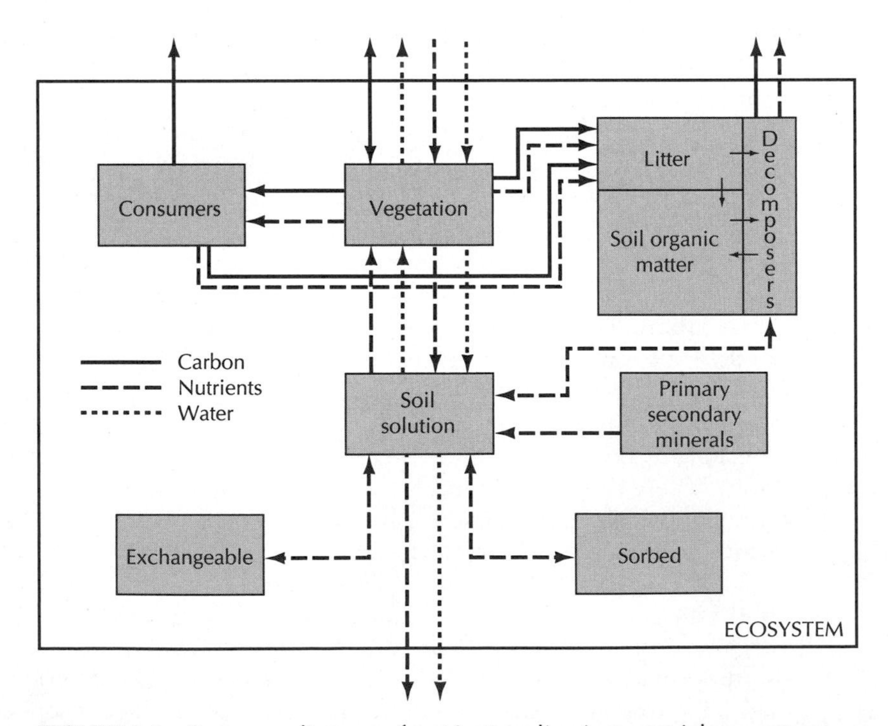

FIGURE 9.1 Summary diagram of nutrient cycling in terrestrial ecosystems.

THE SOIL ENVIRONMENT

Important soil processes occur at extremely small scales and are linked to the environments in moist surfaces. Roots and their microbial symbionts (see Chapter 10) extend long, thin tissues over the surfaces of soil particles to access nutrients. Free-living microbes also occupy these surfaces, exuding enzymes designed to decompose large organic molecules and taking up the smaller products of the resulting reactions. Often microbes and plants compete for the same elements in the same location, with important effects on ecosystem function.

Water passing through the soil travels mainly along these same surfaces in response to concentration (water potential) gradients within the soil, resulting in part from the pull of plant transpiration. As the water passes over these surfaces, the organic and inorganic chemicals dissolved in the water react with those on the soil surfaces. Individual elements are gained and lost, and summary parameters, such as pH (soil acidity) and conductivity (total concentration of dissolved ions), are altered. While general processes of this type can be described, the actual physics and chemistry of particle surfaces in soils are still only partially understood. Sometimes it is even difficult to distinguish which transformations are biological and which are chemical.

Individual elements and ions can move along particle surfaces passively in response to the bulk flow of water, or their movement may be accelerated by gradients in concentration resulting from plant uptake, or retarded by interactions with particle surfaces. Two characteristics are key to all of these processes: (1) the quantity and characteristics of the soil surfaces, (2) how moist they are.

As discussed in Chapter 7, the quantity of soil particle surface present in a given volume of soil is a function of soil texture, which is described as the fraction of separable soil solids of different sizes (Table 9.3). In general, soils with higher fractions of smaller particles (silt and clay) have more total internal surface area than do sandy soils.

When absolutely dry, soils contain a combination only of solids (particles) and air. The total volume of air-filled space is called the pore space. There is generally more total pore space in sandy soils than in finer-textured soils (Figure 9.2). A crucial parameter for both chemical and biological processes in soils is called water-filled pore space, or the fraction of the total pore space that is occupied by water. When water is absent, the moist films along the surface of soil particles disappear and there can be no diffusion of ions or elements or small biological molecules in response to concentration gradients. Enzymes exuded by microbes cannot react with substrates. Nutrients cannot move to roots with the flow of water, and most soil functions cease. Researchers actually make use of this fact when preserving soil samples for long periods of time. Soils that are oven-dried (500°C) and then stored in airtight jars remain unaltered in most respects for decades. In the absence of water, reactions happen very slowly, if at all.

As the water content of soils increases, the fraction of total pore space filled by water increases, and the rates of most soil processes increase as well (Figure 9.2), up to a point. As this fraction approaches 1, passageways between pores for the diffusion of gases begin to close. As respiration continues, oxygen is depleted and cannot be replaced by exchange with the

TABLE 9.3 Classification of Particle Sizes in Soils
(After Roth 1990)

Separate	Diameter (mm)	Number of Particles Per Gram	Surface Area in 1 Gram cm^2
Very coarse sand	2.00–1.00	90	11
Coarse sand	1.00–0.50	720	23
Medium sand	0.50–0.25	5,700	45
Fine sand	0.25–0.10	46,000	91
Very fine sand	0.10–0.05	722,000	227
Silt	0.05–0.002	5,776,000	454
Clay	>0.002	90,260,853,000	8,000,000

atmosphere. Eventually, sites within the soil begin to become anoxic (oxygen-depleted) and many common biological functions shut down. A different set of processes that results in a different set of products (e.g., methane [CH_4] instead of CO_2) become dominant. An anoxic environment also changes chemical processes significantly.

Because coarse-textured (sandy) soils have a high soil pore volume and low particle surface area, they drain very rapidly and are less likely than fine-textured soils to support anoxic sites. Fine-textured soils offer more particle surfaces for reactions and can hold more water but may have less pore space and may become water saturated more easily and therefore support more anoxic sites.

SOIL CHEMICAL PROCESSES AFFECTING NUTRIENT AVAILABILITY

Weathering

Weathering of Primary Minerals – Weathering is the process by which newly created or newly exposed geological substrates are converted into soils and in-

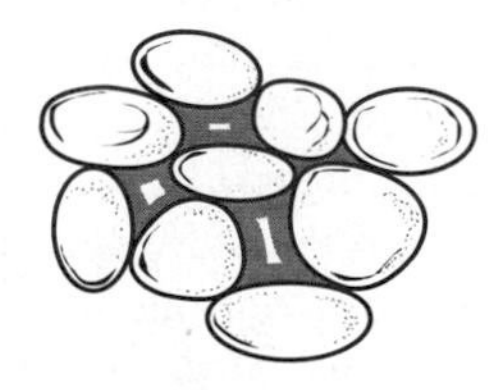

FIGURE 9.2 Diagram of soil solids, pore spaces, and water distribution at different water contents.

volves the physical and chemical alteration of the geological substrate underlying an ecosystem. Weathering is initiated by the retreat of a glacier, the creation of newly hardened lava flows by volcanic activity, or by any other process exposing geological material that has been previously protected from chemical dissolution and physical wear.

Three aspects of the larger study of weathering rates and reactions are important in the function of ecosystems: (1) the rate at which the geological material in the system weathers, (2) the nutrient elements released during weathering, and (3) the type of secondary minerals formed. The first two determine the effect of weathering directly on pools of nutrients available to plants in the system. The third affects the capacity of soils to retain those nutrients against leaching losses and plant uptake.

There are three major types of rocks: (1) sedimentary, (2) igneous, and (3) metamorphic. **Sedimentary rocks** are those formed by deposition of mineral particles eroded from the landscape and deposited in lakes, seas, and oceans. As the sediments into which these particles settle gets deeper, the pressure produced by the weight of overlying material causes the sediment to solidify into rock. The characteristics of these rocks, and the types of soils derived from them by weathering, are largely determined by the type of material deposited. For example, sandstones formed by the compression of sand particles generally weather back into sandy soils, releasing few nutrients in the process. In contrast, limestone is formed by the sedimentation of the remains of aquatic organisms that concentrate calcium and magnesium during life in shells and other structures. Weathering of this type of rock is rapid and releases large amounts of calcium and magnesium.

Igneous rocks are those formed by the cooling of volcanic flows (magma) either at or beneath the surface of the Earth. The properties of these rocks depend on the rate and temperature conditions at which they formed. Igneous rocks consist mainly of iron and aluminum silicates with lesser concentrations of calcium, potassium, magnesium, and sodium, reflecting the relative abundance of these elements in the Earth's crust (Table 9.1).

Metamorphic rocks are either sedimentary or igneous rocks that have been altered by the pressure and heat generated by overlying rocks but have not returned to the melted or magma state. These rocks are modified more in appearance and large-scale structural features than in the mineralogical characteristics that affect weathering.

Figure 9.3 shows the relative content of different minerals in different types of igneous rocks, as well as the relative nutrient contents and rates of weathering. The darker, more easily weathered, and more nutrient-rich rock types are generally formed by rapid cooling after movement to the Earth's surface. Few rock types consist entirely of one mineral. In igneous rocks of mixed mineralogy, the more easily weathered minerals (e.g., olivine and pyroxene families) weather first, leaving a residue of the more slowly weathered minerals, plus secondary minerals (discussed later).

The content of elements other than silicon and oxygen is also affected by conditions during rock formation. The harder rocks tend to have greater substitution of aluminum for silicon and lower contents of calcium and magnesium as "bridging" ions between the crystal structures. Thus, young soils derived from volcanic flows that solidified on the Earth's surface are often rich in nutrient cations (calcium, magnesium, and potassium), as well as

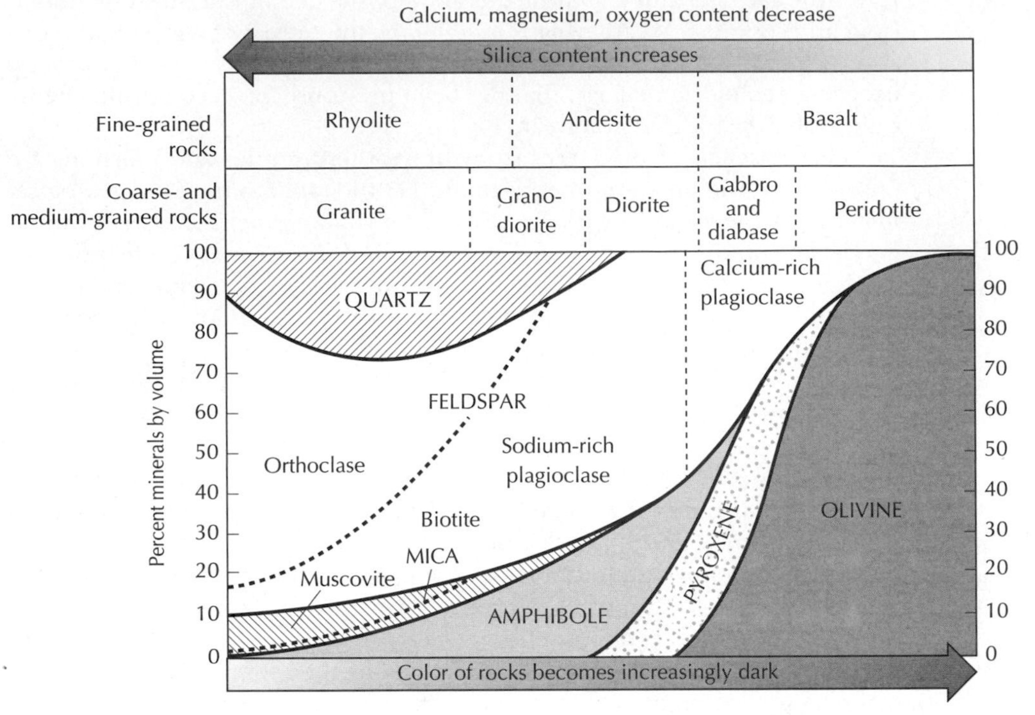

FIGURE 9.3 Types of rocks and the minerals they contain as a function of the conditions under which they formed. The content of plant nutrient elements in the rocks and the rate at which they weather change along this gradient. (After Flint and Skinner 1974)

phosphorus. The macronutrients sulfur and phosphorus are present in trace amounts in many geological formations, usually in minerals that are easily weathered. Nitrogen is absent from all major rock types and is not made available through weathering except in a few, very unusual geological formations.

Weathering rates are increased by high temperatures and high (but not saturated) soil water content. Thus similar rock types are altered more quickly in tropical rainforest conditions than in either cool or dry systems (Figure 9.4). This has the positive effect of liberating the nutrients held in fresh rocks more quickly under tropical conditions and the negative effect of more rapid depletion of weatherable minerals under these same conditions.

Formation of Secondary Minerals – Except in some sedimentary rocks, such as pure limestone, weathering does not occur through complete dissolution of the primary minerals. Rather, secondary minerals are formed by the weathering reactions. These products are dominated by combinations of the major crustal elements, oxides of silicon, iron, and aluminum. Iron and aluminum oxides have properties that are very different from those of silicon oxides and tend to dominate as weathering products under different climatic conditions. Both climate and chemistry can also alter the form that iron and aluminum oxides take as weathering products. These differences result in very different soils.

In hot and humid climates, silicon oxides tend to be more soluble than aluminum and iron oxides and more mobile in soils. They also tend not to

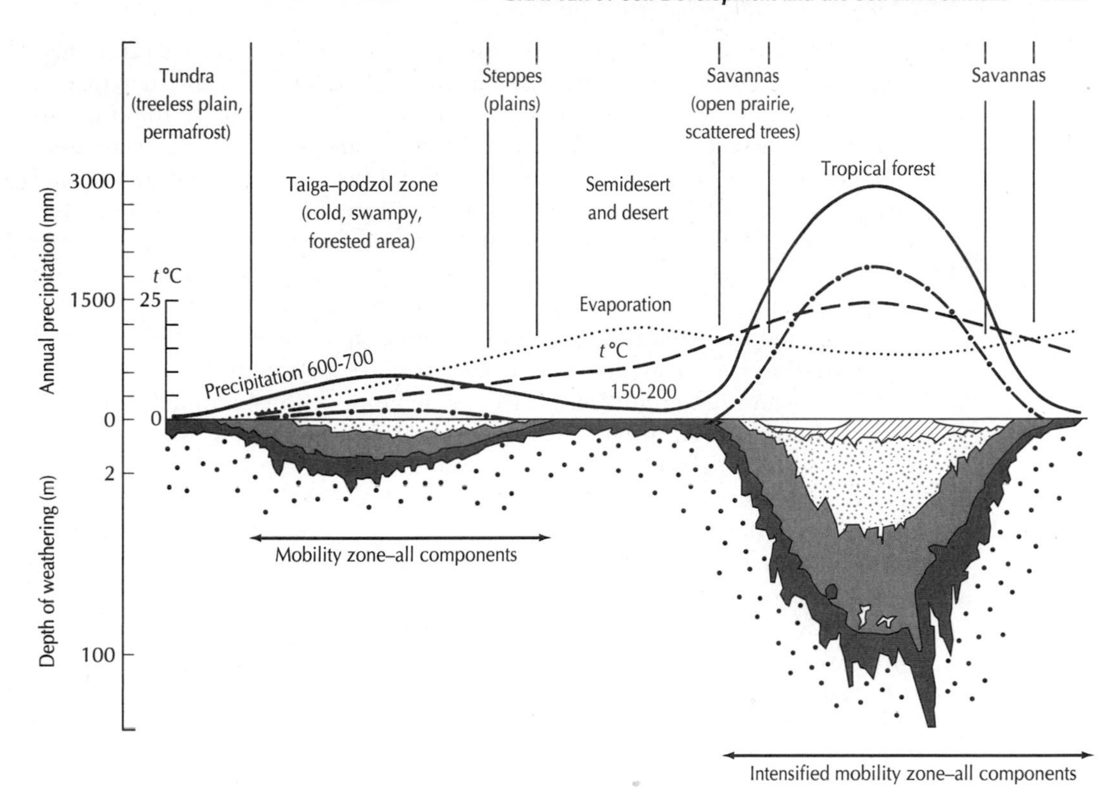

FIGURE 9.4 Schematic view of the effects of climate on the depth of soil weathered. The warm and wet conditions of the tropical forest result in the most deeply weathered soils. Weathering rates are reduced by drier conditions in the semiarid regions and by both reduced precipitation and colder temperatures in the boreal forest and tundra zones (Strakhov 1967). Depth of soil weathering in far northern regions is also reduced by recent glacial epochs that restart the soil development process.

form secondary products, such that all of the original silicon present in fresh substrates tends to be stripped away during the long process of soil formation. Soils then tend to become dominated by iron and aluminum oxides. When this process goes to completion, the residual material is gibbsite (oxides of aluminum) or hematite (oxides of iron), which accumulate in a deep E horizon.

In contrast, under cool and humid conditions, such as in northern forests, the iron and aluminum oxides tend to be more mobile and are leached from the upper soil horizons, often in combination with dissolved organic matter. This results in the accumulation of silicon oxides in the upper horizons. When taken to completion, some pure SiO_2 (quartz sand) occurs in the E horizon. However, unlike the loss of silicon from tropical soils, mobile oxides of iron and aluminum in colder sites tend to reprecipitate in lower soil horizons, forming a B horizon enriched in these compounds.

Silicon, iron, and aluminum oxides represent extreme endpoints of the weathering process and require thousands of years to become dominant soil minerals. Most soils contain a mixture of both fully and only partially weathered minerals. Partially weathered iron and aluminum silicates form a series of secondary products that fall into the clay fraction during physical separation. Two categories of clays are often distinguished based on the structure of the minerals: (1) 1:1 clays and (2) 2:1 clays. The names relate to the interlayering of sheets of aluminosilicates with different structures. The 2:1 clays have alternating layers with different structures that provides spaces in the crystal lattice for the inclusion or substitution of different ions. In 1:1 clays, alternating surfaces of oxygen and hydroxyl provide for a strong bond between layers, allowing little substitution and providing less surface for reactions with other elements. While often placed in a weathering sequence, with the 2:1 clays (e.g., montmorillonite and illite) considered more easily weathered and a precursor to the formation of 1:1 layer clays (e.g., kaolinite), under moderately cool and dry conditions, 2:1 clays make up a considerable portion of the clay fraction even in old soils. Conversely, primary minerals can weather directly to kaolinite under favorable conditions, skipping the 2:1 layer stage.

While moisture and temperature are important in determining weathering rates and the relative solubility of aluminum and silicon, other characteristics of the weathering environment can also be important in determining the type of secondary mineral formed. For example, sandstone substrates, rich in silica and poor in aluminum, can weather to form silicon oxides even under tropical conditions. This occurs because low aluminum concentrations in the substrate does not provide enough soluble aluminum to cause the formation of 1:1 clays or aluminum oxides. The lack of nutrient cations in sandstone also leads to low soil pH and, again, increased mobility of aluminum. As another example, montmorillonite (a 2:1 layer clay) is more stable at near-neutral pH than under acidic conditions, and its formation may depend on the presence of potassium in the soil solution.

Evaporites are formed in arid regions, where water leaching through the soil is minimal. In these areas, easily weathered minerals, such as calcium carbonate, are dissolved in the upper soil horizons during brief wet periods and deposited in lower horizons as percolating water either evaporates or is taken up by plants. Desert subsoils often contain a hard, cementlike layer of deposited calcium carbonate, or "caliche." Halite (sodium chloride, or salt) is another evaporite.

Weathering, Soil Types, and Soil Chemistry

Characteristics of Major Soil Groups – It is not by coincidence that this discussion of weathering has run somewhat in parallel to the discussion of soil-forming factors in Chapter 2. The type of weathering regime active in a soil, and the secondary products formed, are important determinants of soil type. Important secondary products of weathering can be placed on a temperature–precipitation grid similar to the one used in Chapter 2 (Figure 9.5*a*). The tropical rain forest environment produces kaolinite or, in extreme cases, gibbsite and hematite clays associated with lateritic soils. Subtropical and tropical forest soils are rich in the 1:1 clays. The northern temperate and boreal forest environments produce the silicon oxides associated with the E horizon of Spodosols. The intermediate environments of the temperate forests produces a mixture of 1:1 and 2:1 clays that can be leached to the B horizon (lessivage) to create Alfisols. Drier and more neutral soil conditions

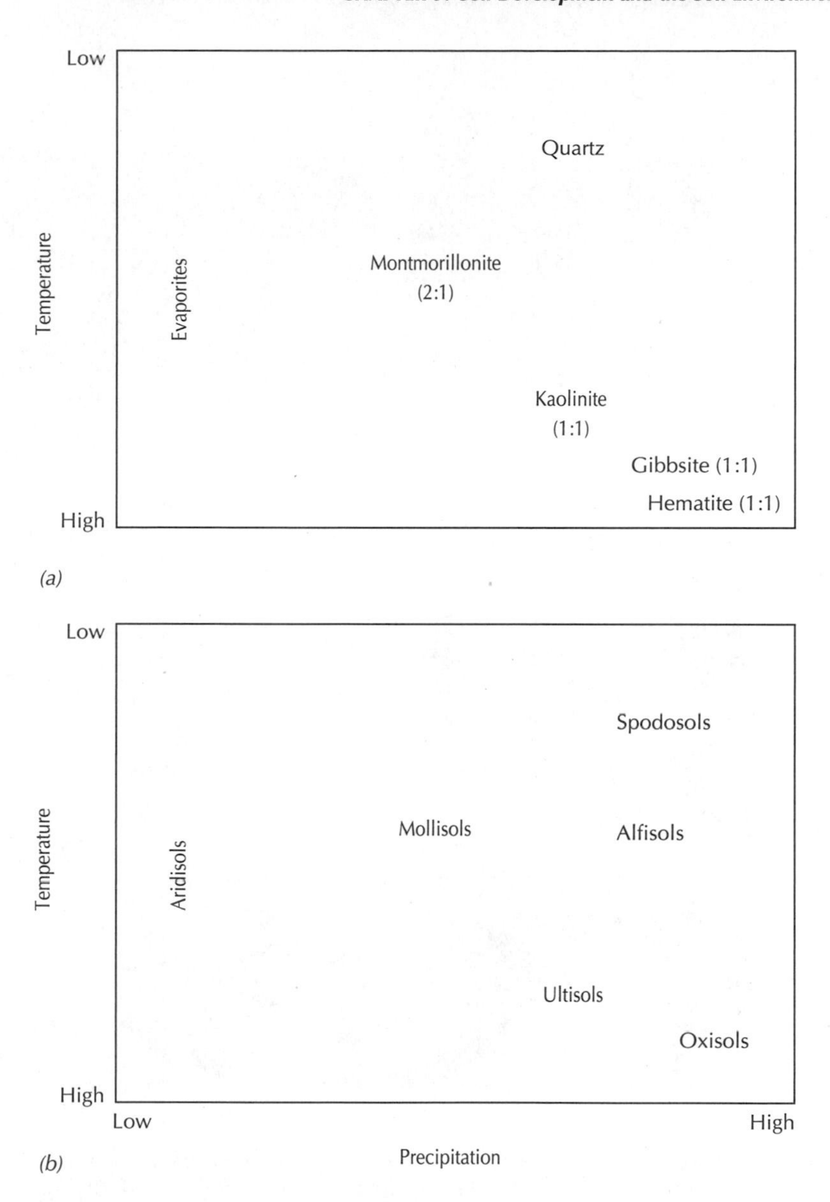

FIGURE 9.5 Effects of climatic conditions on (*a*) the type of secondary minerals formed by the weathering process and (*b*) the distribution of major soil types.

associated with melanization of grassland soils favor 2:1 clays and only slight horizon development (Mollisols). The driest conditions lead to the formation of evaporites that are characteristic of Aridisols. A distribution of major soil types can be overlain on the distribution of major types of secondary minerals (Figure 9.5*b;* the correlation between soil, climate, and vegetation is also presented in Chapter 2).

The different types of secondary minerals formed by weathering confer different physical properties that affect water retention and soil structure.

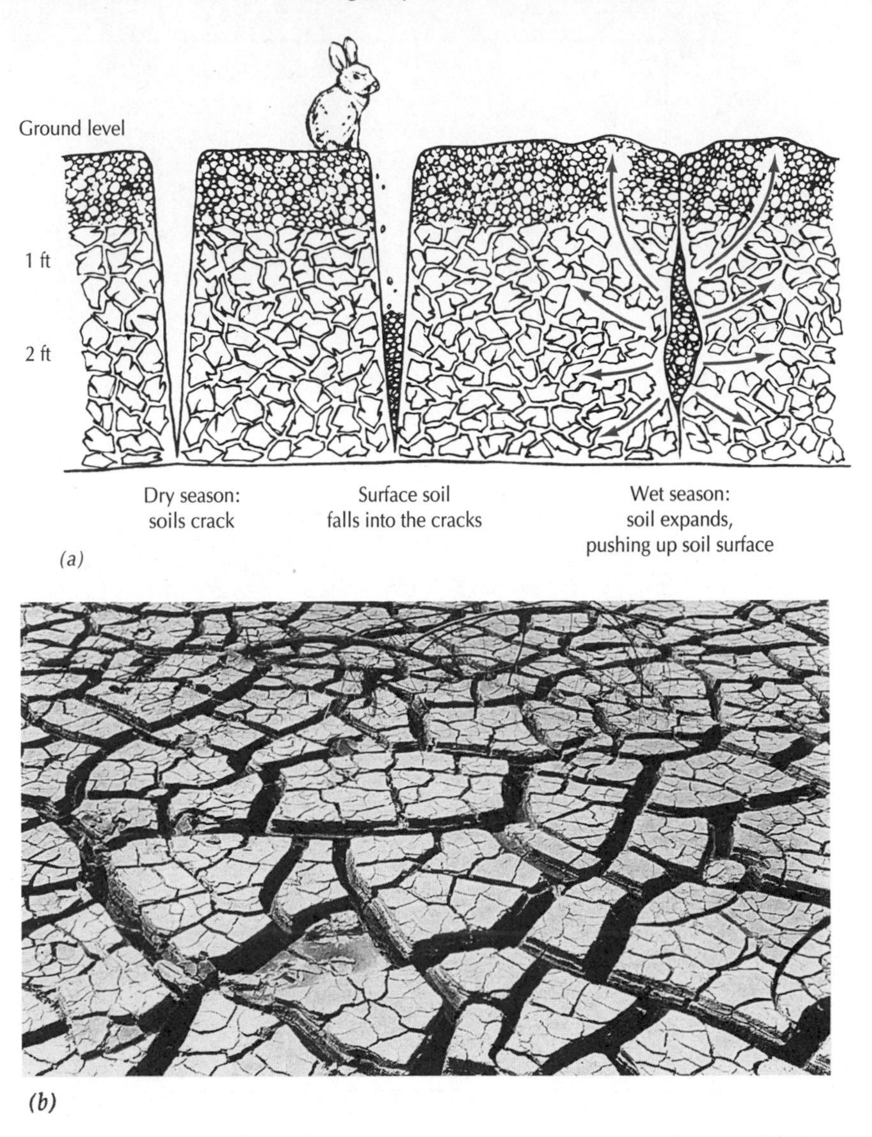

FIGURE 9.6 The effect of extreme drying on the structure of a soil rich in 2:1 layer clays. (*a*) Cross section showing shrinkage and expansion with drying and wetting (Buol et al. 1980). (*b*) Surface cracking during a dry period (H. Armstrong Roberts).

The deposition of evaporites can create a cementlike layer in the soil that restricts water movement down through the profile and further inhibits soil development. The 2:1 clays differ significantly from other products because of their tendency to swell upon wetting and shrink upon drying. This can cause reduced infiltration of water during periods of wet soil conditions and large-scale cracking of surface soils during dry periods (Figure 9.6). Otherwise, the accumulation of clay-sized particles generally increases water retention capacity (Chapter 7).

Soil Colloids and Particle Surfaces – Minerals do not exist in isolation in soils but are generally in close physical and chemical contact with soil organic matter—plant material that has been shed as litter and partially decomposed by microbes (see Chapters 12 and 13). The mineral–organic complexes formed by this combination have been given the name *colloids*. The surfaces that colloids present to the soil solution and to roots and microbes are those on which the major chemical reactions in soils occur. The properties of these surfaces depend on the types of minerals present, the proportion of organic matter, and the soil chemical environment itself, completing a feedback between soil chemistry and mineralogy that defines the distribution of soil types and major soil-forming processes described earlier.

One of the primary characteristics of soil colloid surfaces is the development of an electrical charge. Both positively and negatively charged sites can occur in the same soil, but the ratio of these two varies widely among soils and has major implications for soil chemistry. Ionic forms of the major nutrients (Table 9.2) are the ones most readily available to plants. Soils with a net negative charge increase the content of positively charged ions, such as calcium ions (Ca^{2+}), in soils by holding them against the flow of water. On the other hand, they also inhibit the movement of those same nutrient ions to roots. Soils with surfaces developing a net positive charge have the same effect on negatively charged nutrient ions, such as phosphate ions (PO_4^{3-}). A soil's **ion exchange capacity** is defined as the total ionic charge equivalent of positive (cation) or negative (anion) ions that can be held in a soil. It can be thought of as the total number of negatively and positively charged sites on particles within the soil body. The two broad classes of clays discussed differ significantly in the way this surface charge develops and in ion exchange capacities.

In 2:1 layer clays, one of the major ways in which chemical modification accompanying weathering occurs is through isomorphic substitution, or the replacement of one ion between the major structural layers of the clay with another. For structural reasons, ions of lower charge generally replace those of higher charge. For example, aluminum ions (Al^{3+}) may substitute for silicon (Si^{+4}), or magnesium ions (Mg^{+2}) could replace aluminum. In either case, this creates a net negative charge on the clay, and the colloid of which it is a part. Because this charge is not altered quickly by the acidity or chemistry of the soil solution, 2:1 clays are called fixed charge clays. Because this type of clay predominates in cooler, temperate systems, colloids in these systems tend to have a net negative charge.

In contrast, 1:1 clays do not undergo isomorphic substitution. Rather, they accumulate a net surface charge as a result of the accumulation of either the hydrogen ion (H^+) or hydroxyl ion (OH^-) derived from the ionization of water. Because soils under the humid and warm conditions where 1:1 clays are formed tend to be acidic (more H^+ than OH^- in soil solutions), the hydrogen ions accumulate preferentially on colloid surfaces providing a net positive charge to these surfaces. Also unlike 2:1 clays, the abundance of positive and negative charges on the surfaces of 1:1 clays can change rapidly in response to changes in the acidity of the solution in which they exist. For this reason, these are called variable charge clays. A basic descriptor of colloids or soils containing variable charge clays is the way in which the net charge on surfaces shifts with changing pH, or the relative abundance of hydrogen and hydroxyl ions, in the soil solution. Figure 9.7 is a typical graph of this type and shows the switch from a net positive charge at low pH (high H^+ concentration) to a net negative charge at high pH. The **zero point of charge** is the pH at which this net surface charge is zero.

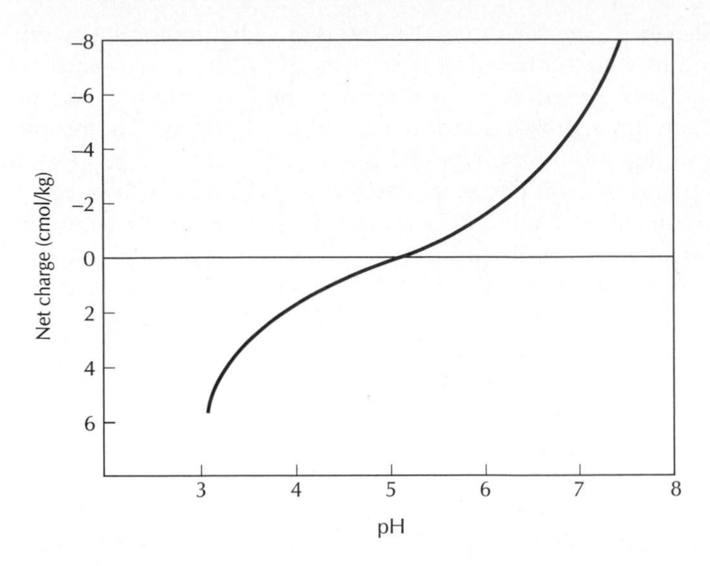

FIGURE 9.7 Change in net charge on 1:1 clays as a function of pH. The pH at which the clay changes from a net positive to a net negative charge is called the isoelectric point. (After Singh and Uehara 1999)

Soil organic matter also contributes to the negative surface charge on soil colloids. Organic acids, alcohols, and other functional groups containing hydroxyl (—OH) groups contribute to particle surface charge by the dissociation of the OH group, creating a negatively charged site (Figure 9.8). The degree to which these functional groups dissociate depends on soil pH. Under acidic conditions, more of this negative sites retain the hydrogen ion, resulting in fewer negatively charged sites. This has been termed pH-dependent exchange capacity. In many temperate zone soils, there is a strong relationship between the amount of soil organic matter present and total **cation exchange capacity** (CEC; Figure 9.10*b*).

In addition to charge characteristics, colloids also present surfaces that can react with ions in the soil solution in less reversible ways. The full spectrum of these reactions define in part the weathering process and are beyond the scope of this book. However, one particular set of these dissolution–

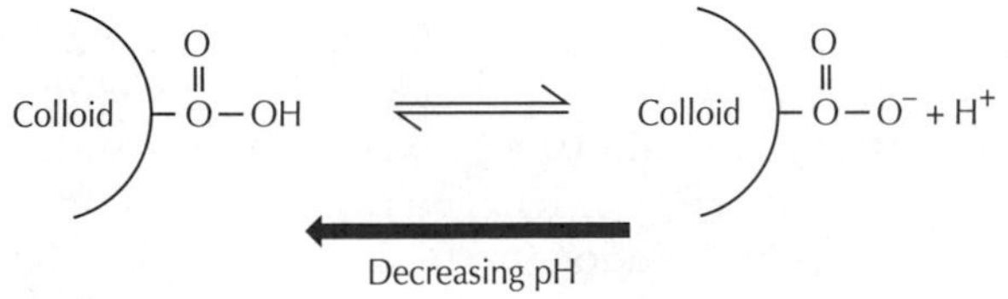

Colloid — O(=O)—OH ⇌ Colloid — O(=O)—O^- + H^+

Decreasing pH

FIGURE 9.8 Charge-dependent cation exchange capacity resulting from organic acids in soils. At high pH (low hydrogen ion content), the organic acid tends to dissociate, or lose a hydrogen ion, creating a net negative charge and increasing cation exchange capacity. At lower pH values, the hydrogen ion is retained and the site is not available for other cations.

precipitation reactions is of major importance in controlling the availability of key elements. Iron and aluminum oxides react strongly with the anions sulfate and phosphate. This process has been termed sorption and creates a strong chemical sink for these ions in soils rich in these minerals (e.g., the E horizons in lateritic soils and B horizons in Spodosols). These reactions are reversible by altering the concentrations of anions in the soil solution immediately surrounding the reaction sites.

Cation Exchange and the Lyotropic Series – In most temperate-zone soils, cation exchange greatly exceeds anion exchange because of the predominance of 2:1 clays and large accumulations of organic matter, both of which tend to develop negatively charged surfaces. The total amount of negatively charged exchange sites, or the total cation exchange capacity, is a basic measure of soil quality and generally increases with increasing clay and organic matter content.

The negatively charged particle surfaces associated with cation exchange attract positively charged ions in the soil solution. However, because these ions also carry a net charge, they also attract water molecules that are oriented with the negatively charged end toward the cation. For this reason, cation exchange sites are occupied by hydrated cations (cations surrounded by water molecules but still bearing a net positive charge).

The relative abundance of different ions on exchange sites is a function of their concentration in the soil solution and the relative affinity of each for the sites. In general, the physically smaller the ion (when hydrated) and the greater the positive charge on it, the more tightly it is held. The ratio of charge to the size of the hydrated ion is called charge density. Thus, aluminum ions (charge of +3) are more strongly attracted than potassium (+1). However, hydrogen ions, even though they carry a charge of only +1, are made up of just a single proton with no occupied electron shells surrounding the proton, so they also carry a high charge density in the hydrated form. The lyotropic series places the major cations in order in terms of decreasing affinity for cation exchange sites:

$$Al^{3+} > H^{+} > Ca^{2+} > Mg^{2+} > K^{+} = NH_4^{+} > Na^{+}$$

Note that the major cations that are also macronutrients (calcium [Ca], magnesium [Mg], potassium [K], and ammonium [NH_4]) are all lower in the lyotropic series than are hydrogen (H) and aluminum (Al) and so tend to be displaced from exchange sites. However, relative concentration on cation exchange sites is a function of not only affinity but also concentration in the soil solution. Thus, the relative concentrations of elements on exchange sites reflect the net effect of several processes that release cations and hydrogen ions into the soil solution (see subsequent discussion).

Total CEC is measured by exposing a soil to a concentrated solution of a single cation such that, at least in theory, this ion occupies all cation exchange sites. The amount of each cation displaced by this one ion can be measured in the solution, and the sum of all of this is the CEC. The term *base saturation* refers to the fraction of cations on exchange sites that are not occupied by the "acid cations" hydrogen and aluminum. Acidic soils (low pH, high concentrations of hydrogen and aluminum in the soil solution) also have low **percent base saturation** and hence low short-term availability of the nutrient cations.

In contrast to temperate zone soils, those of the humid tropics, through a combination of 1:1 clays and acidic soil solutions, produce soils with very low CEC. Low CEC, combined with the extreme leaching of tropical rainforest soils by excessive rainfall and fully weathered substrates, produces soils with very low exchangeable cation content.

Weathering, Cation Exchange, and Soil Acidity – Weathering reactions are an important source of cations. For example, soils developing on limestone receive a continuous input of calcium by weathering. This increases the calcium concentration in the soil solution so that the differences in concentration outweigh the tendencies for hydrogen to displace calcium. Instead, calcium tends to displace hydrogen, base saturation increases, and the soil becomes less acidic. This is precisely what happens when limestone is applied to an acidic soil. The calcium displaces the hydrogen, which then tends to recombine with the carbonate to form bicarbonate or carbonic acid, which may then be leached through the soil and out of the system (Figure 9.9*a*). Many weathering reactions of iron and aluminum silicate rocks also release cations (including iron and aluminum in ionic form) and consume hydrogen ions.

Many physical, chemical, and biological processes counteract the neutralizing effects of weathering and can increase the acidity of soils. Rainfall is somewhat acidic, even in relatively unpolluted areas, because of the formation and dissolution of carbonic acid by reactions between water vapor and CO_2 in the atmosphere (Figure 9.9*b*). In industrialized regions, the formation of nitric and sulfuric acids in the atmosphere due to reactions between water and oxides of sulfur and nitrogen released by the combustion of fossil fuels ("acid rain") can increase acidity 10-fold or more (Chapter 25). As acidified precipitation percolates through the soil, hydrogen ions may displace cations from exchange sites, reducing base saturation (Figure 9.9*b*).

Biological acidification of soils results both from respiration by roots and soil microbes and imbalances in anion/cation uptake by plants. Soil acidification due to hydrogen ion release by plant roots is greatest when nitrogen is taken up as ammonium. Charge balance on the root is then maintained by "pumping" a hydrogen ion out into the soil (Figure 9.9*c*, see also Chapter 10). Soil pH can also be reduced by the release of CO_2 into the soil by respiration. This combines with water to produce carbonic acid, which then dissociates to produce hydrogen ion and bicarbonate (Figure 9.9*c*). The extent to which this dissociation occurs depends on the pH of the soil and is minimal below pH 5.0.

Measured soil pH represents a balance between the processes of acidification due to the acidity of rainfall and the effects of biological activity and the neutralizing effects of weathering. Organic horizons (the O horizon described in Chapter 2) are generally more acidic than are mineral horizons (the E and B horizons) due to higher biological activity and the low levels of weatherable minerals present. However, when weathering rates are low in the mineral soil, pH values may be similar between horizons. We return to the processes controlling soil acidity in Chapter 25, which describes the "acid rain" phenomenon.

Anion Exchange and Sorption – Both anion exchange and sorption affect the availability of negatively charged ions, especially sulfate and phosphate.

(a) Weathering

Colloid –H –K –H –H
K+ (Al+3)
Minerals
H+
Colloid –H –H –H –H
Ca++
CaCO3
CO3--
H2CO3

(b) Precipitation

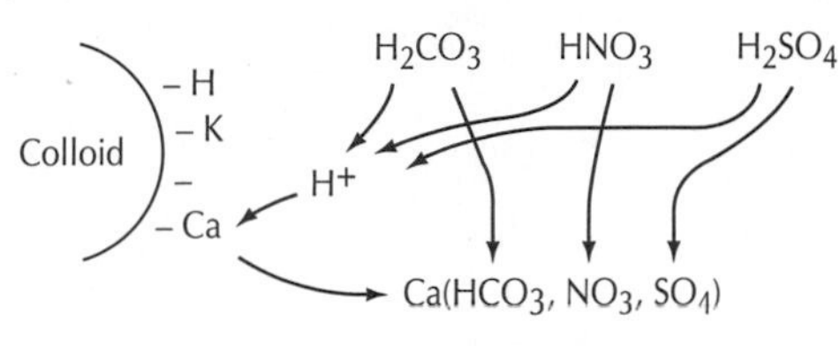

(c) Biological

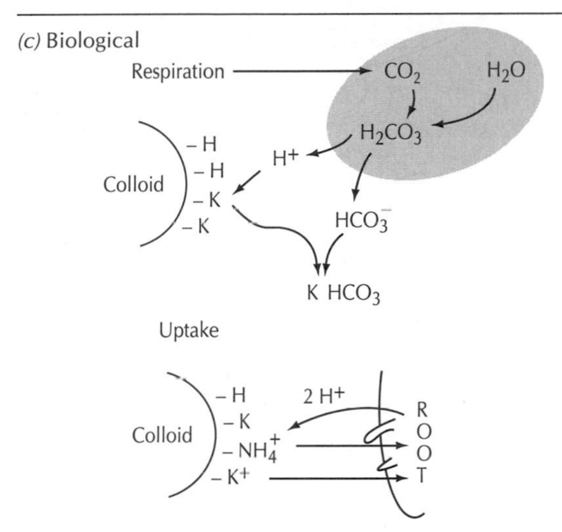

FIGURE 9.9 Processes affecting the relative concentration of different cations on exchange sites in temperate soils.

Because both reactions relate to the presence of iron and aluminum oxides in 1:1 clays, it has been difficult to separate and fully understand these two processes. There is still some uncertainty in the application of these concepts in field soils, although a variety of new techniques are leading to rapid progress in this area.

One of the complicating issues is that anion exchange and sorption are interrelated processes. Figure 9.10 presents this interaction in a conceptual way. Ions tend to be attracted, and held to, the surfaces of soil colloids. Under the acidic conditions occurring in most systems where 1:1 clays are found, hydrogen ions (H^+) occupy more of this surface than do hydroxyl ions (OH^-), resulting in a net positive charge on the colloid surface. A positively charged surface causes a redistribution of charge particles in the adjacent soil solution until the net charge on the soil solution interface is equivalent to that on the colloid surface, but of opposite sign (Figure 9.10*a*).

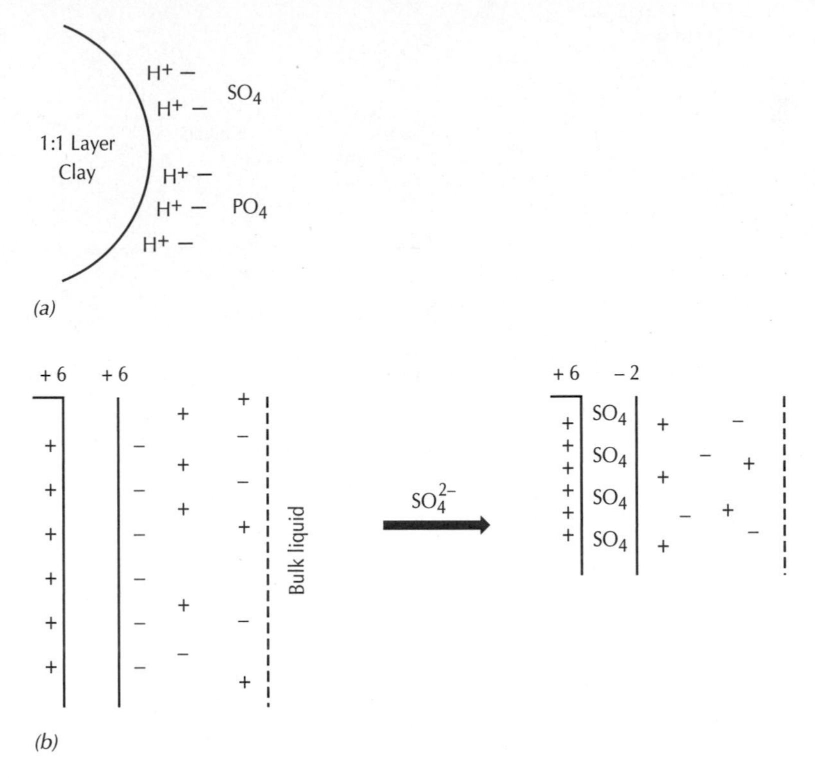

FIGURE 9.10 Anion exchange and anion sorption in soils with 1:1 clays. (*a*) Anion exchange resulting from the selective accumulation of hydrogen rather than hydroxyl ions on colloid surfaces. (*b*) The chemical reaction between clays and anions that removes the anion of the solution and alters the net charge on the clay particle. (After Singh and Uehara 1999)

Because of the positive surface charge, anions are pulled toward the colloid surface and selectively retained by electrostatic forces. This gives rise to anion exchange capacity and creates one mechanism for retention of sulfate and phosphate in soils.

However, both phosphate and sulfate ions can also enter into chemical reactions with clay surfaces (Figure 9.10*b*) by which these ions are removed from the soil solution and the surface charge balance is altered. By this mechanism, sorption of sulfate or phosphate reduces anion exchange capacity. The anions, however, are still held in the soil, but by a different mechanism with different release characteristics.

In the field, it can be very difficult to separate these two processes for anion retention. The primary method for determining the capacity of soils to retain or release phosphate and sulfate has been the development of "sorption" isotherms (Figure 9.11). These describe the quantity of an ion that can be removed from the soil solution and retained by the soil as a function of its concentration in the soil solution. In field soils, a soil horizon that has a lower total quantity of sorbed sulfate or phosphate than the sorption isotherms predicts will remove those ions from the solution, reducing their availability to plants.

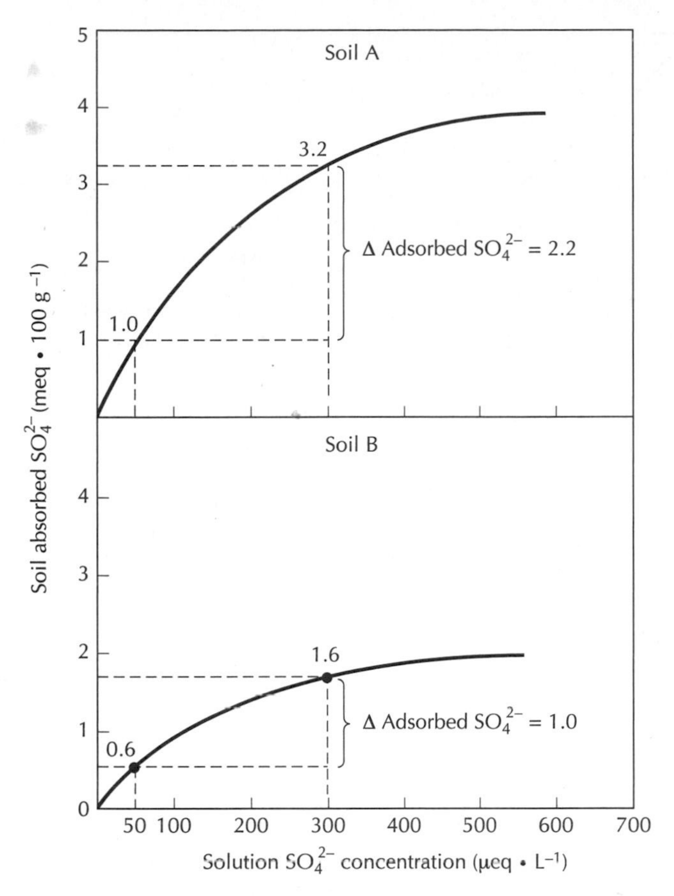

FIGURE 9.11 Hypothetical sulfate sorption curve showing increasing total sorption capacity with increasing concentration of sulfate in solution. (Johnson 1984)

Clearly, this method does not determine how much of this "sorption" is due to electrostatic attraction between anions and colloid surfaces and how much is due to sorption reactions. Still, it does present a basic method for comparing the chemical state of soils relative to anion availability. Not surprisingly, research has determined that pH and the concentration of other ions in the soil solution alters this basic measure of soil "sorption" potential significantly. In addition, these methods have been used to determine that anion retention is positively correlated with iron and aluminum oxides in soils and negatively correlated with organic matter content. The solubility of aluminum in soils, a crucial process in determining the toxic effects of acid water on organisms, is also described by this type of dissolution reaction. In this case, however, Al concentrations increase with decreasing pH (or increases in the concentration of hydrogen ions in solution).

The isotherms describe a set of reactions that are reversible. "Desorption" can occur when the concentration of the anion is reduced in the soil solution (moving to the left on the X axis in Figure 9.11). Biological activity in the soil can play an important role in increasing the availability of sorbed nutrients (see Chapter 10) by this and other mechanisms.

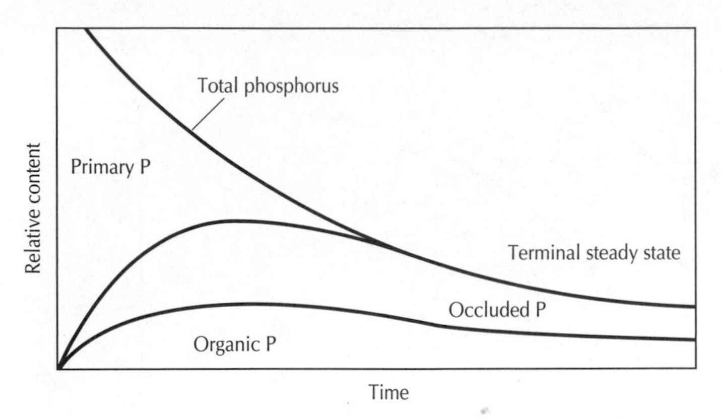

FIGURE 9.12 The generalized effects of long-term weathering and soil development on the distribution and availability of phosphorus (P). Newly exposed geological substrate is relatively rich in easily weatherable minerals, which release phosphorus. This release leads to accumulation in both organic and readily soluble forms (secondary P, such as calcium phosphate). As primary minerals disappear and secondary minerals capable of sorbing phosphate accumulate, an increasing proportion of the phosphorus remaining in the system is held in unavailable (occluded) forms. Phosphorus availability to plants peaks relatively early in this sequence and declines thereafter. (After Walker and Syers 1976)

Interactions of Short-Term and Long-Term Processes: Phosphorus Availability

The processes described here as chemical controls over nutrient availability operate at very different time scales. Ion exchange and sorption/desorption reactions occur very rapidly in response to very short-term changes in soil solution concentrations, but these reactions are constrained by more slowly varying soil characteristics of ion exchange capacity and sorption potential that result from weathering and soil development. Short-term availability of nutrients can be strongly related to a systems position on a very long-term soil developmental sequence. For example (Figure 9.12), phosphorus is present in many primary minerals. A steady supply of phosphate is provided when relatively rapid weathering of young geological substrates occurs. This is taken up by plants and incorporated into organic matter, and decomposition of the organic fraction provides a second relatively constant supply of phosphate to the soil solution. Assuming humid tropical conditions, the weathering process results in the formation and accumulation of secondary minerals (iron and aluminum oxides) that increase the potential for sorption reactions to remove phosphate from the soil solution. Over geological time, erosion and leaching reduce the total amount of phosphorus in the system, and a larger fraction becomes sequestered into secondary minerals. Very old and fully weathered tropical soils contain little primary mineral content to provide new phosphorus from weathering and large accumulations of secondary minerals that provide a strong chemical sink for phosphate. This results in very low phosphorus availability to plants, with much of that recycling through the organic fraction. A classic example of an ecosystem sequence of this type is presented in Chapter 22.

REFERENCES CITED

Buol, S. W., F. D. Hole, and R. J. McCracken. 1980. *Soil Genesis and Classification.* Iowa State University Press, Ames, IA.

Flint, R. F, and B. J. Skinner. 1974. *Physical Geology.* John Wiley & Sons, New York.

Foth, H. D. 1990. *Fundamentals of Soil Science,* 8th ed. John Wiley & Sons, New York.

Johnson, D. W. 1984. Sulfur cycling in forests. *Biogeochemistry* 1:29–44.

Levine, J. S. 1989. Photochemistry of biogenic gases. In Rambler, M. B., L. Margulis, and R. Fester (eds.), *Global Ecology.* Academic Press, Boston.

Salisbury, F. B., and C. W. Ross. 1978. *Plant Physiology.* Wadsworth Publishing, Belmont, CA.

Singh, U., and G. Uehara. 1999. Electrochemistry of the double layer: Principles and applications to soils. In Sparks, D. L. (ed.), *Soil Physical Chemistry,* 2d ed. CRC Press, Boca Raton.

Strakhov, N. 1967. *Principles of Lithogenesis.* Oliver and Boyd, Ltd., London.

Walker, T. W., and J. K. Syers. 1976. The fate of phosphorus during pedogenesis. *Geoderma* 5:1–19.

10

BIOLOGICAL PROCESSES IN SOILS

INTRODUCTION

Soil chemistry plays a large role in determining the availability of nutrients in ecosystems. Biology plays an important role as well. Acting within the constraints set by the basic chemical processes of weathering, sorption, and ion exchange, plants and associated soil organisms can cause significant changes in soil solution chemistry and the density and activity of biological surfaces exposed to the soil environment. With symbiotic relationships between roots and microorganisms, processes not available to either independently can be carried out. All of these require energy and carbon for both structure and respiration, and this establishes tradeoffs in the use of carbon from photosynthesis to increase the uptake of potentially limiting nutrients.

The purpose of this chapter is to present traditional, chemical methods of assessing nutrient availability to plants, as well as the biology of nutrient uptake and methods by which the effective availability of nutrients can be increased. At the beginning of Chapter 9, we stressed that much of what goes on in soils is not clearly understood. This is particularly true for the types of interactions discussed in this chapter. While the processes presented have been shown to occur, it is still very difficult to measure rates of nutrient movements and especially the carbon costs involved. This is currently an area of active research.

MEASURES OF NUTRIENT AVAILABILITY

Static Measures of the Available Pool

Most methods for assessing nutrient availability come from agricultural research and are attempts to mimic chemically the ability of plants to extract nutrients from soils. Each method includes, to a different extent, nutrients in the several soil chemical pools in Figure 9.1.

Total exchangeable nutrient content is a standard measure of soil quality and nutrient availability, especially for cations. The measurement involves saturating a soil sample with a high concentration ionic solution, displacing all cations and anions from exchange sites. Potassium chloride may be used for nitrate and ammonium extraction, and ammonium acetate, for other cations. This approach assumes that only exchangeable nutrients and those already in the soil solution are available to plants and that measuring availability at one point in time is a good indicator of availability over the whole year.

For elements present in excess of plant demand, there may be a fairly good relationship between soil exchangeable pools and plant uptake. High cation retention capacities in temperate soils cause cations present in excess of plant demand to be retained on soil colloids. However, exchangeable amounts of either limiting nutrients or of nutrients that can be either leached from soils or sorbed onto clays (including the often limiting nutrients nitrogen and phosphorus) are generally very low, much less than 1 year's plant requirement for uptake. In general, both nutrient uptake and plant productivity are not related to exchangeable nitrogen or phosphorus, even when experimental additions show significant growth responses. There are two major reasons for this.

First, nutrient availability is a rate phenomenon. Annual uptake is not so much a function of concentration in soils at any one time as it is a rate at which nutrients enter the available pool by organic matter decomposition, weathering, rainfall, etc. In fact, it is a general principle that nutrients that are limiting to production or other ecosystem functions are present in the smallest amounts in the soil solution due to high demand and rapid uptake by plants and microbes (i.e., they have very high turnover rates).

Second, the biological activity of plants and microorganisms can significantly alter availability. There are several mechanisms that involve either the separate function of plants and microbes or their joint function as symbionts. These are discussed later in this chapter.

MEASURES BASED ON THE RATE OF MINERALIZATION FROM ORGANIC MATTER

An improvement on the one-time measurements of nutrient availability are incubation techniques that isolate soil cores for a period of time and measure the rate of release (mineralization) from organic matter. This approach is particularly valuable for elements with cycles dominated by the biological decomposition of organic matter (e.g., nitrogen and sulfur). Incubation techniques involve isolating a soil core either in the laboratory or in the field, allowing it to incubate for some length of time and then measuring the accumulation of the mineral forms of the nutrient (e.g., ammonium and nitrate). The stable isotope ^{15}N has also been used to trace the dynamics of nitrogen metabolism during incubation (see Chapters 12 and 13).

Measuring phosphorus availability is altogether more difficult. High phosphate sorption potentials in many soils (Chapter 9) make it impossible to use the same incubation techniques described for nitrogen. Phosphorus mineralized in the incubating soils can be sorbed quickly and not accumulate in ionic form. The radioisotope ^{32}P has been used with some success, but more frequently used are one of several chemical extraction techniques. Each technique extracts a different proportion of the total phosphate pool (Table 10.1). The best technique would be the one most similar to the ability of plants and microbes to solubilize and take up phosphorus. However, that ability is quite variable. Estimating phosphorus availability and cycling rates remains a challenge.

Sulfur offers some of the same problems as does phosphorus. Decomposition of organic matter is a principal source of sulfate (SO_4^{-2}) for plant up-

TABLE 10.1 Different Methods of Extracting Phosphorus From Soils and the Different form of Phosphorus Removed by each.
(From Olson and Dean 1965).

Extracting Solution	Phosphate Fraction Extracted
Water	P in soil solution
Dilute acid fluoride (HCl + NH_4F)	Easily soluble forms: calcium phosphates plus some iron and aluminum phosphates; considered plant-available
Dilute acid (HCl + H_2SO_4)	A stronger extraction for soils with high P-sorption potentials; considered plant-available
Sodium bicarbonate ($NAHCO_3$)	For extraction of P from neutral to alkaline soils
Perchloric acid (HClO4)	Total P content of soil

P = phosphorus

take, but anion sorption complicates measurement of mineralization by incubation techniques in some soil types. In general, sulfur is rarely a limiting nutrient in natural ecosystems. It has received more attention in relation to acid rain effects (Chapter 25) than as an important nutrient limiting plant growth.

Measuring mineralization rates for cations is also difficult in most soils, but for very different reasons. Unless cations are limiting production (which is generally rare), then they are present in large quantities on exchange sites. Measuring relatively small changes in exchangeable cations due to mineralization can be very difficult.

NUTRIENT UPTAKE AND THE BIOLOGICAL MODIFICATION OF NUTRIENT AVAILABILITY

Even if the problems with measuring rates of nutrient release in isolated soil cores could be solved, this method would still have limited value in predicting rates of nutrient cycling in ecosystems. The reason for this is that eliminating plant activity from the cores removes the crucial interactions between plants and microbes, which can alter nutrient dynamics significantly.

Two related phenomena dominate this interface between plants and soils: (1) symbiotic relationships between plants and microorganisms and (2) the creation of a highly modified environment around this interface: the **rhizosphere**. Symbiotic relationships between plant roots and soil microorganisms are the rule rather than the exception in native ecosystems. Most common is the mycorrhizal (*myco* = fungus, *rhiza* = root) association between plants and different types of soil fungi, although associations with bacteria also occur. These associations provide unique mechanisms for increasing nutrient availability, as is described later. The activities of these symbiotic root systems also create a unique rhizosphere environment immediately surrounding the root. Several critical chemical characteristics are altered in this narrow zone of high biological activity, which in turn alter the functions of free-living microbes in the soil and the availability of nutrients.

The prevalence of symbioses and the tight coupling between plant, symbiont, rhizosphere, and free-living microbes combine to make the study of nutrient cycling in intact soil systems extremely challenging. A major underlying problem is that of accessing this system for measurement without disrupting the critical interactions that control rates of function. Much of what is reported here comes from the study of isolated parts of the system: plant roots, isolated fungi, and bacteria, or symbiotic combinations under simplified laboratory conditions. Much remains to be learned about the actual function of intact soil systems.

We approach the description of these complex systems by first describing mechanisms of nutrient uptake in roots and then three classes of processes that modify soil conditions and, later, nutrient availability. These three classes are: (1) increasing root mass, (2) altering the chemistry and biology of the rhizosphere, and (3) root/microbe symbioses.

Mechanisms of Nutrient Uptake

Nutrient uptake occurs either through active transport processes across root surfaces or by passive movement with the bulk flow of water or in response to electrochemical (charge) gradients. Active uptake is carried out through protein "channels" across membranes that generate strong, local, electrochemical gradients for the transport of specific ions into the root. This is an energy-requiring process and so places a demand on the carbohydrate resources of the plant.

The rate of active root uptake depends on the affinity of roots for different nutrients and their concentrations in the soil solution and is described (Figure 10.1) using a curve that is analogous to our earlier description of the interactions between photosynthesis and light. The key characteristics of this curve are the maximum rate of uptake (I_{max}) and the concentration at which uptake reaches half of this maximum rate (the half-saturation constant K_m; note the similarity to the photosynthetic response curve in Figure 6.2). Shifts in the shape of this curve under different conditions can be attributed to increases in activity of high- and low-affinity uptake mechanisms, also analogous to the potential for photosynthesis in response to light in sun and shade leaves. High I_{max} increases the maximum rate of uptake under conditions of high nutrient availability, while low K_m increases uptake rates at low availability.

Passive uptake can occur as water flows into roots in response to transpirational demand or in response to electrochemical gradients. For example, if nitrogen is taken up as ammonium (NH_4^+) and demands for calcium, magnesium, and potassium (all cations) are high, then the only macronutrients taken up as anions are sulfur and phosphorus (Figure 10.2*a*). Active uptake of cations is required, while passive uptake of sulfur and phosphorus occurs. Any remaining charge imbalance is met by an active pumping of hydrogen ions from roots into soils, resulting in important changes in the chemistry of the rhizosphere. In contrast, if nitrogen uptake occurs as nitrate (Figure

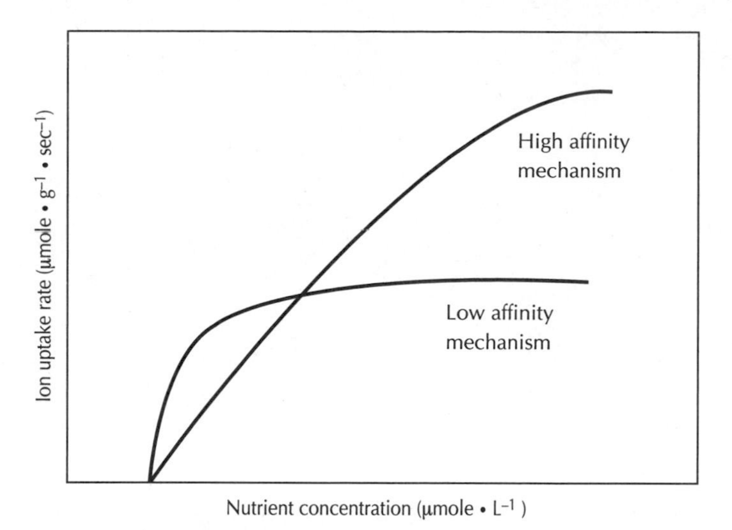

FIGURE 10.1 Example of high- and low-affinity processes for nutrient uptake by roots. (After Lambers et al. 1998)

10.2*b*), then the balance between cation and anion uptake can be nearly equal, and attaining charge balance by pumping H^+ is minimal.

Until recently, most of our information on the physiology of nutrient uptake has come from experiments with agricultural systems, and the accepted paradigm has been passive or active uptake of nutrients in ionic, inorganic form (e.g., NH_4^+ and NO_3^-). Increasing evidence shows that additional

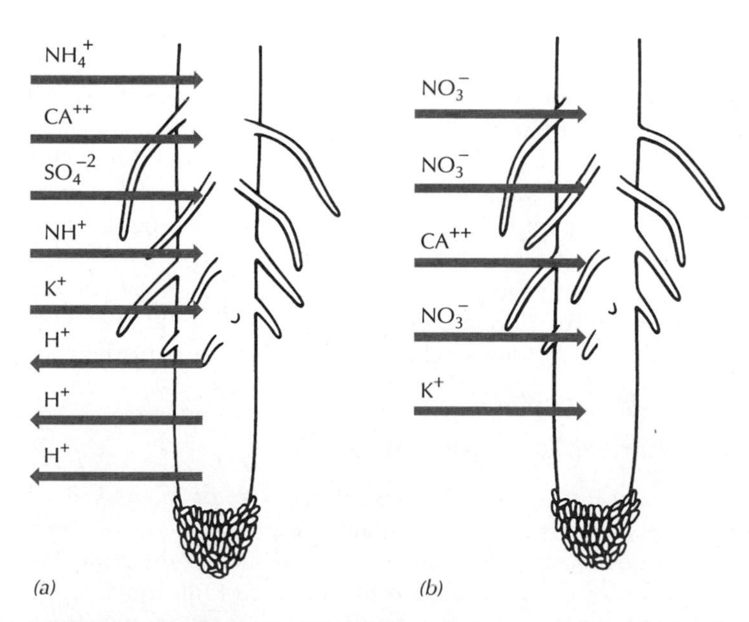

FIGURE 10.2 The effects of nitrate versus ammonium uptake on the movement of hydrogen ions from root to soil.

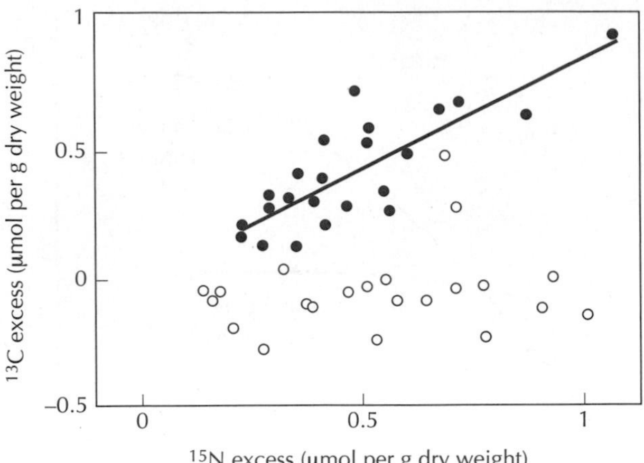

FIGURE 10.3 Direct uptake of organic nitrogen demonstrated by a comparison between the addition of ^{15}N-labeled ammonium (open circles), and amino acids labeled with both ^{15}N and ^{13}C (closed circles). With the ammonium addition, only the ^{15}N content of the root increases. With the double-labeled amino acids, both the ^{15}N and ^{13}C of the roots increase. If the amino acid were decomposed to ammonium before being taken up, then the amino acid addition would show the same response as that of the ammonium addition. The double-label results demonstrate that the nitrogen was taken up as the amino acid. (Nasholm et al. 1998)

mechanisms are active in some natural systems. As is explained in Chapter 13, mineral forms of nitrogen are generated only very slowly in cold, acidic soils, such as those in taiga or tundra systems. This, along with the presence of mycorrhizal fungi, has led to speculation that plants might be able to access organic forms of nitrogen directly.

In one recent experiment carried out in a boreal (taiga) forest, double-labeled amino acids, in which the carbon atoms are labeled as ^{13}C and the nitrogen atoms are labeled as ^{15}N, were applied to soils. Figure 10.3 shows that, for three different growth forms (trees, grasses, and shrubs), both the mobile carbon and mobile nitrogen pools in roots contained the label. This demonstrates that the amino acids were not first decomposed to ammonium before uptake. Had decomposition occurred, only the nitrogen label would be present in the root. In a companion experiment in which $^{15}NH_4$ was added, the root pools showed the expected increase in ^{15}N without any increases in ^{13}C. This potential for "direct cycling" of nitrogen could alter many of our existing ideas about nutrient cycling, especially in very nutrient-poor systems.

Root Density and Nutrient Depletion Zones

The presence of nutrients on exchange sites or in the soil solution and the presence of roots with defined uptake affinities do not ensure uptake by plants. The ion must also move to the surface of the root. The rate of nutrient uptake is frequently limited by this rate of movement.

Movement toward roots can also occur with the mass flow of water or because of concentration gradients within the soil solution caused by nutrient uptake. In temperate zone soils, movement with water is generally limited to anions, such as nitrate, which are not retained by soils either by ion exchange sites or sorption. Mobility of cations in soil is reduced by retention on exchange sites. Phosphate mobility is very low in soils with high phosphate sorption potentials. Thus, the same processes that reduce nutrient losses to groundwater also cause resistance to the movement of ions to the root.

Figure 10.4 shows changes over time in the distribution of a nutrient ion in the soil adjacent to a newly grown root. Initially, the nutrient concentration is constant throughout the soil (t_0). As uptake of nutrients near the root surface begins (t_1), a gradient in nutrient concentration develops within the soil and nutrients begin to move toward the root. At first, uptake is rapid because nutrient concentrations are high, and replenishment of removed nutrients is rapid because the concentration gradient in the soil is steep and the distance over which nutrients move to the root surface is short. As uptake continues, a localized zone of nutrient depletion develops around the root. Nutrients are drawn from further away and the concentration gradient is stretched over a longer distance, reducing rates of flow toward the root (t_2–t_4). Eventually, the concentration gradient is not steep enough to offset resistance to ion movement in the soil, and movement toward the root stops. Again, nutrient uptake has created a very different chemical environment in the rhizosphere.

Depletion zones are of very different sizes for different nutrients. In temperate soils with high cation and relatively low anion exchange capacity, the resistance to movement of ammonium may be 100-fold of that for nitrate. The depletion zone for nitrate will be correspondingly larger. Depletion zones for phosphate are generally much smaller than for ammonium in mineral soil because of sorption potentials but can be quite high in purely organic soil horizons.

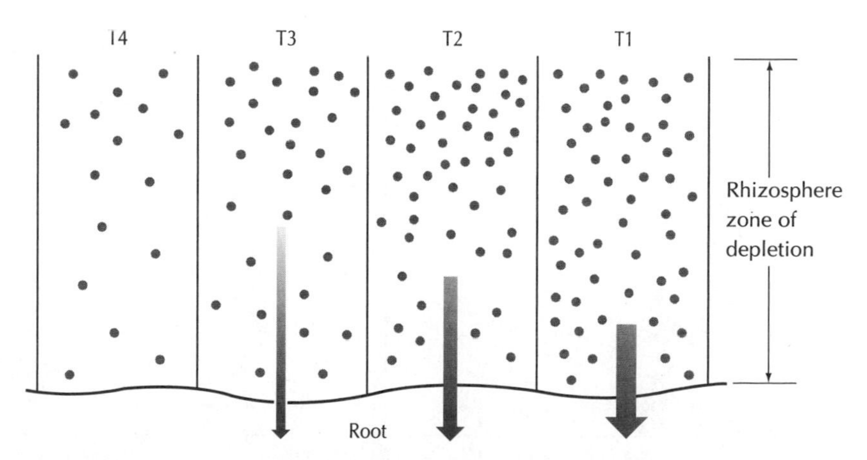

FIGURE 10.4 Development of ion depletion zones in soils as a result of nutrient uptake. As uptake occurs, a concentration gradient develops in the soil, resulting in movement of ions to the root surface. As uptake reduces nutrient concentration in the developing depletion zone, the concentration gradient to the root declines and the uptake rate is reduced. Eventually, the gradient is insufficient to overcome resistance to movement in the soils, and uptake stops.

In the context of the whole soil body, then, not all nutrients that are chemically exchangeable are "available" to plants. Those outside of the rhizospheres of all plant roots or beyond the zones of depletion are not taken up. Under these conditions, a simple way to increase effective nutrient availability to the plant is to increase the root mass or surface area (Figure 10.5).

It follows from this that plant communities on nutrient-poor sites often have higher root masses. In nitrogen limited systems, those communities growing on soils where nitrate is the dominant form of available nitrogen have lower root masses than do similar communities where ammonium predominates (Figure 10.6). It has been hypothesized that root density should be lowest, but rooting depth highest, in ecosystems where plants are limited mainly by water and that root density should increase in systems limited by nitrate, ammonium, and phosphate, respectively.

Altering the Biology and Chemistry of the Rhizosphere

Increasing nutrient uptake by increasing rooting density assumes that there is some excess pool of available nutrients in the soil. It does not increase the rate at which nutrients are transformed from unavailable to available forms. A densely rooted soil ensures complete uptake of all plant-available nutrients, but uptake of nitrogen, for example, cannot continuously exceed the rate at which it is released from organic matter (Chapters 12 and 13). Increasing root mass only ensures that all mineralized nitrogen is indeed taken up and not lost to groundwater leaching. Beyond this, there must be some direct effect of roots on soil processes to foster an increase in mineralization or weathering or desorption of nutrients for root mass to exert an effect on the rate of nutrient availability. There are three broad categories of processes that

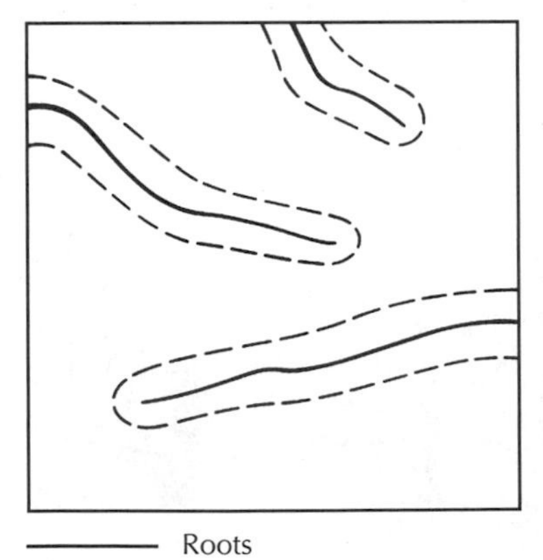

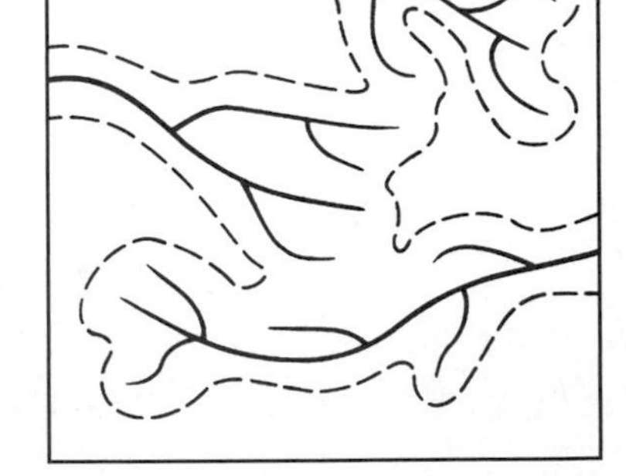

FIGURE 10.5 The effect of increasing root density on the total fraction of soil within the depletion zone of roots.

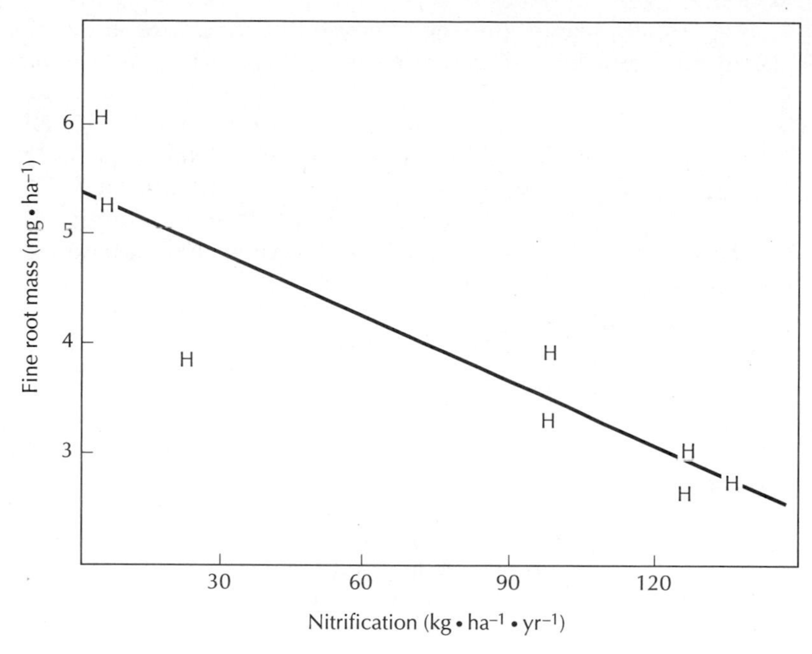

FIGURE 10.6 Relationship between annual availability of nitrate (nitrification) and total fine root biomass for several temperate deciduous forests. (Aber et al. 1985)

are active in plants and that increase the rate at which nutrients become available in the rhizosphere.

First, plants produce enzymes that speed decomposition or even chemical weathering of nutrients. Many plant roots exude enzymes that catalyze the release of phosphate from organic matter (phosphatase enzymes). There are also suggestions that cellulase enzymes (for decomposing cellulose, the primary constituent of plant cell walls, see Chapter 12) may be released, perhaps creating a role for the plant in the decomposition of organic matter. The direct extraction of nutrients, particularly potassium, from fresh minerals has been shown for several species, although the mechanism is not clear.

Second, plants affect decomposition or soil chemistry indirectly by exuding from the root simple organic compounds that stimulate microbial activity and increase organic matter decay. Much of the organic matter in soils is of very poor quality and yields little or no net energy gain to microbes (Chapters 12 and 13). The presence of more easily decomposed carbohydrates from root exudation may provide an energy subsidy that increases the rate of organic matter decay and nutrient release. Organic acids, amino acids, and other growth-enhancing compounds are present in the exudate as well, creating a rich soup of substrates for microbial growth. This interaction can be thought of as a loose symbiosis, with the plant supplying energy and even some nitrogen in amino acids to the microbes to promote more rapid release of nutrients from organic matter.

Finally, plants alter the chemistry of the rhizosphere as a byproduct of nutrient uptake and charge balance. Roots cannot accumulate a significant net electrical charge, either positive or negative. Thus the total net negative charge of the important anions (NO_3^-, SO_4^{2-}, and PO_4^{3-}) taken up must be equal to the total positive charge of cations (NH_4^+, Ca^{2+}, Mg^{2+}, and K^+), or the difference must be made up by movement of H^+ and OH^- between root and soil (see Figure 10.3). Generally, pH is reduced in the rhizosphere compared with the bulk soil outside of this zone, altering base saturation and the solubility of several forms of nutrients. Weathering rates can also be increased. By reducing concentration of ions that are affected by sorption reactions (SO_4^{3-} and PO_4^{2-}), plant uptake can lead to desorption (see Figure 9.13) and increased availability.

Soil Symbioses

This is the most complex and effective method for increasing nutrient availability. Most symbiotic relationships in natural ecosystems are with mycorrhizal fungi and both bacterial and fungal species that fix atmospheric nitrogen.

Mycorrhizal associations are nearly ubiquitous in natural ecosystems. While the species and even the morphology of the symbiosis varies, nearly all plants in nearly all terrestrial ecosystems have mycorrhizal symbionts. There are two major groups of mycorrhizal fungi: (1) ectomycorrhizae and (2) endomycorrhizae (primarily vesicular–arbuscular mycorrhizae, or VAM).

The ectomycorrhizal fungi form large mats called Hartig nets that sheath the infected root tips (Figure 10.7*a*). The sheath may account for 40% of the total weight of the root and the sheath taken together. From this net, hyphae extend both into the intercellular spaces within the root and out into the soil. Thus, the fungi form a bridge between regions of soil nutrient availability and the plant root.

The endomycorrhizae contain a more direct connection between plant and soil. There is no sheath or net in this case. Rather, the hyphae actually penetrate the cells of the root and are in direct contact with cytoplasm. The vesicular–arbuscular mycorrhizae are so named because of the vesicles (rootlike) and arbuscles (saclike) structures that characterize the fungal tissues that grow within plant cells (Figure 10.7*b*). Hyphae then also penetrate out through the root surface and extend into the soil. Fungal weight in the VAM is generally less than 15% of the root/mycorrhizae total.

In both cases, mycorrhizal symbioses are viewed primarily as a means of increasing the surface area of the nutrient-absorbing network. In terms of carbon–nutrient tradeoffs, they do this at a lower carbon cost than root production because the thinner hyphae require less carbon per unit length than do roots. An increased affinity for nutrients, especially phosphorus, may also increase the efficiency of nutrient removal within the depletion zone. The effectiveness of VAM on increasing phosphorus uptake and plant growth can be seen in Figure 10.8.

There are important differences between the two fungal groups. VAM fungi are thought to be obligate symbionts, meaning that they cannot grow actively without the carbon subsidy obtained in the relationship with the

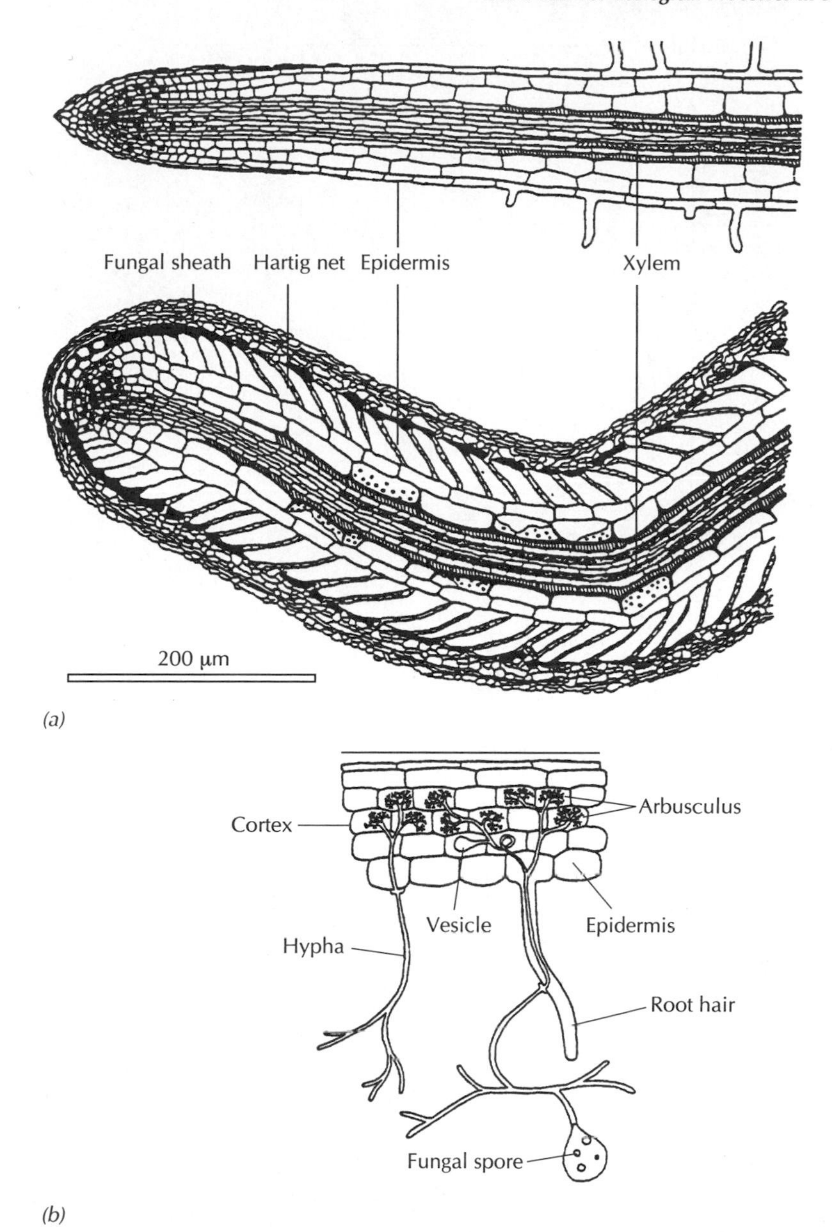

FIGURE 10.7 Structure of the two main forms of mycorrhizal root/fungal symbioses. (*a*) Ectomycorrhizal. (Root shown with and without mycorrhizal association.) (*b*) Vesicular–arbuscular mycorrhizal. (Lambers et al. 1998)

plant. Ectomycorrhizal fungi generally can survive and grow as free-living decomposers in the soil. VAM forms are more generally implicated in increased uptake of phosphate and nitrate, ecto forms in increased uptake of ammonium, and possibly of organic forms of nitrogen (e.g., amino acids) as well.

As discussed earlier, increasing uptake of already available nutrients is of limited value compared with actually increasing availability. However, there are intriguing indications that mycorrhizae may be capable of in-

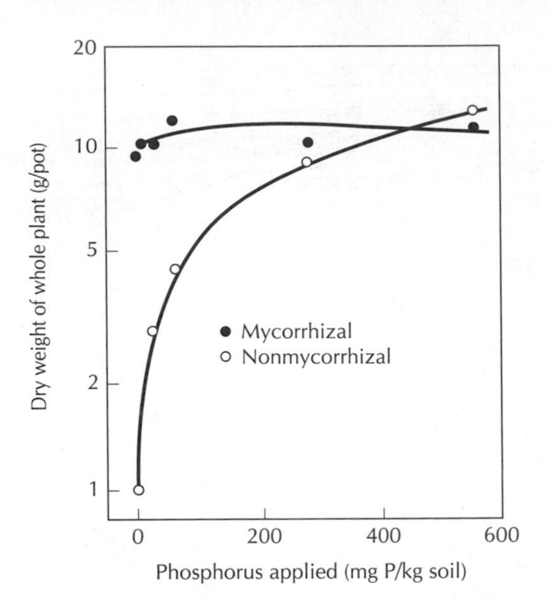

FIGURE 10.8 Effect of mycorrhizal infection on net primary production in Brazilian sour orange. Note that growth rates are high in mycorrhizal plants even with low phosphorus additions, indicating increased phosphorus availability. Nonmycorrhizal plants grow rapidly only with fertilizer phosphorus added. (Abbot and Robson 1985)

creasing availability by (1) increasing desorption of phosphate or decomposing organic forms of nitrogen and phosphorus, (2) increasing the production of enzymes that decay organic matter and mineralize nutrients, and (3) allowing for the uptake of simple organic forms of nutrients (e.g., amino acids). If these processes are found to be more effective with mycorrhizae than for plants growing alone, then the symbiosis can actually increase the rate at which nutrients are made available or the forms in which nutrients can be taken up. The use of plant-derived photosynthate to speed decomposition would also blur the distinction between producers and decomposers.

Symbiotic nitrogen fixation is far less common than are mycorrhizal associations in natural ecosystems, but has received much attention because of its economic importance. Nitrogen is an important limiting nutrient in many natural and managed ecosystems. Producing nitrogen fertilizers requires fossil fuel energy and is expensive. Increasing nitrogen availability and cycling by using solar energy captured by photosynthesis to fix nitrogen from the atmosphere can increase yields in both agriculture and forestry and reduce costs.

The most widely known nitrogen-fixing symbiosis is between leguminous plants and bacteria of the genus *Rhizobium.* This is the combination involved in legumes, such as soybeans, green beans, peas, etc. The legumes are a very large plant family, especially in the tropics, where both herbaceous and tree species are found in abundance. In temperate zones, legumes are less common and generally herbaceous, with only a few tree species.

The legume–*Rhizobium* symbiosis functions through the "infection" of roots of host plants by the bacteria. The infection stimulates the construction of a specialized organ called a root nodule. The specialized environment in the nodule shields the microbial symbionts from soil conditions (acidity and oxygen) that inhibit fixation.

Other plant–microbe symbioses also produce nodulelike systems and the capacity to fix atmospheric nitrogen. At least 20 genera of angiosperms in seven families interact with species of actinomycete fungi (Table 10.2). Most of these are woody shrubs or trees, frequently of wetland habitats. Use of these species in mixed plantations with trees or even in crop-rotation forestry (similar to corn–soybean rotations in agriculture) is currently being tested as a means of increasing site fertility and yield.

TABLE 10.2 Genera of Plants in Which Some Species Form Symbiotic Nitrogen-Fixing Relationships with Actinomycete Fungi
(From Bond 1983)

Genus	Number of Species Recognized in Genus	Present Distribution	Number of Species Recorded Bearing Nodules
Casuarina	45	Australia, tropical Asia, Pacific islands	24
Myrica	35	Many tropical, subtropical, and temperate regions	26
Alnus	35	Europe, Siberia, N. America, China, Japan, Andes	39
Dryas	4	Arctic, mountains of north temperate zone	3
Cercocarpus	20	West and Southwest US, Mexico	4
Purshia	2	N. America	2
Chamaebatia	2	California	1
Cowania	5	Southwest US, Mexico	1
Rubus	250	Many regions, especially north temperate	2
Coriara	15	Mediterranean, Japan, China, New Zealand, Chile, Mexico	16
Colletia	17	Temperate and subtropical America	3
Discaria	10	S. America, New Zealand, Australia	5
Trevoa	6	Andes	2
Talguenea	1	Chile	1
Kentrothamnus	2	Bolivia, Argentina	1
Ceanothus	55	N. America	31
Elaeagnus	45	Asia, Europe, N. America	28
Hippophaë	3	Europe, temperate Asia to Kamachatka, Japan, Himalaya	1
Shepherdia	3	N. America	2
Datisca	2	Mediterranean to Himalaya and central Asia, southwest US	2

Carbon Costs and Tradeoffs in Symbiotic Systems

If there are so many effective mechanisms for increasing nutrient availability, why is there so much evidence for nutrient limitations on plant growth? Why don't plants just subsidize the processes described here until all nutrient limitations are removed, so that they are always growing at the maximum potential rates allowed by light, temperature, and water? Why don't all plants possess nitrogen-fixation symbioses and a high level of phosphatase activity? Using the arguments of natural selection and evolution, why have species that possess those abilities not out-competed and replaced those that do not?

The simplest answer is that each mechanism has a carbon cost. The extra nutrient uptake occurs at the expense of higher root/symbiont tissue construction costs and more respiration. For a plant to realize an adaptive advantage for supporting root symbionts, the increase in photosynthesis due to the added nutrient availability must be greater than the carbon "invested" in supporting the symbiosis. Such is not the case in environments that are already rich in nutrients.

It is generally recognized that the degree of subsidy to root symbionts varies with site quality or nutrient availability. Fertilizing a soybean field leads to a reduction in nitrogen fixation. Forests on richer sites may have less complete infection of roots by mycorrhizal fungi. Natural ecosystems dominated by symbiotic nitrogen fixers tend to be those with a high availability of light and water, and a relatively low availability of nitrogen, including certain frequently disturbed or burned types.

In summary, we can begin to think of plants in ecosystems not as maximizing nutrient uptake but rather as optimizing the allocation of carbon for tissue growth (leaves, roots, etc.) and supportive of physiological mechanisms or symbionts to maintain a relatively constant and proportional acquisition of all the resources (light, water, nutrients) required for growth. The actual measurement of the carbon cost–nutrient gain ratio associated with each of these mechanisms for increasing availability in native ecosystems remains an intriguing and unsolved problem.

REFERENCES CITED

Abbott, L. K., and A. D. Robson. 1985. The effect of VA mycorrhizae on plant growth. In Powell, C. L., and D. J. Bagyaraj (eds.), *VA Mycorrhiza*. CRC Press, Boca Raton, FL.

Aber, J. D., J. M. Melillo, K. J. Nadelhoffer, C. A. McClaugherty, and J. Pastor. 1985. Fine root turnover in forest ecosystems in relation to quantity and form of nitrogen availability: A comparison of two methods. *Oecologia* 66:317–321.

Bond, G. 1983. Taxonomy and distribution of non-legume nitrogen-fixing systems. In J. C. Gordon, and C. T. Wheeler (eds.), *Biological Nitrogen Fixation in Forest Ecosystems: Foundation and Applications*. Martinus Nijhoff/Dr. W. Junk Publishers, The Hague.

Lambers, H., F. S. Chapin III, and T. L. Pons. 1998. Plant Physiological Ecology. Springer-Verlag, New York.

11

RESOURCE ALLOCATION AND NET PRIMARY PRODUCTION

INTRODUCTION

The previous chapters in Part 2 have presented the chemical, physical, and biological factors affecting the availability of resources (water, light, nutrients) required by plants and the physiological mechanism by which they are acquired. We have also discussed the selective advantages in using one resource to increase the availability or uptake of another (e.g., allocating carbon to roots to increase nutrient uptake or nitrogen to leaves to increase carbon uptake). One of the outcomes of the interactions between resource supply and plant demand is the overall allocation of carbon and nutrients to different types of tissues (leaves, roots, stems, etc.). The production of plant biomass, or net primary production (NPP), is perhaps the most frequently measured of all ecosystems processes because it is central to yield of usable products, and, in the context of increasing atmospheric CO_2, to the storage and accumulation of carbon in ecosystems.

While NPP is measured as an emergent ecosystem property, it is the result of all of the resource-gathering processes described in Chapters 6–10 and of the allocation of those resources to plant tissues. It is also the result of the interactions among different plant species that acquire and use resources in different ways. At the plant and species level, high rates of primary production generally increase competitive ability. Faster-growing plants can grow taller and shade out slower-growing ones. They may also have more resources available for the production of seeds. However, plants genetically "programmed" for fast growth or large size may do very poorly in dry or nutrient-poor systems. For example, the tree growth form is not adaptive in semi-arid climates because water availability is too low, and photosynthesis too restricted, to support the high costs of growing and maintaining a tall stem. Natural selection favors species that have resource requirements, growth forms, and life history characteristics that match a site's resource availability profile and seasonal changes in climate.

The purpose of this chapter is to discuss the allocation of plant resources between respiration and biomass production and the allocation of production among foliage, stem, and root tissues. The implications of these allocation patterns for nutrient use efficiency are also presented, and we conclude with an ecological enigma—why trees stop growing.

RESOURCE LIMITATIONS ON PRODUCTION: A SIMPLIFIED VIEW

In 1840, Justus von Liebig first presented the theory that plant growth is limited by a single factor "presented to it in minimum quantities." Stated with true nineteenth century certitude, this became known as Liebeg's Law of the Minimum. It has served as a springboard for discussions of the nature of resource limitations on growth.

What is meant by "minimum quantities"? It is not meant in the absolute sense. Phosphorus can be provided to plants in much smaller quantities than nitrogen, and still nitrogen can be limiting due to a higher demand for this element in the production of new tissues. We have already seen that water is required in much larger quantities than any nutrient. Thus, the expression of "minimum quantities" must be made relative to plant requirements.

In addition, seasonal changes in availability can cause different resources or conditions to be limiting at different times. For example, a temperate forest ecosystem develops a leaf canopy only during the frost-free part of the year so is effectively temperature limited much of the time. What limits the rate of photosynthesis during the growing season? In the morning and evening (and of course, all night long), and also on cloudy days, low light levels are limiting. In the early spring and late fall, low temperatures are limiting. If a midsummer drought occurs, then a lack of soil water causes stomatal closure and photosynthesis is water limited. When temperature, soil moisture, and light are all optimal, then photosynthesis is limited by photosynthetic capacity, which is related to leaf nitrogen content (see Figure 6.3), so nitrogen is limiting. Simple models of photosynthesis can be used to calculate the degree to which each of these actually limits the total amount of carbon fixed by photosynthesis over the course of a year.

So, the answer to "What limits production in ecosystems?" also varies with the time frame for which the question is asked. At an hourly to daily scale, light may be the primary variable. Seasonally, temperature or soil water is important in those systems in which these conditions vary. In the longer term, the ability to produce leaf area and increase photosynthetic capacity depends on the rate of nutrient cycling, which, as we will see in later chapters, changes at the scale of decades to centuries.

RESOURCE POOLS IN PLANTS AND THEIR ALLOCATION

Immediately following uptake or fixation, nutrients and carbon are present in the plant in mobile forms. These resources are allocated to meet demands for respiration and for the construction of new foliage, fine roots, and the structural tissues in stems that support both roots and foliage, allowing for the movement of material between them. Different types of carbon compounds are produced to meet structural and metabolic needs (Figure 11.1). This division, and further allocation for protection against herbivory (Chapter 15), determine the chemical composition of materials that are eventually shed as litter and provide substrate for decomposers. How are these allocation patterns within a plant determined?

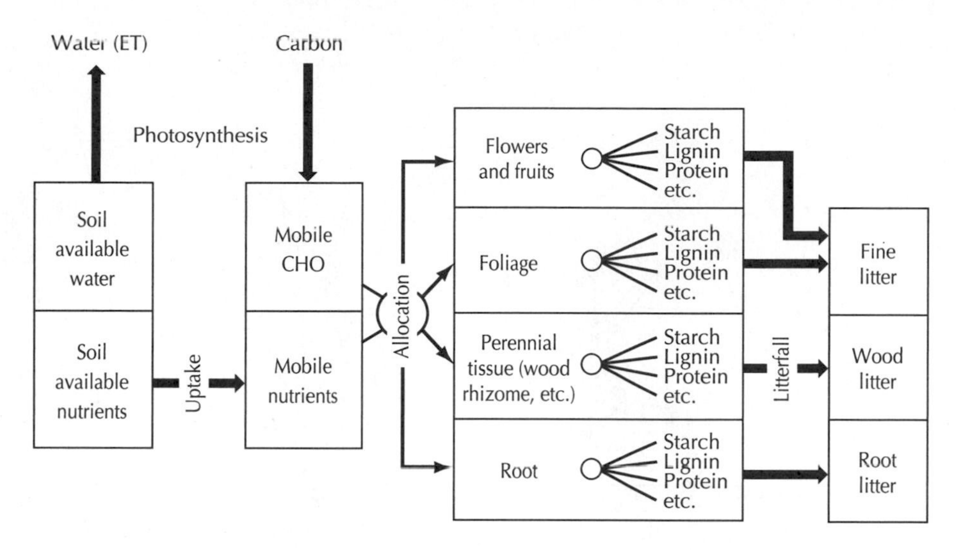

FIGURE 11.1 Schematic diagram of pathways for carbon and nutrient allocation from internal mobile pools in plants. The relative importance of each path depends on the growth form and tissue chemistry of the species involved, as well as the environmental conditions of the site.

A General Framework: Source–Sink Relationships

Patterns of allocation vary with species and stage of growth and are anything but random. A general working hypothesis is that the direction and rate of flow of carbon and nutrients through plants is controlled by the relative demands for resources exerted by different tissues at different times. Areas where demand for carbon is great (including growing leaves) are termed carbon **"sinks."** Mature leaves that are actively photosynthesizing but not growing would be carbon **"sources."** The stronger the sink strength, the greater the allocation to that tissue. Seasonal patterns of allocation in an annual grass plant and a perennial tree (Figure 11.2) illustrate differences in the timing and intensity of sink strength.

In the newly germinated annual grass, early photosynthesis may go initially to roots as the plant establishes an equilibrium between the fixation of carbon and energy from the atmosphere and the uptake of water and nutrients from the soil. The growing top may then become a center for faster growth, creating a greater sink and moving carbon up from established leaves. After the vegetative portion of the plant is fully grown, allocation of carbon to flowers and then to fruits or seeds begins. In annual plants, a large fraction of the annual uptake of carbon and nutrients is directed to seed production, which becomes a very strong sink at the end of the growing season.

In perennial plants such as trees, carbohydrates (starches) stored throughout the plant and nutrients stored in buds, twigs, and branches are mobilized in spring to produce new leaves, as well as some new top and stem growth. The leaves quickly become net sources of carbon, sending carbohydrates back down the plant. Different parts of the tree act as important sinks at different parts of the growing season. Top growth is usually accomplished

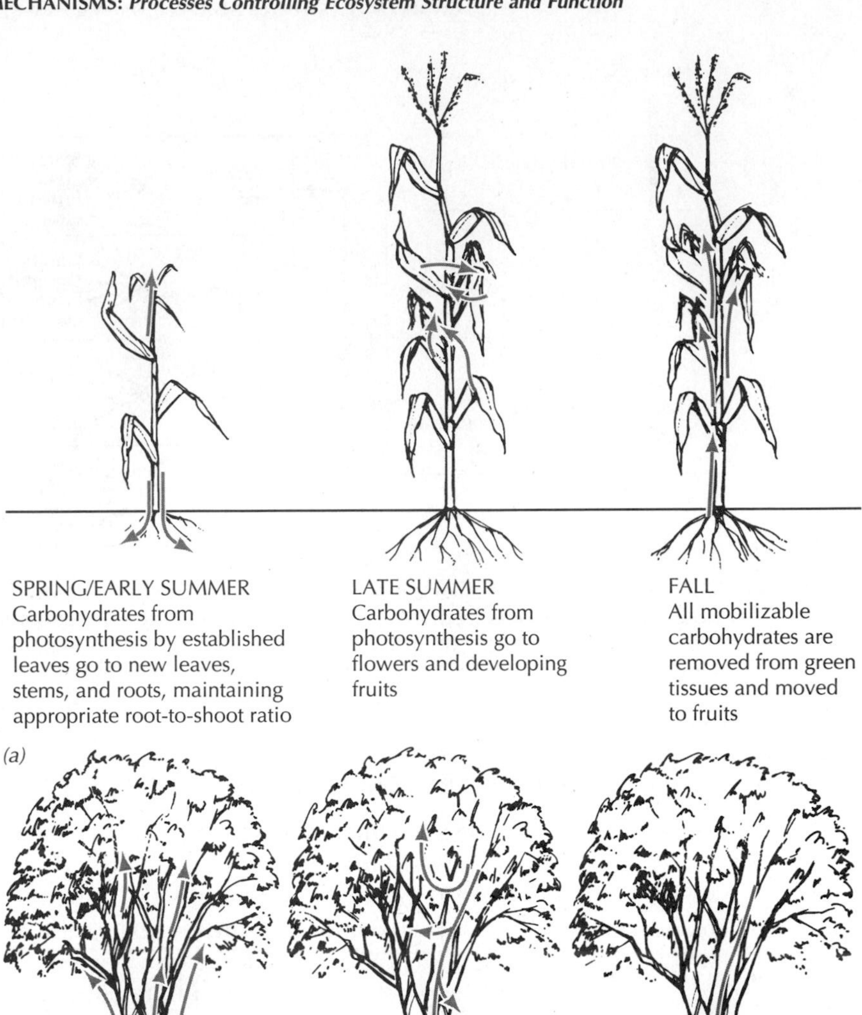

FIGURE 11.2 Seasonal changes in the allocation of carbon in (*a*) an annual grass and (*b*) a tree.

first, followed by stem expansion (radial increment). In many temperate and boreal zone species, these are both completed by the middle of the growing season. Carbon fixed after this period goes either to buds or roots or to storage as starch that is deposited throughout the stems, twigs, and roots. It is a general characteristic of perennial plants that a significant amount of carbohydrate is always held in reserve to buffer catastrophic losses of roots or foliage by insect attack, fire, wind, or other disturbances. This is in strong contrast to annual plants, in which all available carbohydrate reserves are directed to the maturing seeds.

If rapid growth rates of tissues results in the creation of strong carbon sinks, what then controls the stimulation of tops, roots, fruits, or flowers to initiate rapid growth? A proximate or partial answer is that plant hormones (a hormone is a "catalytic" molecule that, when present in small quantities, can cause large changes in plant metabolism) exert control over allocation patterns and that these are produced in response to different environmental conditions in ways controlled by the genetics of the species. Ultimate control resides in the genetic makeup of the plant, which in turn is the result of natural selection of individuals within species for genetically programmed allocation patterns that best fit the environment in which they grow.

Respiration: The Variable Cost of Maintaining Biomass

Respiration and production of new biomass represent competing sinks within the framework described above. In this context, what is the carbon cost of maintaining existing, living biomass? As with photosynthesis, there have been a great many measurements of respiration. Efforts to acquire field measurements from intact ecosystems go back at least to the 1970s (see Figures 3.2 and 3.13).

In general, respiration is controlled in the short term primarily by temperature. The classic relationship between the two is described by the temperature coefficient, or Q_{10} relationship (Figure 11.3), in which the Q_{10} coefficient indicates the logarithmic increase in respiration for every 10°C increase in temperature. An average Q_{10} value might be 2, indicating a doubling of respiration rate for every 10°C increase in temperature, but measured values range from less than 2 to more than 3. Plant tissues can acclimate to different temperature regimes by altering base metabolism rate (Figure 11.3). Still, respiration for support of living tissues does appear to be higher in warmer climates. An empirical relationship developed between mean annual temperature and the amount of carbon allocated to maintaining existing woody biomass in forests (maintenance respiration; Figure 11.4) shows increased carbon losses in warmer areas.

Other factors also affect base metabolism rates, and here another generalized theory has emerged from the large number of measurements taken. As was presented for photosynthesis in leaves, respiration both within a tissue type (Figure 11.5*a*) and across a wide range of tissue types (Figure 11.5*b*) is strongly related to nitrogen concentration. In leaves, this is expressed as an increase in dark respiration rates (negative net carbon flux in the dark), which is proportional to A_{max}, which is in turn related to leaf nitrogen concentration. Woody tissues have a much lower concentration of nitrogen and

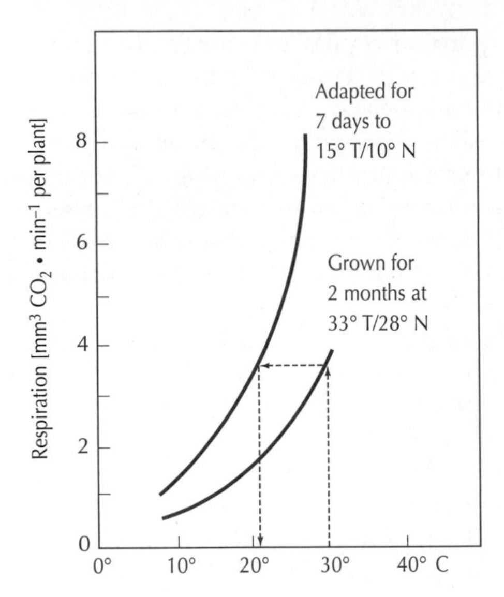

FIGURE 11.3 The short-term and long-term effects of temperature on respiration. Data are for seedlings of the same species grown at different temperatures. In the short term, both sets of seedlings show an exponential increase in respiration with increasing temperature. The long-term acclimation of the seedlings grown at higher temperatures includes a reduction in the basal rate of metabolism or respiration (lower rates of CO_2 production at any given temperature). (Data from Rook 1969 as cited in Larcher 1995)

also show much lower rates of respiration. Fine roots tend to be intermediate between leaves and wood. The generality of this relationship suggests that the rate of metabolic activity in all plant tissues is related to the content of nitrogen. This undoubtedly relates to the high and relatively constant concentration of this element in proteins and enzymes that control rates at which metabolic activity occurs in plants.

The combination of temperature and nitrogen controls on respiration makes it possible to calculate the total carbon cost of respiration for whole ecosystems. It might also be possible to predict how respiration might increase under different scenarios of future climate change. However, Figure 11.3 suggests that down-regulation or acclimation in respiration rates under different temperature regimes alters respiration rates significantly. Understanding the range of acclimation possible within a given plant is an important part of the climate change response prediction.

Allocation and Net Primary Production Above Ground

The production of new plant biomass above ground is one of the more easily measured of ecosystem properties. Clipping of herbaceous growth, measurement of diameter increment in tree stems, and collection of leaf and other litter fall as it hits the ground are well-understood procedures. As a result, there are many estimates for above-ground NPP (ANPP). Predictably, these patterns reflect general relationships between ANPP and climate and are often summarized by biome type (Table 11.1, see also Figure 2.2).

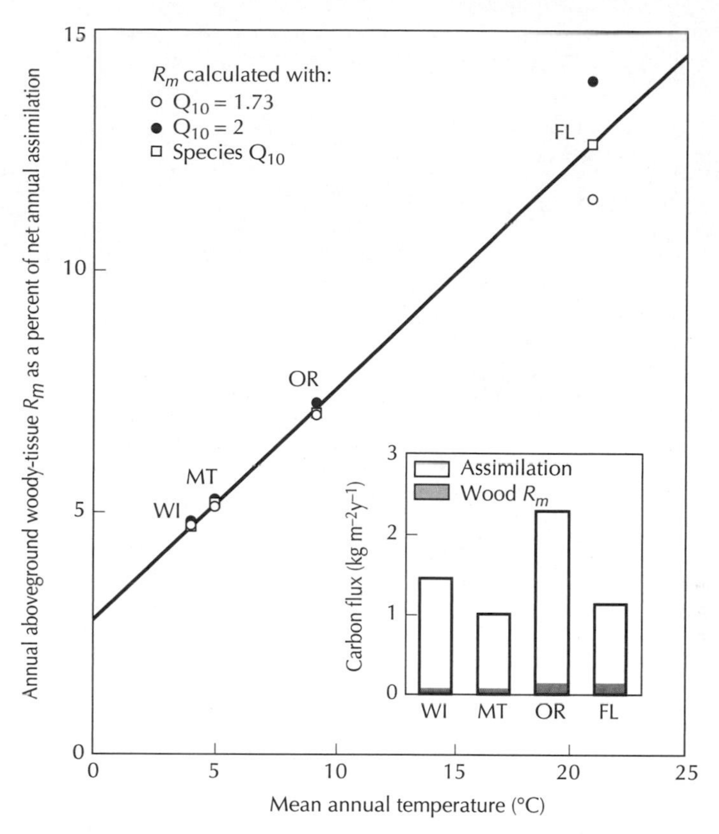

FIGURE 11.4 Changes in total annual respiration from woody stems in four stands of evergreen conifers in very different climatic regimes. (Ryan et al. 1995)

Allocation to growth above ground determines the potential for an individual plant to compete for light. In systems with adequate water and nutrient availability, the development of a full, stratified canopy (Chapter 8) is an expression of this competition. Under conditions of full canopy development, the photosynthetic capacity of the upper layers of the canopy, as determined by foliar nitrogen content, are a good predictor of total above-ground productivity within a climatic zone (Figure 11.6). Because primary production is tied to the uptake allocation and cycling of nutrients, there is also, within biomes, a close relationship between realized ANPP and rates of nutrient cycling (Figure 11.7).

In semiarid zones, large changes in the availability of water over the course of a single growing season or between years can result in large variations in ANPP. Within a single year, spatial differences in rainfall received across a region can drive large differences in ANPP as well. In grasslands, ANPP is basically equivalent to the development of canopy structure and leaf area index (LAI). One regional study of the spatial pattern of ANPP in the Great Plains region of the United States used remote sensing of the spatial pattern in LAI as a method of validating model predictions of grassland ANPP response to variation in precipitation both spatially and between years (Figure 11.8).

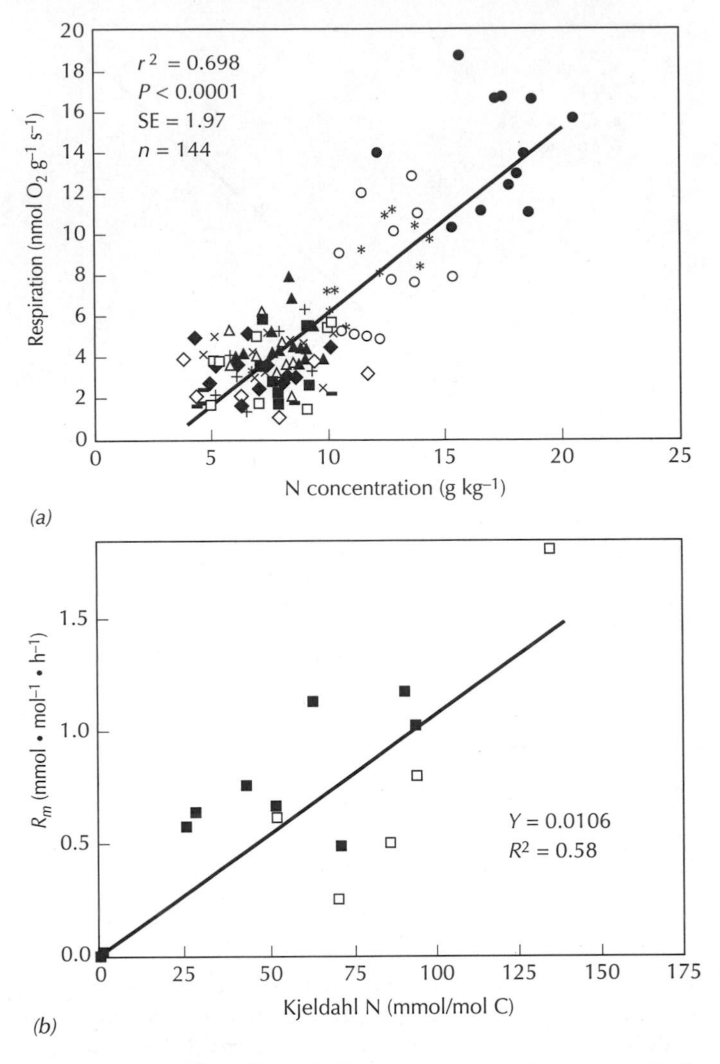

FIGURE 11.5 The effect of nitrogen concentration on respiration rates. (*a*) Relationship for fine roots from different soil horizons and stands in Michigan. (Pregitzer et al. 1998) (*b*) Relationship for a wide variety of tissues ranging from stem wood to foliage in both herbaceous and woody plants. (Ryan 1991)

In mountainous regions, interactions between complex topography and microclimate can also lead to large variations in ANPP. One study approached the estimation of this variability by first using remote sensing to derive estimates of spatial pattern in the amount of foliage present in the evergreen forests that dominated a site in the northern Rocky Mountains of the United States. This was combined with estimated patterns in microclimate with elevation, slope, and aspect as input to an ecosystem model that then predicted spatial patterns in ANPP (Figure 11.9). Similar methods applied at coarser scales have been carried out for all the ecosystems of the Earth. What has been lacking in many of these efforts is a direct and rigorous comparison between model estimates of ANPP and actual measurements on the ground.

TABLE 11.1 Primary Production and Biomass Estimates for the Biosphere
(From Whittaker and Likens 1973)

Ecosystem Type	Area (10^6 km = 10^{12} m^2)	Mean Net Primary Productivity (g C/m^2/yr)	Total Net Primary Production (10^9 metric tons C/yr)	Mean Plant Biomass (kg C/m^2)	Total Plant Mass (10^9 metric tons C)
Tropical rain forest	17.0	900	15.3	20	340
Tropical seasonal forest	7.5	675	5.1	16	120
Temperate evergreen forest	5.0	585	2.9	16	80
Temperate deciduous forest	7.0	540	3.8	13.5	95
Boreal forest	12.0	360	4.3	9.0	108
Woodland and shrubland	8.0	270	2.2	2.7	22
Savanna	15.0	315	4.7	1.8	27
Temperate grassland	9.0	225	2.0	0.7	6.3
Tundra and alpine meadow	8.0	65	0.5	0.3	2.4
Desert scrub	18.0	32	0.6	0.3	5.4
Rock, ice, and sand	24.0	1.5	0.04	0.01	0.2
Cultivated land	14.0	290	4.1	0.5	7.0
Swamp and marsh	2.0	1125	2.2	6.8	13.6
Lake and stream	2.5	225	0.6	0.01	0.02
Total continental	149	324	48.3	5.55	827
Open ocean	332.0	57	18.9	0.0014	0.46
Upwelling zones	0.4	225	0.1	0.01	0.004
Continental shelf	26.6	162	4.3	0.005	0.13
Algal bed and reef	0.6	900	0.5	0.9	0.54
Estuaries	1.4	810	1.1	0.45	0.63
Total marine	361	69	24.9	0.0049	1.76
Full total	510	144	73.2	1.63	829

Allocation to Roots: Using the Ecosystem Context

Are there global patterns in the distribution of below-ground production as there are for ANPP? This is a much more difficult question to answer.

Fine roots, those active in nutrient uptake, and those hosting symbiotic relationships with microbes, remain one of the most difficult, and important, aspects of ecosystems to study. With some effort, estimates of the changes in biomass through time can be obtained. But, unlike foliage that turns color and falls at senescence, fine roots are difficult to separate into live and dead tissues. Most approaches to measuring the growth or death of fine roots modify the soil environment in some way and may invalidate results. Yet obtaining good estimates of fine root production is crucial for calculating total NPP and

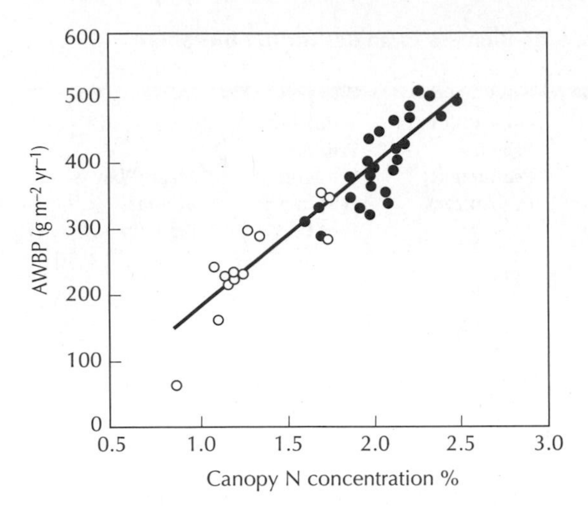

FIGURE 11.6 Relationship between mean canopy foliar nitrogen concentration and total woody biomass production for a series of deciduous and evergreen stands in New Hampshire. (Smith et al. 2000)

total nutrient uptake. Relatively few estimates of production and nutrient requirement of fine roots exist, and most of those are controversial. Because of this, there are disturbingly few measurements of total NPP in terrestrial ecosystems (review Table 3.1 for production terminology and an example of the problems in separating root and soil metabolism). Most productivity values published apply to above-ground tissues only or use simple and possibly inaccurate methods for extrapolating below-ground production from above.

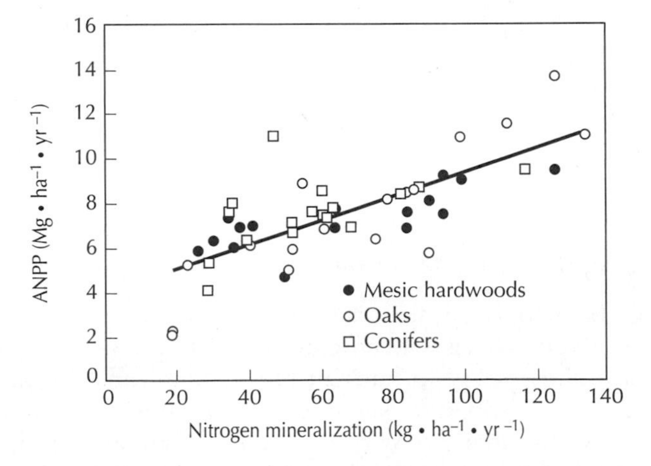

FIGURE 11.7 Net nitrogen mineralization and above-ground net primary production for several temperate forest ecosystems in Wisconsin and Minnesota (Reich et al. 1997). Net nitrogen mineralization is roughly equivalent to the total rate of nitrogen uptake and cycling in this region of low nitrogen deposition.

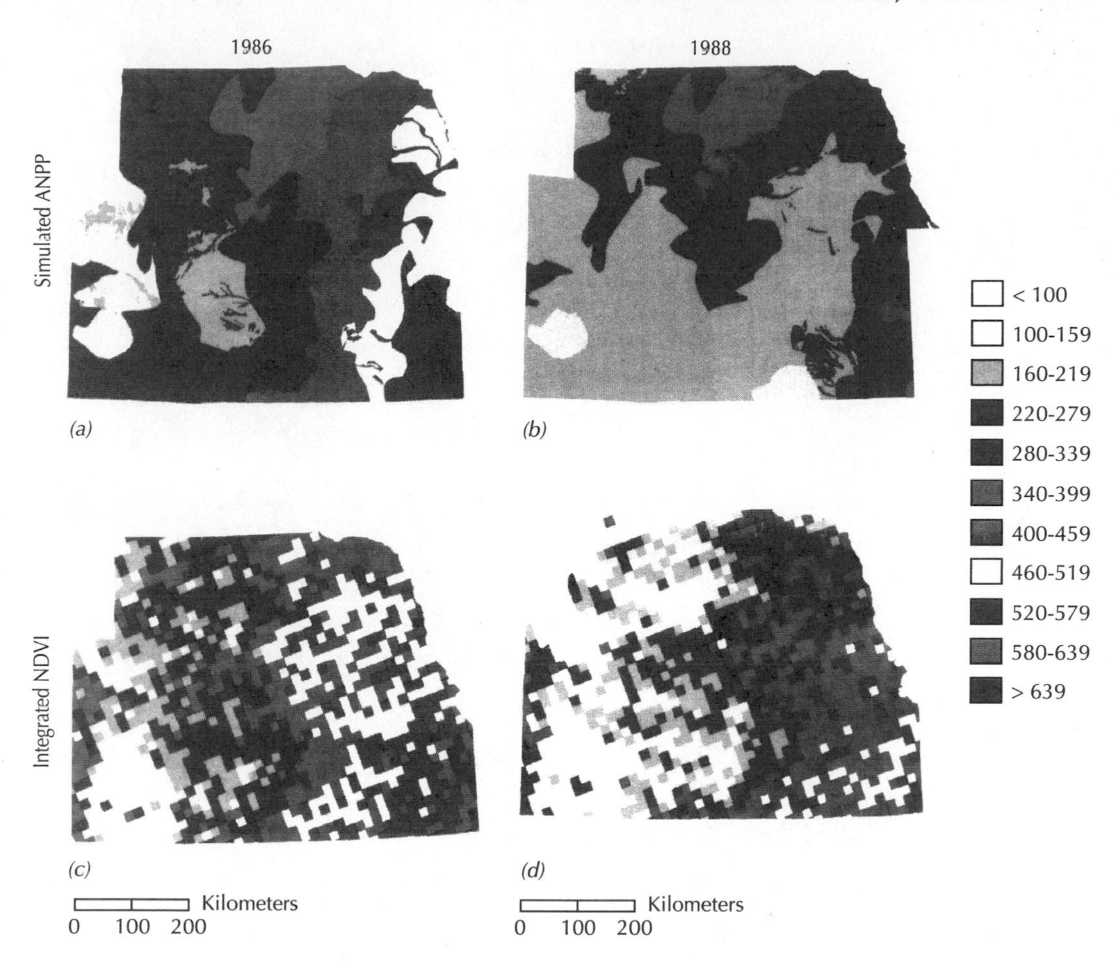

FIGURE 11.8 Predicted above-ground net primary production (ANPP) compared with the Normalized Difference Vegetation Index (NVDI) from remote sensing data for 2 years for the Central Plains region, including Kansas, Nebraska, and eastern Colorado. Differences in precipitation between the 2 years led to very different spatial patterns in ANPP as predicted by the CENTURY model. The strong correlation between predicted ANPP and NDVI suggests that the model predictions are valid. (Burke et al. 1991) **See plate in color section.**

Two divergent views of fine root production have emerged from the research done to date and tend to reflect two different sets of assumptions and approaches to measurement. A third and more recent method attempts actual direct observation of root production and turnover.

Direct Measurements of Fine Root Biomass – The first method uses consecutive measurements of fine root biomass and attributes increases in live or dead root mass to production and mortality. The principal assumption behind this method is that production and mortality occur at different times so that measured differences between sampling periods capture all of the changes in mass related to production. Results from this method often show lower fluctuations in biomass in "rich" sites, those with high ANPP (Figure 11.10*a*). From this, the conclusion has been drawn that root production is low on rich sites and a higher percentage of total productivity is allocated to wood and

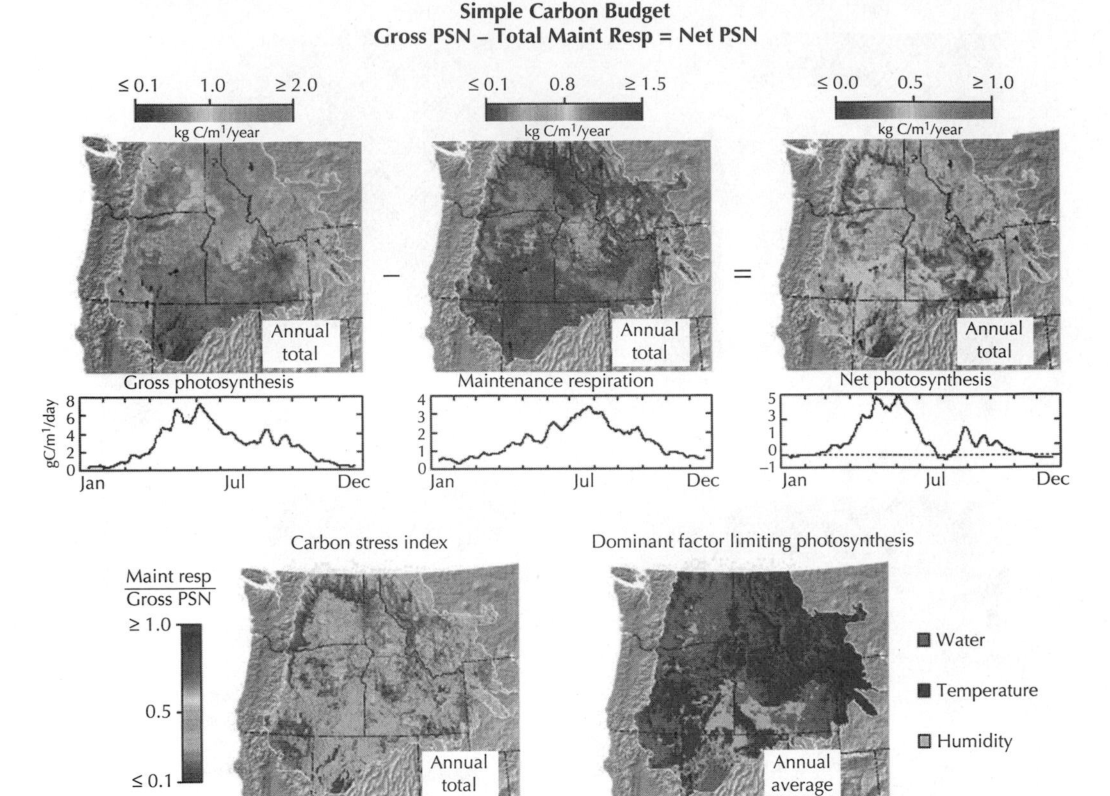

FIGURE 11.9 Extreme topographical variation in this region of the northern Rocky Mountains results in widely different climatic regimes, which are reflected in estimates of leaf area index derived from Normalized Difference Vegetation Index (NVDI) as obtained from remote sensing imagery. Using climate and NDVI as inputs, the Regional Hydroecological Simulation System (RHESsys) model predicts a highly variable pattern of ANPP as well. (Coughlan and Dungan 1997) **See plate in color section.**

foliage (Figure 11.10*b*). This is consistent with the idea of lower allocation of carbohydrate to roots on rich sites, where less may need to be "invested" to obtain adequate nutrient supplies. A problem with this is that fine roots are higher in nutrient content than is senescent foliage and much higher in nutrient content than is wood, so allocating more carbon to roots increases nutrient demand. If "low-quality sites" means low nutrient availability, then there is a logical inconsistency in this approach.

Carbon and Nitrogen Budgets – The second method assumes that root growth and mortality can occur continuously and simultaneously, so that the differences between sites in Figure 11.10*a* could occur by faster and continuous turnover of roots in the richer site rather than by lower root production. Under this assumption, infrequent direct measurements of biomass would not capture root dynamics. Data in support of this assumption has come from ecosystem-level measurements of carbon and nitrogen cycling that establish constraints on the amount of carbon and nitrogen that can flow through roots.

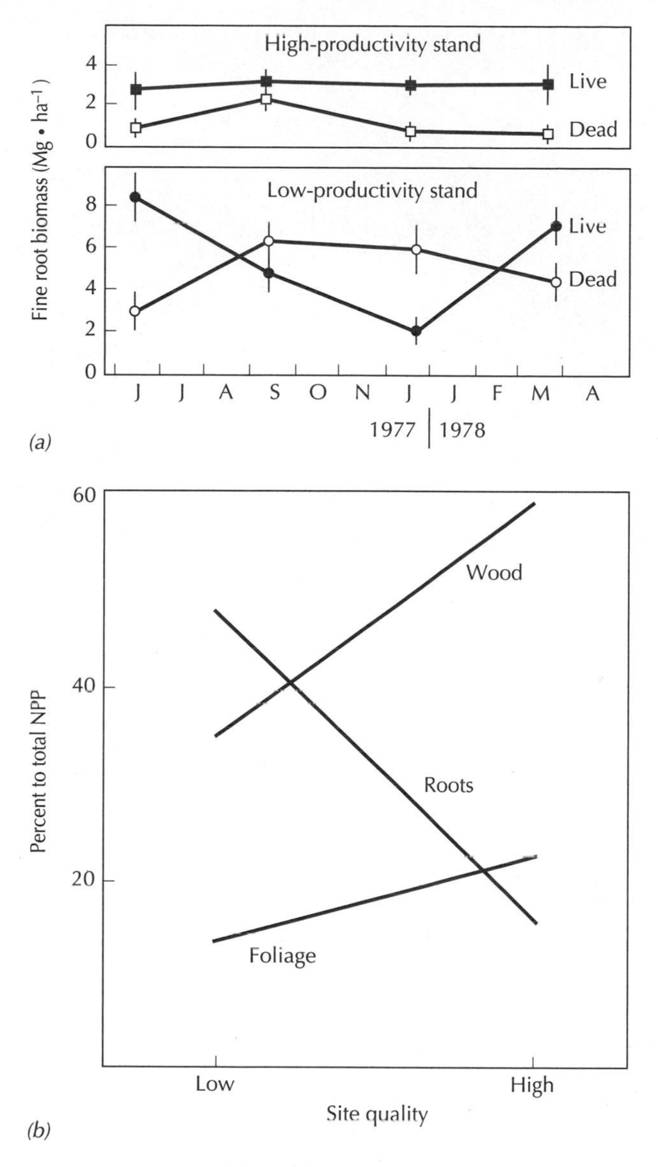

FIGURE 11.10 Carbon allocation between foliage, wood, and fine roots in Douglas-fir on nutrient-rich and -poor sites. (*a*) Measured changes in fine root biomass. (*b*) Estimated changes in allocation of production. (Keyes and Grier 1981)

For example, the amount of carbon allocated below ground by plants can be estimated as the difference between total above-ground litter inputs and soil CO_2 efflux (Figure 11.11; repeated here from Chapter 5). In mature ecosystems, the net change in carbon storage in soils is close to zero, and the CO_2 released from decay of above-ground litter is nearly equal to the carbon deposited in that litter. Under these conditions, subtracting carbon inputs in above-ground litter from total CO_2 release from soils provides an estimate of the carbon allocation by plants to below-ground tissues for both growth and respiration. A survey of data from many systems suggests that a predictable

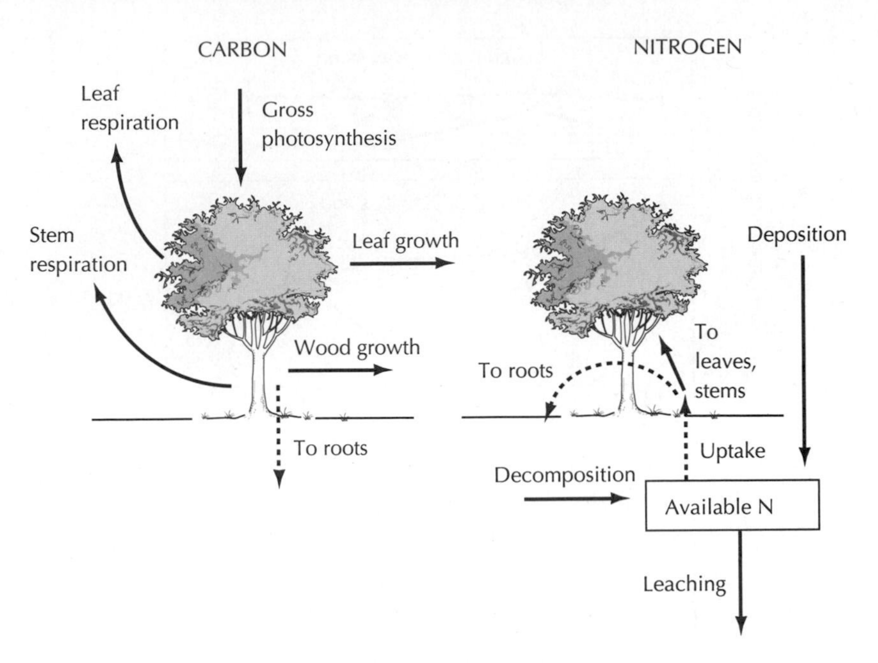

FIGURE 11.11 Examples of estimating root production in ecosystems using the mass balance technique. For carbon balance, direct measurements of gross photosynthesis, leaf and stem respiration, and leaf and stem growth allow for the estimation of allocation of carbon to roots by difference. For nitrogen, uptake is estimated as deposition plus nitrogen released by decomposition of organic matter, minus leaching losses. Root uptake is then estimated as uptake minus allocation of nitrogen to leaves and wood.

fraction of fixed carbon is allocated below ground over a very wide range of above-ground litter inputs (Figure 11.12). The relatively constant ratio (increasing only at fairly low litter inputs) suggests that carbon allocation below ground is a relatively constant fraction of total NPP (Figure 11.12), which in turn supports the notion of lower root biomass through turnover on rich sites.

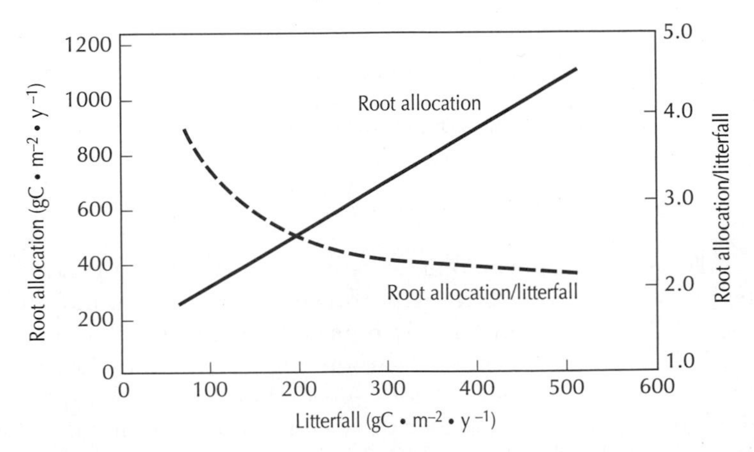

FIGURE 11.12 Summary relationship between above-ground litterfall and total carbon allocation to roots for both growth and respiration. Relationship derived from large number of published studies. (Raich and Nadelhoffer 1989)

The nitrogen budgeting approach begins by measuring total nitrogen availability as nitrogen mineralization plus precipitation inputs minus losses to leaching below the rooting zone (Figure 11.11). The difference is assumed to be plant uptake. By subtracting nitrogen in measured ANPP from total uptake, you have an estimate of nitrogen allocated to roots (and mycorrhizae). Dividing by the concentration of nitrogen in roots provides an estimate of fine root (and mycorrhizal) production. This method was applied to a number of forest sites of varying nitrogen availability, and the conclusions drawn were that fine root turnover does indeed increase with nitrogen availability and that the fraction of total NPP allocated to roots was relatively constant between sites (Figure 11.13). In one boreal system, however, an attempt to use this technique gave results that were very different than either direct measurement or the carbon balance method. In certain soil systems, the measurement of nitrogen mineralization, which is crucial for this method, can be as difficult as the measurement of fine roots.

Direct Observation: Minirhizotrons – Are there methods for the direct observation of roots in soils? Over the last 2 decades, systems involving miniature glass tubes inserted into the ground (minirhizotrons) and coupled with miniaturized camera systems have been developed, improved, and used to image root distributions along the glass interface. In some places, entire glass-walled underground rooms have been constructed to view a larger cross-sectional area of soil. In either case, repeated images at the same location along the glass allow for a direct assessment of longevity (or turnover rate) of roots. Highly magnified imaging systems currently being tested may be able to detect individual fungal hyphae as well, giving an estimate of turnover for this type of tissue that is not detectable by the root biomass methods described earlier.

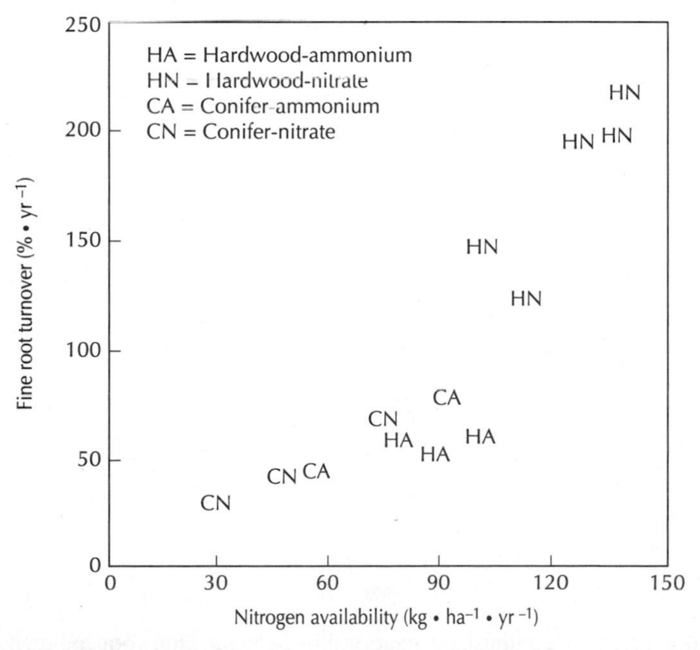

FIGURE 11.13 Estimated rate of fine root turnover in several forests as a function of nitrogen availability. (Aber et al. 1985)

What have these direct observations shown? First, they have shown that the very finest roots (< 0.5-mm diameter), which have the highest nitrogen content and highest respiration rates, also represent 70–80% of total root length and have the highest turnover rate. In northern temperate forests, minirhizotron estimates for turnover rate are in the range of 60% per year, in the same range as those for similar forest types derived from the nitrogen budget method in Figure 11.13. In general, root turnover rates increase with increasing nitrogen availability. Turnover rates also vary significantly with soil depth, with greater life expectancies at greater depths.

Is there a general pattern to root metabolism and longevity that can be drawn from these data? We can assess this by beginning with the hypothesis that fine roots should show the same responses as leaves to increasing nutrient availability: increases in nitrogen concentration and respiration rate and decreases in life span or longevity. Is there support for this theory?

The physiological evidence might say yes. One study of patterns of changes in starch content of roots suggested that new carbon (stored as starch) is allocated to roots only when the root is growing and that, when this reserve is depleted, the root dies. A different study showed that 70% of the variation in root respiration across a range of sites and soil depths was determined by root nitrogen concentration (Figure 11.5*a*) and that this concentration generally increased with nitrogen availability. Thus, under the same climate, on a site with higher nitrogen cycling (a "richer" site) roots have higher nitrogen concentrations and higher respiration rates, leading to a more rapid exhaustion of starch reserves and faster turnover. In one synthesis of existing data, it has been proposed that the pattern of higher root turnover (lower root longevity) is a product of changes in species composition across a nitrogen availability gradient, which counters a trend toward increased longevity for roots of a given species (Figure 11.14).

What is missing most in the study of fine root production patterns is a clear understanding of the tradeoffs involved in shifting allocation from above to below ground. While there are generally recognized trends across biomes (e.g., grasslands in general allocate more production below ground

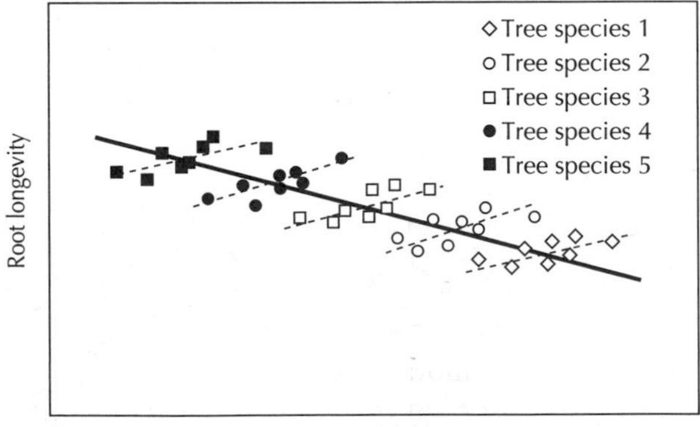

FIGURE 11.14 Hypothesized relationship between fine root longevity and nitrogen availability within forest types (*dashed lines*) and across forest types (*solid line*). (Burton et al. 2000)

than above relative to forests), exceptions to even these simple statements appear in the literature. Existing models of ecosystem production treat roots in very simplistic ways, driving allocation with simple ratios to photosynthesis or above-ground production, such as those in Figure 11.12. Canopies tend to be modeled with some degree of spatial structure and explicit competition for light (or at least changes in light intensity with canopy depth). The costs of leaf production can be estimated, as well as the rates of carbon gain realized. Some models use these to determine optimum canopy development. The study of roots in the field has not progressed much past the measurement of biomass and estimation of turnover. While laboratory studies on root physiology abound, the difficulties associated with measuring the function of the root/mycorrhizal symbiosis in the field have resulted in a lack of data on the basic carbon cost–nutrient gain responses of mycorrhizal function. Lacking these, models cannot begin to describe accurately the carbon–nutrient tradeoffs below ground or determine what optimal root systems would be under different conditions.

Nutrient Use Efficiency: Interactions Between Carbon Allocation and Nutrient Concentrations

In Chapter 6, we defined *water use efficiency* as the ratio of carbon fixed in photosynthesis to water lost or used in transpiration. A similar index can be defined for nutrient use efficiency. Productivity and nutrient demand are linked through the concentration of nutrients in the biomass produced. Nutrient use efficiency is just the inverse of nutrient concentration summed for all types of tissues produced.

However, the actual nutrient requirement for producing a tissue is not its concentration while alive, but the concentration it carries with it at senescence or when it is shed by the plant. Leaves, for example, can lose up to 75% of their nitrogen content during senescence. The "missing" nutrients can be lost by either of two mechanisms: (1) they may be leached from the foliage before senescence or (2) they may be retranslocated back into the perennial part of the plant as a part of the senescence process.

The magnitude of nutrient decline in foliage and the relative importance of leaching and retranslocation are very different for different nutrients. Table 11.2 shows how the decline in foliage content of several macronutrients is partitioned between leaching and retranslocation for a mature northern hardwoods forest. Nitrogen, phosphorus, and potassium content of the canopy decreased substantially during senescence. Magnesium declined to a lesser extent, and calcium actually increased. Both nitrogen and phosphorus were retranslocated strongly, while losses of potassium and magnesium were mostly as throughfall. These results are representative of many temperate and boreal zone systems in that calcium and magnesium are generally considered to be less mobile than nitrogen, phosphorus, and potassium. Of the latter three, nitrogen is often removed from leaves in the largest amounts by retranslocation, with lesser removals of phosphorus. Potassium losses are generally through leaching. This ion is not fixed in organic molecules in the leaf, so it is very mobile and easily leached.

We might reason that nutrient use efficiency is advantageous on nutrient-poor sites and that natural selection and evolution would lead to lower nutrient concentrations in tissues adapted to growing on nutrient-poor sites.

TABLE 11.2 Decline in Foliar Content of Several Macronutrients During Senescence, Partitioned Between Leaching Losses (Throughfall) and Retranslocation in Northern Hardwood Forests of Two Different Ages

	5-Year-Old Forest					55-Year-Old Forest				
	N	P	K	Ca	Mg	N	P	K	Ca	Mg
Foliar contents before senescence (F_1)	63	4.4	37.6	20	7.2	71	5.6	28	20	4.9
Foliar contents after senescence (F_2)	26	1.6	19.5	24	5.5	33	2.1	14	25	4.1
Foliar leaching during senescence (L)	1.4	0.2	9.4	2	0.6	2	0.1	11	2	0.6
Retranslocation (F_1–F_2–L)	35.6	2.6	8.7	–6	1.1	36	3.4	3	–7	0.2

Restricting the analysis to above-ground tissues and assuming that the total amount of nutrients cycling through above-ground litter in a year is related to total nutrient availability, there is indeed a very strong relationship between availability and the nitrogen concentration in above-ground litter (Figure 11.15*a*). For phosphorus (Figure 11.15*b*), the same pattern emerges only for tropical forests, where litterfall concentrations are generally lower than for other forest and shrubland types at a given rate of phosphorus cycling. In contrast, the relationship between calcium concentration and total calcium in leaf litterfall is much less pronounced (Figure 11.15*c*).

Nutrient concentration within tissue types, such as leaves, is one part of the nutrient use efficiency equation. The other is the relative production of different types of tissues. In general, fine roots have higher nutrient concentrations than senescent foliage, and both of these have much higher nutrient contents than wood or stem material. Thus, evergreen forests in which more allocation is directed toward wood production and less to foliage can achieve very high rates of nutrient use efficiency. It should be apparent here that our inability to measure root production accurately increases the uncertainties in calculating nutrient use efficiencies for whole ecosystems.

It should also be apparent that there are tradeoffs involved in the conflicting needs to conserve nutrients on poor sites and to produce enough of the high nutrient content foliage and fine roots required for photosynthesis and nutrient uptake. Increasing nutrient use efficiency by allocating more carbon to wood instead of leaves may not prove to be very adaptive if it seriously reduces leaf area and total photosynthetic capacity. Another mechanism for conserving resources, especially in high-content leaves and roots, is to retain those tissues for a longer period of time. If foliage is retained for 2 years rather than 1, then a significant increase in carbon gain may be realized per unit of nutrient investment in leaf tissue. Indeed, the "evergreen" habit is considered a primary adaptation to low nutrient availability in the temperate forest region, where the benefits of increased nutrient use efficiency offset the respirational costs of maintaining foliage through the winter. Some evergreen species can respond to particularly low site quality by increasing needle retention time even further. For example, the boreal forest species black

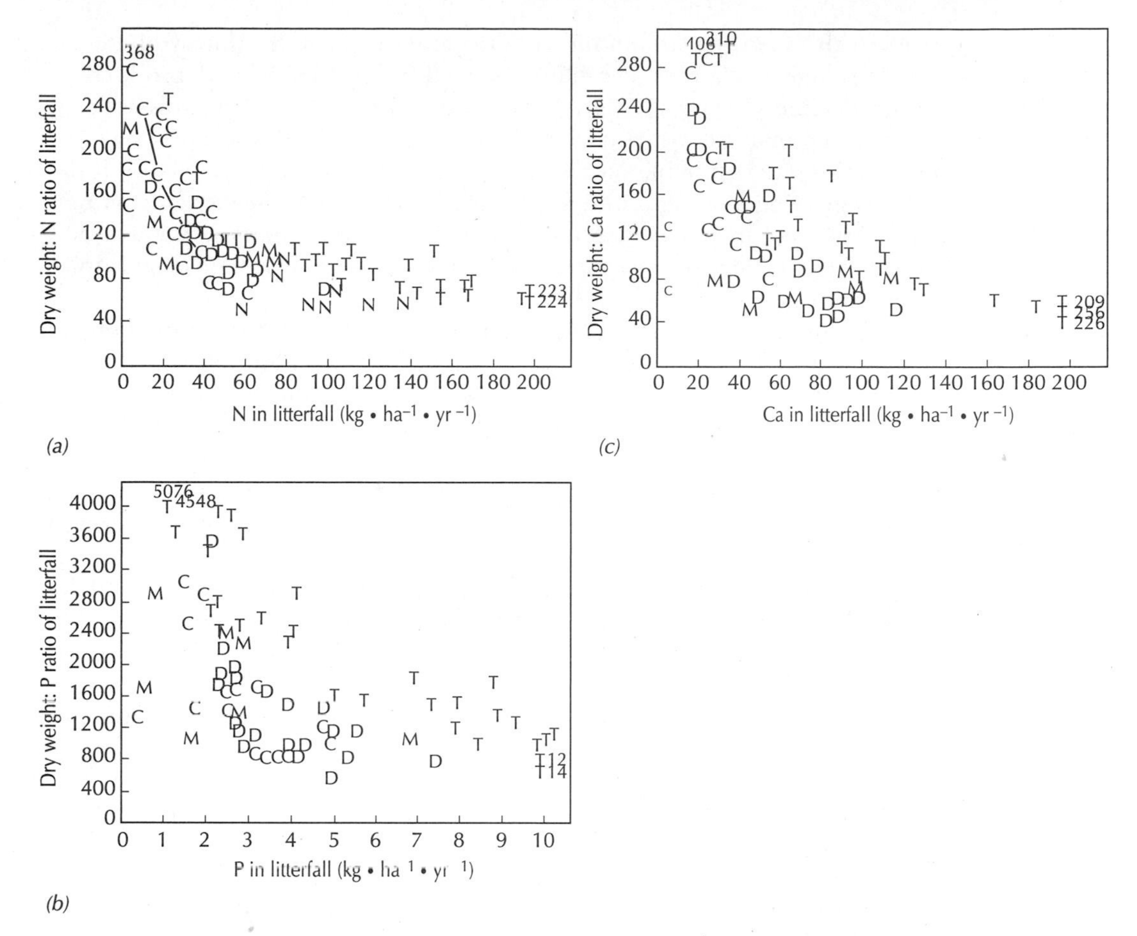

FIGURE 11.15 Nutrient concentration of (*a*) nitrogen, (*b*) phosphorus, and (*c*) calcium in leaf litter, as a function of total nutrient content in litter, for a wide range of forest types throughout the world. Dry weight–to–nutrient ratios are the inverse of nutrient concentration. Letters show types of systems: T = tropical forest, C = coniferous forest, D = temperate deciduous forest, M = Mediterranean shrublands, N = systems dominated by nitrogen-fixing species. (Vitousek 1982)

spruce (see Chapter 19) can retain needles for 10 years and more on the poorest sites.

AN ECOLOGICAL ENIGMA: WHY DO TREES STOP GROWING?

Sometimes asking the simplest of questions leads to very interesting insights into fundamental processes. Here is one that involves several of the processes that we have touched on so far in Part 2: Why do trees stop growing?

It has long been recognized in production forestry, especially of evergreen species, that maximum production occurs somewhere in the middle of stand development and then declines thereafter. For many years, it was assumed that this was due to the accumulation of large amounts of woody biomass in the stem that would increase respiration rates. If photosynthetic capacity remains relatively constant, then NPP, especially wood production, would decline (Figure 11.16*a*). A more critical examination of this concept has stressed that the vast majority of woody biomass in forests has a very low respiration rate as that biomass is made up mainly of nonliving cells. Only the actively growing area right under the bark, in which both new xylem and new bark tissues are produced, has a high metabolic rate. The sapwood, a region of still-living xylem cells nearest the bark, also contains living cells that respire at a lower rate.

If respiration does not change significantly with age, then reductions in stem growth must be linked either to changes in allocation or reductions in photosynthetic capacity. The alternate hypothesis in Figure 11.16*b* predicts increased allocation to fine roots and symbionts (ephemeral tissues that appear as increased litter production in Figure 11.16*b*), while photosynthesis and respiration remain constant at maturity. The other alternative, reduced photosynthetic capacity with constant respiration and allocation, is shown in Figure 11.16*c*.

A recent review has explored these alternative hypotheses. In general, increased allocation to fine roots in older stands is difficult to substantiate for all of the reasons cited earlier. The evidence that exists is contradictory and does not support the alternate pattern in Figure 11.16*a*. Then what might cause photosynthetic capacity to decline in older forests? Three hypotheses have been put forward.

The newest and most intriguing hypothesis is an increase in the resistance to water flow through xylem elements to the tops of trees with age. As older trees begin to decline in growth, there is a greater proportion of smaller, thick walled cells produced late in the season (summerwood) and a lower proportion of larger, thin walled cells (springwood). Resistance to water flow is greater in summerwood cells. Also, as trees grow taller, there is a greater resistance to water flow resulting from the force of gravity. A combination of these two could reduce the rate at which water can be provided to foliage high in the crown, creating more negative leaf water potentials and greater water stress. Thus, the height of trees in different environments would be directly related to the degree of water stress on the site as the trees grow up to a maximum sustainable height.

Older hypotheses for reduced tree growth at stand maturity can also be found. One is that nutrient limitations are the cause, and that this results from increased content of limiting nutrients (especially nitrogen or phosphorus) in plant biomass and nondecomposed litter. As the nutrient capital on a site is removed from the cycle by this accumulation, less would be available for plant growth, leading to lower foliar concentrations and less net photosynthesis. The second is that trees mature through genetic control and that these changes include reduced photosynthetic capacity.

You can imagine a series of experiments that would test these different hypotheses. For example, irrigation and fertilization of an older stand can be carried out to see if removing either of these limitations reverses declines in growth. Such experiments have been proposed, and as the results from these come in, we will have a better understanding of the factors causing tree growth to decline with age and a better understanding of the physiology of forested ecosystems in general.

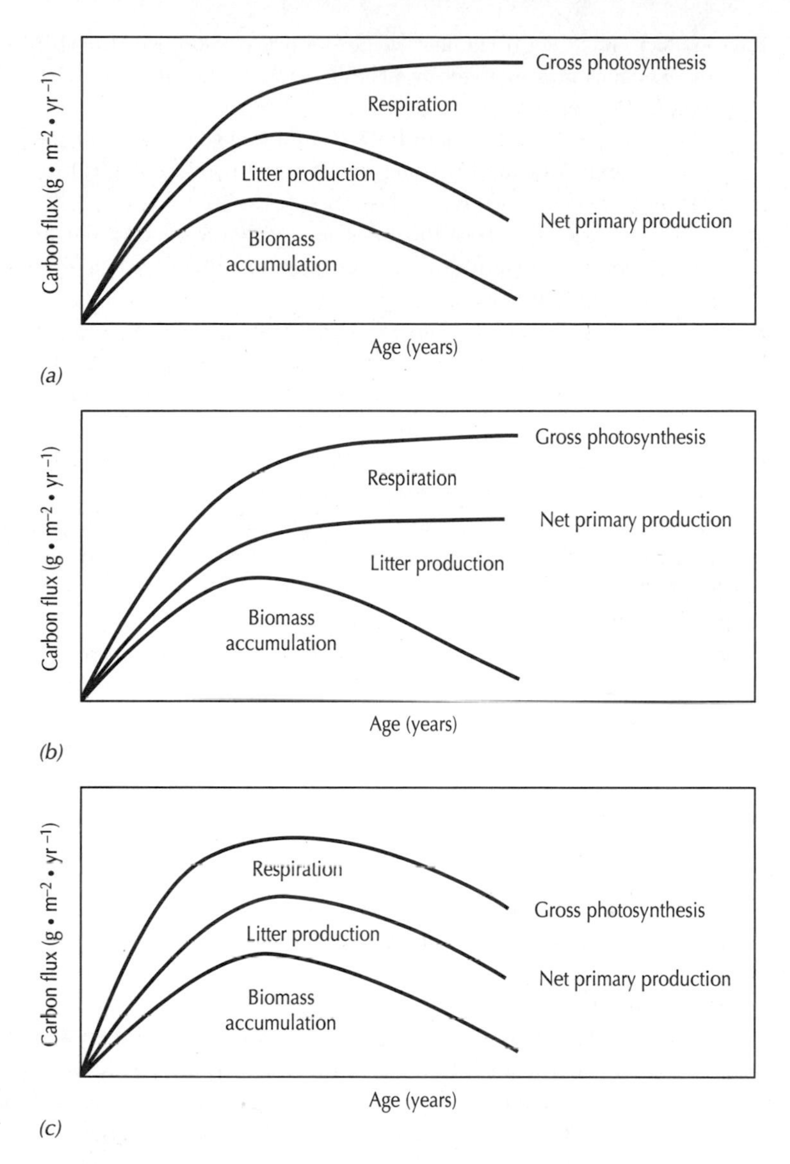

FIGURE 11.16 Three alternate hypotheses for the observed decline in biomass accumulation in mature forest ecosystems. (Ryan et al. 1997, Ryan and Yoder 1997)

REFERENCES CITED

Aber, J. D., J. M. Melillo, K. J. Nadelhoffer, C. A. McClaugherty, and J. Pastor. 1985. Fine root turnover in forest ecosystems in relation to quantity and form of nitrogen availability: A comparison of two methods. *Oecologia* 66:317–321.

Burke, I. C., T. G. F. Kittel, W. K. Lauenroth, P. Snook, C. M. Yonker, and W. J. Parton. 1991. Regional analysis of the central great plains. *BioScience* 41:685–692.

Burton, A. J., K. S. Pregitzer, and R. L. Hendrick. 2000. Relationships between fine root dynamics and nitrogen availability in Michigan northern hardwood forests. *Oecologia.* In preparation.

Coughlan, J. C., and J. L. Dungan. 1997. Combining remote sensing and forest ecosystem modeling: An example using the regional hydroecological simulation system (RHESSys). *Forestry Sciences* 50:135–158.

Keyes, M. R., and C. C. Grier. 1981. Above- and below-ground net production in 40 year old Douglas-fir stands on low and high productivity sites. *Canadian Journal of Forest Research* 11:599–605.

Larcher, W. 1995. *Physiological Plant Ecology,* 3d ed. Springer-Verlag, Berlin.

Pregitzer, K. S., M. J. Laskowski, A. J. Burton, V. C. Lessard, and D. R. Zak. 1998. Variation in sugar maple root respiration with root diameter distribution and soil depth. *Tree Physiology* 18:665–670.

Raich, J. W. and K. J. Nadelhoffer. 1989. Below ground carbon allocation in forest ecosystems: Global trends. *Ecology* 70:1346–1354.

Reich, P. B., D. F. Grigal, J. D. Aber, and S. T. Gower. 1997. Nitrogen mineralization and productivity in 50 hardwood and conifer stands on diverse soils. *Ecology* 78:335–347.

Rook, D. A. 1969. The influence of growing temperature on photosynthesis and respiration of *Pinus radiata* seedlings. *New Zealand Journal of Botany* 7:43–55.

Ryan, D. F., and F. H. Bormann. 1982. Nutrient resorption in northern hardwood forests. *BioScience* 32:29–32.

Ryan, M. G. 1991. Effects of climate change on plant respiration. *Ecological Applications* 1:157–167.

Ryan, M. G., and B. J. Yoder. 1997. Hydraulic limits to tree height and tree growth. *BioScience* 47:235–242.

Ryan, M. G., D. Binkley, and J. H. Fownes. 1997. Age-related decline in forest productivity: Pattern and process. *Advances in Ecological Research* 27:214–262.

Ryan, M. G., S. T. Gower, R. M. Hubbard, R. H. Waring, H. L. Gholz, W. P. Cropper, Jr., and S. W. Running. Woody tissue maintenance respiration of four conifers in contrasting climates. *Oecologia* 101:133–140.

Smith, M. L., S. V. Ollinger, M. E. Martin, J. D. Aber, R. A. Hallett and C. L. Goodale. 2000. Direct prediction of aboveground forest productivity by hyperspectral remote sensing of canopy nitrogen. *Ecological Applications.* In preparation.

Vitousek, P. M. 1982. Nutrient cycling and nutrient use efficiency. *American Naturalist* 119:553–572.

12

CHEMICAL PROPERTIES OF LITTER AND SOIL ORGANIC MATTER

THE DECOMPOSITION CONTINUUM

INTRODUCTION

Carbon allocation and nutrient use efficiency are important processes in plant function. They also control the quantity and biochemical content, or "quality," of dead organic matter (litter) produced. Litters of different quality are decomposed at different rates. For example, fine roots generally decay more slowly than do leaves, so root–shoot production ratios affect not only plant function but also decomposition rates and the development of soil organic matter. These, in turn, affect site quality and future root–shoot ratios.

Why is decomposition important? Litter and soil organic matter contain nutrients bound in complex molecules that are not generally accessible by plants. For most nutrients and in most ecosystems, annual plant demand is far greater than the sum of inputs from deposition and weathering. To use the language introduced in Chapter 5, most nutrient cycles are relatively closed. Plant production, then, depends on the recycling of nutrients within the system, and recycling depends on the decomposition of organic matter and release of the nutrients it contains. In addition, soil organic matter is one of the largest and most important and dynamic reservoirs for carbon in the global carbon cycle. Understanding what controls the carbon balance of this pool is crucial to predicting future CO_2 balance of the atmosphere.

Decomposition rarely goes to completion but rather results in the accumulation of very stable, complex substances collectively called soil organic matter, or humus, which plays a large role in determining soil structure, water-holding capacity, and ion exchange. Interactions between organic and mineral soil materials can also be important. Compounds within soil organic materials that would decay rapidly if exposed to microbes can be stabilized for long periods of time by interactions with mineral particle surfaces.

The purpose of Chapters 12 and 13 is to introduce and discuss the structure and dynamics of soil organic matter in the ecosystem context. This chapter concentrates on chemistry and structure, while describing in general terms the progressive changes that occur as fresh litter decays to humus. Chapter 13 explores the factors that control rates of decomposition and accompanying nutrient dynamics.

ORGANIC MATTER AS A RESOURCE FOR MICROBIAL GROWTH

Organic Matter Quality

Why is it that leaves that fall to the ground in a forest can disappear in a year or two, while a fallen log may last a century or more? Why do certain types of leaves disappear more rapidly than do others, even on the same site? Why is it that such large quantities of organic matter accumulate in soils? An important part of the answer to all of these questions is the "quality" of the different types of organic matter relative to the energy, carbon, and nutrient requirements of the microbes involved in the decay process.

The concepts of resource limitations on production developed for plants apply to microbial growth as well. Litter and soil organic matter are the resources that drive microbial growth, and, as with plants, different resources can be limiting at different times. Freshly fallen leaf litter may have a high proportion of energy-rich compounds on which microbes could grow rapidly. However, if there is little nitrogen in the litter due to retranslocation, then the rate of protein synthesis by microbes is reduced, limiting both population levels and the rate at which the leaf litter is processed. Alternatively, if the microbial population can access available nutrients in the soil or soil solution, then these nutrients can be taken up by microbes, reducing nutrient availability to plants.

Three general characteristics determine the quality of litter materials relative to microbial decay: (1) the types of chemical bonds present and the amount of energy released by their decay, (2) the size and three-dimensional complexity of the molecules in which these bonds are found, and (3) nutrient content. The types of carbon bonds present and the energy they yield constitute the **carbon quality** of the material. Nutrient content, and the ease with which those nutrients can be made available, constitutes the **nutrient quality**. A fourth characteristic, association with mineral particles, is especially important for soil organic matter and is discussed in a later section.

How Decomposition Occurs

Soil microbes cannot "eat" organic matter directly. The vast majority of molecules that make up this substrate are large and complex and cannot be transported across membranes and into cells. Instead, decomposition occurs through the production of enzymes that break the chemical bonds formed during the construction of plant tissues. These enzymes are exuded from the cells into the soil environment (and so are called extracellular enzymes) where they may or may not encounter the desired substrate and so may or may not break the larger molecules into smaller ones that can be taken up directly. Enzymes are expensive to produce in terms of carbon and nitrogen. The costs must be more than offset by energy and/or nutrient gain for the microbial community to grow through the decomposition of a given material.

The three-dimensional structure of large molecules may also offer an additional barrier to decay. Extracellular enzymes cannot be effective unless they encounter an appropriate three-dimensional bond structure. Compounds that do not have a consistent structure, or in which a complex three-dimensional structure impedes enzyme access to reaction sites (see the discussion of lignin and proteins below), decay more slowly than the carbon quality of the material would suggest.

Sorption of organic materials onto the surfaces of mineral soil particles has much the same effect as does large and random molecular structure. By blocking potential sites of reaction with extracellular enzymes, mineral materials reduce decomposition rates significantly and help to stabilize organic matter in soils. Even small molecules with relatively high carbon quality can be preserved in soils for long periods of time by this mechanism.

Nutrient incorporation or release during decomposition depends on the ratio of nutrient to energy content in the material. Microbes growing rapidly on high carbon quality substrates create large nutrient demands for the synthesis of new microbial biomass. If these demands are not met, microbial activity and decomposition are inhibited. Decomposition of low carbon quality organic matter does not support net growth in microbial populations and so results in a net release of nutrients stored in the substrate being decomposed.

BIOCHEMICAL CONSTITUENTS OF LITTER AND THEIR RATES OF DECAY

Glucose (Figure 12.1*a*) and other simple sugars (or carbohydrates) are among the first products of photosynthesis and are very high quality substrates for decomposition. The molecules are small, and the chemical bonds in them are energy rich, yielding much more energy than is required to create the enzymes necessary to initiate the chemical breakdown reaction. They also can be taken into microbial cells and metabolized internally.

In plants, simple sugars not immediately required for respiration and growth may be stored as starch, a carbohydrate polymer formed by bonding of the 1- and 4-position carbons of adjacent molecules (Figure 12.1*b*). Decomposition of starch is somewhat slower than that of simple sugars because the longer molecule must be severed before the sugar units can be metabolized. Still, it is a rapidly decomposed, high energy yielding substrate. Both sugars and starch are present in small amounts in plant litter material (Table 12.1) because they tend to be used for respiration by the plant before senescence is complete.

Cellulose is also a polymer of simple sugars linked by bonds between the 1 and 4 carbons of adjacent molecules. However, the three-dimensional structure of the bond is slightly different (Figure 12.1*c*). This slight difference allows for a totally different function. Cellulose is the main component of primary cell walls in plants. Carbohydrates converted to cellulose cannot be remobilized for respiration or growth by the plant. Respiration of this carbon must be accomplished by microbial decay. Individual polymers of cellulose, consisting of from 2000 to 15,000 sugar units, are further entwined into larger strands or fibers that are laid down in roughly parallel form to create the cell walls (Figure 12.2).

Cellulose is probably the most common molecule in the plant component of terrestrial ecosystems. It is also the source of fiber for paper and paper products, which are nearly pure cellulose. Cellulose is of moderate quality as a decay substrate. Extracellular enzymes are required to cleave the

(a) Simple sugars

glucose

sucrose

(b) Starch

(c) Cellulose

(d) Hemicellulose

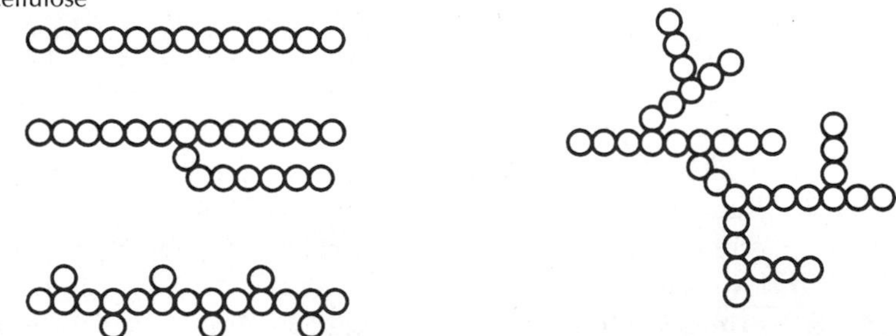

FIGURE 12.1 Chemical structures of several important types of carbohydrates in plants.

TABLE 12.1 Concentration of Major Carbon Compunds in Different Plant Materials
(Data from McClaugherty et al. 1985, Larsson and Steen 1988, Morrison 1980, Hodson et al. 1984)

	Sugars and Starch (%)	Other Solubles (%)	Cellulose (%)	Lignin (%)
Woody plants				
Foliage				
Sugar maple	7.2	37.6	43.1	12.1
Red oak	7.3	25.1	47.4	20.2
White pine	5.7	27.1	44.7	22.5
Fine roots				(Suberin)
Sugar maple	3.9	14.6	47.7	33.8
White pine	5.2	20.0	49.5	25.3
Wood				
Red maple	1.1	5.9	80.5	12.5
Hemlock bark	4.1	16.7	40.3	38.9
Herbaceous plants				
Foliage and stems				
Salt marsh grass				
Tall-form, live		34.4	52.5	13.1
Tall-form, dead		28.9	57.7	14.4
Tall-form, stems		30.3	56.0	13.7
Ryegrass stems				3–9
Leaves				2–6
Timothy stems				5–9
Leaves				3–6
Roots				
Salt marsh grass		36.2	41.6	12.2
Mixed pasture grasses		20	58	22

large polymers into simple sugars that can be taken up and metabolized. Two types of enzyme systems are known; one cleaves bonds within the molecule to create two shorter molecules, and the other separates individual sugar units from the ends of polymers.

Mixed in with cellulose and serving the same basic function are the hemicelluloses. These are also polymers, but they consist of several different basic sugar units combined into both straight and branched chains (Figure 12.1*d*). The branched structure helps bind the long, straight cellulose fibers together in the cell wall. There is little difference between cellulose and the hemicelluloses in rate of decay and energy yield.

A number of compounds containing unsaturated (double) carbon–carbon bonds are very important in plant function and for decomposition. Particularly important are those based on the six carbon phenolic ring. Two classes of such compounds are generally recognized: (1) smaller phenol polymers (polyphenols) made from several phenolic acids (Figure 12.3*a*) and

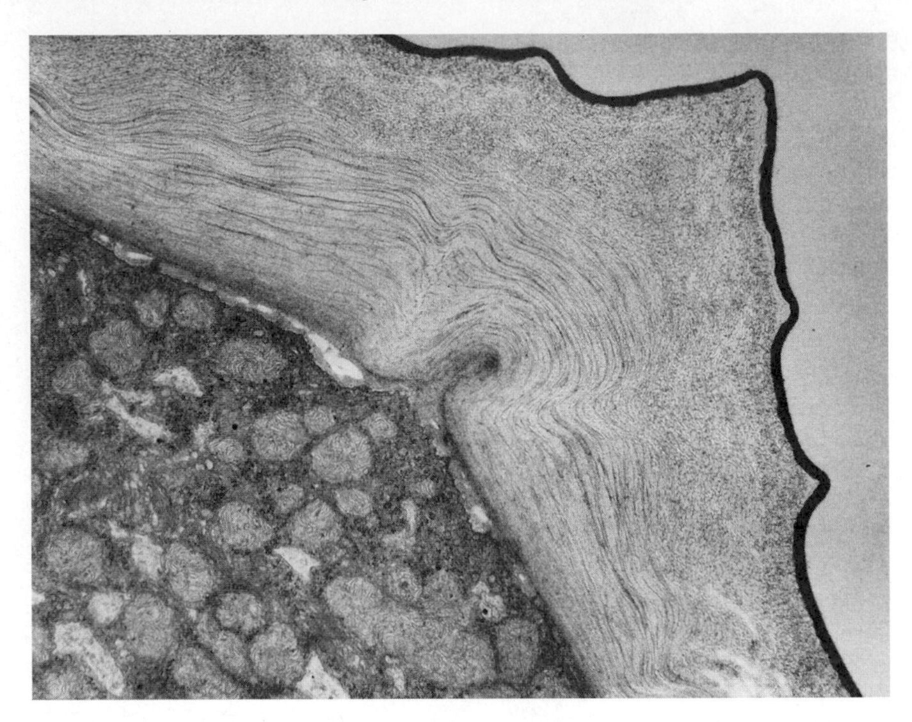

FIGURE 12.2 Arrangement of cellulose fibrils in primary cell wall. (Biophoto Associates/Photo Researchers)

often called tannins, and (2) the larger, amorphous, and very complex compounds, collectively called lignin (Figure 12.3*b*).

The **tannins** are so named because they have historically been extracted from plant tissues rich in these substances (e.g., the bark of some tree species) and used to tan leathers for shoemaking. The tanning process is the combination (or condensation) of polyphenols with proteins present in animal hides to increase the strength and durability of the leather (and to decrease its decomposability). In plants, tannins are thought to be primarily a defense mechanism against animal consumption (Chapter 15) and attacks by pathogenic fungi and bacteria. Many polyphenols are easily extracted from leaf tissues, indicating their potential mobility within plants. Leaching of polyphenol from litter can have important effects on litter dynamics and nutrient balances, as we explain in Chapter 13.

Although the phenolic ring yields less energy than do saturated carbon bonds, the smaller polyphenols can be metabolized. The rate in nature is difficult to determine, however, because of the mobility of the molecules and their capacity to condense with proteins or other polyphenols to form **lignins.** Thus, polyphenols may "disappear" from decomposing litter but may not be decomposed. They may be present as "new" lignin or polyphenol–protein complexes.

The much larger lignin molecules are among the most complex and variable in nature. There is no precise chemical description of *lignin* be-

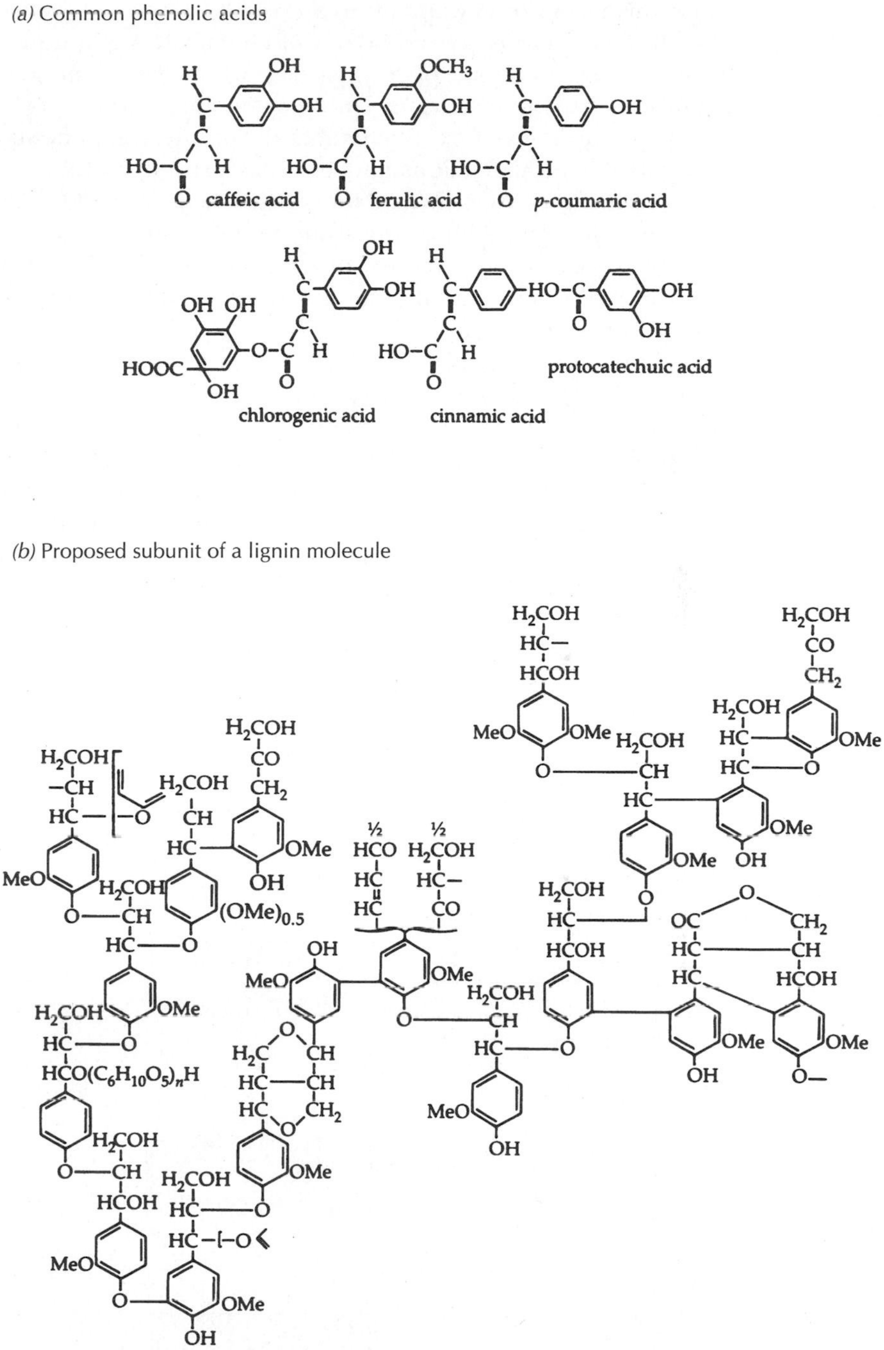
(a) Common phenolic acids
caffeic acid
ferulic acid
p-coumaric acid
chlorogenic acid
cinnamic acid
protocatechuic acid
(b) Proposed subunit of a lignin molecule

FIGURE 12.3 Chemical structure of some phenolic compounds found in plants.

cause this term actually applies to a class of compounds with variable structure. In fact, there is no precise way of chemically separating all of the molecules of lignin from plant material when performing tests of plant chemistry. Rather, a series of "proximate" analyses are performed. Figure 12.4 lists one series of treatments that divide the components of plants into different chemical fractions and that has proven useful in explaining differences in decay rates. When we say that a tissue is 20% "lignin," we are actually saying that 20% of the tissue remains after having been soaked in dichloromethane (a nonpolar solvent) and hot water and boiled in 72% sulfuric acid! There are different sets of proximate analyses, and each gives a different set of results.

It is possible to synthesize specific types of ligninlike compounds in the laboratory or, by chemically degrading a lignin molecule into smaller units, to get some idea of the structure of these subunits. One example of part of a lignin molecule is presented in Figure 12.3*b*. Important characteristics of lignin are the density of phenolic rings and the frequency of side chains, both of which reduce decomposition rates significantly. In addition, these very large molecules can be intricately folded into complex, three-dimensional structures that effectively protect much of the internal portions of the molecule from enzymatic attack.

Lignin is one of the slowest of the common plant components to decay. It yields almost no net energy gain to microbes during decomposition be-

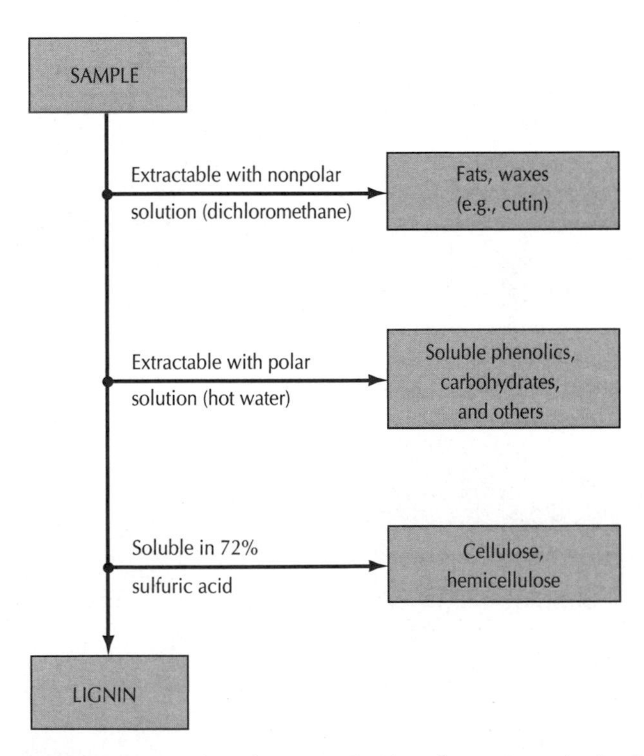

FIGURE 12.4 Flow diagram of series of treatments in the "proximate analysis" of plant tissues.

cause of the large amounts of energy required to initiate its decomposition. Evidence shows that energy derived from the decay of higher-quality substrates may be required to decompose lignin. For example, very complex enzyme systems that simultaneously degrade both lignin and cellulose-like carbohydrate polymers have been identified. In addition, laboratory studies indicate that the release of carbon as CO_2 from lignin can be increased by the addition of high-quality carbon substrates (sugars).

Yet lignin is second only to cellulose in quantitative importance in most plant tissues (Table 12.1). Lignin is what makes wood "woody." It is encrusted on and around cellulose in cell walls to provide rigidity and strength. Paper is easier to bend or tear than thin slices of wood largely because the lignin has been removed in the papermaking process. Lignin can be as concentrated in leaves as it is in wood. Within these tissues, the intricate intertwining of lignin and cellulose may actually cause lignin to "shield" some of the cellulose present from microbial processing, causing some of the cellulose to decay more slowly than would be expected by its biochemical quality.

The hydrocarbons (fats and waxes) are partially saturated, long-chain carbon compounds in which hydrogen ions have largely displaced the hydroxyl (OH) group. They tend to repel rather than interact with water molecules (they are hydrophobic), which, along with their size, makes them intermediate between carbohydrates and phenolics in decay rate.

The most common compound in this category is cutin (Figure 12.5), which is present in the cuticle, or outer waxy layer in plant leaves. Again, no

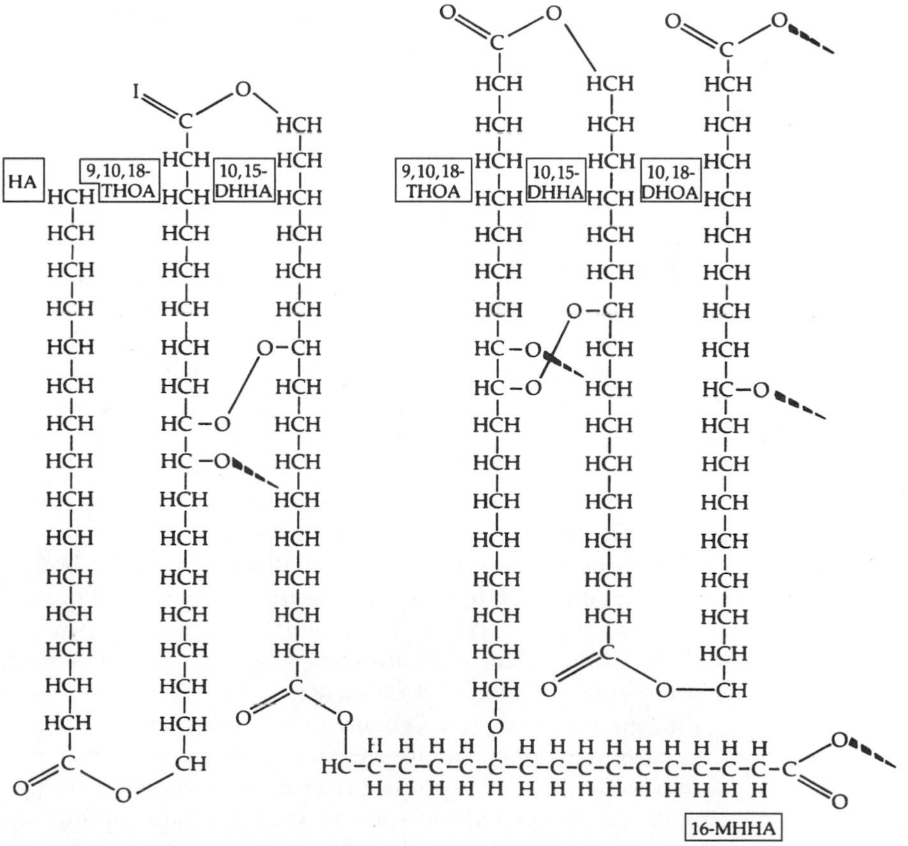

FIGURE 12.5 Chemical structure of cutin. HA, MHHA, DHHA, DHOA, and THOA are all different side chain organic acids.

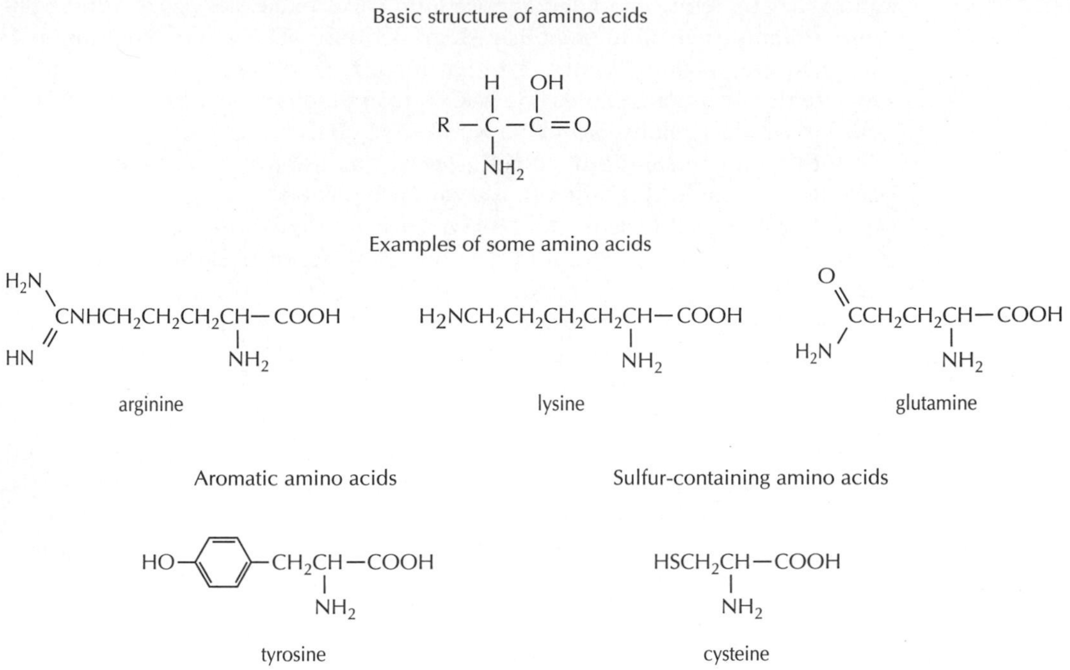

Basic structure of amino acids

$$R-\underset{NH_2}{\overset{H}{C}}-\overset{OH}{C}=O$$

Examples of some amino acids

$H_2N(HN=)CNHCH_2CH_2CH_2CH(NH_2)-COOH$ arginine

$H_2NCH_2CH_2CH_2CH_2CH(NH_2)-COOH$ lysine

$H_2N(O=)CCH_2CH_2CH(NH_2)-COOH$ glutamine

Aromatic amino acids

$HO-C_6H_4-CH_2CH(NH_2)-COOH$ tyrosine

Sulfur-containing amino acids

$HSCH_2CH(NH_2)-COOH$ cysteine

FIGURE 12.6 Basic structure of amino acids and examples of some common amino acids.

strict definition of this substance is known because it is an amorphous combination of constituents into variable structures. Its major role is to reduce evaporation through, and protect, leaf surfaces.

Suberin is an interesting class of molecules composed of roughly half hydrocarbons and half phenolics. This makes it both hydrophobic and of low energy quality, and consequently it is one of the slowest classes of compounds to decompose. Suberin is found in fine roots largely in place of lignin. When fresh, white roots become darker in color and more rigid, they are called **suberized,** because this change is due to the deposition of suberin within the cell walls. Because suberin is of less economic importance than lignin (we do not try to make paper out of root tissues), even less is known about its structure and the compounds from which it is formed.

Proteins are another important constituent of litter both because of their carbon quality and because they contain nitrogen. Nitrogen is present in plants mainly as structural or functional (enzymatic) proteins and the amino acids from which they are constructed (Figure 12.6). The simple amino acids can either be metabolized or used directly in construction of new proteins within the microbial cell. Proteins are polymers of amino acids and can contain more than 10,000 amino acid units. The largest proteins can become resistant to decay by the complexity of their three-dimensional structure, even though the individual units are very energy rich. In addition, proteins are often condensed with, and deactivated by, polyphenols and lignin. These ligno–protein complexes are very resistant to decay, even though they have a high nitrogen content. These condensation reactions may involve the very enzymes released by microbes to effect decomposition! In this way, the polyphenolic content of litter may have an additional negative effect on decay rates.

WHAT IS HUMUS?

At some point in the decay process, litter materials cross a threshold and become soil organic matter, or **humus,** another complex and amorphous form of organic matter in ecosystems. The distinction between litter and humus is often an operational one, based on root penetration or some other visible criterion.

There is general agreement that humus is a series of high molecular weight polymers with a high content of phenolic rings and quite variable side chains. In comparison with plant materials, it is very high in nitrogen and large polyphenolic molecules that are analyzed as "lignin" by the series of proximate analyses listed in Figure 12.4 and low in cellulose and hemicellulose (Table 12.1). Only part of that nitrogen content can be identified with a certain type of compound. As much as 40% of the nitrogen in soil organic matter is neither in protein nor amino acid form. It is unclear how this nitrogen is bound into humus, but a significant fraction may be present as **chitin** (Figure 12.7), which occurs both in the exoskeletons of insects and in fungal hyphae. Another fraction may be in the form of heterocyclic compounds, essentially phenolics combined with amino groups (see subsequent discussion on nuclear magnetic resonance methods).

As with litter, the classes of compounds contained in soil organic matter have traditionally been determined by proximate analyses—specifically by their solubility in different acidic and alkaline solutions. Two common methods of separation are outlined in Figure 12.8*a*. The fractions generated by the first method vary considerably in the weight of the molecules extracted and in their carbon and oxygen content (Figure 12.8*b*), although there are no distinct boundaries between the classes; rather, we can picture the major components of soil organic matter as a sequence of compounds varying in the degree of polymerization (the size of molecules) and in elemental content. In addition to association with mineral particles, humin may also be even more highly polymerized molecules, to the right of humic acids in Figure 12.8.

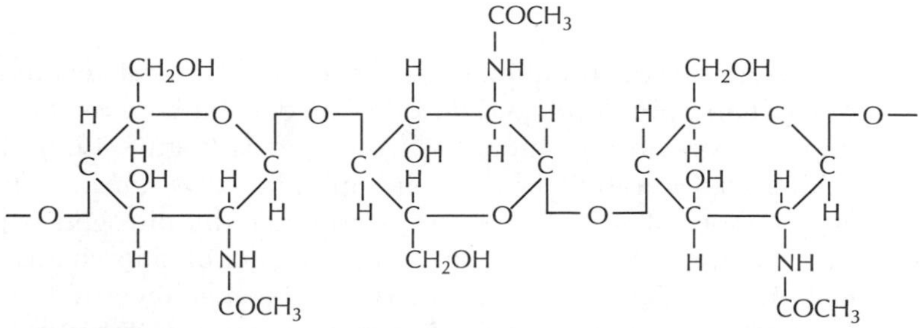

FIGURE 12.7 Structure of chitin.

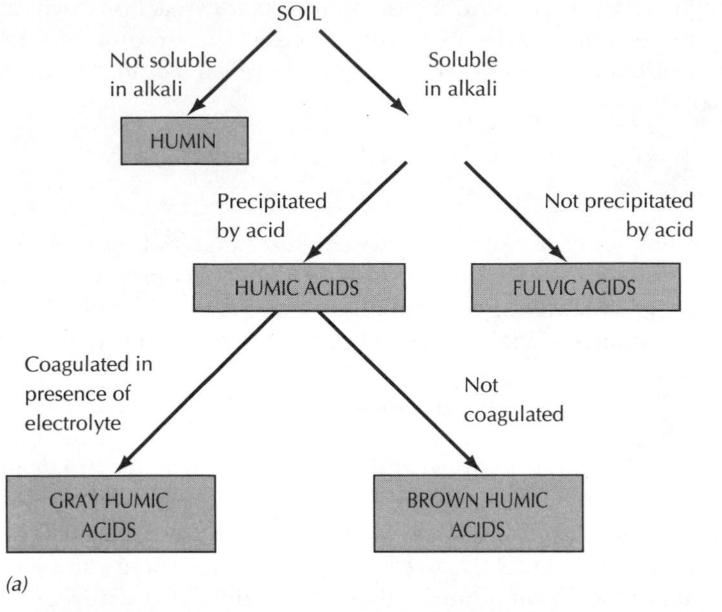

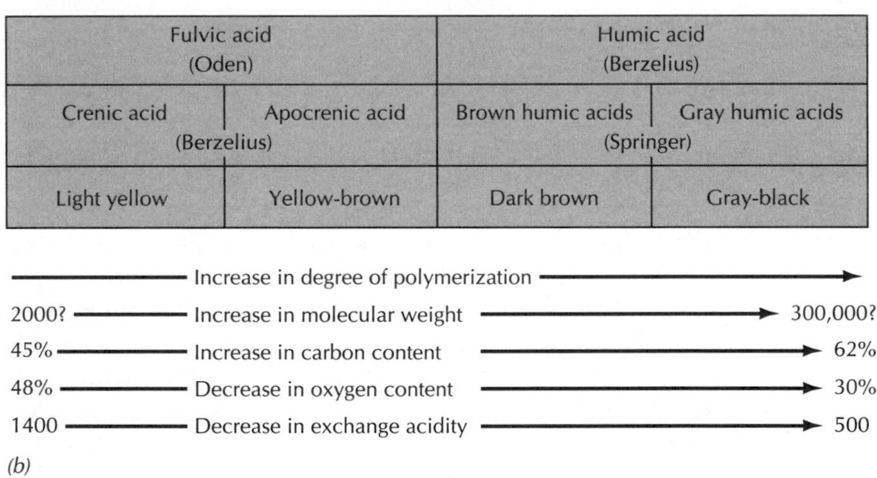

FIGURE 12.8 Classification and characteristics of the components of soil organic matter. (*a*) Methods used to separate humic acids, fulvic acids, and humin. (*b*) Chemical characteristics of different fractions. (Stevenson 1985)

Using several traditional and modern analytical techniques, different researchers have developed different model compounds that may represent the basic structure of humic and fulvic acids (Figure 12.9). These all show a high concentration of phenolic units but vary considerably in the type and number of side chains and in whether the nitrogen is present in the ring structures themselves or as components of side chains. Some do not include nitrogen in the structure at all, assuming that the nitrogen is present as proteins condensed with or associated with the humic and fulvic acids.

(a) Humic acids

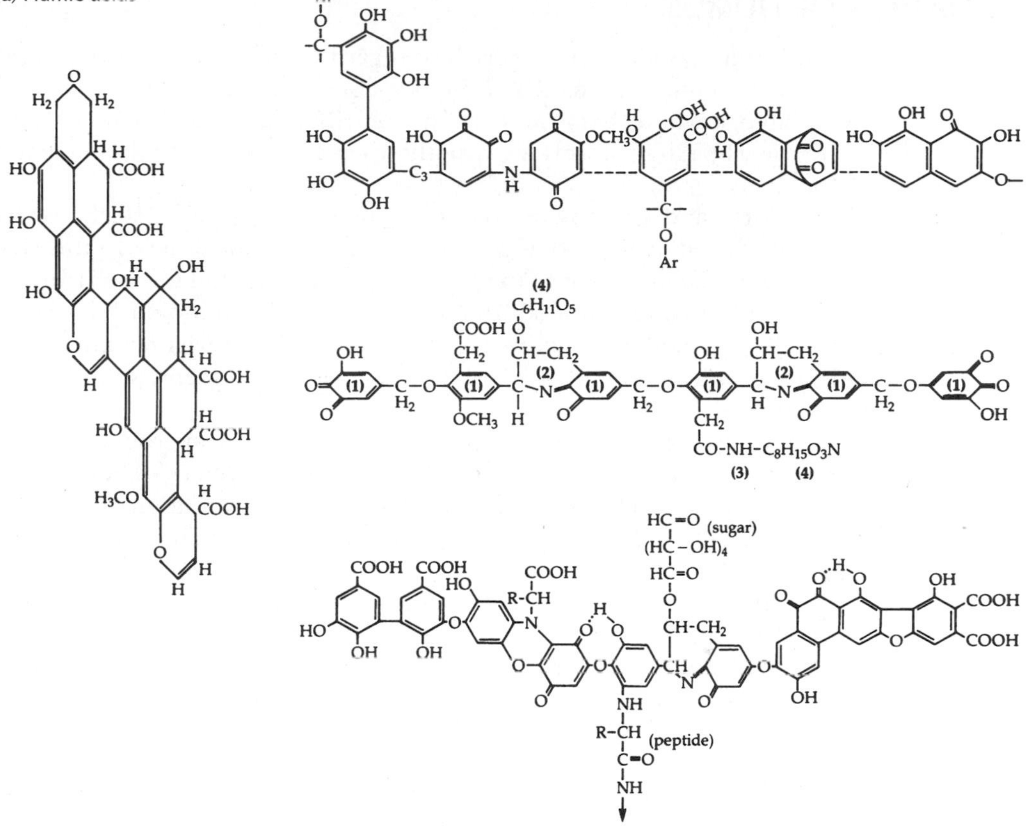

(b) Fulvic acids

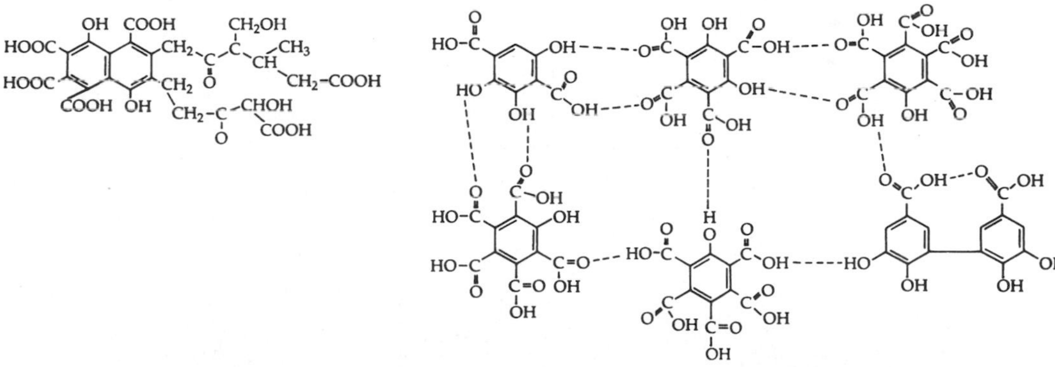

FIGURE 12.9 Proposed chemical structures of humic and fulvic acids. (Cited in Stevenson 1985)

FORMATION OF HUMUS

There is still considerable uncertainty regarding the types of compounds from which humus is formed and the reactions involved. Even the extent to which biological and chemical reactions contribute the generation of humus is unclear. The range of working hypotheses can be summarized as in Figure 12.10.

The first, and perhaps the oldest, hypothesis (Figure 12.10, I), is that humus is formed by the modification of existing plant residues. In this view, lignin is a crucial precursor to the formation of humus, and humus is formed by the slow but continual modification of the initial lignin molecules by microbes. These modifications include condensation into larger and larger molecules, with nitrogen added to the molecule by condensation of lignin with proteins.

The opposing hypothesis (Figure 12.10, III) is that microbes break all large molecules into smaller molecules that then repolymerize chemically to form the high molecular weight humic and fulvic acids. Variations on this theory relate to the extent to which repolymerization occurs within microbial cells.

The third hypothesis is intermediate between the first two. It holds that phenolic substances originating either directly from plant litter or from microbial synthesis are converted first to quinones (Figure 12.10), which then polymerize either with nitrogen containing compounds or other carbon compounds.

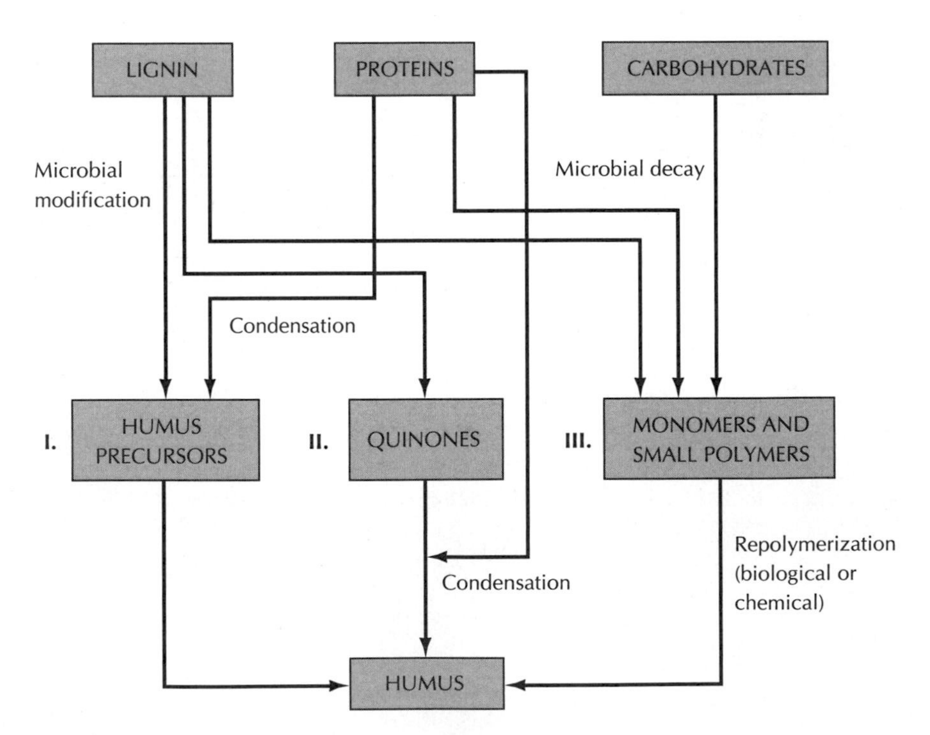

FIGURE 12.10 Summary diagram of hypothesized pathways for humus formation. (After Schnitzer 1978, Melillo et al. 1989)

There is general agreement that all three of the proposed mechanisms may, in fact, be operating simultaneously. For example, there is recent evidence that microbes can synthesize quinines and excrete them into the soil environment. However, two lines of evidence suggest that pathway III may not be the dominant one in most systems. First, this pathway predicts that humic acids are the first substances formed and that the smaller fulvic acids arise from the decomposition of humic acids. Mechanisms I and II would predict that fulvic acids are formed first, with humic acids resulting from continued condensation of the lower molecular weight fulvic acids. Radiocarbon dating (see discussion in Chapter 5) indicates that the carbon in humic acids is generally older than in fulvic acids, suggesting that humic acids are formed from fulvics, rather than the other way around (see Table 13.2).

Second, a 10-year experiment involving the additions of different kinds of plant litter to a sandy soil showed that the accumulation of soil organic matter increased with higher lignin concentration in the added material (Figure 12.11). This also suggests that lignin is the major precursor to soil organic matter, supporting hypotheses I and II.

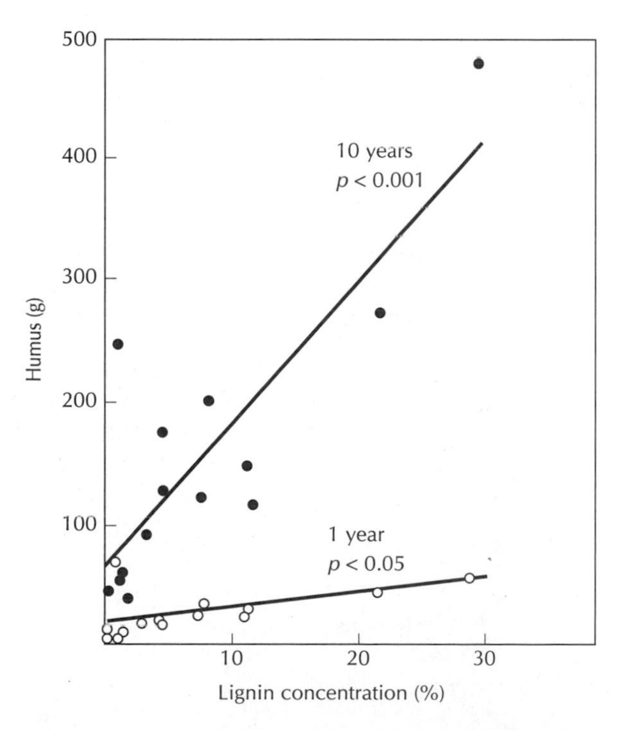

FIGURE 12.11 Effects of 1-year and 10-year additions of plant materials with different concentrations of lignin on the accumulation of soil organic matter in a sandy soil. (de Haan 1977)

DECOMPOSITION AND STABILIZATION OF HUMUS

Humus decomposes very slowly under field conditions. Yet it is generally present in such large amounts that it dominates the chemical and biological dynamics of soils in most terrestrial ecosystems. There are at least two reasons why humus decays slowly.

First, the carbon compounds present are of low biochemical quality. As litter decomposition proceeds, the more easily decayed compounds (e.g., sugars, starches, and cellulose) are selectively degraded. Ligninlike compounds accumulate and may even be produced as a byproduct of the decay of simpler compounds. Thus, the concentration of lignin and ligninlike secondary products increases over time, reducing the potential energy yield and overall carbon quality of the material.

Second, organic matter tends to form colloids with mineral particles in soils. Figure 12.12 is a schematic of the ways in which humus molecules and clay particles can be bound together by metal cations, water, sugars, and other substances. This sort of interaction tends to disrupt the alignment of degrading enzymes with the organic humus molecules and reduces the effectiveness of those enzymes. In fact, enzymes produced by microbes for the purpose of decomposing organic matter can themselves become deactivated, or fixed, by humus molecules and clay particles.

Organic matter combined to different degrees with mineral particles can be separated by a process called density fractionation. This relies on the very different bulk densities (or weight per unit volume) of organic and mineral materials. Water has a bulk density of 1 g $\cdot$ cm^{-3}. Organic matter in soils has a much lower bulk density because of its loose, porous structure. Mineral particles in soils have bulk densities around 2.2. A soil colloid consisting of both

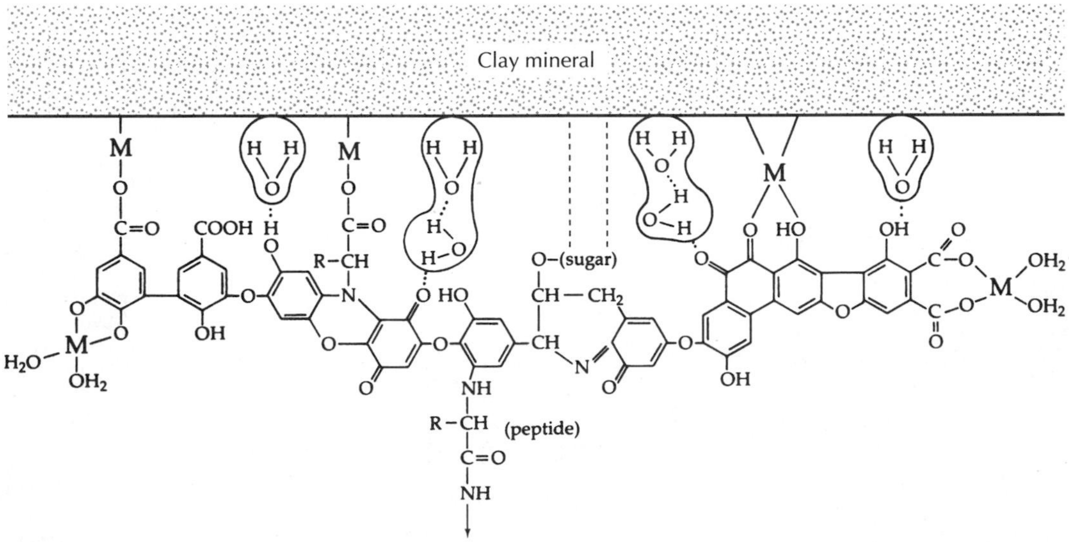

FIGURE 12.12 Schematic diagram of interactions between soil organic matter, clay particles, simpler organic compounds, and metal ions (M) in soils. (Stevenson 1985)

organic and mineral materials combined have a bulk density somewhere between the two, depending on the relative amount of each in the colloid. "Heavy" soil fractions would have higher bulk densities and more mineral matter. "Light" soil fractions would have low bulk densities and more organic matter. The two can be separated by suspending a soil sample in a solution of known bulk density; 1.65 is often used. The light soil fraction floats, the heavy fraction sinks.

In one example, separation of light and heavy soil organic matter has been carried out in several forests in the Pacific Northwest dominated either by conifers or by alder. The light and heavy fractions from these soils had very different carbon–nitrogen ratios and nitrogen release rates (Table 12.2). In general, the light fraction consisted of still-recognizable plant litter with small amounts of mineral material encrusted on it, while the heavy fraction was a combination of mineral particles and more decomposed organic fragments. The light fraction had a lower nitrogen concentration (higher carbon–nitrogen ratio) and much lower rates of nitrogen mineralization as measured by a laboratory incubation method. The light fraction averaged about 25% of the total soil organic matter through the soil profile but was a higher percentage of the total near the soil surface and also changed in total mass by as much as 50% from one season to the next, while the total mass of heavy fraction material remained relatively constant.

The same separation methods were also used on a series of soils of different ages. The sampled soils had developed on mudflows caused by volcanic activity on the flanks of Mt. Shasta in California. In these soils, the total amount of heavy fraction carbon increased continually with age, while the total amount of light fraction carbon tended to level off after about 600 years of soil development.

These characteristics all suggest that light fraction carbon is younger and less processed organic material and that organic matter becomes increasingly associated with mineral particles as decomposition proceeds. It may also suggest that once organic matter is heavily shielded by mineral particles, it is more difficult to decompose and remains in the ecosystem for a longer time.

TABLE 12.2 Carbon-to-Nitrogen Ratios and a Relative Index of Nitrogen Mineralization in Light and Heavy Fractions of Forest Soils Dominated by Conifers or Alder in the Pacific Northwest[a]

(From Sollins et al. 1984)

	Alder		Conifer	
	Light Fraction	**Heavy Fraction**	**Light Fraction**	**Heavy Fraction**
C: N Ratio	24.8	16.7	47.5	23.1
N mineralized (%)	0.71	2.34	0.39	2.91

[a]Values are means for forest type.

The amount of organic matter that can be stabilized in this way increases as soils become finer textured, as indicated by increasing nitrogen and phosphorus contents of soils (Figure 12.13). Smaller soil particles, such as clays, bind more effectively with organic matter and provide more particle surface area for organic–mineral interactions to occur. This form of stabilization is thought to be particularly important in tropical systems in which decomposition of unshielded organic matter is generally very rapid. Physical shielding may be a major reason why tropical soils contain at least as much organic matter as do temperate-zone soils with similar vegetation.

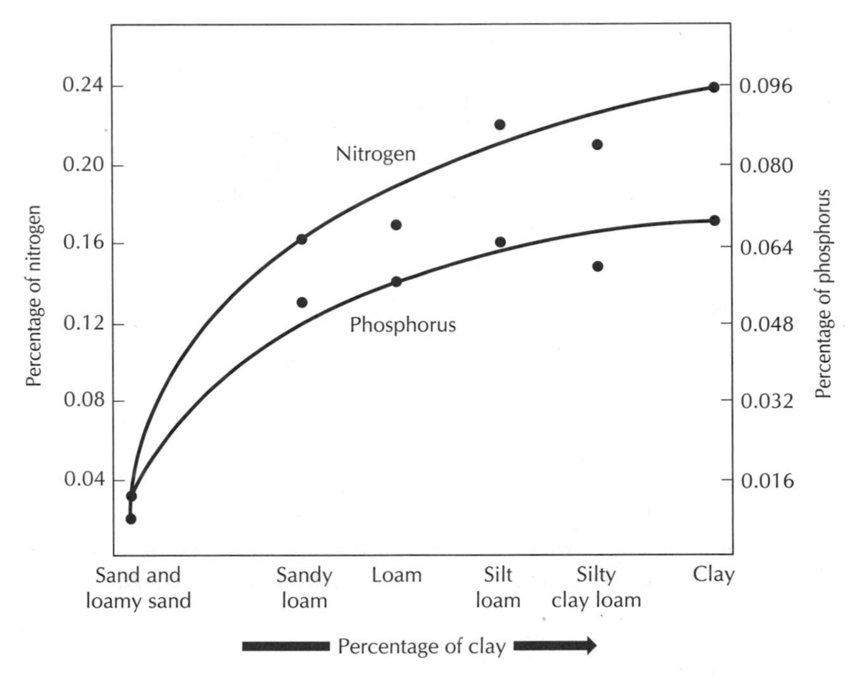

FIGURE 12.13 Differences in nitrogen and phosphorus concentrations in soils as a function of texture for several soils in New York State. (Foth and Turk 1972)

THREE EXAMPLES OF NEW APPROACHES TO "SEEING" THE STRUCTURE AND DYNAMICS OF SOIL ORGANIC MATTER

Humus remains something of a mysterious substance, even after more than a century of study. As new methods of analyzing the structure of organic materials are invented, they are applied to soils in order to provide a new look at this complex and critical material. This chapter concludes with three examples of more modern techniques that have been used to gain new insights into the structure and function of soil organic matter.

Absorption, Reflection, and Emission of Radiation

A host of techniques have been developed that determine the properties of different types of organic matter, or at least distinguish between them, based on interactions with radiation (light). Two classes of responses can be used as examples, absorbance/reflectance and emission or fluorescence.

The reflectance/absorbance characteristics are determined by exposing the material to a known intensity of light in different parts of the spectrum. Ultraviolet, visible, and infrared radiation are all used. The absorbance or reflectance of light is determined by the types of bond structures present. For example, the double, or unsaturated, carbon–carbon bonds common in phenolic compounds have a maximum absorbance at a different wavelength than does a saturated or single carbon–carbon bond. Bonds involving nitrogen and sulfur also absorb selectively in different wavelengths such that this method has been used to measure the concentrations of these elements in organic materials. Because of this, the spectra obtained from different constituents of plant materials, such as lignin, cellulose, and proteins show distinct peaks at characteristic wavelengths (Figure 12.14). These wavelengths can be used to determine the amount of each constituent in a sample of plant material.

Materials can also be induced to emit radiation at a different wavelength than that to which they are exposed, a process called fluorescence (the same process involved in those glow-in-the-dark toys). Energy absorbed by the material is stored temporarily by transitions of electrons to higher energy states. This energy is released over time as radiation. Generally, this release occurs simultaneously with the absorption of energy. In the case of those glow-in-the-dark toys, the energy is released over a much longer period of time.

Fluorescence spectra are generated and used in different ways. The material can be exposed to light in a single wavelength and the distribution of emitted light examined. Alternatively, emissions in a single wavelength can be measured as the exposure wavelength is varied.

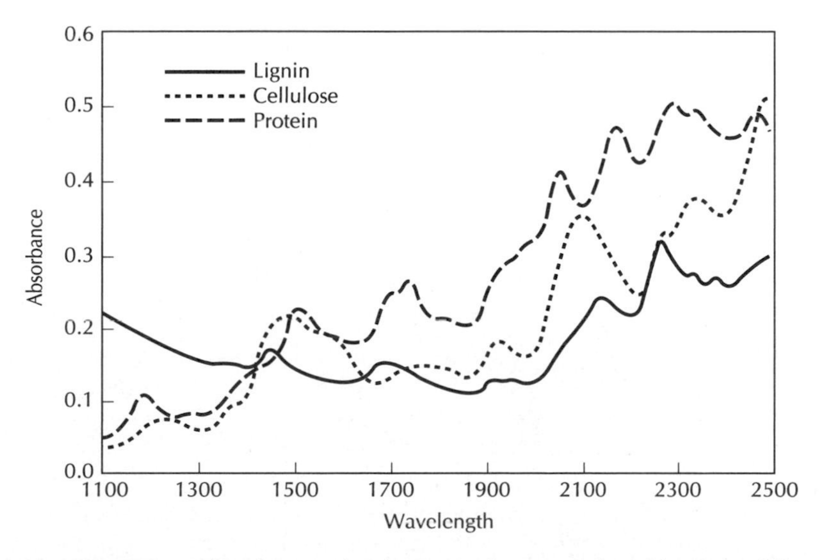

FIGURE 12.14 Absorbance of radiation in the near infrared by three different classes of materials found in plants. (Aber et al. 1994)

^{13}C and ^{15}N Nuclear Magnetic Resonance

Nuclear magnetic resonance (NMR), a new and extremely complex procedure, derives information about the types of molecular structures present in a material by exposing the nuclei of carbon and nitrogen atoms in different materials to a uniform and an oscillating magnetic field. The oscillating field is tuned until resonance is achieved between this field and the elemental nuclei being investigated. Under this condition, the nuclei emit a spectrum that is altered depending on the chemical environment around the nuclei, determined by the distribution of electrons.

For carbon, the spectrum obtained is a quantitative estimate of the importance of different bond types within the material (Figure 12.15). When combined with more traditional chemical procedures described earlier, different parts of the spectrum can be assigned to known bond types. Saturated and unsaturated carbon bonds can be discriminated, as well as finer differences determined.

For nitrogen, ^{15}N NMR has been used primarily to separate two different categories of products: (1) amino acid derivatives that are thought to originate from the incorporation of nitrogen into organic matter during biological processing (decomposition) and (2) heterocyclic compounds that are thought to result primarily from the chemical incorporation (condensation) of ammonium with organic materials. This distinction between biological and chemical immobilization of nitrogen has emerged as a crucial issue in nitrogen cycling in ecosystems.

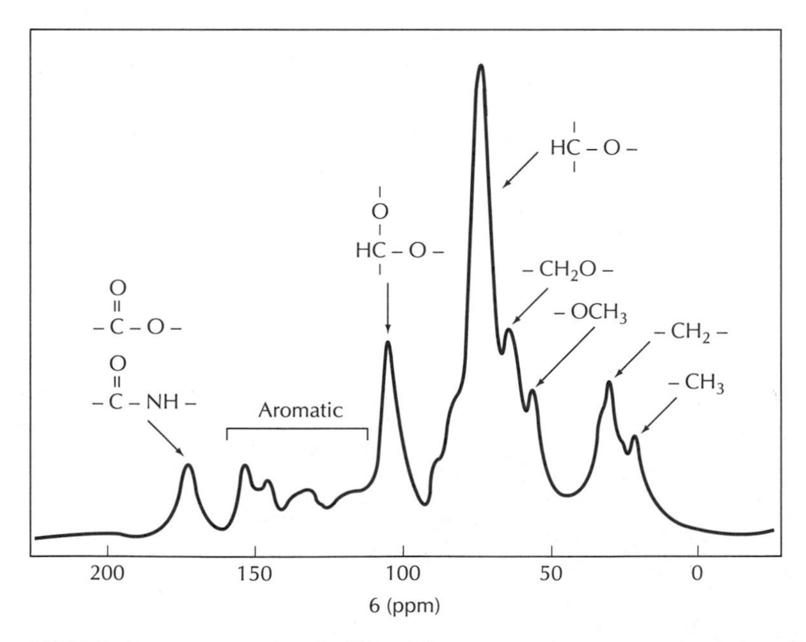

FIGURE 12.15 Example of a ^{13}C nuclear magnetic resonance spectrum for fresh litter material. Different peaks in the spectrum are associated with different types of carbon bonds, and the relative peak heights express the relative frequency of each bond type in the analyzed material. (Clinton et al. 1995)

^{15}N Distribution

Soil organic matter accumulates over very long periods of time, and its current distribution in a soil profile is the result of constant reprocessing by microbes, recombination by chemical reactions, physical movement by soil animals, disturbances such as tree falls, and movement in dissolved form with water. The spatial distribution of nitrogen isotopes in organic matter can yield insights into the relative age of organic matter at different depths.

When reactions involving nitrogen do not use all of the substrate present (for example, if plant uptake uses only some, but not all, of the available ammonium), there is a tendency for the lighter isotope to react more rapidly and to be taken up or combined more rapidly than the heavier isotope (see description in Chapter 5). This process is called fractionation and results in different distributions of lighter and heavier isotopes throughout the system. So, as nitrogen cycles from soils to plants to litter and back to soils, the lighter ^{14}N isotope is preferentially taken up and recycled, while the heavier ^{15}N isotope is left in the soil. As this same process also applies to microbial uptake of nitrogen, the heavier isotope also tends to remain in soil organic matter longer. In humid forests, the combination of this process with the general downward movement of organic matter over time by leaching of dissolved organics results in a distinct distributional pattern of nitrogen isotopes (e.g., Figure 5.2), suggesting that the age of organic matter increases with depth. As plant uptake is an important fractionating process, plants generally contain less of the heavier ^{15}N isotope than do soils.

REFERENCES CITED

Aber, J. D., K. L. Bolster, S. D. Newman, M. Soulia, and M. E. Martin. 1994. Analyses of forest foliage: II. Measurement of carbon fraction and nitrogen content by end-member analysis. *Journal of Near Infrared Spectroscopy* 2:15–23.

Clinton, P. W., R. H. Newman, and R. B. Allen. 1995. Immobilization of ^{15}N in forest litter studied by ^{15}N CPMAS NMR spectroscopy. *European Journal of Soil Science* 46:551–556.

de Haan, S. 1977. Humus, its formation, its relation with the mineral part of the soil and its significance for soil productivity. In *Organic Matter Studies,* volume 1. International Atomic Energy Agency, Vienna.

Foth, H. D., and L. M. Turk. 1972. *Fundamentals of Soil Science.* John Wiley & Sons, New York.

Hodson, R. E., R. R. Christian, and A. E. Maccubbin. 1984. Lignocellulose and lignin in the salt marsh grass *Spartina alterniflora:* Initial concentration and short-term post-depositional changes in detrital matter. *Marine Biology* 81:1–7.

Larsson, K., and E. Steen. 1988. Changes in mass and chemical composition of grass roots during decomposition. *Grass and Forage Science* 43:173–177.

McClaugherty, C. A., J. Pastor, J. D. Aber, and J. M. Melillo. 1985. Forest litter decomposition in relation to soil nitrogen dynamics and litter quality. Ecology 66:266–275.

Melillo, J. M., J. D. Aber, A. E. Linkins, A. Ricca, B. Fry, and K. J. Nadelhoffer. 1989. Carbon and nitrogen dynamics along the decay continuum: Plant litter to soil organic matter. In Clarholm, M., and L. Bergstrom (eds.), *Ecology of Arable Land.* Kluwer Academic Publishers, Dordrecht, The Netherlands.

Morrison, I. M. 1980. Changes in the lignin and hemicellulose concentration of ten varieties of temperate grasses with increasing maturity. *Grass and Forage Science* 35:287–293.

Schnitzer, M. 1978. Humic substances: Chemistry and reactions. In Schnitzer, M., and S. U. Khan (eds.), *Soil Organic Matter.* Elsevier Scientific Publishing Company, Amsterdam.

Sollins, P., G. Spycher, and C. A. Glassman. 1984. Net nitrogen mineralization from light- and heavy-fraction forest soil organic matter. *Soil Biology and Biochemistry* 16:31–37.

Stevenson, F. J. 1985. Geochemistry of soil humic substances. In McKnight, D. M. (ed.), *Humic Substances in Soil, Sediment and Water: Geochemistry, Isolation and Characterization.* John Wiley & Sons, New York.

13

DECAY RATES AND NUTRIENT DYNAMICS OF LITTER AND SOIL ORGANIC MATTER

INTRODUCTION

Chapter 12 introduced the major structural characteristics of litter and soil organic matter and discussed both the relative decay rates of different fractions and mechanisms for stabilization of organic matter over time. These general patterns occur at different rates under different environmental conditions, while litter quality, soil conditions, and other environmental factors also affect the rate at which nutrients are incorporated into, or released from, decaying organic materials.

The purpose of this chapter is to discuss how environmental factors alter decay rates and the end products of decomposition, as well as the nutrient dynamics associated with this process. Because different methods have generally been used to study fresh litter and older soil organic matter, this discussion is separated into these two categories. Two interesting questions arise out of this distinction: When does litter become soil organic matter?; and, What determines the fraction of a given litter material that is transformed into more stable, more slowly decaying soil organics?

LITTER DECOMPOSITION RATES

How Decay Rates Are Measured

Freshly fallen leaf litter is easily collected and sorted, and decomposition rates under field conditions can be measured by enclosing these materials in cloth or fine-mesh nylon bags and replacing them on-site. As a result, there is quite an accumulation of decomposition data from which general conclusions about this process can be drawn.

Patterns of Weight Loss with Time

Decomposition rates for litter have often been described by a constant percentage weight loss per unit time. This yields a curvilinear pattern for the percentage of original weight remaining that can be fit to an exponential equation:

$$\% \text{ original remaining} = e^{-kt}$$

where t = time, and k = litter-specific decay rate constant. This equation should look familiar. It is the same one used to describe light extinction in forest canopies as a function of cumulative leaf area index (Chapter 8, Figure 8.1). Much of the initial work on litter decomposition covered only the first few years of the process, and most of the data presented were summarized using this equation and deriving a statistical estimate of the k value, or decomposition rate, of different materials.

The concept of a constant decay rate for a given litter type is an integral part of the standard approach to litter decomposition but is only indirectly related to the actual decay process. The constant k value assumes that litter decay rates do not change through time. It also implies that all substrates within a litter type are processed simultaneously. This is not the case. Four phenomena demonstrate some of the actual complexity of the decomposition process.

First, significant weight loss can occur by the purely physical process of leaching as rain or melting snow wash over the litter. Low molecular weight sugars, polyphenols, and amino acids can be lost in this way. The total weight loss by leaching depends on the amount of water-soluble material in the litter and the amount of water passing over it.

Second, the constituents that are not water soluble do not all decompose simultaneously. The simpler carbon compounds, which can be degraded rapidly and yield the most net energy, are attacked preferentially. So, when sugars and free cellulose are available, enzymes for the degradation of lignin tend not to be produced. Lignin decomposition does not begin until more easily decomposed materials are nearly exhausted, at least at local microsites, within the material. Because ligninlike substances (ones that appear in the lignin fraction of proximate analyses; see Figure 12.4) are produced as byproducts of the decay process, the total amount of measured lignin in a decaying material can increase in the early stages of decay.

Third, lignin, and probably suberin in roots, is so intimately entwined with the cellulose strands in most materials that it effectively "shields" much of the cellulose from microbial attack. After the nonshielded cellulose is decomposed, lignin must be degraded to allow microbial access to the remaining cellulose.

Finally, both chemical and physical changes in the state of organic matter occurring during the decay process produce modifications in the compounds present and interactions with mineral material that render the remaining material very resistant to decay. Said in another way, decomposition never goes to completion but slows considerably from the initial rates of mass loss predicted by fitting a k value to the first months or years of data. K values continue to be used mainly for short-term decomposition studies.

The effect of these processes can be summarized in a generalized view of the rates of disappearance of different classes of compounds (Figure 13.1). The simple extractable compounds are both leached and decomposed very quickly to low levels. The values never reach zero, however, either because small amounts are stabilized by interactions with mineral or organic materi-

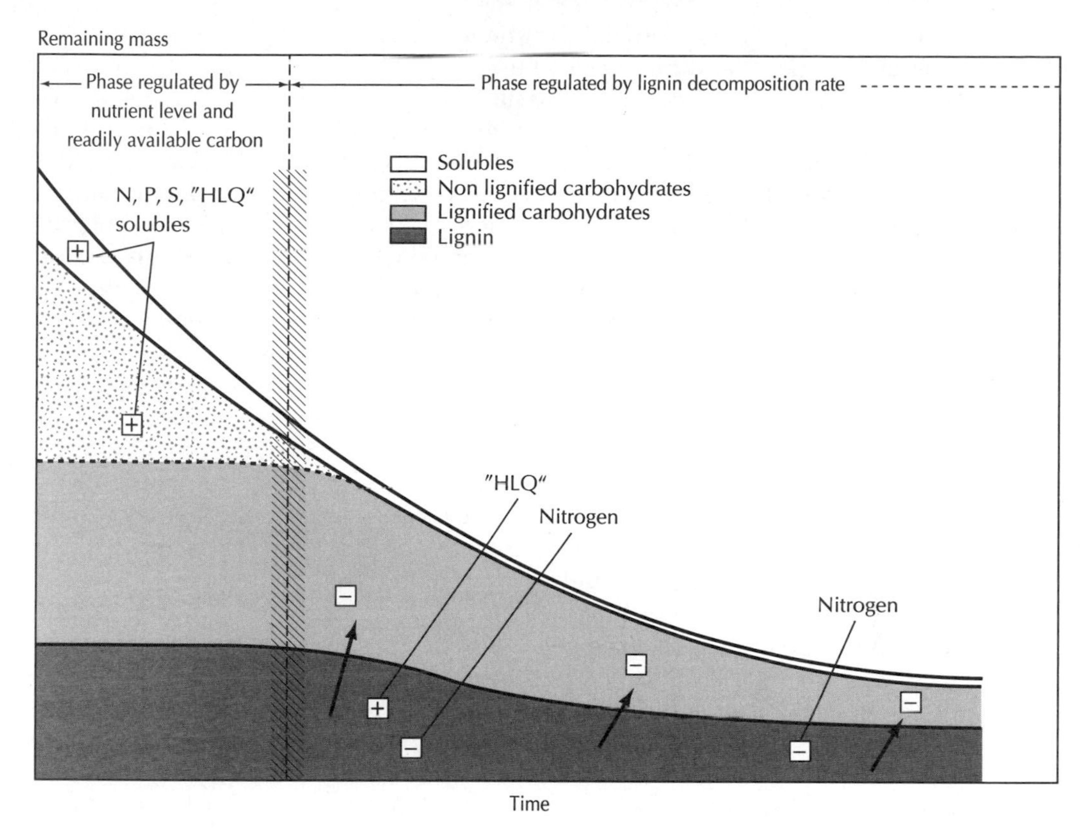

FIGURE 13.1 Generalized pattern of mass loss by carbon fraction during decomposition and associated interactions with carbon quality and nutrient availability. Soluble compounds and nonlignified cellulose decay rapidly, and decomposition is increased by the addition of nutrients (nitrogen, phosphorus, and sulfur [N.P.S.]) and a higher holocellulose–lignin quotient (HLQ), or the ratio of cellulose to cellulose + lignin). In later stages, mass loss is determined by the decay rate of lignin, and the higher the lignin content relative to cellulose (i.e., the lower the HLQ), the slower decomposition occurs. Decomposition is also inhibited by high nitrogen concentrations in contrast to its enhancing effects early on. (Berg 1986)

als or because a small amount is always present within the populations of microbes intimately associated with the decomposing litter. Simple sugars also decay quickly. Nonlignified cellulose, that portion of the total that is not shielded by lignin, decays at an intermediate rate. Up to this point, lignin decay is negligible, and formation of ligninlike compounds by the decay process may cause actual increases in total lignin content. Beyond this point, lignin and cellulose are degraded slowly and simultaneously, and the energy derived from the decay of cellulose is used to attack the associated lignin molecules.

Nutrient Dynamics During Litter Decomposition

The rate of weight loss from decomposing litter is an indication of the amount of energy and carbon available to microbes for growth. As discussed briefly in Chapter 12, microbes also require nutrients in given concentrations to build cells. For microbial growth to proceed at the rate allowable by available energy and carbon in a litter material, nutrients must also be available from either the decomposing substrate or from the surrounding soil solution. Suboptimal nutrient availability to microbes reduces decomposition rates of the litter and may also lead to removal of available nutrients from the soil solution through microbial uptake. Conversely, if nutrients are present in litter in excess of microbial requirements (microbial activity limited by carbon quality rather than nutrient availability), then those nutrients are released as decomposition proceeds.

Figure 13.2 shows weight loss (summarized in an exponential decay curve) and nutrient dynamics accompanying decomposition of Scots pine needle litter in a forest in Sweden. While the needles are of only moderate carbon quality and decay relatively slowly, they are also very low in content of several nutrients. Microbial growth on these needles creates demands for nitrogen, phosphorus, and sulfur that are greater than the supply within the litter material. In Figure 13.2, the net increase in the total content of these nutrients within the decomposing litter results from microbial uptake from outside of the leaf material and incorporation into the litter.

This net increase in the absolute amount (not just the concentration) of nutrients in decomposing litter is called net immobilization. It actually decreases the amount of nutrients available for plant growth. Plants and microbes growing in the same soil can be competing for available nutrients. This is why materials such as fresh sawdust and wood chips can greatly reduce plant growth when used in gardens and planting mixtures; they are rich in high-quality carbon (cellulose) and very low in nutrient concentration.

The rate of nutrient immobilization in the Scots pine litter is highest in the earliest stages of decay, when the most easily decomposed compounds are being degraded and microbial growth is most rapid. After about 1.5 years in the field, the concentrations of nitrogen and phosphorus in the partially decayed litter have increased substantially, and the decomposition rate has decreased. As the carbon and energy yield from decomposition declines, so does the microbial demand for nutrients. Nutrient release from the decaying material, and also from dead and now decaying microbial tissues, is greater than the continuing microbial demand, and the excess nutrients are released in mineral form (e.g., ammonium and sulfate). This nutrient release is called net mineralization and results in a net reduction in the absolute amount of nutrients in the litter and increased availability to plants.

Where do the immobilized nutrients come from? Two sources are most likely. For tissues decaying on or above the soil surface, throughfall is the most likely source. For tissues decaying below ground, microbes have access to nutrients mineralized from older litter and soil organic matter. By taking up these nutrients, microbes compete with plants for what might be an important limiting resource. A third potential source for nitrogen is fixation by free-living microbes living within litter. However, the level of free-living nitrogen fixation (as opposed to symbiotic nitrogen fixation; see Chapter 10) is generally low in most terrestrial ecosystems.

The patterns discussed here for nutrient immobilization and mineralization are fairly typical. Nitrogen and phosphorus tend to be the two nutrients

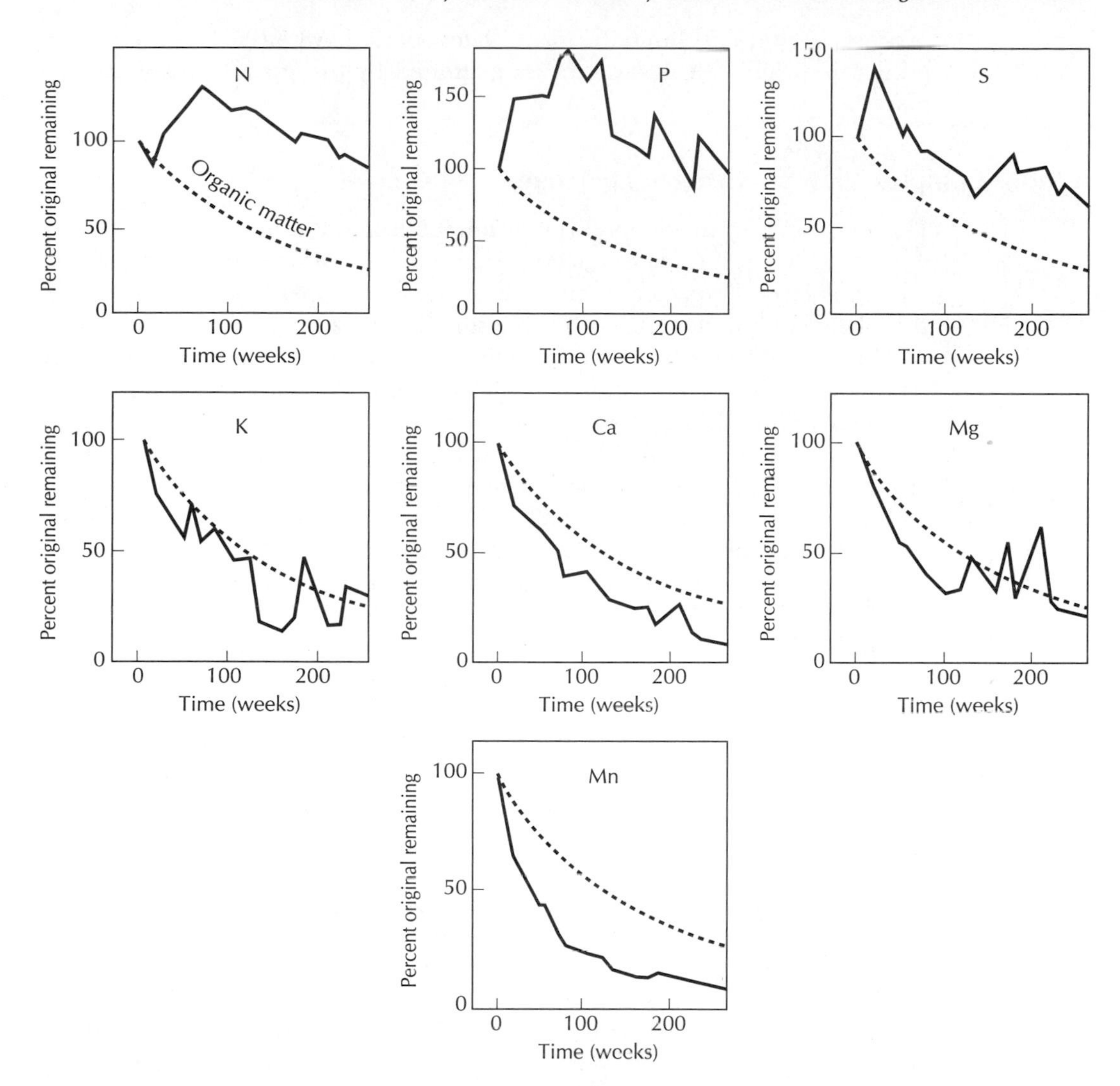

FIGURE 13.2 Rates of nutrient immobilization and mineralization from decomposing Scots pine litter over a 5-year period. (Staff and Berg 1982)

most likely to be immobilized. However, other elements may be immobilized in materials where they are present in low concentrations. The important principle is the supply of the element from decaying litter compared with the nutrient demands of the growing microbes.

When does nutrient mineralization begin? According to our discussion of microbial resource requirements, it should be when carbon quality declines to the point that nutrient availability is no longer limiting to microbial activity. There is evidence that mineralization of nitrogen in nitrogen-limited ecosystems begins at the same time that lignin begins to be decomposed. Both the beginning of nitrogen mineralization and the initiation of lignin decay would indicate exhaustion of relatively high-quality carbon compounds. In Figure 13.2, decay is separated into an initial period, in which immobilization of limiting nutrient(s) occurs and the decay process is nutrient limited, and a later period, when all nutrients are mineralized and the decay process is limited by the carbon quality of the material.

A More Complex View of Nitrogen Dynamics

The net fluxes of nutrients through immobilization and mineralization described earlier result from much more rapid and complex exchanges at the microbial level. Two research projects using ^{15}N isotopes begin to display the complexity of interactions underlying the simple patterns presented earlier.

In the first study, Scots pine needle litter experimentally enriched in ^{15}N was decayed under both laboratory and field conditions. Three insights into nitrogen cycling during decay come from this study. First, the total ^{15}N content of the litter materials changed little during the immobilization phase (Figure 13.3*a*). This shows that nitrogen originally present in litter is used and retained by microbes, while additional nitrogen is acquired from throughfall or the soil solution. Second, the amount of ^{15}N in the acid-insoluble, or "lignin," fraction increased during the immobilization phase, as did total lignin content (Figure 13.3*b*). Third, net release of nitrogen, and the release of ^{15}N as well, was nearly undetectable until the acid-insoluble content of the material began to decrease. This last result suggests that net nitrogen immobilization occurs in the early stages of decay, when high-quality substrates, such as sugars and cellulose, are being decayed, leading to high nitrogen demand in a growing microbial population, while net mineralization occurs when only low-quality substrates remain, and microbial growth is limited by the energy that can be gained from the material, and not by nitrogen.

In the second study, Scots pine litter materials in different stages of decay and having different nitrogen concentrations were placed in a sterile solution with different concentrations of ^{15}N-labeled ammonium. Despite a complete lack of biological processes, the litter material increased significantly in total nitrogen and ^{15}N content (Figure 13.4). The amount of ^{15}N incorporated increased with higher concentrations of ammonium in the solution, and decreased with increasing nitrogen content in the litter material. This result suggests that a significant amount of immobilization can occur through strictly chemical reactions between litter and the soil solution.

An additional study demonstrates the potential importance of soluble organics in controlling nutrient balances. On a set of sites in California supporting pygmy forests growing on extremely poor soils, a relationship was detected between the quantity of soluble phenols (or tannins) in litter and the amount of nitrogen released in mineral (ammonium) versus organic form from decaying litter (Figure 13.5). It has been hypothesized that selective pressure to compete for very small quantities of available nitrogen has led to the presence of plants that produce leaf litter high in tannins, which condense with nitrogen and release nitrogen in dissolved organic form, rather than as ammonium, and that also have mycorrhizal root systems capable of taking up these same dissolved organic compounds. The effect of this is to "short-circuit" the nitrogen cycle, minimizing leaching losses and increasing the fitness of species with both high tannin content in litter and this particular type of mycorrhizal association. Even outside of these very poor systems, a potentially large fraction of weight loss in "decaying" litter can occur as leach-

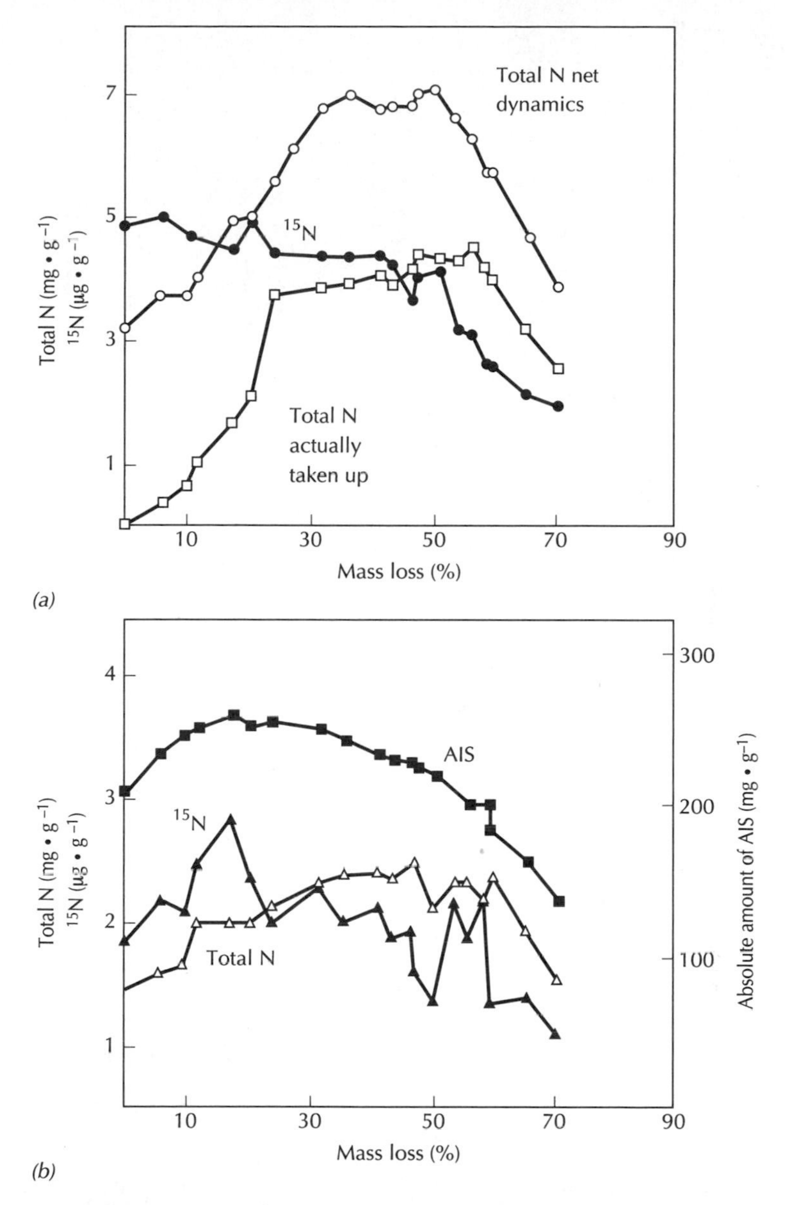

FIGURE 13.3 Gross fluxes of nitrogen in decomposing litter as determined using ^{15}N enriched material. (*a*) ^{15}N losses are minimal during the period of net nitrogen immobilization and increase during net nitrogen mineralization. (*b*) ^{15}N content in the acid-insoluble, or "lignin," fraction increases during the net nitrogen immobilization phase and decreases along with the mass of lignin during the net nitrogen mineralization phase. AIS = Acid Insoluble Fraction. (Berg 1988)

ing of soluble organics (Figure 13.6). What is the fate of this leached material? Does it decay rapidly or become stabilized as long-term organic matter in soils colloids? We do not know the answers to these questions.

Finally, it has become clear that nitrogen content in decaying organic materials plays very different roles at different stages in the decay process. The short-term studies cited earlier support the notion that high initial nitrogen

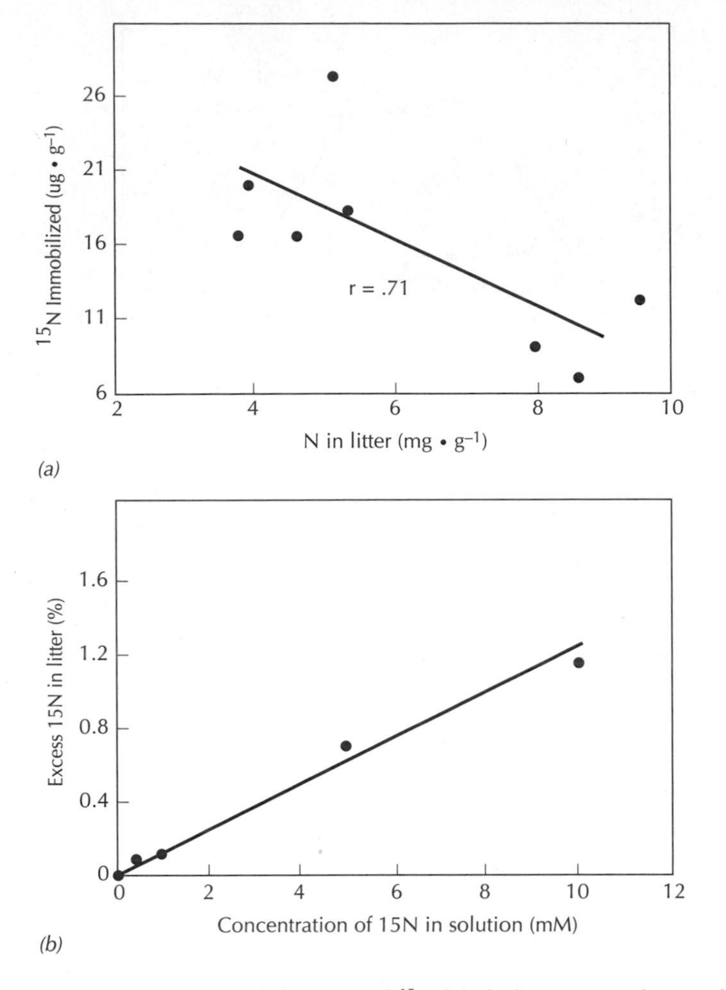

FIGURE 13.4 Immobilization of ^{15}N-labeled ammonia from solution into sterilized Scots pine needle litter in different stages of decay. (*a*) Immobilization decreases with nitrogen content in litter. (*b*) Increases with nitrogen concentration in the solution. (Axelsson and Berg 1988)

content is important in meeting microbial demand for this nutrient and maintaining decay rates. However, studies of CO_2 loss from humus show this to be inversely related to nitrogen content (Figure 13.7*a*), and a field study of leaf litter decay in stands receiving high doses of added nitrogen show depressed rather than increased rates of decomposition over the long term (Figure 13.7*b*). Two theories have been advanced to explain these results. The first is that the addition of nitrogen compounds to organic matter, either by microbial processes or by chemical reactions, randomizes the bond structures in old litter and humus so that the enzyme systems designed to degrade known carbon structures become less effective (see Chapter 12). The second builds on research with specific fungal species that have been shown to shut down the production of lignin and humus degrading enzymes in the presence of large amounts of mineral nitrogen because the requirement for extracting nitrogen from humus to meet fungal nitrogen demands is removed. Both mechanisms may occur.

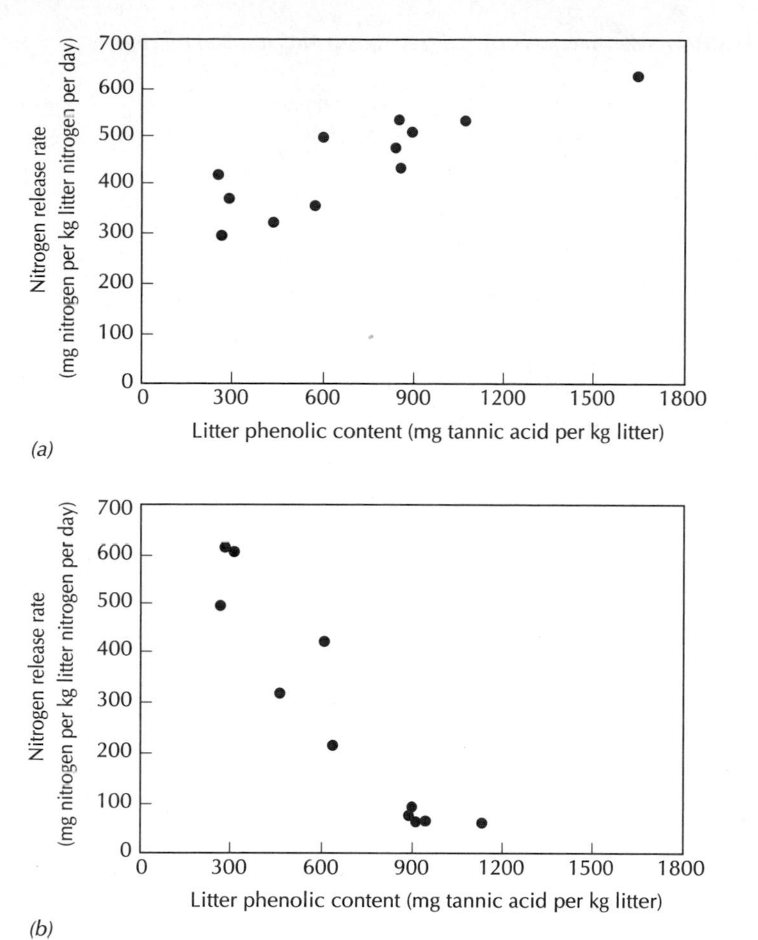

FIGURE 13.5 Changes in the form in which nitrogen is released from moderately decayed pine needles as a function of soluble phenolic ("tannin") content. As phenolic content increases, less of the released nitrogen is in inorganic (NH_4) form and more is as dissolved organic nitrogen. (Northrup et al. 1995)

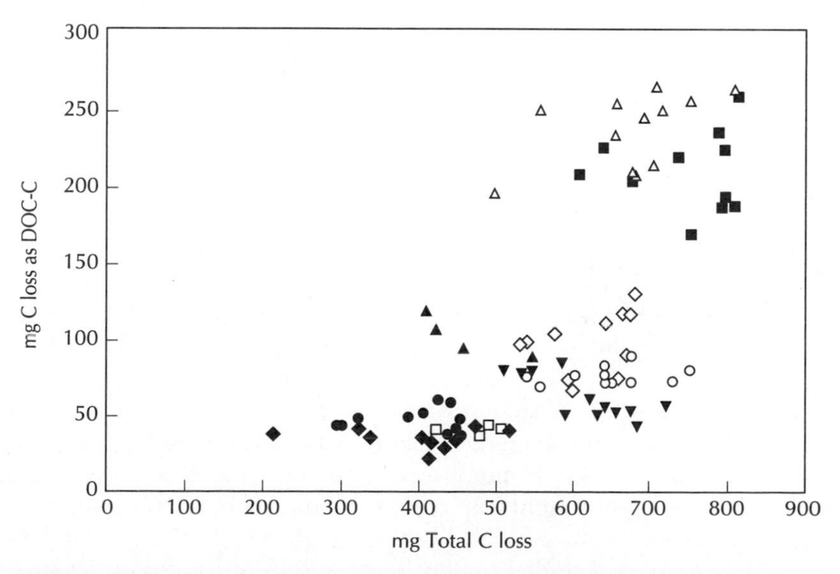

FIGURE 13.6 Dissolved organic carbon loss from decaying leaf litter of several species as a fraction of total carbon loss. The highest values represent nearly 30% of total carbon loss in dissolved form rather than as CO_2. (Magill and Aber 2000)

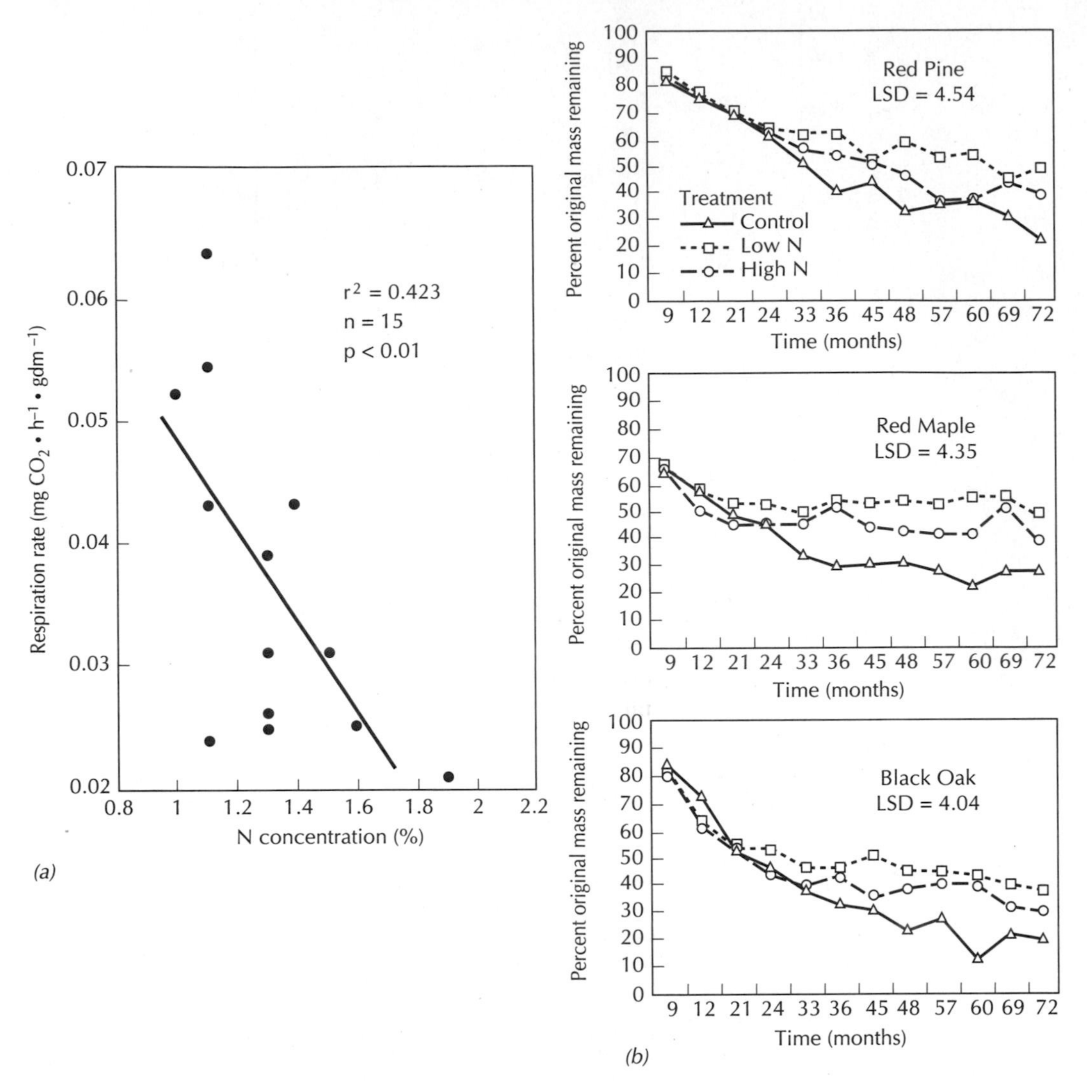

FIGURE 13.7 Effects of nitrogen content or nitrogen availability on decomposition rates. (*a*) Soil organic matter with elevated nitrogen content tends to release CO_2 at a lower rate. (Berg and Matzner 1997) (*b*) Chronic additions of nitrogen to litter decaying in the field reduces total rates of mass loss. (Magill and Aber 1998)

Predicting Litter Decay Rates

Given this understanding of the role of carbon and nutrient quality in litter decay and the large number of field studies completed, can we predict decay dynamics in a general way across different ecosystems? In short-duration studies, decay rates have been related to the initial ratio of lignin to nitrogen in the material (Figure 13.8*a*). In long-term studies, the role of nutrient content is much less important, and rates covering the entire period of rapid weight loss can be predicted by carbon quality alone (Figure 13.8*b*).

Tissues of similar quality decay at different rates under different climatic conditions. The two lines in Figure 13.8*a* represent two study sites: one in

New Hampshire and one further south in North Carolina. It is apparent that litter with similar initial quality as expressed by the lignin–nitrogen ratio decomposes faster at the North Carolina site because of the longer growing seasons and higher mean temperatures at this site.

One of the advantages of litter decay studies is that the materials used can be transported easily between sites. By exchanging different materials between a wide range of sites, the relative effects of climate and litter quality can be distinguished and general relationships developed for use in ecosystem models. One experiment in Europe placed litter materials with a wide range of initial lignin concentrations in sites ranging from the Arctic circle to central Europe. Results showed that Actual Evapotranspiration (AET), as a variable summarizing both cumulative temperature and relative water stress, explained much of the variation in weight loss (Figure 13.9) and that lignin concentration explained much of the variation within a site. Interestingly, lignin became less important as a predictor of decay rates under colder climates. This ability to predict the rate of a central ecosystem process when only a few important variables are known is important to the understanding and prediction of ecosystem function at the global level (see Chapter 26).

Unfortunately, not all materials offer as clear a picture of controlling variables as does leaf litter. In a different cross-site decay experiment conducted in the United States, seven above-ground litter types (including native grassland species and wheat in addition to forest leaf litter) and three root types were decomposed in a tundra site in Colorado. The above-ground litter behaved as described earlier, with decay rates increasing with nitrogen content and decreasing with lignin content. However, the root litter decay showed no effect of lignin and decreased with increased nitrogen content (Figure 13.10). Roots also showed no detectable mass loss after the first year in a 3-year study. In summary, root decay in this study appeared to have more in common with the decay of soil organic matter, as is discussed later, than with the decay of fresh leaf litter.

Decomposition of woody stems in forest ecosystems also varies significantly from the simple models presented earlier. Wood decays much more slowly than would be predicted by its carbon chemistry alone due mainly to the presence of chemical inhibitors in heartwood and to the time required for microbes to fully colonize the large mass of material. For example, the heartwood of temperate-zone conifers can contain any of several carbon-based decay inhibitors. During the tree's lifetime, these protect the structural integrity of the stem. After tree death, they impede decomposition.

Unlike foliage and roots, which offer a large amount of surface area to decomposers relative to their mass, large tree boles may require many years for complete microbial colonization. Figure 13.11 is an example of the effect of stem diameter on the percentage of stem wood colonized with time. Decay is delayed for decades at the center of large boles. Stem size also affects immobilization rates. Wood is very low in nutrient content, yet the interior of a decaying stem is largely isolated from sources of nutrients in either throughfall or soils. Nutrient additions to decaying wood have shown significant increases in decomposition rates.

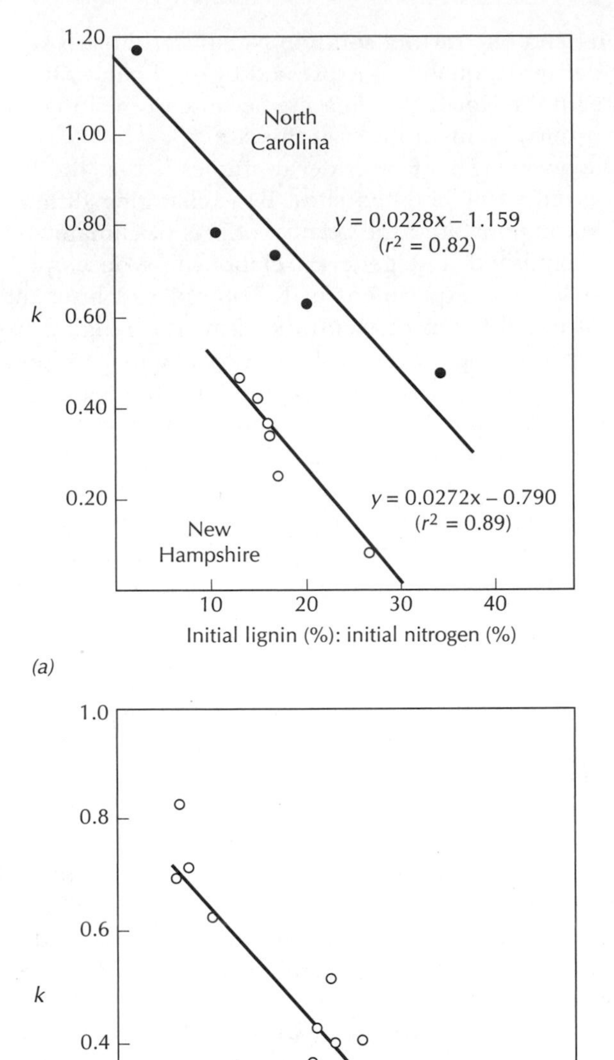

FIGURE 13.8 Decomposition constant k for foliar litter in relation to initial tissue chemistry. (*a*) First-year decomposition as a function of the lignin-to-nitrogen ratio. (Melillo et al. 1982, includes data from Cromack 1973) (*b*) Long-term decay as a function of initial lignin + cellulose concentration. (Aber et al. 1990)

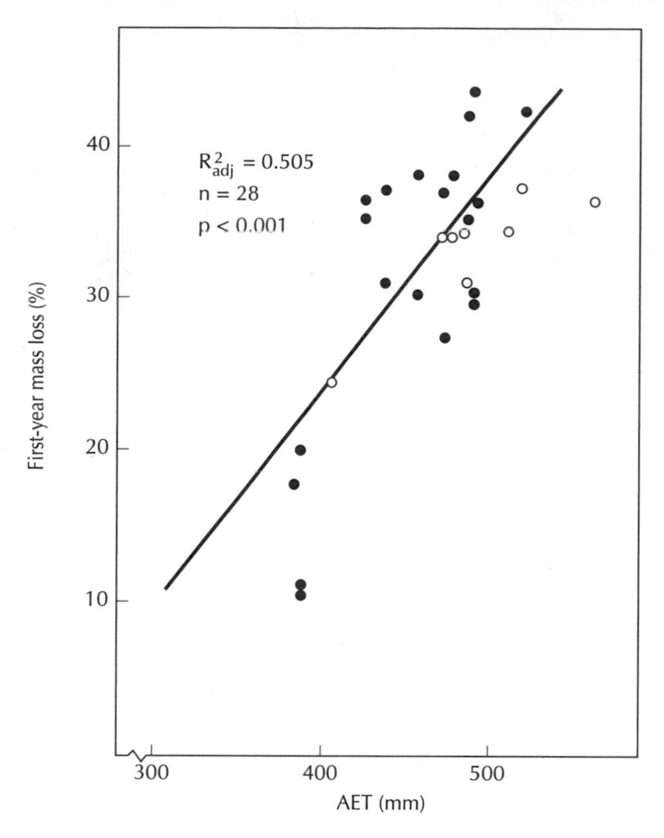

FIGURE 13.9 Summary relationships of first year decay rates as a function of AET for Scots pine litter placed in several locations along a mean annual temperature gradient in Europe. (Johansson et al. 1995)

Effects of Soil Animals on Decomposition

Most of the previous discussion has pertained to forest ecosystems because this is where most of the studies have been carried out. There are indications that things may be even more complicated in the drier ecosystems: grasslands, shrublands, and deserts. Particularly in deserts, much of the small amount of litter produced may be removed from the surface by animals and stored in burrows and tunnels. This greatly complicates the study of the fate and decomposition of litter!

In general, the role in decomposition played by larger animals, including organisms as common as earthworms, is relatively unknown, even in the better-studied humid regions. It is clear that earthworms in particular can move or ingest large amounts of material. Darwin, in an early and classic study of earthworms in field soils, calculated that earthworms deposited more than 24 tons of castings per year on a hectare of English meadow. There is a general relationship between the mass of earthworms present and the rate at

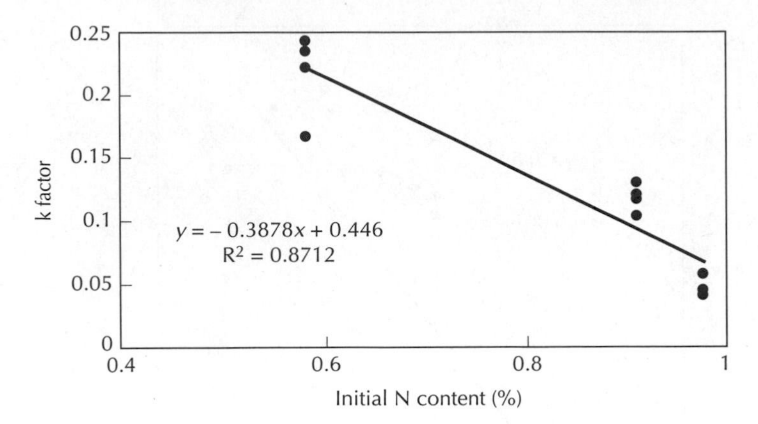

FIGURE 13.10 Decay of three different types of fine root materials in a common site in alpine tundra in Colorado as a function of initial nitrogen concentration. (Bryant et al. 1998)

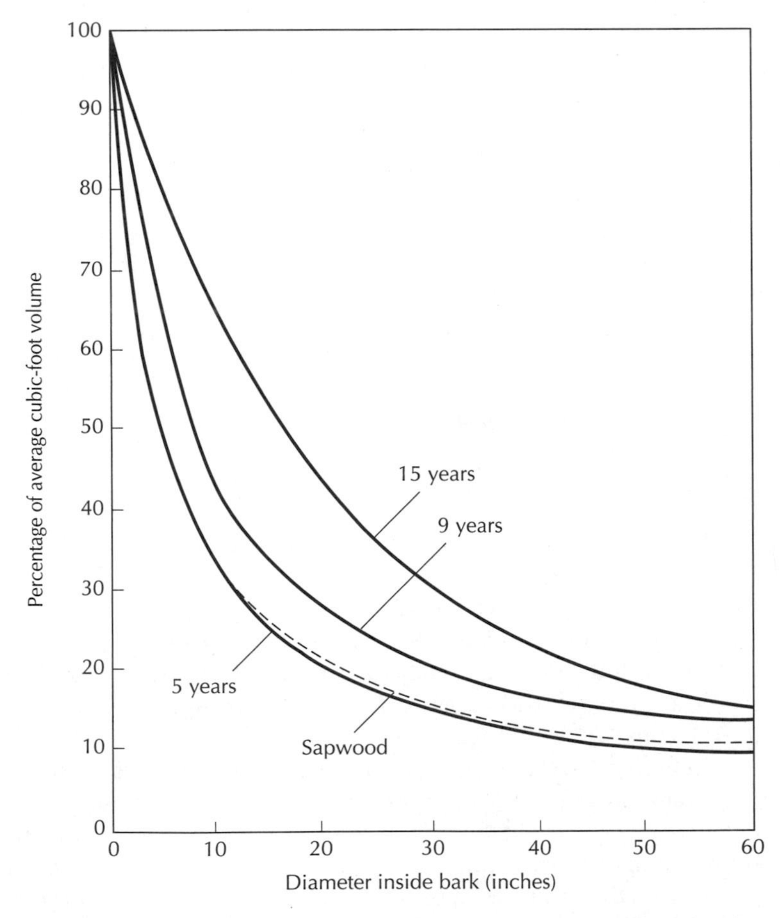

FIGURE 13.11 Percentage of colonization of Douglas-fir logs by decomposer organisms as a function of time and the diameter of the log. (Harmon et al. 1986)

which surface litter "disappears." However, disappearance is not decomposition. A leaf moved from the surface to an earthworm burrow has not changed biochemically, but it is in a very different environment. Whether it then decays differently is unknown. There have been a number of interesting speculations as to the very long-term effect of the biological invasion of the large European earthworm, *Lumbricus,* the species studied by Darwin, on soil structure in North America.

How Much Humus Is Produced by Litter Decay?

We have not yet described the end point of the litter decay process. Litter passes through periods of nutrient immobilization and mineralization, and periods of more and less rapid mass loss, but neither mass loss nor nutrient mineralization proceed to completion. When litter reaches the stage of very slow weight loss and becomes embedded in the surrounding soil matrix, it becomes the conceptual province of soil organic matter research. The boundary between litter and soil organic matter has not been clearly defined, either operationally or intellectually. The definition of this boundary and the transfer of litter to soils constitute a crucial process in ecosystem models, or in any quantitative description of the carbon and nutrient dynamics of soils. Consequently, these questions need to be asked: When does decayed litter become soil organic matter?; and, What fraction of initial litter fall is transferred to this very different organic matter pool?

One emerging characteristic of litter decay is that the process begins with a wide variety of materials of very different chemical quality and produces a much more homogeneous material as a product. For example, several different analyses suggest that the end product of litter decay, regardless of initial chemical quality, has a lignin–cellulose ratio (as measured by the proximate analyses in Figure 12.4) of about 1 : 1. Whether this ratio has real meaning in terms of the fundamental energetics of decomposition or is only a useful empirical result is unknown. The nutrient content of the organic matter produced by this "decay filter" varies considerably depending on the initial content of the litter and the availability of nutrients in the soil solution.

A number of attempts to determine this transfer to humus have been made using long-term litter decay data. In one study, the traditional exponential decay equation was modified to include a residual term or asymptote that specified the amount of very slowly decaying material remaining at the end of the decay process (Figure 13.12*a*). Applying this equation to a large range of litter types yields a wide range in predicted mass remaining when this very low rate of decay occurs. Interestingly, the amount remaining at this limit increased with initial nitrogen concentration in the material, supporting the notion that nitrogen inhibits decay in the later stages of the process (Figure 13.12*b*). This limit was also found to increase (more complete decomposition occurred) with higher initial calcium and manganese contents (Figure 13.12*c*), two elements that are involved in the microbial breakdown of lignin.

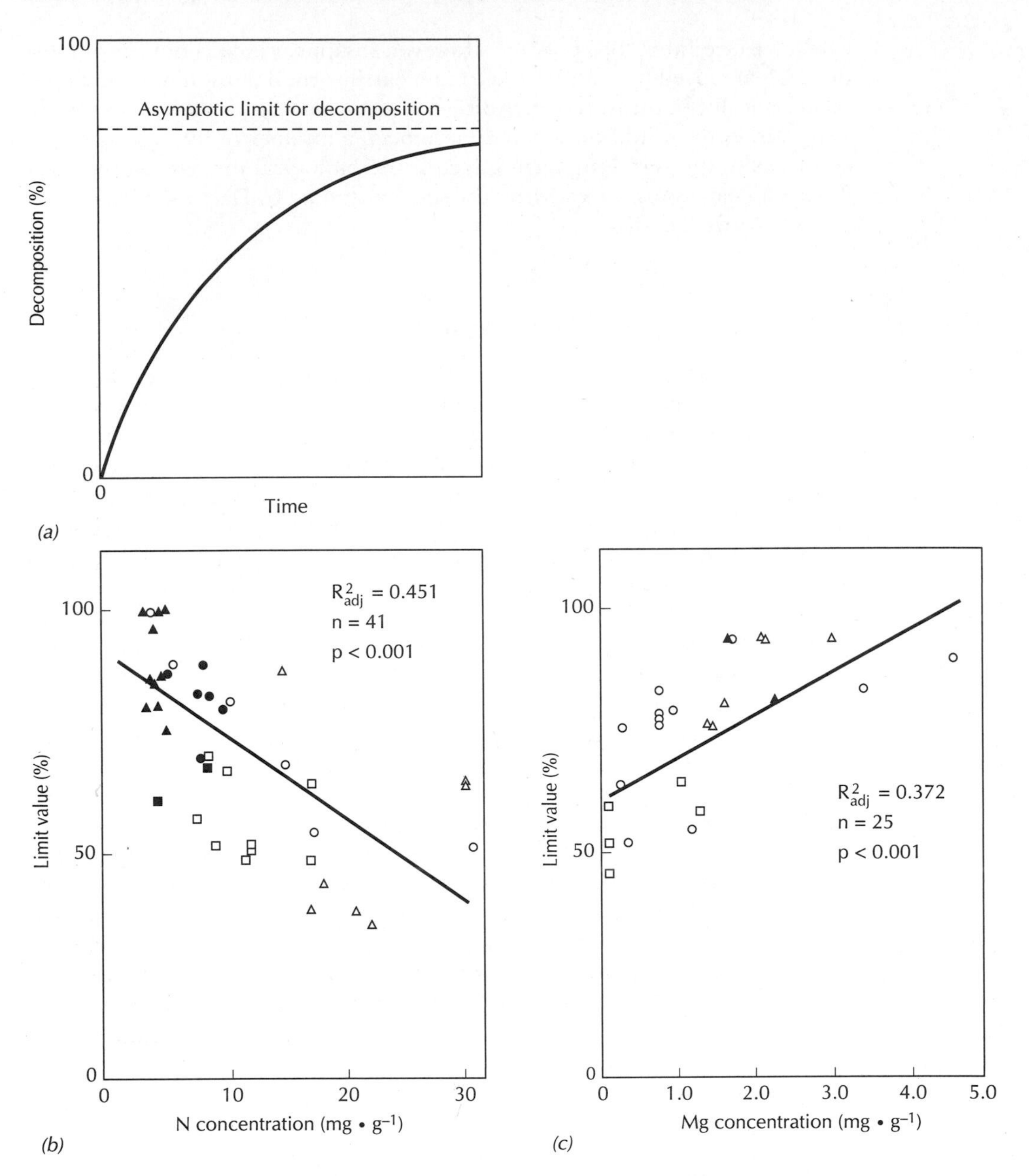

FIGURE 13.12 Predicting the fraction of litter material that becomes soil organic matter, or humus. (*a*) The general form of the asymptotic equation used to predict the limit of decomposition for different materials. (*b*) and (*c*) The relationship between this predicted limit and initial nitrogen and manganese concentration. (Berg et al. 1996)

DECOMPOSITION AND NUTRIENT RELEASE FROM HUMUS

How Are Decay Rates Measured?

Processes such as decomposition are much more difficult to measure in soils than in litter. Above-ground litter is, at least initially, relatively isolated from roots, soil colloids, and the complex matrix of soil systems. After significant decay, the switch to net mineralization, and the invasion of fine roots, the decaying material becomes only one component of this intricate system. The respiration and release of CO_2 by free-living microbes using soil organic matter as a substrate is difficult to separate from that released by mycorrhizae or roots. Nutrients are subject to large and rapid fluxes between colloid surfaces, soil solution, microbes, and roots. We saw in Chapter 10 that the same organisms (mycorrhizae) that act as extensions of the fine root system may also be part of the decomposer community!

The new methods for understanding the chemical structure of humus described at the end of Chapter 12 unfortunately provide little information that might allow us to predict rates of humus decomposition and associated nutrient mineralization or to assign fluxes of carbon and nitrogen to different parts of the soil system. Our inability to predict decay and turnover of humus is in marked contrast to the prediction of litter decay as discussed earlier.

Given this grim initial assessment, what methods have been used to estimate soil organic matter decomposition and nutrient release rates?

One straightforward method is to enclose organic soil horizons in the same type of mesh bag used for fresh litter and examine weight loss over time. This has been attempted in four different forest types in the taiga zone forests in Alaska. There was a distinct difference in decay rates between stands, with the birch and aspen stands having turnover rates of about 3% per year, and the spruce stand forest floors decaying at about 1% per year (Table 13.1A). The low turnover rates suggest that a long time period must be involved in such studies before significant losses in weight can be detected. The problem would become even more difficult in soil horizons with substantial mineral content because weight changes would be an even smaller fraction of total soil horizon mass.

Turnover rates may also be estimated by measuring the net mineralization rates for nitrogen. Field soil incubation techniques have been devised that measure the accumulation (net mineralization) of ammonium and nitrate in isolated soil cores enclosed in gas-permeable plastic bags and replaced in the soil for a given length of time. The accumulation of ammonium and nitrate in these bags provides one estimate of both net nitrogen mineralization (change in nitrate plus ammonium) and net nitrification (change in nitrate alone). Assuming that the ratio of carbon to nitrogen in the soil organic matter does not change during the incubation period, then the percentage of soil nitrogen mineralized during a year is equivalent to the percentage of soil carbon mineralized.

TABLE 13.1 Measured Turnover Rates of Carbon in Several Forest Types in the Taiga Zone of Alaska* and Nitrogen in Several Forests in Wisconsin†

(*From Flanagan and Van Cleve 1983)
(†From Nadelhoffer et al. 1983)

	Turnover Rate (% per year)
A. Taiga forests	
Dominant species	
Aspen	3.1
Birch	3.0
White spruce	1.0
Black spruce	1.3
B. Wisconsin forests	
Dominant species	
Red pine	1.8
White pine	4.9
Sugar maple	3.5
Red oak	7.9
Black oak	2.8
White oak	4.8

Using this technique, a wide range of organic matter turnover rates have been measured in a series of temperate forest ecosystems with similar soil structure and climatic regime but different dominant tree species and past history of disturbance (Table 13.1B). In well-mixed soils, including some of those in Table 13.1B, this method may measure nitrogen turnover for all soil organic matter because fresh litter is rapidly mixed with older litter and the two cannot be separated. This, along with climatic differences, may partially explain the differences in turnover rates between forest soils in Alaska and Wisconsin.

The relatively slow net turnover of carbon and nitrogen in soils suggested by the last two methods may mask much more rapid and dynamic transfers, especially of nitrogen, within the soil system. A method called ^{15}N pool dilution analysis attempts to describe these more rapid exchanges of nitrogen by injecting ^{15}N-labeled ammonium or nitrate into soil and measuring the absolute disappearance of the labeled nitrogen (gross immobilization rate) and the dilution of the remaining mineral ^{15}N pool by ^{14}N (gross mineralization rate). By measuring the rate of carbon respiration at the same time, nitrogen uptake by microbes per unit carbon processed (also called the microbial carbon use efficiency) can be calculated. Results from this method suggest that the gross rate of exchange between the soil solution and microbial pool is as much as 10-fold or more the net rate of mineralization, but questions have been raised as to how much of this rapid turnover is actually chemical rather than microbial and where the energy would come from for microbes to incorporate such large amounts of nitrogen.

One study on a series of forests in Europe with very different rates of carbon respiration in soils revealed that both gross nitrogen immobilization and gross nitrogen mineralization increased with CO_2 release (Figure 13.13). The difference between these two, or net nitrogen mineralization, was predicted to increase slightly because the two lines diverge with higher respiration rates. Ratios of gross to net mineralization vary between 3 : 1 and 4 : 1 in these studies.

A rather crude but large-scale field measurement of the rate of humus decay is the rate of disappearance of total soil organic matter following the initiation of plowing and farming. Agriculture affects soil organic matter content by reducing the annual input of litter as compared with natural systems and because the physical effects of plowing increase decay rates. These effects include alteration of soil structure, increased aeration, and mixing of organic and mineral horizons.

Losses of soil organic matter have often been measured as associated losses in nitrogen content. There is a distinct exponential decrease in soil nitrogen beginning with the initiation of cultivation. Figure 13.14 shows changes in nitrogen content of four different prairie soils over the first few decades of plowing and planting. These changes represent losses of just over 1% per year. The carbon lost from soils in association with this nitrogen represents a potentially important source of CO_2 to the atmosphere (Chapter 26).

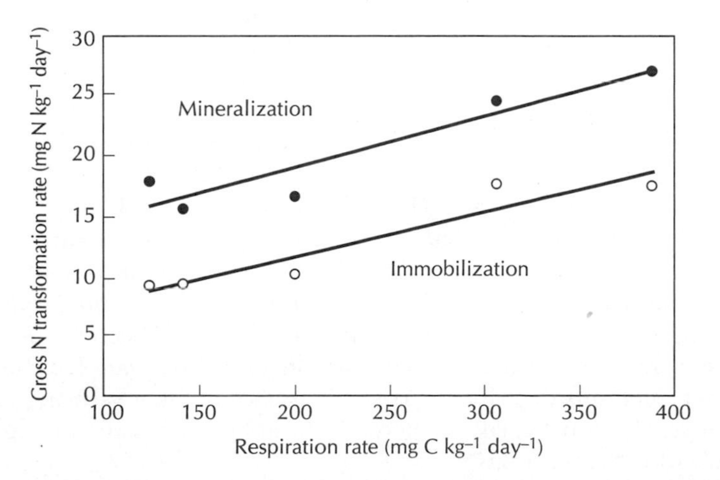

FIGURE 13.13 Rates of gross mineralization and immobilization as a function of soil respiration rate for several European forests with very different rates of nitrogen cycling. (Tietema 1998)

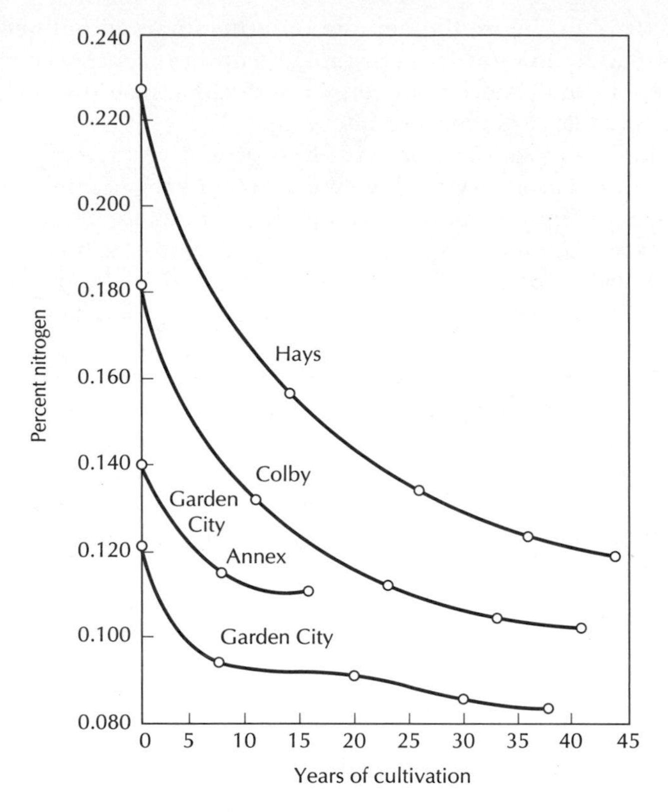

FIGURE 13.14 Changes in soil nitrogen storage in several prairie soil types following the initiation of plowing and cropping. (Cited in Allison 1973)

In contrast to these apparently rapid turnover rates, radiocarbon dating (see explanation in Chapter 5) of extracted soil organic matter (Table 13.2) yields very old mean ages of carbon in the different fractions of organic matter commonly separated from soils. A soil with a turnover rate of 1–4% per year, typical of the values obtained with the CO_2 and nitrogen balance methods mentioned earlier, would have a mean residence time of 25–100 years. Yet even the youngest soil fraction in Table 13.2, the fulvic acids, range from 550–630 years average age. Humic acids were much older.

What this means is that a certain portion of soil organic matter is very inert, decaying at a rate of less than 0.2–0.1% per year. Because radiocarbon methods yield a mean age for a given class of compounds, some of the carbon atoms in that soil fraction have undoubtedly been in place for much longer than 1000 years. The radiocarbon ages also mean that the CO_2 and nitrogen methods described earlier are undoubtedly measuring releases from relatively fresh materials, such as root litter and root exudates, as well as mineralization from humus.

One way of visualizing this wide range in the dynamics of different types of soil organic matter is as a series of different fractions with different turnover rates (Figure 13.15; this is actually the structure of a well-known

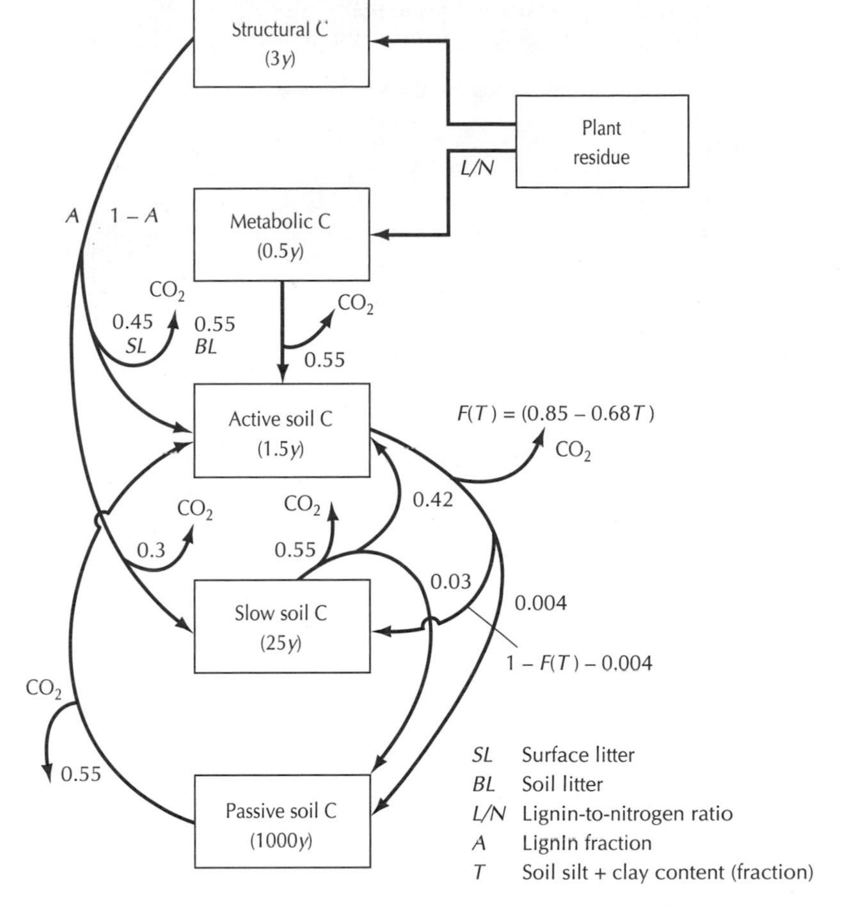

FIGURE 13.15 Compartmentalization of soil organic matter in a computer model (CENTURY) of soil organic matter dynamics in undisturbed and plowed prairie ecosystems. Values in parentheses give the turnover rate for each compartment. Initial distribution between structural (cell wall) and metabolic (cell contents) compartments is determined by the lignin to nitrogen ratio of the material. Further divisions between CO_2 respired and transfers to longer-term organic compartments depend on soil texture, origin of material, and lignin content. This model has successfully predicted patterns of organic matter and nutrient loss from plowed prairie soils similar to those shown in Figure 13.10. (Parton et al. 1988)

computer model of soil organic matter dynamics, the CENTURY model). The distribution of litter among the faster turnover compartments is determined by the initial chemical quality (lignin and nitrogen content) of the plant material. Transfers to the very long-term pools are determined by soil texture, which affects the degree to which organic matter can be stabilized by association with mineral soil particles. The response of a soil to disturbance, such as plowing, depends to a great degree on the distribution of organic matter between these different pools.

TABLE 13.2 Radiocarbon Ages of Soil Organic Matter Fractions in Two Northern Prairie Soils

Soil Organic Matter Fraction	Mean Age (Yr)
Campbell et al. 1967	
Calcium-humates	1400
"Mobile" humic acids	780
Fulvic acids	550
Stevenson 1985	
Humin	1240
Humic acids	1308
Fulvic acids	630

Factors Controlling Rates of Humus Decay

Given the methodological problems in measuring humus decomposition, there is little information available on controls over humus decay rates. In general, the same conditions that affect microbial activity in general also affect humus decay. Thus experiments in temperate zones where soil temperatures are raised experimentally show increases in soil CO_2 release and net nitrogen mineralization. Both experiments with individual soil cores and large-scale summaries of measurements at the regional level (Figure 13.16) show exponential increases in CO_2 efflux from soils with increasing temperature, usually with a Q_{10} value of 2–3 (indicating a two- or threefold increase for every 10°C increase in temperature). Summaries of both laboratory and field studies, normalized for average rates of carbon or nitrogen release, show not unexpected effects of soil moisture and acidity (Figure 13.17). The difficulty arises in describing the causes of the underlying differences in rates of decay once the effects of temperature, moisture, acidity, etc. have been removed.

A major complicating factor is the role that the mineral materials play in stabilizing soil organic matter. We saw in Chapter 12 that fine-textured soils (high in clay content) generally contain more soil organic matter than do sandy or coarse-textured soils. This is due to the role that minerals can play, through sorption or shielding of organics, in disrupting reactions between decompositional enzymes and their intended substrates. Organic material of a given class as separated by chemical techniques (described in Chapter 12) would decay at very different rates if combined or not combined with mineral material.

The essence of this problem, and perhaps also its solution, is in the closed nitrogen cycles in most ecosystems and the tight feedbacks or interactions between nitrogen release from soil and other measurable processes, such as productivity above ground. Models that contain relatively coarse information

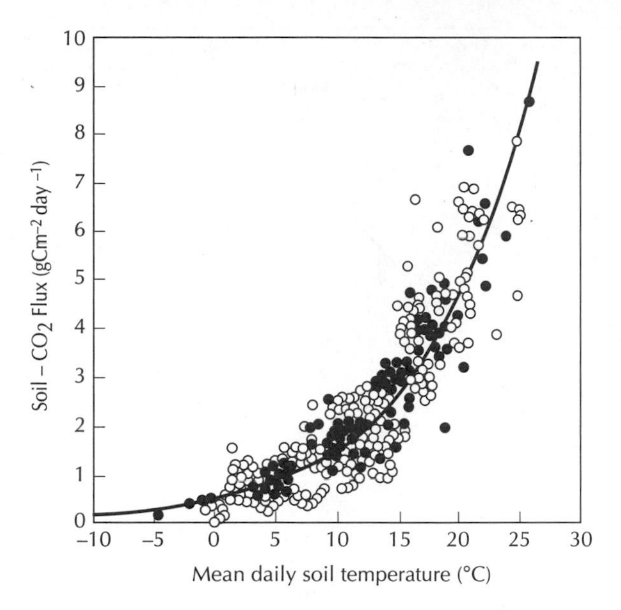

FIGURE 13.16 Effects of soil temperature on CO_2 flux. This figure is a compilation of data from several studies in temperate forests around the world. The Q_{10} value for the regression line through the data is 1.99. (Kicklighter et al. 1994)

about soil organic matter do relatively well in predicting changes in rates of cycling over time. Because nitrogen cycling is to tightly linked to processes of photosynthesis and productivity, rates of change in nitrogen cycling are related to these more easily measured processes. Most models of soil organic matter dynamics use something as simple as the CENTURY structure in Figure 13.15 or an even simpler structure. Prediction about the dynamics of nitrogen cycling are constrained by predicted controls on production, litter production, nitrogen deposition, etc. The nature of the problem is reduced to applying a reasonable number for turnover rates to the one to three soil organic matter compartments described in the model. This controls the rate at which the entire system responds to disturbance or some other change in condition. None of the widely used ecosystem models, at this time, make use of the detailed measurements of soil organic matter structure and chemistry described in Chapter 12.

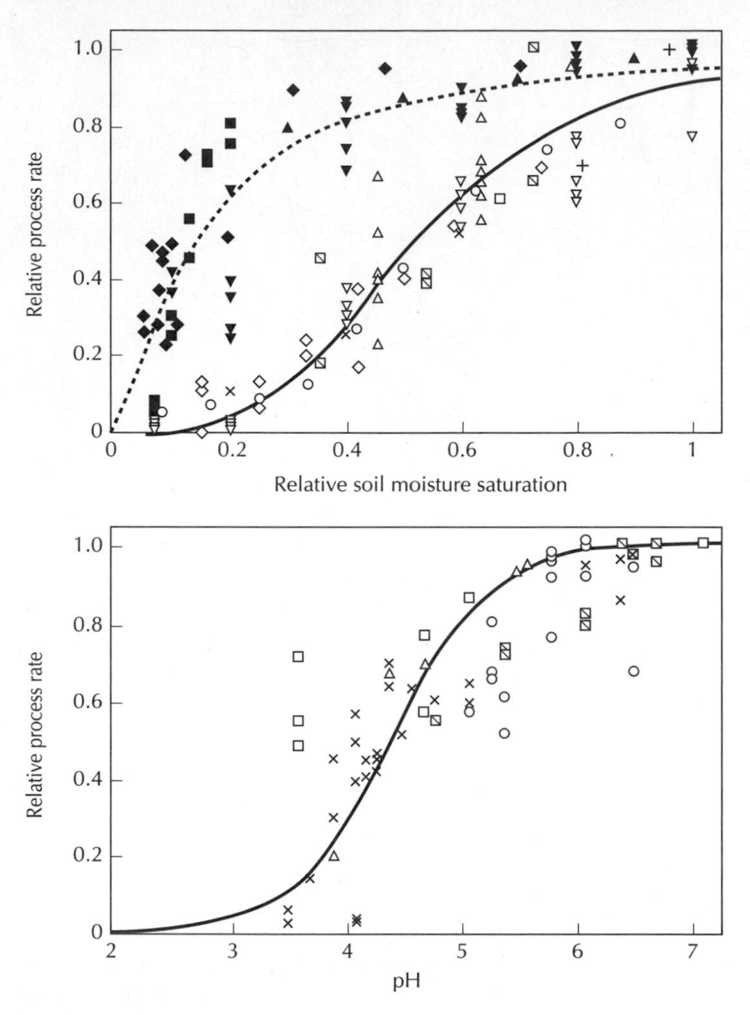

FIGURE 13.17 Summary effects of acidity (pH) and moisture content on relative rates of respiration or nitrogen release. (Walse et al. 1998)

THE IMPORTANT ROLE OF SOIL ORGANIC MATTER

To summarize, soils contain a very wide range of organic materials, combined to different degrees with mineral material, that turnover or decay at very different rates. The net turnover of carbon and nitrogen for the whole soil body represent the combined and averaged effects of all of these different fractions and their different decay rates. Much of our inability to predict the turnover of soil organic matter without measuring it directly results from an inability to accurately quantify the amount of soil organic matter in these different fractions and the degree of association with mineral materials, and also perhaps to understand how disturbing soil structure alters these rates.

Humus represents a large store of essential nutrients. As with other parts of the humus story, the form in which these nutrients occur is unclear. Whatever the form, as the organic matter is decayed, these nutrients are mineralized and made available to plants and microbes.

In an ecosystem context, a large pool of nutrients that becomes available very slowly has the effect of moderating annual fluctuations in nutrient availability. Field measurements in several forest ecosystems have shown that nitrogen mineralization is relatively constant year to year despite significant changes in precipitation and temperature. Humus can be thought of as a slow release fertilizer that acts to preserve site quality or maintain a given level of nutrition. Thus a forest or grassland severely disturbed by fire or insects is "buffered" against drastic nutrient losses (unless the soil is severely eroded as well). Plant growth can begin again immediately after disturbance because a large nutrient reserve is already in place. In fact, as we will see in Part 3, such disturbances often increase the rate of nutrient release from this long-term storage pool, which may increase the rate of ecosystem recovery.

REFERENCES CITED

Aber, J. D. , J. M. Melillo, and C. A. McClaugherty. 1990. Predicting long-term patterns of mass loss, nitrogen dynamics and soil organic matter formation from initial litter chemistry in forest ecosystems. *Canadian Journal of Botany* 68:2201–2208.

Allison, F. E. 1973. *Soil Organic Matter and its Role in Crop Production.* Elsevier Scientific Publishing Company, Amsterdam.

Axelsson, G., and B. Berg 1988. Fixation of ammonia (^{15}N) to *Pinus silvestris* needle litter in different stages of decomposition. *Scandinavian Journal of Forest Research* 3:273–279.

Berg, B. 1986. Nutrient release from litter and humus in coniferous forest soils: A mini review. *Scandinavian Journal of Forest Research* 1:359–369.

Berg, B. 1988. Dynamics of nitrogen (^{15}N) in decomposing Scots pine (*Pinus silvestris*) needle litter: Long-term decomposition in a Scots pine forest: VI. *Canadian Journal of Botany* 66:1539–1546.

Berg, B., and E. Matzner. 1997. Effect of N deposition on decomposition of plant litter and soil organic matter in forest systems. *Environmental Review* 5:1–25.

Berg, B., G. Ekbohm, M. B. Johansson, C. McClaugherty, F. Rutigliano, and A. V. De-Santo. 1996. Maximum decomposition limits of forest litter types: A synthesis. *Canadian Journal of Botany* 74:659–672.

Bryant, D. M., E. A. Holland, T. R. Seastedt, and M. D. Walker. 1998. Analysis of litter decomposition in an alpine tundra. *Canadian Journal of Botany* 76:1295–1304.

Campbell, C. A., E. A. Paul, D. A. Rennie, and K. J. McCallum. 1967. Factors affecting the accuracy of the carbon-dating method in soil humus studies. *Soil Science* 104:81–85.

Cromack, K. 1973. Litter production and litter decomposition in a mixed hardwood watershed and in a white pine watershed at Coweeta Hydrologic Station, North Carolina. Ph. D. dissertation, University of Georgia.

Flanagan, P. W., and K. Van Cleve. 1983. Nutrient cycling in relation to decomposition and organic matter quality in taiga ecosystems. *Canadian Journal of Forest Research* 13:795–817.

Harmon, M. E., J. F. Franklin, F. J. Swanson, P. Sollins, S. V. Gregory, J. D. Lattin, N. H. Anderson, S. P. Cline, N. G. Aumen, J. R. Sedell, G. W. Lienkaemper, K. Cromack, and K. W. Cummins. 1986. Ecology of coarse woody debris in temperate ecosystems. *Advances in Ecological Research* 15:133–302.

Johansson, M. B., B. Berg, and V. Meentemeyer. 1995. Litter mass-loss rates in late stages of decomposition in a climatic transect of pine forests. Long-term decomposition in a Scots pine forest: IX. *Canadian Journal of Botany* 73:1509–1521.

Kicklighter, D. W., J. M. Melillo, W. T. Peterjohn, E. B. Rastetter, A. D. McGuire, and P. A. Steudler. 1994. Aspects of spatial and temporal aggregation in estimating regional carbon dioxide fluxes from temperate forest soils. *Journal of Geophysical Research* 99:1303–1315.

Magill, A. H., and J. D. Aber. 1998. Long-term effects of experimental nitrogen additions on foliar litter decay and humus formation in forest ecosystems. Plant and Soil 203:301–311.

Magill, A. H., and J. D. Aber. 2000. Dissolved organic carbon and nitrogen relationships in forest litter as affected by nitrogen deposition. *Soil Biology and Biochemistry* 32: 603–613.

Melillo, J. M., J. D. Aber, and J. F. Muratore. 1982. Nitrogen and lignin control of hardwood leaf litter decomposition dynamics. *Ecology* 63:621–626.

Nadelhoffer, K. J., J. D. Aber, and J. M. Melillo. 1983. Leaf litter production and soil organic matter dynamics along a nitrogen-availability gradient in Southern Wisconsin (U. S. A.). *Canadian Journal of Forest Research* 13:12–21.

Northrup, R. R., Z. Yu, R. A. Dahlgren, and K. A. Vogt. 1995. Polyphenol control of nitrogen release from pine litter. *Nature* 377:227–229.

Parton, W. J., J. W. B. Stewart, and C. V. Cole. 1988. Dynamics of C, N, P and S in grassland soils: A model. *Biogeochemistry* 5:109–132.

Staff, H., and B. Berg. 1982. Accumulation and release of plant nutrients in decomposing Scots pine needle litter: Long-term decomposition in a Scots pine forest: II. *Canadian Journal of Botany* 60:1561–1568.

Stevenson, F. J. 1985. Geochemistry of soil humic substances. In McKnight, D. M. (ed.), *Humic Substances in Soil, Sediment and Water: Geochemistry, Isolation and Characterization*. John Wiley & Sons, New York.

Tietema, A. 1998. Microbial carbon and nitrogen dynamics in coniferous forest floor material collected along a European nitrogen deposition gradient. *Forest Ecology and Management* 101:29–36.

Walse, C., B. Berg, and H. Sverdrup. 1998. Review and synthesis of experimental data on organic matter decomposition with respect to the effect of temperature, moisture and acidity. *Environmental Review* 6:25–40.

14

PLANT–SOIL INTERACTIONS

SUMMARY EFFECTS ON NUTRIENT CYCLES

INTRODUCTION

The past eight chapters have dealt with interactions among physical climate, the chemistry of soils, and the biological processes of production and decomposition. Much of terrestrial ecosystem research has dealt with the combined effects of these interactions on the movement of nutrients through ecosystems.

In this chapter, we step back and take another look at these interactions from the perspective of comparative and integrated nutrient cycles and using case studies of systems with very different environmental conditions and species compositions. Much of what is presented here repeats the content of earlier chapters, making this something of a midpoint review before going on to discussions of herbivory, fire, and presentations of whole-ecosystem studies and human impacts.

The purpose of this chapter, then, is to synthesize much of the information already presented into a comparative discussion of nutrient cycles in terrestrial ecosystems. Three approaches are taken. First, we return to the generalized diagram of nutrient cycling developed in Chapter 1 and present side-by-side comparisons of the relative importance of each major process for each macronutrient. We then use the movement of water through ecosystems as a means of summarizing where in ecosystems the different processes occur and how those processes might vary under different conditions. We conclude with some examples of the extent to which different species of plants can modify the distribution and cycling of nutrients within ecosystems.

COMPARISONS OF GENERALIZED NUTRIENT CYCLES

The processes that dominate element cycling for the major nutrients differ widely. We have seen that carbon (C) resides in the system only as organic matter and that the atmosphere is the source of available carbon dioxide (CO_2) for plant uptake, through the leaves, during photosynthesis. At the other extreme, potassium is taken up by the root and is never combined in any organic form. It moves through the plant in ionic form and is easily leached from plant surfaces. While these cycles are very different, they can be presented as part of a larger, generalized nutrient cycle similar to the one developed in Chapter 1 (Figure 14.1). The relative importance of each process for each nutrient is presented graphically in Figure 14.2 by the relative thickness of the lines marking transfers between components.

Carbon

Carbon (Figure 14.2*a*) available for plant uptake is in the atmosphere as CO_2. It is fixed by photosynthesis and is then respired or becomes part of the plant biomass. In plants, it is present in the wide variety of compounds described in Chapter 12. Some of these move readily through the plant (simple sugars) and are respired. Some are converted to structural compounds that can no longer be translocated (lignin, cellulose). The immobile compounds eventually become litter either through seasonal senescence, as with leaves, or through the death of whole plants. The litter is decomposed by microbes (see Chapters 12 and 13) and is respired to CO_2 or transformed to more stable organic complexes and, eventually, humus.

Relative to the large gross transfers of C described earlier, only small amounts of organic C enter into the soil solution and are transported to lower soil horizons or groundwater in most systems. However, this dissolved organic material can play an important role in the long-term nutrient balance of a site. Dissolved organic compounds are important in chelating iron and aluminum (Al) and transporting them to lower soil horizons and thus

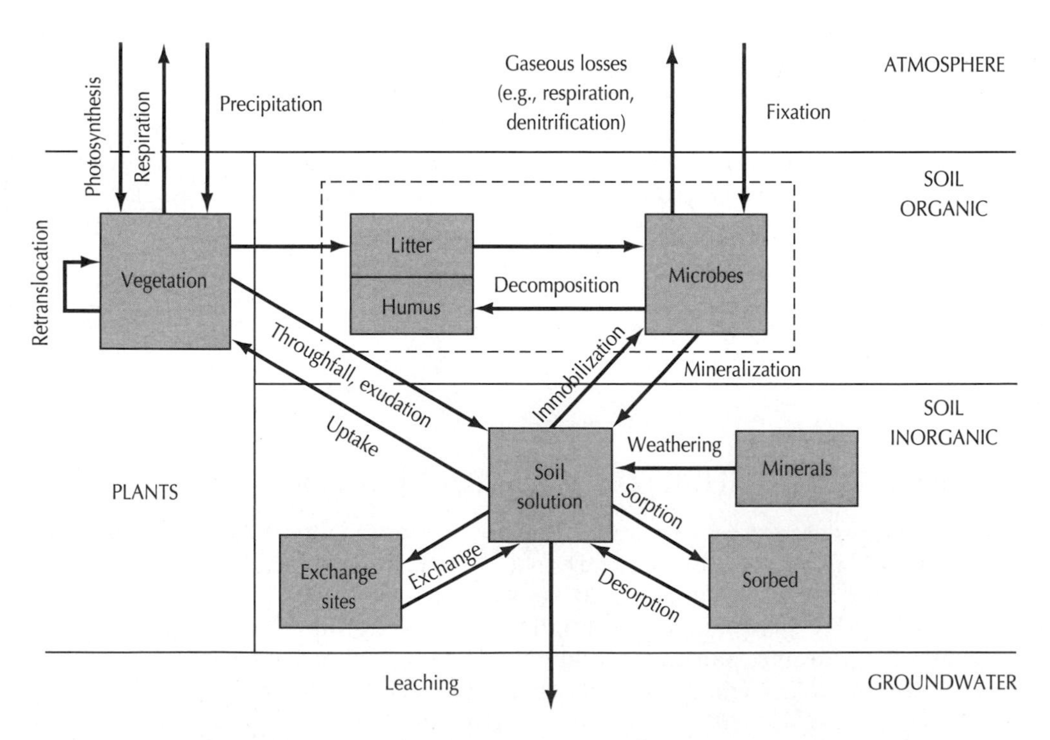

FIGURE 14.1 Generalized nutrient cycle.

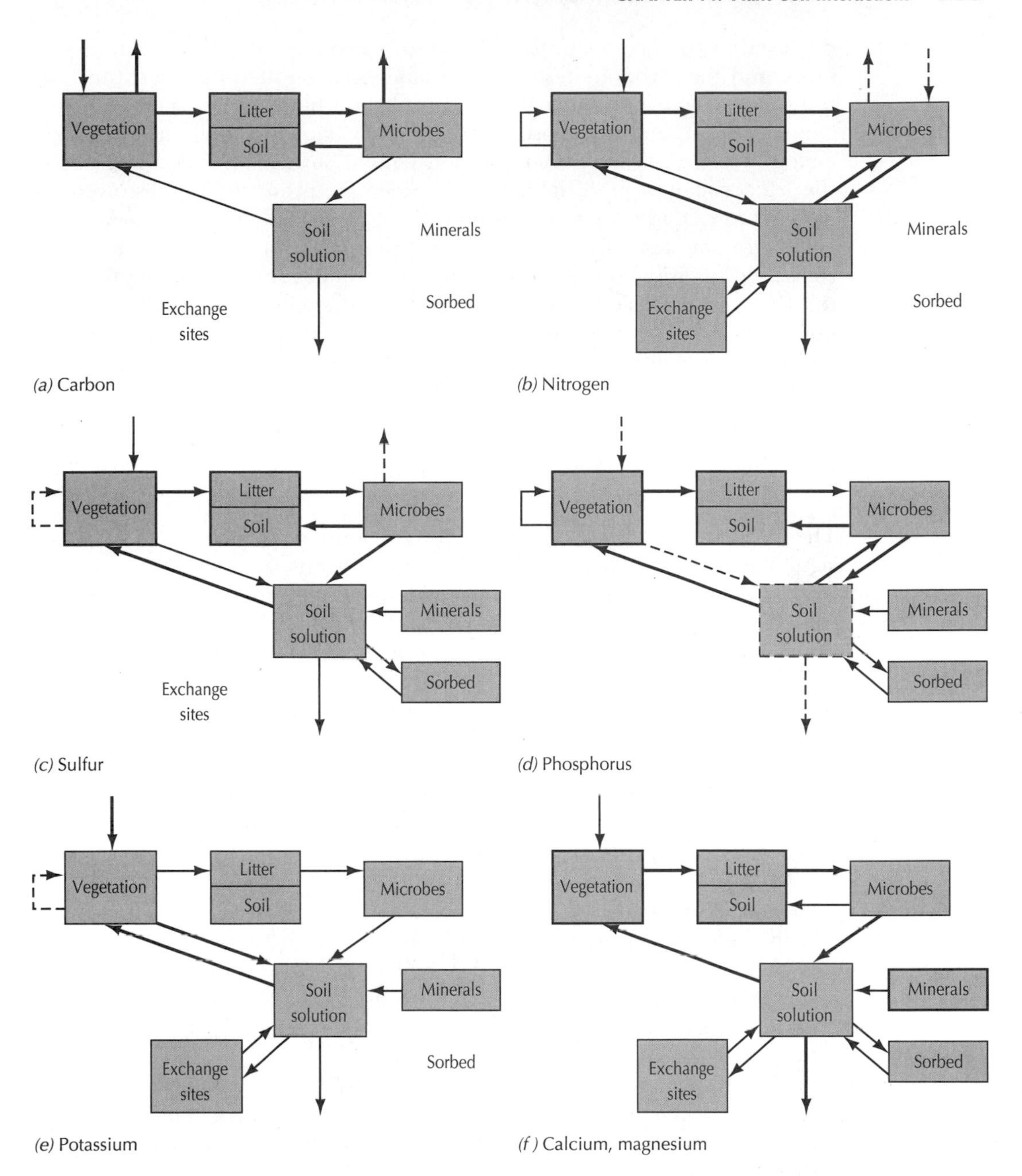

FIGURE 14.2 Comparative cycles for carbon and macronutrients. The width of the arrows indicate the relative amount transferred by that pathway.

play an important role in podzolization. Leaching of dissolved organics into streams can alter both the chemistry and energetics of these aquatic systems. As will be seen in the subsequent discussion of nitrogen (N), the transfer of N to streams in dissolved organic form can play an important role in the N balance of unpolluted systems. We still have an imperfect understanding of the factors that control the production, consumption, and leaching losses of dissolved organic compounds. They absorb strongly to mineral particle surfaces, and the sorption–desorption balance may be the primary control on soil solution concentrations. Sorption isotherms have been described for organics that are similar to those for sulfur (see Chapter 9). Still, the rate of exchange between soil solution and soil colloid surfaces is unknown, nor is there a constant or predictable ratio between the production of dissolved organics and CO_2 in the decomposition process.

Large amounts of C may be present in systems underlain by carbonate rocks, such as limestone. By contributing carbonate ions through weathering, this C can have a major effect on soil pH and other chemical parameters. Carbonate released in this form is mainly lost to streams, and, because this pathway is very different than for C fixed in organic matter, it is omitted in Figure 14.2*a*.

Nitrogen

The N (Figure 14.2*b*) cycle is similar to C in some respects but also differs in significant ways. While the atmosphere is 79% nitrogen gas (N_2), the substantial energy investment required to convert this form of N into a biologically available form limits the amount of N fixation that occurs in most ecosystems. While we have seen that several types of plants have symbiotic relationships with microorganisms that allow N fixation to occur in most natural ecosystems, enough N is available from internal recycling to cause N fixation to be a disadvantageous way of gaining N. Thus, N fixation is relatively low in most, but not all, natural ecosystems. Both N fixation and the complementary process of denitrification (the loss of nitrogen gas [N_2] or nitrous oxide gas [N_2O] to the atmosphere) tend to occur at low and roughly equal rates in most terrestrial ecosystems.

Nitrogen is also present as ammonium (NH_4^+) and nitrate (NO_3^-) in the atmosphere, and these forms are deposited in precipitation. These inputs are usually greater than inputs from fixation and are increasing in many areas due to human activities (Chapter 25). Still, both inputs combined are generally much less than the amount of N available internally through the mineralization of N in soil organic matter. The N cycle is relatively "closed" in most systems.

The internal cycle is dominated by uptake of NH_4^+ and NO_3^- from the soil solution and exchange sites, incorporation into plant tissues, deposition in litter, and decomposition and eventual release to the soil solution. This cycle differs from C in several ways.

First, significant amounts of N can be withdrawn from the leaves of plants as they senesce (retranslocation). Second, decomposition can result in either the release of N from organic matter (mineralization), or the incorporation of available N into the litter (immobilization). Thus, microbes using the litter as an energy source can actually compete with plants for available N. Third, there are three forms of N that plants can take up: (1) NH_4^+, (2) NO_3^-, and (3) organic. The differences in the form of N produced in soils and used by plants has important implications for soil chemistry, as discussed later. Fourth, N is not a significant component of primary or secondary minerals.

The form in which N cycles in soils and is used by plants is an important characteristic of an ecosystem, and one that interacts strongly with several biological processes and has implications for soil chemistry. Figure 14.3 is a simplified diagram of the N cycle, focused on the transformations in soils and uptake by plants. Soil organic matter is the solid phase of humuslike compounds, often rich in N, which are slow to decay and are not available for plant uptake. Traditional thinking has been that N is released from soil organic matter by microbial decomposition resulting in the production of NH_4^+ (net mineralization). In some systems this is followed by the conversion of NH_4^+ to NO_3^- (net nitrification). While there has been some debate about the degree to which different species or growth forms of plants prefer NH_4^+ or NO_3^-, it seems clear that most species can take up either form of mineral N.

There is a distinct difference between the two in terms of charge balances over roots (see Chapter 10). In addition, the production of NO_3^- affects soil chemistry in two ways. First, the nitrification reaction releases hydrogen ions into the soil solution and hence is an acidifying process. Second, NO_3^- is more mobile in temperate-zone soils and can be leached below the rooting zone much more easily than can NH_4^+. This can result both in significant losses of N from the system and further acidification of soils through the concurrent loss of nutrient cations (see Chapter 9). In Chapter 4 we saw that different forest ecosystems responded very differently in terms of NO_3^- leaching losses to clear-cutting. Those systems in which nitrification did not occur could not lose NO_3^- to streams following disturbance, suggesting that understanding the factors that control or suppress nitrification is an important goal.

An early hypothesis was that plants inhibited nitrification by exuding chemicals into the soil that interfered with the process. This is called **allelopathy,** the production and exudation of compounds harmful to other species or their function. This explanation has been refuted in most, but not all, cases.

Current theory holds that nitrification is controlled primarily by a combination of soil pH and NH_4^+ availability. A broad-scale comparison of nitrification rates in natural ecosystems shows that nitrification tends to be high in soils with a pH of more than 5.0 and is very low or nonexistent in soils with a pH of less than 3.5 (see example for temperate-zone forests in Figure 14.4). However, nitrification can occur at very high rates in acidic soils (even those with a pH of 4 or less) when NH_4^+ is produced by mineralization in excess of

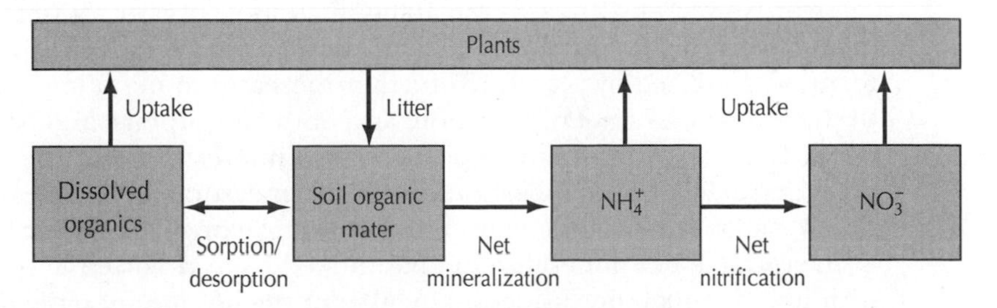

FIGURE 14.3 A generalized and simplified view of the nitrogen cycle.

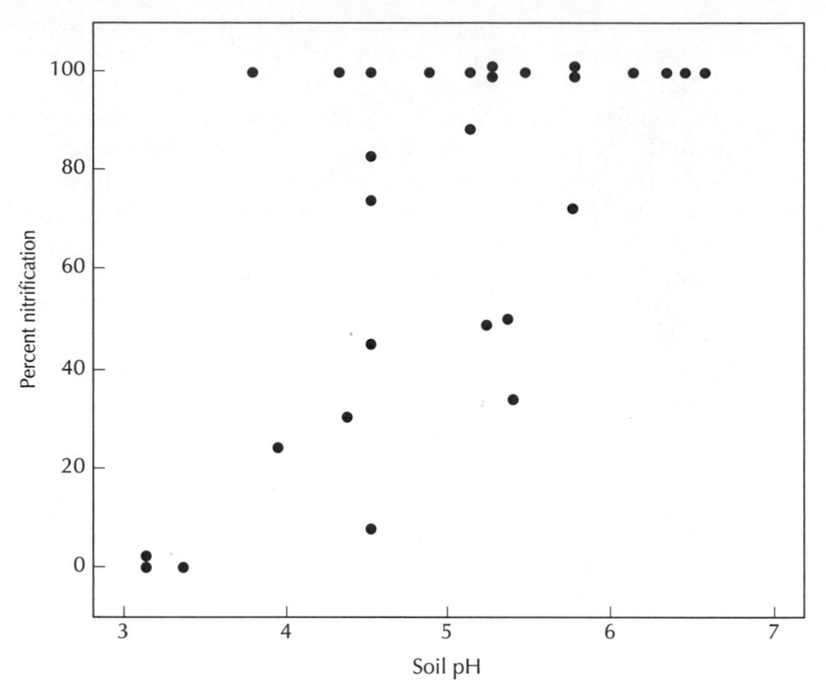

FIGURE 14.4 Change in the rate of nitrification in forest soils as a fraction of soil pH. (Data from Pastor et al. 1984 and Nadelhoffer et al. 1983)

plant demand (see Chapter 25). It can also be low in high-pH soils if the NH_4^+ supply is low and competition for NH_4^+ between nitrifying microbes and plants and their mycorrhizal symbionts is intense.

In addition, several studies suggest that plant roots and mycorrhizae can actually utilize small organic molecules directly, without the organic compounds being converted first to mineral form (see Chapter 13). This direct cycling of organic N appears to be most important in systems with very low rates of N cycling (low rates of net mineralization and plant uptake).

Combining these patterns, we can begin to describe a general theory relative to the form of N produced and taken up in different ecosystems. If we envision a gradient of N availability (or rate of net N mineralization, Figure 14.5*a*) then the following three points appear to be true: (1) the direct uptake of organic N appears to be an important process in systems with very low rates of N cycling; (2) as N availability increases, organic N uptake may decrease, while net nitrification remains minimal, and NH_4^+ is the predominant form of N utilized by plants; (3) further increases in net mineralization lead to the induction of net nitrification, and both NO_3^- uptake and NO_3^- leaching could become increasingly important. It is interesting that this theory predicts that no net mineralization should be measurable in systems with very low N availability because plants release and take up N in organic form. A lack of measurable net mineralization has indeed been reported in several stands with low N availability, using the incubation techniques described in the last chapter. It should be remembered that we are discussing net mineralization and nitrification here and that gross rates of NH_4^+ and NO_3^- production and

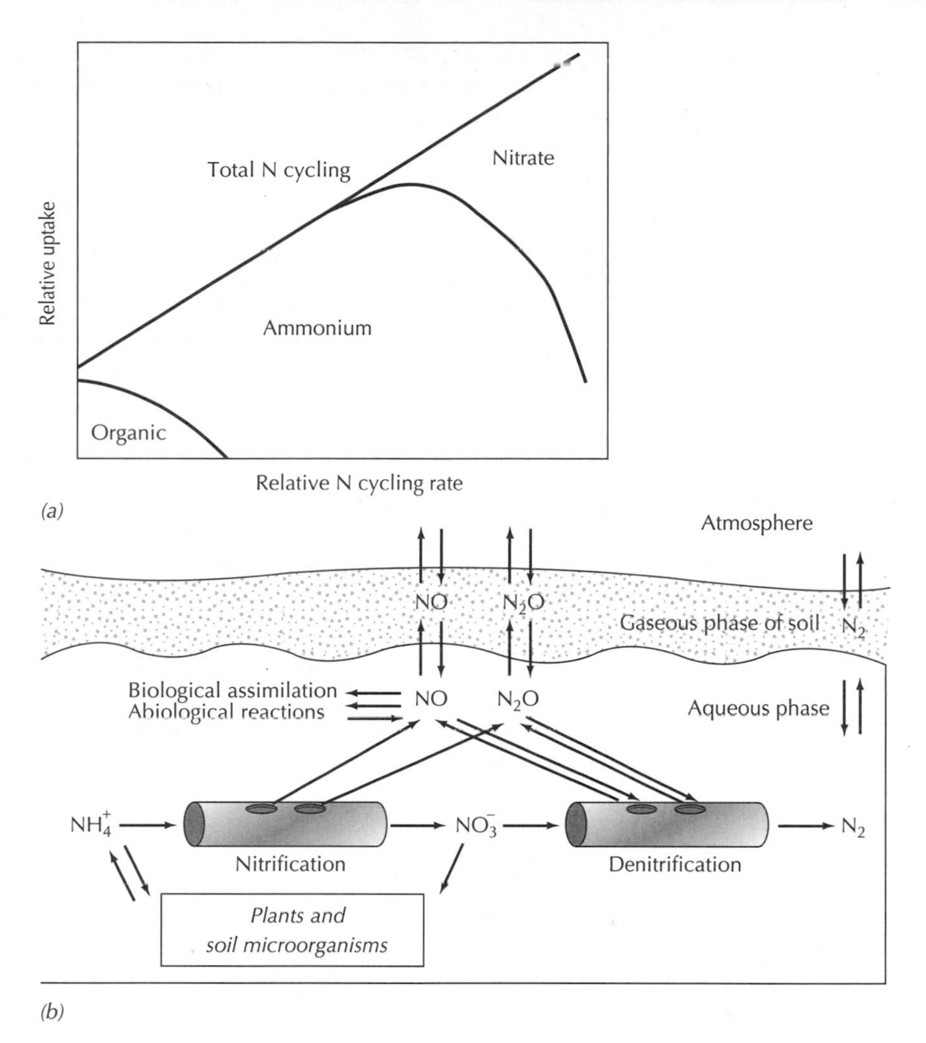

FIGURE 14.5 Changes in nitrogen (N) cycling with increasing N availability. (*a*) Hypothesized changes in the relative importance of organic N, ammonium (NH_4^+) and nitrate (NO_3^-). (*b*) The role of nitrification in the emissions of N trace gases (the "hole-in-the-pipe" theory). NO = nitric oxide, N_2O = nitrous oxide, N_2 = nitrogen gas, NO_3^- = nitrate, NH_4^+ = ammonium. (Davidson et al. 2000)

consumption by microbes could differ from patterns described for net fluxes. We return to this trend in the form of N produced and cycled in our discussion of acid rain in Chapter 25.

The form of N cycled also has implications for the emission of trace gases that can affect atmospheric chemistry and perhaps climate change. The primary gases of interest are N_2O and nitric oxide (NO). N_2O is a relatively long-lived gas that, like CO_2 and water vapor, traps infrared radiation emitted from the Earth's surface and contributes to warming of the atmosphere. Both gases are produced during the aerobic process of nitrification, and both can be produced or transformed during the anaerobic process of denitrification, by which NO_3^- is reduced to nitrogen gas (N_2) or several intermediate compounds, including N_2O and NO (Figure 14.5*b*). In general, emissions of N_2O

are near zero in the absence of nitrification; increase with nitrification; and increase further if conditions such as a fluctuating water table alternately favor nitrification and denitrification.

Sulfur

The sulfur (S; Figure 14.2*c*) cycle is similar to that of N in many ways. Most S cycling occurs through plant uptake of sulfate, deposition in litterfall, and mineralization from soil organic matter. Some important differences are that retranslocation and immobilization are less important because S is generally less limiting in most terrestrial ecosystems. S is also present in certain primary minerals and can be supplied in part by weathering and can also be sorbed strongly in soils with high iron and Al content. Sorption can be increased through soil acidification, and thus the retention of S can be increased in systems experiencing high rates of NO_3^- loss and, hence, some acidification of soils. Gaseous exchanges with the atmosphere do occur, but they are generally small for nonwetland systems compared with precipitation inputs. Sulfuric acid is a major component of the acidity in "acid rain," so sulfate input rates are higher in and near heavily industrialized regions than in remote areas. As inputs increase, leaching losses may also increase unless soil sulfate sorption capacity is high. The leaching of sulfate from soils is a major contributor to soil acidification in regions heavily affected by acid rain (Chapter 25).

Phosphorus

The internal organic cycle of phosphorus (P; Figure 14.2*d*) is much like the N cycle. The vegetation and soil organic pools are major reservoirs of P, with uptake, litter fall, retranslocation, and decomposition as major transfers. Net immobilization into litter does occur in some systems, causing plants and microbes to compete for this element.

There are two important differences between the cycles of N and P. The first is that atmospheric inputs of P in precipitation are much smaller and gaseous exchanges are negligible. Instead, new inputs of P into terrestrial ecosystems come almost exclusively from weathering of minerals. This input is small relative to the annual internal cycle and varies considerably depending on the type of minerals present in the system and the degree to which they have already been weathered. In general, P inputs decrease as the soil profile develops over geological time.

The second is that soils with high concentrations of iron and Al oxides as a result of the weathering process also have high sorption potentials for P. This can severely limit the availability of P for plant growth or increase the energetic cost of P uptake. These competing demands for P cause leaching losses of this element to be minimal from nearly all terrestrial ecosystems, which is one reason why primary production in lakes and streams is often P limited. P sorption potential in soils tends to increase over time as soil development proceeds, especially in the E horizon of Lateritic soils and the B horizon of Podzolic soils. Because of these characteristics of the P cycle, availability to plants tends to be highest on the youngest or most rapidly weathering substrates and decreases over geologic time (see Figure 9.14 and the more detailed discussion in Chapter 22).

Potassium

The cycle for the metal cation potassium (K; Figure 14.2*e*) has little in common with those of C, N, S, or P. It is not bound into any known organic compound but moves through the plant in ionic form. Because it is never bound in organic form, it is susceptible to leaching from leaves and stems as rainfall washes over plant surfaces. In fact, throughfall (water dropping to the soil surface after passing over canopy leaves), and stem flow (water running down the stem), the two avenues for water percolation through plant canopies, play a large role in the K cycle. Both throughfall and stem flow are usually greatly enriched in K as compared to rainfall and together return more K to soils than litter fall in many systems. Retranslocation has been reported for K, but is difficult to detect accurately because of the large leaching losses from leaves in throughfall. The K in litter is also only loosely held and it leaches quickly into the soil solution. Net immobilization is not important, and microbial demand is small. In the soil, K is held only on exchange sites and is not involved in reactions to insoluble products.

Thus, the K cycle is very simple. Inputs are from precipitation and weathering. Plant uptake occurs from exchange surfaces and the soil solution. Leaching occurs from plant surfaces and litter, replenishing the exchange sites. K present in excess of plant demand and exchange site capacity is leached to groundwater.

Calcium

The calcium (Ca; Figure 14.2*f*) cycle is similar to K in that organic cycling, weathering, and cation exchange are important processes. It differs in that Ca is included in stable organic compounds. Some of these compounds do not have a direct physiological use but represent accumulated waste products, such as Ca oxalate in leaves. This may be no less crucial a function than the role of Ca in membrane function, and its accumulation in leaves before senescence may have important effects on soil chemistry (see subsequent discussion on "cation pump" species).

Throughfall and stem flow are less important in the Ca cycle than are litterfall and decomposition. Immobilization is rarely, if ever, reported. Mineralized Ca, along with inputs from weathering (which can vary widely from limestone to nonlimestone substrates) are either taken up, held on exchange sites, or leached to groundwater. In deserts, where leaching is minimal, Ca can accumulate as Ca carbonate (or caliche) in lower soil horizons.

Magnesium

The magnesium (Mg) cycle is very similar to the Ca cycle, with the exception that Mg is generally present in lower concentrations in biological materials. As the central element in the structure of the chlorophyll molecule, imbalances or suboptimal availability of Mg can have important implications for photosynthesis.

CHANGES IN SOLUTION CHEMISTRY IN ECOSYSTEMS

A second way of summarizing and comparing nutrient cycles is to follow changes in nutrient concentrations in water as it passes through an ecosystem. The major types of solutions generally sampled (Figure 14.6) include: (1) precipitation (collected above the canopy), (2) throughfall (collected at the soil surface away from stems), (3) stem flow (collected as it runs down stem surfaces), and (4) leachates from below one or more soil horizons. Changes in concentrations at each step in this process reflect the net effect of all of the major biogeochemical processes occurring in each layer.

Table 14.1 compares the concentrations of macronutrients in solutions collected at several points in this sequence for three pairs of forest ecosystems that differ in important features that can alter biogeochemical cycles and balances. The differences are in dominant species present, geologic substrate, and pollution loading, respectively.

The first comparison is between Douglas-fir and red alder forests in the Pacific Northwest of the United States. Douglas-fir is a long-lived evergreen conifer species that dominates large areas in the western United States and is often associated with low rates of net N cycling and high N use efficiency (low N concentrations in tissues). Red alder, in contrast, is an early successional species that can be dominant following fires or logging operations. The crucial characteristic of red alder for this comparison is that it has root symbionts that can fix atmospheric N.

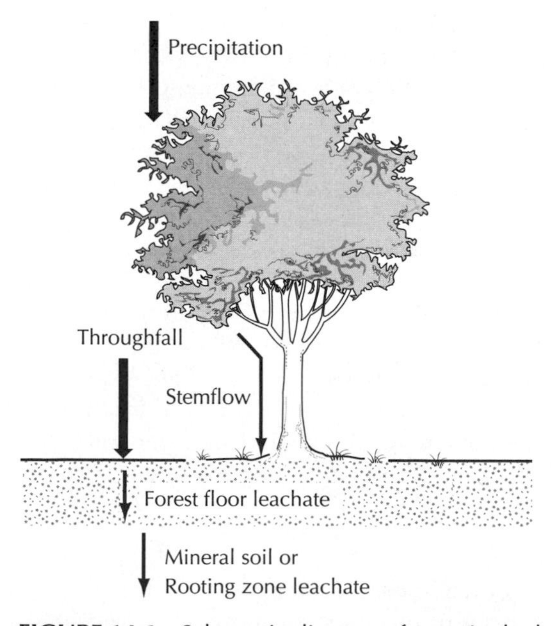

FIGURE 14.6 Schematic diagram of steps in the hydrological cycle where water samples are collected for chemical analysis. The chemistry of the water changes through contact with canopy, stem, and soil surfaces.

TABLE 14.1 Changes in the Element content of Water as it Passes Through Six Different Forest Ecosystems.
The Cedar River sites are in Western Washington State (Johnson and Lindberg 1992).
Coweeta (Johnson and Lindberg 1992) and **Walker Branch** (Johnson and Van Hook 1989) **are in the Southern Appalachian mountains. Ballyhooly and Kootwijk are European sites in Ireland and The Netherlands respectively** (Kreutzer et al. 1998). All values are in g · m^{-2} · yr^{-1}.

	Cedar River		Coweeta	Walker Branch	Ballyhooly	Kootwijk
	Douglas-Fir	Red Alder	Deciduous	Deciduous	Norway Spruce	Douglas-Fir
Nitrogen						
Precipitation	0.48	0.47	0.47	1.30	0.64	1.68
Throughfall	0.12	0.14	0.25	1.70	1.13	4.14
Forest floor	0.02	6.95	0.19	—	0.64	2.69
Rooting zone	0.00	3.89	0.01	0.30	0.14	1.68
Sulfur						
Precipitation	1.02	1.04	1.58	1.80	1.41	2.27
Throughfall	0.93	0.95	1.77	7.20	3.14	5.79
Forest floor	0.67	0.62	1.42	—	2.98	6.56
Rooting zone	0.96	0.23	0.18	1.20	2.08	4.00
Calcium						
Precipitation	0.70	0.70	0.34	1.57	0.64	0.56
Throughfall	0.94	1.13	1.79	2.97	1.64	1.16
Forest floor	7.12	11.40	5.67	—	2.20	3.04
Rooting zone	1.05	7.48	0.12	14.70	0.56	1.16
Potassium						
Precipitation	0.25	0.24	0.11	0.29	0.23	0.12
Throughfall	0.79	1.24	2.25	2.19	2.42	1.48
Forest floor	0.85	4.02	2.99	—	1.05	1.25
Rooting zone	0.13	0.80	0.16	0.71	0.23	0.08

The second comparison is between two deciduous broad-leaved forests in the southeastern United States located on different geologic substrates. The Coweeta site is underlain primarily by granites and biotite, two slowly weathered minerals that release base cations and other nutrients at low rates. In contrast, the Walker Branch site is located over a deep deposit of dolomitic limestone that both weathers rapidly and has high concentrations of Ca and Mg.

The third comparison is between two sites in Europe. Both are plantations of evergreen conifers, but they experience very different rates of atmospheric deposition. The Ballyhooly site in Ireland is relatively free of air pollutants by European standards, while the Kootwijk site in Holland experiences some of the highest rates of N deposition in the world.

Despite these differences, some common patterns can be seen across all sites. For example, in all stands, K inputs in precipitation and leaching below the rooting zone are relatively low, while fluxes in throughfall and from the organic soil horizon to lower soils are quite high. This reflects the mobility of K and its tendency to leach from foliage and stems, and from recently deposited litter, into throughfall and the soil solution. Both cation exchange and high demand for K through root uptake reduce K losses below the rooting zone.

The impact of N fixation can be clearly seen in the first comparison. Red alder increases N input to a forest to several-fold above precipitation inputs in this region. The accumulation of this N leads, over time, to an excess of N availability, which leads in turn to the induction of nitrification and NO_3^- leaching. This can be seen in the N losses below the rooting zone and also in the elevated losses for Ca and K. Leaching losses of the anion NO_3^- must be balanced by cation losses to maintain charge balance in the soil solution. As we saw in Chapter 9, the nutrient cations, such as Ca and K, are leached preferentially. Cation losses at this rate could eventually acidify the soil and lead to nutrient imbalances between N and Mg (see Chapter 25). Differences in the chemistry of the water leaching below the rooting zone in the two forests are not related to nutrient deposition (precipitation inputs are similar) or to bedrock geology.

The dolomitic limestones underlying the Walker Branch ecosystem lead to large differences in cation leaching losses, compared with Coweeta, even though climate and vegetation are similar. Losses are particularly high for Ca (and Mg as well, although this is not shown) because these elements are major constituents in this bedrock material. There are also differences in NO_3^- losses that might be attributable to higher soil pH in the subsoils at Walker Branch but might also be due to previous history and the degree of N limitation, with greater N deficiencies at Coweeta leading to a suppression of nitrification.

Pollution inputs are major drivers of biogeochemical cycles in the most heavily populated and industrialized parts of the world. The impacts of human activity can be seen in the comparison of the Ballyhooly, Ireland, site, which, located in the far west of Europe, lies upwind from most of the major sources of pollution in that region. In contrast, the Kootwijk site is in an area affected both by large background emissions of pollutants and major local sources. As a result, S and especially N inputs are much higher than in Ireland. The cumulative effect of N deposition can be seen in the high mobility of N in soils and the very high N leaching rate (although still lower than that in the red alder stand that also experiences very high inputs of N). The fact that increased N leaching at Kootwijk is not matched by increases in Ca and K leaching suggests that the soil is already very acidic, and that Al and hydrogen are the dominant cations in leachate.

SPECIES EFFECTS ON NUTRIENT DISTRIBUTION AND CYCLING

As a final approach to summarizing nutrient interactions in different ecosystems, we discuss the effects of different species with very different physiological properties, or the spatial distribution of plants, on the distribution and cycling of nutrients. Three examples are presented: (1) N fixation that affects the input–output balance of this nutrient, (2) cation pumping that affects the vertical distribution and cycling of metal cations, and (3) the creation of "resource" islands in arid shrublands resulting from the spatial pattern of shrub occurrence.

Nitrogen Fixation

As described earlier, red alder and Douglas-fir can often be found growing together in young, successional forests in the Pacific Northwest. Alder has often been considered a weedy species that interferes with the natural regeneration and early growth of new Douglas-fir stands. Recently there has been increased interest in the role of this species in augmenting N content of soils and N availability to plants.

Table 14.2 reports data on soil properties for two pairs of stands, both planted to Douglas-fir 23 years before the measurements were made. One site is of low fertility, as judged by growth of Douglas-fir and also foliar nutrient concentrations and soil nutrient availability indices, and one is of high fertility. Red alder seeded naturally into portions of both of these plantations, allowing for a comparison of effects of the N fixer on soils and productivity in rich and poor sites.

In the infertile site, the presence of the alder has increased N and C content of soils (Table 14.2). Levels of extractable Ca and Mg are also higher when alder is present, along with a measured index of N availability. This has in turn increased N concentrations in Douglas-fir foliage, while lowering P concentrations significantly (Table 14.3). Net primary productivity above ground for the Douglas-fir alone is similar in the stands with and without alder, but the added productivity of the alder makes total net primary production more than twice that of the stand without alder (Table 14.3).

In contrast, the presence of alder has had much less effect on soil C and N in the fertile site, and the amount of extractable cations is actually lower. Foliar nutrient content of N and P is unaffected as is total net primary production, but that productivity is split between alder and Douglas-fir when the alder is present. This means an actual reduction in Douglas-fir growth in the presence of alder on the rich site.

These results show that N fixation can significantly alter soil characteristics, but the degree of alteration is related to initial site conditions. On the infertile site, N fixation increases N availability and tree growth. This has the secondary effects of increasing total stand and soil organic matter content but apparently reducing relative P availability. As the N limitation on growth is relieved, P limitation becomes enhanced. How N fixation increases cation availability is unclear, but some of the interactions between nutrient sorption, weathering, and cation exchange discussed in Chapter 9 may be at work. These changes in soil properties would be expected to increase growth of Douglas-fir even long after the short-lived alder has disappeared from the stand.

On the fertile site, soil N storage and N availability are already adequate for Douglas-fir, and the addition of alder does little to alter productivity or soil characteristics. In fact, competition between alder and Douglas-fir reduces Douglas-fir growth. In this stand, the alder does act as a weed, competing with the crop species and reducing its growth.

TABLE 14.2 Differences in Soil Properties in Fertile and Infertile Douglas-Fir Stands Growing With and Without Alder

(Binkley 1983)[a]

Property	Depth (cm)	Infertile Mt. Benson		Fertile Skykomish	
		No-Alder Site	Red Alder Site	No-Alder site	Red Alder Site
Coarse-fragment–free bulk density (kg l^{-1})	0–15	0.48	0.49	0.57	0.55
pH	0–10	4.5	4.4	4.2	4.1
	10–20	4.8	4.9	4.7	4.6
	20–35	4.9	5.0	4.9	4.5*
	35–50	4.9	5.1	4.6	4.6
N (%)	0–10	0.09	0.19****	0.31	0.29
	10–20	0.07	0.11****	0.21	0.25***
	20–35	0.06	0.08*	0.13	0.16*
	35–50	0.05	0.08**	0.09	0.14*
Available-N index ($\mu g g^{-1}$)	0–10	23	77****	82	61*
	10–20	15	52****	29	29
	20–35	39	48*	9	8
	35–50	35	56***	4	5
C (%)	0–10	2.05	3.92****	4.98	4.72
	10–20	1.61	2.13*	3.74	5.03****
	20–35	1.15	1.52*	2.11	2.85
	35–50	1.09	1.44	1.26	2.33**
Extractable Ca (mEq kg^{-1})	0–10	19.9	38.9**	12.50	5.25***
	10–20	7.2	21.0*	3.41	6.35**
	20–35	7.2	16.5*	1.74	1.44
	35–50	6.6	16.8**	2.00	1.29**
Extractable Mg (mEq kg^{-1})	0–10	4.4	6.8	2.10	1.03**
	10–20	1.8	3.9**	0.98	0.99
	20–35	1.5	3.0*	0.78	0.20***
	35–50	1.6	3.9**	1.10	0.43****
Extractable K (mEq kg^{-1})	0–10	2.4	3.7	1.32	1.21
	10–20	2.2	2.5	0.86	1.46*
	20–35	0.9	1.4	0.85	0.92
	35–50	1.0	0.9	0.83	0.77

[a] Asterisks in table body indicate significant differences with and without alder.

TABLE 14.3 Differences in Foliar Nutrient Concentrations and Total Net Primary Productivity by Species in Fertile and Infertile Douglas-Fir Stands Growing With and Without Alder
(Binkley 1983)[a]

	Infertile		Fertile	
	Without Alder	With Alder	Without Alder	With Alder
Foliar N (%)				
Douglas fir	0.93	1.41*	1.54	1.55
Alder	—	3.05	—	2.35
Foliar P (%)				
Douglas Fir	0.22	0.09*	0.14	0.16
Net primary production ($T \cdot ha^{-1} \cdot year^{-1}$)				
Douglas fir	6.9	6.4	23.2	15.5*
Alder	—	9.3	—	7.0
Total	**6.9**	**15.7***	**23.2**	**22.5**

[a]Asterisks in table body indicate significant differences with and without alder.

Cation Pumping

Nitrogen fixation is a well-known example of a process with important implications for nutrient cycling in ecosystems, but there are other, less obvious differences between species that can have similarly impressive effects. One of these is **cation pumping,** a term used to describe the tendency in some species to take up and cycle large quantities of the macronutrient cations Ca, Mg, and K.

The effects of cation pump species can be seen in Figure 14.7. This summarizes changes in the distribution of several nutrients and organic matter following 40 years of growth by four different species planted on old agricultural fields in Minnesota. Four different species were planted: (1) quaking aspen, (2) white spruce, (3) red pine, and (4) jack pine. The first two are known as high metal cation accumulators (cation pumps, white spruce differs significantly from red and black spruce in this respect), while the pines are not.

In Figure 14.7, the total nutrient content in the ecosystem, excluding that present in unavailable mineral forms, is divided into vegetation, forest floor, and mineral soil compartments. Because all four species were planted in a random design in a relatively homogeneous old agricultural field, all four systems had the same initial content of all nutrients. After 40 years, the total content of Ca remains the same for all four. However, under aspen and spruce, considerably more of the Ca has been redistributed to the vegetation and forest floor, while less remains in the mineral soil.

The effect of this redistribution on soil pH can be seen in Figure 14.8. The aspen and spruce stands have more basic forest floors (the O horizons) with higher Ca content, but the A horizons are more acidic. This is because of the removal of Ca from this horizon, which reduces base saturation.

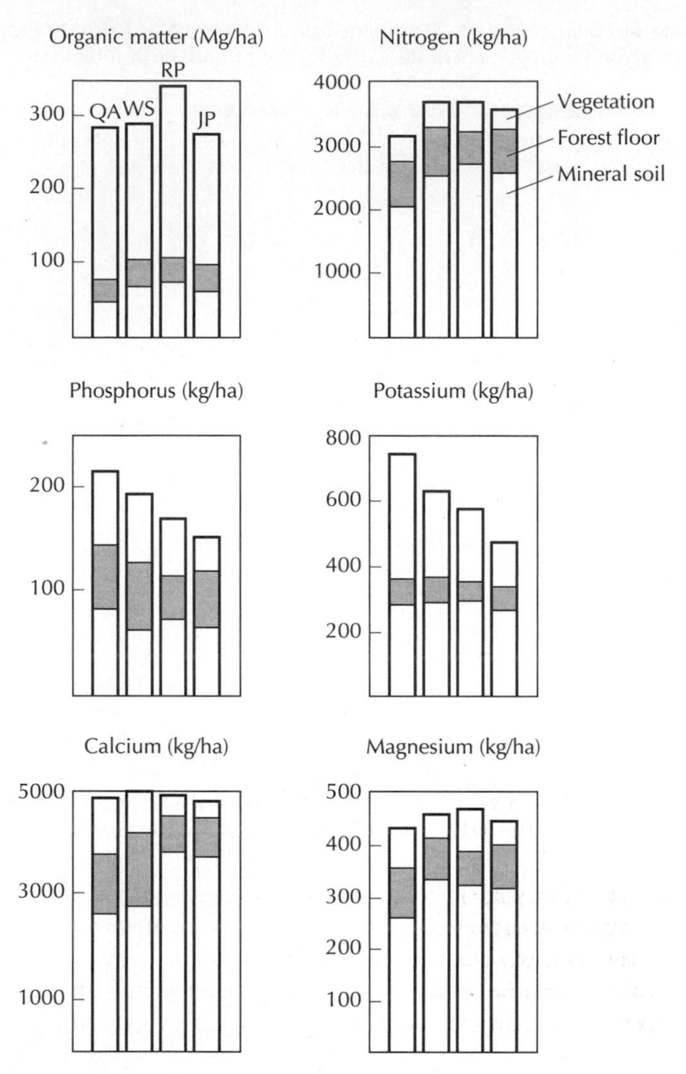

FIGURE 14.7 Distribution of organic matter and nutrients in vegetation and soils of four plantations with different dominant species. RP = red pine, JP = jack pine, WS = white spruce, QA = quaking aspen. (Alban 1982)

The results for K and P are also intriguing. The total measurable amounts of these nutrients have apparently increased in the aspen and spruce stands, relative to the pine stands. This could be a secondary effect of lower soil pH in the A horizon, which could lead to slightly higher weathering rates and the release of these elements from primary and secondary minerals. Other direct physiological processes for biological extraction of these elements have also been identified and may be at work here.

Resource Islands

Plants can also affect the horizontal and vertical distribution of elements. In Chapter 8, we discussed how sparsely distributed shrub canopies in arid shrublands can trap wind-blown particles, enriching the soils beneath (Figure

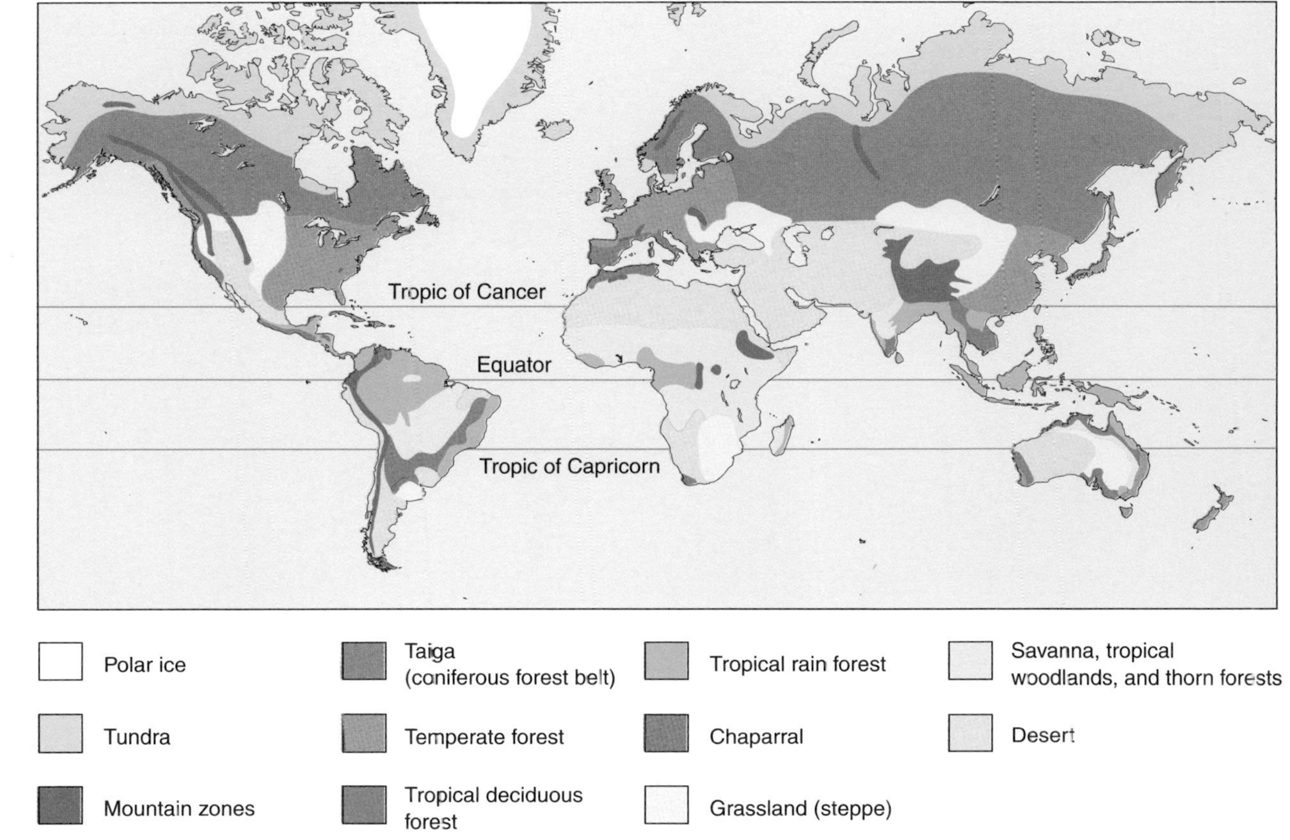

FIGURE 2.1 (b) Geographic distribution of major vegetation types of the world (Solomon, World of Biology, 5th ed, Saunders College Publishing, 1995)

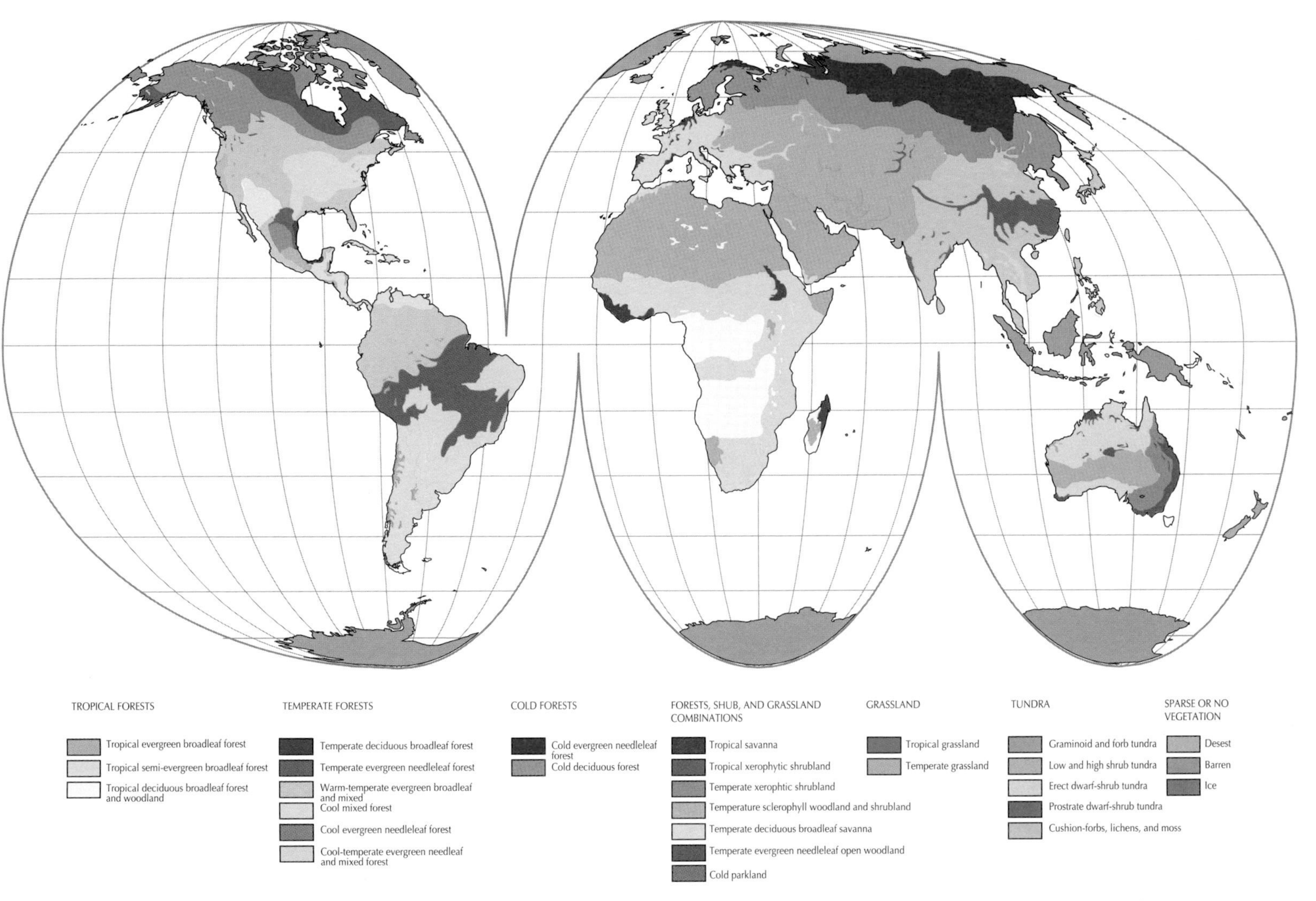

FIGURE 2.3 Comparison of current distribution of major biomes of the world as summarized from field observations. (Haxeltine and Prentice 1996)

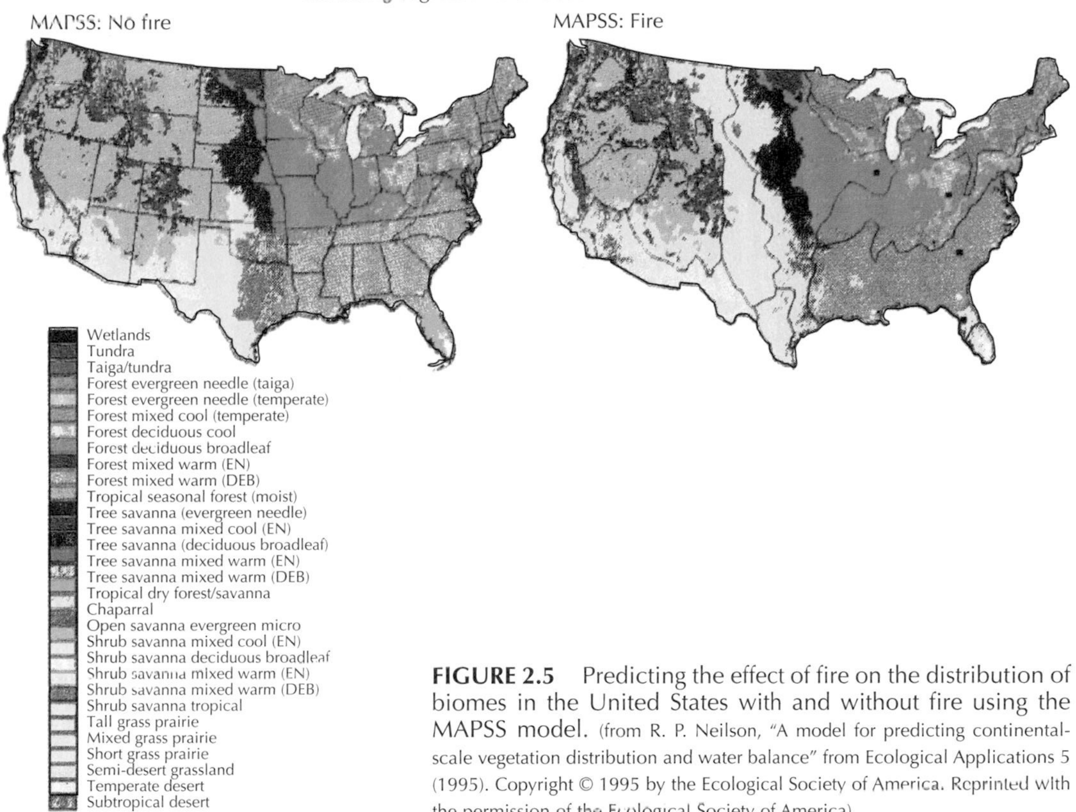

FIGURE 2.5 Predicting the effect of fire on the distribution of biomes in the United States with and without fire using the MAPSS model. (from R. P. Neilson, "A model for predicting continental-scale vegetation distribution and water balance" from Ecological Applications 5 (1995). Copyright © 1995 by the Ecological Society of America. Reprinted with the permission of the Ecological Society of America)

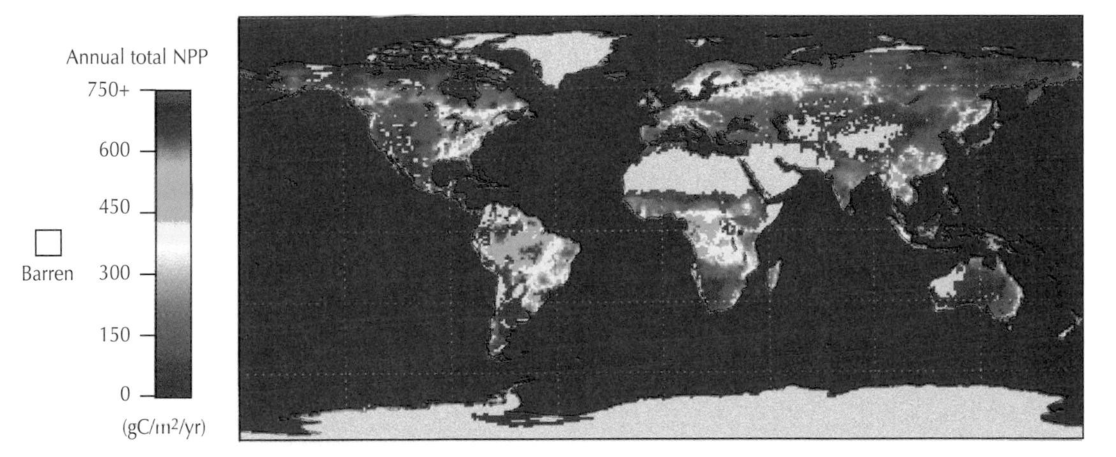

FIGURE 2.8 and Figure 8.14 Predictions of biomass production for ecosystems of the world. (*b*) A global map of net primary production derived using a value for the efficiency of conversion of sunlight to chemical energy by plants based on remote sensing data. (Waring and Running 1998)

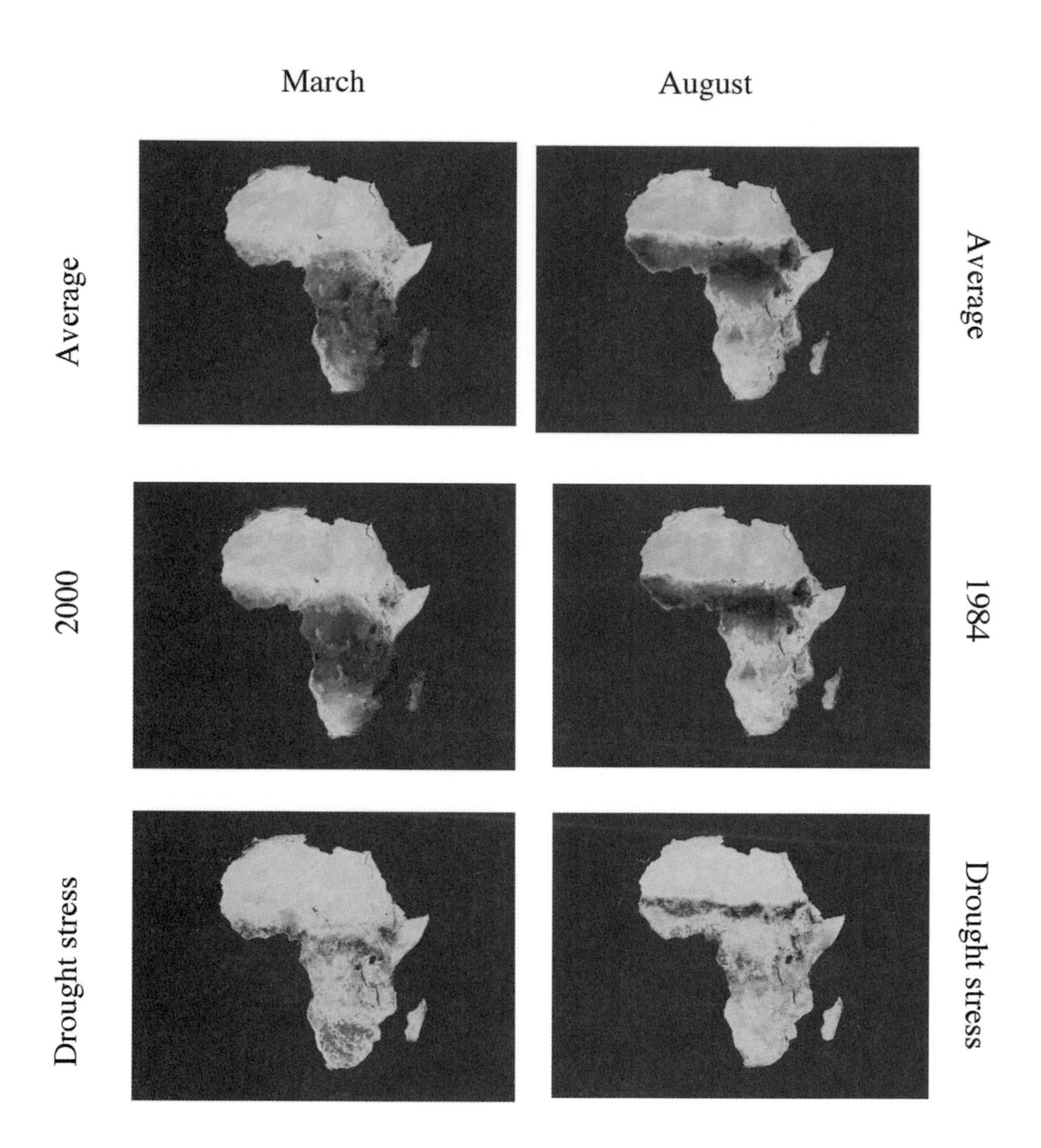

FIGURE 5.9 Mean and anomolous changes in NDVI, or "greenness," of vegetation in Africa. The top two frames show normal seasonal cycles tied to patterns of rainfall and temperature. The center two show maps for the same month, but in years of serious drought. The bottom two are a calculated difference, or anomoly, generated by comparing the two images directly above. (From C.J. Tucker and B.J. Choudhury. Copyright © 1987 by Elsevier Science Limited. Reprinted with the permission of Elsevier Science Limited)

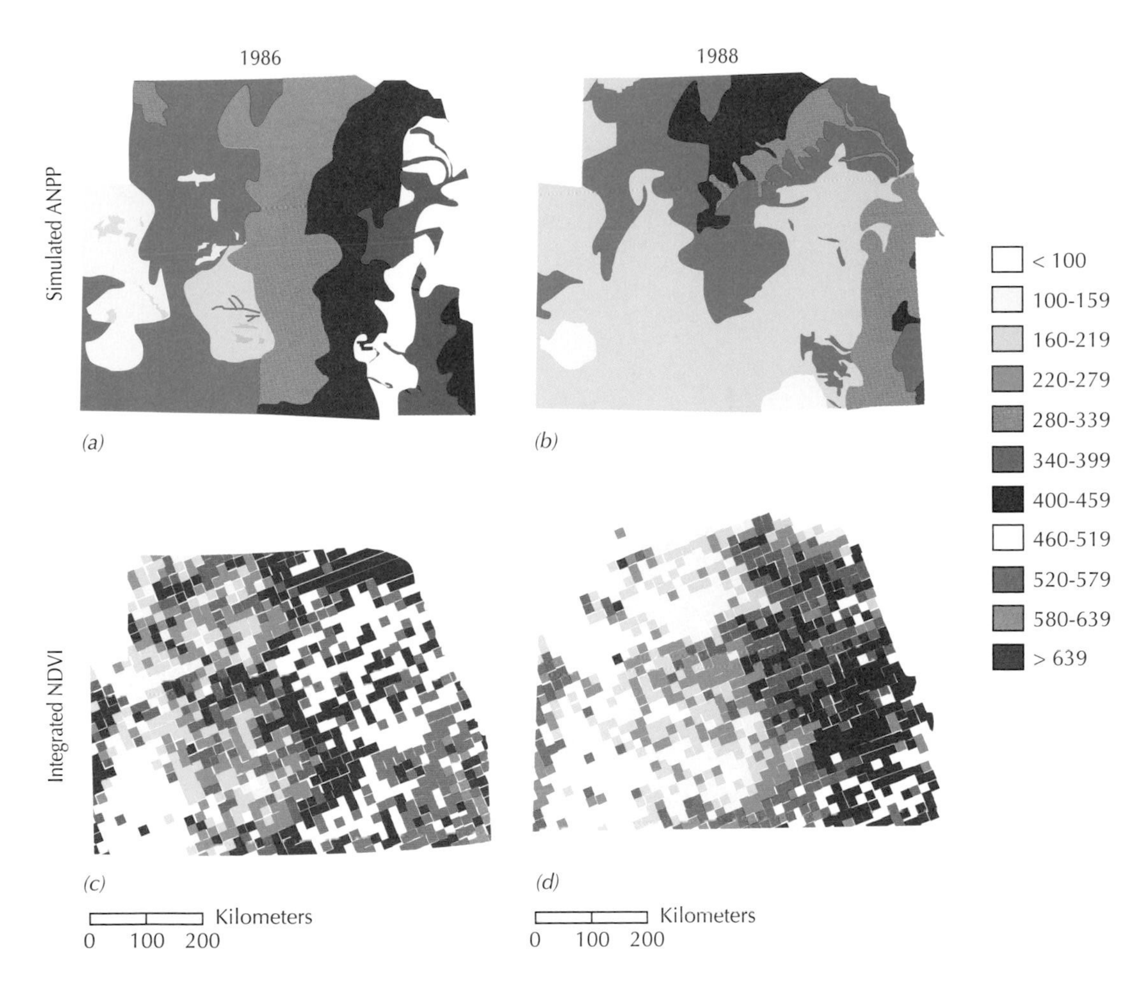

FIGURE 11.8 Predicted above-ground net primary production (ANPP) compared with the Normalized Difference Vegetation Index (NVDI) from remote sensing data for 2 years for the Central Plains region, including Kansas, Nebraska, and eastern Colorado. Differences in precipitation between the 2 years led to very different spatial patterns in ANPP as predicted by the CENTURY model. The strong correlation between predicted ANPP and NDVI suggests that the model predictions are valid. (Burke et al. 1991)

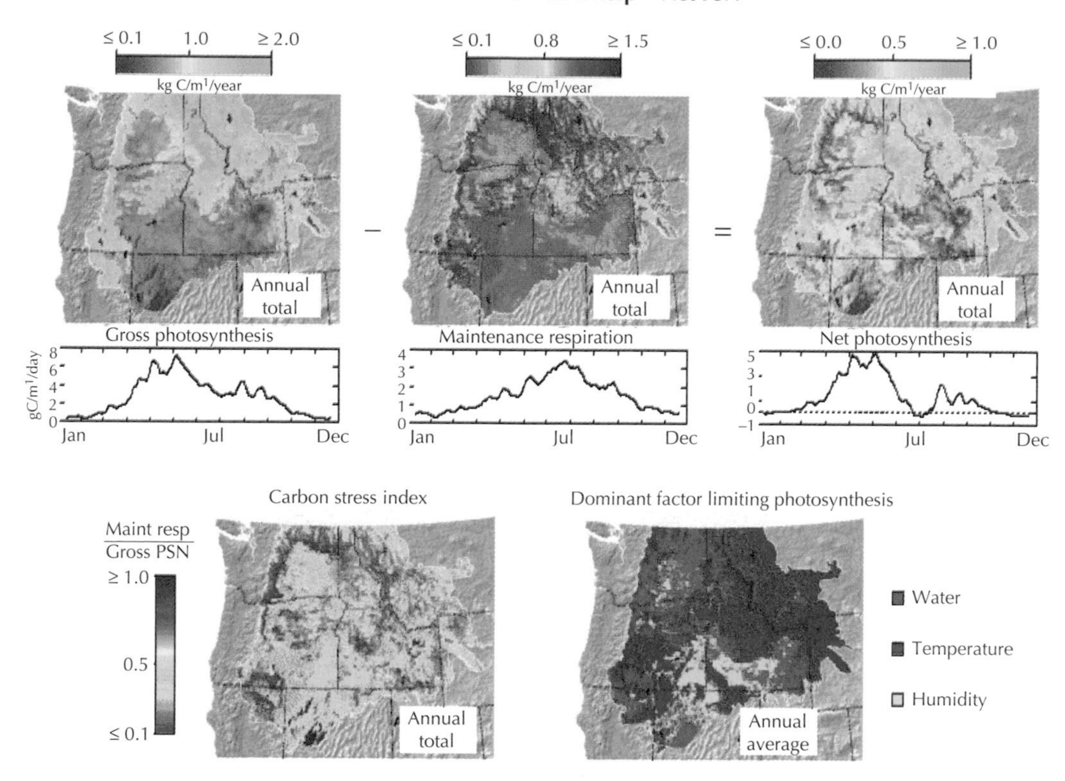

FIGURE 11.9 Extreme topographical variation in this region of the northern Rocky Mountains results in widely different climatic regimes, which are reflected in estimates of leaf area index derived from Normalized Difference Vegetation Index (NVDI) as obtained from remote sensing imagery. Using climate and NDVI as inputs, the Regional Hydroecological Simulation System (RHESsys) model predicts a highly variable pattern of ANPP as well. (Coughlan and Dungan 1997)

FIGURE 17.11 Picture of an intense crown fire in a dry western forest. (Bureau of Management, Alaskan Type One Incident Management Team)

FIGURE 20.4 The short-grass vegetation of the drier plains area of the Serengeti. (Photo courtesy of Julie Newman)

FIGURE 20.5 Acacia woodlands of the wetter northern and western portions of the Serengeti. (Photo courtesy of Julie Newman)

FIGURE 20.7 The three major species of herbivores in the Serengeti: (*a*) wildebeest and zebra. (Photo courtesy of Julie Newman)

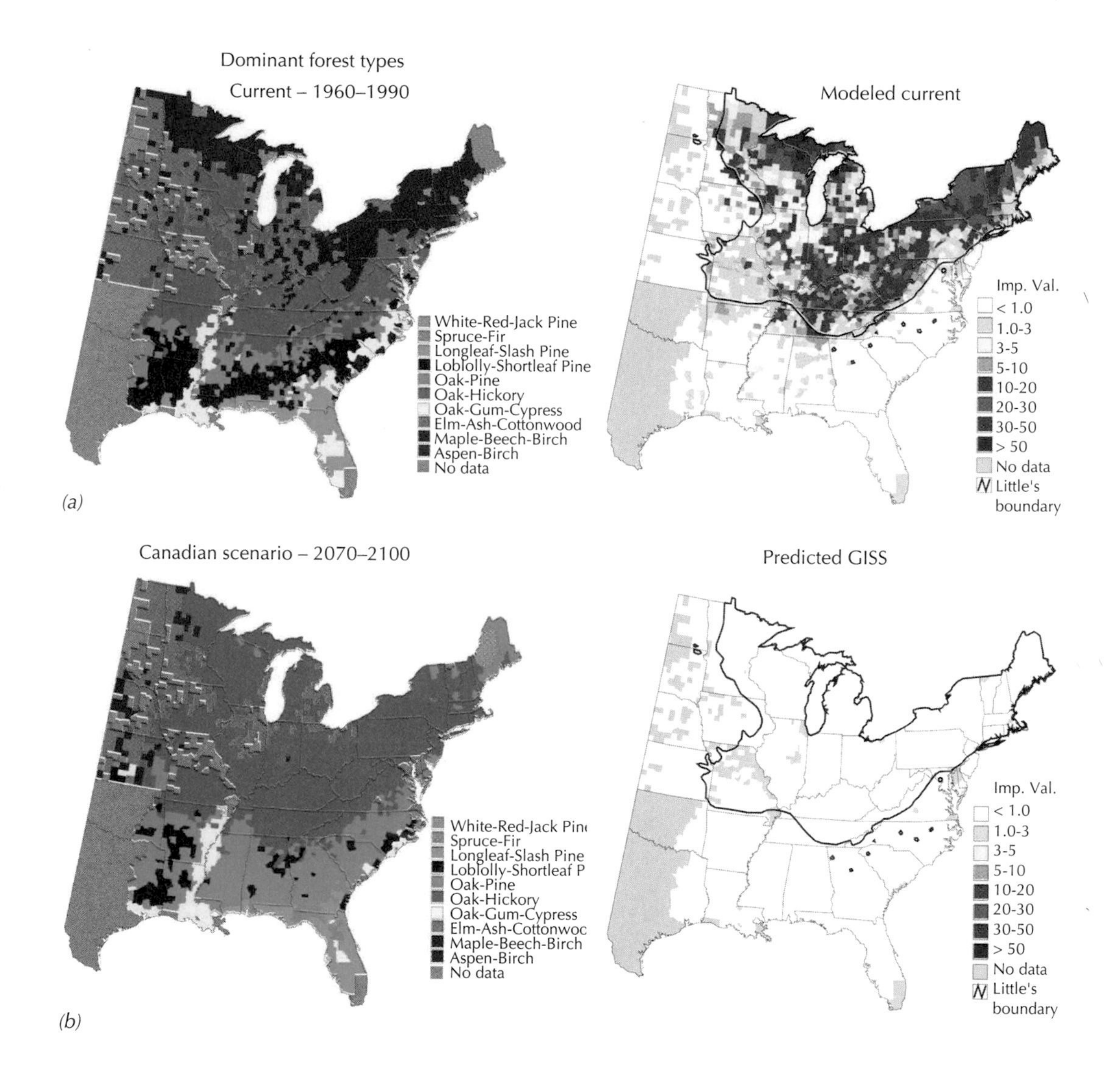

FIGURE 21.19 and FIGURE 24.17 Predicted changes in the distribution of (*a*) major forest types and (*b*) one dominant species (sugar maple) under different climate change scenarios. (Iverson et al., 1999)

FIGURE 24.16 Potential changes in the distribution of major vegetation types in response to predicted changes in climate, as derived from a combined vegetation-fire-biogeochemistry model. (Bachelet et at. 2001, cited in Aber et al. 2001)

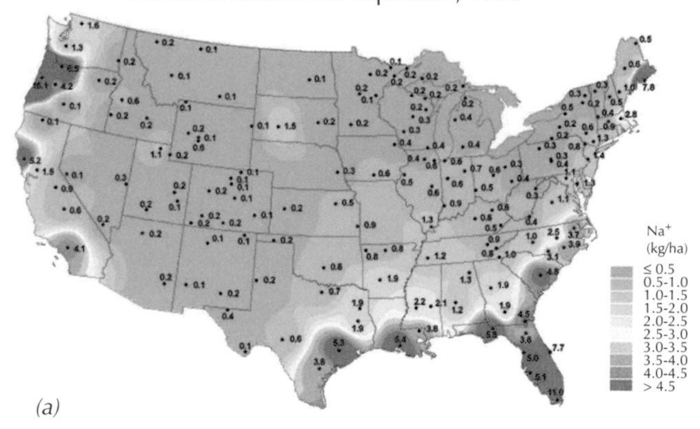

(a)

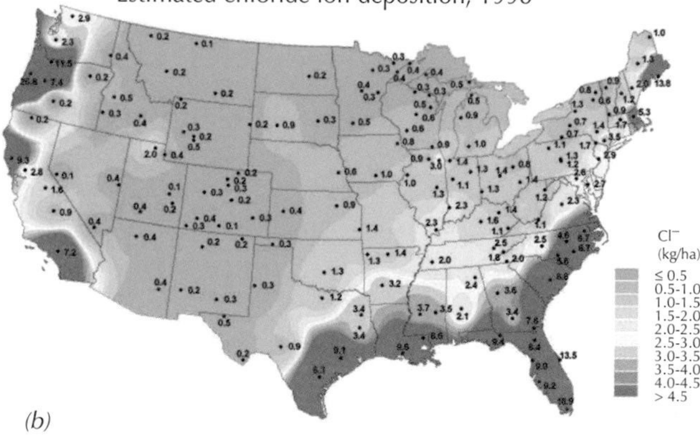

(b)

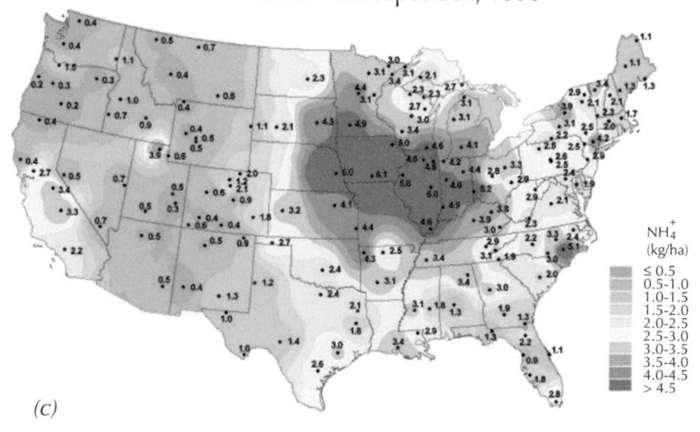

(c)

FIGURE 25.5 (*a–f*) Distribution of total wet deposition of selected chemical species as measured in 1998 through the National Atmospheric Deposition Program. (http://nadp.sws.uiuc.edu/)

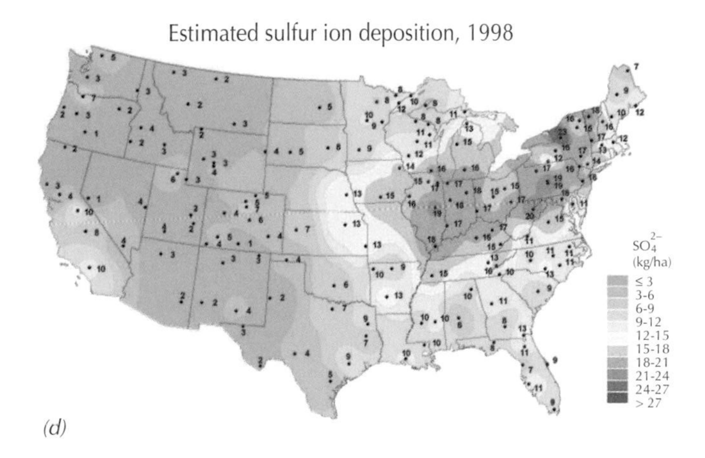

Estimated sulfur ion deposition, 1998
SO_4^{2-} (kg/ha)
≤ 3
3-6
6-9
9-12
12-15
15-18
18-21
21-24
24-27
> 27

(d)

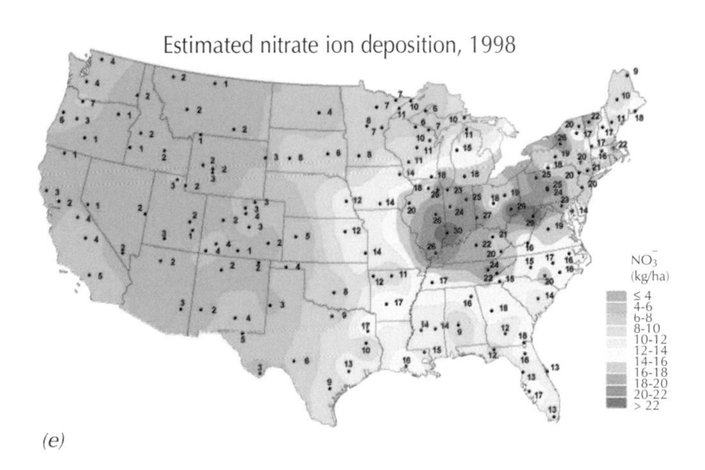

Estimated nitrate ion deposition, 1998
NO_3^- (kg/ha)
≤ 4
4-6
6-8
8-10
10-12
12-14
14-16
16-18
18-20
20-22
> 22

(e)

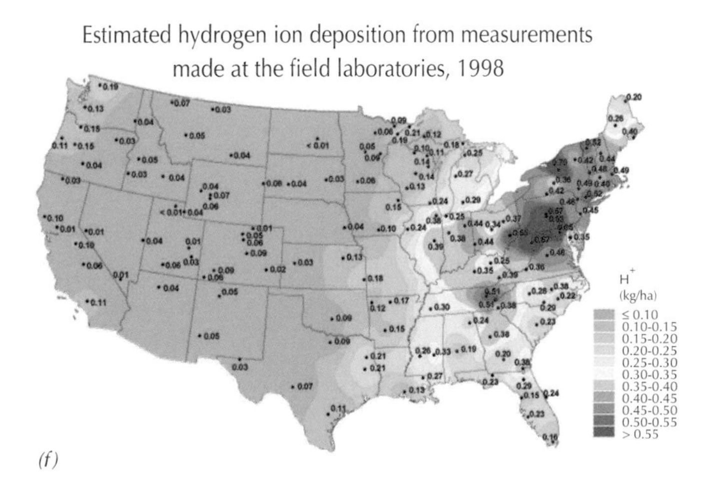

Estimated hydrogen ion deposition from measurements made at the field laboratories, 1998
H^+ (kg/ha)
≤ 0.10
0.10-0.15
0.15-0.20
0.20-0.25
0.25-0.30
0.30-0.35
0.35-0.40
0.40-0.45
0.45-0.50
0.50-0.55
> 0.55

(f)

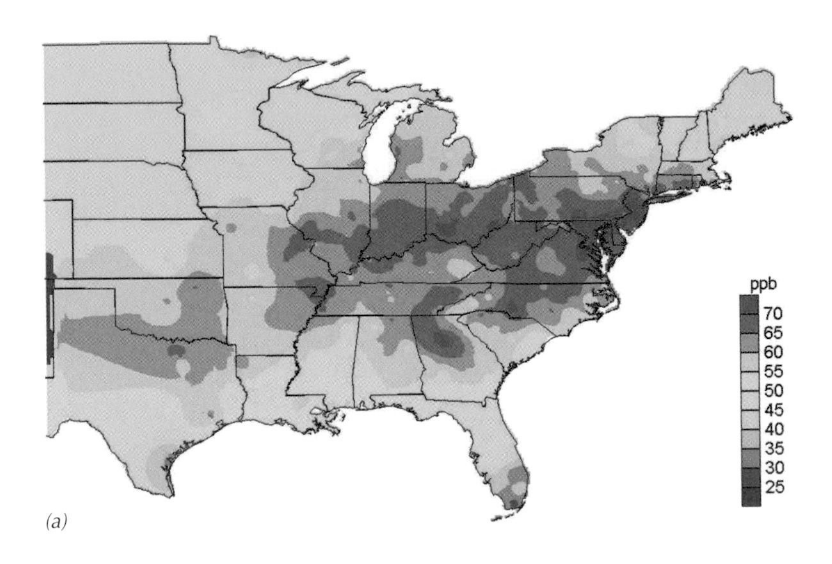

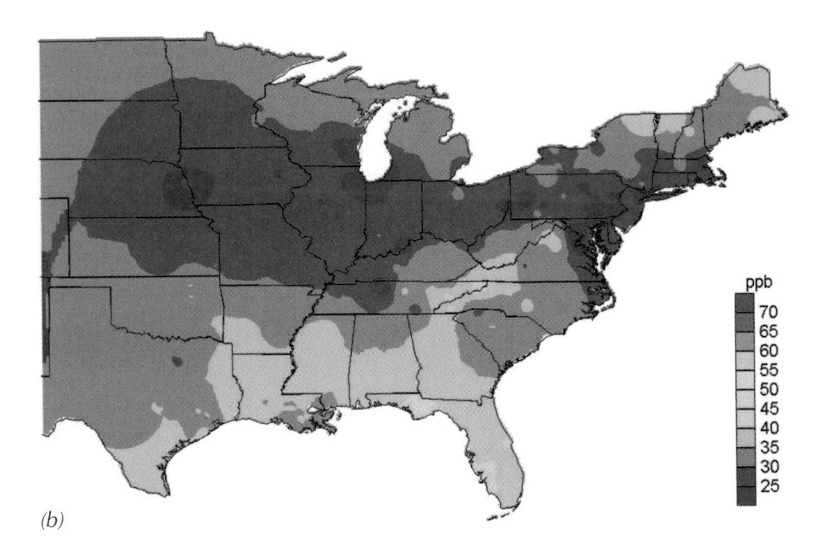

FIGURE 25.6 Distribution of the average daily maximum value for ozone concentration in the eastern United States in (*a*) summer and (*b*) winter. (Heilman et al. 2000)

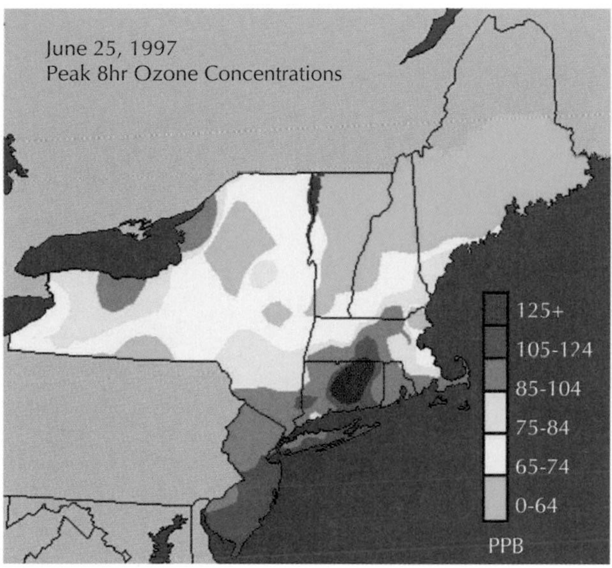

Source: NESCAUM

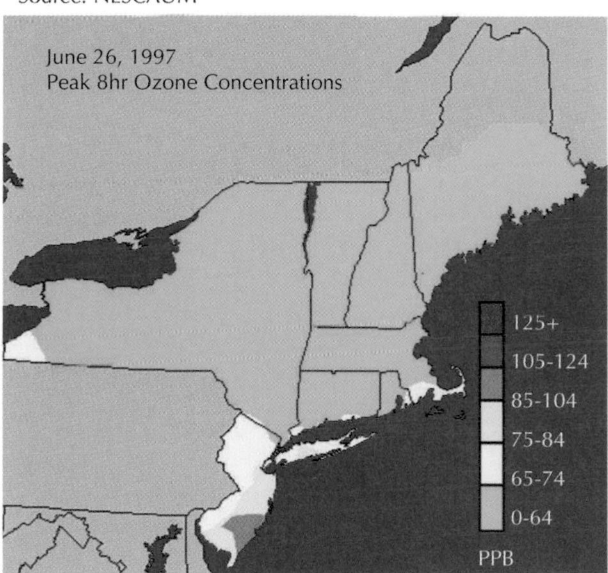

Source: NESCAUM

FIGURE 25.7 Change in maximum daytime ozone concentration on two consecutive days in the northeastern United States. Such large changes generally occur with the passage of a cold front accompanied by decreased temperature and humidity and increased wind speed. (NESCAUM—Northeast State for Coordinated Air Use Management)

FIGURE 25.9 (*b*) Sudbury Ontario and the impacts of three large smelting plants on vegetation. Pictures of vegetation in the different zones. (Photo courtesy of Keith Winterhalder. Reprinted with the permission of Kluwer Academic Publishers)

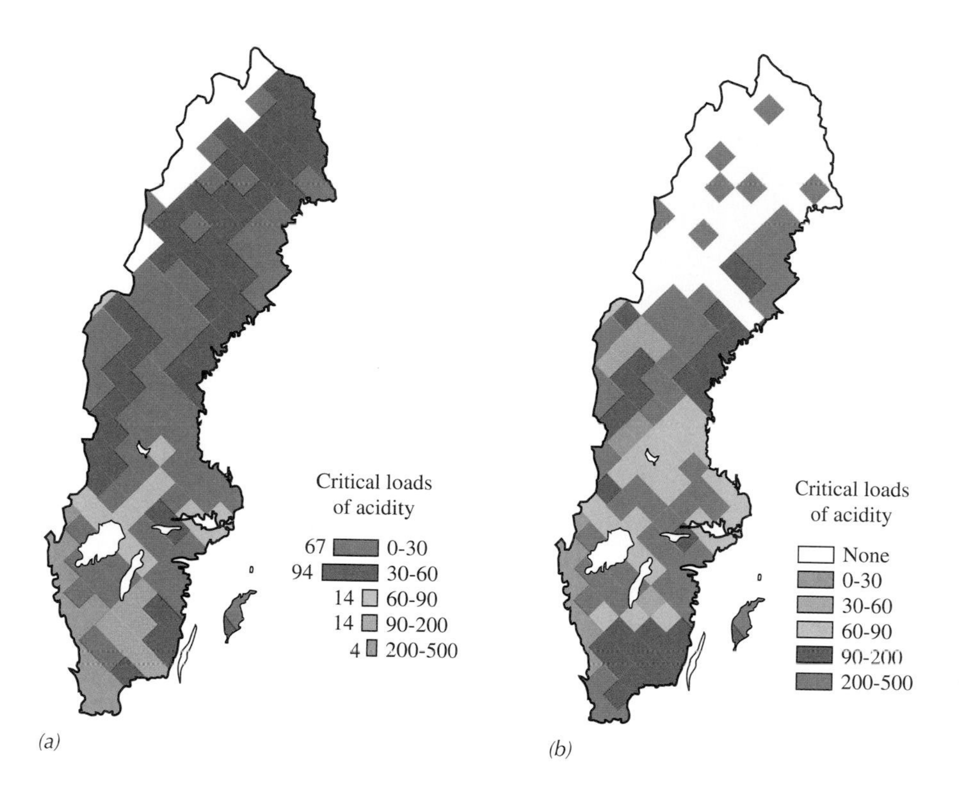

FIGURE 25.17 Calculated critical loads and exceedances for Swedish forests. (*a*) Calculated critical load based on acid neutralizing capacity of different soil types. (*b*) Estimated exceedances-the extent to which current deposition rates exceeds the critical load. (Warfvinge and Sverdrup 1995)

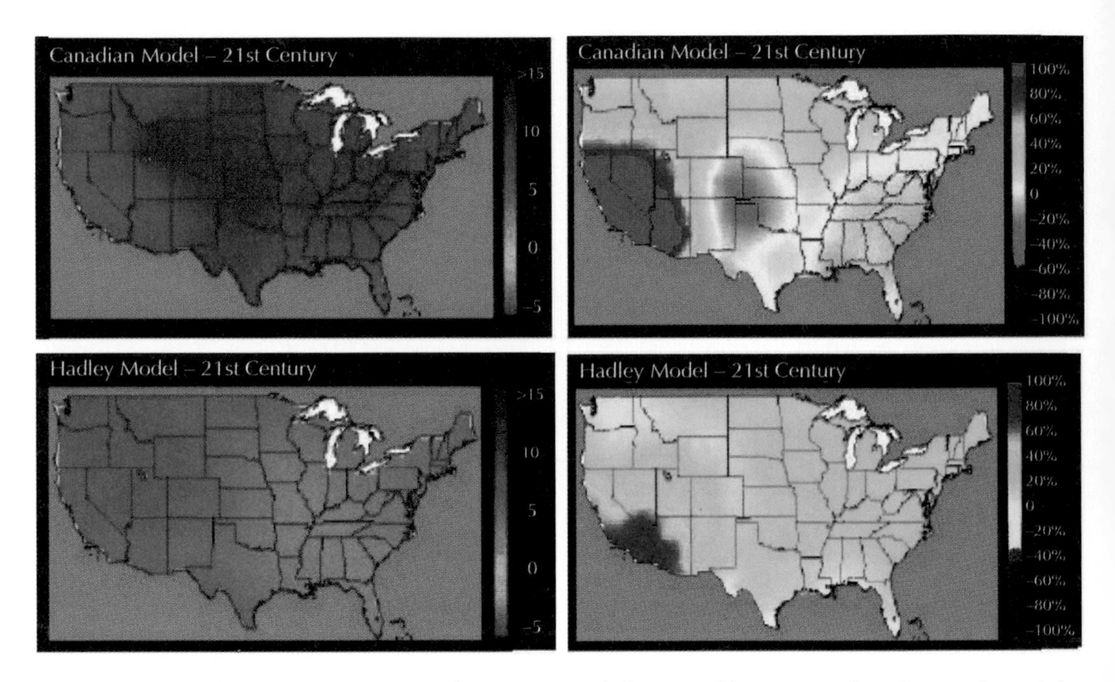

FIGURE 26.5 Changes in mean annual temperature (left top and bottom) and total annual precipitation (right top and bottom) as predicted by two different models. (National Assessment Synthesis Team 2000) Temperature changes are in degrees celcius. Precipitation changes are in percentage of current values.

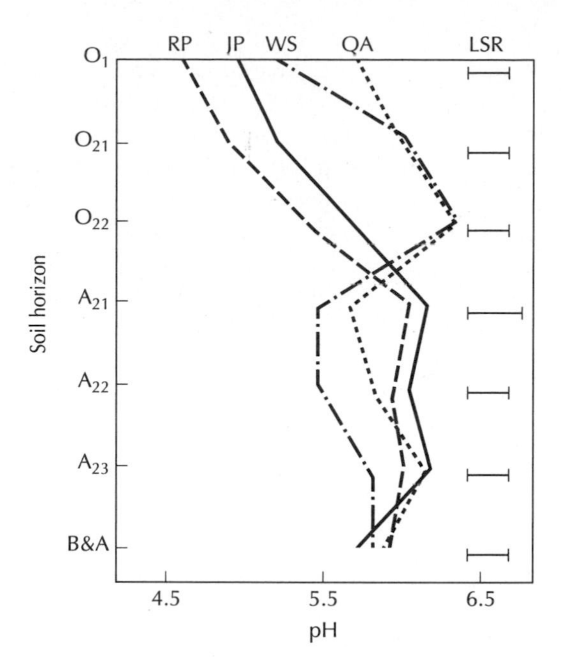

FIGURE 14.8 Differences in soil pH in four plantations with different dominant species. RP = red pine, JP = jack pine, WS = white spruce, QA = quaking aspen. (Alban 1982)

8.6). Big sagebrush is a dominant species in arid shrublands of the intermountain West in the United States. Competition for water in this dry environment leads to a somewhat regular distribution of plants. Root systems of adjacent plants overlap and fully occupy the soil, but large gaps occur between the poorly developed crowns (Figure 14.9).

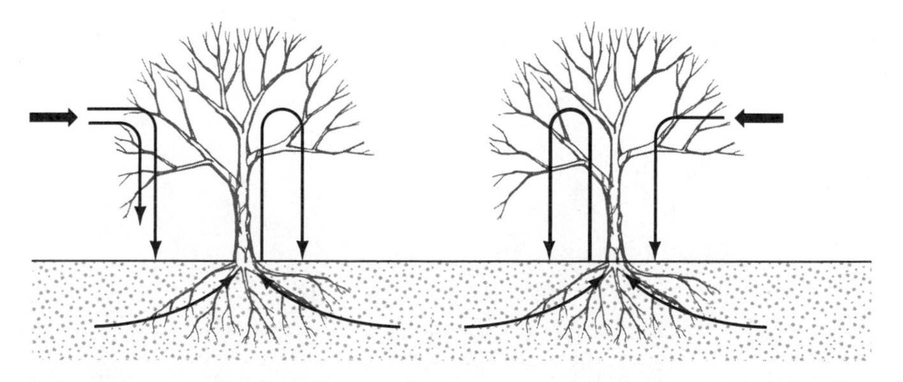

FIGURE 14.9 Schematic diagram of the creation of resource islands by widely spaced shrubs in semiarid systems. Scavenging of particles from the air by reducing wind speed, gathering nutrients with wide-spreading root systems, and localized deposition of litterfall all contribute to the concentration of nutrients and organic matter beneath and around the stems.

Wind erosion is an import process in this system in which the soil surface is exposed to the atmosphere rather than being covered with a protective layer of vegetation and litter. The sparsely distributed sagebrush plants act as miniature wind breaks, reducing wind speed and causing the deposition of dust and particles under the crowns. In addition, nutrients and water taken up from a fairly large area by the root system are used to produce foliage only within the small area occupied by the crown (Figure 14.9). As a result of both of these processes, organic matter and nutrients become concentrated under the crowns and are very unevenly distributed (Figure 8.6). This pattern is reinforced by root growth into these now enriched microenvironments, sometimes called resource islands.

REFERENCES CITED

Alban, D. H. 1982. Effects of nutrient accumulation by aspen, spruce and pine on soils properties. *Soil Science Society of America Journal* 46:853–861.

Binkley, D. 1983. Ecosystem production in Douglas fir plantations: Interactions of red alder and site fertility. *Forest Ecology and Management* 5:215–227.

Davidson, E. A., M. Keller, H. E. Erickson, L. V. Verchot, and E. Veldkamp. 2000. Testing a conceptual model of soil emissions of nitrous and nitric oxides. *BioScience* (in press).

Johnson, D. W., and S. E. Lindberg. 1992. *Atmospheric Deposition and Forest Nutrient Cycling.* Springer-Verlag, New York.

Johnson, D. W., and R. I. Van Hook. 1989. *Analysis of Biogeochemical Processes in Walker Branch Watershed.* Springer-Verlag, New York.

Kreutzer, K., C. Beier, M. Bredemeier, K. Blanck, T. Cummins, E. P. Farrell, N. Lammersdorf, L. Rasmussen, A. Rothe, P. H. B. De Visser, W. Weis, T. Weis, and Y. J. Xu. 1998. Atmospheric deposition and soil acidification in five coniferous forest ecosystems: A comparison of the control plots of the EXMAN sites. *Forest Ecology and Management* 101:125–142.

Nadelhoffer, K. J., J. D. Aber, and J. M. Melillo. 1983. Leaf litter production and soil organic matter dynamics along a nitrogen mineralization gradient in southern Wisconsin (USA). *Canadian Journal of Forest Research* 13:12–21.

Pastor, J., J. D. Aber, C. A. McClaugherty, and J. M. Melillo. 1984. Above-ground production and N and P cycling along a nitrogen mineralization gradient on Blackhawk Island, Wisconsin. *Ecology* 65:256–268.

15

FACTORS LIMITING CONSUMPTION

PLANT–HERBIVORE INTERACTIONS

INTRODUCTION

The transfer from plant to litter can be short-circuited by the consumption of live tissues by herbivores. In ecosystems dominated by woody plants, consumption is generally a much smaller transfer than senescence and litter fall. However, in systems dominated by low-growing, herbaceous species, such as the heavily grazed grasslands of Africa, large populations of herbivores regularly consume a significant fraction of the net annual production above ground. In other systems, such as the tundra, there may be large, cyclical changes in herbivory. Even forests and shrublands may be subjected to irregular irruptions of herbivore populations due to alterations in food webs through the introduction or removal of species or to stagnation (reduced vigor) in the dominant plant species.

The purpose of this chapter and the next is to discuss the processes that control levels of herbivory in terrestrial ecosystems. In this chapter, we will discuss the chemical and physical characteristics that reduce the palatability of plant materials and the evolutionary and ecological context in which these have developed. We will see that many chemical compounds that affect decomposition also affect herbivores and that symbioses between higher animals and microbes have evolved that increase the potential energy gain from ingested plant material. We will also see that the development and the form of herbivore defense employed depends on the physiological state of the plant, including growth rate and nutrition, reflecting the relative availability of carbon (C) and nitrogen (N) to plants. The diversity and complexity of compounds employed is staggering.

This discussion could be extended to include plant defenses against parasites and disease organisms. Many of the same principles apply. However, a full discussion of the chemical interactions at the cellular level that affect the success of disease organisms is beyond the scope of this book.

CONSUMPTION AS A FRACTION OF NET PRIMARY PRODUCTIVITY

Defoliation of large areas of forest by gypsy moths (Figure 15.1), large-scale tree death due to mountain pine beetles, or overbrowsing of forests by deer are conspicuous examples of consumption of a significant fraction of ecosys-

FIGURE 15.1 An example of extreme herbivory in a forested ecosystem: defoliation of a temperate deciduous forest in the northeastern United States by gypsy moth. (Van Bucher/Photo Researchers)

tem net primary production. Yet such examples are newsworthy partly because they do not represent the norm in these forested ecosystems. Herbivory in forests and shrublands generally consumes less than 10% of annual net primary production above ground.

In contrast with these low average levels, irregular irruptions of herbivore populations can completely strip a canopy of all of its foliage. Such large pulses of consumption, even if they occur only rarely, have the potential to alter the species composition and structure of the affected ecosystem. They also demonstrate that the factors generally maintaining herbivory at low levels in forests can, under certain circumstances, be overcome.

When populations of consumers are maintained at artificially high levels for long periods of time, the resulting changes in structure and function can appear permanent. An extreme example of this is the conversion of deciduous forests in central Pennsylvania to brush or grassland by deer. This region contains one of the highest deer densities in the United States because of the elimination of predators, restrictions on hunting, and the presence of an ideal mixture of forest and farmland habitat. When clearcuttings occur in this region, the resulting lush growth of highly palatable species can attract large numbers of deer that browse so completely that all trees are killed or severely cut back. In the most extreme cases, no tree growth occurs and the area eventually becomes dominated by bracken fern and other species less palatable to deer. The effect of deer can be seen in a comparison of a small, fenced plot that excludes deer (exclosure plot) with an adjacent open area (Figure 15.2). Species composition, production, and all other ecosystem processes are significantly different between these two plots.

FIGURE 15.2 Example of the effects of severe deer browsing on the structure of a forest ecosystem. The area outside the exclosure has been subjected to deer activity for several years following a clear-cutting, while the area inside has been protected. This site is in the Allegheny plateau of Pennsylvania. (Marquis and Brenneman 1981)

Unlike forests, where high levels of consumption are unusual, many grassland systems sustain very high levels of herbivory year after year. In these systems, natural selection has favored plant species with morphological and physiological traits that allow for rapid recovery from cropping by herbivores (see Chapter 16). Where herbivory is an important and recurring mechanism for returning plant production to the soil, the **removal** of herbivores causes significant changes in plant species composition and ecosystem structure and function.

STRUCTURAL AND CHEMICAL INHIBITION OF HERBIVORY

One major reason for the generally low level of consumption in nongrassland ecosystems is the chemical and physical complexity of plant materials. Leaves in particular, which have a high N content and could be a high-quality substrate for herbivory, contain a number of compounds that reduce their palatability.

Two general categories of compounds can be identified. The first and least diverse are the **"quantitative,"** or **general feeding inhibitors,** which are present in relatively high concentrations, reduce the potential energy gain from digestion, and so affect large groups of herbivores in similar ways. The second and more complex are the **"qualitative" inhibitors,** toxins that are effective in small concentrations but do not affect all herbivores.

A word of caution before we venture into this rich and complex subject that is tied so closely to evolutionary theory: The existence of a compound that inhibits consumption does not necessarily mean that this compound evolved specifically for this purpose. As we will see, the same compounds that inhibit digestion also affect decomposition. Compounds that serve as qualitative inhibitors can also be waste products from primary metabolism, can protect foliage from ultraviolet radiation, or alter frost hardiness. They can even serve as attractants for species that look very much like herbivores but that also pollinate flowers or spread seeds. Herbivory is only one of many forms of selective pressure that direct the production of secondary metabolites. Amid such complexity, simple and generalizable patterns may be hard to discern and support. Still, there are some emerging theories that attempt to bring order to this somewhat chaotic area of biology.

Quantitative Inhibitors

Not surprisingly, the generalized feeding inhibitors that affect the basic energy gain from digestion are the same compounds described earlier that control decomposition rate. Plant materials that have large amounts of lignin and small amounts of protein and simple carbohydrates make poor-quality food for herbivores large or small. Even cellulose can be difficult to digest for animals that lack the specialized symbiotic relationships with microbes discussed later. Both cellulose and lignin are considered indigestible "fiber" in the human diet.

Because the concentration of these compounds in plant tissues is what determines their effect on energy gain by consumption, they are called "quantitative" feeding inhibitors. When plant tissues are high in quantitative inhibitors, herbivores must take in and process larger amounts of material. This means using more energy for movement to reach the larger food mass required, as well as for the digestive process. In general, consumers of low-quality plant parts require more time to mature, and this increases the chances for predation, disease, or climatic extremes to reduce survival. Under the most extreme conditions, it may become physically impossible to process the volume of material required to extract enough energy and protein. An example is the "starvation" of certain herbivores of the Serengeti Plain in Africa who were found to have stomachs full of very low-quality grass produced because of very dry conditions. There is very little that can be achieved, evolutionarily, to overcome the basic biochemistry of energy and nutrient yield from plants rich in these general digestion inhibitors. One evolutionary path that is available is the development of symbiotic relationships between consumers and microorganisms more closely associated with decomposition.

In general, the digestive systems of insects and mammals cannot produce the enzymes required to degrade polyphenolics or even cellulose. However, diverse groups of animals have evolved a wide variety of symbiotic relationships with microorganisms that can. Perhaps the most familiar example is between ruminant mammals, such as cattle, sheep or deer, and the unique symbiotic microflora of their specialized "stomach," the rumen (Figure 15.3). The ruminant provides a constant, optimal environment in which the mi-

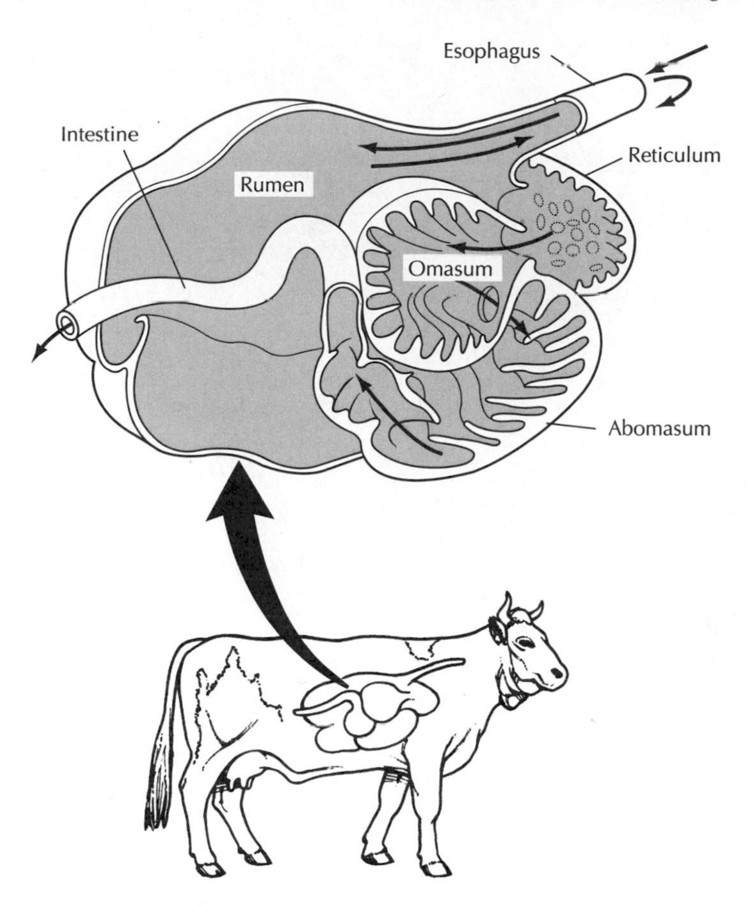

FIGURE 15.3 The structure of the rumen, a specialized digestive system that supports symbiotic microorganisms that assist in the digestion of high-fiber plant material. (Botkin and Keller 1982)

crobes partially decompose the otherwise indigestible fiber. The simple carbohydrates released by the breakdown of cellulose can be absorbed by the ruminant, causing a significant increase in the percentage of energy in the ingested material actually available to the animal.

There are other less pastoral, more exotic examples of this kind of symbiosis. One of the most quantitatively important, especially in tropical ecosystems, is in the termite. Termites support symbiotic microorganisms that both speed the decay of woody or high lignin substances and fix atmospheric nitrogen. The result can be an astonishingly rapid rate of consumption and degradation of even dry, seasoned wood. Termites can become a major factor in the structure and function of certain ecosystems, as indicated by the high density and unique properties of the large mounds that termite colonies produce (Figure 15.4).

In contrast to these physiological, internal, symbiotic relationships, there are interesting external symbiotic associations that rely on the behavior of the host. One example is the "fungal-farming" tropical ant that harvests and predigests poor-quality leaf litter, carries the mash to a specially constructed

FIGURE 15.4 Termite mounds dominate the landscape in many tropical grassland and woodland sites. (Georg Gerster/Rapho/Photo Researchers)

chamber, and regurgitates it into a colony of a specific fungus. The fungus decomposes the material, grows, and is harvested by the ant as its sole food. The chamber is constructed to maintain a constant temperature and humidity (Figure 15.5).

While these three examples are all very different, their effect is the same: to create an optimal micro-environment for the microbiological degradation of poor-quality organic matter. The decay processes themselves are not significantly different from those in litter, but the reactions are controlled for the mutual benefit of host and microbe. These are adaptations that overcome the inherent resistance of the material to degradation. Still, with the exception of the heavy grazing of some grasslands by ruminants, even these complex, highly evolved systems do not allow consumption of a sizable fraction of annual net primary production in most systems.

If producing nearly inedible tissues is such an effective barrier to consumption, why has evolution not led solely to plants with very high fiber and polyphenol content? One important answer is that such products are expensive to synthesize. More than 3 g of glucose is required to produce 1 g of lignin. Coniferous trees may have as much as 25–35% lignin in their leaves and even higher concentrations of similarly complex material (suberin) in roots. This represents a sizable investment of photosynthate that could have gone into height growth, new leaves or stems, or reproduction. All of these would, in the absence of a great deal of herbivore pressure, increase the competitive advantage of one plant over its neighbors. Once again, a tradeoff is involved. A heavy investment in these general consumption inhibitors would confer a selective, evolutionary advantage only if the protected tissues, such as leaves, were thereby able to last longer and so produce enough photosynthate to more than "repay" the "cost" of protection.

This particular tradeoff is expressed clearly at the physiological as well as structural level. The substrate from which polyphenols are produced can also

FIGURE 15.5 Diagram of chambers constructed by the fungal farming ant.

be used as a precursor in protein synthesis (Figure 15.6). The relative demand for this substrate by the two metabolic pathways reflects the relative availability of C and N in the plant and are also thought to reflect genotypic and environmental controls.

Qualitative Inhibitors

It is with the qualitative rather than the quantitative feeding inhibitors that the diversity of the products of evolution is so apparent. The tremendous diversity of compounds arises because each toxin does not affect the basic extractable energy content of the material but rather protects it with a dilute concentration of a toxic substance. For every toxic substance evolved, an herbivore eventually evolves a detoxifying mechanism that makes that plant edible again. This, in turn, increases the selective pressure for the synthesis of a new protective compound. Plants face numerous herbivorous species simultaneously, particularly insects. Each may be capable of detoxifying a wide array of previously evolved substances. Thus, a picture emerges of plants and herbivores in constant evolutionary movement, with herbivores continually pressuring plants employing qualitative inhibitors to develop novel combinations of toxins. Natural selection and evolution continually weigh the advantages of protection against the cost of synthesizing the protective substance.

Detoxification is not the only method for handling qualitative inhibitors. Herbivores can, instead, develop a tolerance for the chemical and then

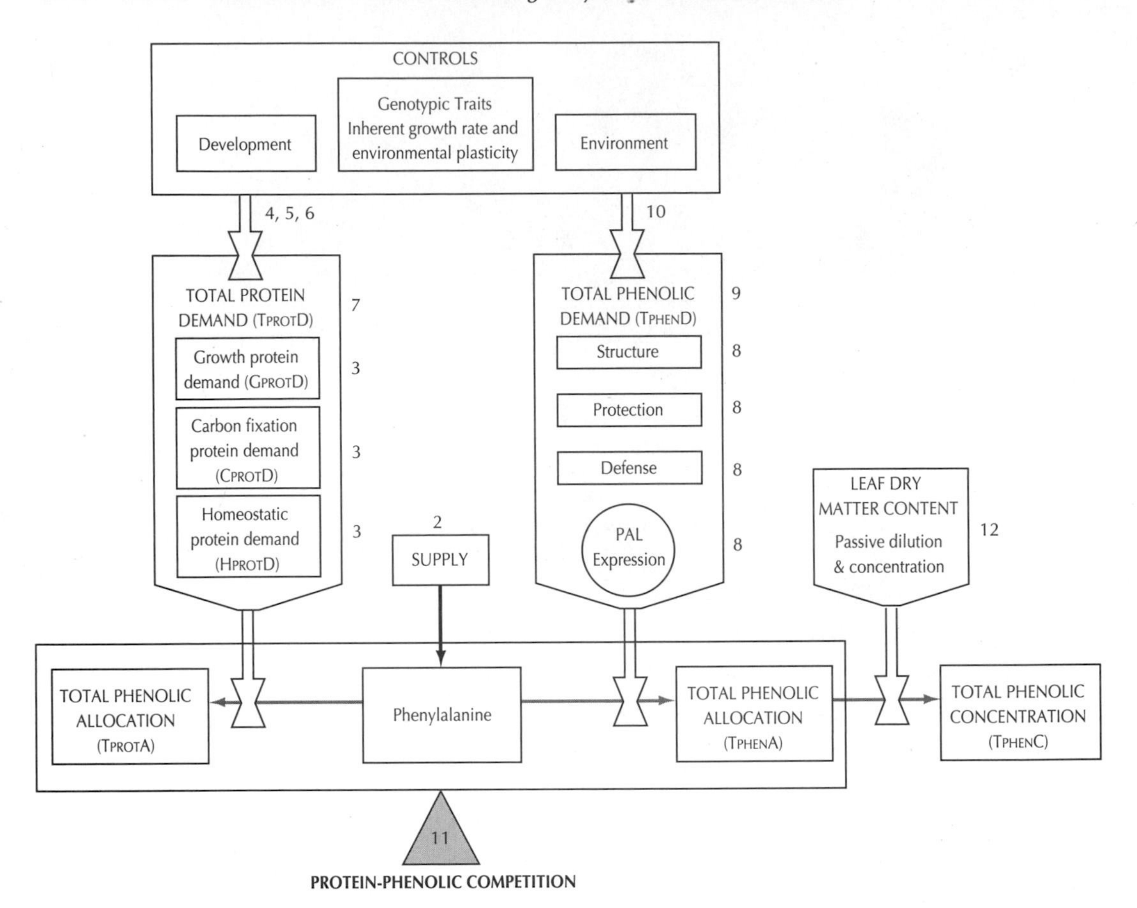

FIGURE 15.6 The protein competition model describing how plant development and environmental conditions alter the supply and demand for phenylalanine, a precursor for the production of both proteins and polyphenols. (Jones and Hartley 1999)

actually incorporate it into their own tissues. The monarch butterfly feeds primarily on different species of milkweed plants. Milkweeds produce a qualitative inhibitor that effectively reduces herbivory by other insects. The feeding monarch pupa (caterpillar stage), however, incorporates the toxin into its tissues, where it is stored and maintained even into the butterfly form, which is then unpalatable to most predators.

There may be more than 20,000 secondary plant compounds that affect the efficiency of herbivory. We will probably never know the actual total. Many have similar basic structures with relatively minor modifications (Figures 15.7 and 15.8). The importance of such minor changes indicates the extreme selectivity and sensitivity of both the biological processes they disrupt and the detoxifying reactions that confer protection.

The vast array of herbivore defense chemicals can be separated into two groups depending on whether or not they contain nitrogen. Carbon-based defense compounds, those without nitrogen, include isoprenoids such as terpenes (Figure 15.7) and sterols, in addition to the quantitative inhibitors

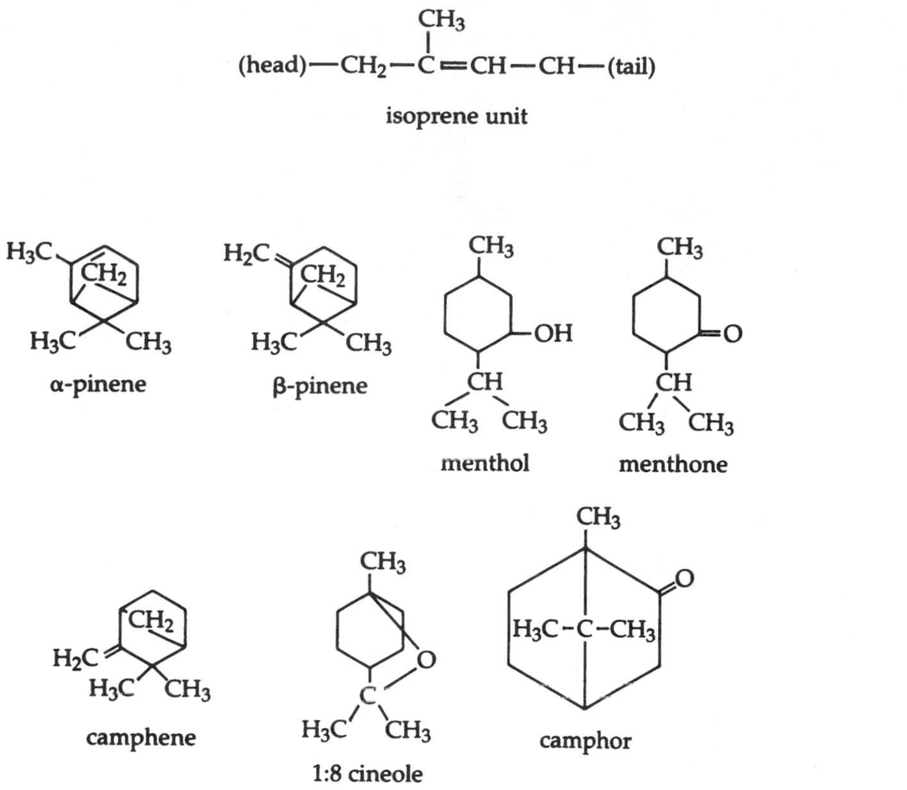

FIGURE 15.7 Generalized structure of isoprenoid compounds and several examples of terpenes, one class of carbon-based, qualitative herbivore defense compounds.

lignin and cellulose. Nitrogen-based defense compounds include alkaloids, among which are several potent medicinal drugs (Figure 15.8), and glucosinates and cyanogenic glycosides.

Structural and Temporal Complexity

The task of finding suitable substrates for herbivory can also be made more difficult by creating complex spatial and temporal patterns of tissue quality. It has been shown that the concentration of herbivore inhibitors can change dramatically across the blade of a single leaf (Figure 15.9), and that the relative proportion of different types of inhibitors presented can vary within a species over time (Figure 15.10). This can be seen essentially as a method for making microscale patches of easily digestible foliage (which should also have higher rates of photosynthesis) less "apparent" (see apparency theory below). It is interesting that the data in Figure 15.10 show a shift from qualitative inhibitors (Figure 15.10*c–f*), and are more important for a short period in the beginning of the year, while the quantitative inhibitors (Figure 15.10*a-b*) increase in importance later on (the accumulation of mineral silicate in

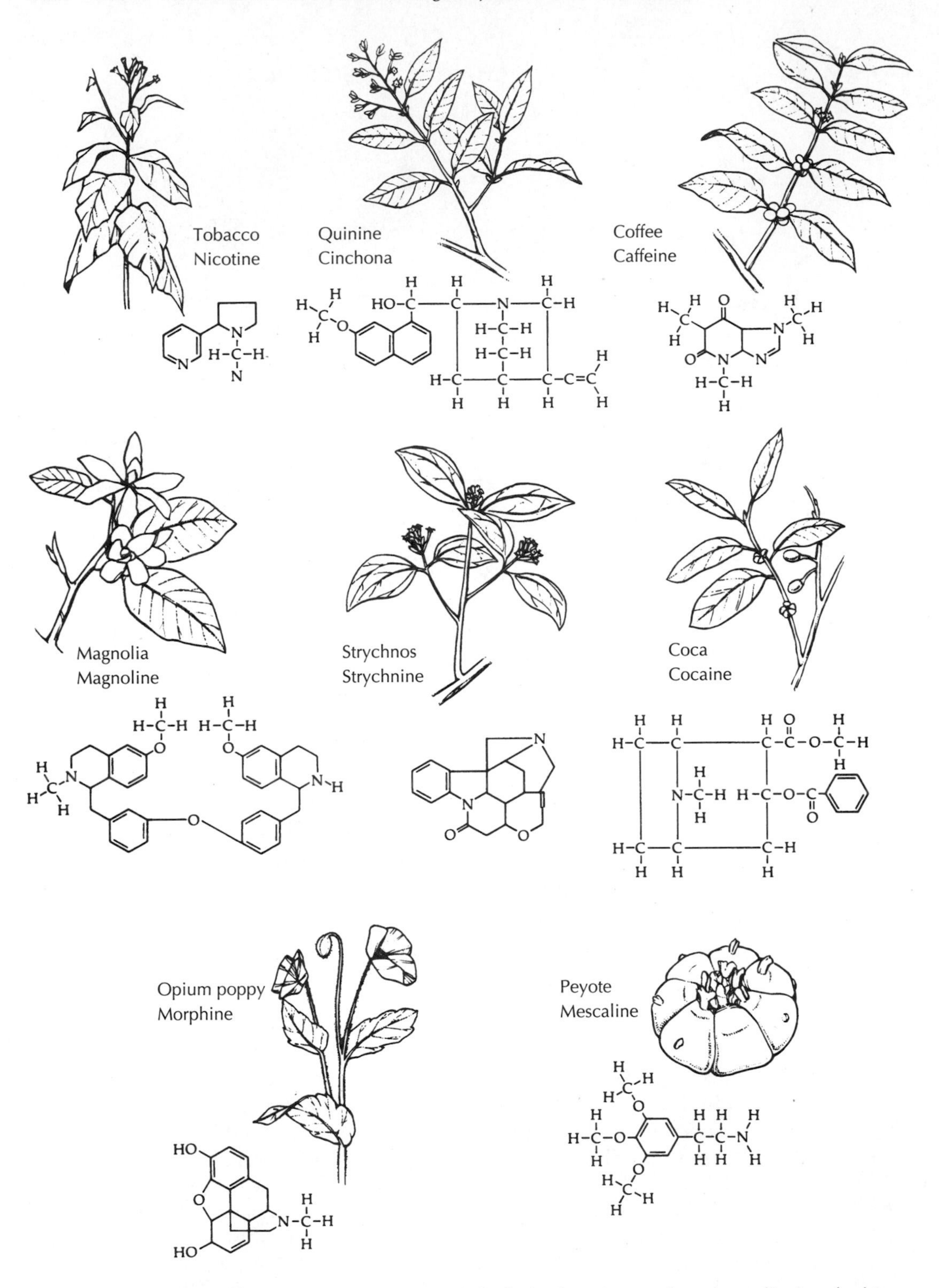

FIGURE 15.8 Structure of selected alkaloids—nitrogen-based, qualitative, herbivore defense compounds—and the plants that produce them.

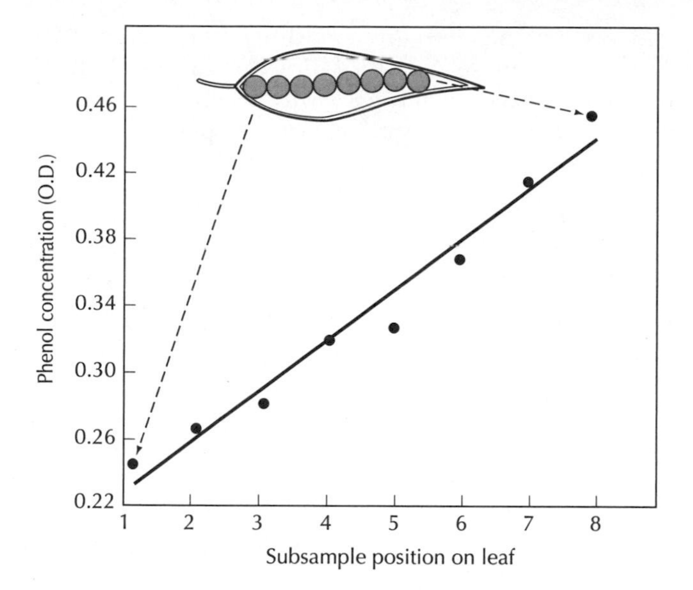

FIGURE 15.9 An example of spatial complexity in the concentration of herbivore inhibitors across a single leaf. The change in phenol concentration with location is shown. (Whitman 1983, cited in Hartley and Jones 1997)

foliage is an important defense strategy in many prairie grasses, and some other plants as well).

PATTERNS OF HERBIVORE INHIBITOR PRODUCTION IN PLANTS

We have now seen that chemical defenses against herbivory can be either quantitative or qualitative, and that the qualitative defenses can be either carbon- or nitrogen-based. There are a number of theories that have been

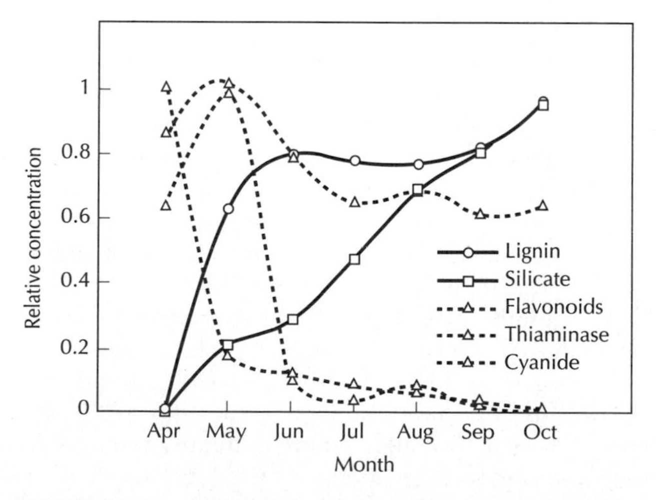

FIGURE 15.10 Relative abundance of quantitative (solid lines) and qualitative (dashed lines) herbivore inhibitors over time in fronds of bracken fern. (After Jones 1983)

developed to predict the diversity and distribution of herbivore defense strategies in plants. Of these, four deal directly with the distinction between quantitative and qualitative defenses. Two explain evolutionary differences between species and two describe adaptations by species growing on different sites.

Evolutionary Theories

Apparency theory suggests that the quantitative inhibitors (lignin, cellulose) will predominate in species that are long-lived or occur in high densities over large areas. This is because of the large numbers of herbivore species that a plant of such a species would encounter during its lifetime. Each of these many species of herbivores would be preselected to detoxify a subset of qualitative inhibitors that might be produced, and together might overcome the entire suite that a species could present. To simultaneously and continuously evolve new sets of qualitative inhibitors effective against all of the potential herbivores could be impossible.

Alternatively, rare species, or those that are important for only short periods during the development of the plant community, gain a measure of defense by being hard to find (less "apparent"). There may be relatively few herbivores adapted to seeking out rare species or the disturbed sites in which many of the less apparent plant species grow. Producing a restricted set of qualitative inhibitors to reduce the efficiency of these few herbivores may be less of an energy drain than producing large amounts of the general compounds, and thus confer a selective advantage in the evolution of these species.

The **resource availability theory** suggests that site quality and leaf longevity, rather than apparency, determine whether quantitative or qualitative defenses are used. Qualitative defenses are less expensive to produce because they are required in such low concentrations. However, they are not chemically stable, and need to be replaced continuously. In contrast, the quantitative defenses are stable and need to be produced only once. As leaf longevity increases, it becomes less expensive to produce quantitative inhibitors once, during leaf development, than to continuously produce qualitative inhibitors (Figure 15.11). If nutrient poor sites are dominated by species with low foliar nitrogen concentrations, then declining site quality should be related to declining nitrogen content and photosynthetic rate, increasing foliar retention time, and greater reliance on quantitative defense compounds. Interestingly, the high contents of foliar lignin implied by the reliance on quantitative inhibitors in poor sites should also reduce decomposition rates and perhaps cause further reductions in site quality.

The resource availability theory also predicts that the use of carbon- or nitrogen-based inhibitors will depend on the relative availability of carbon and nitrogen to plants. Tropical forests in which phosphorus may be more limiting than nitrogen, and in which there may be a significant component of nitrogen-fixing species, have a higher proportion of species that use the nitrogen-based defenses. In systems where light and water are more available than nitrogen, the carbon-based defenses predominate. For example, large amounts of terpenes are produced in many coniferous forest ecosystems.

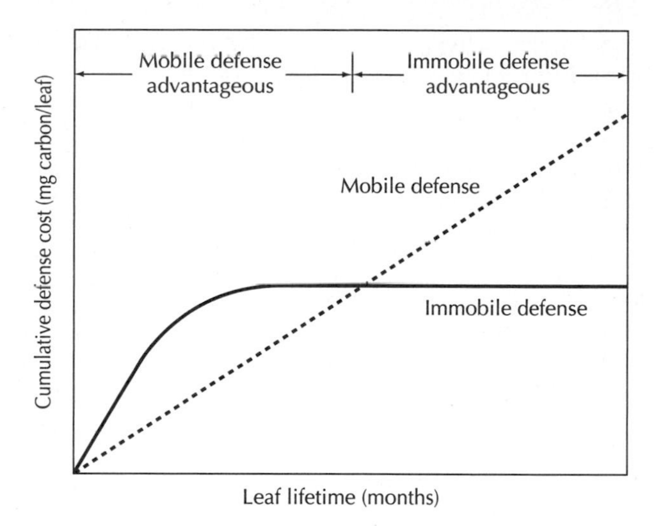

FIGURE 15.11 Predicted relationship between leaf longevity and the type (quantitative or qualitative) of herbivore defense compound employed. (Coley et al. 1985)

In section three, we will discuss ways in which the availability of resources, particularly nutrients, can decrease significantly with time in many ecosystems. This can reduce site quality and plant vigor to the extent that herbivore defense compounds cannot be produced in sufficient quantity. An example of this is in the ponderosa pine ecosystems of the Rocky Mountains. For these systems it has been shown that reductions in tree vigor (measured as the ratio of diameter increment to foliage surface area) significantly increase the probability of mortality due to mountain pine beetle attacks (Figure 15.12). Experimental work has also shown that thinning and fertilizing the stand,

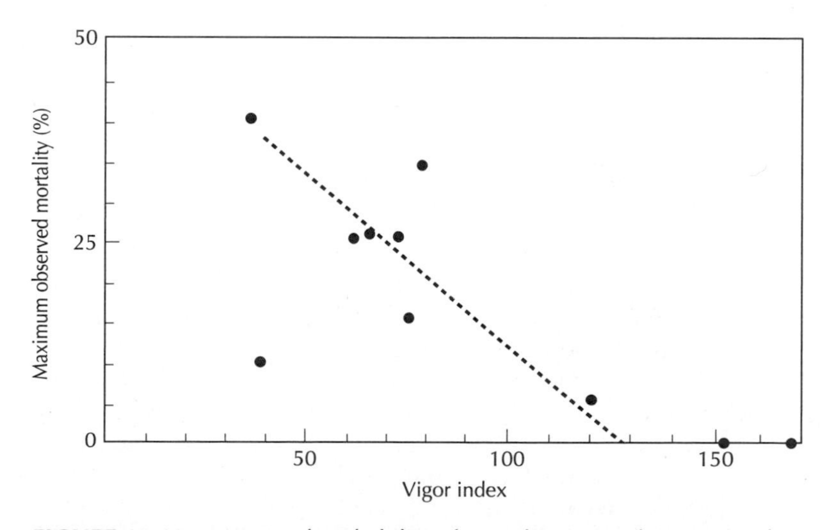

FIGURE 15.12 Measured probability of mortality in ponderosa pine due to mountain pine beetle attack, as a function of tree vigor in the stand. (After Mitchell et al. 1983)

reducing the number of trees, and increasing nutrient availability can increase tree vigor and reduce mortality.

Both apparency and resource availability effects may be at work in natural ecosystems. Since recently disturbed sites tend also to have high resource availability and support rapid plant growth, those early successional species that are less apparent also tend to be those with fast growth rates and fast leaf turnover.

From this discussion it should not be surprising that by far the greatest number of specific defense compounds are found in plants from the tropical rain forests. These systems typically have very large numbers of rare species. As many as 100 species of canopy trees may occur per hectare, with perhaps three or four individuals per species. These same systems also tend to have high densities of nitrogen-fixing species, which increases the reliance on nitrogen-based defenses.

Physiological Theories

Both of the evolutionary theories have ecological or physiological equivalents.

The carbon/nutrient balance theory holds that species can shift from carbon to nitrogen-based defense compounds depending on the relative availability of C and N to plants. For example, a plant growing on nitrogen-poor soil will have an excess of carbon relative to nitrogen. Carbon-based defenses are likely to be more important under these conditions.

The growth/differentiation theory builds from the process of leaf tissue growth. Simple carbohydrates and amino acids in plants can be converted either into new cells for growth, or into specialized cell constituents, including herbivore defense compounds that differentiate cell functions. Greater allocation to growth leaves fewer resources for differentiation and the production of defense compounds. Allocating less to defense frees more resources for faster growth. In both cases, slower growing individuals may have more resources to apply to herbivore defense.

IMPLICATIONS

There are two important implications of the trends discussed superficially in this chapter. The first is the potential importance of the tropical forest regions of the Earth as a source of biochemicals of value to society. Evolution has created a vast repertoire of compounds that can affect the human body in unpredictable ways. Some of these compounds are well known, such as the alkaloids nicotine, morphine, and cocaine. Numerous synthetic drugs are simple modifications of natural products.

Thousands of additional secondary products of plant metabolism remain undiscovered in the tropical forests. The potential for these to reduce human suffering (or increase it?) is a genuine concern of many whose goal is to preserve the species diversity of the tropical rain forest by preserving large areas intact. Because of the low numbers of individuals per species in an area and the large number of species in total, reducing the area in rain forest by conversion to forest plantations or permanent agriculture could result in the loss of hundreds (thousands?) of distinct plant species plus their associated insect species. Each could carry with it the genetic capability of producing a specific compound of great benefit.

A second implication arises from the fact that many of our important crop plants, such as corn, are derived from tropical or semitropical species of disturbed habitats. They are also fast-growing and low in quantitative feeding inhibitors. Under natural conditions, rarity probably conferred some protection from attack, along with qualitative inhibitors for specific pests. We have taken these species, bred much of their natural genetic diversity and energy-consuming defense mechanisms out to increase yield. We plant them in large fields dominating whole regions. It is not surprising that the crop yield gained comes at the expense of substantial investments in pesticides to protect the growing plants. Fossil fuels are used as an energy subsidy so that the metabolic energy from photosynthate in plants may be genetically directed to yield. As fossil fuels become more expensive, society may find it less expensive and more beneficial to make use of natural diversity and the natural defense mechanisms of plants to control herbivory in crops. Alternatively, genetic engineering offers the potential to "program" the production of existing or unique inhibitors into the metabolic machinery of crop plants.

All of this is much removed from those natural ecosystems in which the interplay between chemical defenses and herbivores causes consumption to be generally a small proportion of net primary productivity. Of course, these same ecosystems also produce little food for human consumption! Still, systems with generally low rates of herbivory can experience cyclic or irregular irruptions of herbivore populations that decimate the canopy. In grasslands, herbivory can claim a sizable fraction of above-ground net primary production each year. In Chapter 16, we will examine the characteristics of those systems that support high rates of herbivory, the evolutionary adaptations that have been made by plants in heavily grazed systems, and factors causing occasional outbreaks of herbivores in systems where their effect is usually minimal.

REFERENCES CITED

Botkin, D. B., and E. A. Keller. 1982. *Environmental Sciences: The Earth as a Living Planet.* Merrill Publishing Co., Columbus, Ohio.

Coley, P. D., J. P. Bryant, and F. S. Chapin III. 1985. Resource availability and plant herbivore defense. *Science* 230:895–899.

Hartley, S. E., and C. G. Jones. 1997. Plant chemistry and herbivory, or why the world is green. In Crawley, M. J. (ed.). *Plant Ecology,* 2d ed. Blackwell Science, Ltd., Oxford.

Jones, C. G. 1983. Phytochemical variation, colonization, and insect communities: The case of bracken fern (*Pteridium aquilinum*) (phytophagous insects). In Denno, R. F., and M. S. McClure (eds.). *Variable Plants and Herbivores in Natural and Managed Systems.* Academic Press, New York.

Jones, C. G., and S. E. Hartley. 1999. A protein competition model of phenolic allocation. *Oikos* 86:27–44.

Mitchell, R. G., R. H. Waring, and G. B. Pitman. 1983. Thinning lodgepole pine increases tree vigor and resistance to mountain pine beetle. *Forest Science* 29:204–211.

Marquis, D. A., and R. Brenneman. 1981. The impact of deer on forest vegetation of Pennsylvania. USDA Forest Service General Technical Report NE-65, Washington, DC.

Whitham, T. G. 1983. Host manipulation of parasites: Within-plant variation as a defense against rapidly evolving pests. In Denno, R. F., and M. S. McClure (eds.). *Variable Plants and Herbivores in Natural and Managed Systems.* Academic Press, New York.

16

CHARACTERISTICS OF ECOSYSTEMS WITH HIGH HERBIVORE CONSUMPTION RATES

INTRODUCTION

The chemical defenses discussed in Chapter 15 reduce plant losses to herbivory, especially in systems dominated by woody plants in which insects are the major herbivores. In systems in which herbaceous plants are dominant and in which mammals are the major herbivores, consumption can be very high and the fraction of net primary production consumed can be either fairly constant from year to year or wildly variable, depending on the system.

Irregular irruptions of insects also occur, and when they do, they can cause herbivory to become an important transfer in woodland systems as well. These irruptions are often triggered by the introduction or removal of hosts, herbivores, and predators from the system; by declining vigor in dominant plants, reducing the production of herbivore defense compounds; and by climatic fluctuations.

The purpose of this chapter is to examine the causes and consequences of these herbivore "success stories" and to discuss the mechanisms that introduce and/or sustain high levels of herbivory in certain ecosystems.

UNGULATES AND GRASSES: COEVOLUTION?

The Serengeti Plain of Africa supports the highest density and greatest biomass of large herbivores in the world. Herbivory is such a dominant process that even the extent of the Serengeti system is defined by the movement of herds of the major herbivores in response to wet and dry seasons and the resulting changes in plant growth. Herbivores can locally consume close to 100% of annual above-ground primary production in parts of this region (see Chapter 20). High rates of herbivory are not unique to the Serengeti, however. A compilation of data from grasslands and forests worldwide (Figure 16.1) clearly shows that herbivory is a major pathway for the consumption of leaf growth in all grasslands and is nearly an order of magnitude (10-fold) higher than in forests with similar levels of foliage produced.

At first, you might think that such heavy consumption would represent a very strong selective force favoring the development of effective herbivore de-

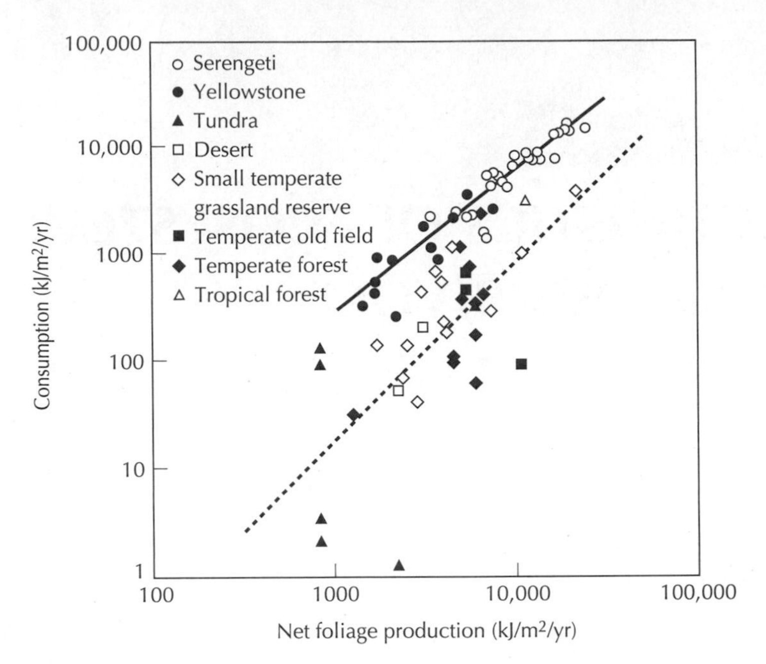

FIGURE 16.1 Average rates of herbivore consumption of foliage as a function of foliage production in grasslands (*solid line*) and other terrestrial (*dashed line*) ecosystems. (Frank et al. 1998)

fense compounds in the plants, and yet the dominant grasses in grassland systems are often **more** palatable than is the foliage of most woody plants. An alternative to the evolution of defenses against herbivory is the development of effective physiological responses to this form of disturbance. If one species of grass could evolve physiological mechanisms that allow for more rapid growth or more successful reproduction following grazing, that species would then have a competitive advantage over others, and the value of discouraging herbivory through production of defensive compounds would be lost. In fact, it would then be advantageous to become increasingly palatable, especially if the herbivores were large and consumed both our "selected-for-grazing" plant and its nonadapted neighbors. It has been noted that the evolutionary appearance and spread of large mammalian herbivores and species of grass are very closely associated. This suggests that the two have evolved in concert and have, together, contributed to the spread of grasslands over geological time.

Grazing is indeed an important selective pressure, and its effects can be seen in differences in plant growth strategies and responses to grazing, even in pairs of closely related species. For example, a species of bunchgrass, *Agropyron desertorum,* has been introduced to the short-grass prairies of the United States from Eurasia, where it evolved under conditions of heavy grazing. A native species, *Agropyron spicatum,* has evolved under lower grazing intensities. In grazed mixtures, the introduced species has been effective in replacing the native species.

Comparisons of responses to simulated grazing (clipping) show that the introduced species can mobilize more carbon and nitrogen for the production of new leaves following this disturbance. Figure 16.2 shows a large difference between the two species in the production of new leaf area and new tillers (root extensions that allow the plant to spread into adjacent areas) following a severe defoliation in May. Total nitrogen content in these tissues is also higher in *A. desertorum*, especially in May and June, resulting in higher

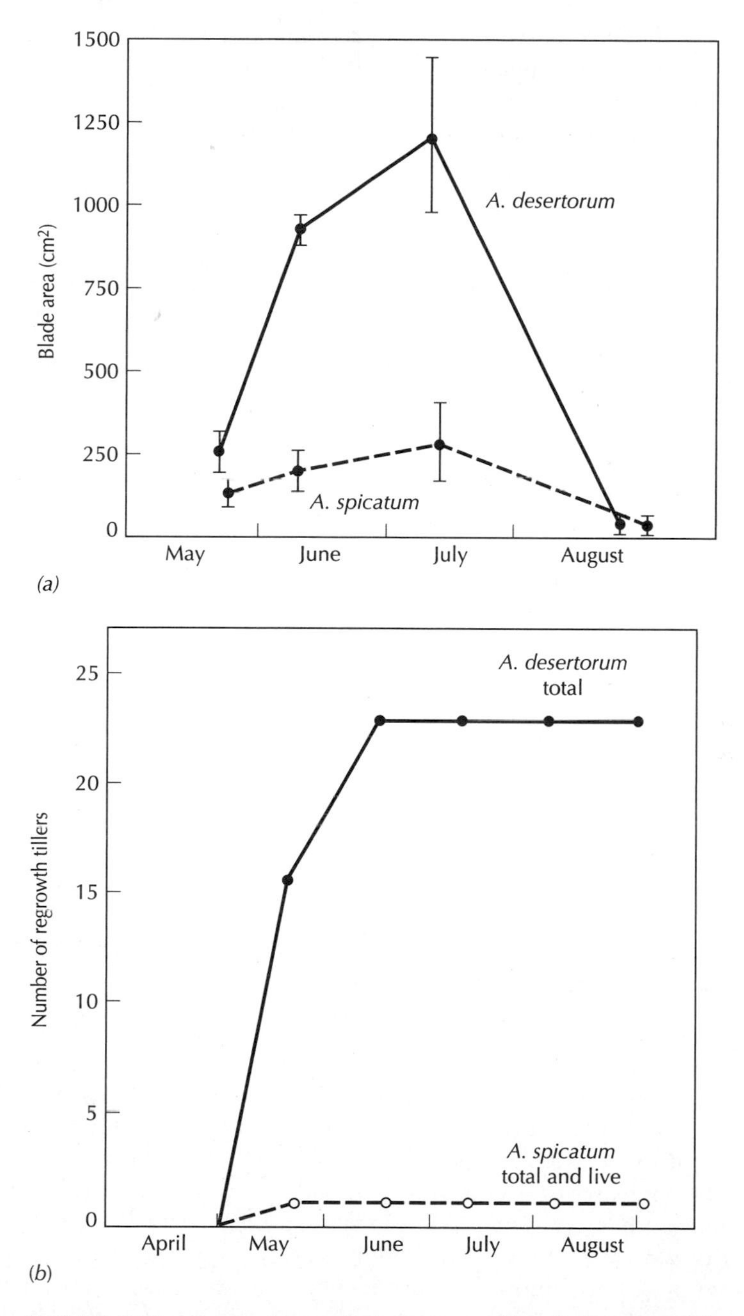

FIGURE 16.2 Comparison of responses to clipping for two similar species of bunchgrass. *Agropyron desertorum* has evolved under heavier grazing pressures, and *Agropyron spicata,* under lower grazing pressures. (*a*) rate of new leaf production. (*b*) Numbers of new tillers produced. (Caldwell et al. 1981)

total net photosynthesis. Taken together, these characteristics create a greater physiological capacity to replace consumed foliage and to colonize adjacent sites that might be disturbed due to grazing or trampling. If grazing gives a species a competitive advantage, then there should actually be a selective advantage in encouraging this form of disturbance to the plant community, perhaps by becoming more, rather than less, palatable. The higher nitrogen content in the grazing adapted species could serve the dual purpose of increasing photosynthesis and increasing palatability.

Very short-term measurements of changes in photosynthesis and carbon allocation in two species of *Panicum* (C4 grasses) from the Serengeti show similar patterns of response. In this experiment, a radioactive isotope of carbon (^{11}C) with a very short half-life (20.3 minutes) was injected into an enclosed atmosphere around the grasses in the laboratory. Two different species found in more and less heavily grazed parts of the region were grown in these atmospheres with and without herbivory by grasshoppers introduced into the enclosures.

Results over a 5-day period showed that the grazing-adapted species maintained higher rates of photosynthesis in the presence of herbivores (Figure 16.3*a*). As a result, this species also had higher total starch content in leaves and roots. After 12 weeks of growth with herbivores, the grazing-adapted species showed less of a reduction in total biomass and a higher shoot–root ratio (Figure 16.3*b*). These responses describe a strategy that combines higher rates of photosynthesis and greater allocation of carbon above ground in the presence of herbivory—a combination of mechanisms for rapidly replacing lost foliage. Tillering, the production of new plants by extension growth of roots into adjacent areas, was also higher in the grazing-adapted species.

Grazing has other effects on ecosystem carbon balance as well. By removing old and partially shaded lower leaves and stems, grazing stimulates the new growth of younger tissues in full sun that will have a higher rate of photosynthesis. Combining all of these responses to grazing, it may not be surprising that grazed grasslands often produce more total above-ground biomass in a year than do ungrazed areas (Figure 16.4).

PLANTS, HERBIVORES, CARNIVORES, AND CYCLIC PATTERNS OF CONSUMPTION IN NONGRASSLAND SYSTEMS

In grassland ecosystems, herbivory can be a large and relatively constant fraction of above-ground productivity. Figure 16.1 suggests that herbivory is relatively low in other systems, but there are important exceptions. Episodic irruptions in herbivore populations may result from migration of a new species into an area or from variations or extremes in climate. Recurring, cyclical increases and crashes may be an intrinsic part of the plant–herbivore–carnivore interaction in systems with relatively simple food webs. In this last case, many of the feedbacks (Chapter 5) between population change and population control mechanisms are positive rather than negative. That is, the mechanisms tend to destabilize rather than stabilize population levels. We will use a few classic examples to demonstrate these interactions.

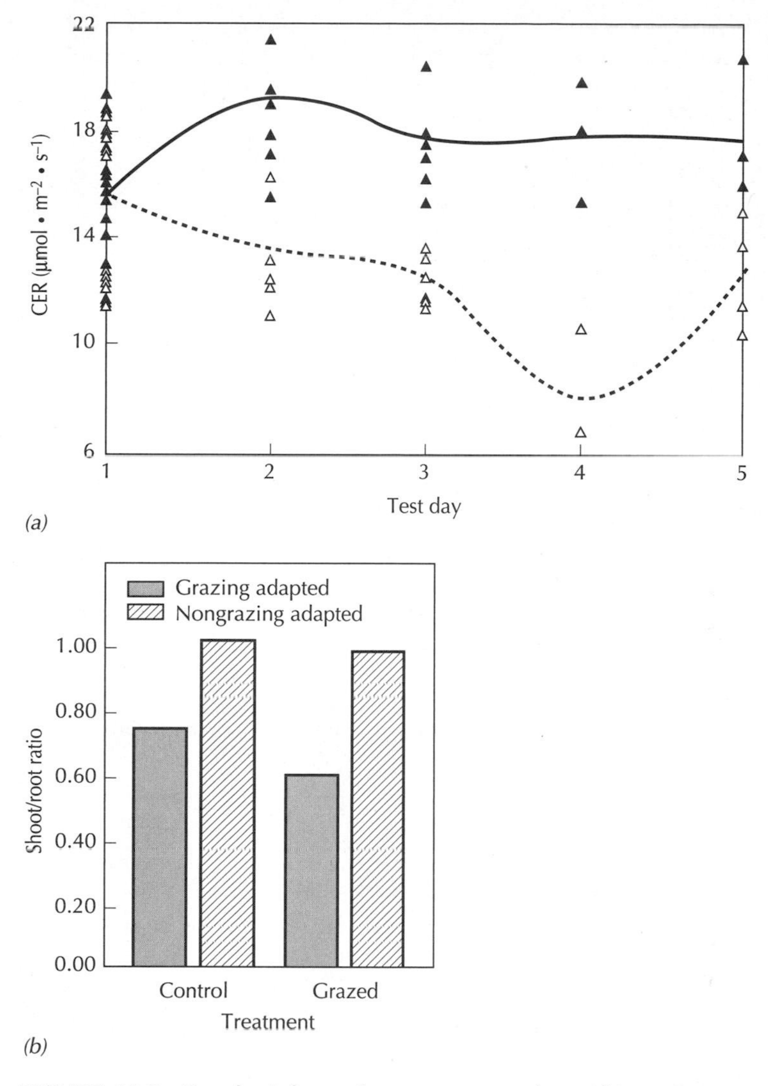

FIGURE 16.3 Ecophysiology of two ecotypes of an African C4 grass, *Panicum coloratum,* grown in the presence of an insect herbivore. (*a*) Over a 5-day period, the grazing-adapted ecotype (*solid line*) maintains a higher rate of net photosynthesis (carbon exchange rate, or CER) than the non-grazing adapted form (*dashed line*) and (*b*) maintains lower shoot–root ratios. (Dyer et al. 1991)

Lemmings in the Tundra

A famous example of drastic, cyclic changes in densities of an herbivore is in the lemming, a small rodent of the arctic tundra. The image of hordes of lemmings rushing into the sea to drown has become part of both ancient folklore and modern social commentary. While actual mass drownings may be rare, very large changes in population density of this species are part of a regular cycle of population explosions and collapse.

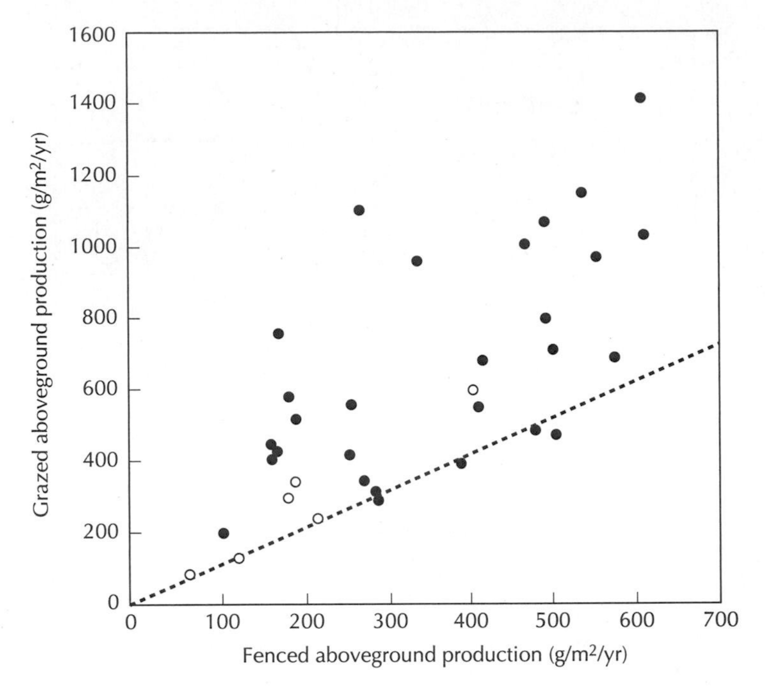

FIGURE 16.4 Total aboveground net primary production for several grazed and ungrazed grassland areas in the Serengeti Plain of Africa and Yellowstone National Park. Grazing is eliminated by the use of large, fenced exclosures. (Frank et al. 1998)

Figure 16.5 shows a 20-year record of lemming population densities at Point Barrow in Northern Alaska. Levels change nearly 100-fold within the 2- to 4-year recurring cycles. Several factors tend to exaggerate, rather than diminish, population fluctuations.

First, the lemmings are the only major herbivore in a very simple food web (Figure 16.6). When lemming populations are high, herbivory severely

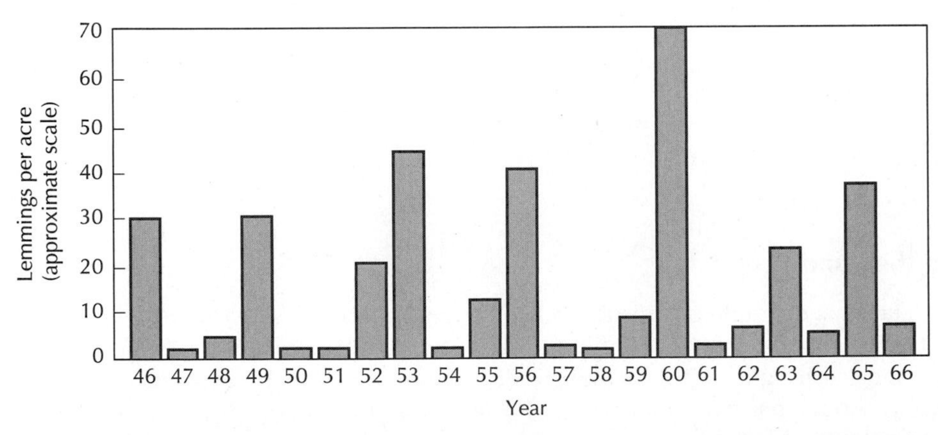

FIGURE 16.5 Changes in population density of lemmings at Point Barrow, Alaska, over a 20-year period. (Schultz 1969)

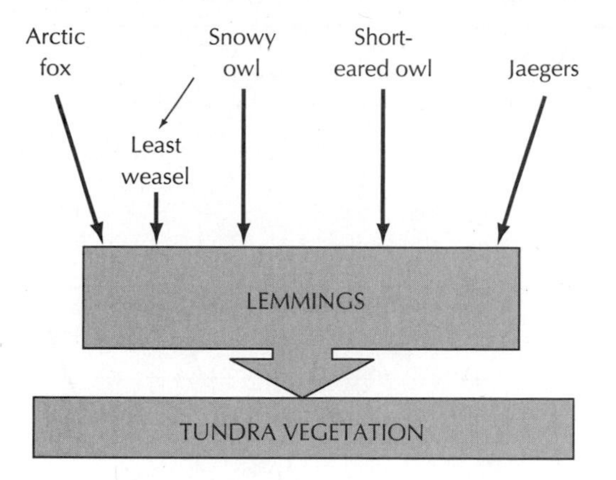

FIGURE 16.6 The food web of the tundra systems in which lemmings are the primary herbivores. (Pitelka et al. 1955)

reduces plant biomass and growth. When lemming numbers crash, vegetation can recover to high levels. Similarly, when lemmings all but disappear at the low point in the cycle, predators have few alternate sources of food and also experience starvation and severe reductions in numbers.

Second, all of the major predators in this system reproduce more slowly than do the lemmings (which can have several large litters in 1 year), so that increases in predator populations lag behind lemming increases. These time lags may actually increase the magnitude of lemming cycles, as predator populations are highest when the rodents are decreasing in number. Heavy predation would then push lemming numbers lower than they would otherwise go, allowing plant growth to rebound more vigorously and stimulating another large increase in lemming numbers.

Third, lemmings lack territoriality and any other type of social interaction between individuals that might reduce reproduction rate to less than the biological potential (see subsequent discussion of moose and wolves). However, lemmings do show evidence of other kinds of social interaction. At high densities, several physiological and behavioral indicators of stress are evident. It has been suggested that reproduction is fast enough, and selective pressure great enough, that the genetics of the population are altered between peak and crash population levels, with individuals tolerant of stress and showing more aggressive behavior becoming more prevalent at high densities. Genetic change is another time-lag factor that tends to increase the amplitude of cycles. Similar changes in genetics, physiology, and behavior have been shown for other cyclic species.

Fourth, fluctuations in lemming numbers induce significant changes in soils and plant nutrition that may also tend to exaggerate population cycles. Tundra plants grow in a thin soil layer that is thawed for only a short time during the summer. The depth to the permanently frozen layer (permafrost) varies with the amount of vegetation and organic litter on the soil surface. Cyclic changes in herbivory alter the amount of insulating organic matter on top of the soil and so affect depth to permafrost, which, along with other interactions, appears to produce cycles in plant element content (Figure 16.7).

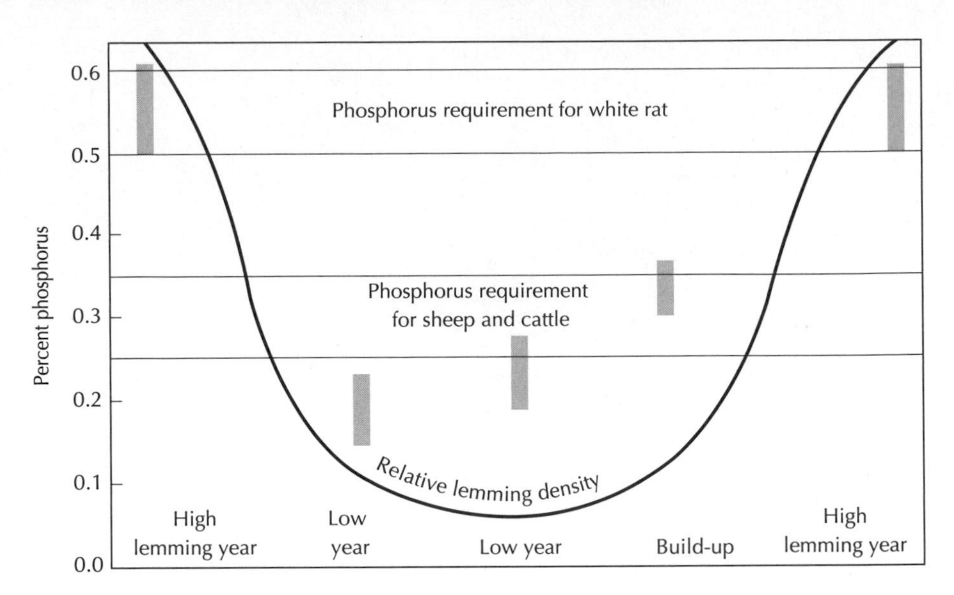

FIGURE 16.7 Changes in the phosphorus content of herbage over the course of a 4-year lemming population cycle in relation to requirement of several species. (Schultz, 1969)

The variable quality of plant forage, in turn, affects lemming nutrition and health.

Figure 16.8 summarizes the set of interactions affecting lemming numbers. There are certainly several mechanisms for limiting population growth. However, as realized for this species in the tundra system, the major checks on lemming population discussed in the previous paragraphs respond slowly to lemming increases and tend more to increase cycle amplitude than to decrease it.

Lemmings also live south of the transition line between tundra and boreal forest, but the forest populations do not show huge cycles in population levels. Rather, rapid reproduction leads to high population levels in the fall, at which time several species of predators, which have fed on other animal species in spring and summer, switch to lemmings. Higher predator numbers and a shorter period of protective snow cover lead to very heavy predation in fall and winter, such that lemming numbers are again very low in the spring.

Hare and Lynx in the Boreal Forest

Other cycles in northern latitudes are well documented. The historical pattern of lynx and hare in the boreal forest zone, as inferred from the number of pelts of each sold to the Hudson Bay Company in Canada, is an often-cited example (Figure 16.9). Here the cycles are repeated at roughly 10-year intervals and, as before, the predator generally lags behind the prey by one or two seasons, suggesting that predation may have a destabilizing effect. There is also evidence that heavy grazing of dwarf birch by hare may induce increased

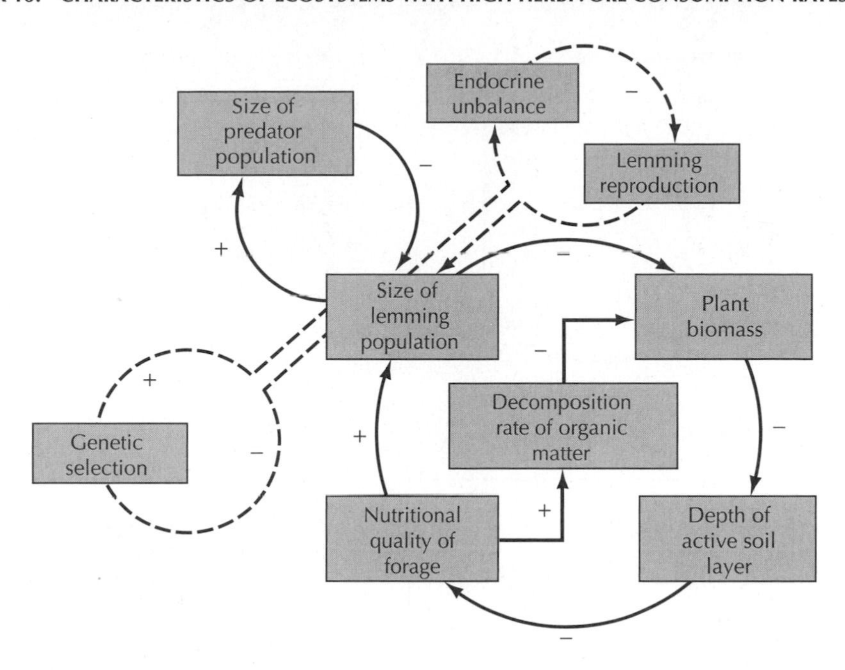

FIGURE 16.8 Interactions of factors controlling lemming populations in the tundra. Note that most of these actually have the effect of increasing the size of the population fluctuations. (After Schultz 1969)

production of resins that reduce the palatability of twigs (Figure 16.10), causing reduced quality of plant material at the same time that populations are falling. This will also tend to increase population fluctuations.

A long-term experiment has been performed with the hare–lynx system in an effort to separate the effects of starvation, predation, and changes in forage quality on the 10-year population cycle. This study first demonstrated that both adult and juvenile hare mortality changed dramatically over the

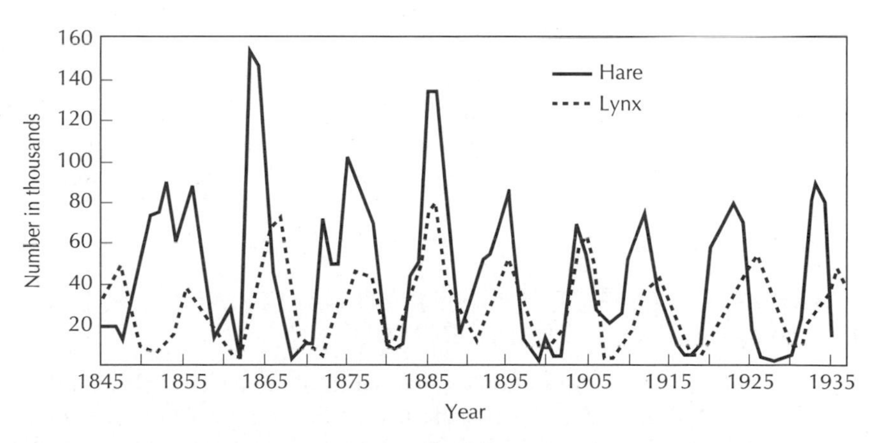

FIGURE 16.9 Changes in sales of pelts of hare and lynx to the Hudson's Bay Company in Canada. Cycles of sales are assumed related to cycles in population levels. (Odum 1971)

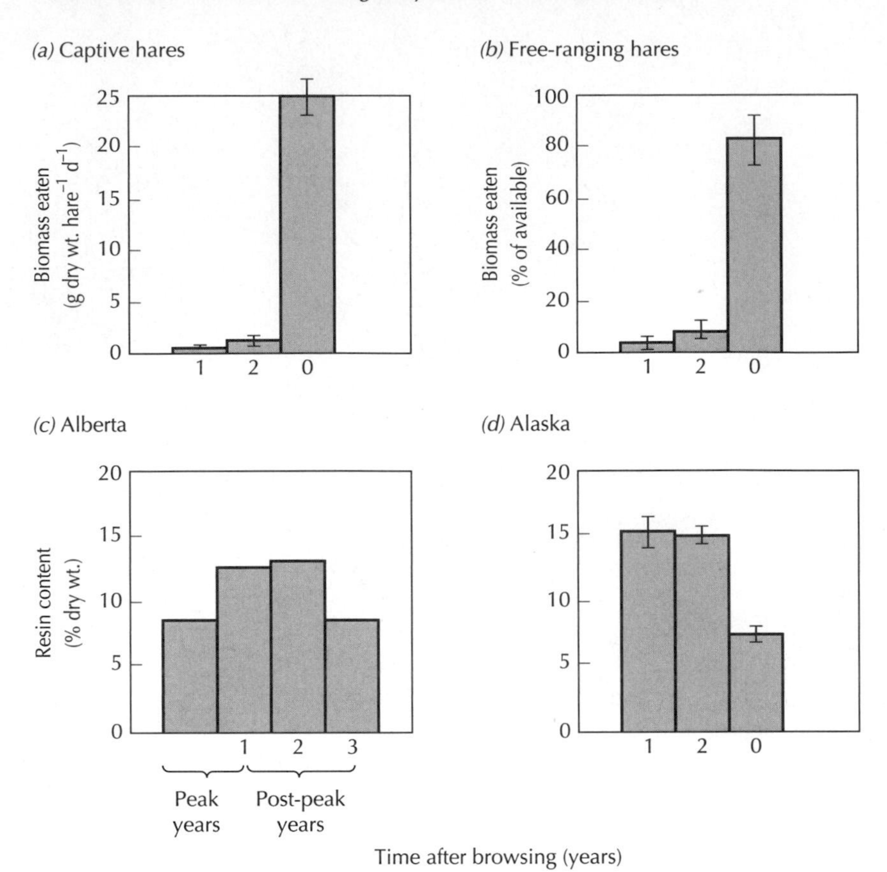

FIGURE 16.10 Changes in the resin content of twigs of dwarf birch in relation to the occurrence of peak populations of hare. (Bryant et al. 1983)

course of a single cycle (Figure 16.11*a*), with increased mortality striking juveniles first. Additional food increased the density of hares, especially during the period of peak numbers (1991; Figure 16.11*b*). The effect of predator exclusion was greater during the period of rapid hare decline, when predator numbers would be at their highest levels (1992–1993; Figure 16.11*b*). Fertilization increased plant growth but did not affect hare density. The interactive effect of feeding and predator exclusion was greater than for either effect alone, suggesting an important three-level interaction between plants, hares, and predators.

THE STABILIZING EFFECTS OF TERRITORIALITY AND PREDATION: MOOSE AND WOLVES ON ISLE ROYALE

Predation need not always have a destabilizing effect. It can also reduce fluctuations in herbivore populations, particularly when the major predators have important social interactions that control their own population density. A rare opportunity to study this process in operation, and to demonstrate the

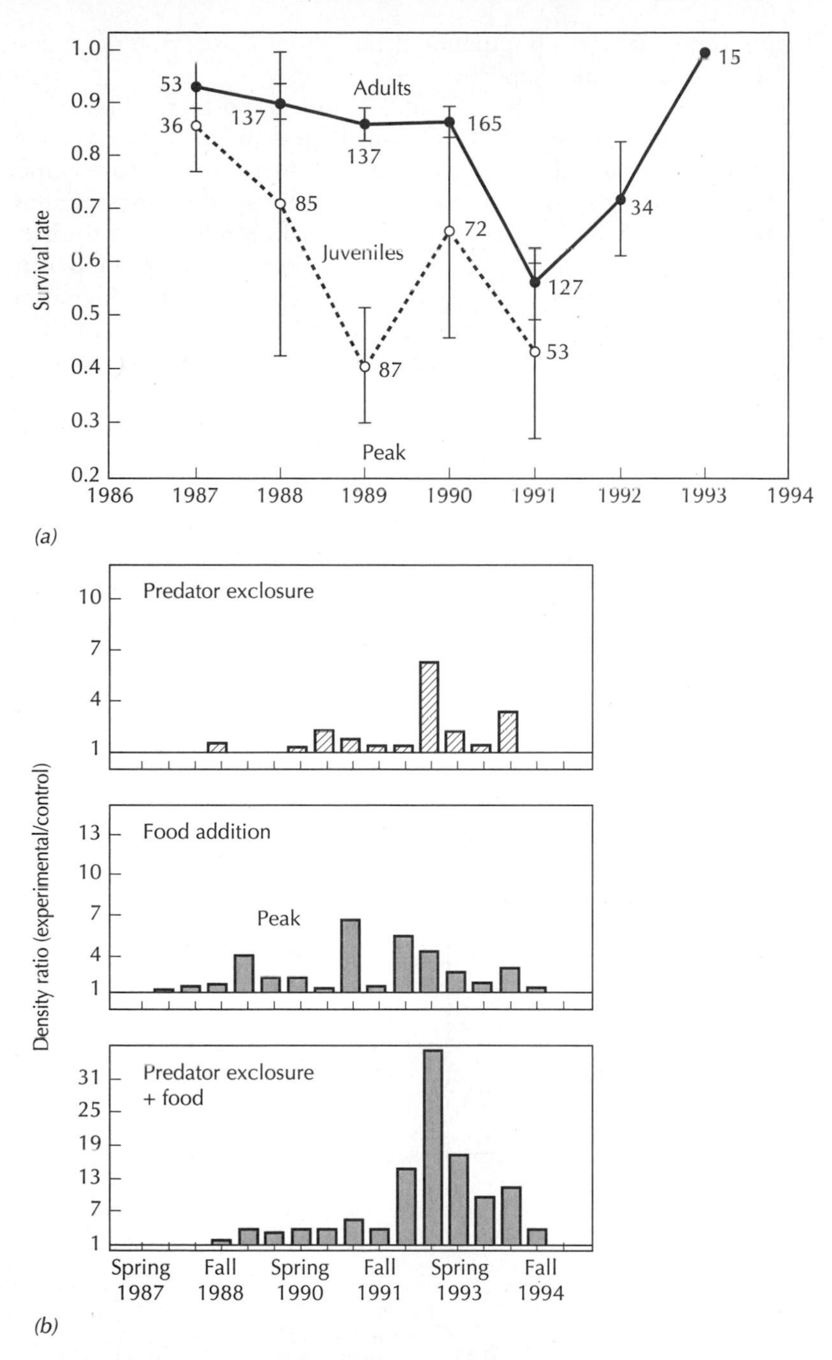

FIGURE 16.11 Mortality rate of hare over an 8-year population cycle and experimental determination of causes of mortality. (*a*) Survival rate of young and adult hare. (*b*) The individual and combined effects of feeding and predator exclusion (note that the combined effect is to increase the hare population by more than 30-fold during the period when mortality would be highest). (Krebs et al. 1995)

inherent patterns of population dynamics in natural systems, has occurred on Isle Royale, in Lake Superior.

Isle Royale is a large (546 km^2) island supporting a mixture of forest and wetland habitats required by moose (Figure 16.12). In 1905, a small number of moose apparently swam from the Canadian side of Lake Superior to Isle Royale, about 32 km offshore. As new arrivals, the moose found abundant food and cover and an absence of their main predator, the timber wolf. From a small beginning, the population grew exponentially (increasing at an increasing rate) for 23 years (Figure 16.13). By 1926–1927, the moose had modified the vegetative structure of the island to such an extent that a shortage of browse—highly palatable leaves and twigs of certain woody species—was developing.

Increasing moose populations and declining browse production resulted in a massive die-off between 1933 and 1937. The moose population declined by more than 90%. A physiologist examining recovered moose corpses determined starvation to be the sole source of mortality.

Low moose densities allowed the vegetation to recover somewhat, a major fire in 1936 increased browse, and another population irruption began. However, the vegetation on the island had been altered during the first irruption, and what remained could support only a smaller moose population. By 1948, moose were again severely reducing the available browse. A second population crash occurred from 1948 to 1950.

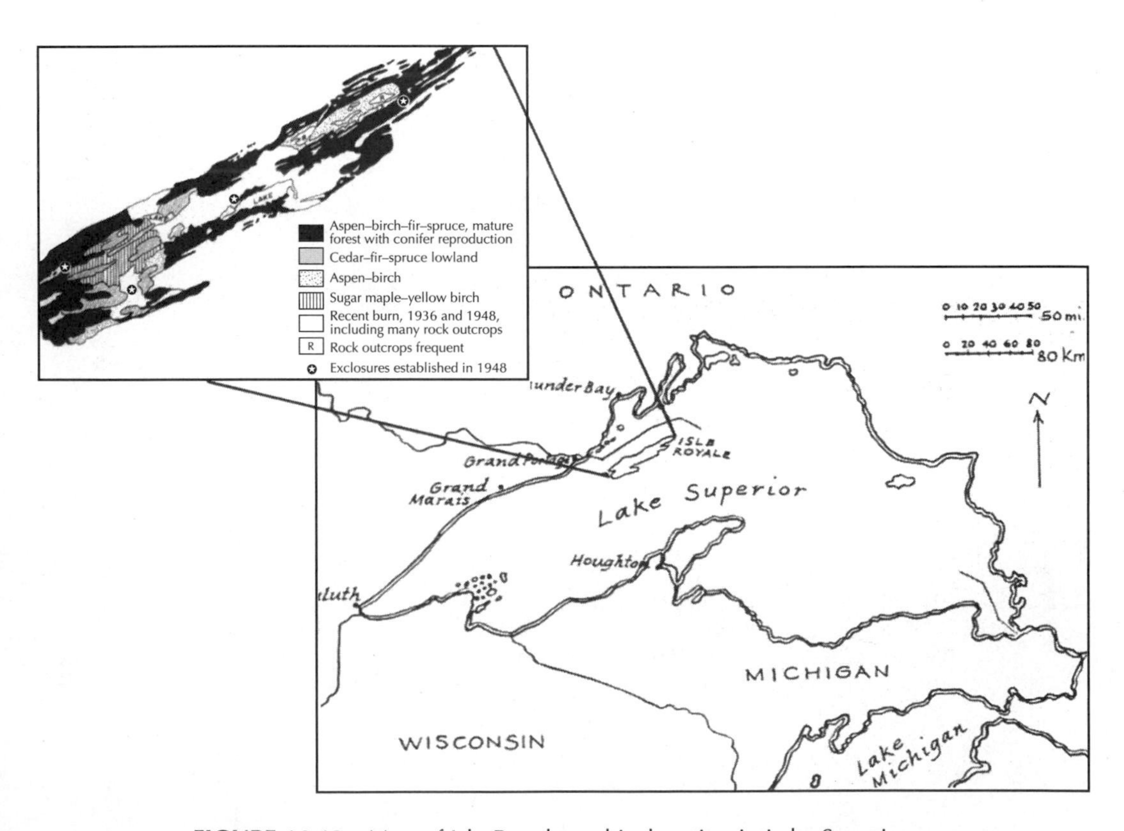

FIGURE 16.12 Map of Isle Royale and its location in Lake Superior.

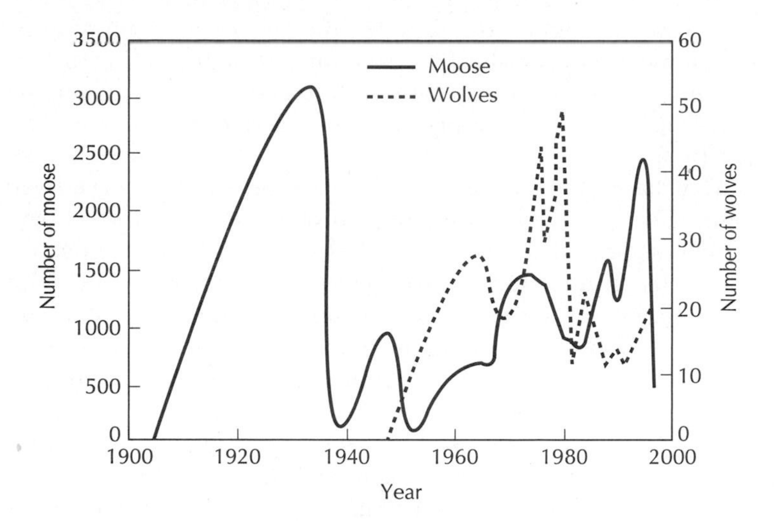

FIGURE 16.13 Changes in numbers of moose and wolves on Isle Royale. (Peterson 1994, 1999)

This pattern is not unlike that described for lemmings, but with a longer periodicity (time between peak populations) due to the slower reproductive rate of moose. Yet such wholesale over-browsing is not common throughout the rest of the moose's range.

By 1956, browse plant species had recovered somewhat, the moose were increasing again, and another cycle of population boom and crash was under-way. However, a new dimension was added with the arrival of wolves across frozen lake ice sometime between 1945 and 1950. The newly arrived wolves found an abundance of moose, and the wolf population began to increase. We might expect that wolves, like predators in the lemming system, would increase the severity of population cycles, and, in a small place like Isle Royale, drive the moose to local extinction. However, wolves have a highly organized social structure that reduces the rate of population increase far below the biological potential. Wolves live in packs, and packs establish distinct territories. Within packs, social interactions limit reproduction mainly to the dominant pair (alpha male and alpha female) in each pack. Thus, the addition of new pups to the pack each year is limited more by social and territorial conditions than biological potential.

The presence of a growing wolf population between 1950 and 1980 could be one of the reasons that the general increase in moose numbers over this period was considerably slower than in the 1920s and 1940s. Decreasing moose numbers in the 1970s may reflect the highest number of wolves that occurred at the same time.

Had this record ended before 1980, we could have comfortably concluded that a stabilized predator population (wolves) had helped stabilize the herbivore population (moose). From 1965–1980, moose populations fluctuated moderately, and browse was never fully depleted as in the late 1920s. Measured fluctuations could be attributed to climatic differences between years and the difficulties of obtaining absolutely accurate population estimates.

However, as Figure 16.13 shows, the record is more complicated than that. An unexpected crash in the wolf population beginning in 1980 led to reduced predation, which may have been linked to the large increase in moose over the next 15 years. Overbrowsing again led to starvation and a moose die-off in 1996.

What happened to the wolves? Two processes may have been most important. The first is a loss of genetic diversity due to inbreeding. A study in the 1980s showed that all of the wolves on the island were descendents of a single maternal ancestor. Lack of genetic diversity can lead to lower reproductive success and increased susceptibility to disease. A second and possibly related event was the arrival on the island of a new disease, canine parvovirus, which was first identified worldwide in 1977 and was present during the crash of 1980–1982.

Thus, both the limited size and geographic isolation of the island and the appearance of a new and invasive species (the parvovirus) may have upset the relative stability of the moose–wolf interaction. It is a general tenet of a facet of ecology appropriately called *island biogeography* (see Chapter 24) that small, isolated populations tend to experience more frequent local extinctions as a result of random events, such as climatic extremes.

Population fluctuations, and the long-term presence of herbivory can produce important changes in vegetation and soils as well. On Isle Royale, annual growth rings in trees small enough to be browsed by moose were significantly reduced around 1975 and 1990, years of high moose numbers. A comparison of ring widths in trees above and below the browse line (Figure 16.14) shows that trees that were out of the reach of moose actually grew much faster in the period of high moose populations after 1980 as competition for water and nutrients with understory plants was reduced, while those within reach were devastated. Overall, browsing decreases total plant growth in a response that is the opposite of what was described earlier for grasslands.

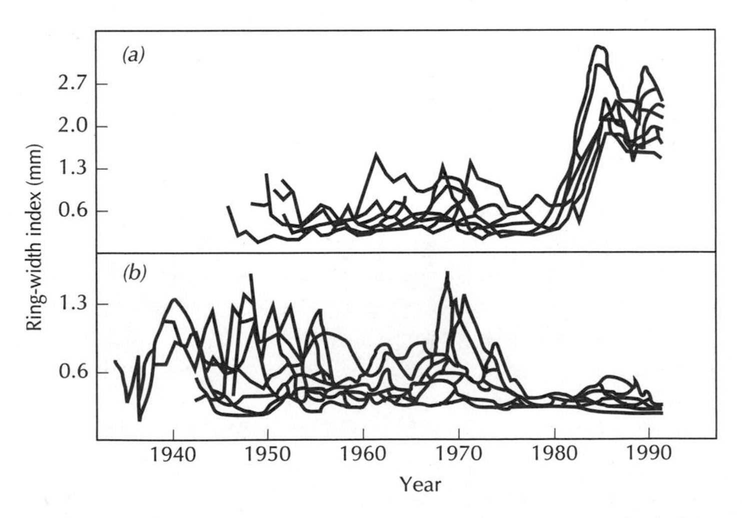

FIGURE 16.14 Comparison of ring width in balsam fir trees that were (*a*) above the browse line and out of reach of moose and (*b*) below the browse line. (McLaren and Peterson 1994)

Moose browsing also has a selective effect on the type and amount of litter that is produced, and this, in turn, alters soil-forming processes. Over 40 years ago, long-term exclosure experiments were initiated on Isle Royale by constructing fences around 101 m^2 areas. Within the fenced areas, foliage was allowed to fall to the forest floor as litter and to decay in place rather than being processed first by moose. In that 40-year period, dramatic differences in vegetation and soils have developed. In the absence of browsing, the plots within the exclosures have both a higher total amount of litterfall and a higher fraction of litterfall as foliage of deciduous species. The soils within the exclosures have a much thicker forest floor (Figure 16.15); contain more organic matter, carbon, and nitrogen; have higher cation exchange capacities; and may have higher rates of net nitrogen mineralization. Moose browsing, at least at the high population levels present on Isle Royale (among the highest in the world, perhaps due to the absence of a second major predator, bears), apparently reduces soil fertility, and may then also reduce future plant growth.

The extreme population fluctuations in moose at Isle Royale are one example of the potential effect of either adding or removing species from an ecosystem. There are many others. Several can be found in Australia and New Zealand, where placental mammals did not exist until relatively recently, and niches generally filled by placental mammals were instead occupied by marsupials, and even by birds. From the introduction of the Dingo from Asia over 3400 years ago through the arrival of a host of European species in the past 300 years, placental mammals have displaced many native marsupial mam-

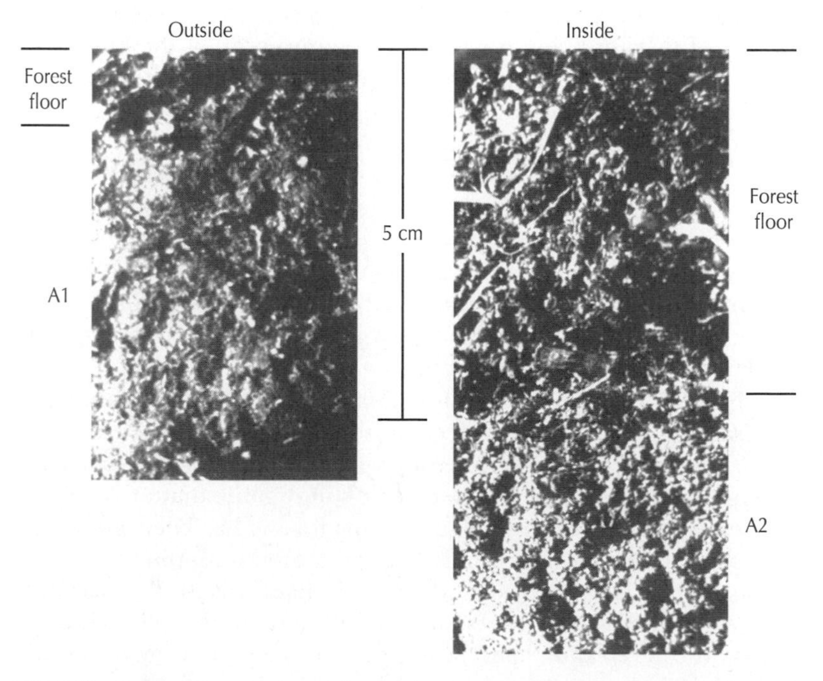

FIGURE 16.15 Comparison of soil structure inside and outside of fenced exclosures on Isle Royale. (Pastor et al. 1988)

mals and often altered the structure of ecosystems. Humans have, in general, greatly accelerated the rate of species extinctions as well as the rate of movement of species between continents. While Isle Royale is essentially a natural experiment in the field of island biogeography, changes in species composition are driven by human actions in most places. We will return to the implications of these phenomena in Chapter 24.

EFFECTS OF VEGETATIVE CHANGE AND CLIMATE ON IRRUPTIONS OF INSECT POPULATIONS

Spruce Budworm

The eastern spruce budworm is an economically important pest of commercial forests in North America. It feeds on mature fir and spruce trees in the boreal forest of the Eastern United States and Canada and has been the object of intense efforts at population control, through pesticides, by forest managers. As with lemmings, irruptions of spruce budworm are also tied to changes in habitat structure. Because the boreal forests change more slowly than tundra, the indications of the cyclic nature of these irruptions has emerged more slowly. An analysis of tree rings in the budworm's range suggests a 30- to 45-year periodicity to irruptions. When irruptions occur, they can spread over thousands of square kilometers in a period of 7–16 years.

The trigger for an irruption of the budworm is the existence of large areas of mature fir and spruce trees, the budworm's preferred food. In this vegetation type, fir increases in dominance with time, replacing the less tolerant birches and other deciduous species that dominate following a disturbance. If succession is allowed to continue, fir tends to be replaced by spruce. However, this succession can be cut short at the point where the concentration of fir is sufficient to stimulate an outbreak of the budworm. This devastates both fir and spruce and reinitiates succession with a new stand dominated by deciduous species.

The human response to budworm irruption has been to spray with insecticides to preserve the economic value of the forest. While this postpones the defoliation caused by the budworm, it also allows further maturation of the forest and increasingly ideal conditions for a major irruption. The situation has been likened to the exclusion of small ground fires from western United States coniferous forests, where they would naturally occur (see Chapter 17). The short-term goal of avoiding economic losses is achieved, but the chance for a major loss increases.

Disturbance by fire, defoliation, or other agents is an intrinsic and necessary part of the function of most terrestrial ecosystems; it is a mechanism for reversing declining rates of nutrient cycling or relieving stand stagnation. The requirement for fire to reverse soil organic matter accumulation and increase nutrient cycling in coniferous forests has been known for some time (see Chapter 17). The ponderosa pine–mountain pine beetle interaction discussed in Chapter 15 is another example of this. In the case of the budworm and the mountain pine beetle, stand breakup, the reinitiation of succession, and the reversal of stand stagnation are facilitated by herbivory rather than fire.

Locusts

Climate may play as large a role as habitat in the irruptions of many insect species. A famous example is the migratory locusts of semiarid, tropical regions in Africa. These are the "plague" species that irrupt three to four times per century and travel thousands of kilometers, increasing their range during the outbreaks from small, isolated patches to large portions of the African continent (Figure 16.16). An irruption may last for 20 years. In actuality, small swarms of locusts form almost every year within the permanent resident areas but only occasionally reach densities high enough to induce large-scale movements. As with the lemmings, swarming is accompanied by large changes in morphology and behavior of individuals.

Population levels and swarming seem to be linked to a series of favorable (wet) years, leading to high densities in the areas of permanent populations. However, changes in habitat and interactions with food source may also be a factor. The vast scale of the locusts' range and the relatively rare irruptions have hampered detailed research.

Thrips

On a less dramatic scale, changes in peak populations of thrips on rosebushes in the gardens of an Australian scientist formed the basis for the theory that populations are generally limited by climate alone. While no longer accepted for most species, climatic regulation can be important in triggering outbreaks, as with locusts, or in limiting increases of insects at the edge of their ranges or in extreme climates.

CONCLUSION

We can conclude that most insect herbivores of woody communities are maintained at low population levels by biochemical alterations in leaf quality generated by plants. Irruptions of these herbivores tend to be related to reduced vigor or increased density of host plants. In these cases, the outbreaks may serve an important "restart" function in the ecosystem. In grassland systems, adaptations between the grasses and mammalian herbivores have lead to continuously high levels of herbivory. In simple systems, such as the tundra, in which the main herbivore (the lemming) reproduces more rapidly than the predators and neither group shows significant territoriality, large, cyclic changes in herbivory can be expected. In more complex systems where predators have several species to prey upon and each herbivore is fed upon by more than one or a few species, or if predators exhibit territoriality or other social interactions that limit reproductive rate, these large cycles may be controlled. In either forests or grasslands, the addition or removal of important herbivores or predators can significantly alter levels of herbivory and change plant species composition. As we will see in the more complete discussion of the Serengeti in Chapter 20, these changes can continue long after the new species has again disappeared from the system.

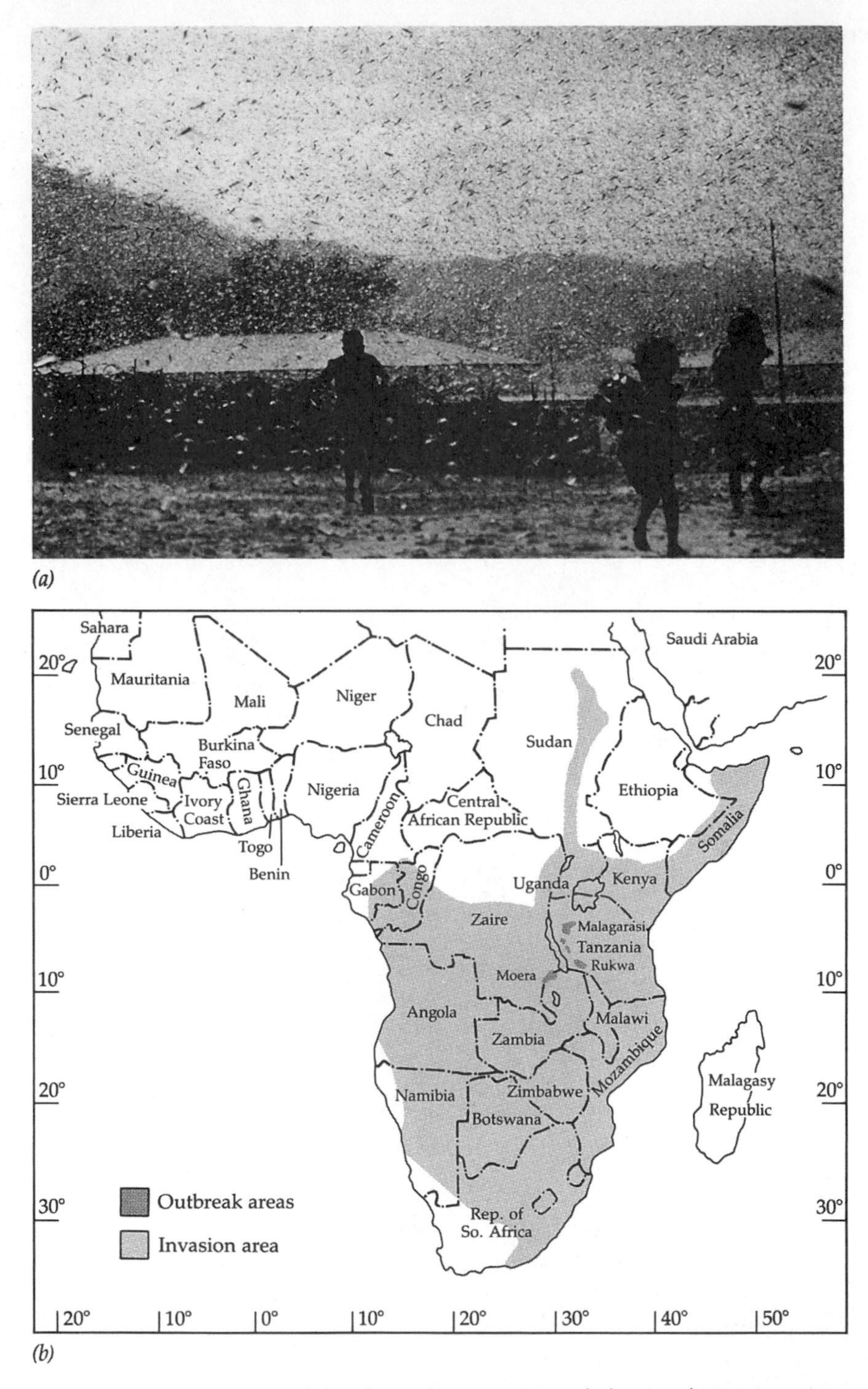

FIGURE 16.16 Swarms of the desert locust. (*a*) Local density during irruptions. (*b*) Map of the normal areas of locust occupation and the much larger area over which they may swarm during population irruptions. (*a*, G Tortoli/H. Armstrong Roberts; *b*, Krebs 1972)

REFERENCES CITED

Bryant, J. P., F. S. Chapin III, and D. R. Klein. 1983. Carbon/nutrient balance of boreal plants in relation to vertebrate herbivory. *Oikos* 40:357–368.

Caldwell, M. M., J. H. Richards, D. A. Johnson, R. S. Nowak, and R. S. Dzurec. 1981. Coping with herbivory: Photosynthetic capacity and resource allocation in two semi-arid *Agropyron* bunchgrasses. *Oecologia* 50:14–24.

Dyer, M. I., M. A. Acra, G. M. Wang, D. C. Coleman, D. W. Freckman, S. J. McNaughton, and B. R. Strain. 1991. Source-sink carbon relations in two *Panicum coloratum* ecotypes in response to herbivory. *Ecology* 72:1472–1483.

Frank, D. S., S. J. McNaughton, and B. F. Tracy. 1998. The ecology of the Earth's grazing systems. *BioScience* 48:513–521.

Krebs, C. J. 1972. *Ecology: The Experimental Analysis of Distribution and Abundance.* Harper & Row, New York.

Krebs, C. J., S. Boutin, R. Boonstra, A. R. E. Sinclair, J. N. M. Smith, M. R. T. Dale, K. Martin, and R. Turkington. 1995. Impact of food and predation on the snowshoe hare cycle. *Science* 269:1112–1115.

McLaren, B. E. and R. O. Peterson. 1994. Wolves, moose and tree rings on Isle Royale. *Science* 266:1555–1558.

Odum, E. P. 1971. *Fundamentals of Ecology.* W. B. Saunders, Philadelphia.

Pastor, J., R. J. Nainman, B. Dewey, and P. McInnes. 1988. Moose, microbes and the boreal forest. *BioScience* 38:770–777.

Peterson, R. O. 1999. Wolf–moose interaction on Isle Royale: The end of natural regulation? *Ecological Applications* 9:10–16.

Peterson, R. O. 1994. Ecological studies of wolves on Isle Royale. Annual Report 1993–1994. School of Forestry and Wood Products, Michigan Technological University, Houghton, MI.

Peterson, R. O. 1977. Wolf Ecology and Prey Relationships on Isle Royale. National Park Service Monograph Series, No. 11, Washington, DC.

Pitelka, F. A., P. Q. Tomich, and G. W. Treichel. 1955. Ecological relations of jaegers and owls as lemming predators near Barrow, Alaska. *Ecological Monographs* 25:85–117.

Schultz, A. M. 1969. A Study of an ecosystem: The arctic tundra. In Van Dyne, G. M. (ed.). *The Ecosystem Concept in Natural Resource Management.* Academic Press, New York.

17

THE ROLE OF FIRE IN CARBON AND NUTRIENT BALANCES

INTRODUCTION

Intense wildfires are among the most spectacular events in nature. In a period of minutes, a green and functioning forest, prairie, or shrubland can be reduced to blackened, dead stems and ash. Plant and animal losses can be important both economically and aesthetically.

However, fire can also play an important, positive role in the dynamics of ecosystems. Organic matter accumulated over long periods of time is oxidized, and the nutrients contained are released. Long-term imbalances in the production–decomposition ratio can be equalized in minutes. Some of the nutrients released from organic matter through combustion can be lost in smoke and gases, but those that remain are generally in forms more available to plants. Cations can be present in the ash in sufficient quantity to increase soil pH significantly. The availability of light and water to the remaining plants, or to the new seedlings and sprouts, is also increased by the removal of some or all of the previous vegetation. All of this can lead to increased vigor and growth of plants.

The human response to fire reflects this confusion between positive and negative effects. Fire has been used frequently by aboriginal peoples throughout the world to reduce pests, prepare land for farming, and increase forage available to grazing animals. Fire remains a very major land-clearing practice in the tropics, with consequences at both the local and global levels. In contrast, forest management practices in the Western world, having developed in regions where fire is a less natural part of the system, have, until recently, stressed complete fire suppression.

The purpose of this chapter is to discuss fire as it affects ecosystem processes. We begin by describing the frequency and intensity with which fires occur in different ecosystem types. We then look at the effects of fire on the availability of resources required for plant growth and how plants respond to the changed conditions, including methods of reproduction to ensure the presence of viable offspring following the fire. Finally, we discuss interactions between fire and herbivory and briefly address some of the important management questions regarding the use of fire in ecosystems.

MAJOR CATEGORIES OF FIRE TYPES

Fires in forest ecosystems can be separated into three major types: (1) surface, (2) crown, and (3) ground (Figure 17.1*a*). **Surface fires** burn only along the soil surface, consuming accumulated litter on the forest floor and part of the organic matter in the forest floor. Shrubs and herbaceous plants in the understory are usually killed, along with seedlings and perhaps saplings of dominant tree species. Surface fires are relatively cool and do not cause

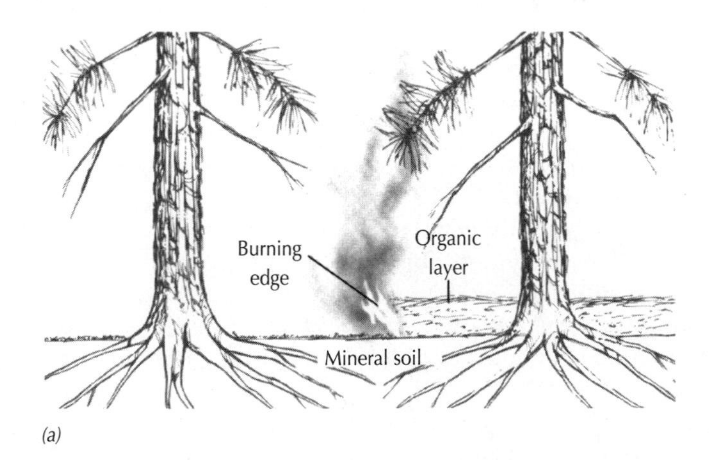

(a)

FIGURE 17.1 Wildfires in different terrestrial ecosystems. (*a*) The three types of fires which occur in forests: surface fire (*top*), crown fire (*middle*), ground fire (*bottom*).

major changes in the visible structure of the forest beyond reduction in forest floor mass and the deposition of ash.

In contrast, **crown fires** reach up into the crowns of the dominant trees, burning live foliage, branches, and stems, as well as the forest floor. Crown fires are much hotter than are surface fires and can kill a large proportion of the dominant trees in the system.

Ground fires occur in forests, bogs, wetlands, or wherever large accumulations of organic matter occur over the mineral soil. When drought dries out the usually wet surface organic mat, ground fires can burn down through this deep organic layer and can "disappear" below the surface for some time before burning back up to the surface in some other location. As a result, ground fires generally burn slowly over long periods of time and can be very effective at consuming accumulated surface organic matter.

In grassland systems, there is no distinction between surface and crown fires. Surface fires also consume the vegetation because there is no discontinuity between the soil surface and plant canopy (Figure 17.1*b*). Shrublands, such as the chaparral (Figure 17.1*c*), also tend to burn completely with each fire.

While crown fires are generally more intense than are ground fires, there are wide variations in intensity within each type. Fire intensity is described in terms of how hot the fire burns in one spot and for how long. The intensity of a fire depends, in turn, on the amount of organic matter present, how it is structured, how moist it is, and on environmental conditions including wind speed and humidity. Even the most intense fires are patchy in nature, and the occurrence of completely burned and completely untouched areas in close proximity is not uncommon.

(b)

(c)

FIGURE 17.1 (*b*) A managed fire in a restored prairie ecosystem. (*c*) A fire in a chaparral system. (Photos by R. Krubner/H. Armstrong Roberts)

FIRE FREQUENCY AND INTENSITY IN DIFFERENT TYPES OF ECOSYSTEMS

Ecosystems vary widely in the frequency with which they burn. Tallgrass prairies may burn every 2–4 years. Pine forests in the Great Lakes region may experience crown fires every 70–100 years. Some of the dry forests of the Rocky Mountains may have frequent surface fires and only rare crown fires. In some systems, there may be important interactions between the frequencies of surface and crown fires because surface fires reduce fuel loads and reduce the chance of more severe crown fires.

Taken together, the frequency, intensity, and types of fires that occur in an ecosystem determine that system's fire regime. A tremendous diversity of fire regimes can result from the interactions between these three factors. Figure 17.2 depicts some of this diversity for ecosystems in North America.

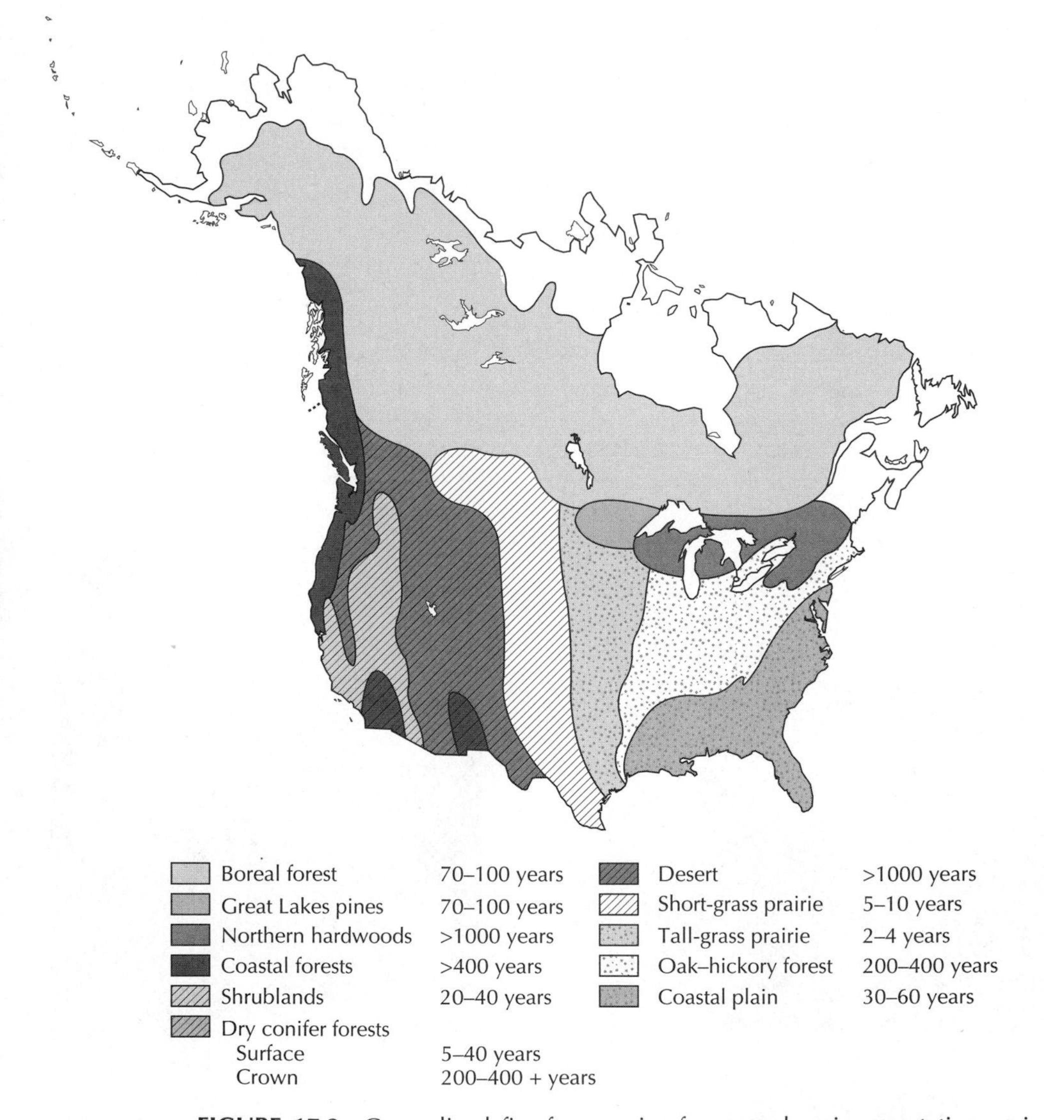

FIGURE 17.2 Generalized fire frequencies for several major vegetation regions of North America. (Heinselman 1973, 1981; Kilgore 1981; Gray and Schlesinger 1981; Christensen 1981; Bormann and Likens 1979)

Fire is least common in eastern deciduous forests. Especially in the northern part of the range, cool, humid climates combine with rapid decomposition of litter to reduce fuel loads and increase moisture content. Fires are not unknown, especially when logging operations, which result in large accumulations of residues (stems and branches, also known as "slash") are followed by unusually dry summers. Some very large fires have occurred in this way.

To the south and east of the deciduous forest zone is a region of generally sandy or poorly drained soils (the coastal plain), where forests are dominated by several species of pine, with an understory of deciduous trees and shrubs. Here, both litter and live biomass can have high concentrations of **pitch**—highly flammable resins that promote, rather than inhibit, combustion. Litter produced by the dominant pines is of low quality and accumulates on the forest floor, increasing fuel loads with time. Needles do not form compact litter layers but rather are loosely structured and dry quickly after a rain. Fire is considered instrumental in reversing nutrient stagnation due to organic matter accumulation in this forest type, and stand-destroying crown fires may occur at intervals of 30–60 years. Some evidence shows that this fire regime has replaced a presettlement pattern of frequent surface fires.

Extensions of the coastal plain forests can be found in the more humid northeast as well, where locally dry sandy soil conditions have led to the development of ecosystems with high fire frequencies. Pitch pine is a dominant species in these pine plain areas, and the frequent recurrence of fire in this system (as often as every 3–4 years under natural conditions) has completely altered the appearance (Figure 17.3) and function of the system.

Fire frequency is somewhat lower in the pine forests of the Great Lakes region. This area was one of the first in which the regular occurrence of natural fires in a forest ecosystem was documented. Sediment cores taken from lakes within the region show distinct layers of increased charcoal input and altered pollen deposition, conditions known to occur following fires. Recognition of this fire frequency as a natural part of the landscape led to one of the earliest theories of ecosystem development in which the **climax**, or end state, is never achieved. Instead, movement toward this end state is continually disrupted and restarted by the occurrence of crown fires (see discussion in Chapter 18).

Fire has always been a regular feature of the midwestern prairie and savannah systems. Early explorers' and settlers' accounts are filled with descriptions of the intensity of these fires and the huge areas that could be covered. The extension of tallgrass prairie into the lake states area, where climatic conditions should favor the growth of forests, has been linked to frequent, tree-killing fires, possibly operating synergistically with grazing by buffalo. Tallgrass prairies require fires at 2- to 4-year intervals to remain vigorous, while the shortgrass prairies of the western plains burn less frequently. In the absence of fire, the buildup of dead grass stems severely restricts light penetration, slowing the warming of the soil in spring and reducing light availability to the sprouting grasses. It also shifts the competitive advantage to taller-growing shrubs and trees. Many forests in the eastern prairie states have developed from savannahs following the removal of fire as a result of the division of the landscape into farms and the plowing of the

FIGURE 17.3 The pine plains area of New Jersey, where repeated fires have severely modified the structure and function of a forest ecosystem within the humid eastern deciduous forest climatic zone.

prairie sod. Bare, plowed fields make excellent firebreaks. These forests still contain the wide-crowned trees that developed in the savannah environment.

In the coniferous forests of the western United States, fires vary from a rare but potentially devastating force in the rain forests of the Olympic peninsula to a dominant factor in the drier forests of the Cascade, Sierra, and Rocky mountains. Several forest types are dominated by very long-lived species (e.g., Douglas-fir, redwood, giant sequoia, and ponderosa pine), which may suggest long periods without catastrophic crown fires, but ground fires may be common and may be critical for reducing fuel loads. In these dry forest types, ground fires may occur at 5- to 10-year intervals, while catastrophic fires may occur only once in several centuries. Interactions between ground fires and crown fires has become a major management issue on federal lands in these dry forest types. Excluding ground fires results in a buildup of fuels that can set the stage for very large and damaging fires, as discussed later.

The dry shrublands of coastal California and the intermountain West are also fire-adapted systems. Relatively low rates of production cause slower rates of biomass accumulation, but decomposition is minimal due to dry and hot conditions. Fires tend to recur at 20- to 40-year intervals. Fire has also been thought to play an additional role in this type of system; that of destroying allelopathic chemicals that accumulate in the soil during succession and that begin to reduce plant vigor over time.

The boreal forest, with its cold climate and low evaporative demand, would seem an unlikely system for fire. But the low mean temperature severely restricts decomposition and, together with the low quality of foliage produced by the dominant conifers, leads to large accumulations of organic matter. As we will see in Chapter 19, fires tend to occur during periods of extreme drought, on a 70- to 100-year rotation, and are crucial to maintaining a functional forest ecosystem.

In contrast, the hot and dry deserts, which would seem to be the ideal climate for frequent fires, burn only rarely. Climatic conditions are so severe that accumulations of organic matter are too low to support extensive burns.

To summarize, fires are rare, and perhaps unimportant, only at the extremes of the climatic continuum across North America. Neither the most humid temperate forests nor the driest deserts experience fire as a major factor. However, between these two extremes, in dry forests, cold forests, grasslands, and shrublands, fire exerts a major control on ecosystem structure and function.

EFFECTS ON SOILS AND PLANTS

The simple act of burning organic matter has some fairly complex effects on soil structure and nutrient availability. The most immediate are the loss of carbon (C) and nitrogen (N) from surface soil organic layers to the atmosphere in gaseous form, and, in the case of crown fires, from vegetation. Most of the other macronutrients (cations and phosphorus [P]) remain in the ash and may be deposited on the soil surface. However, in hot fires, which are often accompanied by strong winds generated by the fire itself, windblown ash may remove P and cations from the site as well. Table 17.1A summarizes losses due to fire in two different experimentally burned systems and a major natural wildfire.

Cation nutrients that remain in the ash, such as calcium, magnesium (Mg), and potassium, tend to be in both mobile and plant available forms (Table 17.1B). When rainfall follows a burn, these mobile cations move down through the soil profile and tend to displace hydrogen ions from exchange sites in the soil, increasing soil pH. Increases can be dramatic in acid forest soils but tend to be minor and unimportant in grassland and shrubland soils, which are nearly neutral even before fires.

The dark color of the ash, plus partial or total removal of plant cover, increases absorption of solar energy at the soil surface, increasing soil temperature. Evapotranspiration may also be lower due to lower foliar biomass, resulting in higher water content in soils. Combined with higher pH, these factors tend to increase microbial activity and may further add to nutrient availability by increasing rates of mineralization of the remaining soil organic matter. This effect may be further augmented by alterations in the chemistry of remaining organic matter by the heat of the fire and also by a narrowing of the C : N ratio. Both higher soil pH and increased availability of ammonium (NH_4^+) may increase the rate of nitrification and the availability of nitrate (NO_3^-). Figure 17.4 shows dramatic increases in the accumulation of both NH_4^+ and NO_3^- in chaparral soils following fire.

It has also been suggested that volatile organic compounds (terpenoids) produced by plants in ponderosa pine and other dry forest and shrubland

TABLE 17.1 Changes in Nutrient Distribution Following Fire: (A) Nutrient Losses from Experimental Surface Fire in a Longleaf Pine Stand,* Experimental Crown Fire in Chaparral,† and Major Wildfire in a Mixed Conifer Forest in Central Washington,‡ (B) Percentage of Soluble Nutrients in Ash From the Longleaf Pine Burns§

	Nutrient Losses in kg/ha (% of Total)			
	N	P	CA	K
A. System				
Longleaf pine	22.3 (66)	1.6 (53)	4.1 (20)	4.2 (40)
Chaparral	146.0 (10)	11.0 (2)	35.0 (.4)	48.0 (12)
Plants + litter only	110.0 (39)	0.2 (—)	9.0 (—)	46.0 (16)
Mixed conifer	907.0 (39)		75.0 (11)	308.0 (35)
B. Nutrient				
N 0.4%				
P 6.7%				
K 58.9%				
Ca 22.4%				
Mg 90.8%				

*From Christensn 1977
†From Debano and Conrad 1978
‡From Grier 1975
§Nutrients that were mobile and available to plants; from Christensen 1977

ecosystems chemically inhibit both mineralization and nitrification. Rapid and immediate increases in net mineralization and nitrification following fire in these systems may result from destruction of the inhibiting compounds by the heat of the fire.

However, the effects of fire on nutrient availability are not always positive. Short-term increases in nutrient availability may be offset by long-term decreases in systems where fire frequency is high and inputs to the system between fires are not high enough to replace losses. Volatilization of C and N in the fire can represent a sizable fraction of the total in plant biomass, litter, and surface soil organic matter (the O horizon; Figure 17.5*a*). Wind erosion of the remaining ash can represent an additional loss, especially for cations that are highly concentrated in ash (Figure 17.5*b*). The actual net effect of wind erosion is difficult to determine because redeposition in rainfall and through settling out of windblown particles can increase apparent inputs to a burned area or into similar systems nearby. The effect of cumulative N losses through fires can be seen as decreased net N mineralization after 20–30 years with increasing fire frequency in both an oak–hickory forest and a tallgrass prairie (Figure 17.6).

These results present an interesting question. Many grassland and forest systems have experienced recurring fires, presumably for millions of years. How have the N losses associated with these fires been offset, especially before the additional pollution-driven inputs of this nutrient (Chapter 25) occurred? Clearly, stability over long periods of time requires that the inputs and outputs of nutrients be balanced.

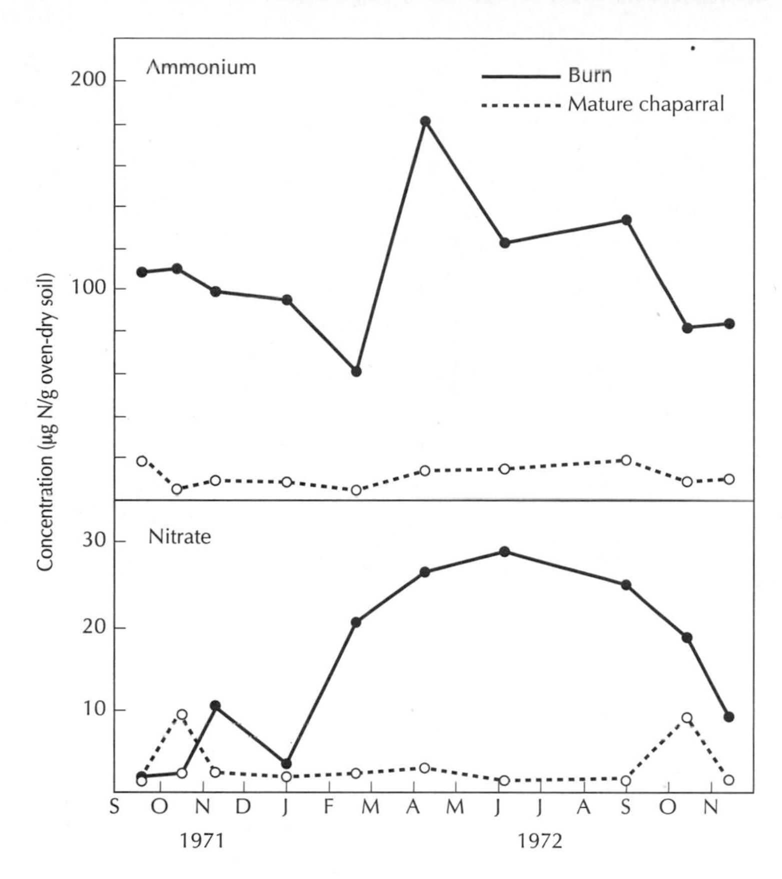

FIGURE 17.4 Concentrations of ammonium and nitrate in burned and unburned chaparral soils. (Christensen 1973)

The simplest answer is N fixation. Several ecosystem types that experience crown fires have N-fixing species as a major component of the postfire vegetation. These include species of alder in the Douglas-fir region, locust in the drier forests of the Southeast and Southwest, and ceanothus in the chaparral. Alder and ceanothus in particular have the capacity to replace N lost to fire within a few years. The red alder example in Chapter 14 is one case in which dramatic changes in soil fertility have resulted from the presence of this early successional species in a forest type thought to experience large, catastrophic fires once every several centuries. Fire increases the importance of species in the legume family in temperate grasslands, but actual rates of N fixation and the effect of these on N balances have not been measured in the field.

Nitrogen fixation can also be carried out by free-living microbes in soils and organic residues. As anoxic and C-rich environments are required for this process, large downed logs are a preferred habitat. One comparison of N fixation rates in burned and harvested pine forests suggested that harvesting could reduce N fixation and alter the long-term N balance by removing large woody debris, which supports most of the N fixation (Table 17.2). The rate of N fixation in soils by free-living microbes can also be limited by low availabil-

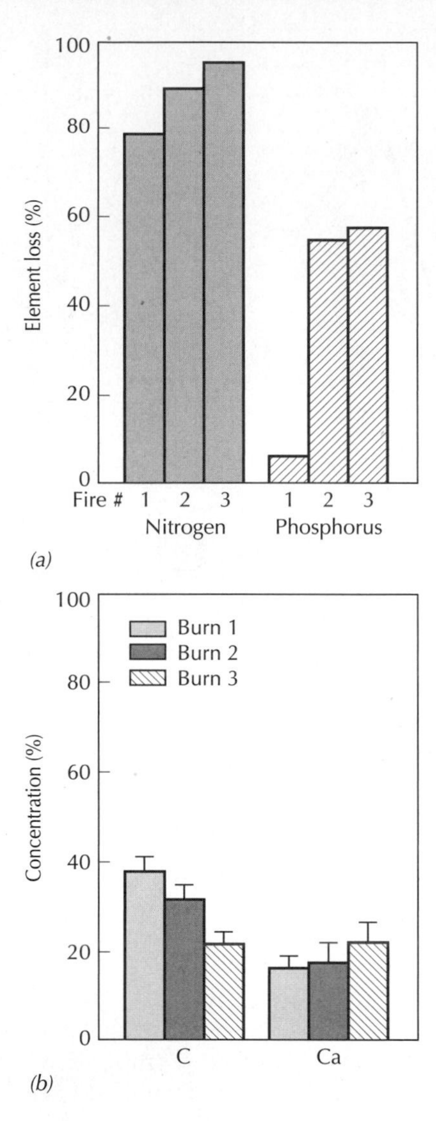

FIGURE 17.5 Loss and redistribution of nutrients following experimental fires in a dry tropical forest. (*a*) Total losses of nitrogen and phosphorus due to volatilization. (*b*) Concentration of carbon and calcium in ash. Note that carbon–calcium ratios are nearly 1 : 1 as opposed to 50 : 1 to 100 : 1 in biomass. (*b,* Kaufman et al. 1993)

ity of Mg and P, both of which would be added to soils in the ash produced in a fire.

To summarize, surface fires tend to increase the pH of acidic soils and to increase the availability of P and cations. Significant amounts of C are lost and the C–N ratio of soils lowered. This may result in increased N mineralization and nitrification, but short-term increases in availability may be offset by longer-term decreases in total soil N content. In very hot ground fires, or the catastrophic crown fires, losses of N can be great enough to reduce N availability significantly in the postfire environment. With high availability of other nutrients, as well as for water and light, species with N-fixing capabilities may have a competitive advantage.

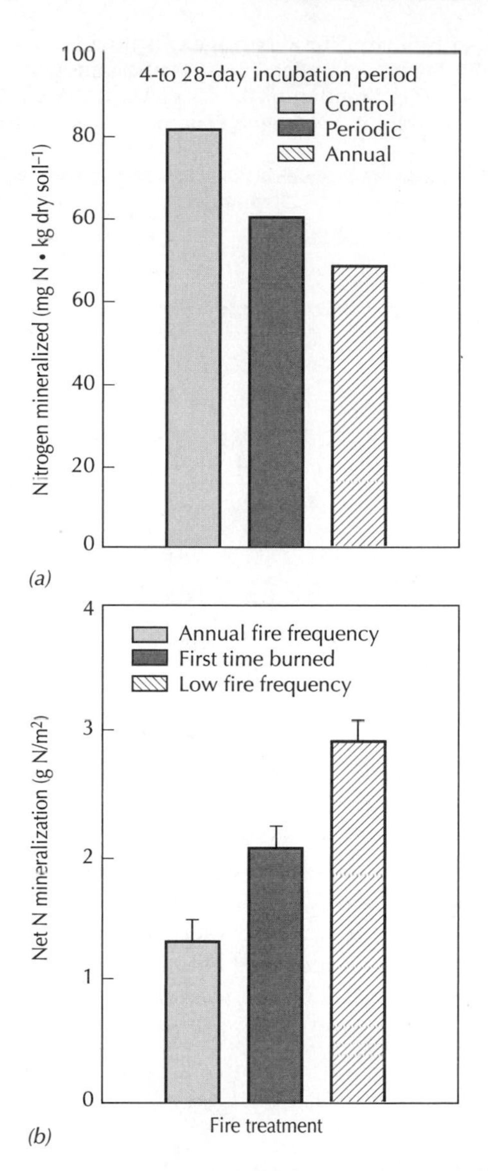

FIGURE 17.6 Long-term effects of surface fires on nitrogen availability. (*a*) Effect of 30 years of annual and periodic (4-year interval) surface fires on net nitrogen mineralization from soils in an oak–hickory forest. (*b*) Effect of annual fires, a first fire, and background or low fire frequency on net N mineralization in tall grass prairies. (*a,* Vance and Henderson 1984; *b,* Blair et al 1998)

Plant response to the postfire environment generally includes increased nutrient concentration in foliage, which may increase photosynthesis and total production. However, these increases may be short-lived as foliar biomass increases and the transient effects of higher soil pH and reduced competition disappear. In grasslands, where rapid and vigorous resprouting of burned vegetation occurs, above-ground productivity is generally increased in the postfire environment, especially in wet years (Table 17.3). While much of this apparent growth is actually the translocation of C and nutrients from

TABLE 17.2 Comparison of Estimated Total Nitrogen Fixation Over a 120-Year Period in a Douglas-Fir Forest Following Forest Harvest and Following Fire. Harvesting Removes a Larger Fraction of the Large Logs, Which Provide the Primary Site for Nonsymbiotic Nitrogen Fixation in this System.
(Wei and Kimmins 1998)

Treatment	Harvest	Burn
Total N Fixation ($kg\ N \cdot ha^{-1}$)	38	70

storage organs below ground, similar to that which occurs following grazing, the reduction of shading by dead material from previous years removes a major impediment to early vigorous growth.

PLANT ADAPTATIONS TO DIFFERENT FIRE REGIMES

As with grazing, fire represents an important selective force in the evolution of species in those systems in which it regularly occurs. As such, several widely distributed reproductive characteristics have evolved in response to fire. These can be summarized in two broad categories: (1) seed dispersal and germination in response to fire and (2) protection or production of vegetative reproductive tissues below ground.

Most of the fire-adapted forest ecosystems in North America are dominated by conifers. Several of the species that dominate these areas have what are called **serotinous cones** (Figure 17.7). Serotinous cones do not open and shed seed in the same year that they are formed, as generally happens with nonserotinous cones. Rather, the scales of the cone are bound together by a resinlike substance that softens only under the conditions of high temperature, which occur during crown fires or intense ground fires. Closed, seroti-

TABLE 17.3 Increases in Above-Ground Productivity in Prairies Following Fires
(Kucera 1981)

Region	Control	Fire
Illinois	3,020	13,210*
	3,610	5,910†
Missouri	5,090	9,330*
	4,820	5,220†
Iowa	3,490	7,500
Eastern Kansas	1,860	3,400
Western Kansas	3,800	1,710

*Wet years. All values are expressed in $kg \cdot ha^{-1} \cdot yr^{-1}$.
†Dry years.

FIGURE 17.7 Serotinous cones are held on the tree and kept tightly closed by a resin-like substance. Fire softens this sealant and results in the opening of the cones soon after the burn occurs. (USDA Forest Service)

nous cones can remain on the tree for up to 15 years in black spruce or 25 years in jack pine. Viable seeds have even been found in cones that have become overgrown by the woody stem or branches of the tree. Serotinous cones cause maximum seed dissemination to occur immediately after a fire, at times when resource availability is high and competition from other plants is low. Seeds within the serotinous cones are generally not damaged by fire unless the cone itself is burned. The cones are resistant to burning and can withstand temperature as high as 300°C for 60 seconds.

An unusual method of protecting buds (or growing apical tips) from fire occurs in longleaf pine, a species that grows in the fire-prone coastal plain of the southeastern United States. Following germination and seedling establishment, this species remains in a "grass stage" (Figure 17.8) for 5 to 20 years. During this time, a large taproot containing large quantities of stored carbohydrates is produced. The apical bud is embedded in the surrounding upright needles, which are resistant to burning and protect the bud. After reaching a critical size, the seedling begins very rapid height growth, up to 1 m/year. This growth pattern is geared to the frequent occurrence of surface fires. The seedling is first protected in the grass stage until sufficient reserves have been accumulated so that rapid height growth carries the growing tip above the level of ground fires in only a few years.

Many tree species of frequently burned areas also have the ability to root-sprout following fire. Viable below-ground buds are present in most oak species, aspen, black spruce, redwood, and even a few species of pine (e.g., pitch pine). It is thought that the ability of oak to survive and sprout following fire is the reason that it dominates many areas in the western portion of the eastern hardwood forest, where reproduction of oak occurs only sporadically now, in the absence of fire.

FIGURE 17.8 Stages in the growth of longleaf pine. (*a*) Seedlings in the grass stage are nearly indistinguishable from real grasses. (*b*) Height growth may be more than 1 m/year after the grass stage is broken. (USDA Forest Service)

Grassland ecosystems show both bud protection and increased reproduction as responses to fire. Perennial grasses and forbs of prairie systems characteristically maintain large root and rhizome systems from which leaves and stems can be produced following either fire or grazing. Many species also show very large increases in flowering and seed production following fires. Again, flowering after fire may occur in response to increased resource availability and is also adaptive in that fire may create openings in the dense prairie sod in which seedling establishment is otherwise very difficult.

In Chapter 16, we presented the idea that successful adaptation to herbivory might cause further selection for plant traits that would encourage this form of "disturbance." Might this also be true for plants that have adapted successfully to fire? Whether plants actually have evolved characteristics that increase fire frequency or intensity has been argued for some time. There are suggestions that the resin content of conifer plant mass and the open, easily dried structure of conifer forest floors are adaptations that increase the frequency and intensity of fire. Similarly, the retention of dried biomass on certain chaparral plants, as well as the high content of volatile and flammable chemical compounds in these species, may be adaptations for increasing fire frequency. The large areas of dried, fine-structured plant mass that occur in prairie systems between midsummer and the following spring certainly offer an optimal physical structure to promote burning.

However, all of these characteristics also affect other aspects of plant function. We have discussed the role of leaf and canopy structure in response to energy and water balances (Chapters 6–8) and the role of secondary plant chemicals as controllers of herbivory (Chapter 15). The extent to which the adaptive advantage of increasing fire frequency has contributed to the evolution of plant structure and biochemistry is still unclear, however.

FIRE–HERBIVORY INTERACTIONS

Inverse relationships have been reported between the intensity of herbivory and the occurrence of fire. For example, fire frequency in the grasslands of the Serengeti region of Africa (Chapter 20) decreased during a period of increasing abundance of major herbivores and higher plant biomass removals through grazing. Mowing, as a proxy for herbivory, decreased the fraction of land surface burned in a tallgrass prairie (Figure 17.9*a*).

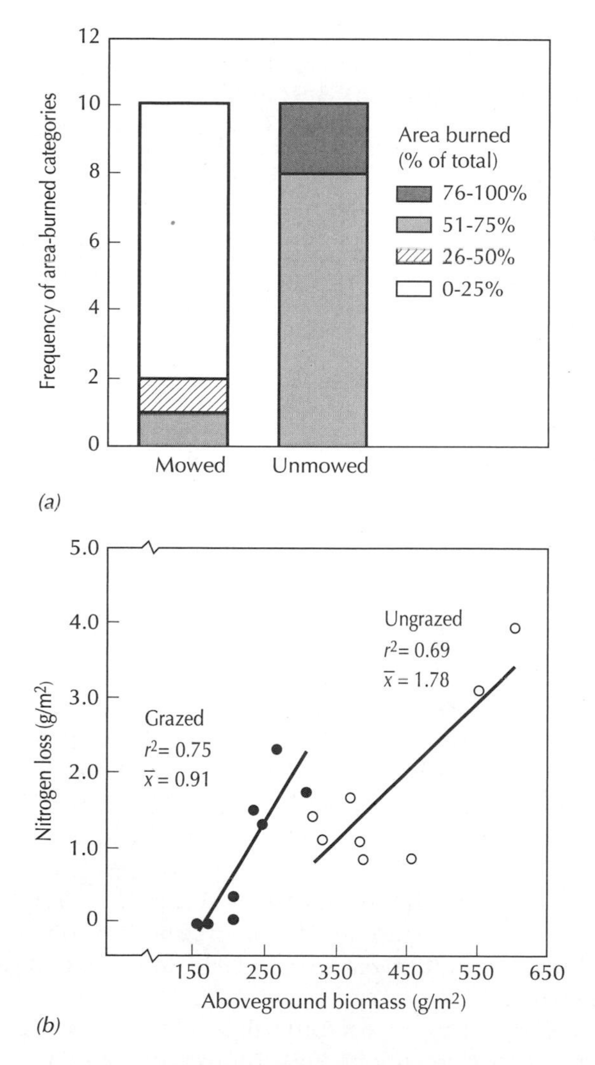

FIGURE 17.9 Effect of simulated grazing (mowing) on fires in tallgrass prairie. (*a*) Fraction of ground burned in mowed and unmowed areas. (*b*) Estimated losses of nitrogen through volatilization in grazed and ungrazed areas. Note that more nitrogen is lost per unit biomass burned in the grazed areas because of higher nitrogen concentrations but that much more biomass is burned in the ungrazed areas. (*b*, Hobbs et al. 1991)

By removing plant biomass, herbivory can also reduce nutrient losses that accompany fire. In a comparison of N volatilization by fires in tallgrass prairies, it was found that while grazing increased the N concentration in plant biomass, reductions in total above-ground biomass due to consumption actually reduced the total amount of N lost from the system due to fire (Figure 17.9*b*).

On the other hand, fire can alter the plant community in ways that increase herbivory. Studies following the major fire in Yellowstone National Park (see subsequent discussion) and other grasslands have shown that large herbivores, such as elk and buffalo, graze preferentially in burned patches, where forage has higher N content and higher C quality. A combination of fire and herbivory is often hypothesized as the reason for the extension of the "prairie peninsula" into the Great Lakes region, where the climate is more suitable for forests.

FIRE AND THE MANAGEMENT OF ECOSYSTEMS

Fire has proven to be a controversial tool in the management of ecosystems. Early attitudes in the United States toward fire were derived from experiences in European and eastern North American forests, where fire was relatively unimportant in ecosystem development and the effect of fire was reduced economic and aesthetic value. Transplanting these ideas into the drier ecosystems of the western United States has caused significant problems.

Fire in the Giant Sequoias

Management of the giant sequoia forests of California offers an example of the complexity of fire management. These groves, which contain the most massive trees in the world, developed under a regime of frequent ground fires. A fire occurred somewhere within the groves in and around Kings Canyon National Park nearly every year. Individual trees experienced minor fire damage, as indicated by the accumulation of fire scars in the stem, on an average of once every 10–40 years. There are indications that many fires were deliberately set by Native Americans.

The ground fires served to reduce fuel loads by burning off accumulations of organic matter on the soil surface and by killing saplings and seedlings. Mature sequoias were not severely harmed by these fires due to their very thick bark, and the chance of a major crown fire was reduced by the tremendous distance between the ground and the lowest branches of the trees.

The incidence of fire was reduced by the displacement of Native Americans by Europeans beginning around 1870. Fire suppression became an official and effective policy in this area around 1900. Fire scars on trees declined markedly after 1875.

Fire suppression led to the accumulation of surface fuels and, perhaps more ominously, to a tremendous increase in the abundance of white fir in the understory. White fir is a relatively shade-tolerant, fire-sensitive species that was kept to low levels in the presettlement forest by ground fires. A marked increase in establishment of stems of both white fir and sugar pine accompanied the reduction in fire frequency (Figure 17.10*a*). The development of this understory altered the appearance of the giant sequoia groves and also provided a potential "fire ladder" for the movement of ground fires into the upper canopy. A fire once started could climb this ladder and "crown," destroying these venerable and majestic trees.

As attitudes toward the potential value of fires in dry forest systems changed in the 1960s and 1970s, studies were begun to assess the potential for reintroducing controlled fires into the giant sequoia groves and to test their effectiveness in reversing the changes caused by fire suppression. Recent results from these studies show that the density of fir has been reduced by about one third in the groves by the use of fire (Table 17.4).

The use of both controlled ground fires (prescribed burns) and partial logging of the understory have been considered as means for reducing the fire hazard and restoring the park to its presettlement condition. However,

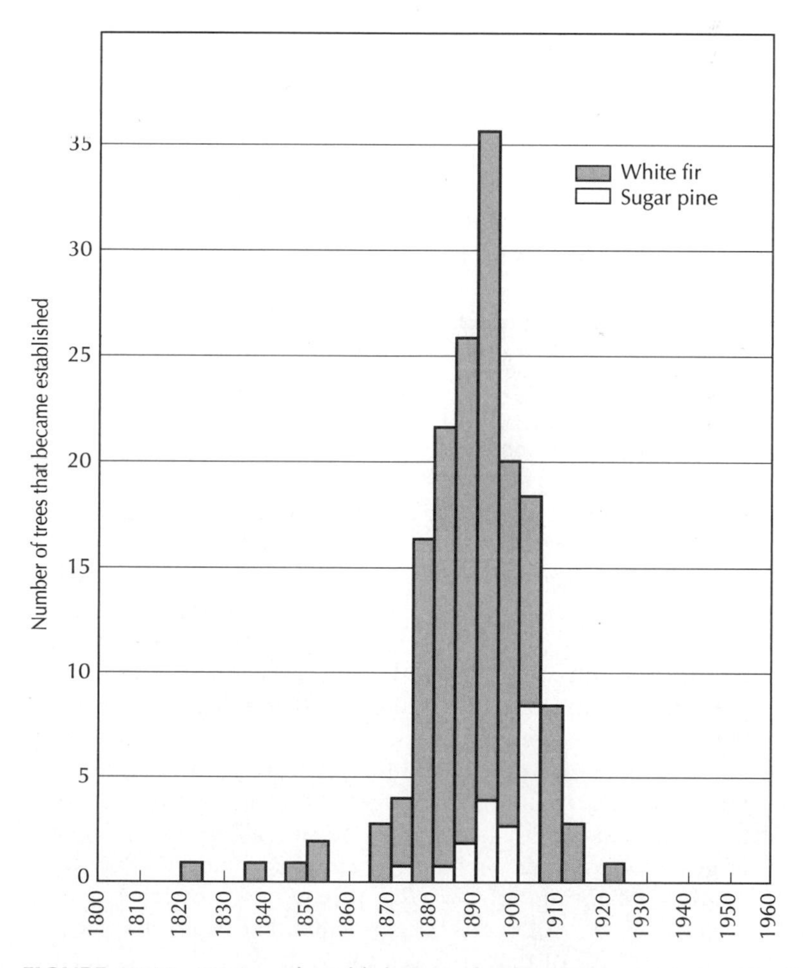

FIGURE 17.10 Timing of establishment of pole-sized understory white fir and sugar pine trees in King's Canyon sequoia stands. Note the peak following fire suppression in the late 1800s. (Kilgore and Taylor 1979)

TABLE 17.4 Changes in Density, Basal Area, and Percentage of Cover of Understory Trees in Giant Sequoia Groves Within Sequoia National Park over a 37-Year Period in Response to a Management Plan Emphasizing Prescribed Burns as an Understory-Control Mechanism.
(Roy and Vankat 1999)

Year	1969	1996
Density (number of stems · ha^{-1})	1,009	618
Basal Area ($m^2 \cdot ha^{-1}$)	955	367
Coverage (% of ground area)	121	88

this type of human intervention in the national parks remains controversial, even though the policy of fire suppression itself represents a major disturbance to the native ecosystem. The dangers inherent in setting fires that get out of control and damage property or threaten lives is one very real consideration that must be factored into any management scheme.

Yellowstone National Park

Late in the summer of 1988, a series of extensive crown fires swept through the oldest national park in the United States, Yellowstone National Park. Following one of the driest summers on record, fires initiated by lightning strikes, and some by human carelessness, spread out of control and covered 45% of the park (Figure 17.11). The social response to this huge set of fires was complex and reflected a growing awareness of the role of disturbance in natural ecosystems and of the limited value of human intervention.

One of the first questions asked was whether human tinkering with the Yellowstone ecosystem had contributed to the severity of the fires. The primary goal of the management of wilderness areas is the maintenance of ecosystems in the same state as would occur if humans had no impact. Were, then, the fires the result of human management? Were they unusual in scope or intensity? Were they so severe that the park's ecosystems could not recover?

In answer to the first question, humans have definitely altered the frequency and intensity of fires. The park was managed under a policy of complete fire suppression, beginning with the inception of the National Park Service in 1916. Even earlier, beginning in 1886, the park was patrolled by the U. S. Cavalry, one of whose duties was fire control. Fire suppression was effective in reducing the percentage of the park burned, the result of which was the accumulation of large areas of late successional forest, those with the largest biomass and fuel loads. In 1972, the National Park Service switched to a policy of fighting only fires that threatened property or human life. Even so, major fires did not occur because climatic conditions were not favorable for many years. All of that changed in 1988.

Was the 1988 fire unusual? Perhaps only in that the burn occurred all in 1 year rather than spread out over a series of years or decades. Figure 17.12 summarizes the fraction of the park burned by decade over a 300-year period.

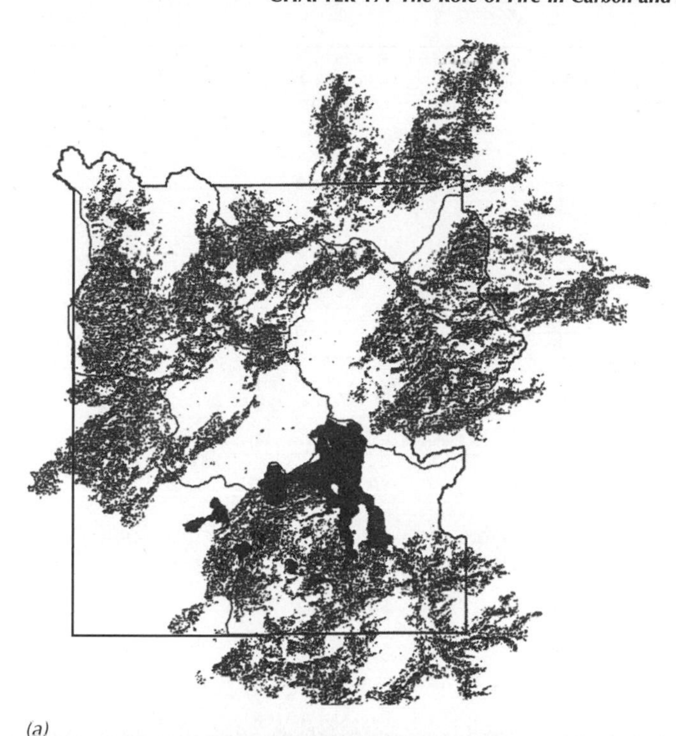

(a)

(b)

FIGURE 17.11 (a) Map of the distribution and intensity of fires in Yellowstone National Park in 1988. (Schullery 1989) (b) Picture of intense crown fire in a dry western forest. (Bureau of Land Management, Alaskan Type One Incident Management Team) **See plate in color section.**

The 1988 fire stands out, but the area consumed in 1690–1730 is much the same as that burned in 1950–1990.

Were the park's ecosystems irreparably damaged? Early studies of the recovery of forests and grasslands suggest that recovery is more rapid than expected. Severely burned forests that were expected to revert to meadows

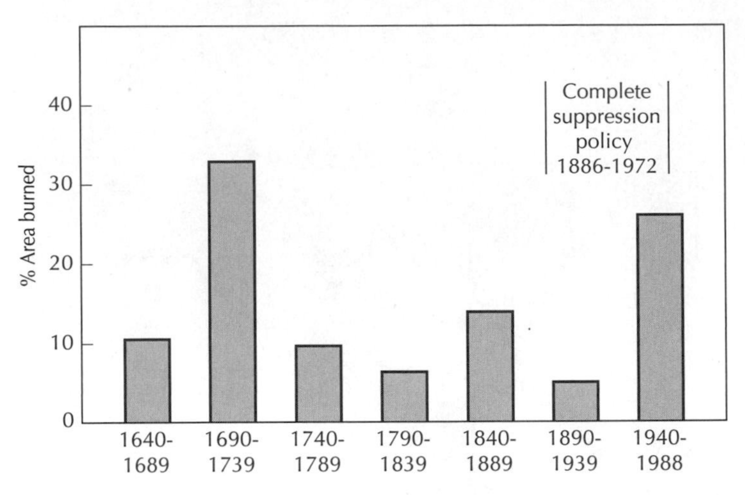

FIGURE 17.12 Estimated fire frequency in Yellowstone National Park from 1640–1988. (Romme and Despain 1989)

for some time instead regenerated to young forests. Grasslands appear to have been less affected by the fire than by the grazing of elk. Wildlife suffered some direct mortality because of the fire, but the quality of forage increased following the fire, and elk grazed preferentially in burned areas.

Can we assume that the fire of 1988 has redressed past management errors and that the park will now be just what it would have been without humans? We can never really answer that question. There is a growing acceptance that change is the norm in natural systems; change that plays out over decades to centuries and has a long memory for disturbance events. Yellowstone may never be just like it was in 1700. But then again, this would probably be true even if humans had never entered the park.

Fire in Commercial Management and Restoration

Prescribed burning is also of interest in the management of commercial forests, both in the national park forests and on private lands. The same concept, that of fire exclusion leading to large fuel accumulation and, in turn, to catastrophic fires, is an important concern in the economic management of forests.

The potential for forest protection through prescribed burns has been illustrated dramatically in the Coconino National Forest in Arizona. As part of a research program, prescribed burns had been carried out in one part of this forest for several years. In 1980, a major crown fire broke out in forest and moved toward the experimental site. However, fuel loads had been reduced in the prescribed burn area to such an extent that the fire could not be supported. Figure 17.14 shows that the fire burned up to and around the experimental area, but not through it.

FIGURE 17.13 Aerial view of part of the Coconino National Forest following a major forest fire. The remaining green (*dark*) area marks the location of an experimental set of prescribed burns that had significantly reduced fuel loads. (Courtesy of Dr. Wallace Covington)

Very early work establishing the necessity of fire in the maintenance of tallgrass prairies was carried out under a very different type of management plan. Efforts to restore or recreate prairies from abandoned farmland in Wisconsin were only marginally successful until fire was reintroduced on a regular schedule. Now the biennial burns are the major management tool used to perpetuate this once nearly extinct type of ecosystem (Figure 17.1*b*).

CONCLUSION

If our view of fire has shifted away from one of suppression at all costs, are there still cases in which fire occurrence has negative effects and is to be reduced? Surely in areas where fire is rare but climatic conditions may occasionally be conducive to fire, and where human activity has greatly increased the ignition of fires due to camping, cigarettes, etc., the argument can be made that fire suppression is a way of reducing human impact on the ecosystem. The prevention of crown fires in the giant sequoia groves may be an example of a case in which suppression is required to reduce the potentially catastrophic loss of those stands until the understory and fuel loads can be reduced.

If there is an emerging principle here, it is that the economically and aesthetically important species in native ecosystems are well adapted to the naturally occurring fire regimes under which they have evolved. Maintaining the integrity and sustainable productivity of these systems may depend on maintaining similar regimes. In areas such as the tallgrass prairie region, where land use patterns now disallow the huge, extensive fires that raced through hundreds of kilometers of continuous grasslands, humans have to reintroduce fire to maintain processes that occurred without our help before we arrived. Direct human intervention is required in this case to maintain remnants of natural systems.

REFERENCES CITED

Blair, J. M., T. R. Seastedt, C. W. Rice, and R. A. Ramundo. 1998. Terrestrial nutrient cycling in tallgrass prairie. In Knapp, A. K., J. M. Briggs, D. C. Hartnett, and S. L. Collins (eds.). *Grassland Dynamics: Long-term Ecological Research in Tallgrass Prairie.* Oxford University Press, Oxford.

Bormann, F. H., and G. E. Likens. 1979. *Pattern and Process in a Forested Ecosystem.* Springer-Verlag, New York.

Christensen, N. L. 1981. Fire regimes in southeastern ecosystems. In *Fire Regimes and Ecosystem Properties.* U. S. Forest Service General Technical Report WO-26. U. S. Department of Agriculture, Washington, D. C.

Christensen, N. L. 1977. Fire and soil–plant nutrient relations in a pine–wiregrass savanna on the coastal plain of North Carolina. *Oecologia* 31:27–44.

Christensen, N. L. 1973. Fire and the nitrogen cycle in California chaparral. *Science* 181:66–68.

Debano, L. F., and C. E. Conrad. 1978. The effect of fire on nutrients in a chaparral ecosystem. *Ecology* 59:489–497.

Gray, J. T., and W. H. Schlesinger. 1981. Nutrient cycling in Mediterranean type ecosystems. In Miller, P. C. (ed.). *Resource Use by Chaparral and Matorral.* Springer-Verlag, New York.

Grier, C. C. 1975. Wildfire effects on nutrient distribution and leaching in a coniferous ecosystem. *Canadian Journal of Forest Research* 5:599–607.

Heinselman, M. L. 1981. Fire intensity and frequency as factors in the distribution and structure of northern ecosystems. In *Fire Regimes and Ecosystem Properties.* U. S. Forest Service General Technical Report WO-26. U. S. Department of Agriculture, Washington, D. C.

Heinselman, M. L. 1973. Fire in the virgin forest of the Boundary Waters Canoe Area, Minnesota. *Quarternary Research* 3:329–382.

Hobbs, N. T., D. S. Schimel, C. E. Owensby, and D. S. Ojima. 1991. Fire and grazing in the tallgrass prairie: Contingent effects on nitrogen budgets. *Ecology* 72:1374–1382.

Kaufman, J. B., R. L. Sanford, J., D. L. Cummings, I. H. Salcedo, and E. V. S. B. Sampaio. 1993. Biomass and nutrient dynamics associated with slash fires in neotropical dry forests. *Ecology* 74:140–151.

Kilgore, B. M. 1981. Fire in ecosystem distribution and structure: Western forests and shrublands. In *Fire Regimes and Ecosystem Properties.* U. S. Forest Service General Technical Report WO-26. U. S. Department of Agriculture, Washington, D. C.

Kilgore, B. M., and D. Taylor. 1979. Fire history of a sequoia-mixed conifer forest. *Ecology* 60:129–142.

Kucera, C. L. 1981. Grasslands and fire. In *Fire Regimes and Ecosystem Properties.* U. S. Forest Service General Technical Report WO-26. U. S. Department of Agriculture, Washington, D. C.

Romme, W. H., and D. G. Despain. 1989. Historical perspective on the Yellowstone fire of 1988. *BioScience* 39:695–699.

Roy, D. G., and J. L Vankat. 1999. Reversal of human-induced vegetation changes in Sequoia National Park, California. *Canadian Journal of Forest Research* 29:399–412.
Schullery, P. 1989. The fires and fire policy. *BioScience* 39:686–699.
Vance, E. D., and G. S. Henderson. 1984. Soil nitrogen availability following long-term burning in an oak–hickory forest. *Soil Science Society of America Journal* 48:184–190.
Wei, X., and J. P. Kimmins. 1998. Asymbiotic fixation in harvested and wild-fire killed lodgepole pine forests in the central interior of British Columbia. *Forest Ecology and Management* 109:343–353.

18

SYNTHESIS: A GENERALIZED THEORY OF ECOSYSTEM DYNAMICS

INTRODUCTION

The past 12 chapters have presented basic information from a tremendous range of traditional disciplines. Taking the central definition of *ecosystem studies* as the theme, we have discussed how plant physiology, soil chemistry, hydrology, soil physics, microbial ecology, population dynamics, plant–herbivore interactions, and fire ecology all affect the fluxes of carbon (C), water, and nutrients through ecosystems. Along the way, we have attempted to stress some generalized patterns and feedbacks that might exist between some of these processes.

Advances in a scientific discipline often proceed from a chaotic set of interesting observations to the development of increasingly generalized theories that encompass those observations. As new observations that violate the existing theoretical framework accumulate, broader or more complete theories need to be established. Writing in the year 2000, it has now been 65 years since the word *ecosystem* was coined and over 40 years since the first large-scale ecosystem-level experiments were begun. Based on the accumulated observations of 4 decades of ecosystem research, can we construct a generalized theory of ecosystem function?

The purpose of this chapter is to review briefly the development of general theories of ecosystem function. Two slightly different approaches are compared: (1) one derived from whole-ecosystem observations and (2) one derived from plant physiology. Conclusions drawn from attempts to reconcile these two approaches are tested in Part 3 against in-depth discussions of four very different types of systems.

SUCCESSIONAL THEORIES OF ECOSYSTEM DEVELOPMENT

Early Theories on Biomass Accumulation and Function

Early theories in terrestrial ecosystems derived from ecological work on succession, especially the classic sequences of old-field succession following the abandonment of farmland in the southeastern United States and primary succession along an age sequence of sand dunes near Lake Michigan. In the mid-1900s, ecological theory viewed succession as an orderly set of changes in species composition that led to a generally recognized endpoint called the **climax community**. The influence of traditional successional thinking can be seen in an early theory of terrestrial ecosystem development (Figure 18.1), which described a continuous and orderly process during which production increased (followed by respiration); nutrient cycles went from open to closed; and species diversity, the complexity of food webs, and perhaps overall stability of the system increased.

This view was heavily influenced by experiences in the eastern United States, where catastrophic disturbances, such as major fires, were not recognized as an important part of ecosystem function and succession was initiated either by human disturbance or relatively small-scale wind-throw events. As we saw in Chapter 17, systems west of the eastern deciduous forest biome do not fit this pattern. Forests of the lake states and the intermountain West, as well as the grasslands in between, demonstrated the important role of continually recurring fires in restarting succession and controlling ecosystem processes.

In response to the evidence from these systems, a more complex theory of "truncated succession" was developed in which the smooth march toward a defined end state (Figure 18.1) was interrupted repeatedly, and at different intervals, by fire (Figure 18.2). An additional implication of this theory is that the hypothesized end state of succession, the climax forest, would be seen rarely, if at all, in the real landscape. Rather, ecosystems would represent a patchwork of stands of different ages, all in the process of recovery from disturbance. Increasing appreciation of the role of both human and natural disturbance in all ecosystems has led to the general adoption of this more complex view.

Taking this one step further, the absolute necessity for fire in tallgrass prairie ecosystems was demonstrated by another set of land-use practices not

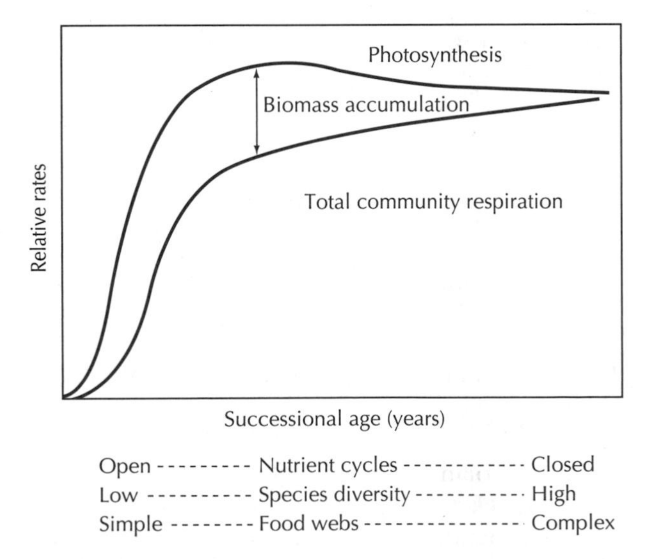

FIGURE 18.1 An early theory of the development of ecosystems over successional time (time since last disturbance). (After Odum 1969)

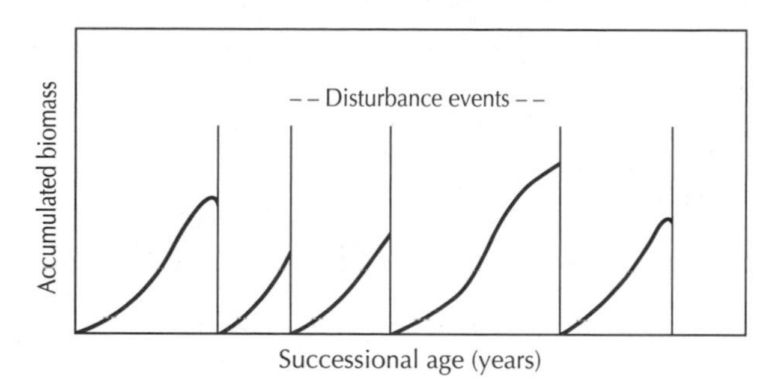

FIGURE 18.2 The concept of truncated succession. Successional accumulation of biomass is cut off by recurring disturbance.

initially conceived as an experiment at all. Prairie restoration is a major environmental initiative in the American Midwest. Native prairie ecosystems were rapidly converted to agriculture in the nineteenth century, with only a few remnants remaining by the early part of the twentieth century. Early attempts to restore abandoned farmland to prairie by importing plants from remnant areas teetered on the edge of failure as reestablished prairies failed to thrive. Large amounts of plant material accumulated over soils, something not seen in native prairies. The shade cast by this material cooled the soil in spring and also shaded out new growth. Eventually, it was discovered that only by the initiation of a regular cycle of prescribed fires that burned off the accumulating plant litter could the restored prairie ecosystem be maintained. In these systems, devastating disturbance was required on a 2- to 4-year cycle (Figure 17.1*b* is a picture of such a fire in progress).

Nutrient Balance Theories

Succession-based theories of ecosystem development pertained mainly to biomass accumulation and community dynamics because data on actual rates of nutrient cycling were sparse. As experimentation with nutrient balances and fluxes increased, theories of disturbance and succession were couched in different terms. A theory developed out of observations on nutrient responses following disturbance in the northeastern United States (see Chapter 4) suggested a more complex series of stages in ecosystem development (Figure 18.3). Biomass loss and accumulation still plays a central role in this theory, acting as a regulator of nutrient losses. Ecosystems at equilibrium or steady state (equivalent in certain ways to the climax system) would have no net accumulation of biomass, and losses of nutrients (in this case, nitrate [NO_3^-] leaching to streams) would be equal to inputs (e.g., in precipitation). A disturbance event that reduced plant uptake results in a pulse loss of nutrients that continues until plant production is reestablished (the reorganization phase). As the rate of plant biomass accumulation increases (the aggrading phase), nutrient availability becomes limiting due to losses in the reorganization phase, and losses in the aggrading phase approach zero. Losses remain very low until net biomass accumulation is reduced by the approach to steady-state or equilibrium conditions.

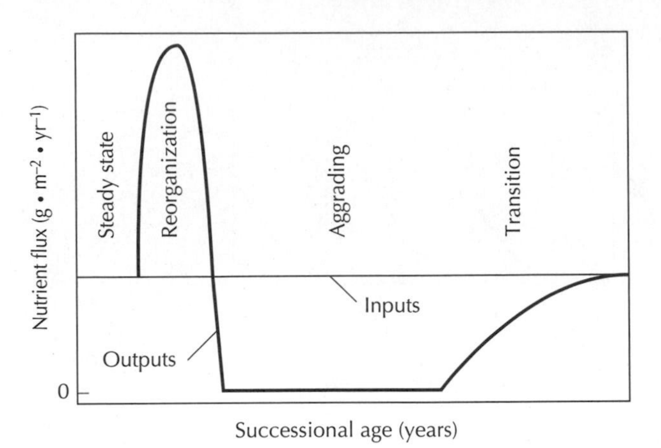

FIGURE 18.3 An early theory of changes in nutrient loss rates over successional time. At steady state, inputs equal outputs. Losses increase following disturbance until vegetation is reestablished (the reorganization phase). Losses of limiting nutrients are zero during the aggrading phase when biomass accumulation is high and increases during the transition to steady state. (After Vitousek and Reiners 1975, and Bormann and Likens 1979)

While presented as a general theory, the ideas captured in Figure 18.3 were at odds with observations in the massive forests of the Pacific Northwest and the boreal forests of Alaska, where nitrogen (N) retention appeared to be nearly complete at all stages of succession and NO_3^- losses did not generally occur, even after clear-cutting of forests. An alternate view of ecosystem development arose in this region in which the accumulation of large amounts of litter and humus over the mineral soil leads to increasing isolation of root biomass and microbial activity from the mineral soil. In this scenario, nutrient cycling decreases over time and is rejuvenated by disturbance (see Chapter 19).

The experiment presented at the end of Chapter 4 was devised in an attempt to resolve the differences between these two patterns of response. Results (Table 4.2) suggested that rates of N availability before disturbance played a role in determining NO_3^- loss responses. This is, of course, only a partial answer, opening the question of what determined rates of cycling before the experiments began. The fact that different ecosystems showed very different responses, even within the same forest type (see again Table 4.2), suggests that longer-term processes can cause significant variation in nutrient cycling over relatively short spatial scales.

One implication of either view of ecosystem development (Figure 18.3 or Chapter 19) is that both types of systems experience significant nutrient limitations on growth over much of each developmental sequence. Both the rapidly aggrading system in Figure 18.3 and the older, maturing system in the boreal example are predicted to show little or no losses of limiting nutrients and suboptimal nutrient availability. Several major ecosystem field and modeling projects in the 1970s were focused on understanding controls over nutrient cycling and on ways of maintaining adequate nutrient availability in forests in the face of increasing harvest removals under intensive management (Chapter 23).

For reasons discussed in earlier chapters, N is the most limiting nutrient in most temperate ecosystems, and the theories discussed earlier generally deal with N limitations or developed out of measurements made on N cycling. In many industrialized areas, the deposition of N to ecosystems has been increased several-fold through air pollution generated by industrial processes and vehicles. This has shifted the paradigm of N cycling studies in forests in industrialized regions from one of maintaining N cycles in the face of forest harvesting and other forms of disturbance to the alternative of minimizing NO_3^- leaching losses and associated acidification of soils and streams in the face of increased N deposition. The focus of N cycling research has shifted from processes that result in NO_3^- losses after disturbance to those that allow for retention of very large amounts of added N. A generalized theory of the progressive changes in ecosystem function as N additions accumulate has been presented (Figure 18.4) and tested in

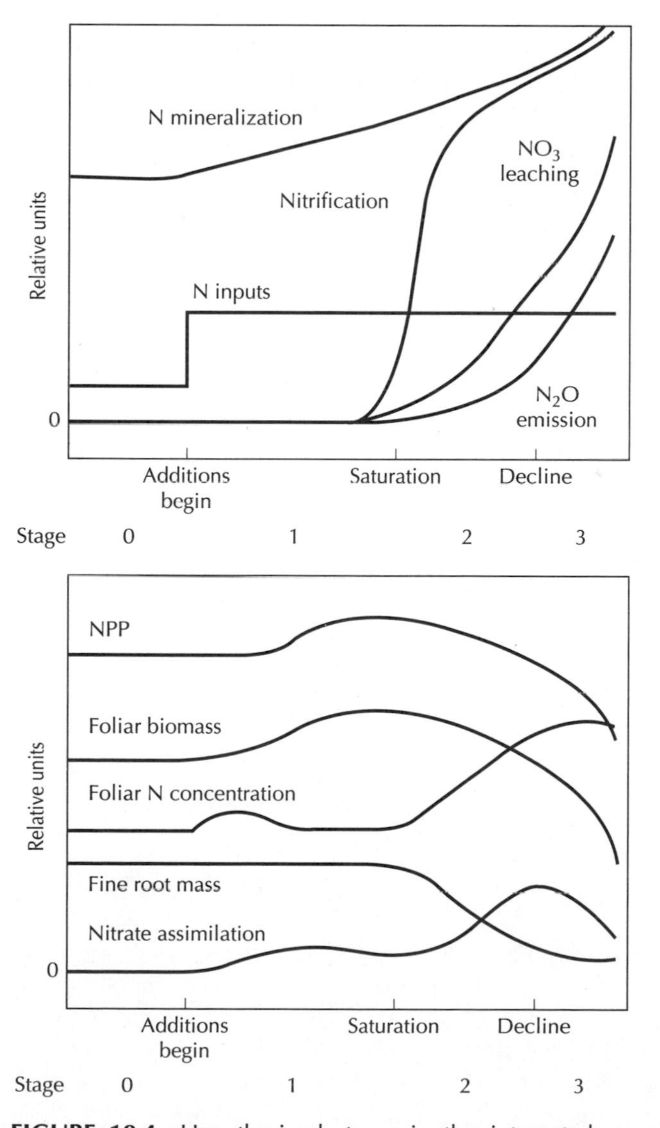

FIGURE 18.4 Hypothesized stages in the integrated response of nitrogen-limited forests to continuing nitrogen additions. (Aber et al. 1989)

both the United States and Europe. One of the crucial results from this set of experiments (see Chapter 25) was that N accumulation capacity is very large in most forests and can occur over long periods of time. An implication of this finding was that land-use events, such as fires or temporary conversion to agriculture, can affect nutrient cycling for well over a century—even longer than the periods of development envisioned in the theories presented in Figures 18.1 and 18.2. Another finding was that N losses from ecosystems can be increased dramatically by direct additions of N without the occurrence of forest maturation.

Recent recognition of dissolved organic losses of N from ecosystems also alters the traditional view of N balances described in Figure 18.3. We saw in Chapter 14 that dissolved organic C (DOC) losses are related to soil C–N ratios across biomes but occur everywhere. Nitrogen is associated with those same organic compounds such that dissolved organic N (DON) is also a small but relatively constant component of drainage water. In heavily polluted regions where NO_3^- or ammonium deposition is elevated, small losses of DON are not an important part of the total ecosystem balance. However, a recent study of the biogeochemistry of old-growth forests in an unpolluted region in southern Chile identified DON as the dominant form of N in stream water (Figure 18.5) and perhaps the dominant form of N loss from the system.

A Modern Synthesis

One recent attempt to create a more generalized theory that can encompass these different patterns of N gain and loss begins with a very simple model of nutrient pools and fluxes (Figure 18.6*a*) and uses this to develop simple extensions of the basic theory presented in Figure 18.3. The model is initialized with relatively low nutrient content and is constrained by a maximum potential uptake rate representing climatic limitations on net primary production.

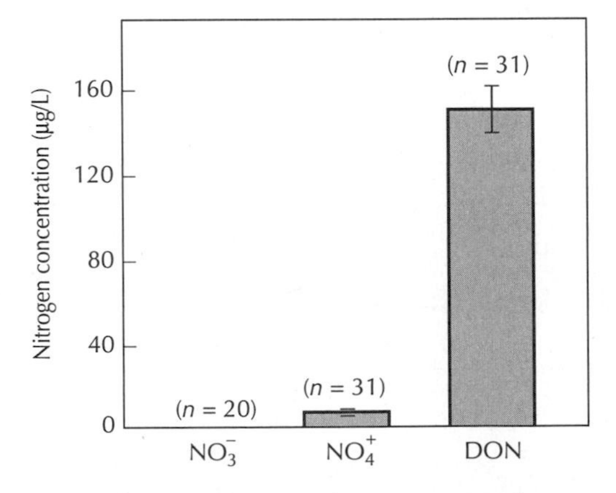

FIGURE 18.5 Relative importance of different forms of nitrogen loss from old-growth forests in an unpolluted region on the southern coast of Chile. (Hedin et al. 1995)

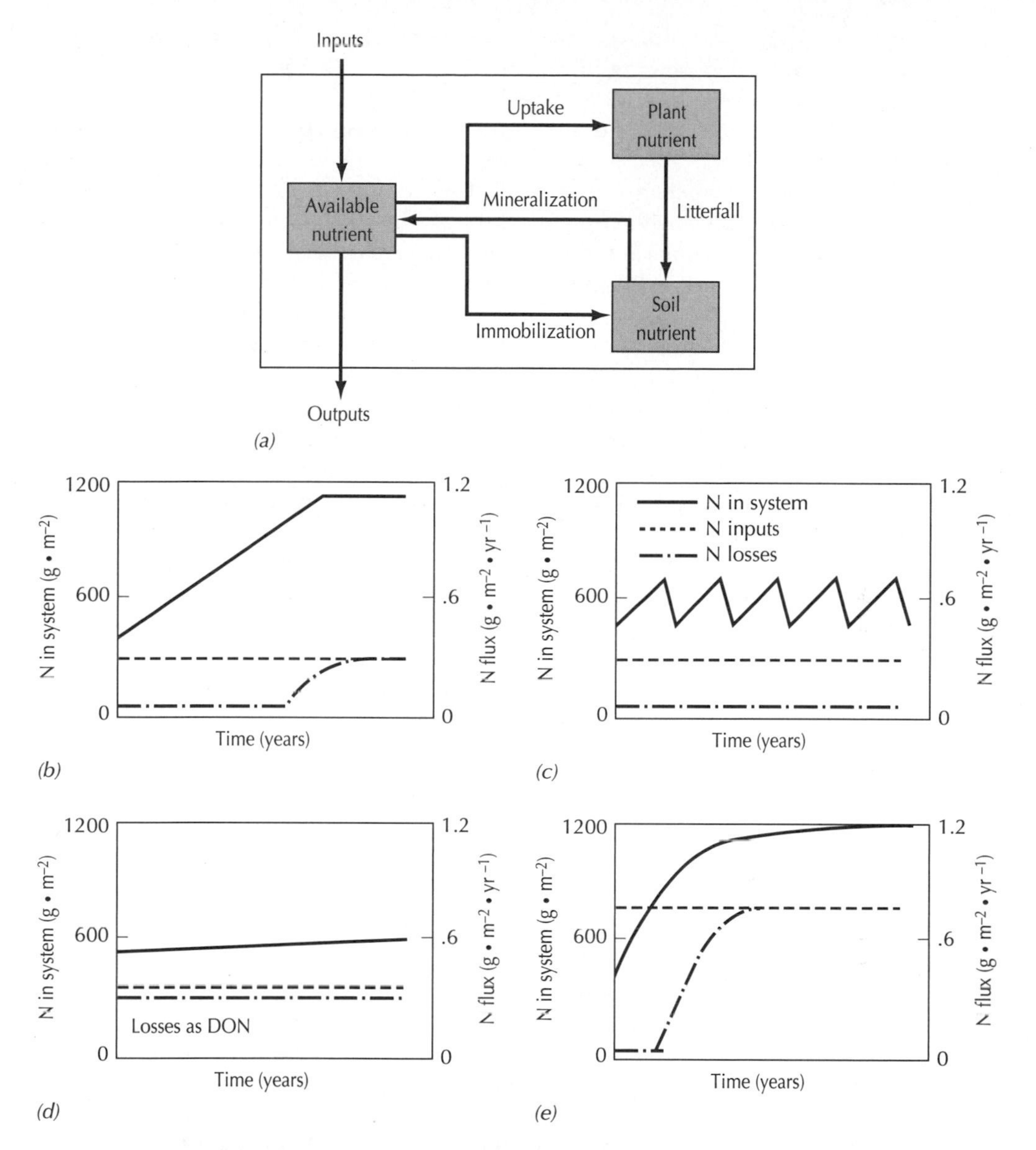

FIGURE 18.6 A generalized theory of nutrient balances over successional time. (*a*) Structure of a simple model used to quantify the hypothesized responses. Also shown is the standard run with: (*b*) no disturbance, no nitrogen losses during nitrogen limitations on plant growth, and relatively high nitrogen inputs; (*c*) recurring disturbance; (*d*) nitrogen losses occurring before nitrogen limitations on plants are removed; and (*e*) greatly elevated nitrogen inputs. DON = dissolved organic nitrogen. (Vitousek et al. 1998)

Turnover rates for transfer of plant nutrients to soils (litterfall) and for net mineralization of the nutrient from organic matter are fixed. Input rates are specified as a variable, and nutrients remaining after plant demand is met are assumed to be lost as outputs. The model is generalized such that outputs could be in dissolved or gaseous forms (e.g., as NO_3^- or as gaseous nitrous oxide [N_2O]). Running the model without additional disturbance produces the general response seen in Figure 18.3 (Figure 18.6*b*). Three alterations to the basic theory are then advanced.

The first alteration is that repeated pulse disturbances that result in nutrient loss can extend the time required to initiate nutrient losses (Figure 18.6*c*). If disturbances are sufficiently frequent or severe, nutrient availability always limits plant production and loss rates always are low. This pattern may predominate in semiarid systems in which fire recurrence is high.

The second is that losses of nutrients can occur before plant demands are met. This addresses the apparently uncontrollable losses of DOC and DON that occur in all humid ecosystems, as discussed earlier and in Chapter 14. Thus, in systems in which N inputs are very low and a constant and unavoidable loss occurs (e.g., DON), accumulation of that nutrient within the system would be small to zero, and nutrient limitations would again be permanent (Figure18.6*d*). The same would hold true for nutrients lost as gases resulting from internal processes within the system (e.g., production and loss of N_2O during nitrification).

The final change relates to cases in which nutrient inputs are elevated due to human activity. Again, N is a good example, as presented in the earlier N saturation discussion. In this case, nutrient limitations can be removed much more rapidly (Figure 18.6*e*) and may not occur at all unless disturbance frequency and loss rates are high.

In summary, these extensions of the basic theory capture the importance of loss rates that are not dictated by plant demand and of the relative rates of external inputs to the system. The basic modeling analysis is an excellent example of how simple models with a few variable parameters can produce a wide range of results; simple theories can explain complex behaviors.

PHYSIOLOGICAL THEORIES OF ECOSYSTEM DEVELOPMENT

The theories presented earlier are essentially based on observations of whole-ecosystem properties over time. Explanations for these patterns at the ecosystem scale are to be found one level down in the environmental hierarchy—that of plant and microbial physiology—and interactions of these with abiotic site conditions. These are the topics we have covered in Part 2 of this book. Is there a set of theories at this physiological level that can be used to explain or substantiate the theories presented at the ecosystem level?

We can begin to answer this question by looking for sets of correlated traits in the physiological response to environmental gradients. In Chapter 6, we saw that the maximum rate of photosynthesis in leaves is strongly related to N concentration and that conductance, the degree to which leaves are open to atmosphere, is related to photosynthesis (Figures 6.3 and 6.8). We also saw that there are a number of leaf characteristics that covary with N concentration, including specific leaf area (leaf thickness or the degree of secondary wall thickening due to cellulose and lignin) and leaf life span, and that this covariation exists in plants all over the world (Figure 6.4). In addition, this suite of leaf-level characteristics is associated with growth rates of plants; plants with high N concentrations also generally have higher relative growth rates (fractional increase in plant mass over time). Natural selection is the force driving the evolutionary development of physiological traits. If the set of co-occurring characteristics listed earlier results from natural selection, then we need a selective force that would reinforce this syndrome and would provide a theoretical framework binding all of the related observations together.

Herbivory as a Selective Force

Surprisingly, the first theory proposed to explain this covariance used herbivory as the proposed agent of natural selection in a positive-feedback loop involving N as the nutrient most limiting to ecosystem function (Figure 18.7). Plants growing on nutrient-rich sites have high concentrations of N in foliage and thus higher rates of net photosynthesis. As a result, leaves repay the C cost of leaf construction rapidly and so do not have to be maintained on the plant for a long period of time. Rather, an aging leaf with declining rates of C gain can be shed and a new one produced. As we discussed in Chapter 15, a short-lived leaf is less likely to be found and attacked by herbivores, so less of the plant's resources are required to protect that leaf from consumption. Expensive, generalized defenses against herbivory, such as low N concentration and high fiber content, are not adaptive in this case, so selection would lead to higher N content and a lower specific leaf weight.

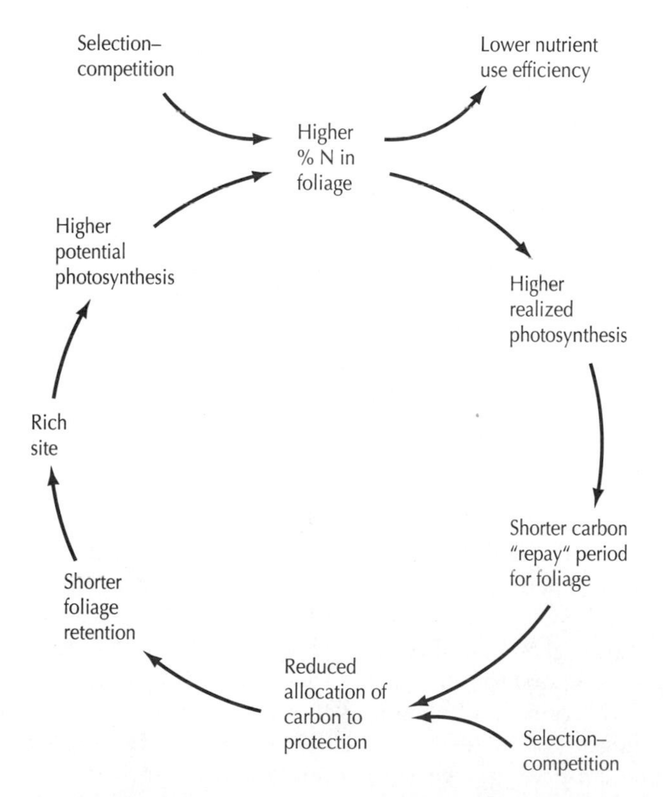

FIGURE 18.7 Hypothesized positive feedbacks among site quality, foliar nitrogen content, leaf longevity, and investment in herbivore defense. (After Mooney and Gulmon 1982)

In contrast, species adapted to low-quality sites would generate foliage with low N concentration and low rates of C gain. Leaves would have to be maintained for a longer period of time, and, therefore, have to be protected more effectively against herbivory. This reinforces the value of low foliar N content and would also select for high fiber content, especially the concentration of lignin in secondary cell wall material.

Our discussion of decomposition and humus formation in Chapters 12 and 13 completes the cycle of positive feedbacks inherent in this theory (Figure 18.7). Fresh plant litter with higher N content and lower fiber decay more rapidly and release more N through decomposition. Thus, the same interactions that lead from high nutrient availability to rapid growth and short leaf longevity also tend to increase N cycling even further. This theory also predicts a set of positive feedbacks leading to decreasing nutrient cycling rates on initially nutrient-poor sites. In this case, low N content in leaves leads to slower growth, longer leaf life span, lower-quality litter, and decreasing nutrient availability.

Generalized Response to Stress

A second and related theory developed at about the same time as the herbivory theory discussed earlier ascribed differences in plant function on rich and poor sites to a generalized response of plants to stress. Borrowing from an ecological theory that divided plants into those that exploit rich habitats and those that tolerate poor ones, a set of physiological responses to low nutrient availability was described (Figure 18.8). The key selective force in this case was not herbivory explicitly but rather the adaptive nature of growing slowly and accumulating reserves of resources on poor sites. Increased leaf longevity and low leaf nutrient content were seen as mechanisms for conserving nutrients by reducing the demand resulting from frequent replacement of nutrient-rich foliage. Thus, the evergreen habit is a nutrient conservation mechanism with the unavoidable consequence of low rates of photosynthesis.

Several processes at the cellular level that reinforce this theory have been discussed. The growth–differentiation and the C–N (nitrogen) balance theories presented in Chapter 15 provide mechanisms by which plants can respond to reduced nutrient availability or other forms of stress. Changes in the production of growth-regulating compounds (e.g., abscisic acid and cytokinins) are a generalized response to stress, with generalized effects on plant growth, form, and differentiation.

Competition and Niche Differentiation

An alternative to these two theories is based on processes of competition between plants rather than herbivory or stress tolerance. This idea applies particularly to humid and fertile systems in which competition for light is extreme. Here, openings in the forest canopy created by the death of dominant individuals provide short-lived opportunities for understory plants and new seedlings to grow rapidly in full sunlight and to produce seed. A suite of characteristics related to tolerance of shade, rate of growth, age at production of first seeds, N content of foliage, and longevity of stems are all related to this gradient in light availability (Table 18.1). Species that are well adapted to exploiting these openings are short-lived and fast growing, reproduce at a young age, and have high N content in foliage and low root–shoot ratios. One theory suggests that such plants are short-lived because the shoot outgrows the ability or the root system to provide physical support as well as nu-

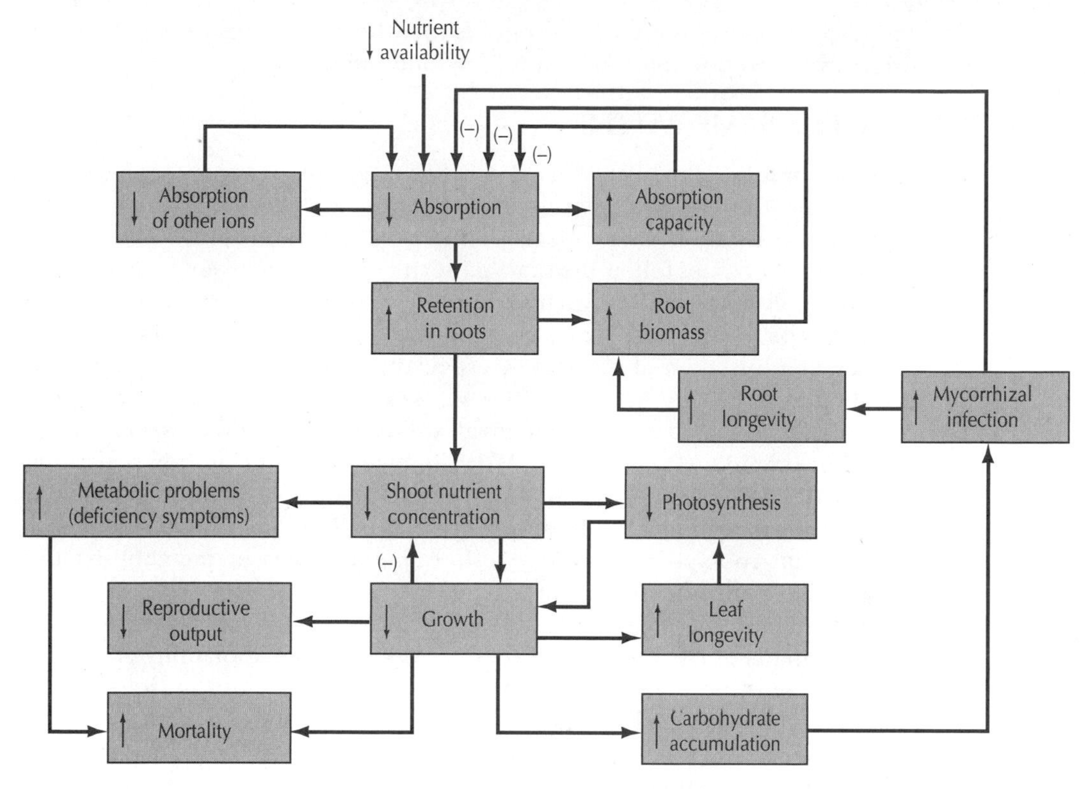

FIGURE 18.8 Direct responses of plants to nutrient deficiencies and feedbacks that both decrease growth rate and increase uptake of additional nutrients. (Chapin 1980)

TABLE 18.1 Correlated Patterns in the Growth Strategies of Early and Late Successional Species in the Northern Hardwoods Forest Type in New England (see Chapter 21 for a more complete discussion)

(After Marks 1975, Bormann and Liken 1979)

	Characteristic				
Species	**Age at Maturity**	**Foliar Nitrogen(%)**	**Longevity (years)**	**Height Increase (cm/yr)**	**Root–Shoot Ratio**
Late succession (shade tolerant)					
Beech	40	2.2	300–350	25–40	0.36
Sugar maple	30	2.1	300–400	25–40	0.37
Mid succession (intermediate)					
Yellow birch	40	2.6	200–300	45–50	—
White ash	20	—	250–300	50–60	—
Early succession (shade intolerant)					
Pin cherry	4	2.8	30–60	60–100	0.19
Paper birch	15	—	140–200	50 +	—

trients and water. Another theory holds that there has simply been no selective pressure to develop traits leading to long life spans.

DO ROOTS FIT THE SAME PATTERN?

Changes in foliar chemistry have direct effects on plant function and herbivore success. As we have seen, the production and decomposition of fine roots can be as important as above-ground processes in the cycles of C and N. Whether root tissues follow the same pattern as leaves represents one method for testing the generality of what we have discussed earlier.

It is worth repeating that far less is known about the function and dynamics of roots than about leaves, especially concerning root and mycorrhizal herbivory. In Chapter 11, we saw that roots with higher N concentration should exhibit higher rates of respiration and do seem to turn over more rapidly (Figure 11.14). We also saw that both C and N budgeting approaches to estimating fine-root production suggest faster turnover and lower biomass in more fertile sites (Figures 11.12 and 11.13). These results suggest that roots as well as leaves show higher rates of metabolism and shorter life spans on enriched areas.

What about other aspects of tissue chemistry? While there are very few measurements of the C quality of roots across resource availability gradients, one study did find a significant increase in N concentrations in fine roots in relation to total availability of NO_3^- across a very wide range of temperate forest conditions (Figure 18.9). Lignin concentrations did not change across this gradient, such that lignin–N ratios declined, suggesting faster rates of decomposition on richer sites.

Taken together, these results suggest that roots do have shorter life spans on richer sites and may release N more rapidly during decomposition. These results have been combined with Figure 18.7 to present a more complete set of hypotheses on the interactions between site quality and the physiological responses of plants (Figure 18.10).

COMPARING SUCCESSIONAL AND PHYSIOLOGICAL THEORIES

This brief comparison of successional versus physiological models of ecosystem development reveals a very basic difference between the two. In general, successional models have emphasized homeostatic, or self-correcting, mechanisms that tend to move a disturbed system back toward an equilibrium condition. They are supposed to be driven by negative feedbacks. In contrast, the physiological models are driven by positive feedbacks, leading to divergence from an average, or mean, condition of site quality. Can these two be reconciled?

The positive-feedback loops described in Figures 18.7 and 18.10 cannot continue to operate endlessly. While nutrient cycling rates can be increased by these mechanisms, eventually some other resource that cannot be altered by plant activity will become limiting. Total radiation, precipitation to soil water holding capacity, and length of growing season are characteristics of sites that plants can do little to change.

The negative feedbacks operating on poor sites pose a different problem. Theoretically, nutrient cycles can be slowed to the point at which they approach zero. If litter quality becomes poor enough, mineralization can be-

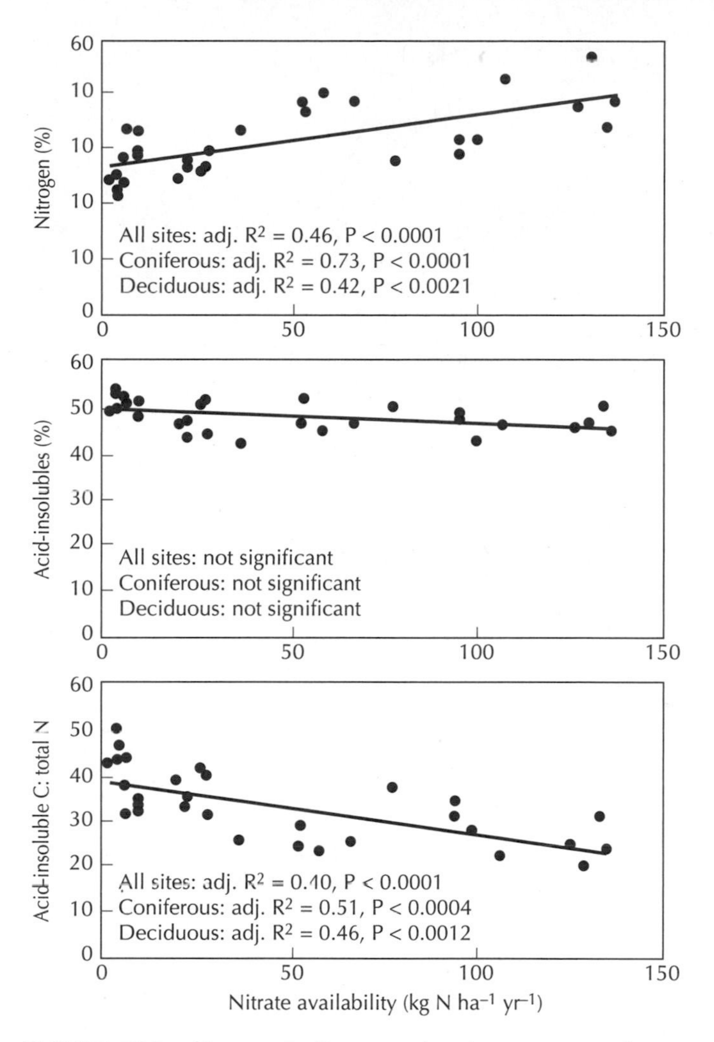

FIGURE 18.9 Changes in fine-root chemistry across a nitrate availability gradient in forest ecosystems. (Hendricks et al. 2000)

come almost undetectable and production can approach zero. Where is the limitation or control on this cycle? Rejuvenation through disturbance is one mechanism. As we saw in Chapter 17 and will see in Chapter 19, fires can release large amounts of nutrients from recalcitrant organic matter and initiate succession, with more productive species having higher-quality litter. This effectively reverses the positive feedbacks, directing the system toward a more productive, rich site condition.

External nutrient inputs can also redirect the physiological feedbacks. Figure 18.4 describes a series of responses to elevated N deposition in which N contents in foliage increase with increased N availability. This response would initiate the site-enriching feedbacks in Figure 18.10, perhaps reversing the feedbacks moving toward lower N cycling. Enhanced N deposition in heavily industrialized areas of the world may have completely altered the course of ecosystem development in these regions. The pattern in Figure 18.3

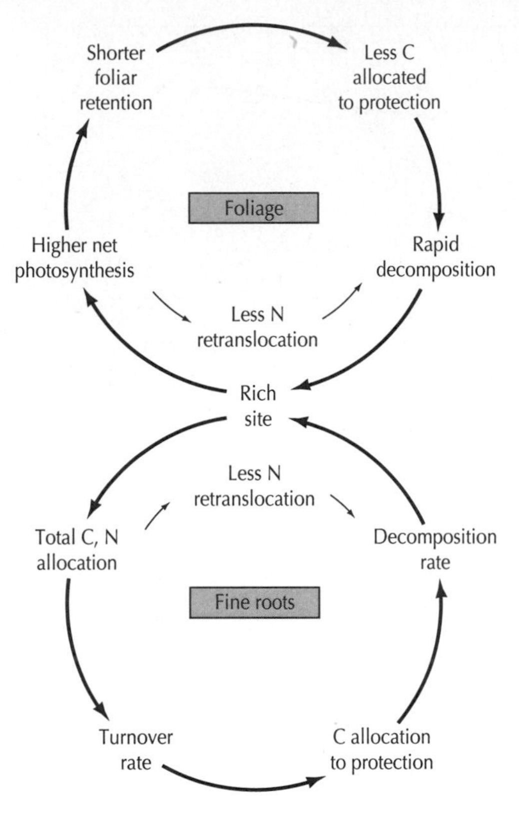

FIGURE 18.10 An extension of the theory in Figure 18.7 to include fine-root dynamics. (Hendricks et al. 1993)

may be a result of this broad-scale disturbance rather than a basic pattern in natural systems.

To summarize, the two sets of theories can be reconciled in that the positive feedbacks predicted by the physiological models operate within a set of larger-scale ecosystem-level constraints that limits how far nutrient availability can be altered by plant–soil interactions.

It must be noted here that the majority of examples from which these theories have been drawn deal with N in temperate-zone forests. There is currently a tremendous amount of research going on in tropical and nonforest ecosystems that will undoubtedly require additional extensions of these generalized explanations of ecosystem responses. In this case, field research and analysis are ahead of the theory.

REFERENCES CITED

Aber, J. D., K. J. Nadelhoffer, P. Steudler, and J. M. Melillo. 1989. Nitrogen saturation in northern forest ecosystems. *BioScience* 39:378–386.

Bormann, F. H., and G. E. Likens. 1979. *Pattern and Process in a Forested Ecosystem.* Springer-Verlag, New York.

Chapin III, F. S. 1980. The mineral nutrition of wild plants. *Annual Reviews of Ecology and Systematics* 11:233–260.

Hedin, L. O., J. J. Armesto, and A. H. Johnson. 1995. Patterns of nutrient loss from unpolluted, old-growth temperate forests: Evaluation of biogeochemical theory. *Ecology* 76:493–509.

Hendricks, J. J., J. D. Aber, K. J. Nadelhoffer, and R. D. Hallett. 2000. Nitrogen controls on fine root substrate quality in temperate forest ecosystems. *Ecosystems* 3:57–69.

Hendricks, J. J., K. J. Nadelhoffer, and J. D. Aber. 1993. Assessing the role of fine roots in carbon and nutrient cycling. *Trends in Ecology and Evolution* 8:174–178.

Louchs, O. L. 1970. Evolution of diversity, efficiency and community stability. *American Zoologist* 10:17–25.

Mooney, H. A., and S. L. Gulmon. 1982. Constraints on leaf structure and function in reference to herbivory. *BioScience* 32:198–206.

Odum, E. P. 1969. The strategy of ecosystem development. *Science* 164:262–270.

Vitousek, P. M., L. O. Hedin, P. A. Matson, J. H. Fownes, and J. C. Neff. 1998. Within-system element cycles, input–output budgets, and nutrient limitation. In Pace, M. L., and P. M. Groffman (ed.). *Successes, Limitations, and Frontiers in Ecosystem Science.* Springer-Verlag, New York.

Vitousek, P. M., and W. A. Reiners. 1975. Ecosystem succession and nutrient retention: A hypothesis. *BioScience* 25:376–381.

Part 3

SYNTHESIS

Dynamics of Selected Ecosystems

In Part 1, we introduced concepts and patterns related to the study of ecosystems and then, in Part 2, took those systems apart to look at the specific mechanisms that drive ecosystem function. In Part 3, we put the pieces back together again through the analysis of specific ecological systems.

Unfortunately, it is not possible to present here even a cursory review of all of the major terrestrial ecosystem types. Instead, we have selected four systems that differ widely in terms of climate, vegetation, dominant biogeochemical processes, and disturbance regime. We present a synthetic view of function in (1) a fire-dominated system (the taiga forests of interior Alaska), (2) an herbivory-dominated system (the Serengeti Plain of Africa), and (3) a system in which these two forces are relatively minor factors (the northern hardwood forests of eastern North America). In addition, we present a unique study of the development of forest ecosystems over geological time in the Hawaiian islands. In each case, we focus on integrated, large-scale, ecosystem-level studies carried out in the system of interest.

Over most of the Earth, human activity, a relatively new factor in terms of geologic or evolutionary time scales, now produces significant changes in the productivity and nutrient cycles of ecosystems worldwide. We address these impacts directly in Part 4, concentrating in this section mainly on the functions of relatively remote and undisturbed systems. However, we will see that, even in these relatively "natural" systems, important processes have been altered by human actions.

19

A FIRE-DOMINATED ECOSYSTEM

THE TAIGA FORESTS OF INTERIOR ALASKA

INTRODUCTION

The taiga, or boreal, forest covers a large, circumpolar area of the northern hemisphere (Figure 19.1). Extremely cold winters and a short growing season are the dominant environmental factors. Major tree species present include spruce and pine, with deciduous species, such as birch and aspen (poplar), important in disturbed, warm, or geologically young sites. Larch, or tamarack, a deciduous but needle-leaved tree, dominates much of the boreal forest in Siberia. In newly exposed areas created by the retreat of glaciers or the shifting of riverbanks, nitrogen (N)-fixing species of alder can play an important role. Upland soils are generally poorly developed Inceptisols, tending toward Spodosols because of the cold, humid climate. Much of the region contains flat to gently sloping terrain, and small differences in slope and elevation can be associated with large changes in soil water content. Permafrost (permanently frozen soil) occurs under much of the boreal forest (Figure 19.1), and the depth to which soils thaw in the summer can be both an important determinant of the rate at which processes occur and, in turn, determined by the amount of organic matter accumulated over mineral soils. Permafrost and the flat, low-lying topography lead to large areas of boggy or saturated soils, such that Histosols (totally organic soils) are also common. In parts of the boreal forest, fire plays an important role in correcting the imbalance between production and decomposition due to the slow decay rates caused by low temperatures and low-quality litter.

The purpose of this chapter is to present results from a long-term, detailed study of ecosystem dynamics in the taiga of interior Alaska as well as recent studies on carbon (C) balance and response to a trend of increasing temperatures. We examine how both site conditions and the occurrence of fire affect the distribution of tree species and operation of important ecosystem processes in natural conditions and discuss implications for management and response to a changing climate.

THE TAIGA FORESTS OF INTERIOR ALASKA

The taiga system of interior Alaska is representative of the larger boreal forest region. It lies between the Alaska and Brooks mountain ranges (Figure 19.2) and is quite distinct from the more productive forests of the southern

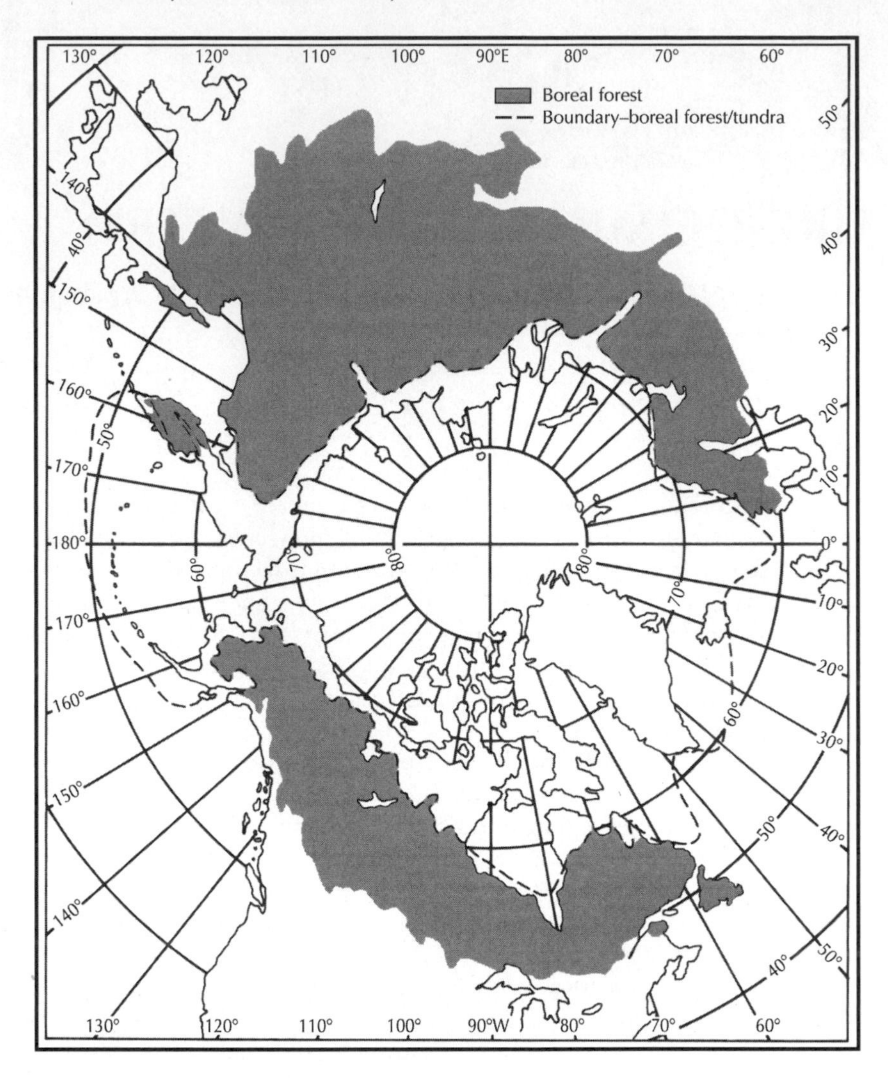

FIGURE 19.1 Circumpolar view of the distribution of taiga forests. (Van Cleve and Dyrness 1983)

coast and the tundra systems to the north. The climate in the taiga zone is continental, meaning that the moderating effect of marine air masses is minimal and that temperature extremes are pronounced. At Fairbanks, the mean January temperature is −25°C. In July, it is +16°C. Average rainfall at Fairbanks is a relatively low 28.6 cm, but the low average annual temperature reduces total annual evaporative demand such that water availability is still sufficient to support the growth of forests. A full 90% of the young, slowly weathering soils are Entisols or Inceptisols, 7% are Histosols, leaving only 3% that show significant profile development.

Within the Alaskan taiga, topographic location plays a large role in determining the relative importance of the four major tree species through interactions with soil temperature and forest floor depth (Figure 19.3).

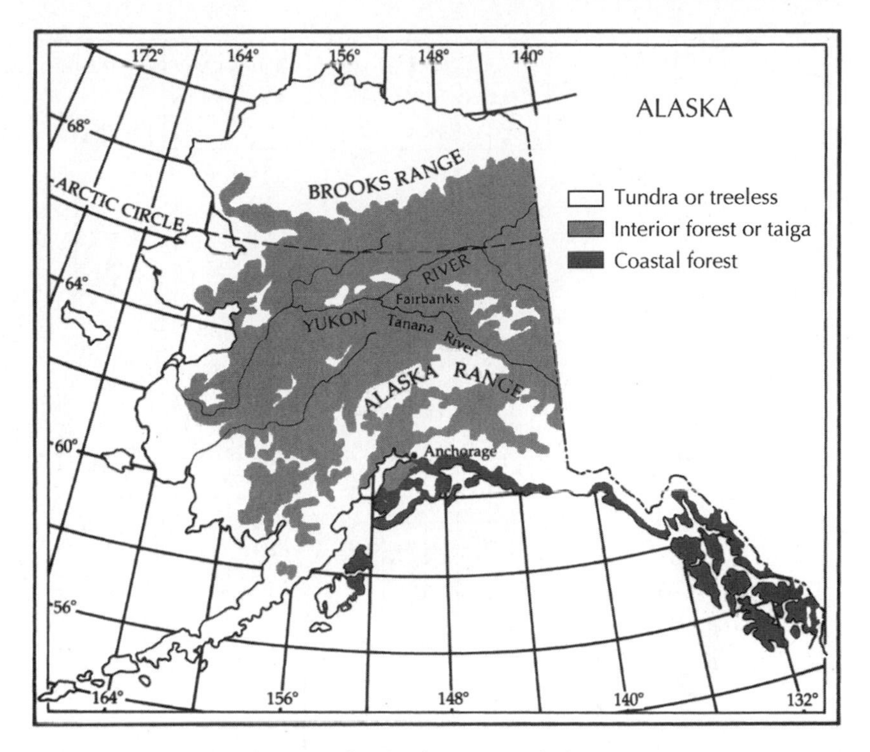

FIGURE 19.2 Distribution of taiga forests in Alaska. (Van Cleve et al. 1983, © 1983 by the American Institute of Biological Sciences)

Ridgetops and south-facing slopes with well-drained soils and warmer microclimates are dominated by the deciduous species quaking aspen and paper birch. These stands may be successional to white spruce on deeper, midslope soils. Poorly drained sites are dominated by black spruce with an understory of feathermoss. On very poorly drained sites, black spruce stands become open and a continuous cover of sphagnum moss develops over the forest floor. In the oldest and most open of these stands, the productivity and nutrient cycling through the moss can be greater than through the spruce trees. Floodplains adjacent to rivers are dominated by the deciduous balsam poplar.

Above-ground net primary production varies widely between these forest types (Figure 19.4). The deciduous forests are roughly equal in productivity and tend to be somewhat higher than white spruce, which in turn is much higher than black spruce. Currently, over 40% of the interior taiga forest in Alaska is dominated by black spruce.

There are some very strong correlations between the species that dominate a site and important environmental and ecosystem parameters. Black spruce stands have both higher mean water content in soils and lower mean soil temperature, measured as total degree days above 0°C at 10-cm depth (Figure 19.5; a degree day is the mean temperature for that day minus the index temperature [in this case, 0°C]. A day with mean soil temperature of 5°C yields 5 degree days. Three consecutive days at that temperature give a

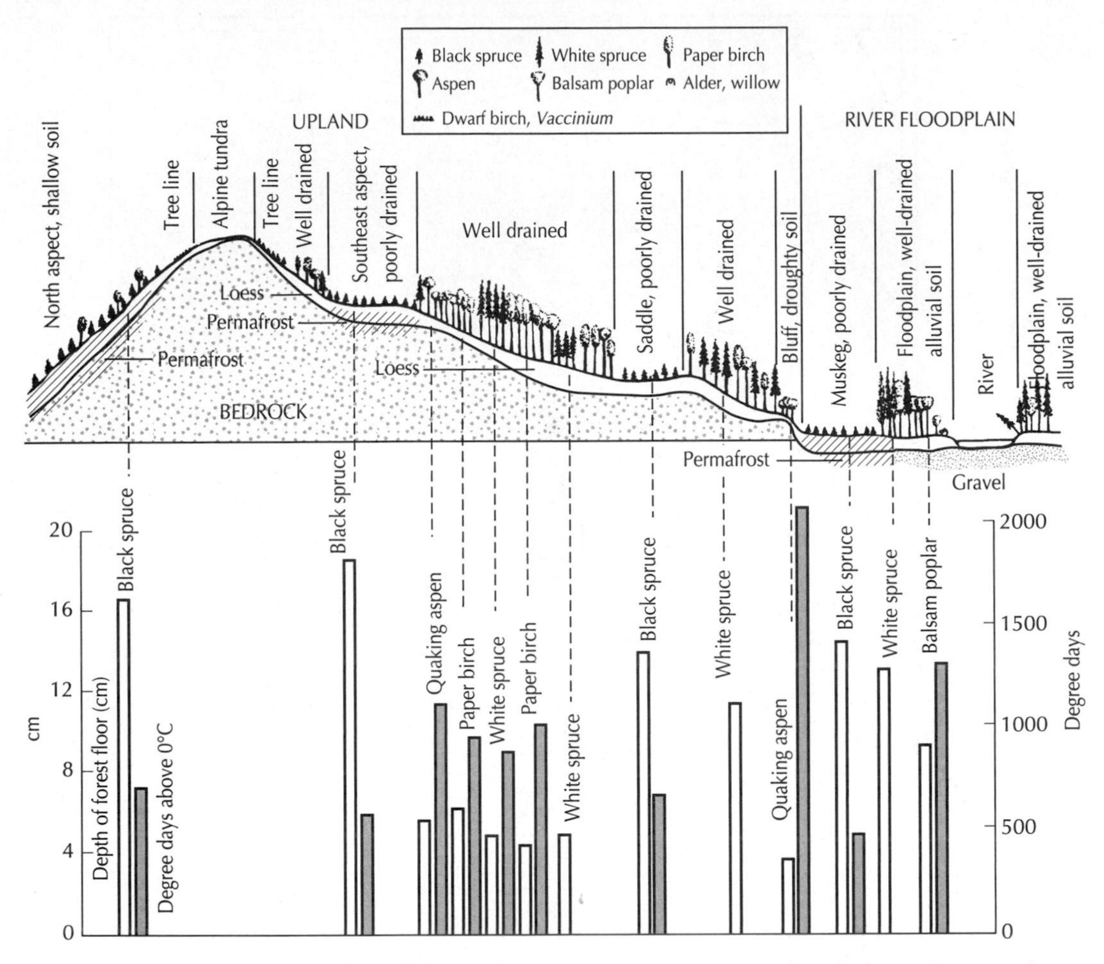

FIGURE 19.3 Distribution of major vegetation types in the Alaskan taiga zone in relation to topography, forest floor depth, and soil degree days above 0°C. (Van Cleve et al. 1983, © 1983 by the American Institute of Biological Sciences)

total of 15 degree days, etc.). There is, in turn, a strong correlation between soil degree days and both mean soil water content and the depth of the forest floor (Figure 19.6). A final important factor is that low average soil temperature generally relates to a shallower depth to permafrost during the growing season.

It could be argued from this either that black spruce is adapted to cold and wet sites and so grows in places with low soil temperature and high water content or that, because black spruce litter decays slowly, it causes accumulation of a thick forest floor and thus creates low-temperature, high-water-content soils. This type of cross-correlation is the rule rather than the exception in ecosystem studies and makes the separation of cause and effect very difficult using only observations of existing stands. A more complete understanding of the interactions between species and site can usually be obtained by some sort of experimental modification—manipulating or disturbing the system in some way and then measuring its response.

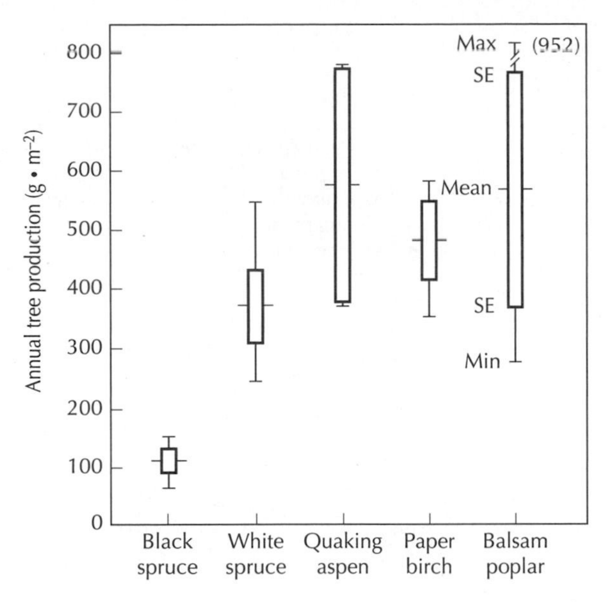

FIGURE 19.4 Range in above-ground net primary production values for the four major forest types in the Alaskan taiga. (Viereck et al. 1983)

FIRE AND SUCCESSION IN TAIGA FORESTS

Fires provide a natural disturbance that temporarily alters the close correlations between species and site in taiga forests. Studying the changes in burned stands through time allows a clearer look at the dynamics that lead to the final distribution of mature stands and their related environmental conditions.

Fire is a common component of the interior taiga forests of Alaska.

> It would be hard to overestimate the importance of fires in shaping the vegetation pattern on the upland of interior Alaska. The distribution of aspen and paper birch can often be traced to patterns of past fires. In the mosaic of burned conditions, the hardwoods generally invade areas where the entire forest floor has been consumed, whereas black spruce replaces itself in areas where a less intense fire left the forest floor intact. This mosaic of highly diverse conditions in the wake of a fire makes it difficult to generalize about the effects of fire on the ecosystem. However, due to increased soil temperature and rates of nutrient cycling, the net effect of periodic fire in black spruce ecosystems is a warmer, more productive site, at least for a period of 10 to 20 years after the fire. (Van Cleve et al. 1983)

Some forest types experience fire on a 30- to 50-year cycle, while the average return interval over the whole region is approximately 100 years. Forests over 170 years old are rare. As discussed later, the area of Alaskan taiga forest burned each year has been increasing since 1970.

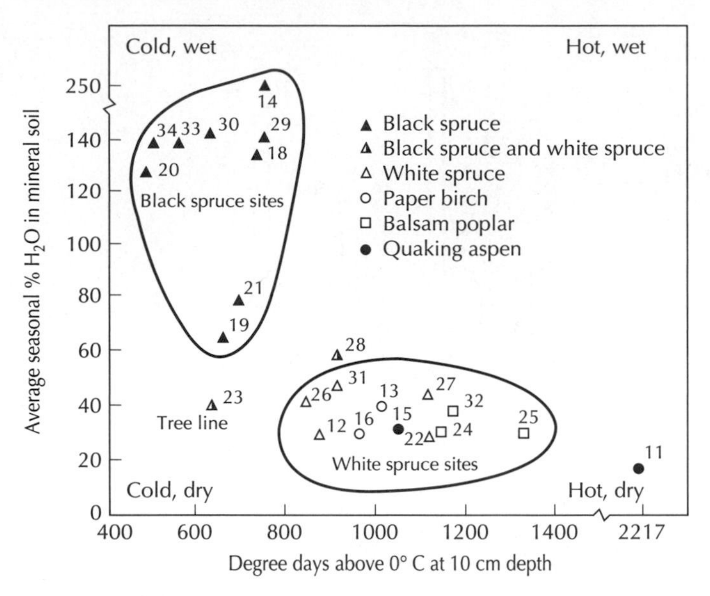

FIGURE 19.5 The distribution of white and black spruce sites with respect to average annual soil water content and total soil degree days. (Viereck et al. 1983)

Figure 19.7 shows the integrated changes in species composition, forest structure, nutrient availability, forest floor and moss biomass, primary productivity, soil temperature, and depth to permafrost for a hypothetical 150-year succession following a crown fire in a mature black spruce forest on poorly drained soils.

Immediately following the fire, the forest floor and moss layers have been greatly reduced by burning and the shading effect of the evergreen spruce canopy is removed. More sunlight reaches the forest floor, increasing soil temperatures and causing a retreat of permafrost to lower levels during the growing season. This increase in available soil volume interacts with higher temperatures to increase mineralization rate of soil organic matter, further increasing nutrient availability to more than that already caused by the deposition of ash.

Fire also affects species composition. While black spruce is present immediately after fire because of seed input from serotinous cones, there is an increase in deciduous species that both sprout vigorously and seed in from outside of the burned area. The enriched soil conditions interact with the higher growth potential of aspen and birch to favor dominance by these species.

This shift in resource availability and species composition sets in motion a series of changes in ecosystem structure and function. The deciduous species do not provide leaf cover in spring, allowing for greater light penetration and earlier soil warming. In addition, the leaf litter produced is higher in N and lower in lignin content, so the forest floor thus increases in N content and in substrate quality for decomposition. These factors lead to increases in mineralization rates (Figure 19.7). Faster mineralization and

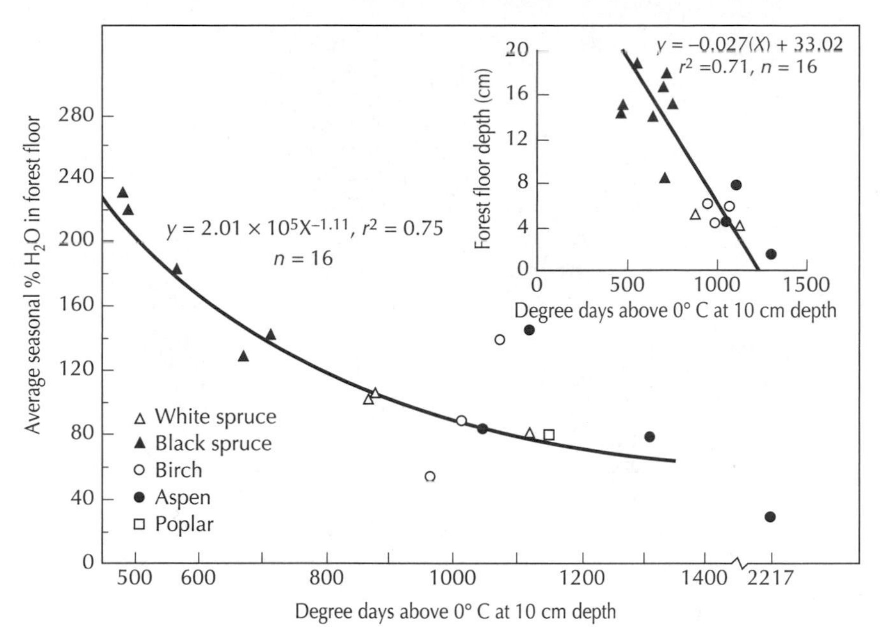

FIGURE 19.6 Correlation between total soil degree days above 0°C and both average seasonal soil water content and forest floor depth. (Van Cleve et al. 1983, © 1983 by the American Institute of Biological Sciences)

higher total N mineralization rates are, in turn, linked to greater leaf (Figure 19.8) and total tree net primary production.

The system is now in the type of positive feedback condition described in Chapter 18 (Figure 18.7). Disturbance has increased nutrient cycling rates and altered species composition in ways that, through increased productivity and higher litter quality, would be expected to lead to further increases in cycling. However, the resulting increase in production also provides the mechanism for a negative feedback that eventually returns the system to the prefire condition.

One result of increased leaf production for deciduous species is a more completely shaded forest floor during the growing season. Black spruce is more tolerant of shade than aspen and birch and increases in importance through successional time (Figure 19.7). As it does, its evergreen foliage decreases average annual soil temperatures by shading the soil surface in spring and fall. The foliage of black spruce is also high in lignin content, so, as its litter becomes an increasingly important component of the total litterfall, decay rates decline. Declining decay rates lead to increasing forest floor depth and organic matter content, lower soil temperatures, and slower mineralization rates. As nutrient availability declines, the species in this system increase nutrient retranslocation from senescing foliage (Figure 19.9), leading to lower nutrient content in litterfall, still lower decomposition rates, and further reductions in nutrient availability. The system is now in a positive-feedback condition, but this time heading toward a state of very low nutrient cycling and productivity.

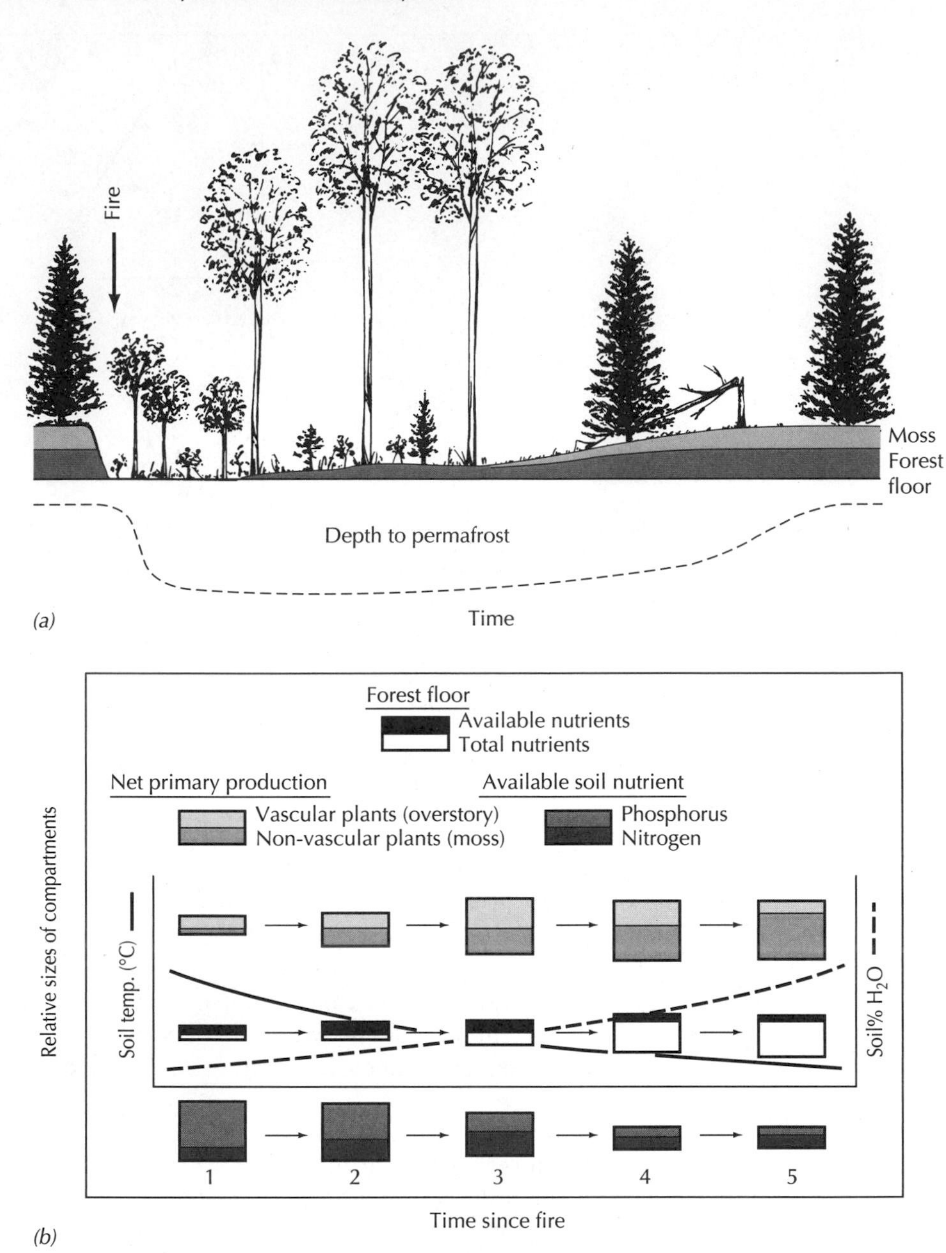

FIGURE 19.7 Integrated response of taiga ecosystems to fire and succession following fire. (*a*) Changes in species composition and ecosystem structure. (*b*) Changes in environmental conditions, resource availability, and new primary production. (Van Cleve et al. 1983, © 1983 by the American Institute of Biological Sciences)

As soil temperature declines, the maximum depth to permafrost also declines. This both restricts access to nutrients frozen in permafrost and also restricts drainage of water from the soil. Higher water content increases the energy required to raise soil temperatures, so mean soil temperature continues to decline. Finally, high soil water content favors the establishment and

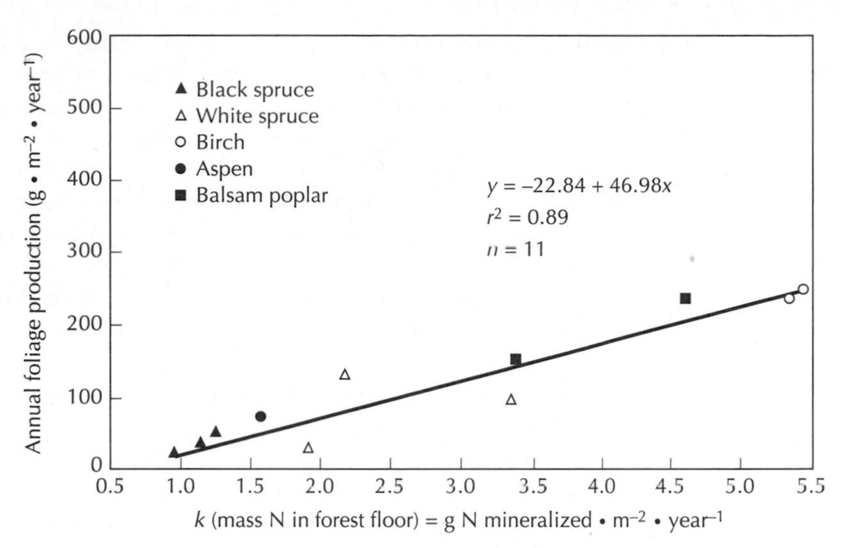

FIGURE 19.8 Net annual production of foliage as a function of estimated annual nitrogen mineralization in taiga forests. (Flanagan and Van Cleve 1983)

spread of first feathermosses and then sphagnum moss, further insulating the forest floor and reducing soil temperatures. The mosses can actually compete effectively with black spruce for the scarce nutrient supply, reducing spruce growth and leading to a "stand" dominated functionally, if not structurally, by moss. The very large accumulations of organic matter over the mineral soil surface that occur at this stage make the system very susceptible to fire during extended periods of drought, which leave both the moss and tree components tinder dry.

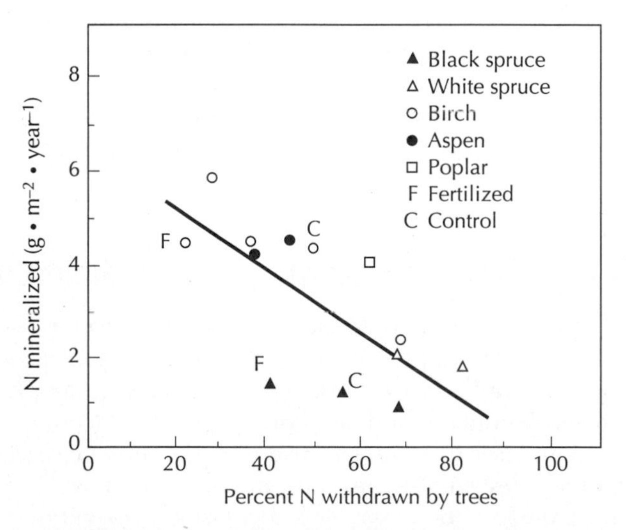

FIGURE 19.9 Percentage of foliar nitrogen withdrawn from foliage as a function of annual nitrogen mineralization rate in taiga forests. (Flanagan and Van Cleve 1983)

From this description it is still unclear where the "cause" is and where the "effect." This is because ecosystems, as all systems, are not controlled by linear, one-way interactions. Rather, they are a series of interactions between biological, physical, and chemical processes. In the taiga example, interactions between climate, soils, and species tend to lead the system toward a decadent end stage of low nutrient cycling and low productivity, which then is conducive to fire. Fire is the restart mechanism in the taiga that increases rates of several processes simultaneously but which also reinitiates succession to the low-productivity end stage. Forest floor organic matter depth, soil water content, and soil temperature are crucial environmental parameters in these interactions.

Underlying site conditions can alter the scenario described earlier. On south-facing, sloping sites with good internal soil drainage, the deciduous forest stage may be prolonged, appearing permanent on extreme sites. On moderate sites, the more productive and nutrient-demanding white spruce may predominate over black spruce. In this case, succession to the decadent, moss-dominated end stage may be retarded and the moss stage may never be reached before the next fire event. There is a clear distinction between sites dominated by white and black spruce in terms of soil temperature and mean soil water content (Figure 19.5). On wetland, muskeg, or bog soils, the importance of deciduous species in the scenario mentioned earlier is reduced, movement away from the moss end stage is less, and succession toward it is more direct.

EXPERIMENTAL MODIFICATION OF TAIGA ECOSYSTEMS

While natural fires offer an uncontrolled look at system response to disturbance, direct experimental manipulations may also help break up the correlations that occur, for example, between soil water content, forest floor accumulation, and soil temperature. An experimental heating of the forest floor in a black spruce stand was carried out in conjunction with the other studies summarized here. Heating cables were embedded in the forest floor and used to raise the temperature of the lowest layer in the forest floor by 9°C above control levels. The heating resulted in a 20% reduction in the forest floor biomass and significant increases in ammonium and phosphorus availability. Black spruce foliage showed a roughly 40% increase in N and phosphorus (P) concentration.

SUMMARY OF INTERACTIONS AND RELATION TO GENERAL THEORY

Figure 19.10 is a summary view of the interactions between dominant plant species, environmental conditions, and rates of ecosystem processes in the Alaskan taiga. It reemphasizes the role of species changes in altering the nature of biomass produced and affecting rates of decomposition. These changes then feed back to reduce nutrient availability; increase retranslocation; and favor species with low nutrient demands, such as black spruce and the eventual dominance by moss. This successional sequence represents an excellent example of the generalized theory of ecosystem development (Figure 18.7) presented in Chapter 18. Aspen and birch represent the rich-site

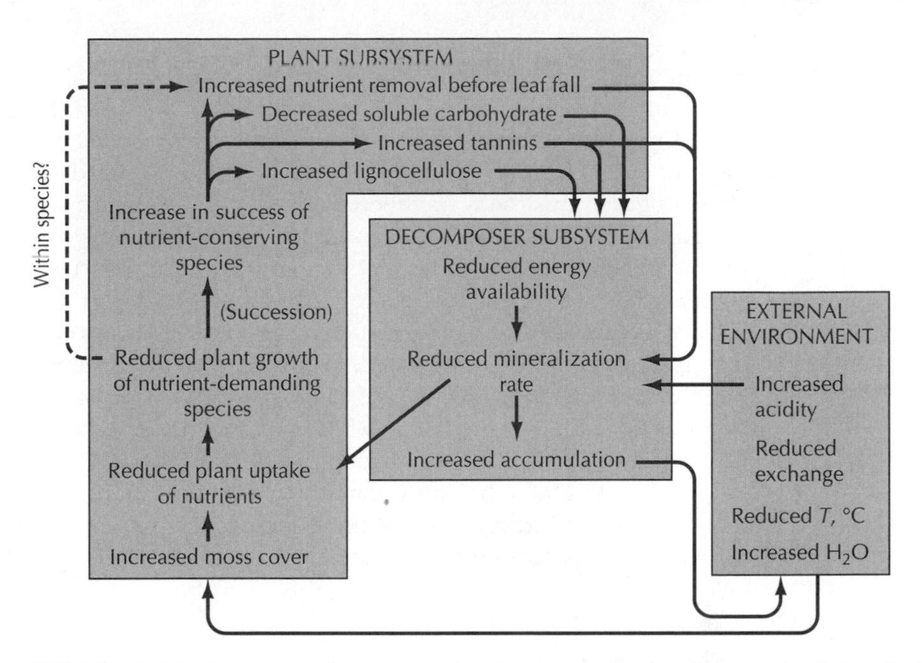

FIGURE 19.10 Summary of processes that interact to lead to low-productivity black spruce stands in Alaskan taiga forests. (Flanagan and Van Cleve 1983)

species, and black spruce, the classic poor-site species. Foliar nutrient levels are higher in the deciduous species, and lignin levels are lower. Foliage is replaced annually in birch and aspen, while the evergreen spruce can hold foliage for up to 20 years, with retention time generally increasing on lower-quality sites. This sequence also fits the generalization that nutrient availability and production are higher in recently disturbed areas and decrease with succession. As is typical of fire-dominated ecosystems, succession leads to the generation of stagnant systems that then are ripe for fire. Fire restarts the cycle of enrichment, succession, and stagnation.

IMPLICATIONS FOR HUMAN USE OF THE TAIGA

Present human use of the Alaskan taiga zone is minimal but increasing.

> Large-scale development is just beginning in the taiga of interior Alaska. The trans-Alaska oil pipeline extends through a broad band of taiga. . . . The pipeline has brought roads and settlements. There are plans for a gas pipeline and increased mineral, coal and oil exploration are underway. Forestry and agriculture in the taiga are in the initial phases of development. We must learn more about the ecosystem in which much of the future development in Alaska will take place. (Van Cleve et al. 1983)

Only 7% of the total forested area in this zone is considered commercial, with the rest consisting mainly of open, low-productivity black spruce forests. The results of the intensive study summarized here suggest that the judicious use of fire could serve to increase the productive potential of some of the remaining area. However, if timber production is to become an important use in the taiga, it must be remembered that logging and fire are very different in terms of their effects on the system. While both forms of disturbance remove vegetation and increase soil temperature, fire returns most of the nutrients in biomass to the soil, while logging removes them.

The actual manipulation of such large natural areas requires consideration of many other factors as well. Wildlife values and recreational and aesthetic values are important in the management of the landscape, as well as the value of wilderness lands, in which natural processes are allowed to continue to operate. If, for example, the taiga region had been heavily settled and modified before the study summarized in this chapter was undertaken, we would have no idea as to how the dominant ecosystem type in this region functioned in its natural state.

BOREAL FORESTS AND GLOBAL CHANGE

Boreal forests play a crucial role in the emerging debate on global climate change (Chapter 26) for two reasons. First, this forest type, with extensive areas of deep organic soils, represents one of the largest reservoirs of organic C in the global system. Alterations in the C balance of boreal forests could have significant effects on the global C budget, either accentuating or buffering the rapid increase in carbon dioxide in the atmosphere. Second, current models of climate change predict that warming will occur more rapidly over far northern regions than in tropical or temperate zones.

How would the C balance of taiga forests change in a warmer environment? We can envision that warmer temperatures and a longer growing season would translate into increased forest growth and more C storage. On the other hand, higher temperatures could also mean higher rates of decomposition of the very large pool of stored soil organic C, leading to increased release of C to the atmosphere. The issue is further complicated by drought. Do drier soils mean reduced or increased soil respiration, and how does this interact with fire frequency?

We can begin to answer these questions with data already in hand. Mean annual temperatures in Alaska showed a steady upward trend in the second half of the twentieth century, increasing by as much as 2°C at some stations (Figure 19.11*a*). A simple water balance index, calculated as a function of annual precipitation minus summer temperature, suggests a long-term increase in drought stress dating back to the 1920s (Figure 19.11*b*). Three major indicators of ecosystem function have changed in concert with temperature.

First, three characteristics of annual tree rings in white spruce from relatively productive stands all show increased drought stress and reduced growth rates, which should be associated with reduced C gain. Actual ring width, related to tree diameter increment and total wood production, has declined since 1930 and is highly correlated with declines in the climate index described earlier (Figure 19.12*a*). Increased discrimination against the heavy C isotope (^{13}C) in the wood produced correlates with the temperature trend (Figure 19.12*b*) and, as discussed in Chapter 5, suggests increased drought stress. Finally, wood density in tree rings has also increased over time (Figure 19.12*c*). Slower growth and reduced production are often associated with increased density of the wood produced. We might keep in mind that these data are from white spruce stands that should occur on the warmest and driest sites. Responses in colder and wetter sites may be different.

Second, estimates of the total area of boreal forest burned in North America show a significant increase in the second half of the twentieth cen-

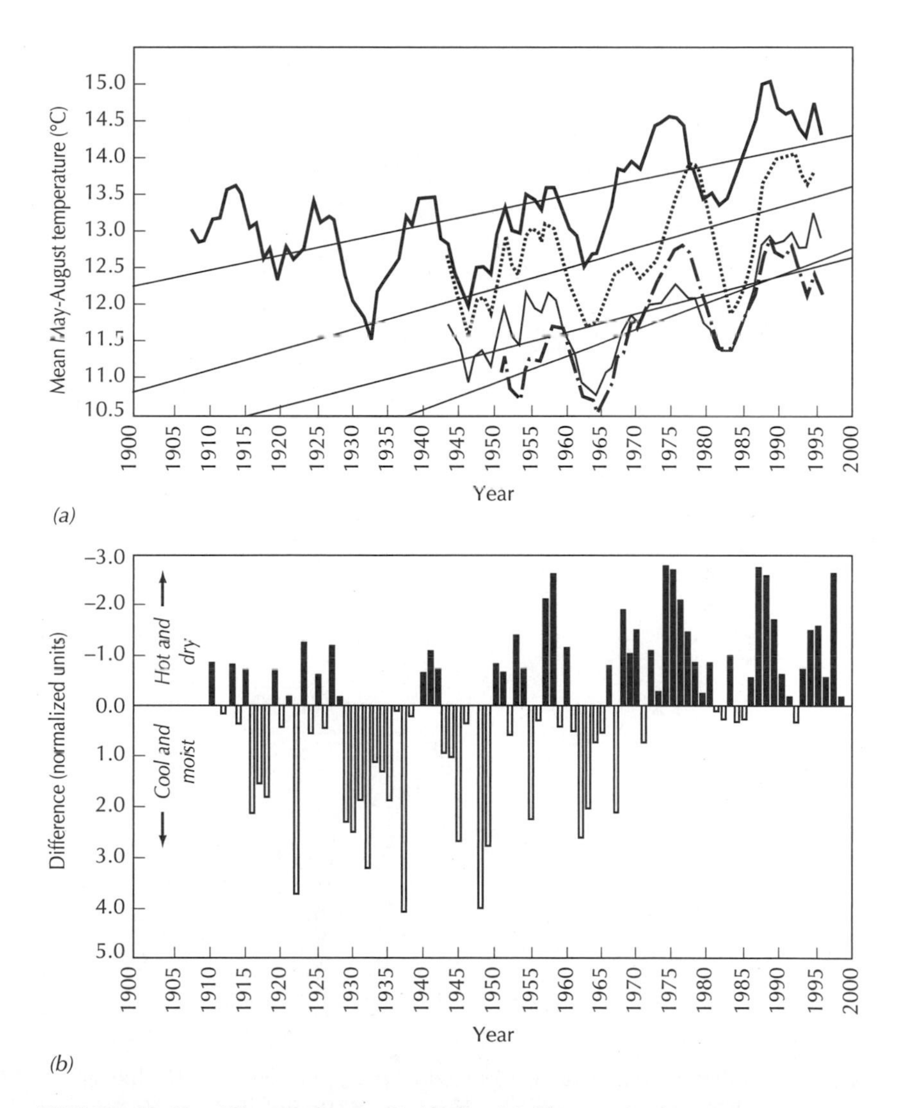

FIGURE 19.11 Climate change in Alaska. (*a*) Changes in the 5-year running average for mean annual temperature at four weather stations. (*b*) Changes in a drought index that includes both temperature and precipitation (higher values indicate greater drought stress). (Barber et al. 2000)

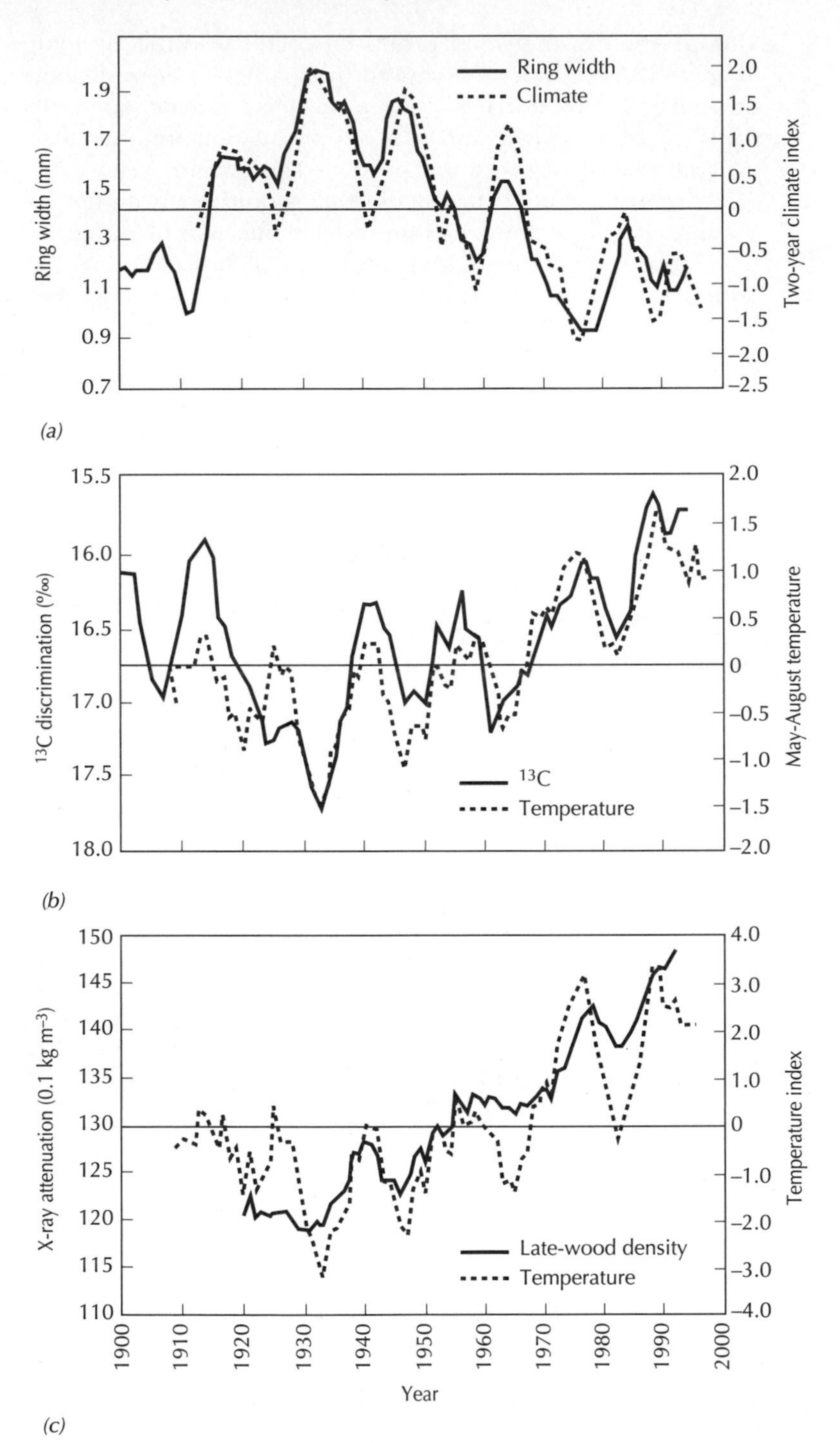

FIGURE 19.12 Changes in characteristics of tree rings in white spruce trees in Alaska in response to climate (all data are 5-year running means). (*a*) Average ring width compared with the drought index in Figure 19.11. (*b*) Relative discrimination against ^{13}C compared with temperature. (*c*) Density of wood produced in midsummer. (Barber et al. 2000)

tury, driven largely by 3 very large fire years in the 1980s and 1990s (Figure 19.13). Fires restart the successional sequence described in this chapter and may stimulate increased forest production, but during the transition time, while fire frequency is increasing, total C storage within this forest type would decrease.

Third, permafrost temperatures have been increasing with long-term increases in air temperature (Figure 19.14). In certain areas, permafrost is disappearing, causing soils to lose structural integrity and collapse. Forests perched on this disappearing substrate can collapse as well. In addition, exposing substantial amounts of previously frozen soil organic matter could also lead to increases in C loss from taiga soils.

While the rest of the world wonders about interactions between climate change and the C balance of taiga forests, residents of the region see the issue from a very different perspective.

> Alaska is vulnerable to climate change. . . . If present climate trends continue, . . . impacts . . . will hit the state's strongly resource-dependent economy hard. . . . A 20% increase in growing degree days [has] benefited both agriculture and forestry. Both the expansion of forests and their increased vulnerability to fire and pest disruption are expected to increase. . . . Subsistence livelihoods [of native peoples] are already being threatened. (Alaska Regional Assessment Group 1999)

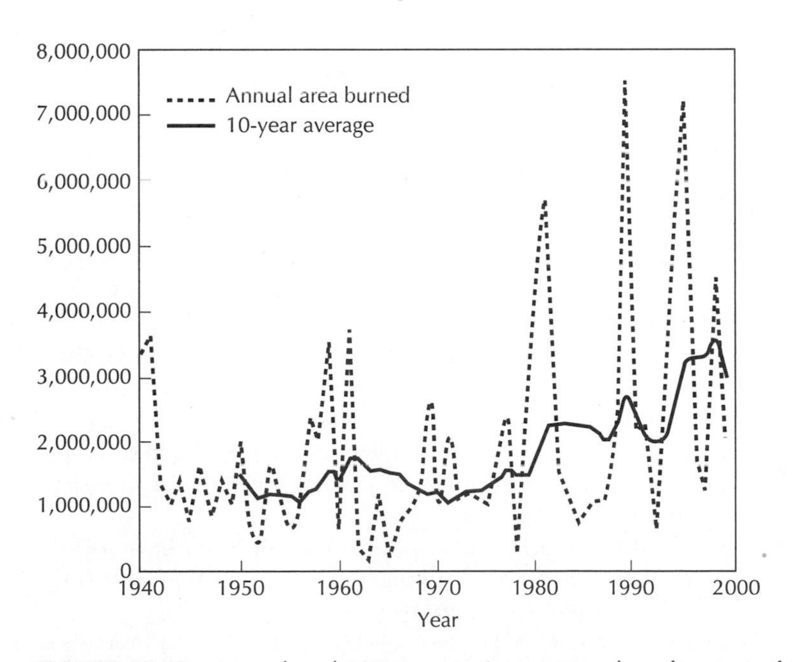

FIGURE 19.13 Annual and 10-year running mean values for area of boreal forests burned in North America. (Kasischke and Stocks 1999)

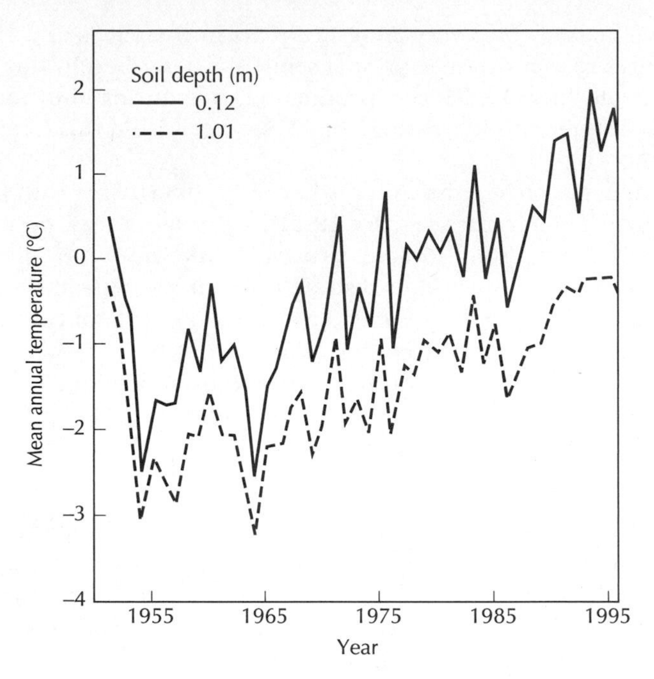

FIGURE 19.14 Estimated changes in mean temperature of permafrost at different depths around Fairbanks. (Alaska Regional Assessment Group 1999)

Social and economic systems are always challenged by rapid change. The predicted rate of change in climate in the boreal zone will eliminate or render uneconomic many traditional practices while creating opportunity in new areas. Perhaps this region, more than any other, will face the challenge of managing and responding to rapid environmental change.

REFERENCES CITED

Alaska Regional Assessment Group. 1999. *Preparing for a Changing Climate: The Potential Consequences of Climate Variability and Change.* Center for Global Change and Arctic System Research, University of Alaska, Fairbanks.

Barber, V. A., G. P. Juday, and B. P. Finney. 2000. Reduced growth of Alaskan white spruce in the twentieth century from temperature-induced drought stress. *Nature* 405:668–673.

Flanagan, P. W., and K. Van Cleve. 1983. Nutrient cycling in relation to decomposition and organic matter quality in taiga ecosystems. *Canadian Journal of Forest Research* 13:795–817.

Kasischke, E. S., and B. J. Stocks. 1999. Introduction. In Kasischke, E. S., and B. J. Stocks (eds.). *Fire, Climate Change and Carbon Cycling in the Boreal Forest: Ecological Studies.* Springer-Verlag, New York.

Van Cleve, K., and C. T. Dyrness. 1983. Introduction and overview of a multidisciplinary research project: The structure and function of a black spruce (*Picea mariana*) forest in relation to other fire-affected taiga ecosystems. *Canadian Journal of Forest Research* 13:695–702.

Viereck, L. A., C. T. Dyrness, and K. Van Cleve. 1983. Vegetation, soils, and forest productivity in selected forest types in interior Alaska. *Canadian Journal of Forest Research* 13:703–720.

Van Cleve, K., C. T. Dyrness, L. A. Viereck, J. Fox, F. S. Chapin, and W. Oechel. 1983. Ecosystems in interior Alaska. *BioScience* 33:39–44.

20

THE SERENGETI

AN HERBIVORE-DOMINATED ECOSYSTEM

INTRODUCTION

No two terrestrial ecosystems could be more different than the boreal forests of the taiga and the grasslands and plains of the Serengeti region of Africa. Cold temperatures and wet soils are key controls on ecosystem function in the taiga, and fire is the major disturbance factor reinvigorating nutrient cycles and tree growth. In the Serengeti, the severity and timing of seasonal drought determines the amount and timing of plant growth, which determines the carrying capacity and seasonal distribution of the large populations of herbivores. Herbivory, in turn, is a major force controlling structure and productivity of the plant community. It is the tremendous diversity and abundance of both large mammal herbivores and their predators that make the Serengeti such a focal point for ecological research and biological conservation efforts.

The purpose of this chapter is to introduce some of the interactions among climate, vegetation, herbivores, and predators that affect the dynamics of the Serengeti ecosystem. Because of the biological significance of, and human interest in the herbivore–predator interactions, much more information is available on this aspect of the system than on the plant–soil–nutrient cycle interactions emphasized in other chapters. Still, important interactions between animal and plant populations occur, and will be presented.

ENVIRONMENT OF THE SERENGETI REGION

The Serengeti region includes a variety of vegetation and habitat types all of which are linked by the grazing activities of the dominant mammalian herbivores. The geographic extent of the Serengeti ecosystem is defined operationally as the area covered by the large migratory herds, especially of wildebeest (Figure 20.1). While the image of the Serengeti may be one of endless plains teeming with wildlife, the area covered by both the migratory herds and the Tanzanian National Park created to protect them is not large, extending less than 200 km at both its longest and widest points. The area contains a diversity of plant communities, ranging from short-grass systems in the east, to mid- and tall-grass systems in the central plains area, to open and occasionally closed woodlands in the far north and west.

Seasons are determined by the timing of rainfall in this tropical region, rather than by changes in temperature. The average annual climatic regime

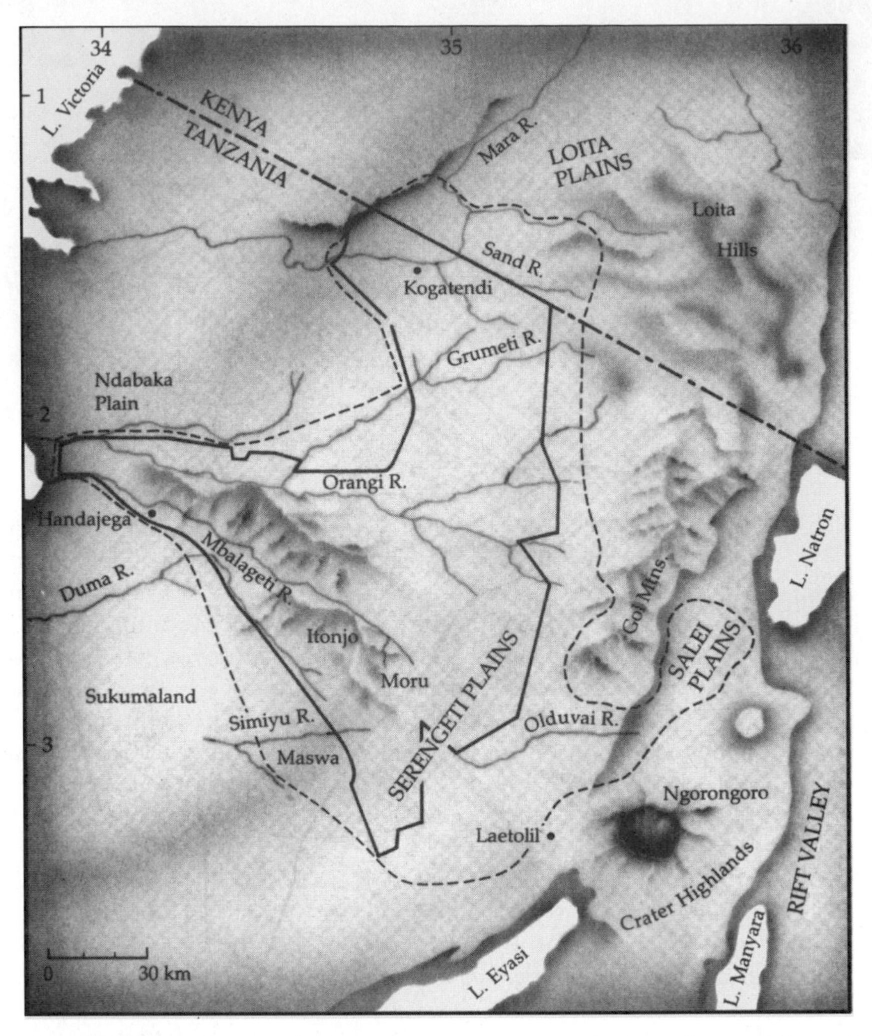

FIGURE 20.1 Map of the Serengeti region showing the boundaries of the park (*solid line*) and the range of migration of the wildebeest herd (*dashed line*). (Sinclair 1979a)

includes a marked dry season from June to October, followed by light rains in November through February, and a major wet season from March to May. Rain arrives on winds from the southeast. The Ngorongoro highlands at the southeast corner of the Serengeti area, rising to 3000 m in elevation, create a strong rain-shadow effect, so that rainfall in both the dry and wet seasons increases from southeast to northwest across the area (Figure 20.2). Rainfall varies significantly year to year as well as season to season. The average values in Figure 20.2 mask important interannual differences that have had major impacts on population levels.

The geological substrate of the region includes mainly ancient granitic rock formations, overlain in the southeast by ashfall from past eruptions of Ngorongoro, which ceased to be active about 2 million years ago, and both older and younger volcanoes to the east of the area. One nearby active volcano, Lengai, erupted four times between 1917 and 1966, with some ashfall reaching the Serengeti during one event.

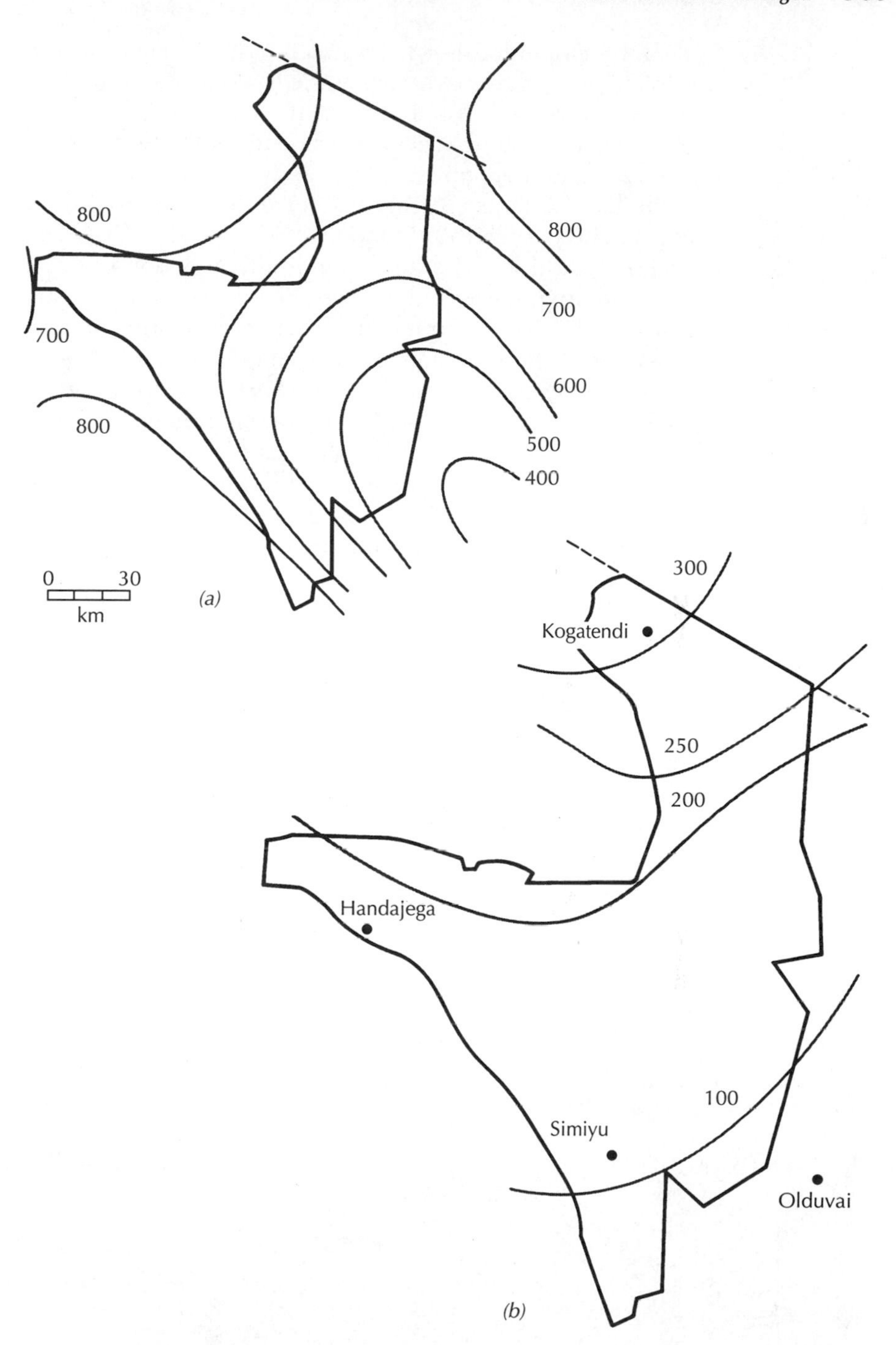

FIGURE 20.2 Distribution of rainfall over the Serengeti region in the (*a*) wet and (*b*) dry seasons (values are expressed in kilometers). (Sinclair 1977)

Soil development shows a regional pattern expected from the range in precipitation (Figure 20.3). Soils at the eastern end of the region are derived from alkaline volcanic ash. Lower rainfall has resulted in minimal soil weathering and development. Calcium carbonates have been leached only from the top horizon and reprecipitated at a depth of about 1 m. This forms a hardpan (caliche layer; Chapter 9) that further reduces downward water movement and maintains high soil alkalinity and pH (> 8). All of these characteristics are typical of an Aridisol. The vegetation in this area is mostly sparse grasses and sedges, and productivity and biomass are relatively low (Figure 20.4).

Moving to the west, rainfall increases, increasing both soil weathering and leaching. The calcium carbonate hardpan diminishes as alkalinity and pH decline. Vegetation grades from short grass to tall grass. Soils are described as Vertisols, meaning that they are rich in expandable 2 : 1 clays that cause soil cracking under dry conditions (see Chapter 9).

In the far northwest, greater rainfall support a transition to woodlands and savannahs. Species of *Acacia,* a genus often associated with symbiotic nitrogen fixation, are a dominant part of the woodland vegetation (Figure 20.5). In adjacent areas of eastern Africa, there is a very close relationship between the alkalinity of the soil and the density of trees (Figure 20.6), suggesting that soil development and climate may play a large role in determining those areas in which trees can grow. We shall see that the extent of woodland development in the Serengeti area is greatly affected by interactions between herbivores, fire, and vegetation.

In general, the transition from dry to subhumid environments described here repeats the trends presented in Chapter 2 for tropical regions. This same transition also occurs on a more local scale within the region along

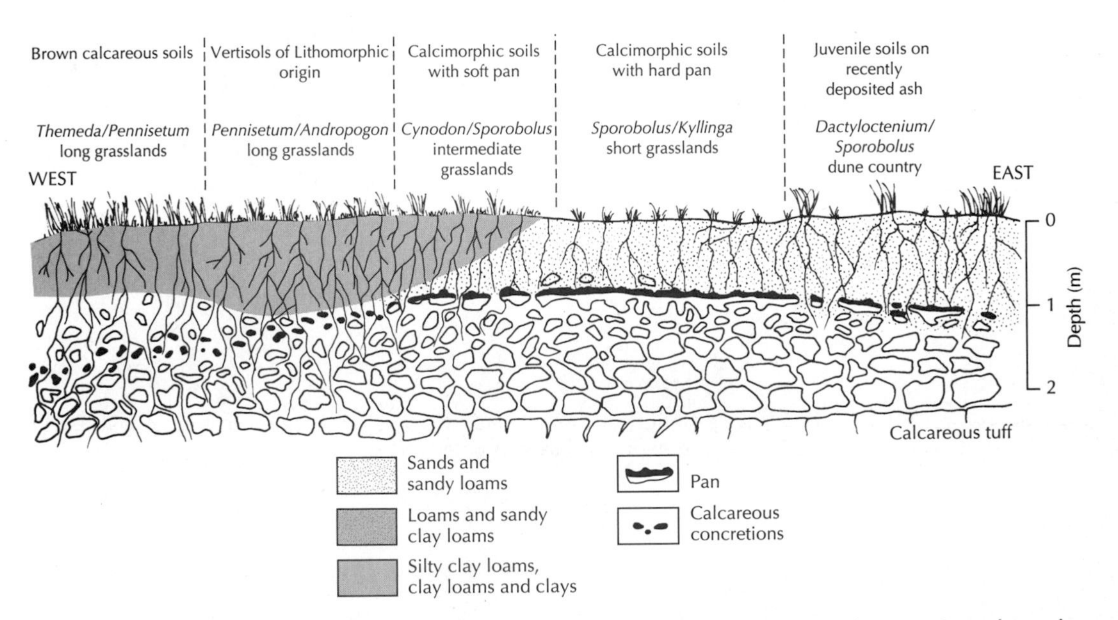

FIGURE 20.3 Schematic view of changes in soil structure and vegetation along the east–west gradient of total annual rainfall in the Serengeti. (Anderson and Talbot 1965)

FIGURE 20.4 The short-grass vegetation of the drier plains area of the Serengeti. (Courtesy of Julie Newman) **See plate in color section.**

topographic sequences called **catenas.** Even the gently rolling topography of the Serengeti can produce significant upper- to lower-slope erosion of fine soil particles. Lower-slope soils tend to be enriched in silt and clay and have a higher water retention capacity. Grass growth is greater in lower-slope positions.

FIGURE 20.5 Acacia woodlands of the wetter northern and western portions of the Serengeti. (Courtesy of Julie Newman) **See plate in color section.**

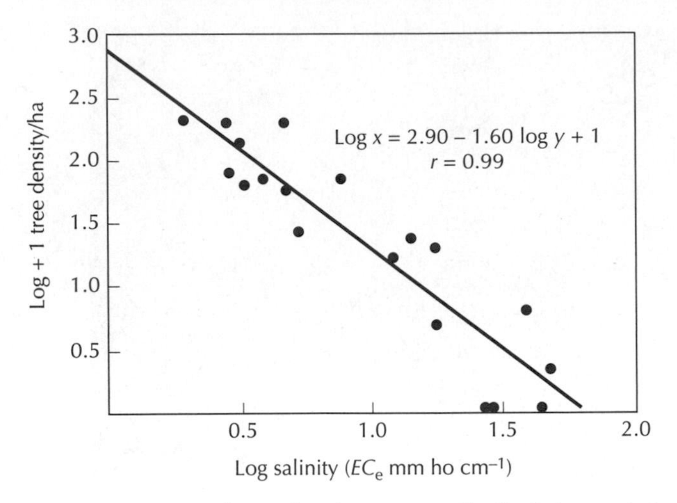

FIGURE 20.6 Relationship between soil alkalinity and density of trees in eastern Africa. (Western and Van Praet 1973)

RESOURCE PARTITIONING AND USE BY HERBIVORES

The grasslands and savannahs that dominate the Serengeti region appear remarkably uniform. Niche theory (Chapter 5) suggests that no two species can occupy the same physical or ecological range unless there is some distinction between them in how the resources in the area are used. Long-term studies of the Serengeti as an ecosystem has revealed some of the ways in which this apparently uniform environment is divided or partitioned by herbivores.

> The Serengeti's ungulates impress the visitor by their huge numbers, the vast extent of the country they occupy, and the number of species that occur together. These combined impressions pose the classic ecological question: how do so many individuals of so many species coexist in such an apparently homogeneous plant community? Studies have now provided some evidence that interspecific competition is taking place and is likely to be the process leading to the pattern of resource partitioning shown by the ungulates. . . . Thomson's gazelle, wildebeest and zebra . . . all face the same problems of obtaining sufficient quantity of good quality food, and they overcome these problems by methods dictated by body size, metabolic rate, and shape of mouth parts. (Jarman and Sinclair 1979)

There are three major migratory species in the Serengeti: zebra, wildebeest, and Thomson's gazelle (Figure 20.7). These three tend to use the grasslands of the region in different ways and at different times, reflecting the distinct growth forms and stages of development of the plant resource that they can process most efficiently. The timing of use is related to the nutritional quality of the grasses in different stages of growth.

Young, short, rapidly growing grass has the highest content of protein and nutrients and the lowest content of cell wall material or fiber. In the wet season, slower grass growth on the upper-soil catena positions of the generally drier areas in the southeastern part of the park prolongs the period of short grass availability, and all three of the major herbivores concentrate here. Food is abundant during the wet season, and competition and niche separation are less stringent.

As the dry season approaches, zebra move down the soil catena first, into the taller grass at the foot of the slopes. Zebra are adapted to processing large amounts of relatively poor-quality herbage. They selectively remove stems and sheaths, opening up the vegetation. The remaining material is higher in leaf biomass, which is preferred by wildebeest. The very large number of wildebeest in the single herd that ranges over the Serengeti removes up to 85% of the vegetation, leaving a short sward that is relatively enriched in low-lying, broad-leaved, herbaceous plants.

Thomson's gazelle avoids tall-grass areas, selecting the high-quality herbs made more apparent by the removal of grass by zebras and wildebeest. In addition, the removal of the tall, mature grass vegetation actually stimulates further grass production (Figure 20.8), increasing the total above-ground productivity of the grassland (see discussion in Chapter 16). Two additional species with smaller populations, Grant's gazelle and eland, appear to utilize the longer-lived leaves of woody species (browsers) as well as grasses, thus prolonging their stay in the upper-catena areas.

This topographic movement is repeated on a far grander scale with the migration of the dominant herbivores across the whole of the Serengeti region in response to shifting patterns of grass growth driven by the wet season–dry season cycle. More than 1 million wildebeest comprise a single herd, which moves much like a prairie fire across the open grasslands of the plains in the wet season and the savannahs and woodlands in the dry season. While the wildebeest are not dependent on the zebra to modify the grass cover to allow grazing, it appears that grazing by wildebeest does have a positive effect on further grazing by Thomson's gazelle.

The timing of mortality indicates that total grass growth and availability, particularly during the dry season, is a major control on herbivore population levels (Figure 20.9). The quality and abundance of dry-season grass is directly related to the rate at which wildebeest lose the weight put on during the wet season. This weight loss, in turn, relates to malnutrition and mortality, which occur mainly during the dry season.

In addition to these migratory species, there are several nonmigratory herbivores, including water buffalo, topi, impala and other gazelles, giraffes, and elephants, which reside in the more humid western and northern reaches of the Serengeti area. During the wet season, when the migratory species are in the plains area to the southeast, the resident species suffer little competition. In the dry season, the migrants return and divide the existing plant resources more finely. Buffalo favor the woods and grasslands of riverine habitats, affected only by the trampling of wildebeest through these areas in search of water. Impala switch to browsing leaves of woody species in addition to grazing of herbs. Topi are adapted to the partial use of long-grass resources, putting them between zebra and wildebeest. The giraffe represents

(a)

(b)

FIGURE 20.7 The three major species of herbivores in the Serengeti: (*a*) wildebeest and zebra, and (*b*) Thomson's gazelle. (*a, Courtsey of Julie Newman; b,* Kruuk 1972) **See plate in color section.**

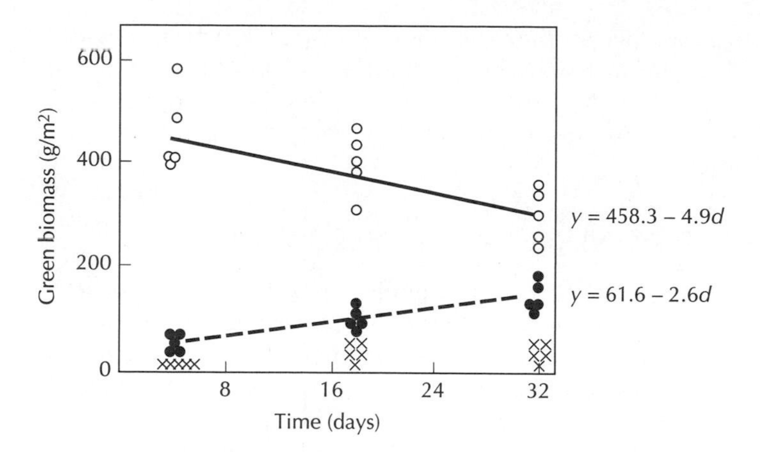

FIGURE 20.8 Comparisons of changes in plant biomass in grazed and ungrazed areas. The open circles describe a loss in total plant biomass over time in ungrazed plots due to respiration and mortality in excess of growth. Grazed plots (*solid circles*) show a net increase in green biomass due to continuing plant growth. (McNaughton 1976)

an extreme adaptation for browsing that confers a unique niche space not directly affected by other herbivores. However, indirect effects of herbivory on fire and tree reproduction can severely alter tree density and availability of food for giraffes. Elephants, with their penchant for knocking down small trees to obtain browse (Figure 20.10), can be particularly effective at altering vegetation structure.

This finer division of niches puts greater pressure on food resources in the dry season. Competition is at its peak, as is mortality, and animal populations are most sensitive to unfavorable shifts in climate. For example, when total annual rainfall in the Serengeti is low, the wildebeest herd spends a greater proportion of the year in the tall-grass and woodland areas of the

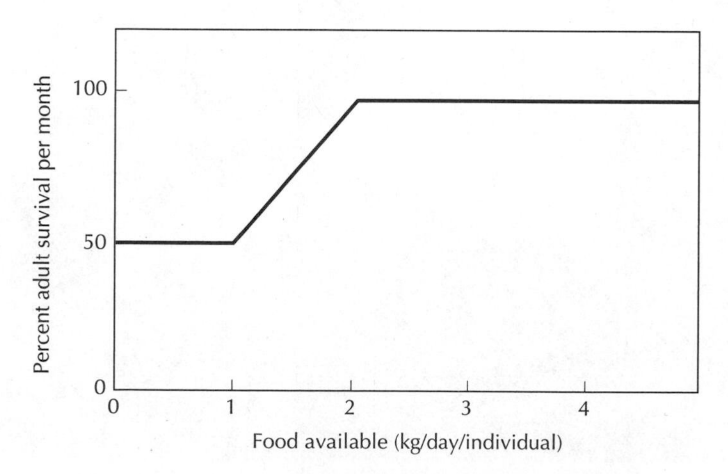

FIGURE 20.9 Assumed relationship between monthly mortality rates of wildebeest and availability of food per individual. (Hilborn and Sinclair 1979)

north, and less in the drier, short-grass plains. This increases the pressure on the northern areas and reduces food availability to resident herbivores as well as other migratory species. As is discussed later, populations of herbivores in the Serengeti have responded markedly to historical changes in dry-season rainfall.

RESOURCE PARTITIONING AMONG PREDATORS

Limitations on niche overlap apply to predators as well as herbivores. The five major predators in the Serengeti—lions, hyenas, leopards, cheetahs, and wild dogs—should also have to divide the herbivore resource by size, condition, location, or season of use. Table 20.1 suggests that this is the case. The major species have different preferences for habitat type, time of day, species taken, and method of hunting. The closest competition between any two pairs of these five species appears to be between hyenas and wild dogs. Recent increases in the hyena population have been linked to reductions in wild dogs, which are nearly extinct in the Serengeti.

Lions account for over half of the predation in the Serengeti. They are territorial and nonmigratory, living in groups, or prides, consisting of 4–15 adults plus a number of subadults and cubs, all interrelated. As the dominant predator, they feed on the dominant, migratory herbivores during their passage through the territory and on the less numerous, resident herbivores at other times of the year. Much like the wolf populations described in Chapter 16, lion populations are stabilized by social interactions within the pride.

In contrast, neither leopards nor cheetahs have large social groupings. They are solitary hunters that feed on the smaller and swifter herbivores, particularly gazelles, which they catch by speed. This is in contrast to the group

FIGURE 20.10 Elephants destroying acacia trees in a Serengeti woodland. (Sven-Olof Lindblad/Photo Researchers)

TABLE 20.1 Differences in Hunting and Social Behavior Between Major Predator Species in the Serengeti
(Bertram 1979)

Characteristic	Cheetah	Leopard	Lion	Hyena	Wild Dog
Approximate weight of adult (kg)	40–60 kg	35–60 kg	100–200 kg	45–60kg	17–20 kg
Number in Serengeti ecosystem	220–500	800–1200	2000–2400	3000–4500	150–300
Habitat in Serengeti	Especially plains, but woodlands too	Woodlands only	Mainly woodlands, but plains too	Mainly plains	Plains and woodlands
Hunting—time of day	Entirely by day	Mainly at night	Mainly at night	Night and dawn	Mainly by day
Number of animals hunting together	1	1	1–5	1–3 (for wildbeest & gazelles), 4–20 (for zebras)	Whole pack, i.e., 2–19
Method of hunting	Stalk, then long fast sprint	Stalk to close range, then short sprint	Spread out; stalk then short sprint	Long-distance pursuit; others join in	Long-distance pursuit
Distance from prey when chase starts	10–50 m 50–70 m	5–20 m	10–50 m	20–100 m	50–200 m
Speed of pursuit	Up to 95 km/hr	up to 60 km/hr	50–60 km/hr	Up to 65 km/hr	Up to 70 km/hr
Distance of pursuit	Up to 350 m	Up to 50 m	Up to 200 m	0.2–3.0 km	0.5–2.5 km
Measured success rate	37–70%	5%	15–30%	35%	50%–70%
Commonest species taken (in order of frequency in diet)	Mainly Thomson's gazelle, also hare, Grant's gazelle, impala	Impala, Thomson's gazelle, dik dik reedbuck, many others	Zebra, wildebeest, buffalo, Thomson's gazelle, warthog, others	Wildebeest, Thomson's gazelle, zebra	Thomson's gazelle, wildebeest, others
Health of prey	Healthy	Healthy	Healthy	Sick and healthy	Sick and healthy
Age and sex of prey	Especially small fawns	All ages; only the young of topi, wildebeest, zebra	All ages, but disproportionately more young than old	Especially males and young of wildebeest and gazelles; female zebra	Esp. gazelle adult males; wildebeest calves; zebra adult females
% of kills that are partially or wholly lost to other carnivores	10–12%	5–10%	Almost none	5% (20% in Ngorongoro)	50%
% of diet obtained by scavenging	None	5–10%	10–15%	33%	3%
Important interference competitors	Hyenas; possibly lions	Possibly lions	None	Possibly lions	Hyenas

hunting of the larger wildebeest practiced by lions, hyenas, and wild dogs. These latter two also show pack or group organization, with greater organization in the wild dogs (Table 20.1).

VEGETATION–HERBIVORE–PREDATOR INTERACTIONS

There are important constraints in resource acquisition that affect the size, distribution, and abundance of both predators and their prey, as demonstrated in the Serengeti. The efficiency of digestion of grasses by herbivores increases with body size because food ingested is held for a longer period of time. This means that the larger herbivores have an advantage over smaller ones in areas where grasses predominate. In contrast, smaller species with less efficient digestion and also a higher metabolic rate (higher respiration per unit body weight) are constrained to consume higher-quality plant material. Thus, it is the gazelles that specialize by eating the broad-leaved herbs and new, short-grass growth that has the highest nutritional value.

High-quality plant material is a more dispersed resource, requiring greater area coverage per animal. This, in turn, dictates smaller herd size. Herding is an effective mechanism against predation for any individual animal. The absence of large herds places increasing value on alertness, speed, and protective coloration as predator defenses.

Large herbivore size and large herd size both increase the value of large size and group hunting behaviors in predators. Thus, the predators on zebra and wildebeest are large and/or hunt in groups. In contrast, the dispersed nature of the herbivores feeding on the dispersed, high-quality food source, requires stealth and speed in would-be predators. The cheetahs and leopards that hunt the smaller and less abundant species also exhibit camouflaging coloration, speed, and solo hunting. The risk in losing a captured prey to a larger group of lions or hyenas is answered in the leopard by carrying the capture up into a tree. Cheetahs do not have this advantage.

One could see, in this division between the generalist and specialist herbivores, patterns similar to those described for apparent and less apparent plant species in relation to protection from herbivory. The generalist or dominant species rely on generalized defenses (lignin and cellulose for plants, body size and herding for herbivores). The specialists, or less apparent species, show a variety of defenses (alkaloids in plants, the wide variation in coloration and behavior in animals).

PREDATION VERSUS FOOD AS LIMITING FACTORS IN HERBIVORE POPULATIONS

Predation is a highly visible process in the Serengeti, but is it the dominant factor determining herbivore population levels? Apparently not. About one third of the mortality of herbivores in the Serengeti is due to predation, with the rest due largely to malnutrition during the dry season. Two mechanisms may interact to cause this: (1) territoriality of predators and (2) the patterns of migration of the major herbivores.

The dominant herbivores in the Serengeti are migratory, while the major predators are resident and territorial. Roughly two thirds of the total herbivore biomass is in the migratory zebra, wildebeest, and Thomson's gazelle. Transitory movement of large numbers of herbivores through stationary predator territories results in a very uneven availability of prey throughout the year.

As the migratory herds pass through a territory, food abundance is very high for a short period of time. However, just as the herbivore population levels are determined by food availability during the dry season, resident populations of predators in territories are determined by resident game populations when the migrating herds are gone. Thus, predator levels throughout the Serengeti are lower than they would be if the herbivores were nonmigratory. Conversely, predator pressure on the migratory herbivores is less than if the herbivores were nonmigratory, and starvation becomes a more significant cause of mortality. Some evidence shows that herbivore populations are lower and predation is higher in other, smaller areas of eastern Africa, where such seasonal migrations of herbivores do not occur.

Herding and migration are two aspects of a larger strategy known as "swamping" of predators—the synchronized appearance and disappearance of prey. Another example of this in the Serengeti is the highly synchronized birth of wildebeests during the wet season on the plains. Not only is this the period of lowest competition for food but also the plains afford the clearest long-range vision for the detection of predators. The synchronized birth produces large quantities of young at one time, numbers far too large to be captured by predators. Giving birth within this period results in a reduced chance of early mortality for the young and so creates strong selective pressure.

PERTURBATIONS, SUCCESSION, AND THE DYNAMICS OF THE SERENGETI SYSTEM

The description of static patterns of distribution, abundance, and function do not allow for an understanding of the factors that control those patterns. Experimentation is required, but the value and uniqueness of the Serengeti, as well as its size, rule out many potential experimental perturbations. However, the region has experienced two major uncontrolled perturbations over the last century, in addition to changes in the degree of human use and modification. These "natural" experiments afford an opportunity to observe the interaction of important processes.

The first perturbation was the outbreak of rinderpest, a virus that strikes both wild and domesticated cattle. Rinderpest was introduced to northern Africa in domestic cattle in the 1880s. It appeared in eastern Africa in 1890, and, by 1892, 95% of buffalo and wildebeest had died, along with most of the domesticated cattle of the pastoral and nomadic tribes in the region. The resulting loss of human life and depopulation of the area is a tragic tale of human misery. Such apparently disassociated phenomena as the disappearance of the tsetse fly and the appearance of man-eating lions in the region can be traced to the decimation of the ruminant species.

Reduced human population led to abandonment of agricultural fields, and previously cultivated areas in the more humid areas returned to brush and woods. As some wildebeest became immune to rinderpest and the population stabilized, the tsetse fly, which uses wildebeest as a host, returned to the area. Increased brush coverage also served to increase the tsetse fly population, and malaria, carried by the fly, returned with a vengeance to the remaining human population.

In the 1930s, mechanical brush clearing and a vaccination against rinderpest reversed these trends, and both human and cattle populations have increased since. Wholesale inoculation of all domestic cattle surrounding the Serengeti area eliminated rinderpest from the population, at least temporarily, by 1963. The establishment of the Serengeti National Park has limited, if not eliminated, direct human impacts on the region since that time. Still, increasing human populations in the surrounding area have caused some shifts in animal abundance, as we will see later.

The effects of rinderpest and its eradication have been profound. Populations of wildebeest and buffalo have increased tremendously, while zebra, a nonruminant unaffected by rinderpest, has remained constant (Figure 20.11). Computer simulations suggest that rinderpest combined with predation were active in keeping the wildebeest population at around 300,000. With rinderpest removed, the population exploded and now swamps predator demand.

The second "natural" perturbation was an increase in dry-season rainfall between the 1967–1971 and 1972–1976 periods (Figure 20.12). Dry-season rainfall delays the movement of herds into the woodland systems of the north and west and also increases total grass production throughout the region (Figure 20.13).

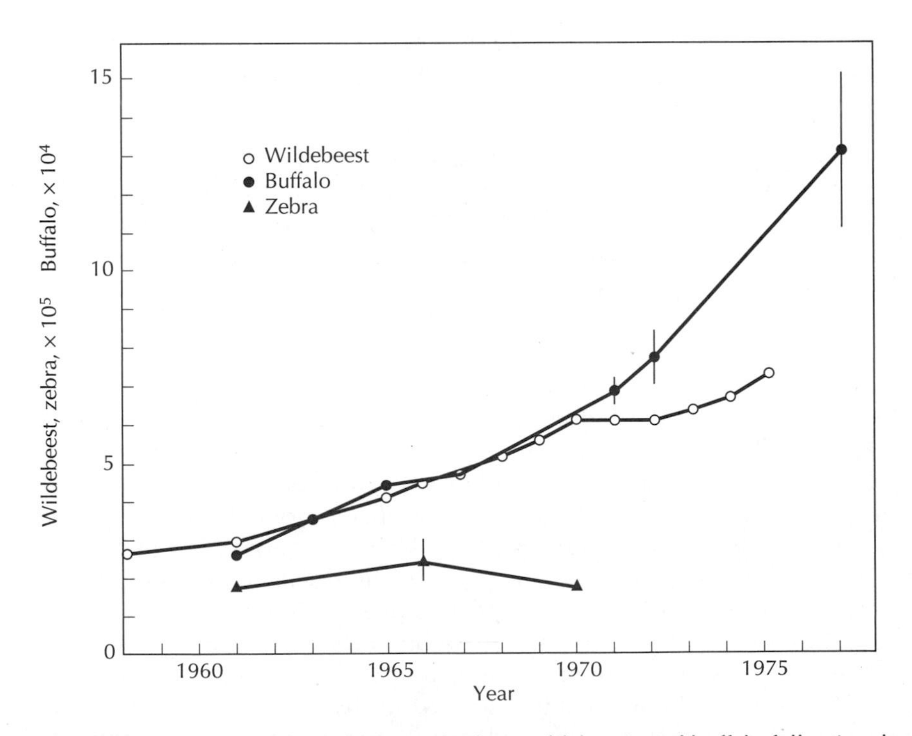

FIGURE 20.11 Population changes of zebra, wildebeest, and buffalo following the removal of rinderpest from the Serengeti. (Sinclair 1979b)

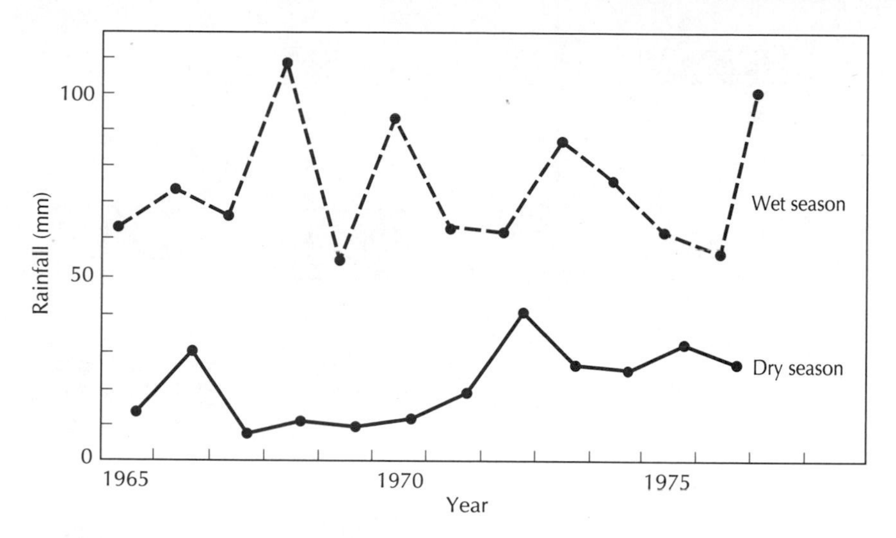

FIGURE 20.12 Changes in the amount of rainfall in the Serengeti between 1965 and 1977 (note the increase in dry-season rainfall). (Hanby and Bygott 1979)

Rinderpest control and increased rainfall fostered an increase in numbers of wildebeest through the 1960s and 1970s. The increase in wildebeest has had important secondary consequences throughout the park. The ways in which the populations of different large mammals have responded are summarized in Figure 20.14. This summary diagram is the distillation of years of research in the Serengeti. It represents a first attempt at a systems analysis of this complex ecosystem. We will step through parts of this diagram to explain briefly the effects of perturbations on the Serengeti system.

Heavier rainfall means more grass production in the plains and longer residence there by the migratory herds. Heavier grazing here alters the relative abundance of grass and herb species. This is in comparison with the

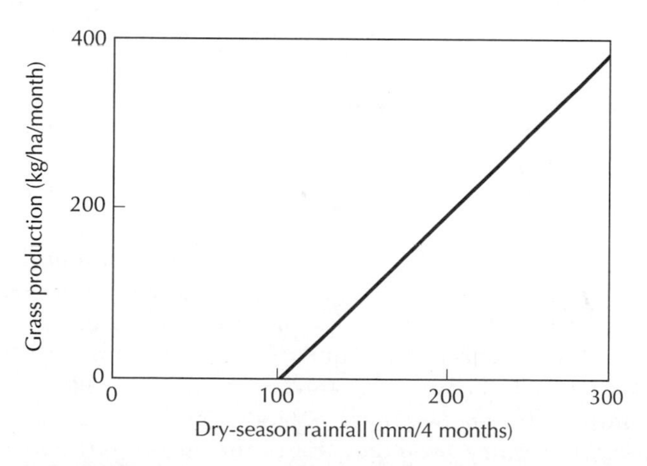

FIGURE 20.13 Summary relationship between dry-season rainfall in the Serengeti and dry-season grass production. (Hilborn and Sinclair 1979)

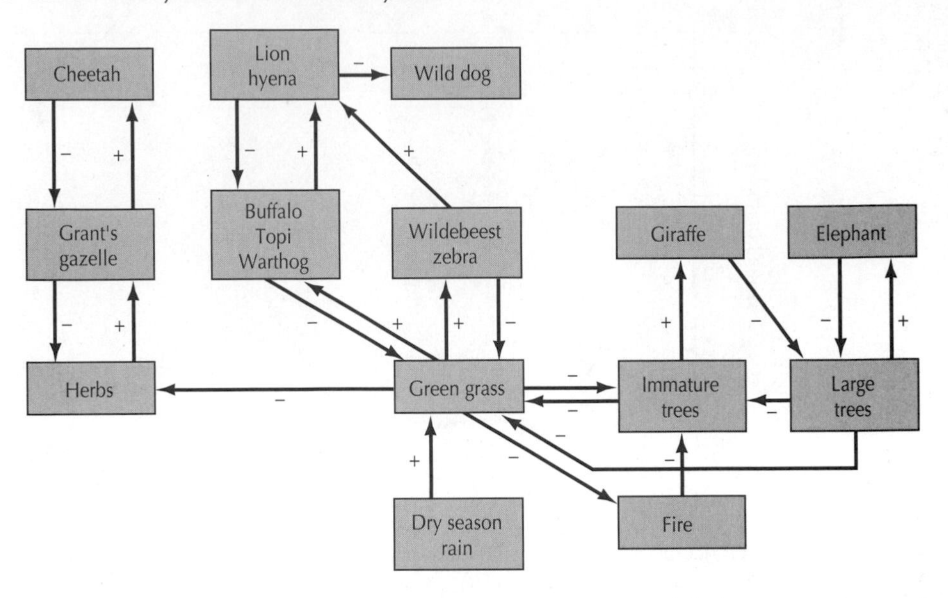

FIGURE 20.14 Schematic model of interactions among rainfall, fire, vegetation, and major species of herbivores and predators in the Serengeti. (Sinclair 1979a)

northern and western areas grazed during the dry season, when plants are senescent and consumption is not a selective force. Greater concentrations of herbs leads to increases in populations of Grant's gazelle, which feeds on herbs in the plains for most of the year. This may, in turn, have caused an increase in numbers of cheetah, which feed on the gazelle. In contrast, feeding and trampling by wildebeest of grasses in the buffalo's riverine habitat has reduced the increase in that species due to the removal of rinderpest. The increased dry-season rainfall has also increased the populations of relatively rare species, such as topi, kongoni, and warthog, that live on the plains year-round.

At first glance, an increase in wildebeest would be expected to cause an increase in the numbers of lions. However, both the territoriality of lions (see Chapter 16) and the migratory nature of the wildebeest herd minimize the impact of this dominant herbivore in determining population levels of the dominant predator. To the extent that growing numbers of wildebeest compete with the year-round prey on which lions subsist when wildebeest are absent, increasing wildebeest numbers could reduce availability of prey for lions.

Both the increase in wildebeest through greater consumption of grass in the woodlands and increased precipitation by maintaining green grass for a longer period reduced the incidence of fire throughout the park (Figure 20.15). Fire and consumption are similar in that they both reduce storage of carbon in plants and soils and recycle nutrients, but they differ in their effects on competitive balance between grasses and trees. By selecting against herbaceous growth, herbivory promotes the occurrence of trees. Conversion of grasslands to arid shrublands, or deserts, through overgrazing has been a general occurrence in many parts of the world. In contrast, fire kills many

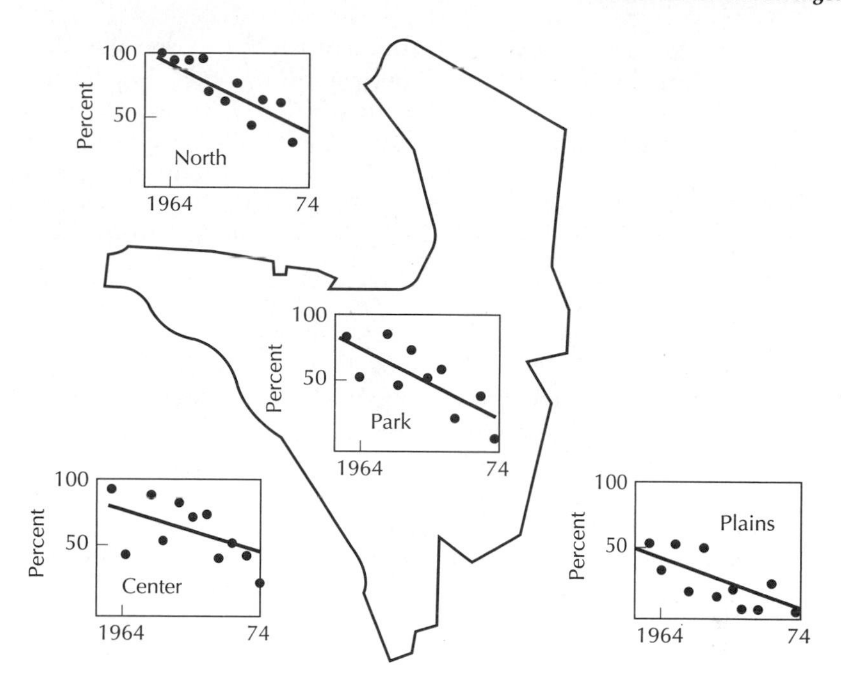

FIGURE 20.15 Changes in the percentage of area burned in different areas of the Serengeti from 1964 to 1974. (Norton-Griffiths 1979)

woody species, or at least suppresses them relative to the rapidly sprouting grasses that are adapted to dying back to ground level each year.

The presence or absence of trees affects browsers, particularly giraffe. The presence of immature trees increases the number of giraffe, which in turn tend to kill mature trees. In the absence of fire, this further increases the importance of immature trees. The giraffe–tree interaction has been affected by the increase in human use around the park. In the 1970s, this forced large numbers of elephants into the woodlands. Elephants are very effective at removing mature trees (Figure 20.10) and can reduce regeneration of young trees, degrading giraffe habitat.

HUMAN USE AND CONSERVATION CONCERNS IN THE SERENGETI

External factors continue to intrude into the Serengeti and initiate rapid changes in populations. While the human vacuum created by the rinderpest tragedy made the establishment of the Serengeti Park possible, human population density in the area surrounding the park is now increasing by as much as 15% per year. Political change and instability have also worked against management and conservation initiatives. In particular, closure of the border between Kenya and Tanzania in 1977 reduced tourism by 85%, reducing revenues for management and antipoaching measures.

> Political events have also had a major impact on the ecosystem . . . In 1977 the international border between Kenya and Tanzania was

> closed . . . and the main tourist route . . . remains closed. The combined effects of human population increase and the marked drop in anti-poaching patrols [resulted in] a 52% loss of rhinoceros in . . . the first year of border closure. By 1980, the population was virtually extinct. . . . Some 50% of the [elephant] population disappeared during 1984–1986, . . . By [1984] the northwest of Serengeti National Park was devoid of buffalo. (Sinclair 1995)

Poaching has reduced the elephant population to one sixth of its former level, and the small population of rhinoceros in the park has been hunted to extinction. This, combined with an anthrax epidemic in resident impala in the woodland areas of the northwest, has led to a rare period of rapid regeneration by acacia, increasing shrub cover by 20%. An outbreak of canine distemper in lions in the mid-1990s reduced the population within one long-term study area by nearly 50%. The appearance of rabies in the rapidly dwindling population of wild dogs has raised concern about the remaining population that, in the early 1990s, numbered less than 60. Through all of these changes, the numbers of wildebeest and zebra have remained relatively stable (Figure 20.16).

Clearly, several serious problems confront the managers of the Serengeti National Park and its surroundings. Attempts to preserve a large area for native species within a region in which the human population experiences poverty and harsh living conditions will lead to continuing conflicts. Political restrictions also place limitations on access to the park for research purposes and limit the amount of information available to managers.

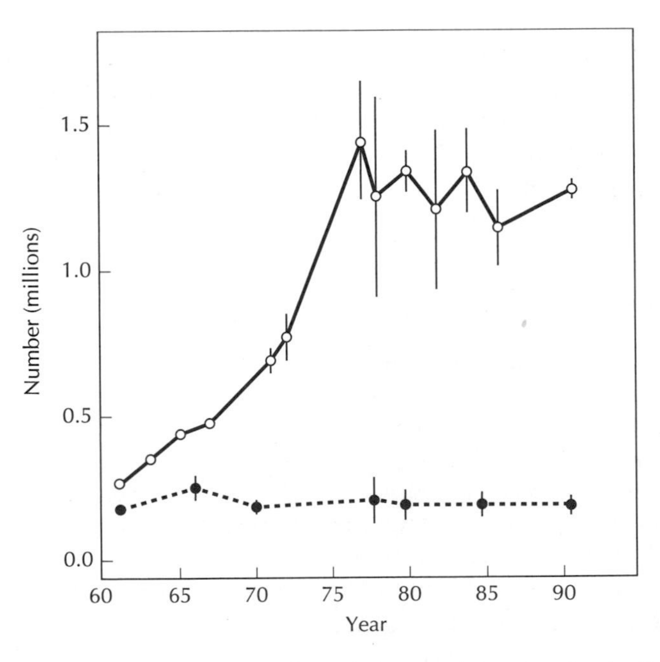

FIGURE 20.16 Historical trends in population levels of wildebeest (*open circles*) and zebra (*closed circles*). (Dublin et al. 1990, cited in Sinclair 1995)

Within the park, managers face the challenge of maintaining an ecosystem that is still recovering from several major disruptions and is continually exposed to new external forces. Large shifts in species abundance might be expected in such a system, and it is difficult to know exactly what the management goals should be. Management for population stability is probably not possible and could be harmful.

> Because ecosystems must be considered dynamic, one must resist the temptation to manage them with a view toward maintaining some arbitrary status quo. The present evidence shows that there are natural negative feedback mechanisms operating between some components in the system. . . . Provided these negative feedbacks are strong enough, the system can absorb disturbances without management. . . . Droughts . . . could produce rapid declines in wildebeest and other herbivores, but these species will build up again when the food supply increases after the drought is over. Such declines are desirable and should be expected. (Sinclair 1979a)

The kind of systems analysis outlined in Figure 20.14 is a first step toward understanding the dynamics of such a system, assessing the implications of changes that may occur in fire frequency, population levels, and other state variables, and developing management plans that work with the processes active in the system. However, any scientifically based management plan will always be affected by the political and economic realities of the countries hosting this global gem.

There is a more general concern demonstrated in the Serengeti but that applies to conservation reserves worldwide: the increasing isolation of these treasured places from similar, adjacent systems that provide an ecological buffer against dramatic shifts within the parks. At the heart of this concern is the question of the size of reserve required to ensure persistence into the distant future. It is axiomatic in ecological studies that rare species can become locally extinct due to random perturbations in climate, disease, or other factors (Chapter 16). The absence of moose and wolves from Isle Royale could be one example of this (see discussion of island biogeography in Chapter 24). It is unlikely that both species never existed on the island before 1900. More likely is that small populations of both were driven to extinction in this small area. Given the recent, large fluctuations in populations of both species, local extinction could happen again.

However, Isle Royale is embedded in a larger region supporting similar species and ecosystems. Recolonization from the mainland would be likely (even if humans chose not to intervene). Can the same be said for the Serengeti? The tenuously low populations of wild dog is another example of a species that may be driven to local extinction by "random" events. Would recolonization be possible in this case?

The upshot of this discussion is that reserves must be large enough to support target species through periods of very low population levels. If the reserve must rely on periodic recolonization of rare species from surrounding areas, then the integrity of the system being preserved is in doubt. For this reason, increasing attention is being given to the management or acquisition of areas surrounding many existing reserves. Research allowing for a clearer definition of the size of reserves is required to ensure the persistence of endangered ecosystems.

REFERENCES CITED

Anderson, G. D., and L. M. Talbot. 1965. Soil factors affecting the distribution of the grassland types and their utilization by wild animals on the Serengeti Plains, Tanganyika. *Journal of Ecology* 53:33–56.

Bertram, B. C. R. 1979. Serengeti predators and their social systems. In Sinclair, A. R. E., and M. Norton-Griffiths (eds.). *Serengeti: Dynamics of an Ecosystem.* University of Chicago Press, Chicago.

Dublin, H. T., A. R. E. Sinclair, S. Boutin, E. Anderson, M. Jago, and P. Arcese. 1990. Does competition regulate ungulate populations? Further evidence from the Serengeti–Mara ecosystem. *Oecologia* 82:283–288.

Hanby, J. P., and J. D. Bygott. 1979. Population changes in lions and other predators. In Sinclair, A. R. E., and M. Norton-Griffiths (eds.). *Serengeti: Dynamics of an Ecosystem.* University of Chicago Press, Chicago.

Hilborn, R., and A. R. E. Sinclair. 1979. A simulation of the wildebeest population, other ungulates and their predators. In Sinclair, A. R. E., and M. Norton-Griffiths (eds.). *Serengeti: Dynamics of an Ecosystem.* University of Chicago Press, Chicago.

Jarman, P. J., and A. R. E. Sinclair. 1979. Feeding strategy and the pattern of resource partitioning in ungulates. In Sinclair, A. R. E., and M. Norton-Griffiths (eds.). *Serengeti: Dynamics of an Ecosystem.* University of Chicago Press, Chicago.

Kruuk, H. 1972. *The Spotted Hyena: A Study of Predation and Social Behavior.* University of Chicago Press, Chicago.

McNaughton, S. J. 1976. Serengeti migratory wildebeest: Facilitation of energy flow by grazing. *Science* 191:92–94.

Norton-Griffiths, M. 1979. The influence of grazing, browsing, and fire on the vegetation dynamics of the Serengeti. In Sinclair, A. R. E., and M. Norton-Griffiths (eds.). *Serengeti: Dynamics of an Ecosystem.* University of Chicago Press, Chicago.

Sinclair, A. R. E. 1995. Serengeti past and present. In: Sinclair, A. R. E., and P. Arcese (eds.). *Serengeti II: Dynamics, Management and Conservation of an Ecosystem.* University of Chicago Press, Chicago.

Sinclair, A. R. E. 1979a. Dynamics of the Serengeti ecosystem: Process and Pattern. In Sinclair, A. R. E., and M. Norton-Griffiths (eds.). *Serengeti: Dynamics of an Ecosystem.* University of Chicago Press, Chicago.

Sinclair, A. R. E. 1979b. The eruption of the ruminants. In Sinclair, A. R. E., and M. Norton-Griffiths (eds.). *Serengeti: Dynamics of an Ecosystem.* University of Chicago Press, Chicago.

Sinclair, A. R. E. 1977. *The African Buffalo.* University of Chicago Press, Chicago.

Sinclair, A. R. E., and M. Norton-Griffiths. 1979. *Serengeti: Dynamics of an Ecosystem.* University of Chicago Press, Chicago.

Western, D., and C. Van Praet. 1973. Cyclical changes in the habitat and climate of an east African ecosystem. *Nature* 241:104–106.

21

A GAP-REGENERATION SYSTEM

THE NORTHERN HARDWOOD FORESTS OF THE UNITED STATES

INTRODUCTION

The northern hardwood forest region of North America is a less extreme environment than either of those presented in Chapters 19 and 20. Located primarily in a band along the border between the northeastern United States and southern Canada (Figure 21.1), the area is neither as cold as the taiga nor as dry as the Serengeti. The climate is a combination of cold winters with significant snow accumulation and warm summers. The current physical and chemical climate of the region combine with characteristics of the major species to produce a balance between production and decomposition that does not lead to the continuous buildup of organic matter seen in the taiga. Herbivory is dominated by insects rather than mammals and is generally much less of a factor in ecosystem function than in the Serengeti. With fire and herbivory as minor factors, the population dynamics of the dominant tree species, along with occasional pulse disturbance due to windstorms and hurricanes, largely determines the structure and dynamics of the ecosystem.

The purpose of this chapter is to describe the dynamics of vegetation and the changes in ecosystem processes that occur in the eastern portion of the northern hardwood forest. We discuss disturbance patterns, how disturbance alters resource availability, and how the major plant species vary in their response to disturbance. Combining these processes yields an integrated view of the function of this ecosystem type.

THE NORTHERN HARDWOOD ECOSYSTEMS OF NEW ENGLAND

The northern hardwood forests in northern New England are representative of the eastern portion of this ecosystem type. They occur mainly at mideleva-tion in the major mountain ranges of the area, the northernmost extension of the Appalachian Mountains. The climate of the region is determined by a mixture of continental and maritime influences, providing fairly wide temperature extremes and abundant precipitation. A typical mean July air temperature is 19°C; for January, it is −9°C. The frost-free season lasts about 160 days, and precipitation occurs fairly evenly throughout the year.

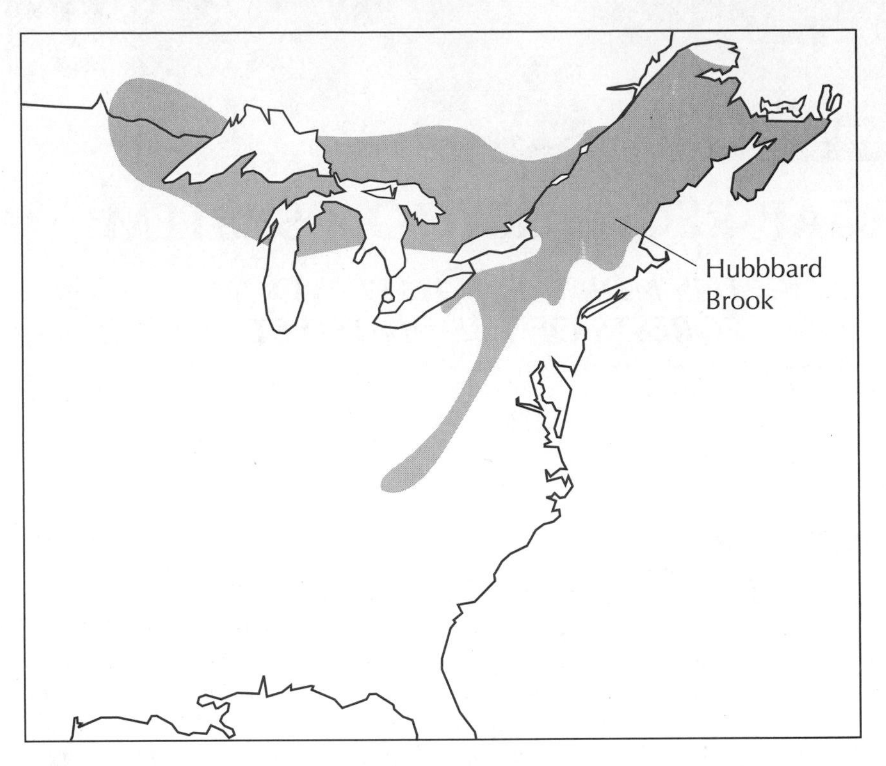

FIGURE 21.1 The distribution of northern hardwood forests in North America and the location of the Hubbard Brook Ecosystem Study. (Bormann and Likens 1979)

The region supports closed forests dominated by sugar maple and yellow birch, along with beech in the eastern portion and basswood in the western portion of the range. Evergreen conifers, such as hemlock and pine, also are present. At the northern edge and at high elevations, boreal species, such as spruce and fir, mix with the hardwoods. At the southern edge and lower elevations, oaks, red maple, white ash, and other species of the oak–hickory forest region are present. Early successional species include the fast-growing, deciduous species pin cherry and aspen, and the oldest stands can contain a thick understory of slow-growing, low-stature shrub species or an increasing abundance of the tolerant conifers, such as spruce and hemlock.

The entire northern hardwood region was stripped by glaciers in the last ice age, and soils have developed on materials left behind or deposited since the glacial retreat. In the east, soils have developed mainly on thin glacial till on mountain slopes. This is in contrast to the Midwestern portion of the northern hardwood region where deposition of windblown loess has produced deeper and richer soils. In both cases, soil depth and the texture of the material (sand, silt, clay, and larger rocks) play a major role in determining site quality and productive potential. In the east, the cool, humid climate and shallow, coarse-textured soils favor podzolization as the dominant soil-forming process. Soils are somewhat older than in the boreal zone and subject to greater weathering intensity (Figure 21.2). In the west, deeper, less

FIGURE 21.2 Spodosol developed under a northern hardwood forest. (Likens et al. 1977)

acidic, and finer-textured soils favor melanization and lessivage as soil-forming processes, and Alfisols are also an important soil type.

Although the Spodosols include an organic surface horizon (Figure 21.2), this need not indicate that litter decomposition is too slow or incomplete to maintain adequate rates of nutrient cycling. Rather, this layer of humus indicates only the lack of mixing of this end product of active decomposition into the lower soil horizons. Unlike the taiga system, moderate temperature and moisture conditions, along with the high quality of litter produced by the deciduous species that currently dominate the area (see subsequent discussion), drive relatively rapid decomposition. This means that fire is not required to maintain nutrient cycles or to balance the production–decomposition ratio. Herbivory is also only rarely important in altering forest structure and ecosystem processes. As we will see, human influences may have played a role in determining both the species composition and the steady-state nature of the forest floor.

When fire and herbivory are not major forces shaping ecosystems, changes in structure and function are driven more by processes of natural tree death and by wind and storm damage—and the interactions between species in the competition to fill the gaps these create (Figure 21.3). Much of the research into vegetation dynamics in northern hardwoods has examined processes that cause these gaps, those that favor the entrance of the different species into different-sized gaps, and those that determine the relative competitive abilities of the different species in gaps of different sizes.

FIGURE 21.3 A gap created by the loss of a large canopy dominant in a northern hardwood forest. (Bormann and Likens 1979)

Important characteristics related to the movement of species into gaps include distance of seed dispersal and requirements for seed germination (both related to seed size) and the extent of vegetative reproduction—the sprouting of new stems from root systems of mature trees and shrubs. Related life-history phenomena include maximum tree size and age, and the age at which seed production begins. Maximum tree size also affects the size of gaps created by tree death.

Success in the gap environment depends mainly on the rate of height growth that can be achieved and the ability to tolerate partial or deep shade. Secondary effects of nutrient requirements and alterations in nutrient cycling rates by the quality of litter produced may also be important.

PATTERNS OF DISTURBANCE IN NORTHERN HARDWOODS AND EFFECTS ON RESOURCE AVAILABILITY

How frequent is disturbance in a forest in which fire- and herbivory-related tree mortality are rare? The occurrence of disturbance in such systems is tied to the life expectancy of the dominant species and to the occurrence of major windstorms and blow-downs or other damaging agents, such as ice storms. Small-scale disturbances caused by individual tree death occur regularly and randomly over the landscape, while the large-scale blow-downs take place in major storms and occur less frequently. Taking a plot of a given size—for example, a half-hectare—there is an inverse relationship between the frequency of disturbances and the percentage of trees in the plot that are removed (Figure 21.4). Disturbances removing less than 20% of the trees occur, on average, every 200 years. This size and frequency of disturbance would be tied to the loss of individual trees. A disturbance removing two thirds of the trees in our example half-hectare, such as a major windstorm blow-down, occurs only once every 1000 years. Within this 1000-year return time for catastrophic disturbance, the stand could go through several cycles of "gap-phase replacement."

Blow-downs in the northern hardwood forest are similar to disturbances in other ecosystems in that plant demand for resources is reduced and availability is increased. This is particularly true for elements that are made available mainly through decomposition and mineralization, such as nitrogen (N). Decay is limited more by temperature than by moisture in this area. Removal of the forest canopy increases the penetration of sunlight to the forest floor, increasing soil temperatures and decomposition rates. Because of the rapid decay of litter produced by most northern hardwood species and the high nutrient demand of these deciduous species, N cycling rates are high even before disturbance.

Combining all of these factors, we can understand how the large-scale removal of vegetation from the watershed at Hubbard Brook (Chapter 4), located in the northern hardwood region, resulted in very large increases in the loss of N and other nutrients to streams. Similar measurements of N loss from northern hardwoods, using trenched plots over which the canopy remained intact (also Chapter 4), suggest very high availability in small-scale disturbances as well. In the terminology presented in Chapter 5, northern hardwood systems have very low resistance to disturbance, as measured by

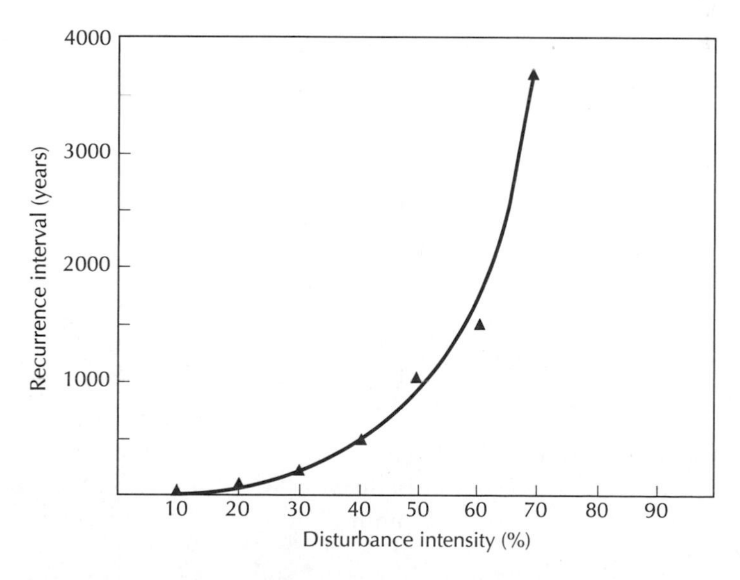

FIGURE 21.4 Relationship between the frequency of disturbance in northern hardwood forests in the great lakes region and the percentage of trees removed. (Lorimer and Frelich 1989)

rates of N loss. In terms of resource availability to plants, the disturbed areas are highly fertilized and irrigated (because of reduced transpiration) sites with increased light availability as well.

SPECIES ADAPTATIONS TO THE DISTURBANCE GRADIENT: REPRODUCTIVE AND LIFE HISTORY STRATEGIES

Gaps come in a wide variety of sizes. We can think of gap size in terms of a gradient of resource availability (Chapter 5). The larger the gap, up to a certain size, the greater the enrichment. Larger gaps also take longer to close, and so the duration of the enriched environment varies with gap size. The major northern hardwood species show a wide variety of reproductive and growth characteristics, combined in different ways (Table 21.1). According to niche theory (Chapter 5), each combination should have adaptive value for a particular set of environmental conditions—in this case, related to disturbance and recovery.

The northern hardwood forest in New England is also called the beech–maple–birch forest type because of the predominance of beech, sugar maple, and yellow birch in mature stands (Figure 21.5). Of these three, beech is considered the most tolerant of shade. It also has large, heavy seeds and active vegetative reproduction. Sugar maple is somewhat less tolerant of shade and has lighter seeds and no vegetative reproduction. Yellow birch has still

TABLE 21.1 Seed Dispersal and Growth Characteristics of Six Major Northern Hardwood Species
(After Bormann and Likens 1979, Marks 1974, Melillo et al. 1982, Likens and Bormann 1970, USDA 1974)

Species Characteristic	Beech	Sugar Maple	Yellow Birch	Pin Cherry	Aspen	Striped Maple
Seed size (mg/seed)	283	65	1.0	32	0.15	41
Seed dispersed by (A = animal, W = wind)	A	W	W	A	W	W
Soil requirements (U = undisturbed, D = disturbed, M = mineral soil)	U	U	DM	D	DM	?
Vegetative reproduction	+	−	−	−	+	+
Age at first seed production	40	30	40	4	10	15 (tree form)
Maximum longevity	350	350	250	45	80	30
Maximum growth rate (cm height/year)	30	35	40	100	100	?
Shade tolerance (T = tolerant, I = intermediate, N = intolerant)	T	T	I	N	N	T?
Leaf chemistry and decomposition						
% Nitrogen (green)	2.2	2.1	2.8	2.8	—	—
(litter)	0.9	0.9	1.1	1.2	—	—
% Lignin	24.0	10.0	15.0	19.0	—	—
Decomposition rate (%/year)	8.0	22.0	30.0	30.0	—	—

FIGURE 21.5 A typical second-growth northern hardwood stand dominated by sugar maple, beech, and yellow birch. (Bormann and Likens 1979)

lighter seeds, is less tolerant of shade, and also does not reproduce vegetatively. All three species have long life spans.

These three species are thought to form a cyclic pattern involving continuous recovery from small-scale disturbance within the mature forest (Figure 21.6). The creation of a gap by the fall of an individual large tree creates increased light availability on the forest floor and also usually creates patches of bare mineral soil around the upturned root mass (Figure 21.7*a*). The small, light seeds of yellow birch can be dispersed 100 m or more by wind. In addition, some seeds are shed during the winter onto snow cover, over which they may be blown for even greater distances. The small seeds have very limited energy reserves for building the initial root and stem and can suffer drought damage if they germinate on the seasonally dry forest floor. They survive better on mineral soil and frequently grow on top of the old root mass of a fallen tree, sending roots down to and into the soil. As these roots mature and become woody, and the old root mass on which the seedling became established erodes away, a stilt-rooted tree characteristic of yellow birch (Figure 21.7*b*) may develop.

The partial shade and developing forest floor produced by yellow birch foliage inhibits the further establishment of yellow birch seedlings but allows for the germination and establishment of the heavier-seeded sugar maple and beech. However, sugar maple produces many more seeds (as many as $10{,}000{,}000 \cdot ha^{-1} \cdot yr^{-1}$) than beech. The seeds are also lighter and are contained in a samara (Figure 21.8), a winged structure that catches the wind,

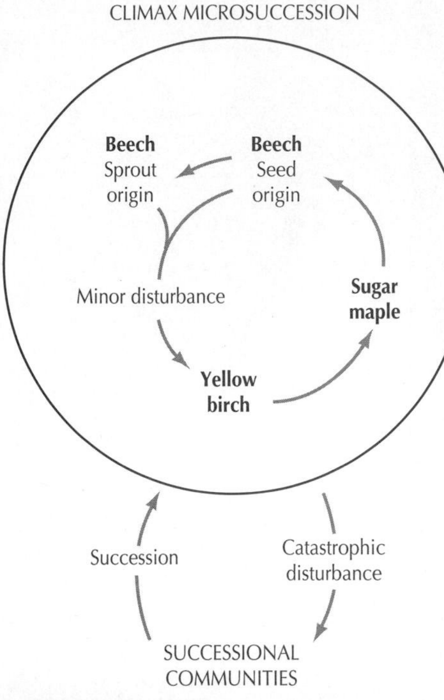

FIGURE 21.6 Hypothesized cyclic successional sequence of yellow birch, sugar maple, and beech in northern hardwood forests. (Forcier 1975)

(a)

(b)

FIGURE 21.7 Early and late stages of colonization of a large tip-up mound. (*a*) A recently created mound that will soon support tree seedlings, including yellow birch. (*b*) A yellow birch that seeded onto a tip-up mound now grows on "stilt roots" after the mound has eroded away. (Bormann and Likens 1979)

makes the entire structure spin like a helicopter blade, slows the descent of the seed, and results in greater dispersal distances. All of these factors tend to favor more rapid colonization by maple under the birch.

Sugar maples live and grow slowly under the birch, moving to the canopy after the death of the canopy tree. Under the darker shade of the maple, beech root sprouts, which receive photosynthate from the parent tree, have a competitive advantage, and tend to take over the understory along with several species of very tolerant shrubs. The heavy beechnut reduces the distance that seeds are moved by wind, although limited animal dispersal occurs. The large energy reserves in the seed allow for seedling establishment in shade and on a thick forest floor.

At any time in this sequence, the death or blow-down of a large, individual tree can reset the pattern by creating an opening large enough to allow for the reentry of yellow birch. The continued occurrence of these small-scale disturbances and the relatively long life span of yellow birch are what allow for the continued existence of this relatively intolerant "gap-phase" species even in mature forests.

Large-scale disturbances, such as those caused by major windstorms, hurricanes, and commercial clear-cutting, bring a different set of actors to the stage. Pin cherry is a rapidly growing, short-lived species that sprouts from a population of dormant seeds stored in the forest floor. These seeds may have been present in the soil since the last major disturbance and may survive for 60 years or more and still remain viable and capable of germi-

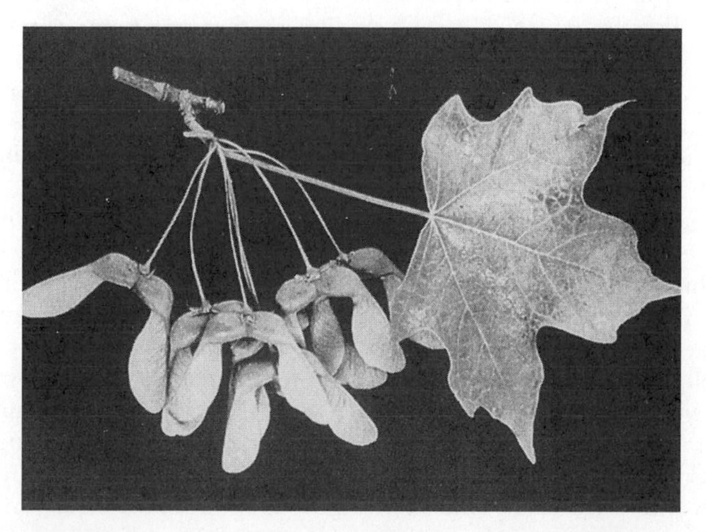

FIGURE 21.8 Sugar maple leaf and samaras. The winglike structure of samaras cause the seeds to whirl in the air, slowing their descent to the ground and allowing the seeds to travel greater distances from the parent tree. (Fowell 1965)

nating after disturbance. What factor triggers germination is unclear, although increased levels of soil nitrate have been implicated in certain fertilizer experiments.

Aspen can also occur in these same areas, but its role is much less important here than in the Alaskan taiga, or even in the western end of the northern hardwood range, perhaps due to the reduced importance of fire in the eastern portion of the northern hardwood region. Aspen does not have a buried seed strategy but rather relies on prolific production of very light, wind-borne seeds to reach disturbed areas. Even a minor surface fire could destroy much of a buried seed crop, decreasing the adaptive value of that strategy and increasing the value of aspen's "fugitive," or light seed, characteristic.

Several species of shrubs can dominate the understory in mature stands and contribute to the vegetative response to disturbance. One of the most interesting is striped maple, which can persist through vegetative reproduction and slow growth for decades under a dense canopy, but it grows to tree height, flowers, and fruits only following a major disturbance. After 10 to 20 years, the main stem dies back, and the "tree" reverts to vegetative reproduction and a shrub growth form.

There are also species that respond to special conditions of resource availability. At low elevation and on rich sites, white ash can be more important than yellow birch as the "gap-phase" replacement species. Red maple becomes more important on very wet or very dry sites and also seems to occur in place of sugar maple on sites where nitrification does not occur.

From this description, we can summarize the different "niches" occupied by the major northern hardwood species in terms of forest conditions under which they reproduce and grow most effectively. Both maple and beech are tolerant, dominant species that maintain both mature individuals capable of producing seed and shade-tolerant seedlings and saplings as advanced regeneration. However, there is a clear distinction between these two species. Beech vigorously reproduces vegetatively by root sprouts, while maple has more effective seed dispersal because of the samara structure.

Yellow birch is the gap-phase species that relies on maintaining long-lived, reproductive individuals in the overstory; these trees can disperse seed to small openings as they occur. Striped maple is also present in the mature forest, but as a low-growing shrub, and also reproduces in gaps, but by converting from long-lived shrub form to short-lived tree form. Pin cherry has the buried seed strategy, while aspen has the "fugitive" strategy.

Successful seed dispersal to disturbed areas is accomplished in all of these species. What differs is the way in which the different species survive or avoid the highly competitive environment between disturbances. Maple and beech are tolerant of these conditions, surviving as seedlings and saplings in the understory. Yellow birch survives in the overstory by being long-lived. Striped maple is also long-lived but grows in the understory, only reaching into the overstory in the first 2 decades after disturbance. Pin cherry and aspen are not present in the closed forest as live stems. Pin cherry is present as dormant seeds. Aspen relies on the presence of disturbed sites within the larger area to provide a constant source of highly mobile seeds. Both pin cherry and aspen require larger gaps to compete successfully.

Certain life history and growth characteristics are strongly related to mechanisms of seed production and dispersal. Table 21.1 shows that the most intolerant species tend to be short-lived, grow rapidly, have low root-to-shoot ratios, and reproduce at a younger age. All of these traits are consistent with selective pressure to keep foliage above the general level of the developing canopy and to take the fullest advantage of the temporary nature of the enriched environment created by disturbance. Table 21.1 also shows that intolerant species tend to have higher N concentration in green foliage and in litter. All of these trends are in keeping with the general theory developed in Chapter 18, which suggests that enriched, disturbed environments tend to be filled by rapidly growing, short-lived species with high nutrient concentration in foliage.

INTEGRATION OF PLANT AND BIOGEOCHEMICAL RESPONSES TO DISTURBANCE

While all disturbed areas show increased resource availability, the degree and duration of those increases depend on the size of the gap created. Small gaps surrounded by intact forest close quickly as roots and branches from adjacent trees expand into the opening. The central areas in large gaps are unaffected by surrounding plants; experience this enrichment for a longer period; and tend to be revegetated by intolerant species, such as pin cherry. Intermediate-sized openings favor the establishment and growth of the moderately tolerant and moderately fast-growing gap phase species, such as yellow birch.

In the largest openings resulting from windstorms or from commercial clear-cutting, there is usually, but not always, a remarkable and dramatic response to the enriched environment created. Germination of previously dormant pin cherry seeds can be as dense as 25–40 stems · m^{-2}. This rapidly growing intolerant species can create a fully closed canopy within 3–4 years after disturbance (Figure 21.9). With the additional productivity of other intolerant and some remaining and new stems of tolerant species, annual net primary production can actually be higher in a 5-year-old stand than in a mature forest (Figure 21.10). Pin cherry also has very high nutrient content in all tissues, such that at age 4 years, a vigorous pin cherry stand can contain a significant fraction of the nutrients that were lost from the devegetation experiment at Hubbard Brook. Thus, pin cherry acts as a major sink for nutrients following disturbance and plays an important role in reducing leaching losses.

Pin cherry leaf litter may also play an important role in minimizing losses. It is high in both N and lignin content, such that rates of decomposition are moderate and considerable N immobilization occurs. In the fourth year of regrowth, immobilization into pin cherry litter may remove 17 kg · ha^{-1} · $year^{-1}$ from the N cycle, reducing the amount available for leaching losses. While some of this immobilization may be temporary, with mineralization occurring in the following 2–3 years, the high lignin content of pin cherry leaf litter suggests that a significant fraction of the immobilized N is converted into long-term soil organic matter (Chapter 13).

FIGURE 21.9 A 5-year-old pin cherry stand showing the complete recreation of canopy biomass and rapid growth that is typical of this species. (Bormann and Likens 1979)

Pin cherry thus contributes substantially to the resilience (Chapter 5) of northern hardwood stands. While initial increases in nutrient losses are high following disturbance, the rapid colonization of disturbed sites by pin cherry soaks up much of the available nutrients. By the fourth year, nutrient losses from disturbed sites in which regeneration occurs is nearly back to predisturbance levels. Other system processes, such as primary productivity and internal nutrient cycling, are also nearly restored.

A Hypothesis of Homeostasis – We propose that severe stress initiated by clear-cutting not only accelerates the activity of the mechanisms responsible for biotic regulation [in the intact forest], but also calls into action another set of mechanisms largely quiescent during that phase. . . . The cutover ecosystem responds to the conditions of increased resource availability by a burst of primary production, not only by species that characterize the precutting forest but also by a group of species not part of the predisturbance forest (e.g., buried seed species) that may have evolved specifically to fill a niche created by this type of disturbance. . . . The coupling of heterotrophic processes to autotrophic processes is far from perfect. Considerable leakage of nutrient from the ecosystem occurs during the first few years, even in immediately revegetating systems. . . . This suggests that the

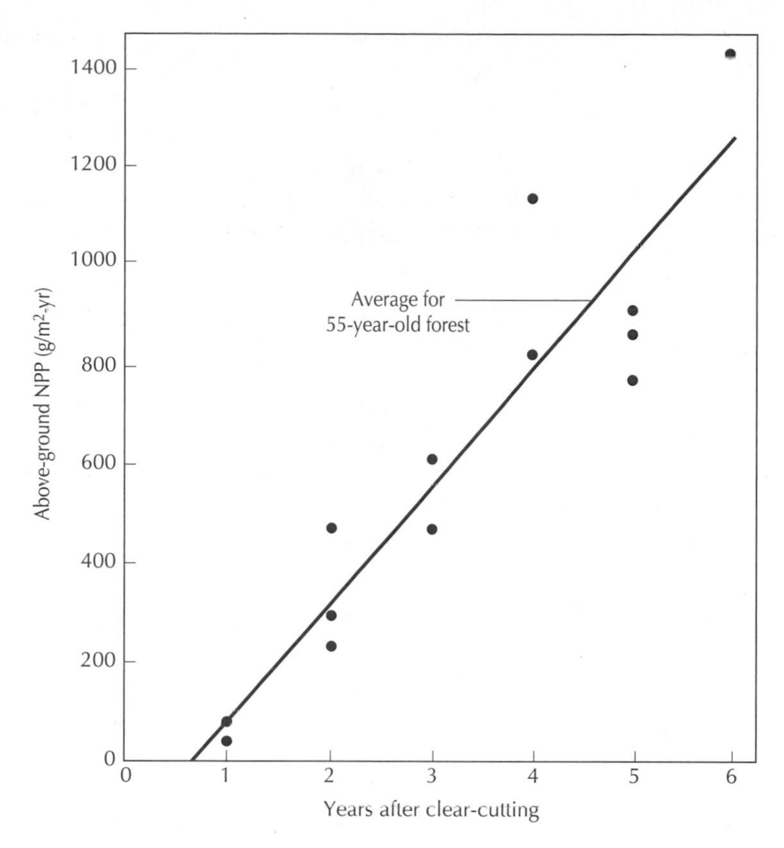

FIGURE 21.10 Changes in total above-ground net primary production (NPP) for the first 6 years following clear-cutting in northern hardwoods. (Bormann and Likens 1979)

rapid increase in productivity that follows disturbance is relatively inefficient activity and is costly in terms of nutrient and biomass storage within the ecosystem. However, the sacrifice of efficiency results in accelerated production, which may be considered an effective strategy of ecosystem stability since it forestalls a still-greater sacrifice in biotic regulation of erosion. (Bormann and Likens 1979)

Pin cherry appears crucial to the northern hardwood ecosystem in terms of minimizing the impact of disturbance and maintaining long-term ecosystem function. Yet its buried seed reproductive strategy does not ensure vigorous regeneration on all sites. If pin cherry trees have a life expectancy of 30 years, and buried seeds can remain viable for 60 years, then this strategy is effective only if disturbance occurs roughly once every 90 years. The disturbance frequency–intensity values in Figure 21.4 suggest that certain areas are likely to remain undisturbed for much longer periods of time. In this case, pin cherry regeneration depends not on seed dropped in place by previous populations, which occurs at very high densities, but rather on seed dispersed into the area by birds, which eat the fleshy fruits and drop the seeds. This mechanism results in many fewer viable seeds per unit area. There is, then, both a high density and low density pin cherry regeneration cycle (Figure 21.11).

Do other species replace pin cherry where it is absent and provide this resilience function? Apparently not. There is a large difference in the rate at which the canopy is recreated in the absence of pin cherry (Figure 21.12). This would suggest that nutrient losses should be greater and more prolonged following disturbance if pin cherry is present only in low densities.

ALTERNATE ENDPOINTS FOR SUCCESSION: SPECIES–SITE INTERACTIONS

The presence or absence of pin cherry in recently disturbed areas creates different pathways of recovery for northern hardwood systems. Do species characteristics affect the end stage of succession as well?

While we have depicted sugar maple as somewhat less tolerant than beech, there are many areas within the northern hardwood region, especially in the western part of the region, and on less acidic soils, where sugar maple is the predominant species. In addition, soils under maples tend to be less acidic than soils under beech or hemlock. This poses another classic ecological question: Are the soils different because of the characteristics of the species that grow there, or are the species there because of inherent differences in soils?

This question can be very difficult to answer in the absence of experimentation, and successional experiments can require decades for completion. Another approach is to relate species distributions not to soil properties that can

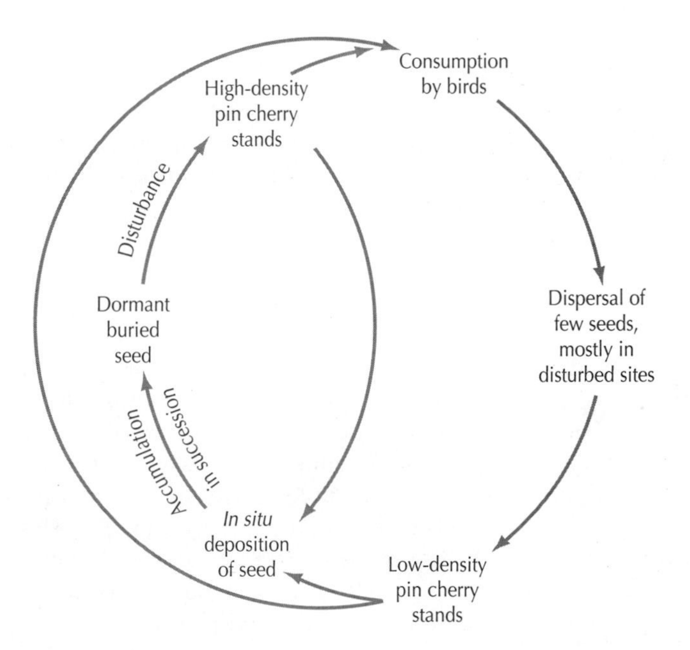

FIGURE 21.11 Representation of the high- and low-density pin cherry cycles active in northern hardwood forests. The low-density cycle depends on dispersal of seed by birds and results in much lower density of pin cherry stems in disturbed areas. (Marks 1974)

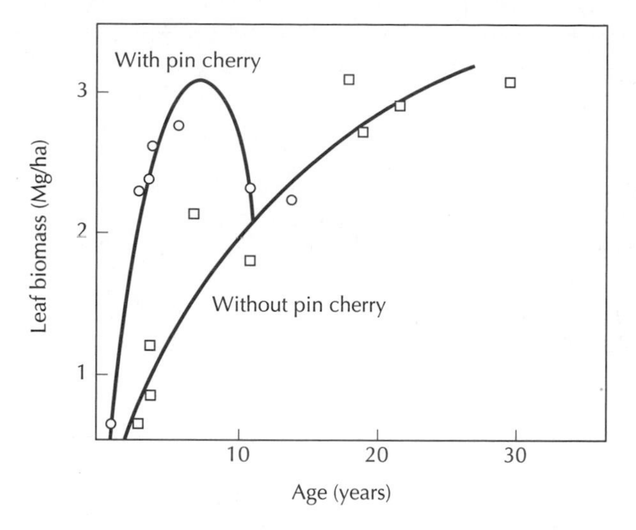

FIGURE 21.12 Differences in successional changes in total foliar biomass in young northern hardwood stands with and without pin cherry. (Covington and Aber 1980)

change over successional time but rather to those that do not. One such study has been carried out in an old-growth forest area in western New England. On this site, there was significant spatial variation in the amount of calcium, phosphorus, and other elements in the bedrock, a characteristic that cannot be altered by plants. There were also differences in the fine-scale distribution of major tree species. Red and sugar maple, as well as white ash, were found preferentially on sites with high calcium and phosphorus substrates, while beech, hemlock, and red oak virtually disappeared (Figure 21.13).

For most of the sites studied, there was also a strong relationship between total mineral calcium and soil exchangeable calcium (Figure 21.14). However, there are a few outliers in this figure with significantly higher exchangeable calcium. Most (but not all) of these are sites dominated by sugar maple and white ash, two cation pump species that apparently magnify the effects of rock calcium content by retaining and recycling this element. The cation pump species have the expected effect of increasing pH, especially in the forest floor and upper mineral soil (Figure 21.15). Assuming that these species actually realize competitive advantages when growing in the types of environments they create (an important assumption), then the two groups of species (cation pumps and noncation pumps) provide positive-feedback mechanisms by which their own abundance should be increased. However, it should be noted that the relationships presented here are relatively weak (see Figure 21.13, for example), suggesting that this is just one of many interactions that actually control species importance and soil characteristics.

Other aspects of ecosystem function can also be affected by species characteristics. The same study cited earlier also showed significant differences between the two groups of species in total net N mineralization and net nitrification (Figure 21.16). This effect has also been seen in a comparison of both net N mineralization and net nitrification rates under stems of sugar maple and beech over a wide N deposition gradient occurring across New

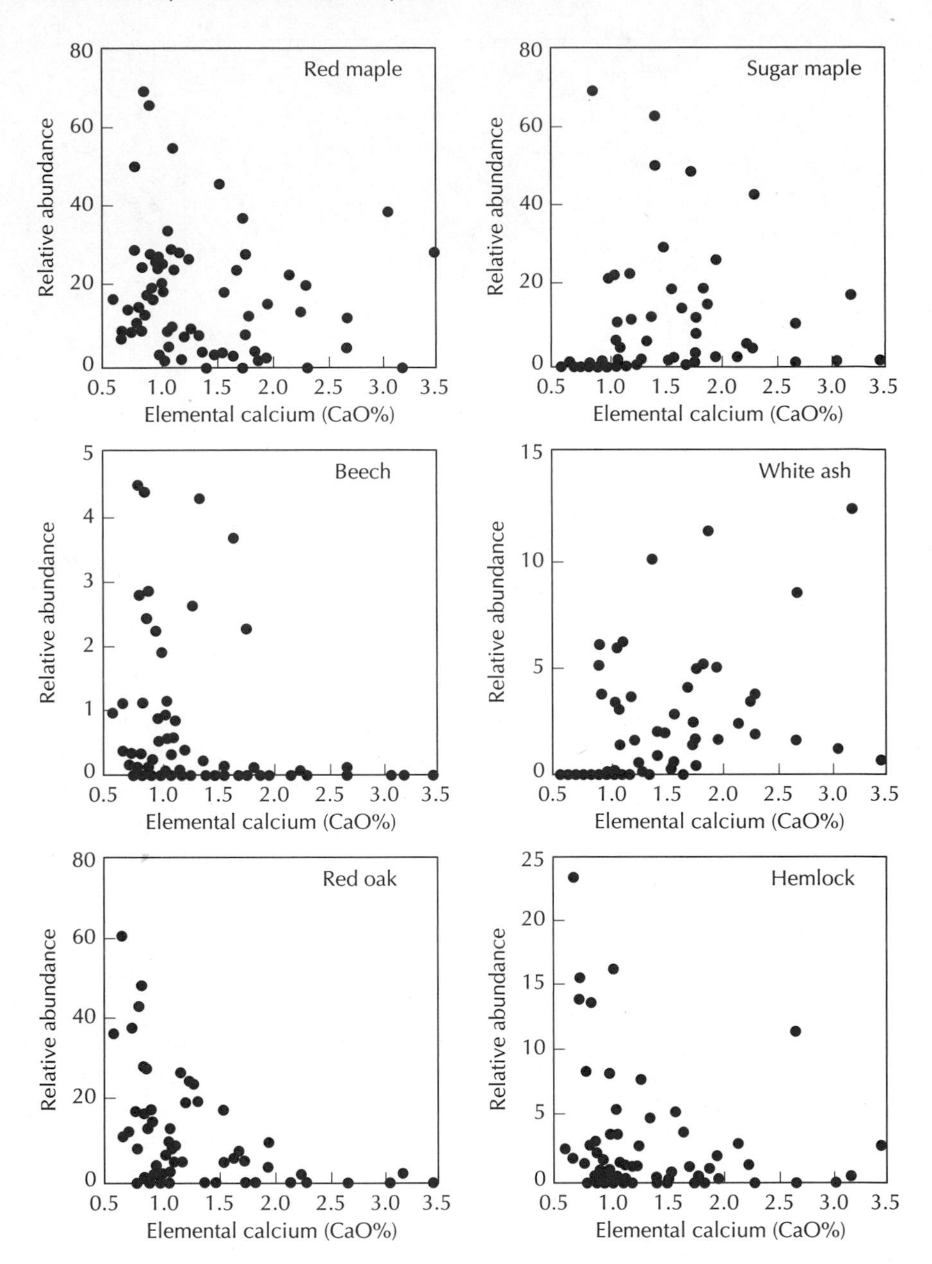

FIGURE 21.13 Relative contribution of different species to total forest basal area as a function of total elemental calcium content in soils in two old-growth forest sites in western New England. (van Breeman et al. 1997)

England (Figure 21.17). In this case, N deposition has had the effect of increasing both net mineralization and net nitrification, as is expected under maple but not under beech.

We have seen (Chapter 9) that plant uptake of N as ammonium or nitrate alters the charge balance over the root and affects mechanisms of uptake for other elements. Uptake of N as the nitrate anion requires increased uptake of cations to achieve charge balance over the root (or reduces the metabolic

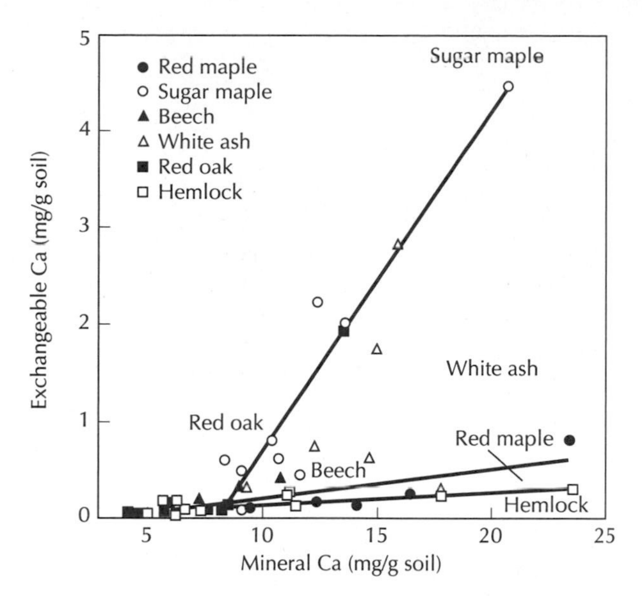

FIGURE 21.14 Relationship between total mineral calcium and total exchangeable calcium under crowns of different species within two old-growth forest sites in western New England. (Finzi et al. 1998a)

cost of cation uptake by allowing it to occur by diffusion along the charge gradient). All of this suggests that the form of N taken up and the associated patterns of cation uptake may also provide a mechanism for niche separation. "Cation pump" species (Chapter 14) may also be selected for acquisition and use of nitrate rather than ammonium. The availability of calcium in soils or

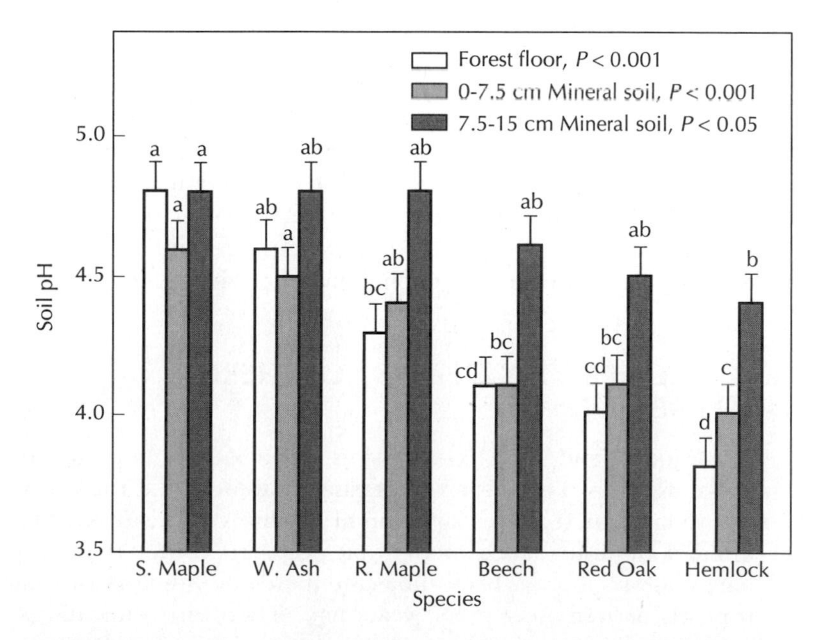

FIGURE 21.15 Soil pH under different species within two old-growth forests in western New England. (Finzi et al. 1998a)

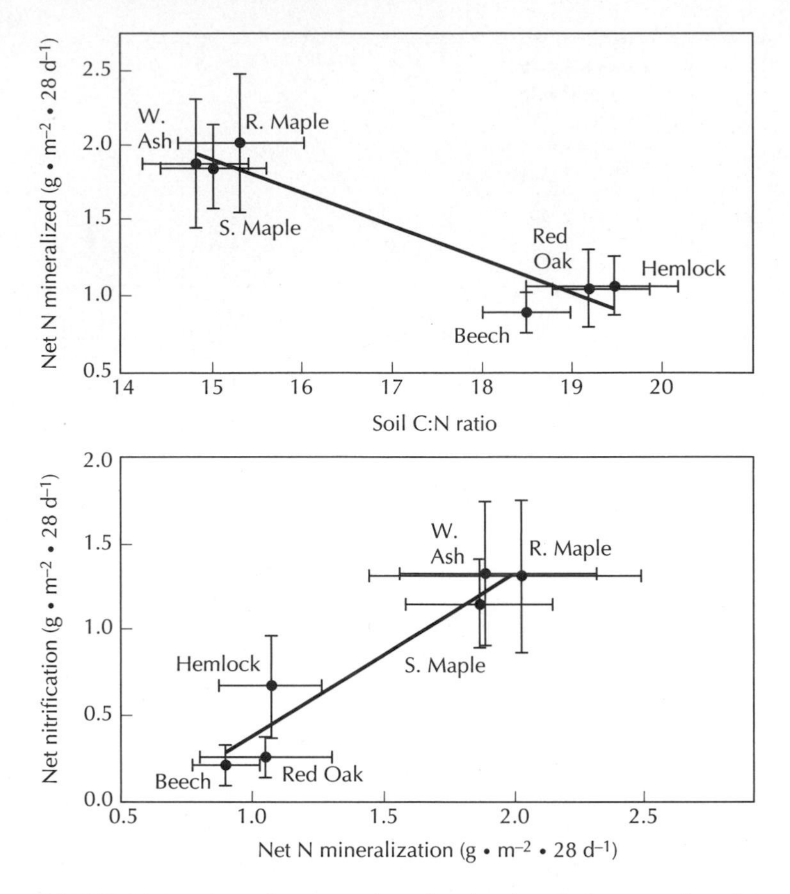

FIGURE 21.16 Net nitrogen mineralization in relation to soil carbon–nitrogen ratio and net nitrification in relation to net mineralization in soils under six different species in two old-growth forests in western New England. (Finzi et al. 1998b)

preexisting soil pH may play a role in determining whether sugar maple or beech dominate the later stages of succession. The effect of the two species on sites where they occur may reinforce those initial differences. Again, the fact that beech and sugar maple co-occur in so many locations suggests that this is a relatively weak or secondary response and that the correlation between site and species is weakened by other factors.

HUMAN USE AND HISTORY OF THE NORTHERN HARDWOODS REGION

Another end state to succession may also have been more important in previous times. There is an increasing awareness that the ecosystems that we see today and that we think of as relatively undisturbed may have been molded by disturbance events that occurred centuries ago. The northern hardwood region has been subject to increased human disturbance since Europeans arrived nearly 400 years ago. There are indications in historical records that both hemlock and spruce were far more important components

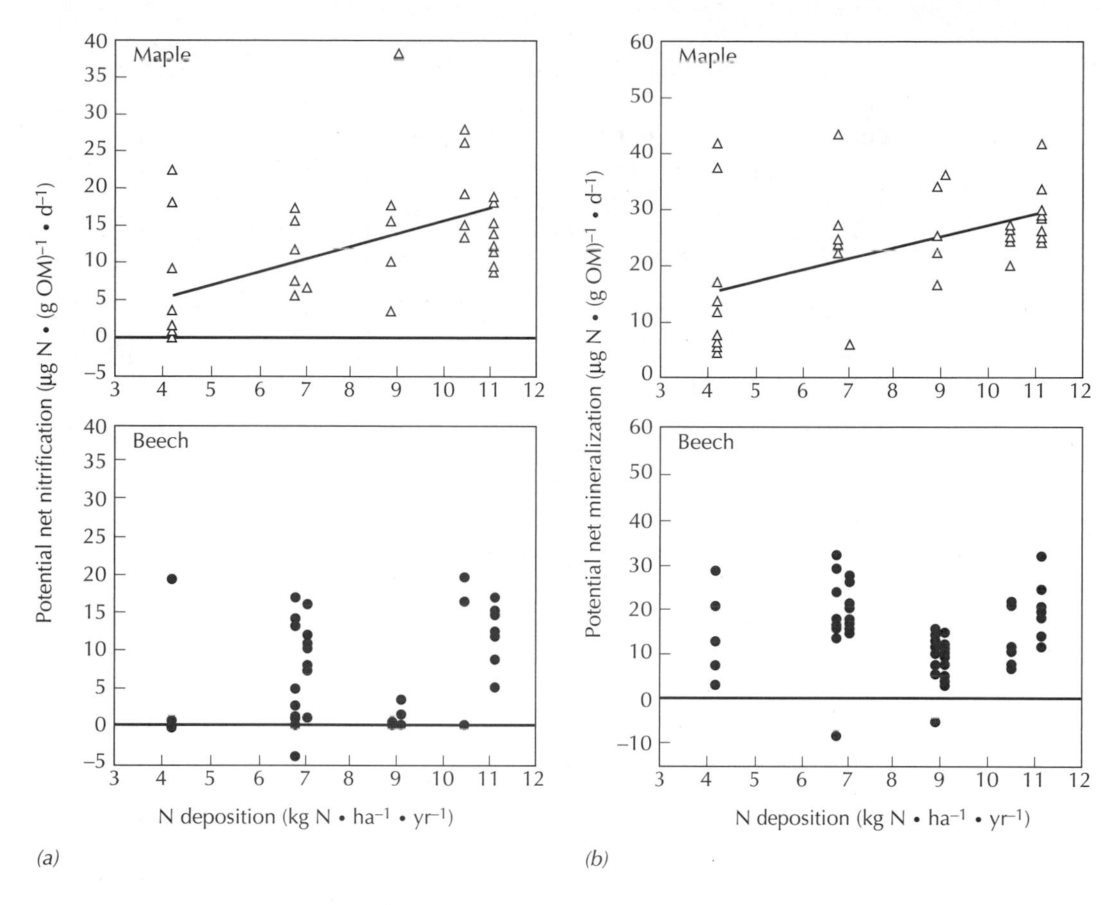

FIGURE 21.17 Changes in net nitrogen cycling along a nitrogen deposition gradient in New England under sugar maple and beech. (*a*) Changes in net nitrogen mineralization. (*b*) Changes in net nitrification. (Lovett and Rueth 1999)

of this forest type 200 years ago than they are today. The first wave of forest harvesting for timber in most of this region was for spruce. In areas in New York State and southern New England, where the manufacture of leather was a major activity, hemlock was cut heavily for its bark, which is rich in the tannins used in the tanning of animal hides.

Both spruce and hemlock are late successional species with extreme evergreen habits (very long needle retention time) and produce small amounts of low-quality litter each year. Both are associated with the generation of thick, very acidic forest floors. Hemlock in particular has been described as creating an understory environment in which almost nothing else can grow. Some of the well-developed podzolic soils to be found under the relatively nutrient-rich hardwood species in this region may well be relics from an era in which hemlock or spruce were more important. Historically, late successional stands may have been dominated by neither beech nor maple but rather by long-lived conifers that reduced rates of nutrient cycling and production.

Historically, then, northern hardwood forests may have shown a pattern of development that was similar to, if less extreme than, the boreal forests to the north, with rapidly growing, early successional species, such as aspen and pin cherry, yielding to midsuccessional sugar maple and beech (with the differences in pH and nitrate cycling occurring between stands), which, in turn, gave way to hemlock and spruce.

One aspect of the nutrient cycling theory of ecosystem development presented in Chapter 18 (Figure 18.6) was that the end stage could be altered by increasing nutrient inputs. The northern hardwood region in the United States is in an area of greatly increased N deposition. Could it be that this has altered, or is altering, the patterns of succession that occurred in this forest type in precolonial times? If the preferential harvesting of hemlock and spruce created a more purely deciduous forest in the eastern part of this region and initiated increases in N cycling, will N deposition maintain both N cycling rates and the deciduous dominance and reduce the importance of spruce and hemlock in later stages?

Today, the northern hardwood forest type is adjacent to some of the most densely populated areas in North America (Figure 21.18). While wood harvesting continues in the national forests and in the relatively small areas owned by wood-products companies, most of the land still in forest is owned by individuals who tend to value the forest more for recreation and aesthetics than for wood production or economic return. Direct human use now involves the conflicting demands of wood production, recreation, and wildland preservation. Hotly contested practices, such as clear-cutting on the national forests, tend to bring groups representing very divergent populations into court to settle differences. The large number of small landowners may be very important to the environmental stability of the region because they inhibit rapid, large-scale shifts in land use.

Perhaps more important at present are the indirect influences of air pollution and climate change.

FIGURE 21.18 Satellite view of eastern North America at night. Bright areas represent urban centers. Note that the northern hardwood forest region (Figure 21.1) lies amid the region with the highest density of urban settlement. (Bormann 1976)

The natural ecosystems of the United States, using solar energy, help to provide the stable and predictable environment on which we all depend. Yet, they are being subjected to increasing stress and destruction as our population and economy grow, and as our profligate use of fossil energy continues. As these natural ecosystems are degraded or destroyed, they lose their capacity to carry on basic biogeochemical functions [which control] climate, hydrology, circulation of nutrients, and the cleansing of air and water. (Bormann 1976)

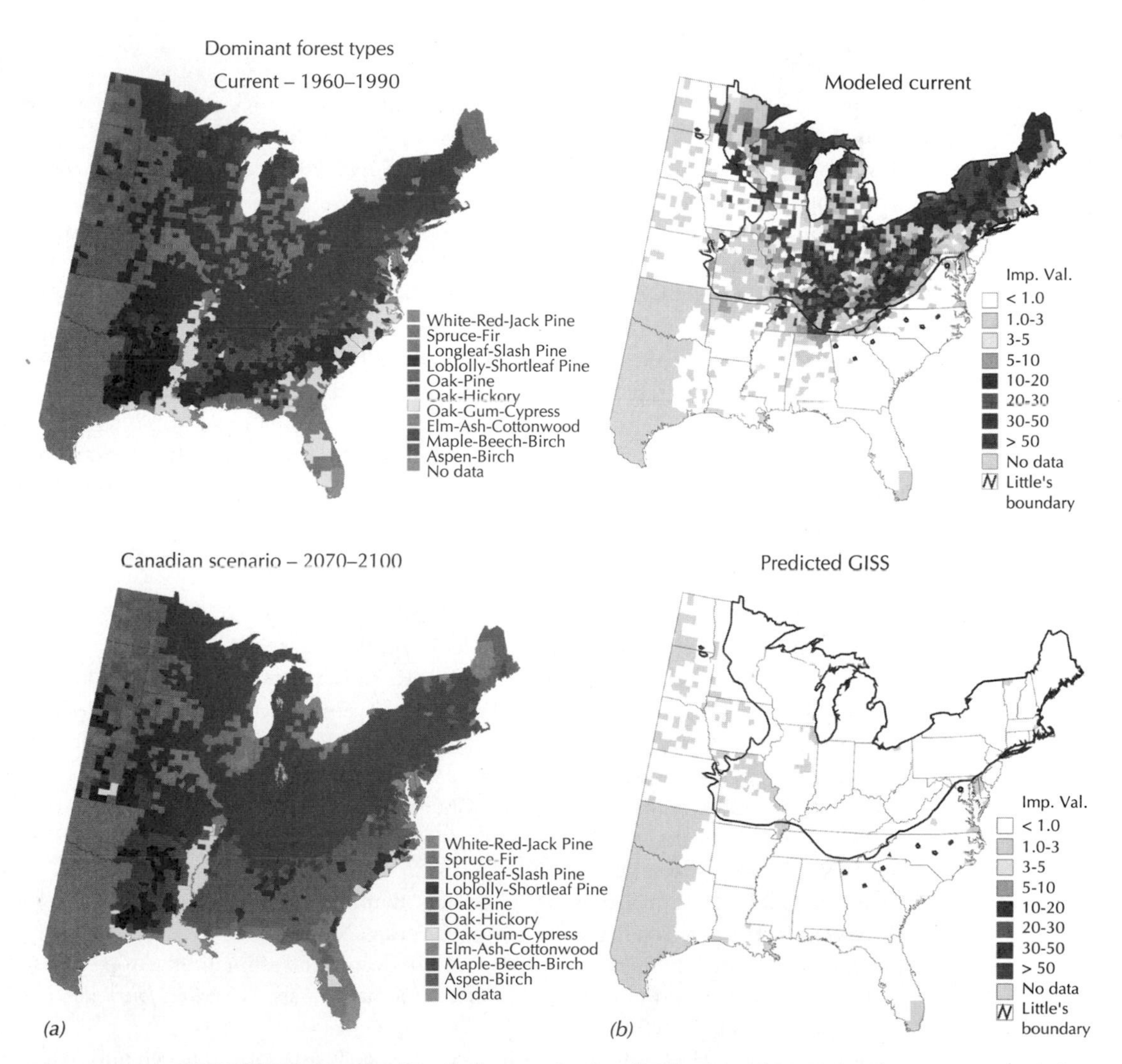

FIGURE 21.19 Predicted changes in the distribution of (*a*) major forest types and (*b*) one dominant species (sugar maple) under different climate change scenarios. (Iverson et al., 1999) **See plate in color section**

Air pollutants, such as acid rain and ozone, can reach into these largely rural areas, presenting a novel and important form of stress to the ecosystem. The integrity of these systems and their ability to perform the services of air and water purification, which we generally take for granted and on which we rely, may be jeopardized (see Chapter 25). Climate change may induce even more fundamental changes. Using current predictions for climate change over the twenty-first century as inputs to models predicting the distribution of major forest species and forest types suggests that the northern hardwood forest, which occupies a relatively small region, will be shifted to the north, moving completely out of the United States and into Canada (Figure 21.19). The physical area we now study as northern hardwood forest may become dominated by a completely different set of species.

REFERENCES CITED

Bormann, F. H. 1976. An inseparable linkage: Conservation of natural ecosystems and the conservation of fossil energy. *BioScience* 26:754–760.

Bormann, F. H., and G. E. Likens. 1979. *Pattern and Process in a Forested Ecosystem.* Springer-Verlag, New York.

Covington, W. W. 1981. Changes in forest floor organic matter and nutrient content following clear cutting in northern hardwoods. *Ecology* 62:41–48.

Covington, W. W., and J. D. Aber. 1980. Leaf production during secondary succession in northern hardwoods. *Ecology* 61:200–204.

Finzi, A. C., C. D. Canham, and N. van Breeman. 1998a. Canopy tree–soil interactions within temperate forests: Species effects on pH and cations. *Ecological Applications* 8:447–454.

Finzi, A. C., N. van Breeman, and C. D. Canham. 1998b. Canopy tree–soil interactions within temperate forests: Species effects on soil carbon and nitrogen. *Ecological Applications* 8:440–446.

Forcier, L. K. 1975. Reproductive strategies and the co-occurrence of climax tree species. *Science* 189:808–810.

Fowells, H. A. (ed.). Silvics of forest trees of the United States. In *Agricultural Handbook,* No. 271. U. S. Department of Agriculture, Washington, DC.

Iverson, L. R., A. M. Prasad, B. J. Hale, and E. K. Sutherland. 1999. Atlas of Current and Potential Future Distributions of Common Trees of the Eastern United States. Northeastern Research Station General Technical Report NE-265. U. S. Department of Agriculture, U.S. Forest Service, Washington, DC.

Likens, G. E., and F. H. Bormann. 1970. Chemical analyses of plant tissues from the Hubbard Brook ecosystem in New Hampshire. *School of Forestry Bulletin* No. 79. Yale University, New Haven, CT.

Likens, G. E., F. H. Bormann, R. S. Pierce, J. S. Eaton, and N. M. Johnson. 1977. *Biogeochemistry of a Forested Ecosystem.* Springer-Verlag, New York.

Lorimer, C. G., and L. E. Frelich. 1989. A methodology for estimating canopy disturbance frequency and intensity in dense temperate forests. *Canadian Journal of Forest Research* 19:651–663.

Lovett, G. M., and H. Rueth. 1999. Soil nitrogen transformations in beech and maple stands along a nitrogen deposition gradient. *Ecological Applications* 9:1330–1344.

Marks, P. L. 1974. The role of pin cherry (*Prunus pennsylvanica L.*) in the maintenance of stability in northern hardwood ecosystems. *Ecological Monographs* 44:73–88.

Melillo, J. M., J. D. Aber, and J. F. Muratore. 1982. Nitrogen and lignin control of hardwood leaf litter decomposition dynamics. *Ecology* 63:621–626.

National Assessment Synthesis Team. 2000. *Climate Change Impacts on the United States: The Potential Consequences of Climate Variability and Change.* U. S. Global Change Research Program, Washington, DC.

U. S. Department of Agriculture. 1974. Seeds of woody plants of the United States. In *Agricultural Handbook,* No. 450. U. S. Department of Agriculture, Washington, DC.

van Breeman, N., A. C. Finzi, and C. D. Canham. 1997. Canopy tree–soil interactions within temperate forests: Effects of soil elemental composition and texture on species distributions. *Canadian Journal of Forest Research* 27:1110–1116.

22

ECOSYSTEM DEVELOPMENT OVER GEOLOGIC TIME

THE TROPICAL FORESTS OF HAWAII

INTRODUCTION

In the last three chapters, we have presented the dynamic aspects of very different ecosystems and how those dynamics are shaped by ecological processes and disturbance regimes. The successional changes presented varied on time scales of decades to centuries, the return time of major disturbance agents, or the lifetime of 1–3 generations of dominant plants. Ecosystems also exist within a matrix of factors that change over much longer time scales. The processes of weathering and soil development discussed in Chapters 2 and 9 imply very slow changes that produce cumulative differences only at the scales of thousands to millions of years. The advance and retreat of glaciers during our own geological era drive a cycle of soil development and species migrations with return times of hundreds of thousands of years. The relatively young soils in both the northern hardwood and taiga systems described in earlier chapters are the product or slow and continuing weathering and development processes acting since the last glacial retreat just 12,000–15,000 years ago.

You can imagine the difficulty inherent in trying to study and measure changes in ecosystem development over several million years. In general, studies of ecosystem processes over long periods of time are accomplished with "space-for-time" substitution by examining several different sites of different ages since the last disturbance of interest. The problem with this approach is the difficulty inherent in locating a set of sites that are identical in every respect except for age. Intervening disturbances (including human use of the land), differences in species composition, and differences in geology and soils can all invalidate the comparison. Where, then, could we hope to find a sequence of sites that would allow a relatively clear view of processes operating over millions of years?

THE HAWAIIAN ISLANDS

This island chain in the central Pacific (Figure 22.1*a*) offers a truly unique set of conditions for examining changes in ecosystem development over geologic time. The islands are formed by the movement of the ocean floor (part of

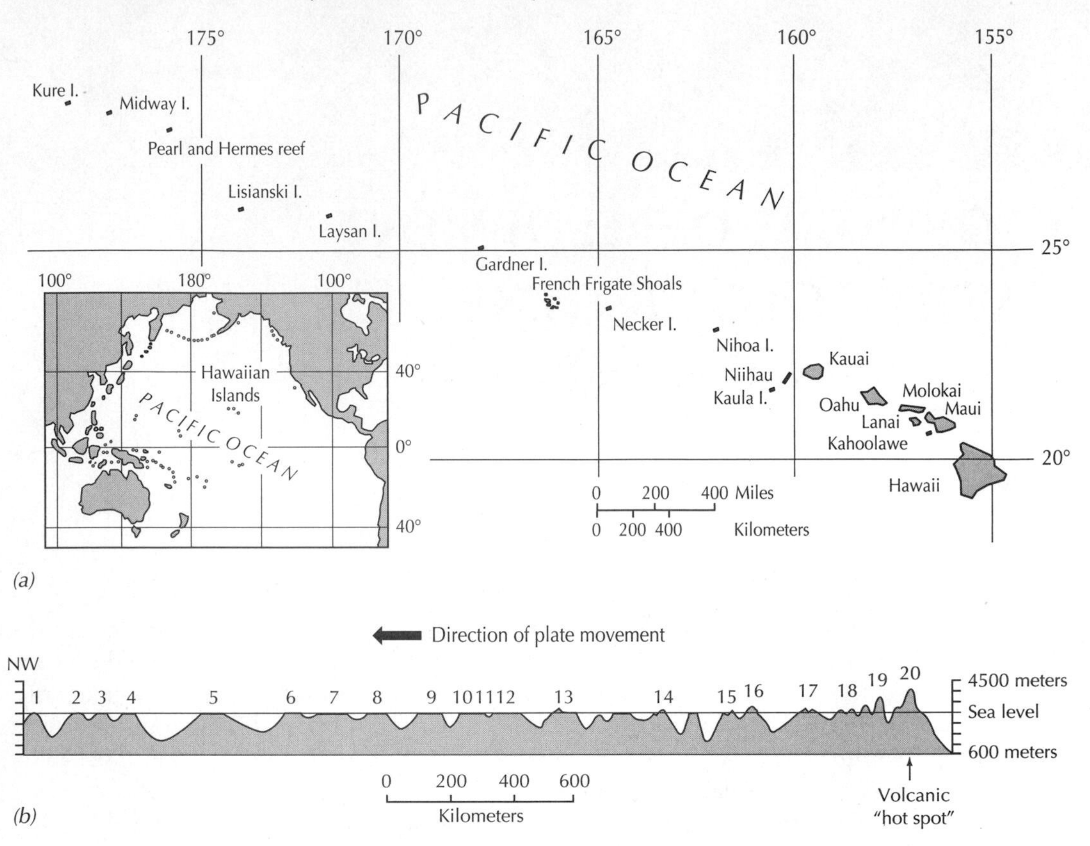

FIGURE 22.1 Geography of the Hawaiian archipelago. (*a*) Distribution of islands, reefs, and atolls. (*b*) Elevation showing pattern of formation over a stationary volcanic "hot spot" and erosion during movement away from lava source over millions of years. (MacDonald et al. 1983)

plate tectonics that drives the slow migration of continents over the Earth's surface) over a constant, deep source of fresh volcanic lava (Figure 22.1*b*). This volcanic "hot spot" builds first an underwater volcano and then eventually a volcanic island over a period of about 600,000 years. As the island is carried off to the northwest, a new volcano and eventually a new island is formed. Without fresh lava flows, the moving islands age, weather, and erode with the passage of time. Distance from the hot spot equates then to geological age, and 4–5 million years is required to erode each island down to a small residual reef or atoll maintained near sea level by reef building organisms. In the map in Figure 22.1*a,* "islands" numbered 1–15 represent very small residual islands, shoals, and reefs.

The larger Hawaiian islands, then, are a series of land masses that increase in age from southeast to northwest, forming a clearly defined age sequence covering over 4 million years. The constancy of the lava source over millions of years causes the initial minerals from which soils develop to be relatively constant as well, although differences in rate of cooling and in the admixture of ash with lava cause differences in substrate that must be controlled in establishing a sequence of sites.

Other potential sources of noise in an intersite comparison are also controlled by the location and history of the islands. Their small size produces only local orographic effects (increases in rainfall due to lifting of air masses over tall mountains). Otherwise, the climate is dominated by oceanic influences and is nearly identical between islands. Migration of species to the islands over evolutionary time has been very limited because of the distance to either the Asian or American continents such that the number of dominant species present is much lower than in mainland tropical forests. In fact, stands in which a single species (*Metrosideros polymorpha*) contributes over 80% of the total basal area of trees can be found across the entire sequence of islands. Human migration to the islands has also been limited (until recent times), and this, along with problems of accessibility to remote parts of the islands due to extreme topography, has limited the impact of human intervention over much of the island chain.

The Hawaiian islands, then, provide a very wide range in geological age with relatively constant climate, vegetation, and initial geological substrate. By controlling for type of lava flow and topographic position and avoiding areas with human impact, a series of study sites can be identified for which the long, slow processes of weathering and soil formation are the major source of variation. The result is a more than 4-million-year time series (Figure 22.2) for the study of these processes.

SOIL DEVELOPMENT, SOIL CHEMISTRY, AND NUTRIENT AVAILABILITY

As we saw in Chapter 9, geological substrates formed from the cooling of lava at the Earth's surface are generally rich in nutrient elements and weather rapidly. Under the humid, tropical conditions found in Hawaii, laterization is the major soil-forming process, with the preferential dissolution and leaching

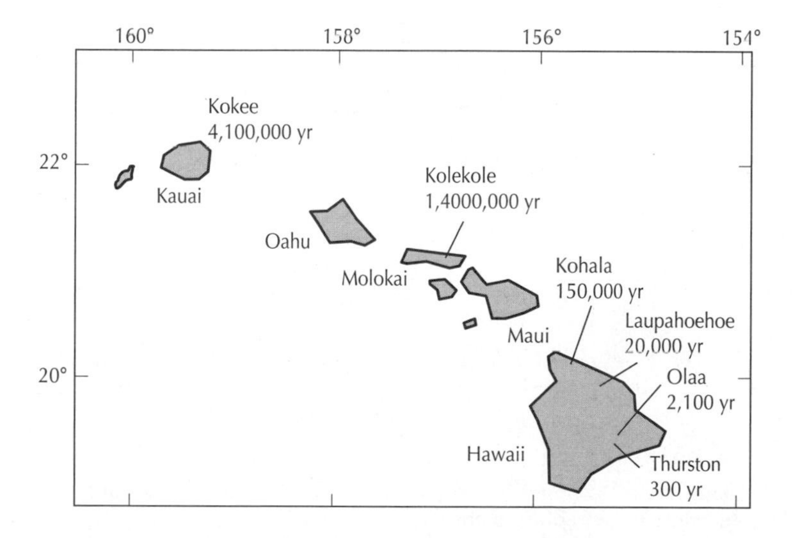

FIGURE 22.2 Distribution and age of one set of study sites used in the research described in this chapter. (Crews et al. 1995)

of silicon and nutrient cations, an accumulation of iron and aluminum (Al) in the A horizon, and no substantial B horizon formed. As an indication of the completeness of the weathering process in the oldest sites, as much as 10 m of original lava depth can be reduced to less than 1 m of residual soil across the 4-million-year time sequence, in the absence of any erosional losses.

With age as the only major variable across the island sequence, the relative rates of weathering and removal of each element can be calculated by comparing the quantity of that element in the residual soil, after correcting for the total mass of soil lost, with that in the freshly cooled lava. Figure 22.3 shows these relative changes in content for five elements and the calculated rates of weathering for phosphorus (P) and calcium (Ca). In keeping with our description of *laterization*, Ca, magnesium, and silicon are all released at the same rate, and none are reprecipitated or retained in the system in solid-phase mineral form.

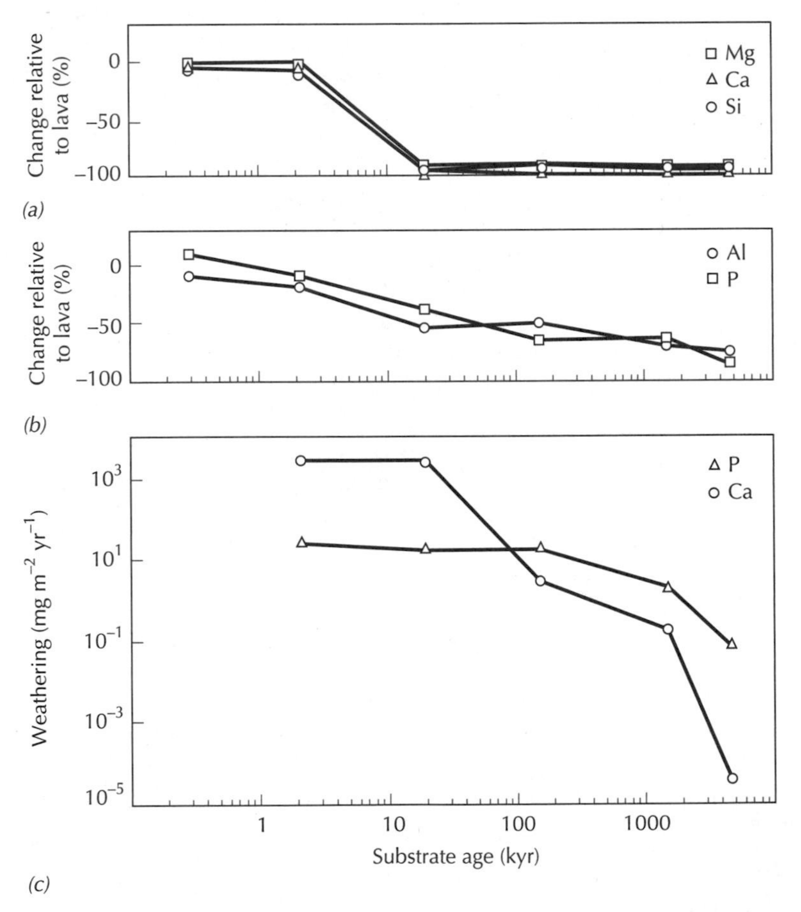

FIGURE 22.3 Changes in the content of five elements in volcanic substrates and estimated rates of phosphorus and calcium weathering rates during 4 million years of soil development in the Hawaiian islands (note log scale for *x* axis). Si = silicon. (Chadwick et al. 1999)

In contrast, Al content declines more slowly, being retained in the system through the formation of secondary minerals. Total mineral P content also declines slowly, in concert with Al. Recall that the phosphate ion reacts with iron and Al oxides and that this process both retains P in the system and reduces its availability to plants. Total release rates (Figure 22.3*c*, *bottom panel*) decline much more rapidly for Ca than for P (note the logarithmic scale on this figure).

In Chapter 9 we described a long-term sequence for the change in the fraction of total soil P in different pools (Figure 9.14), with more easily weathered mineral forms declining over time while "occluded" or relatively unavailable forms increased. Results from the Hawaiian gradient confirm this pattern (Figure 22.4). Organic P also increases initially and declines later on over this sequence of sites.

Nitrogen (N) is not an important element in the freshly formed lava, so N content and availability are very low in the youngest substrates. In contrast to the other elements described so far, N accumulates in developing soils over time through the slow accumulation due to atmospheric deposition and free-living N-fixing species. There are no native forest species in Hawaii that support the actinorrhizal symbiotic N-fixation mechanism (Chapter 10) common to forest environments (but see discussion at the end of this chapter), so rates of N fixation are very low (about 0.1 $g \cdot m^{-2} \cdot year^{-1}$). In comparison, rainfall currently provides about 0.5 $g \cdot m^{-2} \cdot year^{-1}$.

Changes in plant available forms of nutrients follows trends in weathering rates and soil development (Figure 22.5). Initially, availability of cations is high due to rapid release through weathering, while N availability is low. As

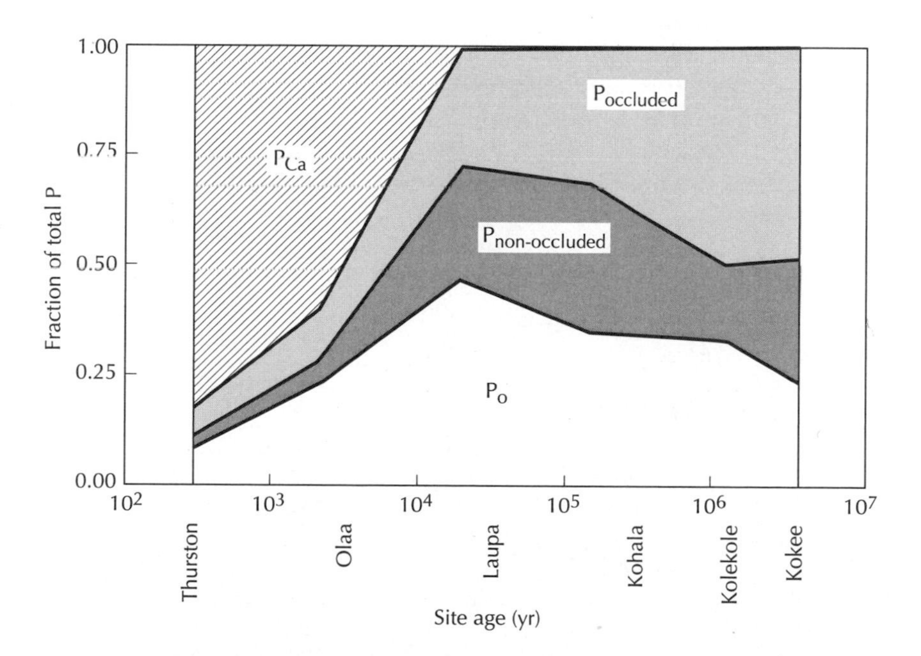

FIGURE 22.4 Changes in the relative concentration of different forms of phosphorus across the Hawaiian age gradient. (Crews et al. 1995)

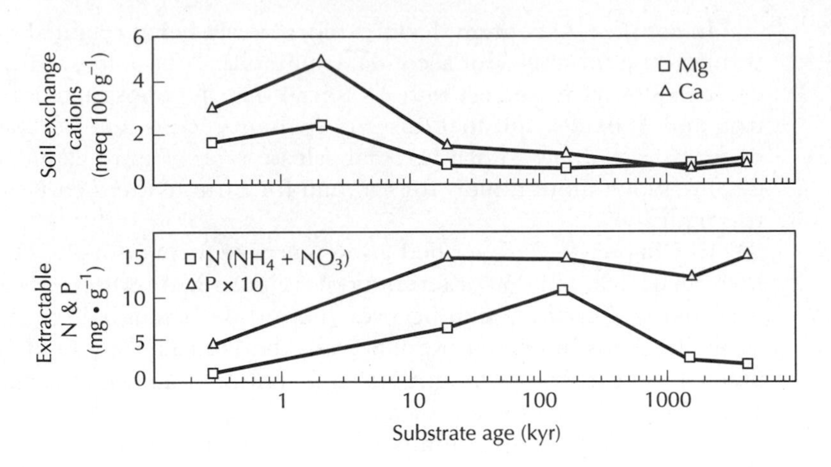

FIGURE 22.5 Changes in exchangeable cations and extractable nitrogen and phosphorus along the Hawaiian age gradient. (Chadwick et al. 1999)

cation exchange capacity increases with the accumulation of soil organic matter, total exchangeable cation content also rises over the first 1000 years. As N accumulates, both total soil N and readily available extractable forms (ammonium and nitrate) also increase. Extractable P increases through the first 100,000 years, but as iron- and Al-rich clays accumulate, extractable P declines.

Foliar nutrient contents have often been used as indicators of the relative degree of limitations on plant function by different elements. Foliar chemistry along the Hawaiian gradient (Figure 22.6) mimics the patterns for availability described earlier. Cation contents are very high initially and decline rapidly through the first 10,000 years. Nitrogen and P contents increase synchronously over this same 10,000-year period and then decline through the

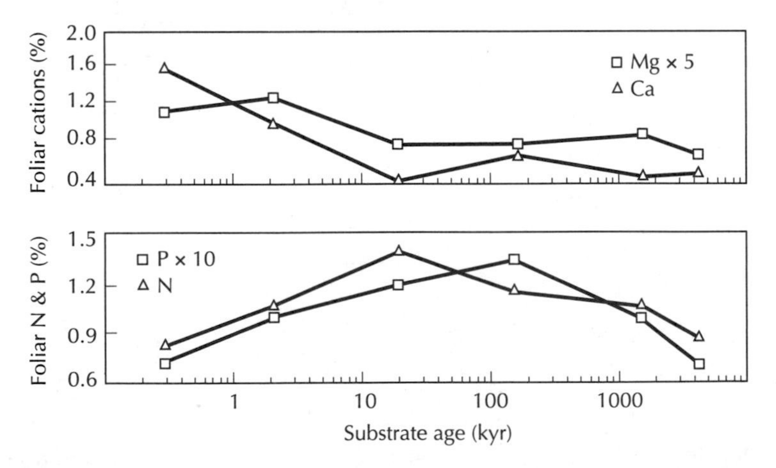

FIGURE 22.6 Changes in foliar nutrient concentrations across the time sequence. (Chadwick et al. 1999)

oldest stands. The lowest N and P concentrations along the Hawaiian gradient are among the lowest seen in any tropical forests.

Measured rates of net N mineralization and net nitrification provide another indicator of the degree of N limitation on plant growth. In agreement with other measures, net N mineralization increases with substrate age (Figure 22.7), suggesting increasing availability. More interesting is the delayed increase in net nitrification. Nitrification is controlled to some extent by plant demand for ammonium and so tends to be inhibited in N-limited sites (Chapter 14). Temporal trends in net nitrification reinforce the foliar nutrient values, suggesting that N availability peaks in the middle of the age sequence and is lower in both the early and late stages.

While foliar nutrient content and soil mineralization rates can suggest relative levels of nutrient availability over time, they cannot, by themselves, be used to determine which nutrient is most limiting to plant function. The direct method for assessing nutrient constraints is through fertilization trials, and such experiments have been performed at the youngest and oldest ends of the age sequence. In none of the sites was there a response to cation additions. Rather, additions of N and/or P were found to be the only fertilization treatments that elicited growth increases. In the youngest sites, N tended to

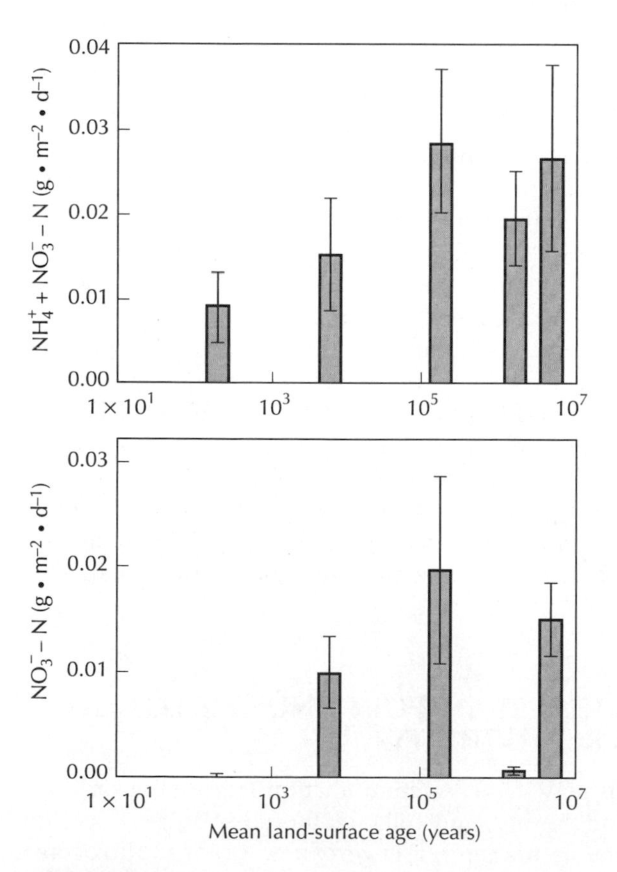

FIGURE 22.7 Changes in net nitrogen mineralization and net nitrification with time along the Hawaiian age gradient. (Riley and Vitousek 1995)

TABLE 22.1 Change in Mean Tree Height Across the Hawaiian Age Sequence
(Crews et al. 1995)

Age (years)	Maximum Canopy Height (m)
300	16.5
2,100	20.0
20,000	24.7
150,000	11.6
1,400,000	8.2
4,100,000	13.7

be the most limiting element, although colimitation by P was detected in the most slowly weathering lavas. Phosphorus alone limited growth in the oldest sites. Data on mean tree height across the gradient (Table 22.1) suggests further that total productivity peaks between 10,000 and 100,000 years and declines thereafter, such that limitations by N and P are strongest in the youngest and oldest stands.

FEEDBACKS BETWEEN PLANT LIMITATIONS AND NUTRIENT CYCLING

In Chapter 18, we presented a generalized theory of physiological responses to site quality (Figure 18.7) that proposed a set of positive feedbacks accentuating either the fertility or sterility of sites. A series of sites with widely varying nutrient availability all dominated by a single species would seem to be a perfect place to test this theory. Is there evidence of plant responses that would reinforce relative site conditions?

For sites at five different ages, there is a definite pattern in the quality of leaf litter and resulting decomposition rates that is in keeping with the theory in Chapter 18 (Table 22.2). Concentrations of both N and P in leaf litter follow the same pattern as in green foliage (Figure 22.6), peaking in the midpoint of the gradient. Carbon fractions in litter also change such that lignin–N ratios are lowest at the midpoint and higher in stands on both younger and older substrates. The changes in quality affect litter decay rates as expected, with decomposition being slowest in the youngest and oldest sites.

LONG-RANGE NUTRIENT TRANSPORT AND THE LONG-TERM MAINTENANCE OF PRODUCTIVITY

These results may suggest a contradiction to you. Cations are removed most rapidly by weathering, and fully weathered soils, such as the oldest soils in this sequence, generate nearly zero new cations. Nitrogen and P, on the other hand, accumulate over time, although P is increasingly locked in un-

TABLE 22.2 Data on Litter Quality and Decomposition Rate (k) for Foliar Litter from Five Sites of Different Geological Ages
(Crews et al. 1995)

Age (years)	% N	% P	Lignin–N Ratio	k
140	0.36	0.026	88	—
300	0.40	0.026	65	0.3
20,000	0.80	0.053	45	0.8
150,000	0.74	0.054	34	1.3
4,100,000	0.37	0.022	99	0.2

available forms. Why, then, does NPP not become cation limited, especially in the intermediate and oldest soils? At the end of the sequence, why does P not cease cycling entirely and limit NPP even more? The solution to this contradiction lies not in the landscape itself but in the surrounding ocean and, for P, on continents thousands of kilometers away. Documenting this solution involves the use of both direct measurements of inputs and isotope ratios within the systems for elements with chemical activities similar to nutrient cations.

Evaporation of foam and spray over the ocean generates atmospheric aerosols that contain Ca, magnesium, and potassium. Winds bring these aerosols to the islands, where they can be deposited either in rainfall or as dry deposition to plant and soil surfaces. While weathering inputs of Ca to Hawaiian ecosystems declines across the age sequence, atmospheric inputs (rainfall and dry deposition) are relatively constant over time. Because of this, the fractional input of these elements from deposition and weathering changes significantly across the 4-million-year sequence (Figure 22.8). While forests on young substrates experience excess cation availability through high weathering rates, those on the oldest sites rely on continuous inputs from atmospheric deposition and closed cycling within the systems to meet cation nutritional needs.

There is a possible flaw in this reasoning. Cations deposited from the atmosphere might not come from the ocean but from dust generated on the islands themselves. If this is the case, then deposition does not represent a net input to the island ecosystems but rather is a redistribution of existing nutrients already present, but perhaps transferred from drier to wetter systems. This idea was tested using the element strontium (Sr) as an analog for Ca. Previous research had demonstrated that the biological and chemical processes that control the cycling of Ca do not distinguish between this element and Sr. Thus, small amounts of the much less common Sr are carried along with, and show the same relative rates of cycling as, Ca. Strontium also occurs in two stable isotopes (^{86}Sr and ^{87}Sr), and the ratio of these two forms is different in seawater (0.7092) and in the lava from Hawaiian volcanoes (0.7035). By measuring the ratio of these two isotopes in soils and foliage from forests across the age sequence (Figure 22.9*a*), it is possible to deter-

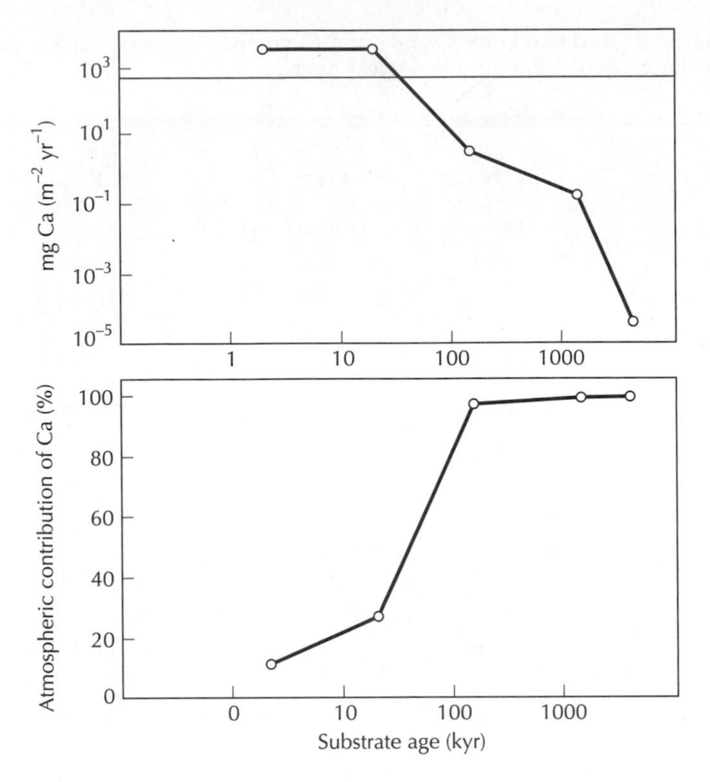

FIGURE 22.8 Estimated absolute and fractional input of calcium to soils from weathering of minerals and from atmospheric deposition. The straight line is atmospheric deposition (assumed to have been constant across all sites). The other line captures the changes in weathering rate over time. (Chadwick et al. 1999)

mine the fraction of Sr that originated in lava and that which entered the system as sea salt (Figure 22.9*b*). The results verify that there is a major shift in the source of Sr, and presumably of Ca, from lava to sea salt across the age sequence.

Marine aerosols, however, do not contribute substantially to the availability of P in older systems. The uptake of P by open ocean plankton is high relative to supply, such that P content in seawater, and hence in foam and sea spray, is very low. Is there another source? One of the more interesting connections between ecosystems separated by very long distances is through the transportation of dust. In particular, dust storms in desert regions inject particles into the atmosphere that may be transported thousands of kilometers. As minerals weather slowly and incompletely in the dry desert environment, minerals in the dust may be highly weatherable under humid conditions.

Once again, the distinct chemistry of the lava flows that form the Hawaiian islands allow the use of isotopes of rare elements to estimate the input of P by dust deposition. By comparing the isotope ratios of several elements, total external dust inputs were determined. Multiplying this value by an average concentration of P in dust provides an estimate of the total delivery of P to the islands by this mechanism. As with Ca, the fractional input of P to the islands' forest systems from the atmosphere increases with time (Figure 22.10).

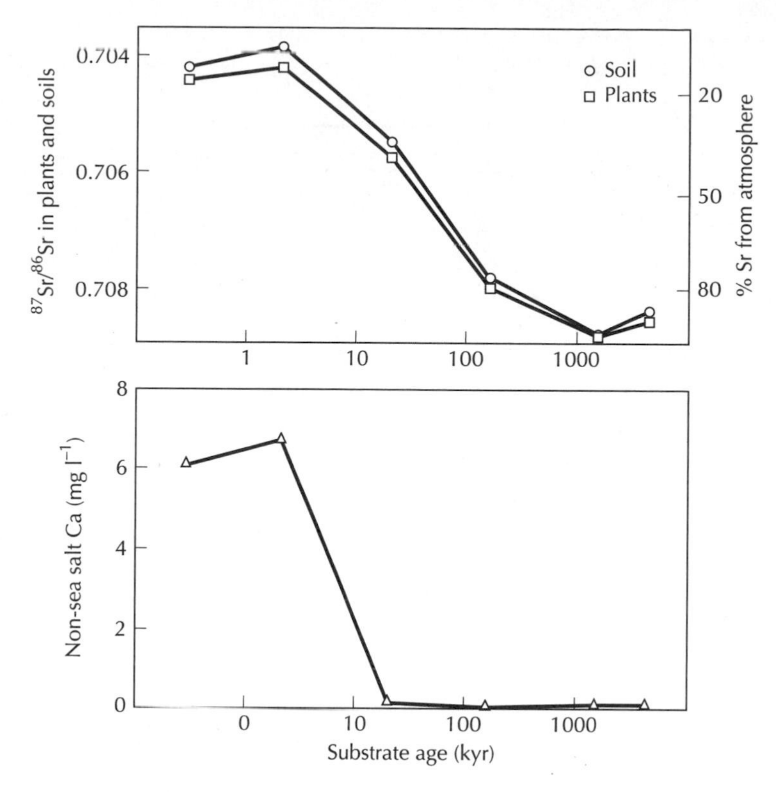

FIGURE 22.9 Ratio of strontium-87 (^{87}Sr) to strontium-86 (^{86}Sr) in plants and soils across the Hawaiian age gradient. The ^{87}Sr–^{86}Sr ratio is 0.7035 in fresh lava and 0.7092 in seawater. Data in the top panel are used in comparison with these "endpoint" ratios to determine the amount of strontium, and therefore calcium, derived from lava and seawater. (Chadwick et al. 1999)

This description of controls on nutrient cycles in the Hawaiian islands is a classic example of our increasing understanding of the linkages between very different types of ecosystems, even those that are separated by long distances. It is curious to think that P limits tree growth on the oldest islands in Hawaii in part because the plankton in the surrounding ocean are P limited rather than cation limited, such that the aerosols generated by sea spray are also depleted in P. This, in turn, leads to an apparent dependence of forest production in the oldest islands on the rate at which P is blown in with dust from Asia, which, in effect, adds new weatherable minerals to the fully weathered soils. A similar role for dust deposition has provided a link between the Sahara Desert and the rain forests of the Amazon.

HUMAN INFLUENCES AND CHANGES IN ECOSYSTEM FUNCTION

A dominant characteristic of the Hawaiian islands over geological and evolutionary time has been their isolation from the species and environments of the Earth's major continents. This has determined the limited flora and fauna of the islands and has minimized human intervention.

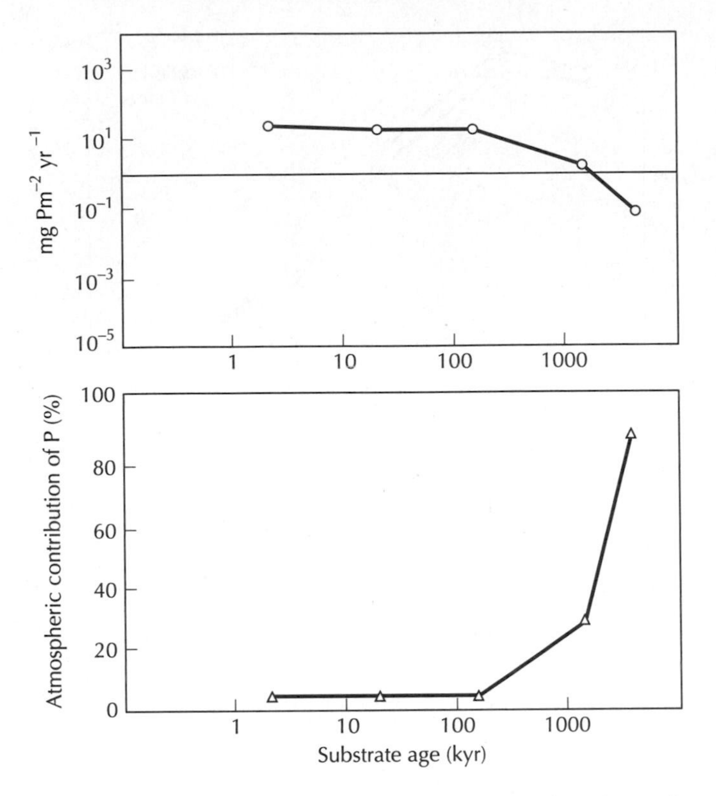

FIGURE 22.10 Absolute and relative inputs of phosphorus from weathering and atmospheric deposition across the Hawaiian age sequence. The straight line is atmospheric deposition that is assumed to have been constant over the past 4 million years. The other line captures the changes in weathering rate over time. (Chadwick et al. 1999)

Absolute isolation is no longer characteristic of any part of our world. The islands do remain separated from most of the localized forms of air pollution indigenous to densely populated or highly industrialized regions, and the importance of marine influences and the tropical location may minimize the influences of climate change. However, humans now travel freely, and Hawaii has become a major tourist destination. Land use will be a continuing source of conflict, and preservation of native ecosystems both for intrinsic values and as a resource for education and renewal of visitors will remain an important cause.

The combination of evolutionary isolation and recent heavy human use may pose one of the most serious problems for these islands. We mentioned earlier that migration by species to the islands has been limited and that species diversity is low as a result. This is a common characteristic of isolated regions and often means that the indigenous species do not exploit resources as efficiently or divide niches as finely as those from other, larger landmasses. This creates a strong potential for the displacement and extinction of local species through migration of more aggressive exotics from distant lands. In addition to causing extinctions, some species may cause significant shifts in ecosystem function, leading to further extinctions of native species.

TABLE 22.3 Total Inputs of Nitrogen to Very Recent Volcanic Sites With and Without the Exotic, Nitrogen-Fixing Shrub *Myrica faya*. All Values in kg · ha^{-1} · year^{-1}.
(Vitousek et al. 1987)

Source	Without *Myrica*	With *Myrica*
Rainfall	5.0	5.0
Nonsymbiotic fixation	0.5	0.5
Myrica	0.0	18.0
Total	5.5	23.5

We will return to this topic in Chapter 24, but one Hawaiian example will make the point here. In the nineteenth century, *Myrica faya*, an exotic, low-growing, N-fixing shrub species native to the Azores and Canary Islands, was introduced into Hawaii. It first appeared in Hawaii Volcanoes National Park in 1961. By 1985, it covered 12,200 hectares and was still spreading rapidly. *Myrica* aggressively invades pastures, native forests thinned by lava flows, and open-canopied forests, but also grows on fresh cinder cones and even in mature forests. Before *Myrica* arrived, there were no symbiotic N-fixing species in Hawaii present in the youngest geological substrates that dominate the national park.

As with other N-fixing species, *Myrica* alters the N economy in sites where it grows. In one study, N fixation greatly exceeded deposition rate and increased total N inputs into the system by more than fourfold (Table 22.3). In response, extractable ammonium and nitrate increased from 0.7 μg N · g^{-1} soil under a native species to 2.6 μg N · g^{-1} soil under *Myrica*. Net N mineralization was undetectable under the native species but averaged 5.6 μg N · g^{-1} soil · month^{-1} under *Myrica*.

Beyond displacing native species and altering soil conditions, *Myrica* might also facilitate invasions by other exotics. The Hawaiian flora inhabiting the youngest substrates on the most recent islands are adapted to living under conditions of severe N limitations. Most exotics are not. By altering the rate at which N accumulates on young sites, *Myrica* can prepare the ground for a new community of species that will displace the preexisting one. Soil development might occur more rapidly, and NPP might increase with N limitations modified, but the community carrying out this development will not be the same as the one that has been in place for millennia.

REFERENCES CITED

Chadwick, O. A., L. A. Derry, P. M. Vitousek, B. J. Huebert, and L. O. Hedin. 1999. Changing sources of nutrients during four million years of ecosystem development. *Nature* 397:491–497.

Crews, T. E., K. Kitayama, J. H. Fownes, R. H. Riley, D. A. Herbert, D. Mueller-Dombois, and P. M. Vitousek. 1995. Changes in soil phosphorus fractions and ecosystem dynamics across a long chronosequence in Hawaii. *Ecology* 76:1407–1424.

MacDonald, G. A., A. T. Abbot, and F. L. Peterson. 1983. *Volcanoes in the Sea: The Geology of Hawaii,* 2d ed. University of Hawaii Press.

Riley, R. H., and P. M. Vitousek. 1995. Nutrient dynamics and nitrogen trace gas flux during ecosystem development in montane rain forest. *Ecology* 76:292–304.

Vitousek, P. M., L. R. Walker, L. D. Whitteaker, D. Mueller-Dombois, and P. A. Matson. 1987. Biological invasion by *Myrica faya* alters ecosystem development in Hawaii. *Science* 238:802–804.

ADDITIONAL REFERENCES

Herbert, D. A., J. H. Fownes, and P. M. Vitousek. 1999. Hurricane damage to a Hawaiian forest: nutrient supply rate affects resistance and resilience. *Ecology* 80:908–920.

Hobbie, S. E. and P. M. Visoutek. 2000. Nutrient limitation of decomposition in Hawaiian forests. *Ecology* 81: 1867–1877.

Part 4

APPLICATION

Human Impact on Local, Regional, and Global Ecosystems

In Part 3 we discussed the function of relatively undisturbed ecosystems—those in which we could glimpse the long-term geological, ecological, and evolutionary forces that guided the development of soils and vegetation—and system-level responses to disturbance. These systems are increasingly rare. Over much of the globe, conversion of native ecosystems to agriculture or intensive forestry has created an entirely new set of ecological systems. While plants and soils in managed systems are constrained by the same basic processes as those acting in native systems, inputs, outputs, disturbance regime, and species composition have all been altered to increase yields of food and fiber for human use. Increasingly, we are changing the fundamental genetic makeup of the plants we grow to combat diseases in crop plants, reduce herbivory, and increase yield.

In addition, human activity in highly industrialized or intensively farmed areas creates pollutants that can be carried by weather systems over long distances. Deposition of these pollutants to both native and managed ecosystems can also alter element budgets and ecosystem function. At the largest scale, increased production of gases that can alter the energy balance of the atmosphere, even when present in very small concentrations, and can bring the prospect of entirely new climate regimes over large parts of the Earth.

In summary, human use of the landscape at local to regional scales is altering the biogeochemistry of the entire planet. The purpose of Part 4 is to present an overview of human impact in four domains. Chapter 23 deals with traditional uses for the production of food and fiber by looking at the management of forest and agricultural ecosystems. Chapter 24 deals with one of the major effects of large-scale land conversion: the loss of biodiversity through the extinction and invasion of species. In Chapter 25, we present the regional effects of air pollution on ecosystems in a discussion of "acid rain," or the more general problems of the deposition of air pollutants in systems distant from the source of those pollutants. In Chapter 26, we move up to the global level, using concerns over global climate change due to increases in carbon dioxide and other trace gases in the Earth's atmosphere as a vehicle for discussing the interactions between ecosystem dynamics and human use of the Earth.

23

ECOSYSTEMS MANAGED FOR FOOD AND FIBER

INTRODUCTION

For all of the advances in human technology, we still share with our earliest ancestors a dependence on productive ecosystems for food and, to some extent, for energy, shelter, and clothing. While synthetic materials and inorganic products, such as steel and glass, are widely used, a sustainable flow of products from forests and fields are crucial to supporting the 6 billion people who currently (year 2000) inhabit the Earth.

As the human population increases and the reach of our industrialized society is extended, new lands are converted from native systems to production. On existing production lands, genetic manipulation and site enrichment through fertilization and irrigation are expanding. Two major concerns lie at the heart of the issues raised by rapid expansion and industrialization of production from domesticated systems: (1) Can production keep ahead of population? and (2) What is the cost in terms of degradation and loss of native systems? The bottom-line question being asked with increasing frequency is that of sustainability: Can the current levels of natural resource and industrial production and the increases that are foreseen for the near future be sustained in the long term?

The purpose of this chapter is to discuss the production side of this question. We begin with a general discussion of the race between population increase and agricultural production. We then present the concept of a gradient of intensity of management and conversion of native ecosystems and then continue with a discussion of several very different practices currently in place for the extraction of food and fiber from ecosystems.

MALTHUS AND THE RACE BETWEEN POPULATION GROWTH AND INCREASED AGRICULTURAL PRODUCTION

In 1798, Thomas Malthus issued his historical essay on the essential dilemma of producing enough food for an incessantly growing human population from a fixed land base. Envisioning the impossibility of increasing food production at the required rate, he foresaw an inescapable cycle of overpopulation, malnutrition, poverty, and famine.

Famine has been no stranger at the local and regional levels at many different times in human history. Major episodes on every continent are linked especially to periods of natural disasters, crop failures, epidemics, social unrest, or large-scale warfare. Still, the essence of Malthus' predictions have not been met. Over the last half of the twentieth century, despite very rapid increases in population, per capita grain production has more than matched population growth (Figure 23.1). Increases have been particularly high in regions of highest population density and increase in southern and eastern Asia.

What lies behind this tremendous increase in food production? To date, concentrated breeding and selection programs for the major grain species, especially for rice, linked with increased industrialization of agriculture, including huge increases in fertilization, irrigation, and mechanization, have driven increases in production. In one analysis, strong relationships were found between increased grain yield globally and simultaneous increases in the use of nitrogen (N) and phosphorus fertilizers and irrigation. Increases in the total amount of land in cultivation were minimal over the same period (Figure 23.2).

The human population has crossed the 6-billion threshold and is currently increasing at a rate of 1.4% per year. That rate of increase has fallen by one third from a maximum value of 2.1% per year in 1968. A crucial question is whether it is possible to continue to increase the production and distribution of water and fertilizers at the rate seen in the recent past such that food production can keep pace with the demand created by population growth. The environmental burden placed on the Earth's ecosystems by such increases can already be seen in loss of species and pollution loading to non-

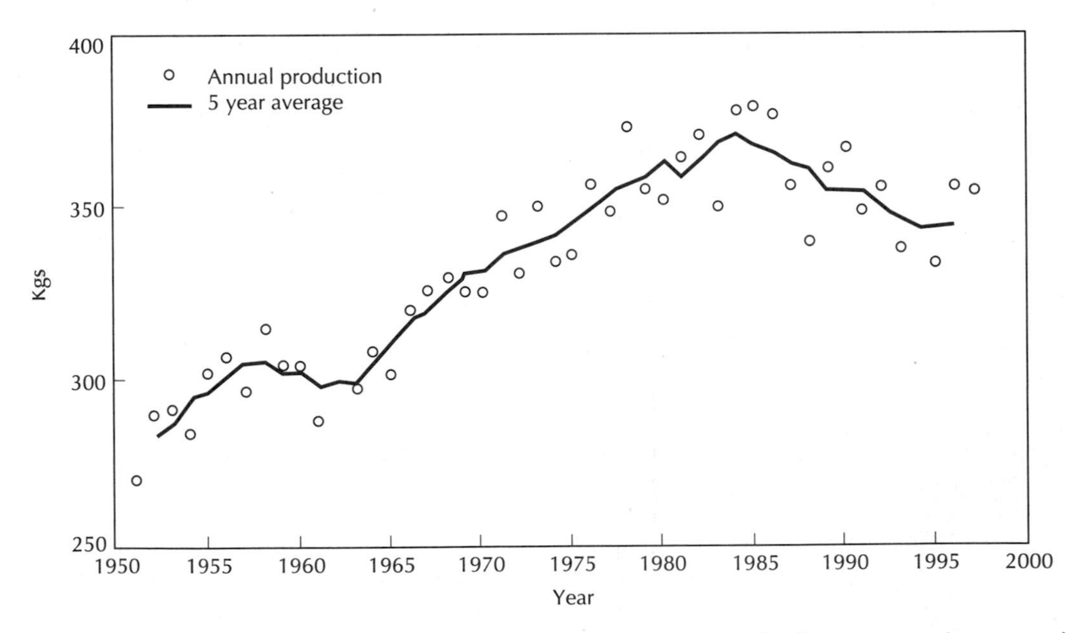

FIGURE 23.1 Change in per capita grain production for the main cereal crops totaled for the world. (Dyson 1999, see also Pimentel and Wightman 1999)

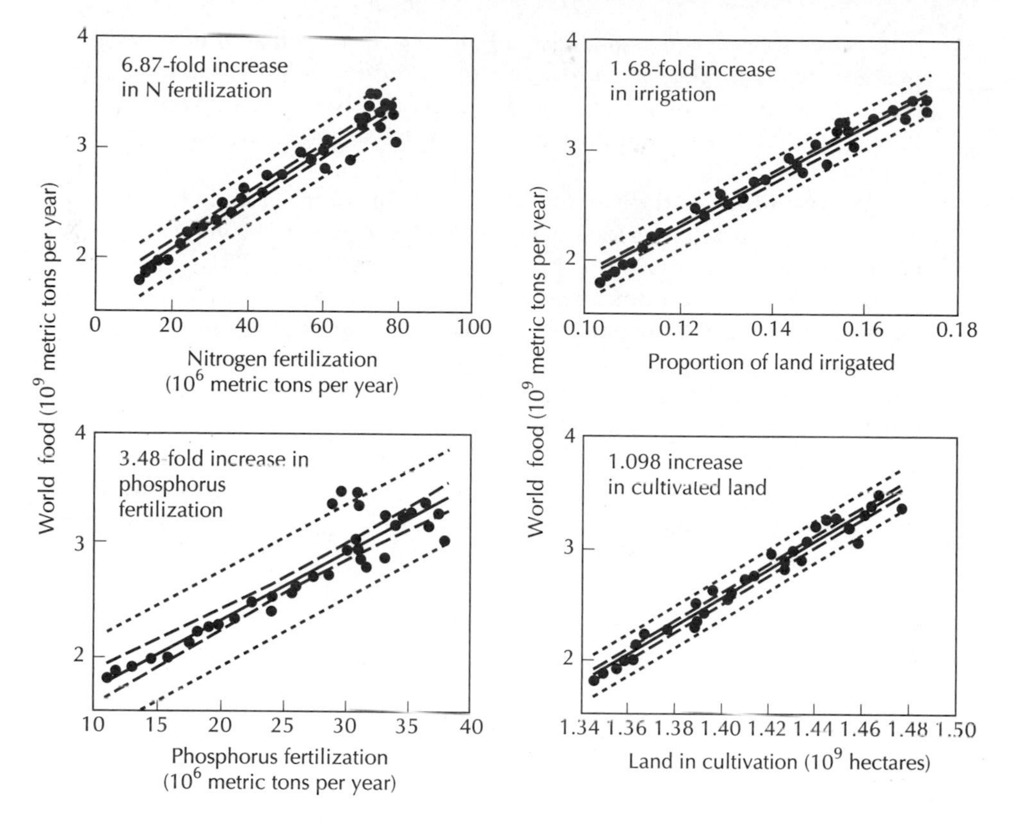

FIGURE 23.2 Relationship between several production factors and total world food production. Note the very different ranges for the different resources. Total land in cultivation has changed little over the time period in the data set (1960–1996). (Tilman 1999)

agricultural systems. On the other hand, the implications of failing to meet that demand are unthinkable. Further intensification of management of those lands currently in production offers the best hope of feeding the world's population and preserving some of the bounty of our natural systems until global population increases are brought under control. Emerging technologies, such as genetic engineering, offer hopeful, if controversial, methods for achieving these ends.

A GRADIENT IN THE INTENSITY OF MANAGEMENT OF ARABLE LAND

We may think of forestry and agriculture as two fairly distinct activities. Managed forests are systems dominated by long-lived perennial plants that are harvested once every couple of decades for fiber and other products derived from the woody stems. Agriculture, on the other hand, generally involves planting short-statured annual plants and harvesting the edible portions all within a single growing season. However, differences may be more a matter of degree. In both systems, disturbance regimes are altered and products are removed in a way that is unlike natural disturbances. In lightly managed forests, recurring disturbances accompanying harvests may alter species composition, while in plantation forestry or agriculture, species are selected and planted to meet management goals. In pasture systems, the imposition of grazing at very high intensities alters both the structure and species composition of the system.

We can think of the tremendous range in intensity of management in ecosystems as variations along a single gradient that trades native ecosystem or conservation values against production values (Figure 23.3). At each step along this sequence, intensification of management increases the disparity between native and managed systems, while also increasing production. How far we need to move along this gradient and over how much of the Earth's land surface depends in large part on the size of the human population and our total demand for food and fiber.

Using Figure 23.3, we can examine the characteristics of managed ecosystems at several points along this intensification gradient. We begin with the least intensive management of forests and follow on to the most intensively managed agricultural systems.

MANAGEMENT OF NATIVE FORESTS FOR TIMBER AND FIBER

In the upper left corner of Figure 23.3 are those native forest ecosystems that are occasionally harvested for timber and fiber. Native species generally are allowed to regenerate in the cutover area and something similar to natural succession occurs. Within this category, however, there are degrees of intensity of management that are the cause of serious concern and debate.

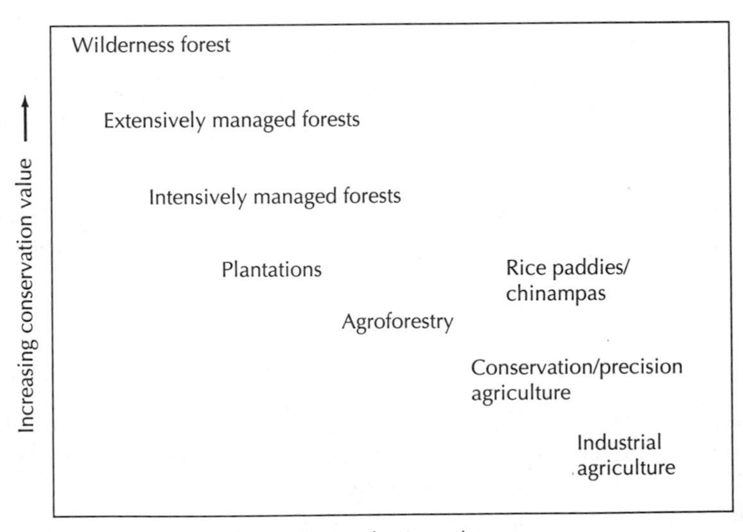

FIGURE 23.3 Tradeoff between the production and conservation value of land-use practices. Conservation value includes both use of native species for production and the long-term conservation of soil and nutrients. (After Nambiar 1996 and Noble and Dirzio 1997)

At one extreme are the unmanaged wilderness or national park areas in which human impact is kept to an absolute minimum. Increases in intensity in management can be described in terms of the amount of biomass removed at each harvest and the frequency at which those harvests occur (also known as the rotation length). Selection cuttings are those in which only certain crop trees are removed. While this practice minimizes nutrient losses and disturbance to the forest floor, it can also cause degradation of the stand if only the most valuable species or better formed or faster-growing trees are taken. This practice, called **high-grading,** has become a major concern in the American tropics, where very high species diversity means that there are very few stems of any given species in a particular area. Selective removal of high-value species, such as teak and mahogany, may lead to local or even regional extinction by removing potential "parent" trees that would be the source of the next generation.

In clear-cutting, all of the stems above a certain diameter limit are removed, regardless of species. This practice is widely used in temperate-zone industrial forests. The visual impact immediately following cutting are extreme, and nutrient losses in both dissolved form and as surface erosion can be much higher than in selection cutting. However, clear-cutting does avoid the problems of high grading and is a preferred method for regenerating forests with a high content of fast-growing early successional species.

One aspect of the intensification of forest management over the last several decades is that tractor-based machines have largely replaced the chainsaw as the major harvesting tool (Figure 23.4). With these devices, whole trees can be clipped off at the base, laid in windrows, dragged to a staging area, and reduced to chips for transport to a mill by truck. Most of the wood-products industry is now based on chips, rather than large pieces of lumber, so that the biomass of small trees, as well as branches, twigs, and even leaves of larger ones, can contribute to the yield from a site. This makes it possible to harvest smaller trees or to harvest whole stands at a younger age. "Whole-tree harvesting" was seen at first as a way of increasing the efficiency of logging operations by removing a larger fraction of the total biomass on a site and also of returning to the site sooner for a second harvest.

Maintaining nutrient stocks and nutrient cycles in the face of repeated harvests is one of the principal concerns in the intensive management of forests. Increased rates of biomass removal, shorter rotations, and the harvesting of whole trees can greatly accelerate rates of nutrient removal. Branches, twigs, and leaves have much higher nutrient concentrations than does stem wood (Table 23.1). By adding these components of stand biomass into the harvest, the overall removal of nutrients is increased by far more than is the yield of usable biomass. These removals must be offset by inputs to the system from precipitation and weathering or severe nutrient limitations on forest growth could develop.

Modifications of the whole-tree harvesting method can be made to minimize nutrient losses (e.g., harvesting deciduous stands in winter after leaves have already become a part of the forest floor). However, this approach fails to make use of the fact that trees concentrate biomass in nutrient-poor, physiologically inactive heartwood in the stems, which effectively separates most

(*a*) (*b*) (*c*)

FIGURE 23.4 Machines used in whole-tree harvesting operations. (*a*) Feller-buncher for harvesting trees. (*b*) Skidder for removing trees to the loading area. (*c*) Chipper that reduces entire tree to chips for transport to the processing plant.

of the harvestable fiber from the nutrient-rich, physiologically active foliage, roots, and bark. Stem-only harvesting takes advantage of this characteristic of tree growth by removing primarily wood and leaving most of the nutrients on site.

One simple approximation of the sustainable harvest from forests can be obtained by tabulating the inputs to the forest from precipitation and weathering and comparing these with losses occurring through removals in harvested biomass as well as leaching and erosional losses resulting from the harvest (Figure 23.5). Accumulation in biomass is a nonlinear function due to the concentration of growth in nutrient-rich foliage, twigs, and branches in the first few years, followed by accumulation of heartwood later on. Nutrient inputs, on the other hand, are assumed to remain the same over time and so accumulate linearly. The sustainable rotation length, then, is the point at which the two lines cross (less than 50 years for a stem-only harvest in this example, and about 90 years for whole-tree harvesting). So, one of the proposed benefits of whole-tree harvesting—the ability to shorten rotation time through removal of a larger fraction of the biomass in a stand at each harvest—is contradicted by the need for longer rotation lengths to replenish lost nutrients.

TABLE 23.1 Comparisons of Biomass and Nutrient Removals in Stem-Only and Whole-Tree Harvests for Selected Forests*

(From Weetman and Weber 1972, Jorgensen et al. 1975, Marks 1974, Pastor and Bockheim 1981, Hornbeck 1977)

Stand Type and Age	Component	Twigs and Foliage	Branches	Stemwood and Bark	Total
Northern hardwood (pin cherry/aspen, age 6)	Biomass	3.7 (11)	4.5 (14)	24 (75)	32.2
	N	75 (50)	14 (9)	62 (41)	151
	K	48 (59)	5 (6)	28 (35)	81
	Ca	26 (24)	16 (15)	68 (61)	110
Aspen/maple (age 63)	Biomass	2.5 (2)	19 (13)	125 (85)	146.5
	N	45 (17)	78 (28)	152 (55)	275
	P	6 (16)	12 (32)	20 (52)	38
	K	22 (7)	56 (18)	235 (75)	313
	Ca	35 (5)	192 (28)	465 (67)	692
Northern hardwood (age 90)	Biomass	3.6 (2)	54 (27)	145 (71)	202.6
	N	81 (17)	195 (40)	207 (43)	483
	Ca	25 (5)	229 (42)	293 (53)	547
Loblolly pine (age 16)	Biomass	8 (5)	23 (15)	125 (80)	156
	N	82 (32)	60 (23)	115 (45)	257
	P	10 (32)	6 (19)	15 (49)	31
	K	48 (29)	28 (17)	89 (54)	165
Jack pine (age 65)	Biomass	8 (7)	10 (9)	89 (84)	107
	N	65 (38)	31 (19)	71 (43)	167
	P	16 (38)	6 (14)	20 (48)	42
	K	30 (36)	13 (15)	41 (49)	84
	Ca	73 (26)	41 (15)	163 (59)	227

*Biomass values in Mg/ha, nutrient value in kg/ha, numbers in parentheses are percentage of total.

More complex predictions of the impact of intensive harvesting on internal nutrient cycling and forest production can be obtained from computer models of ecosystem function. In fact, because the detrimental effects of increased nutrient removal are cumulative over time and may require 2 or 3 rotations to become evident, the use of models to predict long-term effects may forewarn forest managers and forestall the degradation of forest lands. In essence, models allow us to run the experiment and see the conclusions rapidly rather than waiting the 90–100 years for results to be evident in the field, at which time remedial action may be expensive or difficult.

Several models predicting the impact of intensive management on temperate-zone forests were produced during the 1970s and 1980s, when whole-tree harvesting first became feasible. The predictions of three of these are summarized in Figure 23.6. Even though the models were structured very

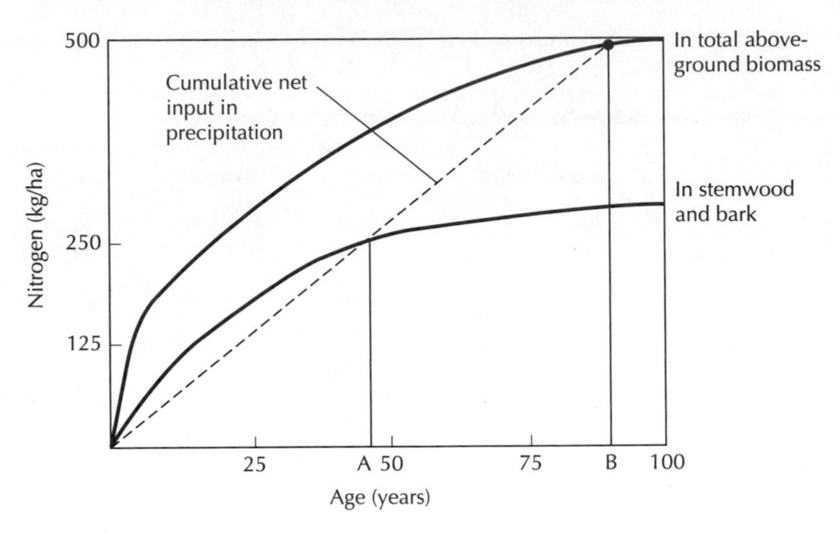

FIGURE 23.5 Graphic representation of the time required to replace nitrogen removed in stem-only versus whole-tree harvesting. Upper curve is nitrogen accumulation in biomass of the whole tree. Lower curve is nitrogen accumulation in stem only. Dashed line reflects constant accumulation of nitrogen inputs in precipitation.

differently and were applied to different forest types, they all agreed that significant losses of soil organic matter would occur with the switch from clear-cutting to whole-tree harvesting. Two of the models used also predicted forest production and did not agree. One suggested that decreases in NPP would be offset by increases in the fraction of biomass harvested, such that wood harvested would change little. The other predicted significant declines in wood harvested.

There is now a general recognition that intensive harvesting also requires other aspects of intensive management, such as fertilization. As forestry moves farther into intensive management (Figure 23.3), it becomes increasingly like agricultural crop management, and many of the principles derived from agricultural sciences become increasingly relevant.

PLANTATION FORESTRY

At the high-intensity end of the gradient in forest management are the commercial plantations of the southeastern United States; Europe; New Zealand; and, increasingly, in tropical America. In these stands, a single species, generally an evergreen conifer of genetically selected stock, is planted at regularly spaced intervals over a large area. Managed as a monoculture, these stands can be treated with herbicides, mechanical weeding, or managed fires to reduce competition from other species that invade the understory. After the first rotation, nutrient removals reduce nutrient cycling and availability, and fertilizers can be applied, usually right after planting or near the harvest date. Whole-tree harvesting, or at least whole-stem harvesting with minimal removal of branches, is carried out at the end of the rotation, which can vary from 20–30 years in the southeastern United States to over 100 years in parts

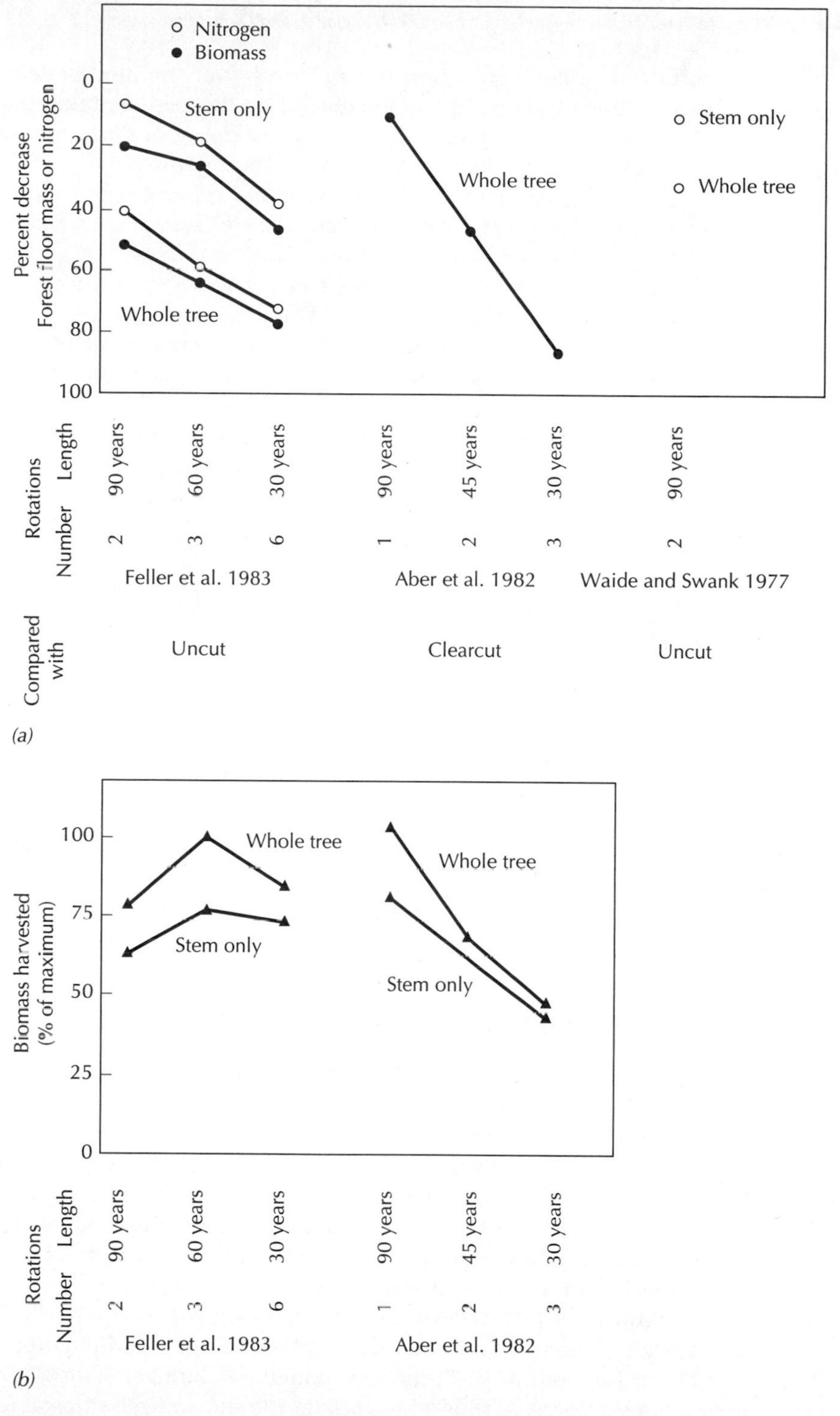

FIGURE 23.6 Comparison of predictions of whole-tree harvesting effects on forest ecosystems. (*a*) Forest floor biomass or nitrogen content in three different forest types by three different models, expressed as a percentage of reduction from uncut or clear-cut stands. (*b*) Comparison of predictions of whole-tree harvesting effects on yield (expressed as a percentage of the maximum value). Note the importance of rotation length. (Waide and Swank 1977, Feller et al. 1983, Aber et al. 1982)

of Europe. Site preparation before planting of the next generation of trees can include windrowing and burning of slash (residue of plants left on site by the harvesting operation), mechanical chopping of slash, mounding of soils for planting, and other agricultural-like treatments.

Two systems that are still experimental represent the ultimate in intensively managed forests. The "sycamore silage" system is envisioned for major river floodplains, where resprouting stumps of sycamore, a fast-growing deciduous species, on floodplain soils in the southern United States would be "mowed" every 2–4 years for fiber. This system makes use of continuous nutrient deposition in sediments in floodplain areas and realizes very high fiber yields without the risk for loss to flooding run by agricultural use of this land type. Experimental energy plantations of fast-growing hybrid poplars (aspen trees) are both fertilized and irrigated for maximum yield—a type of forest management that approaches standard agricultural practices, except that the "crop" grows for more than 1 year, and the harvested product is fiber rather than food.

CONSERVING FOREST RESOURCES

While the intensification of forest management offers potential gains in yield, most of the world's forests remain in the seminatural state. On the other hand, only a small fraction of the world's remaining forests are in conservation holdings and are managed in ways designed to sustain native ecosystems. Forests are a major reservoir of biodiversity (see Chapter 24), and pressures to increase yields are met by counter pressures to preserve the integrity of forest communities and the ability to provide clean water and air purification (see Chapter 25). The future of this global resource will be a major part of the response of the world community to the challenges of overpopulation, economic development, and quality of life.

LOW-INPUT, LOW-YIELD AGRICULTURE: TRADITIONAL PRACTICES IN THE HUMID TROPICS

A first step from forest management to agriculture is the traditional, extensive model of **shifting cultivation** practiced in humid tropical regions. Also called **"slash-and-burn"** or **swidden-fallow agriculture,** this approach has been alternately despised by temperate-zone observers as a great waster of the tropical forest, or held up as the ultimate model of sustainable, low-resource demand agriculture. To the native peoples of the tropics, it has provided subsistence for thousands of years.

Shifting cultivation in the tropical forest zone (Figure 23.7*a*) consists of a relatively long cycle that begins with the cutting and burning of the forest. This is followed by short-term cropping and a longer "fallow" period, during which the forest is allowed to reclaim the site and rebuild soil fertility.

This process has many features that are in keeping with the tropical environment. In Chapter 2, we discussed the relatively nutrient-poor nature of many tropical soils and the relatively large proportion of rapidly cycling nutrients that are held in the vegetation. Clearing and burning the forest releases these nutrients in a pulse. This pulse of high nutrient availability supports a brief period of agricultural production. However, rapid leaching by excessive rainfall, and reduced total plant growth in crops compared with the original forest, cause a steep decline in the productivity of cleared land over a 2- to 3-year period (Figure 23.7*b*). After the area has lost much of its productive value, it is let go to fallow and a secondary forest invades rapidly. A fallow period of 20–30 years is usually required to rebuild the nutrient stores in the vegetation and soil.

Figure 23.8*a* depicts a generalized pattern for nutrient retention and loss over the first 4 years. Nutrient content in trees is reduced initially by burning and continues to decline as decomposition of the remaining material proceeds. Content in soils increases initially as some of the nutrients stored in vegetation are transferred by decomposition and ash deposition. It then declines due to leaching losses and removals in the crop. Once the field is abandoned to fallow, plant regrowth begins rebuilding the soil until preclearing levels are reached (Figure 23.8*b*).

Shifting cultivation has other advantages in addition to nutrient self-sufficiency. In the first year of cultivation, weeds and insect pests are present in relatively low numbers, having been absent from the closed forest environment present before clearing. Invasion increases in years 2 and 3, reducing yield and increasing the amount of work required to bring a crop to harvest.

Considering that crop plants are generally of high nutrient and protein content and relatively low in general feeding inhibitors, shifting cultivation can been seen as a managed system that mimics the environments in which such species would naturally grow. Looking back to Chapter 15, we discussed the rapid growth, short life span, and prevalence of "qualitative" defense strategies in plants that occupy disturbed areas. These plants grow rapidly in the temporary, nutrient-enriched environments created by disturbance and invest less in herbivore defenses. Crop plants tend to be derived from wild species native to disturbed sites. Shifting cultivation is a system by which disturbance frequency is increased, and specific high-value crop plants are introduced to the disturbed area.

As a low-input, extensive form of agriculture, shifting cultivation is not highly productive. The total wet-weight yield of manioc over a 3-year cultivation period may average only 280 $g \cdot m^{-2} \cdot year^{-1}$. Dry-weight yield may be half that amount (compare with corn production in Table 23.2 and consider that corn grows for less than half of the year). That 280 $g \cdot m^{-2} \cdot year^{-1}$ becomes only 2.8 $g \cdot m^{-2} \cdot year^{-1}$ when the production of manioc is spread over the entire 30-year cropping and fallow rotation. As such, this system is suitable only for low population densities. As population density increases, the fallow cycle must be shortened, giving the soil and vegetation less time to recover. This can result in long-term declines in productivity and an actual reduction in total crop yield (the similarity to the case of whole-tree harvesting should be clear).

AGROFORESTRY: INCREASING YIELDS BY INTERCROPPING AND MANAGING THE FALLOW FOREST

If temperate-zone forestry is becoming more like agriculture, agriculture is, especially in the tropics, becoming more like forestry. The term **agroforestry**

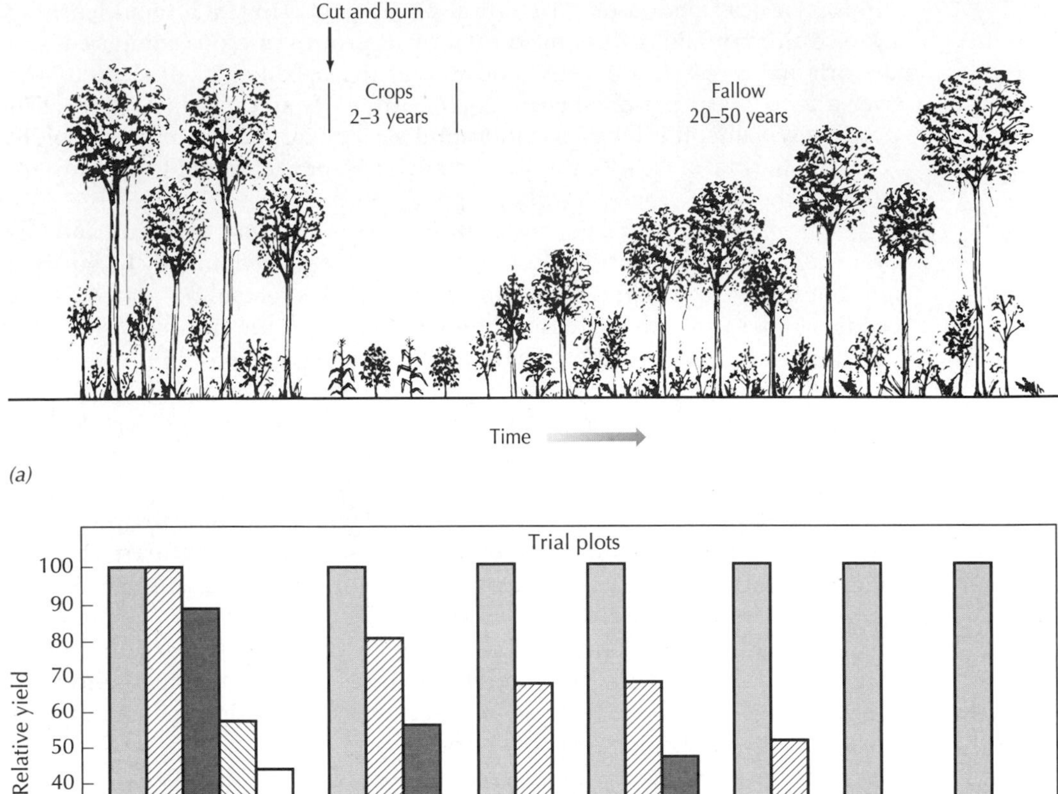

FIGURE 23.7 Shifting cultivation in the humid tropics. (*a*) Schematic drawing of the structure of different stages of the cycle. (*b*) Relative changes in yield in different years following cutting and clearing. (Cox and Atkins 1979)

has been coined to describe either permanent agriculture using woody plants as crop producers, or a combination of woody and herbaceous vegetation in complex systems designed to mimic the structure and function of natural systems.

Two potential advantages of agroforestry systems have been put forward. The first involves the differing characteristics of resource use by trees and herbaceous annuals (Figure 23.9). An idealized agroforestry system would combine physical protection from wind and rain as well as acquisition and recycling of nutrients from deep within the soil provided by trees, with the high productivity of herbaceous crop species. The second is that

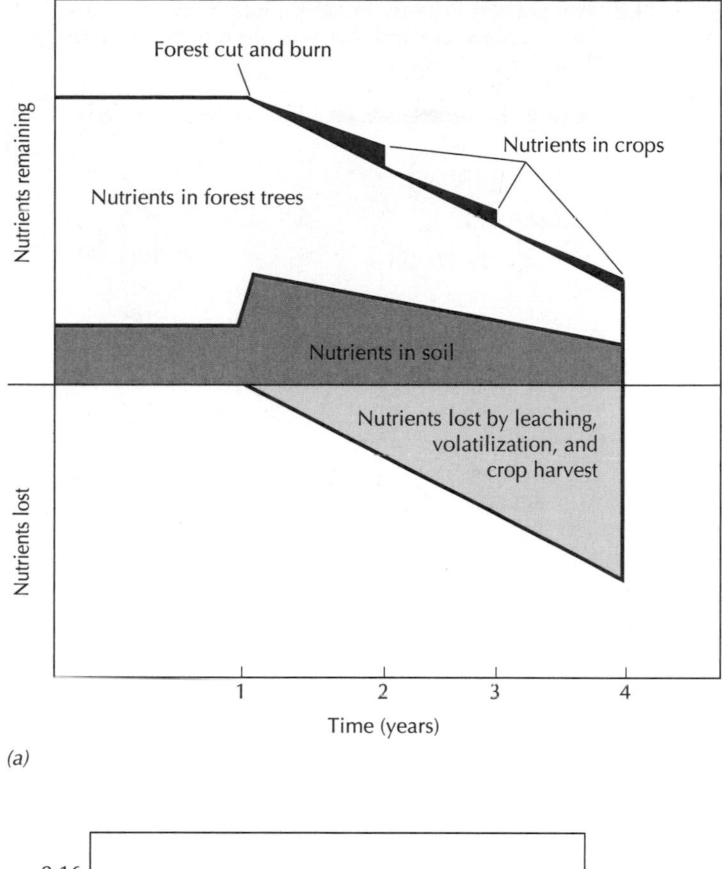

(a)

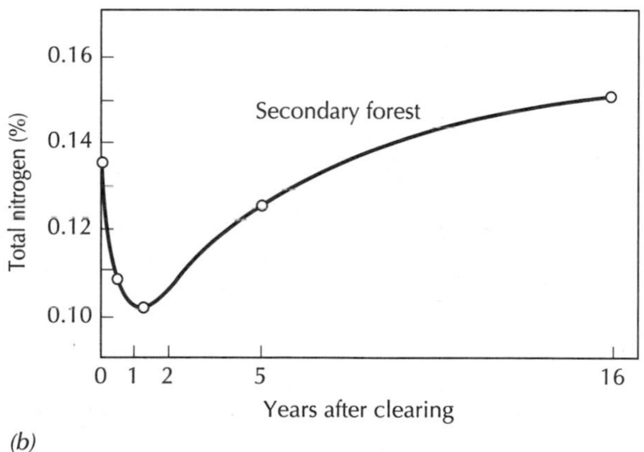

(b)

FIGURE 23.8 Effects of shifting cultivation on nutrient content of humid tropical ecosystems. (*a*) Generalized pattern of the distribution of nutrients during the first 4 years of the shifting cultivation cycle. (*b*) Changes in nitrogen storage over a full 16-year cycle. (*a*, Uhl et al. 1983; *b*, Sanchez 1982)

the complex structure and species composition provides protection from herbivory by reducing the "apparency" of the crop plants (Chapter 15). In an experiment designed to test the effects of species diversity on herbivory, recently cut areas in Costa Rica were allowed to regenerate naturally, or were modified to contain more and fewer species. As species richness in

TABLE 23.2 Annual Net Primary Production in a Cornfield and a Tall-Grass Prairie Ecosystem and the Allocation of Production to Different Plant Parts
(Boyer 1982, Sims ad Singh 1978)

Plant part	Prairie ($g \cdot m^{-2} \cdot year^{-1}$)	Corn ($t \cdot ha^{-1} \cdot year^{-1}$)
Above ground		
Vegetative	346 (39%)	800 (53%)
Reproductive	—	700 (47%)
Below ground		
Perennial	286 (32%)	0
Annual	256 (29%)	?
Total	**888**	**1500 + roots**

these areas increased, losses to herbivory declined and became less variable (Figure 23.10).

Stepping from swidden-fallow agriculture to agroforestry can be as simple as managing the secondary forest in the fallow period. This can be accomplished by selecting and fostering useful species that reseed naturally into the fallow forest or by planting tree crops into the fallow at the end of the

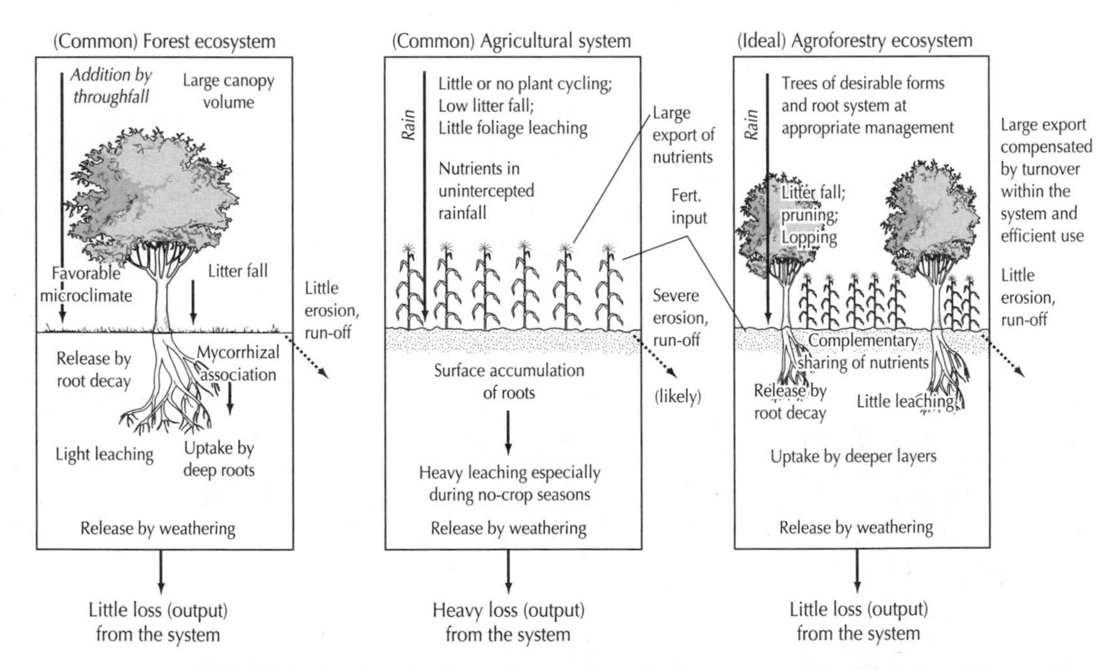

FIGURE 23.9 Deeply rooted trees tap nutrients and water in different soil layers than do annual, herbaceous crops. An "ideal" agroforestry system maximizes both production in the two types of crops and minimizes losses of soil and nutrients. (Nair et al. 1999)

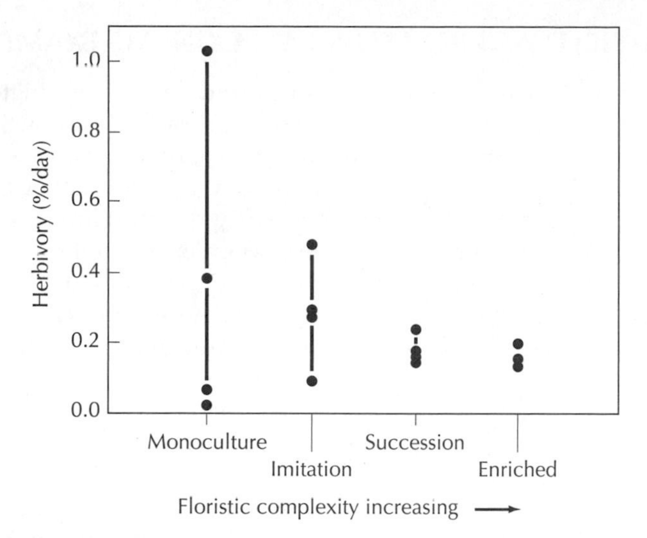

FIGURE 23.10 Comparison of herbivore consumption of leaves in early successional tropical forests in Costa Rica in which species diversity was experimentally manipulated. Plots with higher species diversity had lower herbivore consumption rates. (Brown and Ewel 1987)

cropping period and tending them as the secondary forest develops. While the species actually involved are as varied as the sites and floras available, combinations of food, fiber, and fuel species are used. The goal of these substitutions is often to replicate the structure of the natural secondary forest (Figure 23.7*a*) while substituting crop species for noncrop species. There is evidence that keeping the structure of the fallow vegetation while substituting economic species for ones that do not produce a useful product does not affect the soil building processes of the fallow period.

More complex than tending the fallow is the permanent establishment of perennial tree crops such as rubber, tea, oil palm, coffee, and banana. As tree crops occupy the site constantly, they offer some protection against erosion and leaching of nutrients. Root systems and understory plants provide a constant, if reduced, supply of organic matter to soils. The yield from such systems is low, but the economic value of the product is relatively high. With shade-loving plants such as coffee, selection of N-fixing species for the overstory of shade trees can improve yields and increase sustainability. However, remember that methods of agriculture cannot be separated from the socioeconomic systems of which they are a part. The conversion of land from subsistence-shifting agriculture to export-oriented cash crops such as the tree crops listed here, has profound social implications for the regions involved.

More intensive still is the combination of cropped areas with windbreaks and/or hedgerows that provide tall, woody borders, reducing wind stress and erosion. In environments with intense solar radiation levels, the partial shade provided by woody borders may also reduce heat and water stress without excessive reductions in energy for photosynthesis. Significant increases in yield in tropical agricultural systems using hedgerows and windbreaks is a general occurrence.

PERMANENT HIGH-YIELD AGRICULTURE: AN EXTREME EXAMPLE

Residents of the northern temperate zones find permanent, high-yield agriculture far more familiar than those described earlier (although apple orchards or pecan groves could be described as agroforestry, especially if the grass between the trees is harvested and used as feed for animals).

A good example of an agricultural system at the most intensive end of the management spectrum is corn production as practiced in the American Midwest. The Corn Belt occupies a large expanse of flat to rolling land centered on the states of Iowa, Illinois, and Wisconsin. Highest productivity is found on deep and highly organic soils formed by millennia of true or tall-grass prairie growth and the deep and extensive root systems formed by the dominant grasses. Corn cultivation has converted the system from a very diverse mixture of mostly perennial grasses and forbs to a monoculture of a highly productive annual, corn (Figure 23.11*a*). While the original prairie system and the new corn system may appear similar in terms of late summer height of vegetation, they are very dissimilar in nearly every other ecosystem parameter.

At the heart of agricultural practice is the maximization of edible yield. Total net annual primary production above ground is increased greatly over native tall-grass prairie systems through fertilization and sometimes irrigation. Production is also allocated very differently between roots, stems, and seeds in the two systems (Table 23.2). In prairies, a large part of the annual productivity is allocated to root growth and to perpetuating a large root and root-crown system. This is partly in response to the occurrence of droughts but more as an adaptation for survival and regrowth following fires and grazing. In contrast, corn is an annual plant which, even in the wild state, allocates much of its growth to seed production to ensure successful reproduction in the following year. Genetic selection has accentuated this tendency, producing plants in which over 45% of the total above-ground primary production can be harvested as grain.

Much of the remaining production is also harvested for use as animal fodder or silage (the conversion of low-quality plant residues to high-quality material by partial fermentation). A conspicuous part of this type of cultivation is the bare-soil winter aspect produced by the complete removal of all plant residues (Figure 23.11*b*). Large reductions in organic matter inputs to soils also occur, which in turn have led to the long-term reductions in soil organic matter levels discussed in Chapter 13. In prairies, much of the aboveground plant material is removed by frequent fires, but the large fraction of production going below ground is protected from burning. The accelerated removal of plant materials and the associated loss in soil organic matter under cultivation are very similar in principle to the discussion of intensive forest management presented earlier in this chapter.

The high yield of grain in modern corn culture comes at the cost of very high inputs of nutrients and protective chemicals, as well as intensive manipulation of soils through plowing and cultivation. Table 23.3 compares the inputs and outputs of nutrients in cornfields and native prairie systems. Fertilizer application is well in excess of the total uptake of N by the growing corn plants. In prairies, only 10–20% of the annual N uptake comes from outside of the system through precipitation and dust, with the remaining fraction being supplied by mineralization of soil organic matter.

(a)

(b)

FIGURE 23.11 Structure of cornfield ecosystems in comparison with the native tall-grass prairies they replaced in the American Midwest. A cornfield in (*a*) summer and (*b*) winter is shown. (*a* and *b*, H. Armstrong Roberts)

The application of high doses of fertilizer also increases nutrient losses to groundwater, as availability is too high, even for the high demands of the growing corn. Such losses are very low to negligible in prairies. In general, conversion of native prairies to industrial farmland converts the dominant form of nutrient cycling in the landscape from "closed" to "open."

(c)

FIGURE 23.11 (*c*) A prairie in summer is shown. (Tom Riles)

Of even greater concern in the long term is the loss of soil through erosion under standard agricultural practices. Spring plowing and the absence of plant or plant residue cover during those periods when the crop is not growing leave the soil vulnerable to severe erosion. Images of the dust bowl in the American Midwest in the 1930s, featuring the specter of worn-out and deeply eroded farms, are an extreme example of the type of damage that can occur. Currently, losses of topsoil in the American Midwest are estimated to be 35 $t \cdot ha^{-1} \cdot yr^{-1}$. While soil organic matter can be rebuilt by changes in tillage practices and harvesting practices, the chemical and physical weathering of soils to produce topsoil is a much, much slower process.

The overriding concern with the kind of modern agriculture described here is: Is it sustainable? This system is vulnerable in at least two respects. First, the high demands for fertilizers and pesticides and mechanical cultivation assume abundant and relatively inexpensive fossil fuels. As the cost of these fuels increases, so will the price of food and the proportion of the national economic effort required to produce it. A continuous supply of fertil-

TABLE 23.3 Nutrient Input and Output Budgets for a Cornfield and for a Tall-Grass Prairie. Inputs are Precipitation and Fertilizers, Outputs Are Leaching Below the Rooting Zone or to Streams. All Values in $g \cdot m^{-2} \cdot year^{-1}$.

(Frissel 1978, Gilliam 1987, Woods et al. 1983, Timmons and Dylla 1983, Blair et al. 1998)

	Prairie			Corn		
Element	**Precip. In**	**Stream Out**	**Fire Out**	**Fert. In**	**Stream Out**	**Harvest Out**
Nitrogen	1.5	0.03	3	16.0	3.5	6.0
Calcium	3.2	—	—	19.0	4.7	1.0
Phosphorous	0.03	—	—	3.0	0.3	1.2
Potassium	0.02	—	—	7.5	1.5	1.3

izer and fuels to all parts of the globe requiring these inputs also requires a stable worldwide social order and smooth operation of international trade. Second, the continuous loss of topsoil is beyond the ability of any economic system to repair. How the decline in topsoil will affect productivity is not fully known. One can imagine that the effects will be greater if inputs of fertilizers need to be reduced for economic reasons. If we face a future of higher fossil fuel costs and lowered availability, then conserving topsoil as a means of ensuring future agricultural productivity seems prudent at least.

There has been a long-standing concern over attempts to transplant highly productive methods for crop production from temperate to tropical zones. Differences in climate, biota, and particularly in soils defeated early attempts. However, permanent agriculture is not impossible in the tropics. The region contains a wide variety of soil and climatic conditions, some of which can be made permanently productive if the methods applied reflect the realities of the tropical environment. An ongoing demonstration of permanent agriculture in Peru, using moderate chemical inputs, has demonstrated continuous productivity and economic viability (Table 23.4). The systems developed include constant **crop rotation** and considerable human labor. They represent a compromise between the high intensity, continuous cultivation practices of the temperate zone, and the realities of the tropical environment. On sites with young soils, such as those described on the volcanic islands of the Hawaiian chain (Chapter 22), productivity can be very high and sustainable over human time scales.

METHODS FOR IMPROVING SUSTAINABILITY

What we have described so far is an extreme form of Corn Belt agriculture. There are several practices discovered long ago and that are already widely used to reduce environmental impact and maintain productivity under permanent agriculture.

Methods for reducing erosion include a system of crop rotation that involves the establishment of bands of continuous plant cover, such as a perennial hayfield, between bands of plowed ground. A particularly conservative

TABLE 23.4 Crop Yields for Different Types of Continuous Agriculture Near Yurimaguas, Peru
(From Nicholaides et al. 1985)

Production System	Crop Rotation											
	Corn–Peanut–Corn			Total	Peanut–Rice–Soybean			Total	Soybean–Rice–Soybean			Total
Traditional	2.44	1.10	1.77	5.31	0.97	1.91	1.34	3.53	1.43	1.91	1.15	4.49
Improved, no lime or fertilizer	3.81	1.36	2.75	7.90	1.22	3.56	1.98	6.76	2.09	2.25	1.89	6.23
Improved, with lime and fertilizer	5.12	1.62	4.66	11.40	1.49	4.53	2.75	8.77	2.73	2.53	2.22	7.48

rotation might be 1 year of corn followed by 1 year of wheat, followed by 3 years of hay. In a prairie region, this resembles a short-rotation form of shifting agriculture. Crop rotation can be combined very easily with **contour tillage** (Figure 23.12), a method of plowing around rather than up and down slopes. This reduces the overland flow of water and, hence, surface soil erosion. When alternating strips of grass and crops are included, an efficient barrier to erosion is created.

Crop rotation can also be used to increase nutrient availability. A common rotation now in the Corn Belt is between corn and soybeans. Soybeans are legumes and, as such, can fix significant amounts of N from the atmosphere. This can reduce the need for N fertilizer while producing a crop of higher protein (and N) content than corn. The soybean rotation can increase N availability for the next crop as well. However, this contribution is diminished if heavy N fertilization is used during the soybean part of the rotation. High N availability inhibits N fixation.

No-till forms of agriculture do not remove all crop residue from fields after harvest. Rather, residues are left on the soil surface, forming the equivalent of a litter layer over the soil that reduces water runoff and erosion. This layer can then be plowed under in the spring to add to stores of soil organic matter. Mulches and wet manures have also been used as a soil cover to reduce erosion outside of the growing season. Of course, the ancient practice of returning animal manures to the fields has been used for centuries as a method for ensuring soil fertility.

Any of these practices can reduce soil erosion significantly (Table 23.5). All can be added to conventional farming systems with relatively little cost. In-

FIGURE 23.12 Contour tillage with interspersed strips of cropland and perennial grass. This combination both reduces erosion and provides a fallow period for cropland. (H. Armstrong Roberts)

creases in the total productivity of agricultural fields through fertilization and crop breeding, along with the increasing use of methods for preserving or augmenting soil organic matter, have actually begun to reverse the decline in total carbon storage in agricultural fields (Figure 23.13). This suggests that soil organic matter contents have begun to increase as well, reversing decades of decline and raising prospects for longer-term sustainability of modern temperate-zone practices.

The off-site effects of erosion and nutrient loss can be reduced by landscape-level management practices. For example, forest borders along streams running through an agricultural area can act as important sinks for nutrient exported from cropland (Figure 23.14). While it is not yet clear what the capacity of these riparian-zone forests is for absorbing nutrients, it is clear that some are still effective after decades of outside inputs. Methods of nutrient loss from these forests that are not always well measured, such as denitrification (loss of gaseous N_2 and N_2O to the atmosphere), may be important in keeping nutrients out of the stream. Of course, they then represent an input to the atmosphere and may have implications for global climate change (see Chapters 25 and 26).

A practice called **precision farming** is a recent development in the effort to both increase the profitability of agriculture and reduce environmental impact. One example of this was the use of research and an increased under-

TABLE 23.5 Losses of Cropland to Erosion Under Different Agricultural Practices
(From Pimentel et al. 1987)

Technology	Treatment	Soil Loss ($t \cdot ha^{-1} \cdot year^{-1}$)	Slope (%)	Country
Rotation	Corn–wheat–hay–hay–hay–hay	3.0	12	U.S.
	Continuous corn	44.0	12	U.S.
Contour planting	Potatoes on contour	0.2		U.S.
	Potatoes, up and down hill	32.0		U.S.
Rotation plus contour planting	Cotton on contour and grass strips	8.0		U.S.
	Continuous cotton planted up and down hill	200.0		U.S.
Terraces	Peppers on terraces	1.4	35	Malaysia
	Peppers on slope	63.0	35	Malaysia
Manure	Corn with 36 $t \cdot ha^{-1}$ of wet manure	11.0	9	U.S.
	Corn without manure	49.0	9	U.S.
Mulch	Corn planted on land with 6t $\cdot ha^{-1}$ of rice straw	0.1	5	Nigeria
	Continuous corn	148.0	5	Nigeria
Grass cover	Grass	0.08	10	Tanzania
	Plowed	13.6	10	Tanzania
No-till	Corn	0.14	15	Nigeria
	Conventional corn	24.0	15	Nigeria
Ridge planting—crop residues left in trenches on land surface	Corn	0.2	2	U.S.
	Conventional corn	10.0	2	U.S.

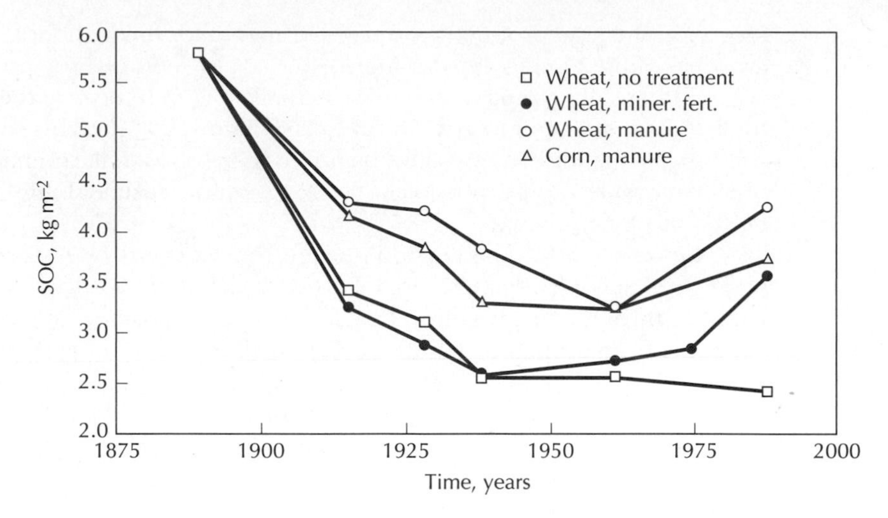

FIGURE 23.13 Changes in total soil organic matter content under four different long-term management practices in the American Midwest. (Buyanovsky and Wagner 1998)

standing of the effect of fertilization at different times of the year to reduce N inputs to wheat cultivation in Mexico. By moving the application of fertilizers to a later point in the growing season and reducing the amount added, economic returns to the farmer were increased, and losses of the important trace gases nitric and nitrous oxide to the atmosphere were reduced (Figure 23.15).

Methods are also being developed to reduce dependence on chemical pesticides. Integrated pest management is a rapidly growing field that employs a combination of biological and chemical methods for reducing pest damage. One problem with chemical pesticides is that they generally affect both the pest and the predators and parasites that help control the pest. Reintroduction of these predators and parasites can be effective in reducing pest problems. General examples of this approach include the introduction of parasitic wasps and bacterial pathogen sprays that attack specific groups of organisms and so selectively reduce pest species while leaving natural predator and disease organisms still viable in the system. In practice, chemical pesticides are also used in integrated pest management to ensure success, but the quantities used are generally lower than in chemical-only programs.

A final method of reducing input demands for modern agriculture is to alter the genetic makeup of the plant. This can be either by the standard practices of genetic selection or by genetic engineering. Breeding programs are currently under way to produce strains of the major crop species that are more tolerant of water and nutrient stress and are resistant to disease while still maintaining high rates of production. This approach is essentially fine-tuning previous selection efforts to meet the changing conditions.

These breeding programs require that the widest possible degree of genetic variability be available to the breeders. There has been growing concern that rigid selection for high yield, along with the disappearance in the

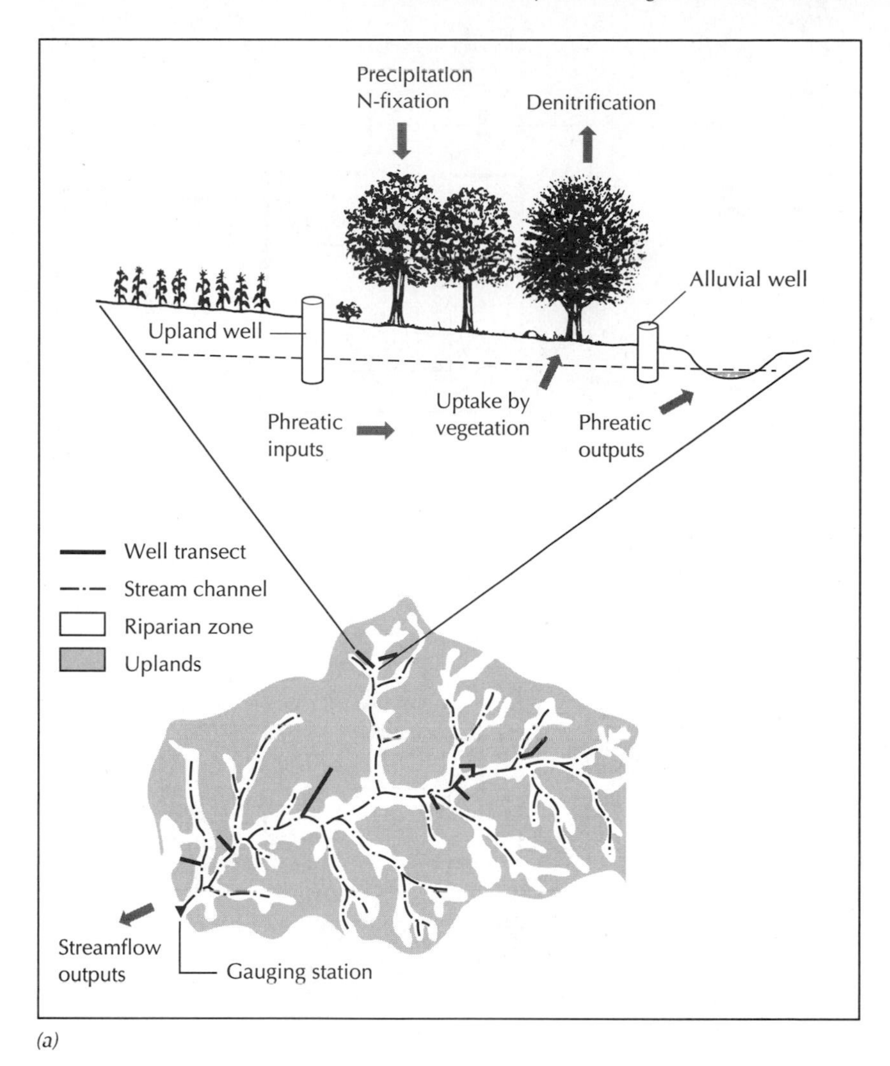

(*a*)

FIGURE 23.14 The role of riparian forests in reducing nitrogen losses to streams in agricultural regions. (*a*) Schematic diagram of Watershed N, a subcatchment of the Little River watershed in Georgia (note the presence of forests adjacent to stream channels). (*b*) The nitrogen, phosphorus, and calcium balances over the riparian forest zone. The riparian zone is a net sink for all three nutrients. Including denitrification, the riparian forest removes 37.9 kg of nitrogen per hectare from the water entering the adjacent stream. (Lowrance et al. 1984, © 1984 by the American Institute of Biological Sciences)

wild of the native species from which today's crop plants have been bred, will severely limit our ability to breed new varieties. In response to this, there are now efforts under way to preserve both the genetic information held in the seeds of wild plants and the ecosystems in which these plants naturally grow.

Genetic engineering offers the potential to make extremely precise changes in the genetic makeup of crop species. Experimental approaches include the insertion of genes that will produce chemicals toxic to potential

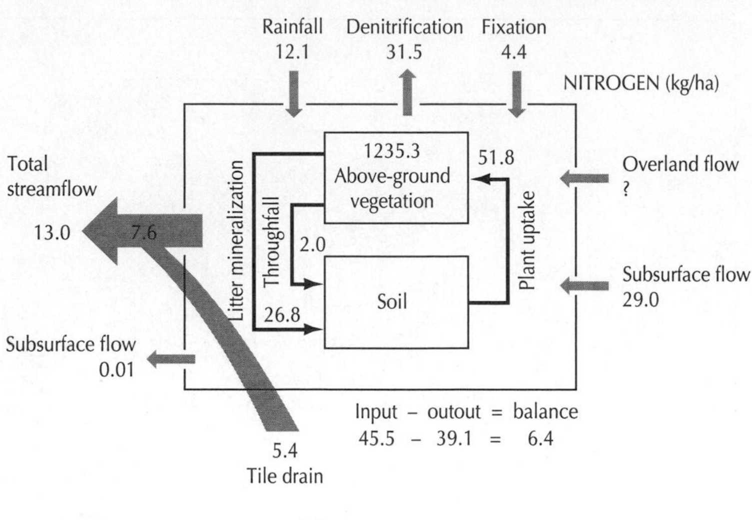

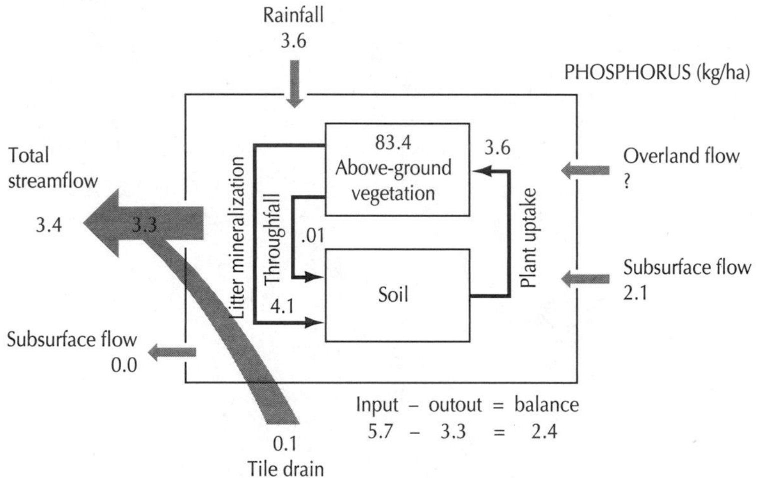

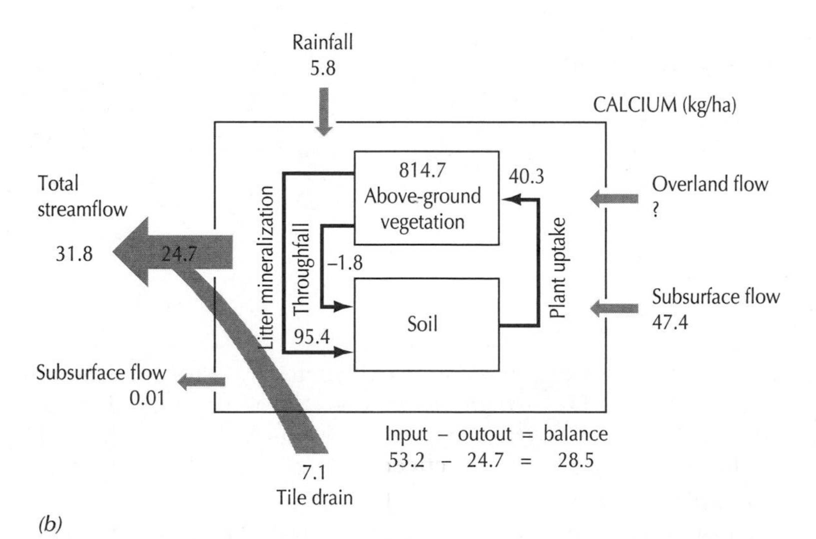

FIGURE 23.14 ***(continued)***

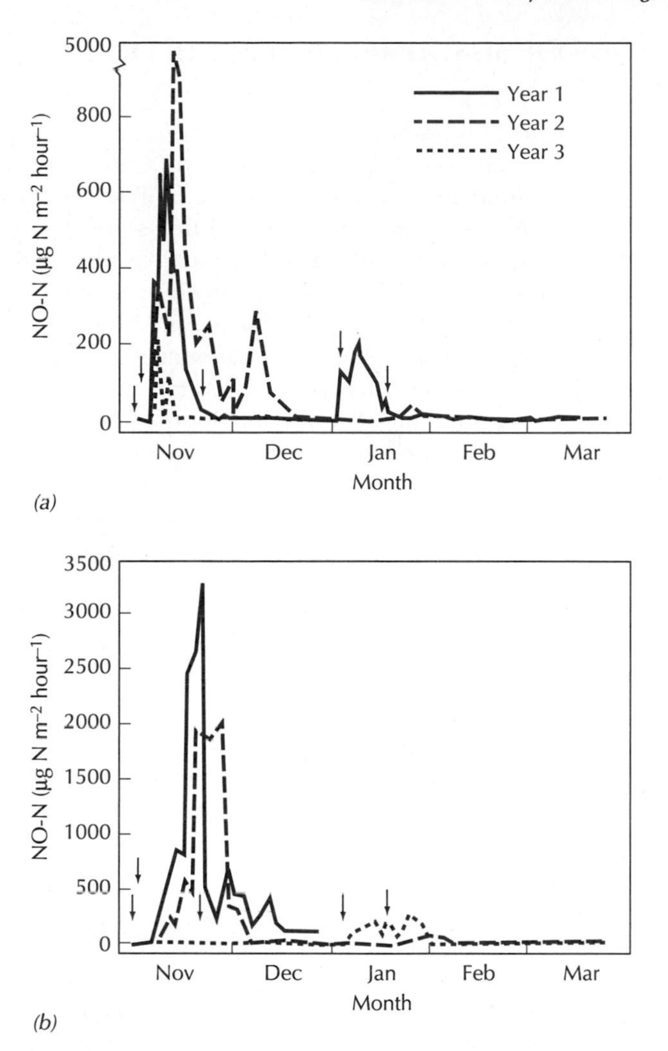

FIGURE 23.15 Changes in fluxes two nitrogenous gases [(*a*) N_2O–N and (*b*) NO–N] important in greenhouse warming and atmospheric chemistry from a wheat field in Mexico with and without "precision farming" methods for the timing of fertilizer applications. In each panel, the 2 years with high losses are for 2 years with regular practices; the lower year is with precision farming practices. (Matson et al. 1998)

pest species (effectively introducing human intervention into the predator–prey chemical dance described in Chapter 15). While the ability to manipulate genetic material is increasing rapidly, consumer resistance to the purchase of manipulated plants is rising as well.

Other, more experimental approaches to increasing the productivity and sustainability of agriculture in the temperate zone exist. From the total exclusion of all chemicals, as practiced by a growing number of organic farmers, to the introduction of totally new species of crop plants, to the total enclosure of the food production and waste treatment system in a recycling life support system—these all contribute to the goal of understanding the potentials for sustaining a viable food-production system for a growing human population.

HISTORICAL METHODS OF SUSTAINABLE AGRICULTURE IN THE TROPICS

As a counterpoint to these modern methods, there are ancient precedents for continuous crop production, even under conditions considered most challenging, for example, nonvolcanic tropical soils. The best known is paddy rice production. Paddy rice is grown in flooded fields using very labor-intensive methods of planting, cultivation, and harvest (Figure 23.16*a*). The field can be on level ground in floodplain areas, or even on steeply sloping hillsides that have been terraced to produce flat areas (Figure 23.16*b*). The flat and flooded paddies provide excellent protection against erosion and can actually trap sediments from the water used in flooding and result in a buildup of soils. The presence of blue–green algae in the stagnant water is also thought to provide N inputs to the system through N fixation from the atmosphere. The combination of sediment trapping, reduced erosion, and N fixation reverses the usual pattern of nutrient loss from cultivated land. Some paddy systems in Asia have been in operation for thousands of years.

A New World version of the rice paddy is the chinampas system, developed in pre-Columbian Mexico and still in use today in low-lying wetland areas (Figure 23.17). Chinampas consist of raised plots surrounded by moats or canals filled with water. The moat traps runoff and erosion and reduces these losses. Plant growth, which is luxuriant in the moat, is harvested regularly and mounded up onto the plot, along with trapped sediments. This system provides a constant supply of organic matter and nutrients to the soil within the plot, including what would have been lost in runoff. Productivity in these plots is very high.

CHARACTERISTICS OF SUSTAINABLE AGROECOSYSTEMS

It should be apparent from this brief review that sustainable, high-productivity agriculture, no matter where it is practiced, involves both a balance between gains and losses in soils and nutrients and a battle against damage inflicted by pests and diseases. Intensive temperate-zone systems replace nutrients lost in harvest and erosion through chemical fertilization and mechanical cultivation of soils, and replaces natural herbivore defense compounds in plants with chemical pesticides. Thus, we are living to a certain extent on the historical forces that created rich prairie soils and on the largesse allowed by our current availability of fossil fuels. It has been said that all of the most productive soils in the world are on transported material (i.e., the loess soils of the American Midwest and the volcanic soils of both tropical and temperate regions). In the temperate zone, erosion is rapidly retransporting those soils off the farm.

Sustainable upland systems in the tropics that rely on natural processes for rebuilding soils require long fallow periods and thus produce low yields per unit area. (There are plenty of precedents in the temperate zone for the exhaustion of soils by preindustrial agriculture, the most dramatic of which might be the cultivation of tobacco in colonial America). Intensive tropical systems that have proven sustainable for centuries minimize losses, increase inputs from outside the system, or both.

FIGURE 23.16 Paddy rice culture in Asia. (*a*) Preparing plots with water buffalo and human labor. (*b*) Continuous rice production in terraced hillside slopes in the Philippines. (H. Armstrong Roberts)

This discussion has been limited to crop production systems and has not dealt with animal systems. Many of these, such as moderate-intensity pasture systems and extensive grazing systems, are highly sustainable. As long as grazing does not reduce plant cover to the point at which erosion becomes important, or palatable species are reduced and replaced by unpalatable ones, they can be permanent systems requiring only low inputs. However, the quantity of food produced is relatively low. The inefficient conversion of plant mass to animal mass (a maximum of 10%, and often less) is something of a luxury, except in those areas where grazing animals can convert plant mass indigestible by humans into meat. Efficient systems would include primarily

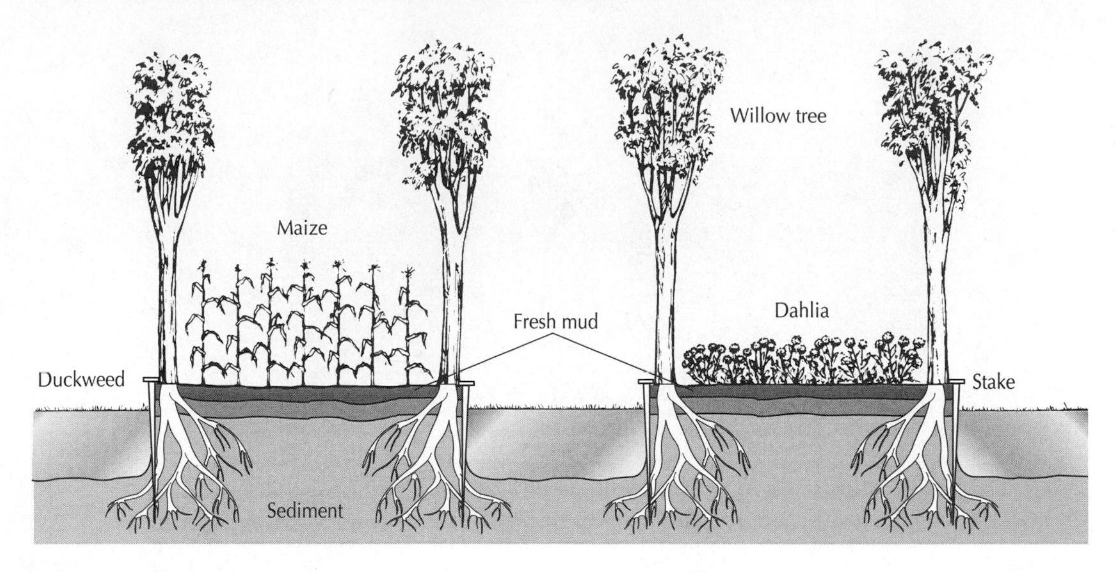

FIGURE 23.17 The Chinampas system of agriculture in Mexico. (Cox and Atkins 1979)

extensive grazing systems in semiarid grasslands and certainly would not include the highly centralized, grain-based feedlots found in developed countries. In general, human domestication of the pasture-grazing system (management of the plant community, removal of predators other than humans) has been found to increase the production of animal biomass by 10-fold over that seen in natural systems (Figure 23.18).

RELATION TO CONSERVATION OF NATIVE ECOSYSTEMS

The management of agricultural ecosystems cannot be separated from discussions of the conservation of wild ecosystems. A growing world population means increasing demands for food. This demand can be met by increasing the intensity of management on existing agricultural lands or by increasing the amount of land in agriculture. Both of these can have negative environmental implications. Expanding the base of agricultural land will mean a drain on the small stock of remaining area in native ecosystems. This can even affect areas nominally protected in parks and reserves, as indicated by the increasing poaching pressure in the Serengeti Park (Chapter 20). More intensive agriculture can bring increased pollution due to increased use of agricultural chemicals and fossil fuels unless new systems less dependent on these inputs are developed and applied.

Three avenues seem open: (1) controlling world population growth; (2) increasing the intensity of agricultural production, particularly in the tropics; and (3) increasing the amount of land in agriculture. A fourth option—failing to provide enough food to meet the human demand—is very real but unthinkable. Conservation of natural ecosystems means choosing one of the first two options.

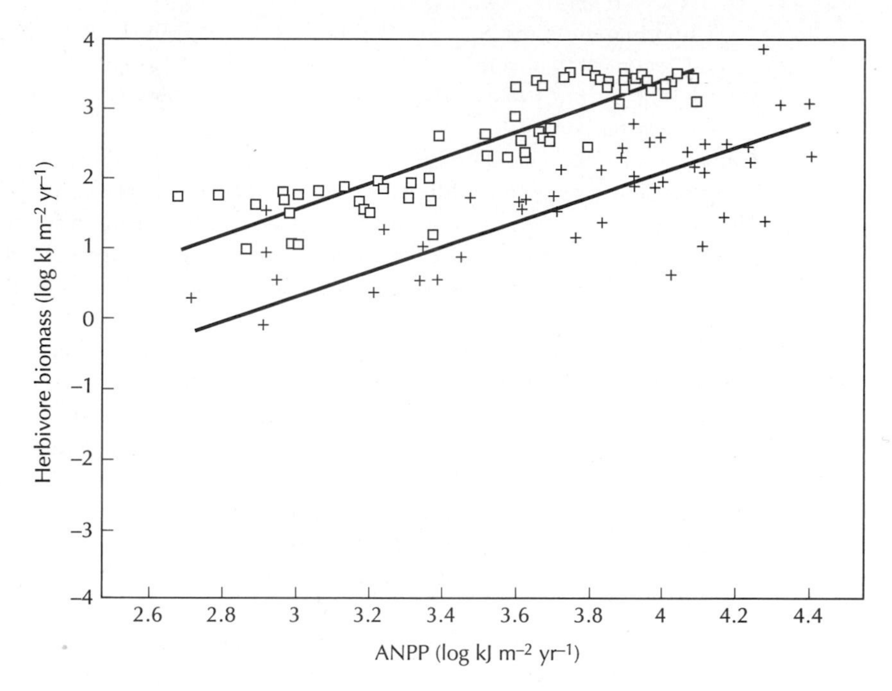

FIGURE 23.18 Relationship between net above-ground primary production (ANPP) and herbivore biomass for natural (+) and agricultural (□) ecosystems. Note that both the x- and y-axes are log scales. An increase of 1 unit on the y-axis equals a 10-fold increase in herbivore biomass. (Oesterheld et al. 1992)

REFERENCES CITED

Aber, J. D., J. M. Melillo, and C. A. Federer. 1982. Predicting the effects of rotation length, harvest intensity and fertilization on fiber yield from northern hardwood forests in New England. *Forest Science* 28:31–45.

Blair, J. M., T. R. Seastedt, C. W. Rice, and R. A. Ramundo. 1998. Terrestrial nutrient cycling in tallgrass prairie. In Knapp, A. K., J. M. Briggs, D. C. Hartnett, and S. L. Collins (eds.). *Grassland Dynamics: Long-term Ecological Research in Tallgrass Prairie.* Oxford University Press, New York.

Boyer, J. S. 1982. Plant productivity and environment. *Science* 218:443–448.

Brown, B. J., and J. J. Ewel. 1987. Herbivory in simple and complex tropical successional ecosystems. *Ecology* 68:108–116.

Buyanovsky, G. A., and G. H. Wagner. 1998. Changing role of cultivated land in the global carbon cycle. *Biology and Fertility of Soils* 27:242–245.

Cox, G. W., and M. D. Atkins. 1979. *Agricultural Ecology.* W. H. Freeman & Co., San Francisco.

Dyson, T. 1999. World food trends and prospects to 2025. *Proceedings of the National Academy of Sciences* 96:5929–5936.

Feller, M. C., J. P. Kimmins, and K. A. Scoullar. 1983. FORCYTE-10: Calibration data and simulation of potential long-term effects of intensive forest management on site productivity, economic performance, and energy benefit/cost ratio. In IUFRO

Symposium on Forest Site and Continuous Productivity. U. S. Forest Service General Technical Report PNW-163. U. S. Department of Agriculture, Washington, DC.

Frissel, M. J. 1978. *Cycling of Mineral Nutrients in Agricultural Ecosystems.* Elsevier Scientific, Amsterdam.

Gilliam, F. S. 1987. The chemistry of wet deposition for a tallgrass prairie ecosystem: Inputs and interactions with plant canopies. *Biogeochemistry* 4:203–218.

Hornbeck, J. W. 1977. Nutrient: a major consideration in intensive forest management. In Proceedings of the Symposium on Intensive Culture of Northern Forest Types. U. S. Forest Service General Technical Report NE-29. U. S. Department of Agriculture, Washington, DC.

Jorgensen, J. R., C. G. Wells and L. J. Metz. 1975. The nutrient cycle: Key to continuous forest production. *Journal of Forestry* 73:400–403.

Lowrance, R., R. Todd., J. Fail, O. Hendrickson, R. Leonard, and L. Asmussen. 1984. Riparian forests as nutrient filters in agricultural watersheds. *BioScience* 34:374–377.

Marks, P. L. 1974. The role of pin cherry (*Prunus pennsylvanica L.*) in the maintenance of stability in northern hardwood ecosystems. *Ecological Monographs* 44:73–88.

Matson, P. A., R. Naylor, and I. Ortiz-Monasterio. 1998. Integration of environmental, agronomic and economic aspects of fertilizer management. *Science* 280:112–115.

Nair, P. K. R., R. J. Buresh, D. N. Mugendi, and C. R. Latt. 1999. Nutrient cycling in tropical agroforestry systems: Myths and science. In Buck, L. E., J. P. Lassoie, and E. C. M. Fernandes (eds.). *Agroforestry in Sustainable Agricultural Systems.* CRC Press, Boca Raton.

Nambiar, E. K. S. 1996. Sustained productivity of forests in a continuing challenge to soil science. *Soil Science Society of America Journal* 60:1629–1642.

Nicholaides, J. J., D. E. Bandy, P. A. Sanchez, J. R. Benites, J. H. Villachica, A. J. Coutu, and C. S. Valverde. 1985. Agricultural alternatives for the Amazon basin. *BioScience* 35:279–285.

Noble, I. R., and R. Dirzio. 1997. Forests as human-dominated ecosystems. *Science* 277:522–525.

Oesterheld, M., O. E. Sala, and S. J. McNaughton. 1992. Effect of animal husbandry on herbivore-carrying capacity at a regional scale. *Nature* 356:234–236.

Pastor, J., and J. G. Bockheim. 1981. Biomass and production of an aspen-mixed hardwood Spodosol ecosystem in northern Wisconsin. *Canadian Journal of Forest Research* 11:132–138.

Pimentel, D., and A. Wightman. 1999. Economic and environmental benefits of agroforestry in food and fuelwood production. In Buck, L. E., J. P. Lassoie, and E. C. M. Fernandes (eds.). *Agroforestry in Sustainable Agricultural Systems.* CRC Press, Boca Raton.

Pimentel, D., J. Allen, A. Beers, L. Guinand, R. Linder, P. McLaughlin, B. Meer, D. Musonda, D. Perdue, S. Poisson, S. Siebert, K. Stoner, S. Salazar, and A. Hawkins. 1987. World agriculture and erosion. *BioScience* 37:277–283.

Sanchez, P. A. 1982. Nitrogen in shifting cultivation systems of Latin America. *Plant and Soil* 67:91–103.

Sims, P. L. and J. S. Singh. 1978. The structure and function of ten western North American grasslands: III. Net primary production, turnover and efficiency of energy capture and water use. *Journal of Ecology* 66:573–597.

Timmons, D. R., and A. S. Dylla. 1983. Nitrogen inputs and outputs for an irrigated corn ecosystem in the northwest corn belt. In Lowrance, R. R., R. L. Todd, L. E. Asmussen, and R. A. Leonard (eds.). *Nutrient Cycling in Agricultural Ecosystems.* College of Agriculture Special Publication No. 23. University of Georgia, Athens, GA.

Uhl, C., C. F. Jordan, and F. Montagnini. 1983. Traditional and innovative approaches to agriculture on Amazon Basin tierra firme sites. In Lowrance, R., R. Todd, L. Asmussen, and R. Leonard (eds.). *Nutrient Cycling in Agricultural Ecosystems.* College of Agriculture Special Publication No. 23. University of Georgia, Athens, GA.

Waide, J. B., and W. T. Swank. 1977. Simulation of potential effects of forest utilization on the nitrogen cycle in different southeastern ecosystems.

Weetman, G. F., and B. Weber. 1972. The influence of wood harvesting on nutrient status of two spruce stands. *Canadian Journal of Forest Research* 2:351–369.

Woods, L. E., R. L. Todd, R. A. Leonard, and L. E. Asmussen. 1983. Nutrient cycling in a southeastern United States agricultural watershed. In Lowrance, R., R. Todd, L. Asmussen, and R. Leonard (eds.). *Nutrient Cycling in Agricultural Ecosystems.* College of Agriculture Special Publication No. 23. University of Georgia, Athens, GA.

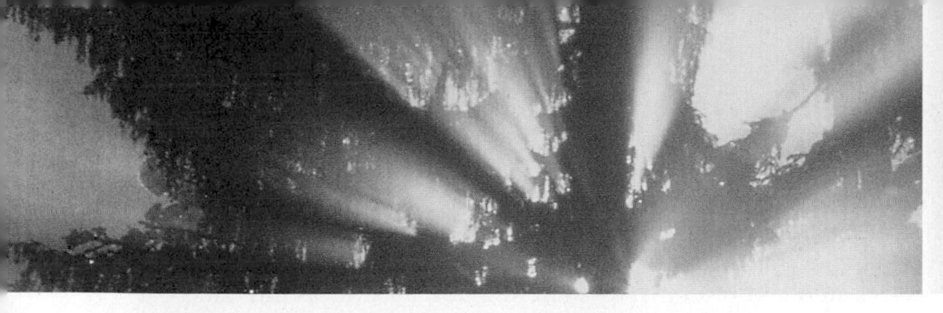

24

BIODIVERSITY AND ECOSYSTEM FUNCTION

INTRODUCTION

Most forms of human activity alter the distribution and abundance of plant and animal species.

Agricultural development generally entails the wholesale substitution of crop species for native plants. While recent advances in agroforestry and other complex, multispecies forms of agriculture incorporate some of these species, the domestication of the landscape invariably means a reduction in the number of species present. When new agricultural development occurs in areas where plants have not been preadapted to succeed in the face of the repeated disturbances that are part of agricultural management, then invasive "weedy" species from other regions or continents can gain a foothold and spread rapidly, becoming an important part of disturbed or even undisturbed communities in the new region. In North America, for example, as much as one third of the species in old fields and pastures can be of European origin.

Global transport of goods and products and regional to global distribution of pollutants are now defining characteristics of our world. Bulk shipments of grains, wood, and other commodities can bring with them seeds, spores, or live specimens of new invasive species, new insect herbivores, or plant and animal pathogens against which the native species have no preevolved defenses. Where these new species become established, the composition of the community can be changed significantly. Regional to global changes in atmospheric chemistry and climate, as described in Chapters 25 and 26, can alter the environment, shifting the competitive balance between species and altering composition.

Biodiversity was initially coined as an ecological term describing the number of species in a study area or the distribution of numbers of organisms among species. It has now come to hold a broader meaning in common usage, encompassing processes of evolution and extinction, especially as affected by human activity. Biodiversity has become a keyword for the integrity of the biological world, and the loss of biodiversity, as a primary indicator of the impact of human society on that world.

Biodiversity is a very rich and mature field—the subject of many more volumes than is ecosystem studies—and we cannot begin to present a full view of it here. However, a much smaller body of literature deals explicitly with the interactions between diversity and ecosystem function. We concentrate here specifically on the role that the gain and loss of particular species, or the overall number and diversity of species in an area, has on the dynamics of ecosystems. The interactions are presented in both directions: (1) as the effects of species gain and loss on ecosystem function and (2) as the effects of natural and altered ecosystem function on the diversity of the plant and animal communities.

In examining the role of biodiversity in ecosystem studies, the question is often posed in this way: Does the presence of one species rather than another, or the gain or loss of a species, cause a shift in ecosystem function? We have already addressed this question several times in previous chapters. In Chapter 14, we looked at the effects of cation pump and nitrogen (N)-fixing species on soil properties and nutrient cycling, of which the invasion of the exotic *Myrica fava* into young lava flows in Hawaii (Chapter 22) is an example. In Chapter 21, we examined the likely course of succession following natural or managed disturbances in the presence or absence of pin cherry. Several examples of large herbivores or predators, which can alter ecosystem structure directly or indirectly, were presented in the second chapter on herbivory (Chapter 16), including deer, moose, wolves, locusts, and lemmings. The role of elephants in restructuring the woodlands in the Serengeti was presented in Chapter 20, and the complete restructuring of soils in tropical systems by termites in Chapter 15.

The purpose of this chapter is to focus the discussion on the role of species in directing ecosystem development, which has been presented piecemeal throughout the book. We will discuss the interaction between biodiversity in its broadest meaning and ecosystem function.

DEFINING BIODIVERSITY

Biodiversity has traditionally been defined in one or both of two ways: (1) species richness and (2) equitability.

Species richness is simply an exhaustive listing of all of the species of a given life from (e.g., vascular plants, mammals, or birds) that exist in an area. Data collected can be presented in terms of the number of species in total or divided into taxonomic groupings. While this is simple in concept, in areas with a large number of species, individuals of rare species may be encountered only if the search area is very large. Thus, the larger the area, the longer the species list, making comparisons difficult. This has been overcome in part by the development of species–area curves, which describe how the number of species encountered increases with the area searched. In Figure 24.1, the number of species encountered with increasing search area within tropical forests increases much more rapidly in the moist, evergreen systems than in drier or disturbed areas and has a much higher level of biodiversity.

Species equitability is a more difficult characteristic to measure. This requires not only a list of all of the species present but also a measure of the importance of each (total number of individuals or total biomass per species have both been used). With these data in hand, a calculation is made in which the derived number increases both with the number of species present and when the relative importance of different species is more equitable. One

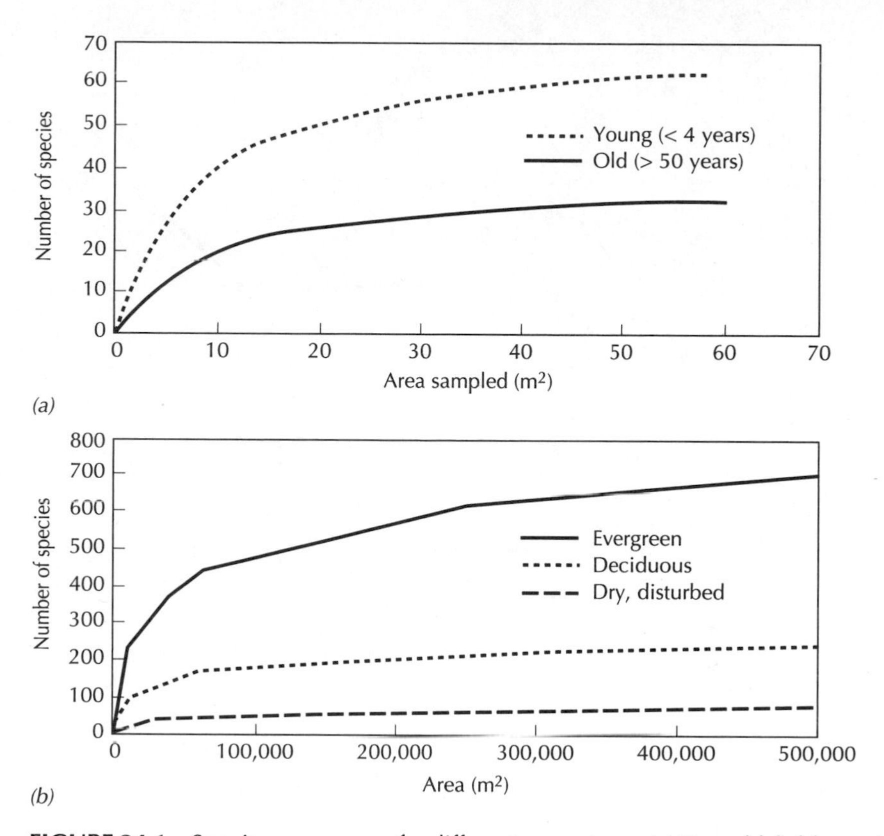

FIGURE 24.1 Species area curves for different ecosystems. (*a*) Two old-field grasslands of different ages in Minnesota. (*b*) Three tropical forest/woodland systems: a moist evergreen forest, a seasonally dry deciduous forest, and a dry forest subject to harvests, fires, and elephant damage. Note the much larger area required to capture all species in the forest systems dominated by larger individual organisms (trees versus grasses). (*a*, Inouye 1998; *b*, Condit et al. 1996)

of the first equations used for this calculation was borrowed from information sciences and was developed as a measure of the information content of an electronic signal.

BIODIVERSITY AT THE GLOBAL SCALE: EVOLUTION AND EXTINCTION

What we can see and measure as the current assemblage of species and their distribution over the Earth is the product of the ongoing processes of evolution and extinction of species. In the roughly 4-billion-year history of the Earth, land plants did not appear until approximately 400 million years ago. Over the geological ages since that time, the total number of land species has increased nearly linearly as the terrestrial environment has become more structurally and biologically complex (Figure 24.2). This increase does not result from a continuous emergence of new species along with continuation of the old, but involves both evolution and extinction.

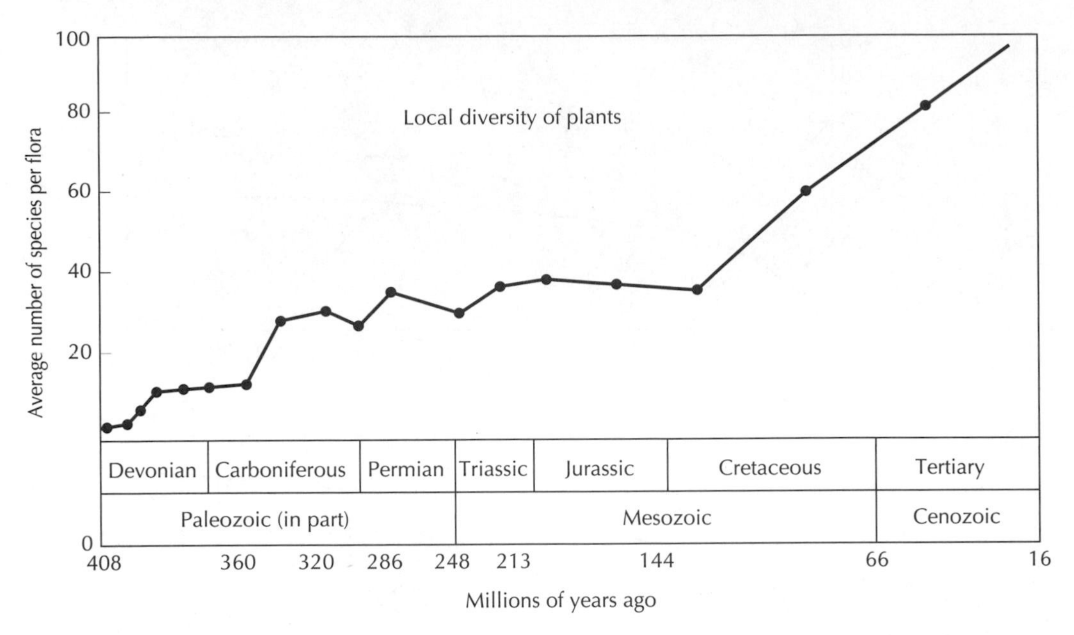

FIGURE 24.2 The average number of plant species in local assemblages has increased steadily since the first appearance of land plants 400 million years ago. (Wilson 1992)

One species can evolve into two whenever separate subpopulations are isolated from one another long enough for random genetic events, different environments, and natural selection to lead to the accumulation of genetic and behavioral differences that preclude successful breeding. If the ranges of these two species come to overlap once again, one may displace the other in all or part of the shared range, depending on the fitness of the two forms within the environment. If partial displacement occurs, then there are two species where there was one before, and biodiversity has increased. If one species is totally displaced by the other, it will become extinct, leaving a single, presumably better-adapted species in place, and biodiversity is unchanged.

PATTERNS OF BIODIVERSITY IN TERRESTRIAL ECOSYSTEMS

Two classic questions in the study of population ecology are: (1) Why are there so many species? and (2) What controls the pattern of species richness across the Earth's ecosystems? Answering these questions lies well beyond the scope of this book, but some general patterns of importance to the rest of this chapter are presented.

In a very general way, species richness tends to increase from the poles to the tropics (Table 24.1) and from low-productivity ecosystems to high. One characteristic of the increase in diversity with production is the greater structural complexity of vegetation in more productive systems. For example, within a region, the diversity of bird populations tends to increase with the diversity in the structure of the canopy (Figure 24.3). Because the accumulation of species through evolution is a very long-term process, the age of an

TABLE 24.1 Estimated Increases in the Number of Species in Major Taxonomic Groups from Polar to Equatorial Regions in the Northern Hemisphere Found in Roughly Equivalent Sampling Areas.
(Data compiled from Dobzhansky 1950, Kusnezov 1957, Wilson 1974)

	Group		
Region	**Ants**	**Birds**	**Mammals**
Arctic/Boreal	10	50	40
Temperate	75	100	100
Tropical	150	1500	150

ecosystem type and the stability of that type, or the existence of predictably repeatable disturbance regimes, may also increase diversity.

A key characteristic linked to species diversity is the concept of **niche packing,** or the degree to which individual species can divide existing environmental gradients and still maintain sufficient population size to ensure persistence. In evolutionarily old systems such as tropical forests, theory holds that longer periods of evolution have led to more finely divided niche spaces and higher species diversity, while in recently glaciated systems such as boreal and many temperate forests, recreation of the system from bare soil and a greatly diminished and dispersed biota resets the evolutionary accumulation of species and the packing of niches. In this case, we should think of the

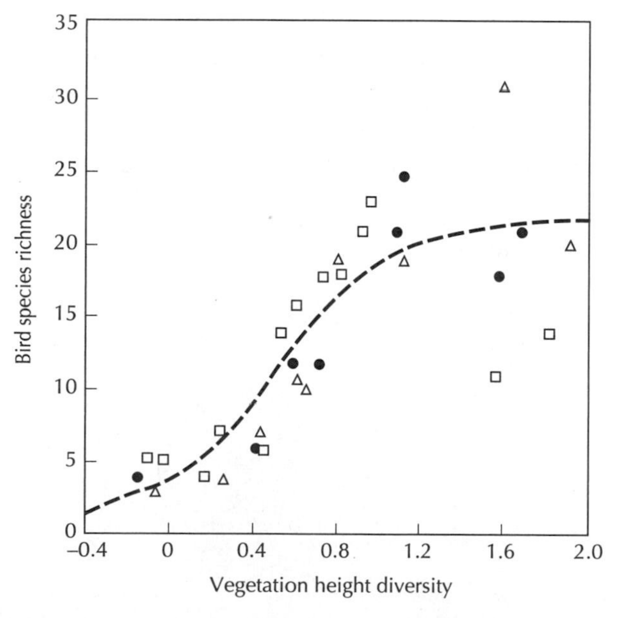

FIGURE 24.3 The relationship between the diversity in canopy structure of semiarid shrubland habitats and the number of bird species present. (After Cody 1975)

niche defined not strictly by availability of resources for growth but also as the complex interactions among species. Much of the differentiation in tropical systems leading to high species richness results from interactions between plants and pollinators, plants and seed dispersers, plants and herbivores, herbivores and predators, and all of these with pathogens.

There is another classic area of community ecology related to both species–area curves and biodiversity: island biogeography. Early naturalists often noted that islands had both a higher percentage of endemic species (those existing in no other place) and a lower total number of species in any one taxonomic group than similar habitats on continental mainlands. Twentieth-century resurveys of earlier lists of island populations, such as those conducted for bird species on the Channel Islands off the coast of Southern California (Table 24.2), would show a similar number of total species present but a large amount of turnover in those species, with extinctions and immigrations occurring at approximately the same rate.

These trends were elaborated into a general theory in which the introduction of new species was related to distance to mainland sources and number of species already present, and extinctions to the size of the island and, therefore, the potential for random fluctuations in populations to cross the threshold of local extinction (Figure 24.4). One prediction of this theory is that larger islands should support a greater total number of species—a trend that has been measured in several cases (e.g., Figure 24.5). An experimental test of the theory involved the removal of all arthropods from four small mangrove islands just a short, but variable, distance off the coast of one of the Florida Keys. In this case, recolonization was rapid, and the total number of species present after 300 days was similar to pretreatment levels, with the closer islands supporting a larger number of species. The actual species

TABLE 24.2 Changes in Species Richness and Species Composition in the Bird Community on the Channel Islands Off the Coast of Southern California Between 1917 and 1968. Extinctions Are Species Present in 1917 and not in 1968. Additions are Species Present in 1968 and not in 1917, and Include a Limited Number of Game Birds Introduced by Humans. Note that the Number of Species Changes Little When the Turnover Rate Is High. Turnover Is Calculated as the Mean of Extinctions and Additions Divided by the Mean of the Number of Species in 1917 and 1968.
(After Diamond 1969)

	Number of Species				
Island	**1917**	**1968**	**Extinctions**	**Additions**	**Turnover (%)**
Los Coronados	11	11	4	4	36
San Nicolas	11	11	6	6	55
San Clemente	28	24	9	5	27
Santa Catalina	30	34	6	10	25
Santa Barbara	10	6	7	3	63
San Miguel	11	15	4	8	46
Santa Rosa	14	25	1	12	45
Santa Cruz	36	37	6	7	18
Anacapa	15	14	5	4	31

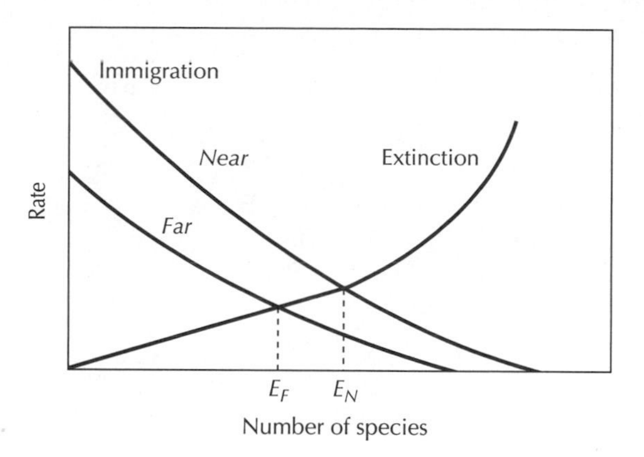

FIGURE 24.4 Hypothesized trends in rates of introductions of new species to islands and extinctions of existing species as a function of the number of species present and the size of the island. (MacArthur 1972)

present were not the same before and after the treatment; only the total number was similar. It should be noted that this rapid recolonization is from existing and nearby populations and does not occur for species that become extinct globally.

The theory of island biogeography can be extended well beyond literal islands. Isolated habitats on mountaintops, in small wetlands within a matrix of upland sites, or patches of woodlands within a grassland system all represent habitat islands. On a larger scale, many of our treasured ecosystems preserved within national parks and other reserves represent islands of native ecosystems surrounded by managed or otherwise disturbed areas. The theory of island biogeography can help to predict the size of the preserve required to maintain viable populations of key species.

HUMAN EFFECTS ON BIODIVERSITY AND CONSEQUENCES FOR ECOSYSTEMS

Humans are now the major force controlling rates of change in biodiversity over the planet. Through modification of the land surface for agriculture, forestry, and urbanization, significant portions of the land area of the Earth have been modified (see Chapter 23). As part of these same activities, we move tremendous quantities of plant material (e.g., grains, wood, ornamental plants, pets, and domestic animals) over tremendous distances. Ecosystems in various parts of the world support unique assemblages of plants and animals in large part because of barriers to migration presented by the intervening oceans. Human activity has done much to remove these barriers, with the result that invasions of species into new areas has been greatly accelerated. This can have important effects on biodiversity and also on ecosystem function. Biodiversity can also be impacted indirectly by modification of the environment that alters the fitness of species for a given land area. We discuss each of these—extinction, invasion, and environmental modification—in turn.

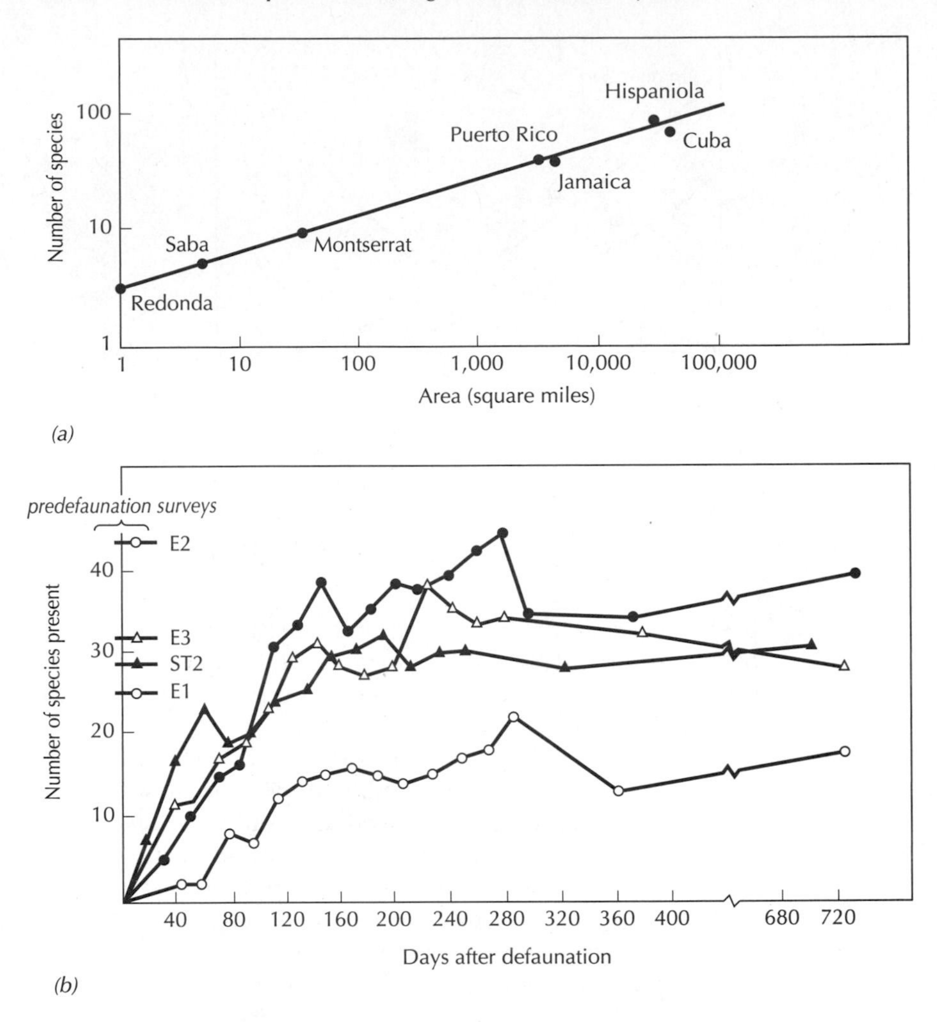

FIGURE 24.5 Tests of the theory of island biogeography. (*a*) Relationship between island size and the number of amphibians and reptiles present for several Caribbean islands. (*b*) Timing and extent of arthropod species recolonization of fumigated mangrove islands off the Florida Keys. (*a,* MacArthur and Wilson 1967, cited in MacArthur 1972; *b,* Simberloff and Wilson 1970)

Extinction, Loss of Biodiversity, and Ecosystem Function

Causes and Rates of Extinctions – We mentioned earlier that extinction has always been a part of the evolution of biodiversity. However, rates of extinctions are not constant over time. While low levels of species turnover occur at all times, there have been several epochs in which huge pulses of extinctions have occurred (Figure 24.6). The recognition of these epochal events, which mark the boundaries between major geologic eras in the history of the Earth, has occurred relatively recently, and the causes remain a topic of heated scientific debate. Explanations range from shifts in the distribution of continents by continental drift, to drastic changes in climate, to the evolution of new life forms, to the impact of meteorites. Clearly, these four causes are not mutually exclusive. One plausible explanation for the end of the Cretaceous Period, identified in part by the disappearance of the dinosaurs, includes a very large meteor strike that would have injected huge masses of dust into the atmosphere, generated volcanoes and tidal waves, altered the chemistry of the atmosphere, and changed the climate regime sufficiently to make the Earth uninhabitable for a wide range of plants and animals. The follow-on or feedback effects of drastic shifts in dominant life forms could further extend the string of extinctions over many millennia. Indeed, the fossil record suggests that massive extinctions occurred over a period of about 1 million years. The time required for evolution to return biodiversity to pre-extinction levels ranges from 25 to 100 million years.

It appears that humans as hunters have long been a cause of extinctions, especially of large, hunted animals. Significant megafauna extinction events in Australia and North America co-occur with the migration of humans to these continents (Figure 24.7). In modern times, hunting continues as a cause of the loss of aesthetically important species, but the combined effects of agriculture and other practices that alter land use and reduce species habitat have become the dominant threat (Table 24.3). The species–area curve (Figure 24.1) and island biogeography give us some insight into this process. Rare species are those encountered only when a large area is searched. When the total area of a habitat type is reduced, the tail of this distribution, the rare species, are lost. When large areas of intact habitat become fragmented into smaller islands within a matrix of disturbed or managed lands, the diversity of each patch is reduced in proportion to the size of the patch and the distance to adjacent patches.

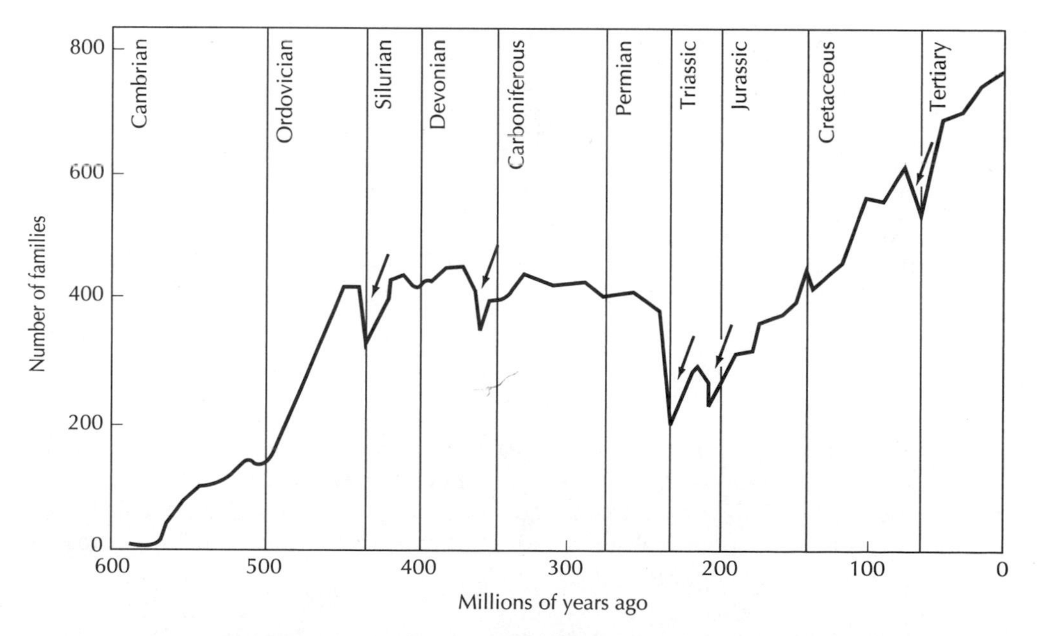

FIGURE 24.6 The timing of major extinction events over the geological history of the Earth. Note that each event marks the transition to a new geological age. Data are for families of marine organisms. (Wilson 1992)

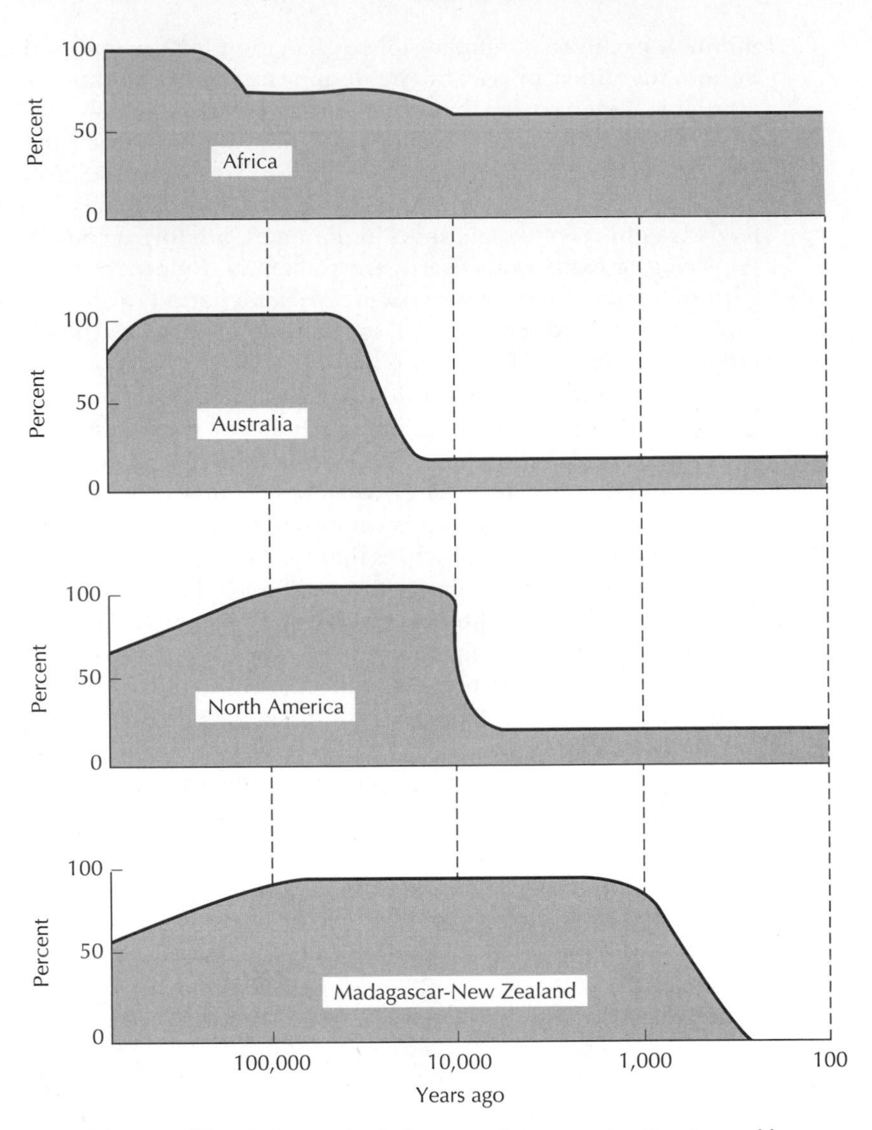

FIGURE 24.7 The timing and relative completeness of extinctions of large mammals and flightless birds in three areas of the world. In each case, the period of rapid loss of species coincided with the arrival of humans. (Wilson 1992)

Because the total number of plants and animals on the Earth today is not well known, rates of extinctions at the species level also cannot be clearly enumerated. Still, it is clear that a new cause for extinctions, human activity, is at work, and the potential for creating an extinction event similar to those in Figure 24.6 is very real.

Relationship Between Species Richness and Ecosystem Function – It should be made clear, before we head into this discussion of the diversity–function interaction, that ecosystem function is only one, and not the most important,

TABLE 24.3 The Reasons for Population Reductions That Have Resulted in Species Designations as Threatened or In Danger of Extinction in the United States.
(Czech et al. 2000)

Cause	Number of Species
Interactions with non-native species	305
Urbanization	275
Agriculture	224
Outdoor recreation and tourism development	186
Domestic livestock and ranching activities	182
Reservoirs and other running water diversions	161
Modified fire regimes and silviculture	144
Pollution of water, air, or soil	144
Mineral, gas, oil, and geothermal extraction or exploration	140
Industrial, institutional, and military activities	131
Harvest, intentional and incidental	120
Logging	109
Road presence, construction, and maintenance	94
Loss of genetic variability, inbreeding depression, or hybridization	92
Aquifer depletion, wetland draining or filling	77
Native species interactions, plant succession	77
Disease	19
Vandalism (destruction without harvest)	12

of several reasons for maintaining biodiversity. The moral and ethical imperatives of conserving a planet and a biota created over billions of years carries tremendous weight on its own.

Three types of responses in ecosystem function to changes in species richness have been hypothesized (Figure 24.8). In a type 1 response, the measured function (production, nutrient cycling, etc.) would increase continuously with increasing species richness. A type 2 response would be asymptotic, increasing with new species added, but at a decreasing rate. A type 3 response would suggest that only one species of each functional type would be required to maximize ecosystem function. A type 2 response has been hypothesized as the most common pattern.

At the heart of these alternate hypotheses is the concept of redundancy. In a system with high species richness, is every species obtaining the basic resources required for life in slightly different ways, such that the presence of each additional species increases the efficiency of resource use and, hence, productivity or nutrient cycling (the type 1 response in Figure 24.8)? Knowing what we do about the physiology of resource acquisition (Part II), it seems unlikely that each of the 200–300 species in a hectare of tropical rain forest is seeing a different resource base. The extent to which different species perform the same functions in the same way suggests that physiological redundancy in species-rich systems would minimize the effects of the loss of any one

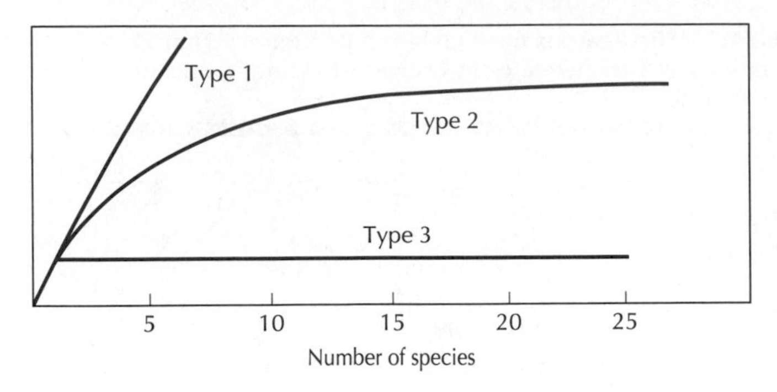

FIGURE 24.8 The alternate hypotheses for the relationship between species richness and ecosystem function. (Vitousek and Hooper 1993)

species. The type 3 response would suggest that all species do things in the same way and redundancy is very high. Two tests of the hypotheses in Figure 24.8 can be offered.

Grassland ecosystems modified so as to produce stands with very different numbers of species present have been established and studied in Minnesota. Two different measures of utilization were used to describe the efficiency of resource use: (1) total canopy cover (a measure of the completeness of light use), and (2) nitrate concentration in the rooting zone (a measure of the efficiency or completeness of N uptake). For both resources, a type 2 response was detected, with about 80% of resource use occurring in the presence of 5–10 species (Figure 24.9).

In a similar experiment in a tropical rain forest, cleared land was maintained as bare ground, was planted with a single species, and was allowed to regenerate naturally or was augmented with extra seed to increase species richness. Results after several years were similar to the grassland experiment (Figure 24.10), with the monocultures supporting soil organic matter and nutrient contents in soils that were generally closer to the full and enriched successional plot values than to the cleared plot values. The lack of plots at intermediate levels of species richness makes it difficult to describe the shape of the responses accurately.

These results suggest that redundancy relative to resource acquisition may be very high in most ecosystems. However, the diversity of species may play another and more significant role in maintaining the integrity of the ecosystems by increasing the resistance and resilience of the systems in response to disturbance. Two examples can be given.

In the Serengeti systems (Chapter 20), resilience was measured as recovery in above-ground biomass following heavy grazing (expressed as a percentage of increase in green biomass per month). There was a positive relationship between one common index of species diversity and the resilience of the system (Figure 24.11).

Similarly, a severe drought in the Minnesota grasslands discussed earlier resulted in a major reduction in production and loss of species in all test plots. However, the extent of the reduction in biomass was strongly related to

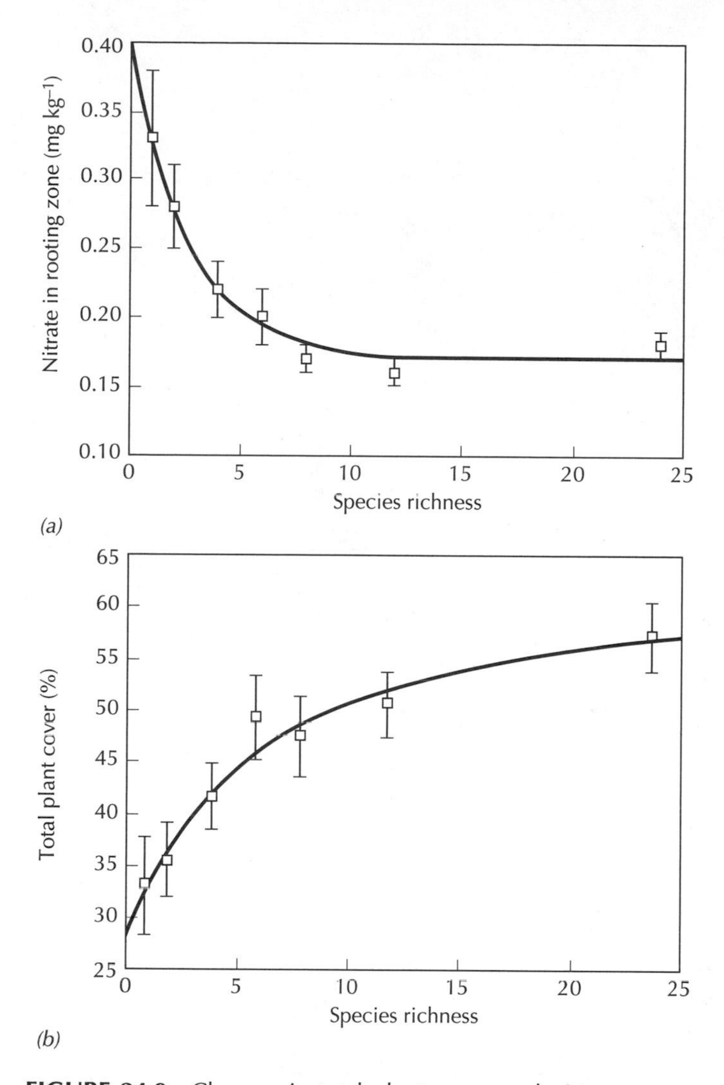

FIGURE 24.9 Changes in total plant cover and nitrate concentration in the rooting zone for a series of grasslands in Minnesota manipulated to achieve different levels of plant species richness. (Tilman et al. 1996)

the number of species present before the drought (Figure 24.12). In effect, higher species richness dampened the effect of the drought on production, providing increased resistance to this form of disturbance.

In Chapter 23 (Figure 23.10), we presented data on herbivory in the manipulated tropical succession example described earlier, which also supports the idea that greater diversity increases resistance to disturbance. In that example, the fraction of foliage consumed by herbivores was reduced on average, and was much less variable, in the plots with higher species numbers compared with the monoculture stands.

These examples suggest that a crucial characteristic of communities with high biotic diversity is the ability to either resist disturbance or to recover rapidly following disturbance. Considering the value of rapid recovery or

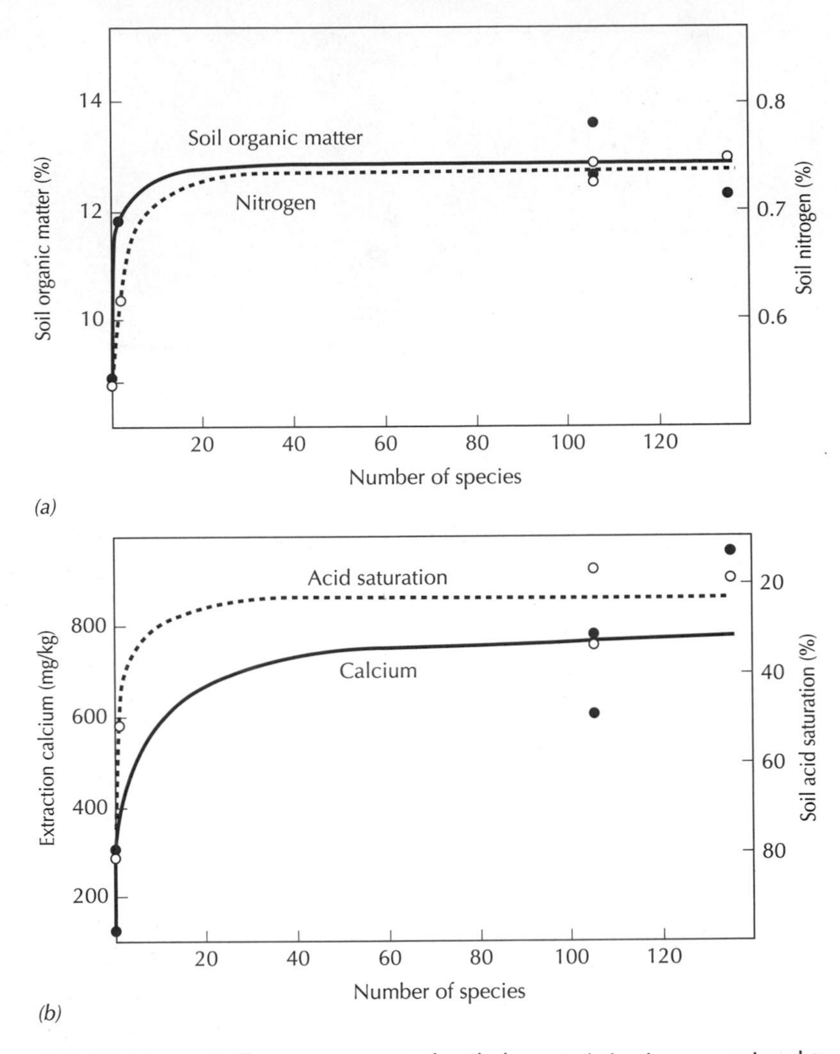

FIGURE 24.10 Differences in several soil characteristics in successional tropical forest systems following 63 months of treatments achieving different levels of species richness. (Ewel et al. 1991, as cited in Vitousek and Hooper 1993)

continued vegetative cover on nutrient retention and erosion control, these responses can have long-term value in maintaining ecosystem function.

All of these examples treat all species alike in terms of their importance to maintaining function. While this makes sense when experimenting with species-rich assemblages such as plants, there are many examples in which one or several species play a very large role in controlling function, and their addition to or removal from the system leads to major alterations in structure or function. Such species are called **keystone species.**

Many examples of keystone species include large herbivores or top predators in systems with multilevel food webs. We have already discussed the importance of elephants in modifying the structure of the Serengeti woodland–grassland system (Chapter 20) and the role of wolves in modifying cycles of moose and vegetation dynamics at Isle Royale (Chapter 16). Removal

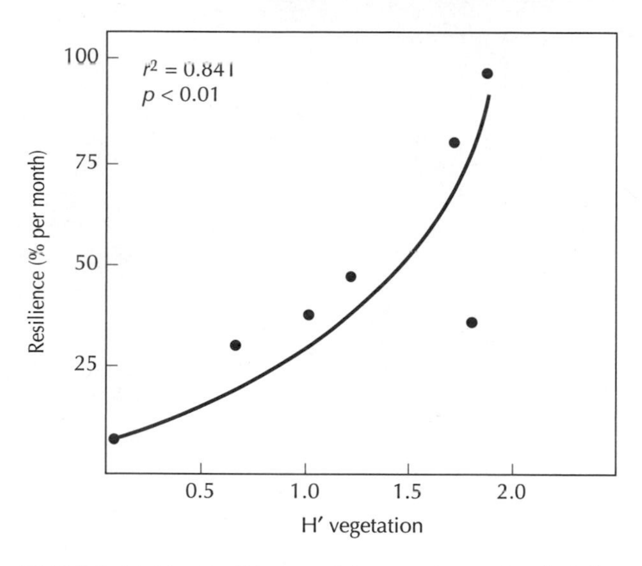

FIGURE 24.11 Resilience of Serengeti grasslands following grazing, estimated as a percentage of recovery in green biomass per month, as a function of a standard measure of species diversity. (McNaughton 1993)

of predators on deer in the eastern United States has led to sustained high population levels and the overbrowsing of regenerating forests into fern barrens (Figure 15.2). Animals that modify soil structure, from ants and termites (Figure 15.4) to burrowing rodents, or that alter hydrology, such as beavers in North America, can engineer large changes in the structure and function of systems.

Keystone plant species are harder to identify, outside of the examples of aggressive invasive species discussed in the next section. However, it has been argued that mycorrhizal fungi may play the keystone role in controlling the distribution of woody plants at the edges of forested regions, with invasion of shrubs and trees into grasslands inhibited by the lack of mycorrhizal symbionts.

INVASIVE SPECIES AND INTRODUCTIONS

While most human-induced extinctions have been the inadvertent result of overhunting or habitat destruction, we have specialized in the transportation and introduction of novel species over most of the land surfaces of the Earth. In addition, many other species have "come along for the ride" and become established in regions far from that in which they evolved. Unlike the consistent patterns in response to general levels of diversity described earlier, the effects of introductions are nearly impossible to generalize and predict. In essence, we do not know enough about the particular physiology, immunology, and behavior of most species to predict the likelihood of survival in a new habitat or the effect of the new species on the target ecosystem. As a result, most of the responses to introductions and invasions are reported as examples, with generalizations hard to come by.

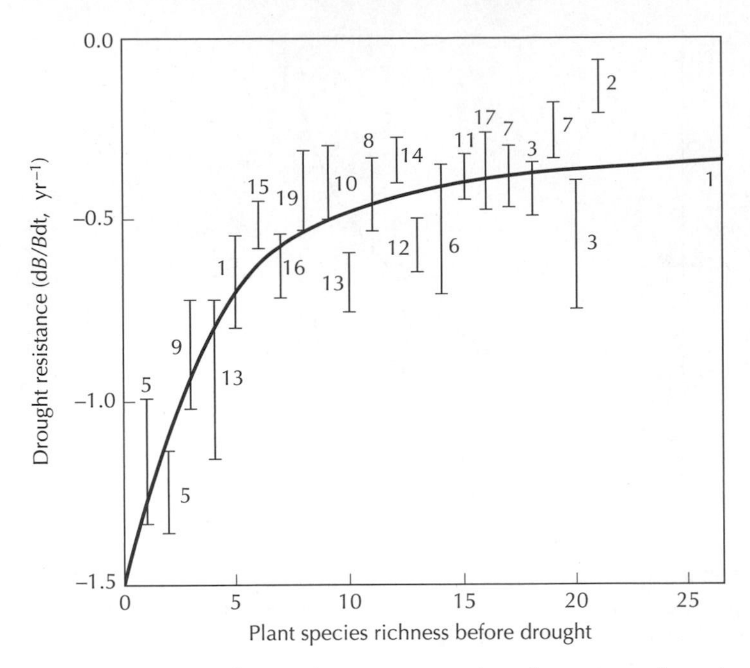

FIGURE 24.12 Differential resistance to drought, estimated as change in maximum standing biomass between a dry and a normal year, in grasslands with different species richness. (Tilman and Downing 1994)

It is likely that most plant species introductions and invasions do not result in drastic shifts in community structure or large-scale extinctions. The many hundreds of plant species of European origin that now exist in the more disturbed and European-like landscape of North America are a case in point. The ability of plant species to fundamentally alter ecosystem function through invasion of existing communities most generally results from the introduction of new life forms or physiological functions. Examples include the invasion of *Myrica fava* into Hawaii, the escape of planted North American pines into the Eucalyptus forests of Australia, and the introduction of rapidly growing vines (e.g., *Kudzu*) into North America. Some fraction of the hardiness and vigor of these species is probably due to escape from pests and pathogens that did not travel with the plant. Planned introductions of such pests and pathogens, intended to control the invading species, can have additional and unpredictable effects on native plants.

The movement of pests and pathogens themselves can have dramatic effects. If we take the deciduous forests of eastern North America as an example, the importation of chestnut blight, Dutch elm disease, and hemlock wooly adelgid have greatly reduced (or are currently reducing) the importance of these three species. Added to this is the spread of the introduced gypsy moth, which out-migrated from a single release point in eastern Massachusetts, causing a series of complete defoliations of many deciduous species that were not preadapted to this herbivore. In general, pathogens can restructure a community and eliminate a species more rapidly than can any other alteration in the biota. The human counterpart, the spread of deadly European diseases to the native peoples of the Americas and the islands of the Pacific is a tragic tale of massive die-off and only slow and partial recovery.

The importation of favored animal species can also have a bigger effect than for plants. Again, introductions of life forms not already present can have the greatest effect. The displacement of marsupial mammals in Australia by the introduction of placental mammals is a source of ongoing concern. The release of wild boars into the forests of the Great Smoky Mountains of the southeastern United States resulted, by the boar's habit of rooting through soils, in a total restructuring of soils. Effects of introductions to freshwater systems are generally even more profound, but are beyond the scope of this book.

ENVIRONMENTAL CHANGE AND BIODIVERSITY

In addition to the direct movement of species from place to place, the indirect effects of an altered local or global environment can lead to changes in species richness and the nature of the dominant species. We present four examples operating at very different scales.

In discussing general patterns of biodiversity, we suggested that ecosystems with higher NPP could support more species per unit area and so would exhibit higher biodiversity. This comparison was across large climatic gradients and between major biome types (e.g., tropical forests more diverse than tundra). Within a given vegetation type, the opposite trend may be seen.

Nitrogen fertilization was another part of the experimental design in the Minnesota grassland ecosystem research described in Figure 24.12. Eight levels of N additions were carried out for several years. After several years of treatment, a single species (*Agropyron repens*) showed a substantial increase in the fraction of total biomass with increasing N additions, coming to dominate almost exclusively in the highest addition plots (Figure 24.13). Total species richness declined markedly with treatment (Figure 24.14*a*), and in particular, plants using the C4 photosynthesis pathway (Chapter 6), which increases photosynthetic rates at a given foliar N content, effectively disappeared at the highest N loadings (Figure 24.14*b*).

An analysis of the distribution of central European plant species along an N availability gradient concluded that much larger numbers of species occur in N-poor areas than in N-rich sites (Figure 24.15) and that there was a preponderance of threatened species in the N-poor habitats. Overall, the fraction of species within an N availability class that was threatened decreased markedly with increasing N availability. One explanation for this would be that N-poor sites support a lower biomass, less complete plant cover, and a greater microscale diversity in resource availability, leading to the presence of a wider range of plant types. Very nutrient-rich sites lead to high biomass and complete canopy coverage, resulting in a loss of, for example, shade-intolerant species and a lower overall biodiversity level. This finding gains importance with the high level of N deposition over central Europe and the increases in N availability that are predicted to ensue (see Chapter 25).

At the other end of the spatial scale for deposition, the extreme pollution loadings around large point sources, such as the smelters described in Chapter 25, result in nearly complete denudation of the landscape in close prox-

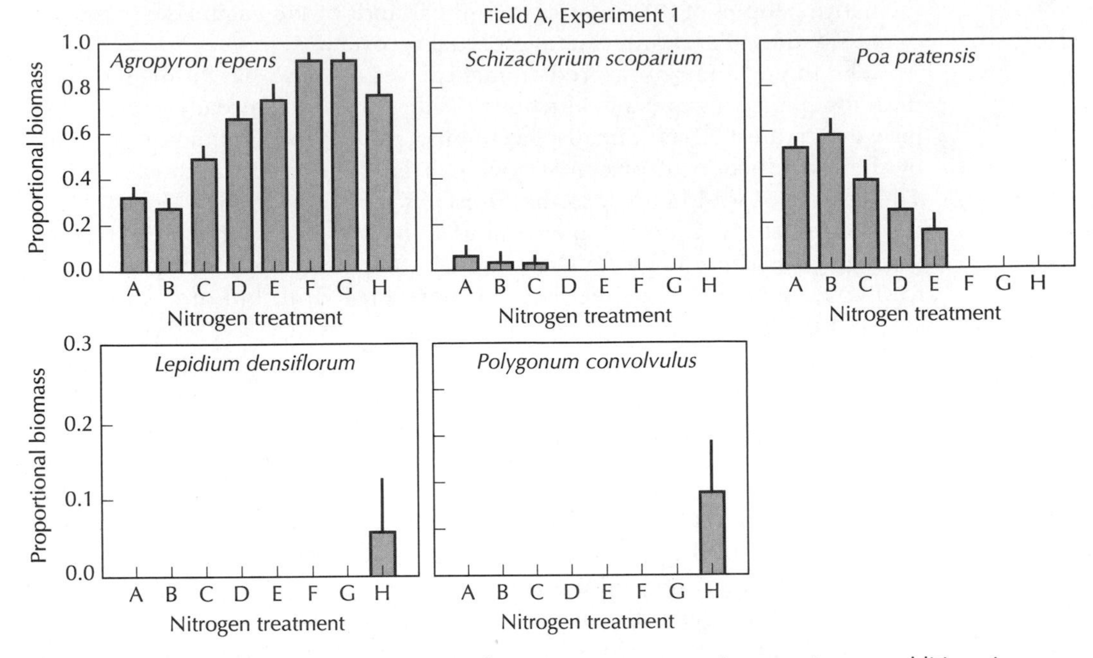

FIGURE 24.13 Response of major species to continuous nitrogen additions in experimental grasslands in Minnesota. Treatments increase in nitrogen added from A to H. Response is measured as change in the fraction of total biomass present. (Inouye and Tilman 1995)

imity to the source, with reduced biomass, species richness, and structural complexity (Figure 25.9). One of the first demonstrations of this general trend in the simplification of ecosystems through biotic impoverishment in response to chronic stress was in a temperate forest in New York exposed to continuous fields of ionizing radiation in a demonstration of what could follow on the heels of a nuclear war.

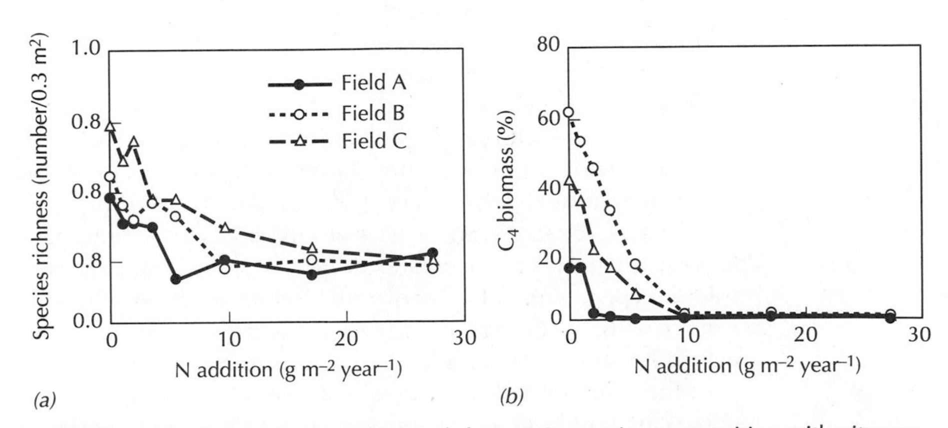

FIGURE 24.14 Summary expressions of changes in species composition with nitrogen additions in experimental grasslands in Minnesota. (*a*) Total species richness. (*b*) Percentage of total biomass in C4 species. (Wedin and Tilman 1996)

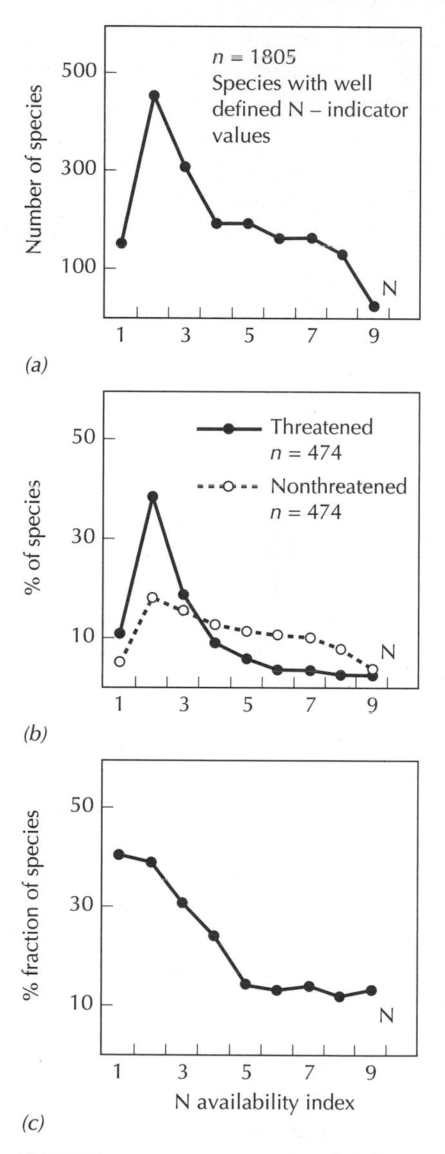

FIGURE 24.15 Nitrogen availability and species richness in central European forests. (*a*) Number of species found at each relative level of N availability. (*b*) The total number is divided between threatened and nonthreatened species. (*c*) The fraction of all species in the threatened category. (Ellenberg 1987)

At the global scale, predicted changes in climate in response to increased carbon dioxide (CO_2) concentration in the atmosphere may result in the wholesale movement of species and ecosystems across the landscape.

At the physiological level, CO_2 enrichment can have effects similar to overfertilization with N. In a CO_2-rich environment, the selective advantage of the C4 pathway, which reduces CO_2 concentrations within the leaf, increases photosynthetic rate, and increases water use efficiency, may be lost. For example, C4 grasses in warm, semiarid grassland systems may be replaced by C3 species.

In Chapter 26, we discuss in greater detail the implications of a CO_2-enriched environment on global and regional climate (Figure 26.5). In this context, it is enough to say that significant changes are predicted on the order of 5–8°C in northern temperate landscapes over the next century. Variation in mean climatic conditions is a primary controller of the spatial distribution of species and community types (see Figure 2.1*a*). Some very large changes in these distributions are predicted in response to estimated changes in climate.

Figure 24.16 shows one estimate of the redistribution of vegetation types in the United States predicted using two climate-change scenarios and a combined model of biome distribution and biogeochemical cycling. Two climate change scenarios are used: the Hadley model predicts less warming and more rainfall, while the Canadian model calls for larger increases in temperature and less of an effect on precipitation (Figure 26.5). The combined ecosystem model predicts a contraction in desert (arid land) distribution in the southwestern United States as both increased rainfall and the effects of CO_2 reduce drought stress (Figure 24.16). In the Canadian scenario, the boreal forests effectively disappear from the far north, shifting entirely into Canada. While all vegetation types move north in both scenarios, greater drying in the southeastern United States is predicted under the Canadian scenario, as well as an increased incidence of fire that begins to create grasslands in place of forests and even introduces arid lands in the far South.

Another model based on statistical relationships between climate, topography, soil conditions, and species distributions taken from an extensive forest inventory network draws similar conclusions at both the forest type and species levels. In Figure 24.17, the top two panels show the current distribution of major forest types in the eastern United States, as well as the current distribution of one species—sugar maple. The two bottom panels are predictions as to the distributions of both forest types and sugar maple under different climate scenarios. Both this species, and the forest type of which it is a part, are predicted to move entirely out of the United States and northward into Canada.

These last two predictions assume that species can move as quickly as the climate changes and that barriers to migration will not be a problem. This is not guaranteed, however. In many areas, the interspersing of native vegetation with agricultural and urban areas may restrict movement. There is also the problem that any one species may not be able keep pace with the rate of climate change. Heavy-seeded species, or those that rely on vegetative reproduction, may be particularly stressed.

Finally, what is the implication of all of this for those cherished ecosystems that have been preserved in major parks throughout the world? If Isle Royale is a rare natural laboratory for the study of plant–herbivore–predator interactions in the boreal forest, what is the implication of a shift in climate that puts the island outside the climatic range for this forest type?

One long-time student of the dynamics of the Serengeti and Yellowstone parks doubts the ability of highly integrated ecosystems, in which soils and topography, as well as plants and animals, play an important role: to move as a unit over the distances required by climate change and into a new and suitable location.

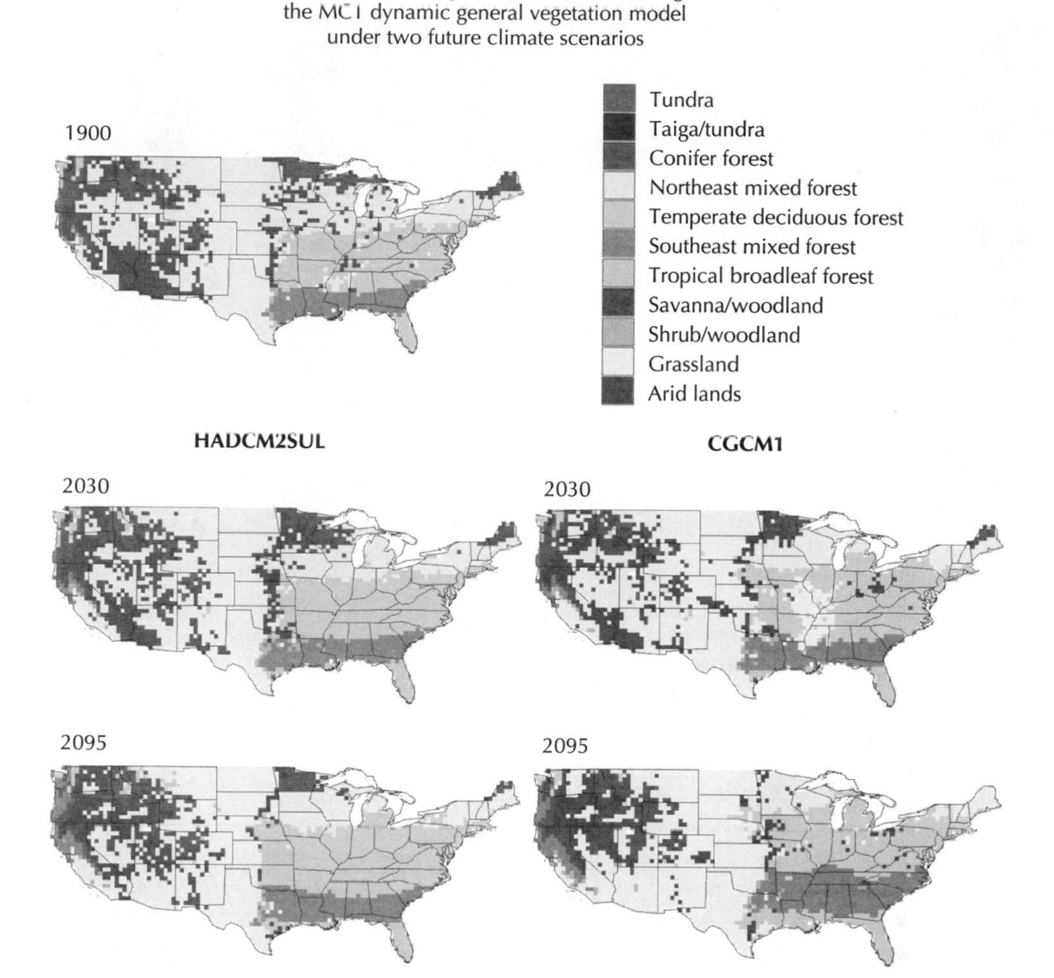

FIGURE 24.16 Potential changes in the distribution of major vegetation types in response to predicted changes in climate, as derived from a combined vegetation-fire-biogeochemistry model. (Bachelet et al. 2001, cited in Aber et al. 2001) **See plate in color section**

Consider the spatial contexts of the Yellowstone and Serengeti ecosystems. Both are essentially, like most conserved, native systems, isolated islands surrounded by land subject to various multiple uses. . . . Particularly implausible is that either of these ecosystems would "migrate" geographically in response to climate change. Because of the unique combinations of edaphic and climatic properties that define the Yellowstone and Serengeti grazing ecosystems, both would be destroyed if climatic conditions within which they currently occur shift spatially into entirely different geological and edaphic contexts. (McNaughton 1993)

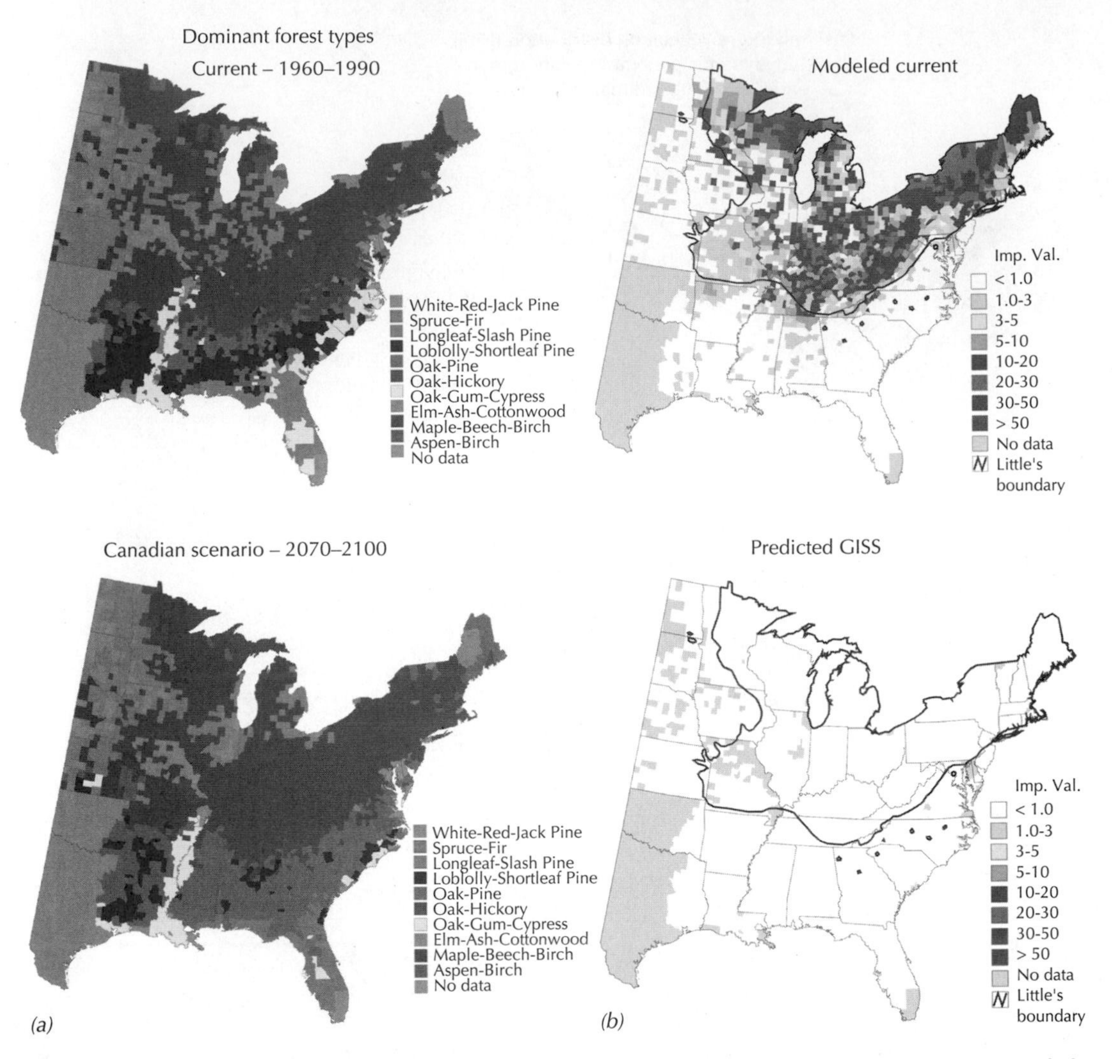

FIGURE 24.17 Predicted changes in the distribution of (*a*) major forest types and (*b*) one dominant species (sugar maple) under different climate change scenarios. (Iverson et al., 1999) **See plate in color section**

REFERENCES CITED

Aber, J. D., R. Neilson, S. McNulty, J. Lenihan, D. Bachelet, and R. Drapek. 2001. Forest processes and global environmental change: Predicting the effects of individual and multiple stressors. *BioScience* (in press).

Cody, M. L. 1975. Towards a theory of continental species diversity. In Cody, M. L., and J. M. Diamond (eds.). *Ecology and Evolution of Communities*. Belknap/Harvard University Press, Cambridge, MA.

Condit, R., S. P. Hubbell, J. V. Lafrankie, R. Sukumar, N. Manokaran, R. B. Foster, and P. S. Ashton. 1996. Species–area and species–individual relationships for tropical trees: a comparison of three 50-ha plots. *Journal of Ecology* 84:549–562.

Czech, B., P. R. Krausman, and P. K. Devers. 2000. Economic associations among causes of species endangerment in the United States. *BioScience* 50:593–601.

Diamond, J. M. 1969. *Avifaunal equilibria* and species turnover rates on the Channel Islands of California. *Proceedings of the National Academy of Science* 64:57–63.

Dobzhansky, T. 1950. Evolution in the tropics. *American Scientist* 38:209–221.

Ellenberg, H. 1987. Floristic changes due to eutrophication. In Asman, W. A. H., and H. S. M. A. Dierderen (eds.). *Ammonia and Acidification: Proceedings of the Symposium.* National Institute of Public Health, Bilthoven, The Netherlands.

Ewel, J. J., M. J. Mazzarino, and C. W. Berish. 1991. Tropical soil fertility changes under monocultures and successional communities of different structure. *Ecological Applications* 1:289–302.

Inouye, R. S. 1998. Species–area curves and estimates of total species richness in an old-field chronosequence. *Plant Ecology* 137:31–40.

Inouye, R. S., and D. Tilman. 1995. Convergence and divergence of old-field vegetation after 11 years of nitrogen addition. *Ecology* 76:1872–1887.

Iverson, L. R., A. M. Prasad, B. J. Hale, and E. K. Sutherland. 1999. *Atlas of Current and Potential Future Distributions of Common Trees of the Eastern United States.* General Technical Report NE-265. USDA Forest Service, Radnor, PA.

Kusnezov, M. 1957. Numbers of species of ants in faunas of different latitudes. *Evolution* 11:298–299.

MacArthur, R. H. 1972. *Geographical Ecology.* Harper & Row, New York.

MacArthur, R. H., and E. O. Wilson. 1967. *The Theory of Island Biogeography.* Princeton University Press, Princeton, NJ.

McNaughton, S. J. 1993. Biodiversity and function of grazing ecosystems. In Schulze, E. D., and H. A. Mooney (eds.). *Biodiversity and Ecosystem Function.* Springer-Verlag, Berlin.

Simberloff, D. S., and E. O. Wilson. 1970. Experimental zoogeography of islands: The colonization of empty islands. *Ecology* 50:278–296.

Tilman, D., and J. A. Downing. 1994. Biodiversity and stability in grasslands. *Nature* 367:363 365.

Tilman, D., D. Wedin, and J. Knops. 1996. Productivity and sustainability influenced by biodiversity in grassland ecosystems. *Nature* 379:718–720.

Vitousek, P. M., and D. U. Hooper. 1993. Biological diversity and terrestrial ecosystem biogeochemistry. In Schulze, E. D., and H. A. Mooney (eds.). *Biodiversity and Ecosystem Function.* Springer-Verlag, Berlin.

Wedin, D. A., and D. Tilman. 1996. Influence of nitrogen loading and species composition on the carbon balance of grasslands. *Science* 274:1720–1723.

Wilson, E. O. 1992. *The Diversity of Life.* Belknap/Harvard University Press, Cambridge, MA.

Wilson, J. W. 1974. Analytical zoogeography of North American mammals. *Evolution* 28:124–140.

Woodwell, G. M. Ionizing radiation and terrestrial ecosystems. *Science* 2001.

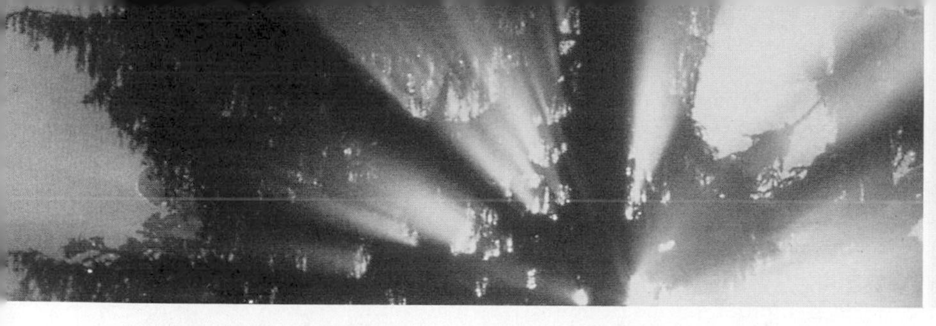

25

EFFECTS OF AIR POLLUTION ON TERRESTRIAL ECOSYSTEMS

INTRODUCTION

Changes in the landscape and in ecosystems due to conversion to agriculture or to the loss and introduction of species tend to be very visible and easily recognized. The landscape looks very different. Important species are present or absent. The effects are apparent in the same location as the conversion or introduction occurred. In contrast, the effects of industrial pollutants may be subtle and difficult to identify, especially in the early stages of response, and may occur far from the source. Pollutants also occur in a wide variety of forms, and interactions between them may be important.

Many primary and secondary forms of pollution affect the concentration of biologically important elements in the atmosphere and their deposition to the surface. The purpose of this chapter is to discuss the origins and distribution patterns of some important forms of air pollution and their potential effects on terrestrial ecosystems. We will concentrate on those components that are most likely to change ecosystem function. We will draw on information presented in several previous chapters because most core ecosystems, processes are affected in some way by some form of air pollution.

AIR POLLUTION SOURCES

Burning fossil fuels is a major component of human intervention into natural elemental cycles. The injection of tremendous amounts of carbon (C) into the atmosphere as CO_2 is one major result of combustion and is the topic of Chapter 26. Because of the importance of nitrogen (N) and sulfur (S) in the biological materials from which they are derived, by extracting the energy stored in fossil fuels, we also release N and S into the atmosphere. Combustion, especially in high-compression automobile engines, can also convert atmospheric N (N_2, which constitutes 79% of the atmosphere) to reactive oxides of N (NO_x). Trace amounts of heavy metals (e.g., zinc, lead, cadmium, nickel, and mercury) are also released by combustion, but even larger amounts of these toxic metals are released by the smelting of ores into pure metals and alloys for industrial use.

The major forms of air pollution affecting terrestrial ecosystems are summarized in Table 25.1. They are grouped into three categories: (1) oxidant gases, (2) "acid rain," and (3) heavy metals. Sulfur dioxide (SO_2), NO_x, and ozone are all oxidant gases. Sulfuric and nitric acids, when dissolved in precipitation, are the principal causes of acidity in "acid rain." Lead, cadmium, zinc, nickel, mercury, and others are grouped as heavy metals.

Pollutants are also categorized as primary or secondary products of industrial activity. Primary products, such as SO_2 and NO_x, are the immediate result of combustion. Carbon dioxide (CO_2) is another primary product, as are a number of other partially oxidized hydrocarbons that are important in the formation of ozone. Secondary products, such as ozone and nitric and sulfuric acids, are formed in the atmosphere from the primary products or as a result of their effects on atmospheric chemistry. Secondary pollutants may be distributed more widely than primary pollutants because of the time required for the reactions to occur (see Figure 25.4).

Nitric and sulfuric acids are formed by reactions between water in the atmosphere and NO_x and SO_2, which can be generalized as:

$$SO_2 + H_2O \rightarrow H_2SO_4 \leftrightarrow H^+ + HSO_4^- \leftrightarrow 2H^+ + SO_4^{2-}$$

$$NO_x + H_2O \rightarrow HNO_3 \leftrightarrow H^+ + NO_3^-$$

Nitric and sulfuric acids are "strong acids" in that they tend to dissociate completely (these reactions go strongly to the right), releasing hydrogen ions (H^+). Thus, clouds, fog, and eventually rain and snow are increased in H^+ concentration; they are acidified.

Acidity is expressed as the concentration of H^+ in a solution and is summarized by the term *pH,* which is:

$$pH = -\log \text{ (concentration of } H^+)$$

Two things are important in this equation. First, it is a logarithmic scale, so every unit change in pH means a 10-fold change in H^+ concentration. Sec-

TABLE 25.1 Major Components of Air Pollution Thought to Affect Terrestrial Ecosystems

Air Pollutant Type	1°/2°*	Compounds	Ecosystem Effects
Oxidant gases	1° 1° 2°	Sulfur dioxide (SO_2) Nitrogen oxides (NO_x) Ozone (O_3)	Reduction of net photosynthesis, damage to leaf cell membranes, formation of toxins (SO_2)
Dissolved acids	2° 2°	Sulfuric Acid (H_2SO_4) Nitric acid (NHO_3)	Soil acidification, metal mobilization, cation nutrient deficiencies, nitrogen saturation
Heavy metals	1°	Lead (Pb), nickel (Ni), copper (Cu), zinc (Zn), cadmium (Cd), others	General interference with biochemical reactions

*Specifies whether the compound is primary or secondary production of industrial activity.

ond, it is inversely related to H^+ concentration, and a lower pH value denotes increased acidity. Distilled water has a pH of 7 (it is neutral). Anything below pH 7 is considered acidic; anything above pH 7 is considered alkaline. Figure 25.1 shows representative pH values for different kinds of solutions.

Unpolluted rainfall is not neutral. The presence of CO_2 in the atmosphere leads to the formation of carbonic acid:

$$CO_2 + H_2O \rightarrow H_2CO_3 \leftrightarrow H^+ + HCO_3^- \leftrightarrow 2H^+ + CO_3^{2-}$$

At current levels of atmospheric CO_2, carbonic acid should produce rainfall with a pH of about 5.65. However, there are other naturally occurring chemicals that can also lower the pH of rainfall. An extensive study of the acidity of rainfall in places remote from industrial activity (Figure 25.2) suggests that background pH levels may actually be closer to 5.2. However, the same study emphasized the magnitude of change in precipitation chemistry in the northeastern United States brought about by human activity, where the annual mean rainfall pH may be as low as 4.0 and individual storm events may reach pH 2.5 (several hundred times the H^+ content of "normal" pH 5.2 rain).

While *acid rain* has become the most generally recognized term in popular discussions of the effects of industrial pollutants on forests and other ecosystems, it is actually only one form of air pollution. Over the years, scientific terminology has changed *acid rain* to *acid precipitation* (which includes snow, fog, etc.) to *acid deposition* (with the realization that acidic or acidifying compounds can be deposited directly on surfaces in dry form as dust or aerosols [small airborne particles]) and finally to *atmospheric deposition* (which includes nonacidifying pollutants, e.g., heavy metals, lead, zinc, etc.). This still fails to include the gaseous pollutants, such as SO_2, NO_x, and ozone (O_3). In dealing with the effects of air pollution on terrestrial ecosystems, we have to address the potential effects of each of these components as well as their potential interactions.

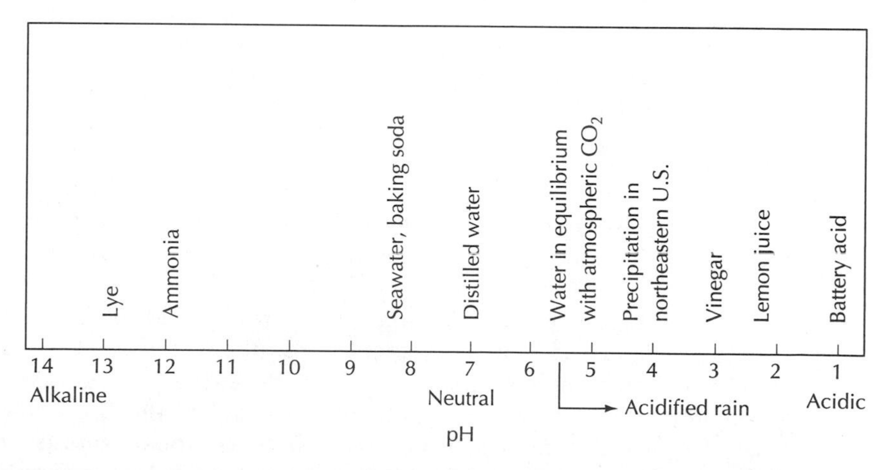

FIGURE 25.1 The pH scale and pH values of some common substances compared with current values for rainfall in the northeastern United States.

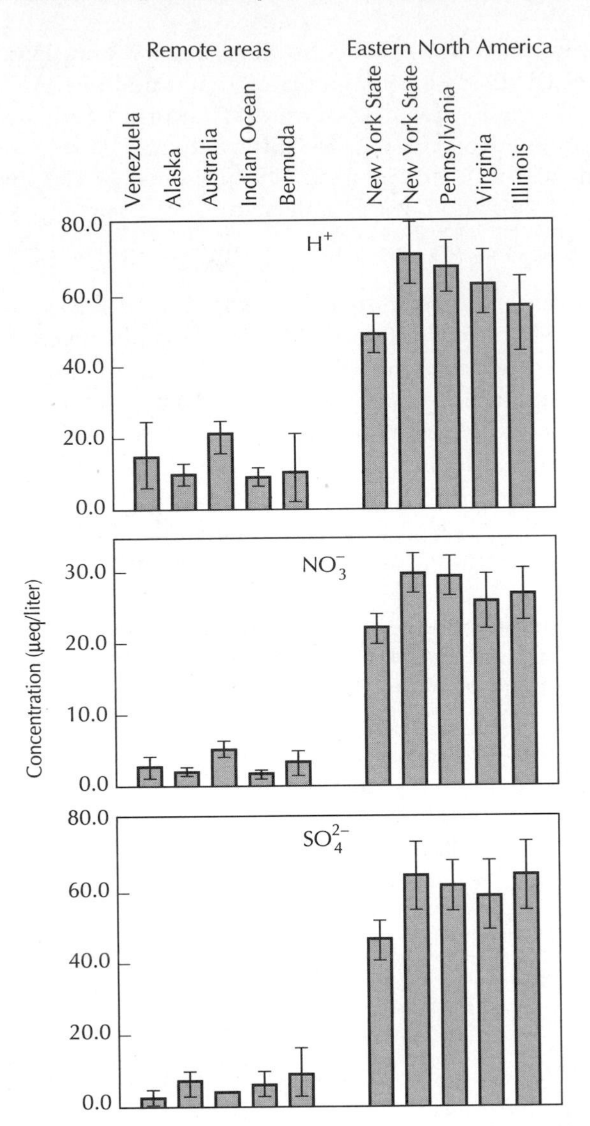

FIGURE 25.2 Differences in the concentrations of hydrogen ion, nitrate, and excess SO_4^{2-} (in addition to that provided by sea salt from evaporated sea spray) between five remote locations and five locations in eastern North America. (Galloway et al. 1984)

Ozone is increasingly recognized as an important component of air pollution, with negative effects on both human health, and vegetation. Ozone is a highly reactive gas formed in the lower atmosphere (or troposphere) by a complex set of reactions involving NO_x, partially oxidized hydrocarbons, and oxygen (Figure 25.3) in the presence of sunlight. This process and its consequences are completely separated from problems and processes associated with the "ozone hole" phenomenon that involves the breakdown of a thin layer of ozone at the top of the stratosphere, or upper atmosphere. The rate of ozone accumulation in the lower atmosphere can be limited either by NO_x or hydrocarbons, and determining which is the limiting factor in different regions is an important step in establishing successful control strategies. Fur-

PHOTOLYTIC NO_2 CYCLE

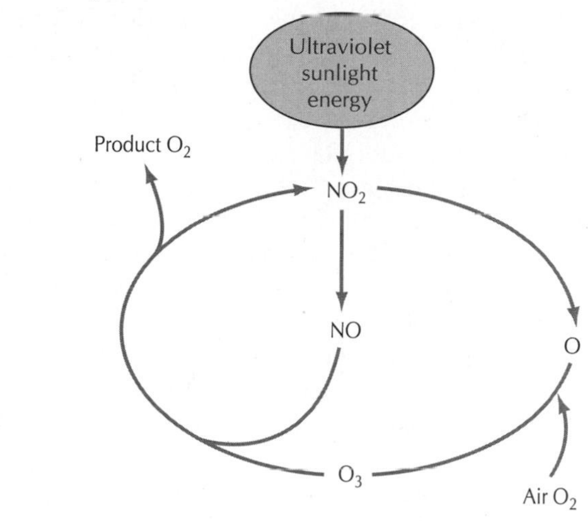

(a)

DISRUPTED PHOTOLYTIC CYCLE

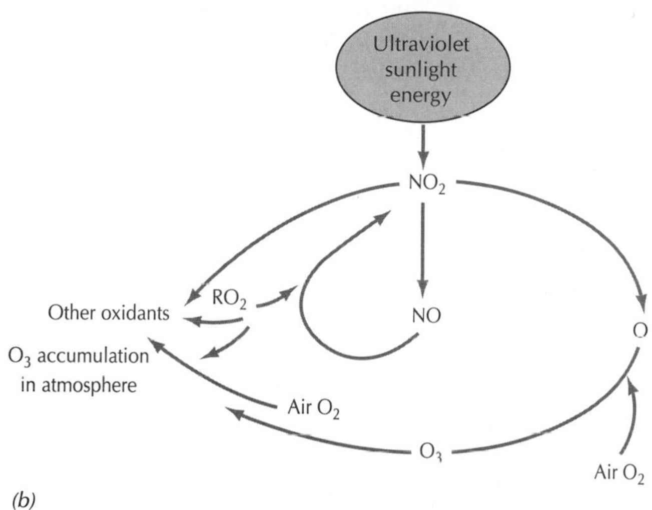

(b)

FIGURE 25.3 The production and breakdown of ozone in the lower atmosphere (*a*) under normal conditions and (*b*) in the presence of partially oxidized hydrocarbons (RO_2).

ther complicating this picture is the production of hydrocarbons or volatile organic carbon (VOC) compounds by green plants. In certain regions, hydrocarbon concentrations in the atmosphere are determined largely by plant activity, and this can affect whether VOCs or NO_x limits the rate of ozone accumulation.

DISTRIBUTION OF AIR POLLUTANTS

As early as the mid-1800s, early atmospheric chemists were beginning to measure the differences in the chemical content of air samples taken near industrial areas and in remote areas. In the late 1800s and early 1900s, exceedingly high concentrations of SO_2 in urban areas was linked to serious direct human health effects. A series of "killer fogs" in London in this era resulted in part from the large number of very inefficient home and industrial heating sources burning "soft" coal, a form particularly high in S content. The frequency of fogs was also increased, as aerosols of both soot and sulfate (SO_4) provide condensation nuclei that catalyze the formation of clouds (or fog) in humid air. Killer fogs increased both elevated hospitalizations and mortality, especially for those with weak respiratory systems.

Centralized energy systems and power plants with much higher chimneys (or "stacks") were a strategy for reducing the concentrations of primary air pollutants in urban areas. Increased efficiency in the centralized systems reduces emissions per unit of energy generated and helps to reduce total environmental loading. The taller stacks, however, only spread the pollution load over a wider area, increased residence time in the atmosphere, and increased acid rain downwind (Figure 25.4).

Widespread and systematic measurement of the chemistry of precipitation has occurred only within the past 30 years. In the early 1970s, initial, isolated measurements indicating significant changes in the chemistry of precipitation even in remote and rural areas led to the establishment of nationwide and continent-wide networks of precipitation collection stations, using standardized methods of analysis. By summarizing the tremendous amount of information gathered by these networks, large-scale patterns in precipitation chemistry can be determined.

Figure 25.5 summarizes patterns of element deposition in precipitation (rain and snow) across North America and emphasizes the role of emission sources, residence time, and regional wind patterns.

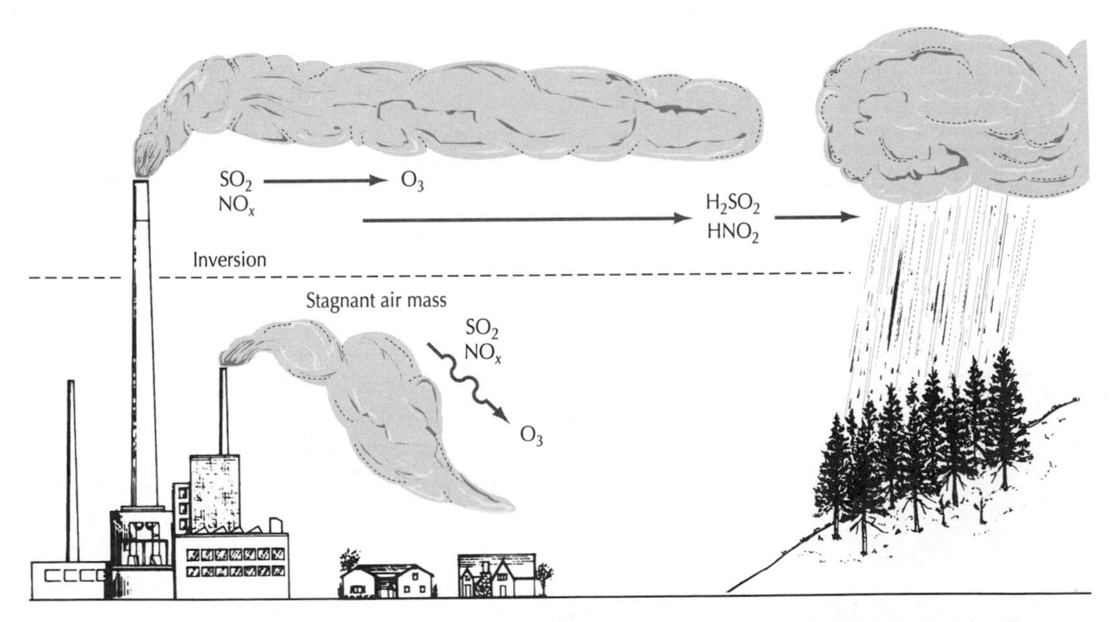

FIGURE 25.4 The effect of taller smokestacks on the distribution of emitted pollutants. A taller stack increases the likelihood that emissions will escape lower, possibly stagnant layers and be carried away from the source area.

For example, sodium and chloride, the elements present in common salt, are among the most common ions in seawater. Aerosols of these ions are injected into the atmosphere over the oceans by the evaporation of ocean spray and are carried by winds over land, where they are either deposited as dry particles or washed out of the air in precipitation. As a result, coastal areas receive large amounts of these two elements (Figure 25.5*a* and *b*). The high concentrations right along the coast and the rapid change in deposition away from the coast suggest that these elements do not travel great distances before leaving the atmosphere. Similarly, the deposition of calcium (Ca) can be higher in areas where limestone is a common geological substrate or where agricultural practices such as liming and plowing can release Ca-carrying dust for deposition in other areas.

Human activities dominate the distribution of N deposition as both ammonium (NH_4^+) and nitrate (NO_3^-). The application of urea- or NH_4^+-based N fertilizers to agricultural fields can result in the volatilization (loss to the atmosphere in gaseous form) of ammonia (NH_3). This is converted to NH_4^+ in the atmosphere and deposited in rainfall somewhere downwind (to the east in Figure 25.5*c*) of the major agricultural regions. Feedlots can also be an important regional source of ammonia to the atmosphere, and some of the highest rates of N deposition in the world are adjacent to large livestock-raising areas in western Europe.

As NO_3^- and SO_4^{2-} are formed in the atmosphere from NO_x and SO_2, their deposition rates are much higher downwind of the heavily industrialized areas of the upper Midwest and the Northeast (Figure 25.5*d* and *e*). These two strong acids control the acidification of rainfall, so there is a general correlation between the deposition rates for NO_3^- and SO_4^{2-} and the H^+ content of precipitation (Figure 25.5*f*).

A crucial feature of the deposition distributions for NO_3^- and SO_4^{2-} is that they can be transported hundreds of kilometers before coming back to Earth. Thus, industrial activity in the Great Lakes region of the United States affects rainwater quality in eastern Canada. Similarly, pollution produced in central Europe falls on the relatively rural areas of Scandinavia. The political complexity of dealing with a problem of this type should be apparent.

The distribution of ozone is also determined by the interaction of emissions of precursors (NO_x and hydrocarbons) with climatic conditions and patterns of wind movement. Ozone levels tend to be particularly high on still, hot, summer days when stagnant air masses reside over industrialized regions, accumulating the precursors that result in the accumulation of ozone in the presence of high temperatures and sunlight. They are generally much lower in winter than in summer (Figure 25.6). Ozone is a very reactive gas and dissipates quickly once the conditions leading to rapid accumulation are lost. As a result, ozone concentrations can change dramatically from one day to the next (Figure 25.7).

Most heavy metals enter the atmosphere as aerosols and do not travel the large distances indicated for NO_3^- and SO_4^{2-}. Areas of highest deposition have historically been concentrated around large smelters. However, the use of leaded gasoline through the 1970s proved to be a very efficient method of distributing large amounts of lead generally throughout urbanized regions. Growing concerns over the human health effects of lead prompted a mandated shift to lead-free fuels, resulting in lower deposition, especially in urban

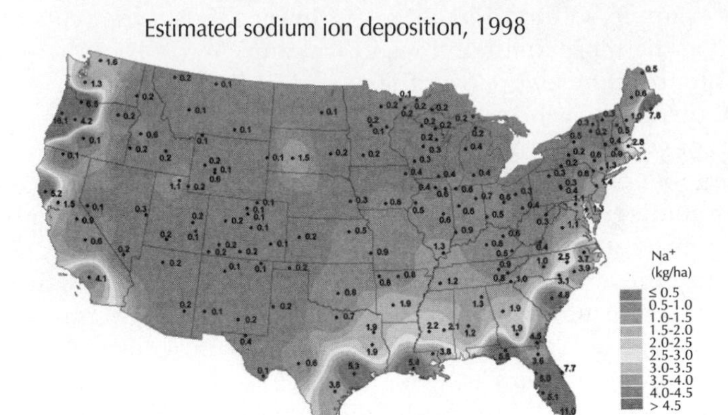

(a)

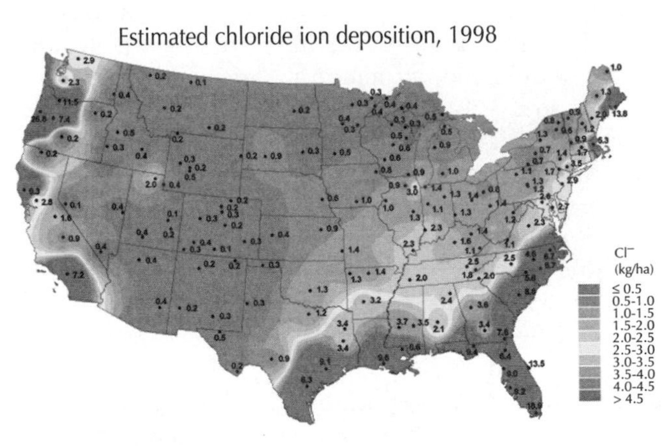

(b)

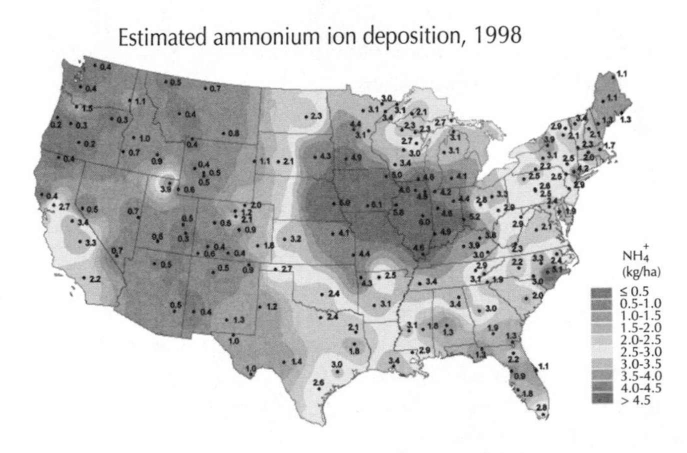

(c)

FIGURE 25.5 (*a–f*) Distribution of total wet deposition of selected chemical species as measured in 1998 through the National Atmospheric Deposition Program. (http://nadp.sws.uiuc.edu/) **See plate in color section**

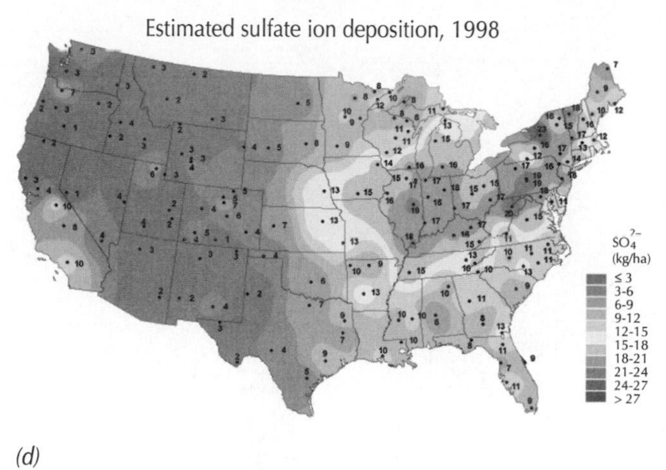
Estimated sulfate ion deposition, 1998
SO_4^{2-}
(kg/ha)
≤ 3
3-6
6-9
9-12
12-15
15-18
18-21
21-24
24-27
> 27

(d)

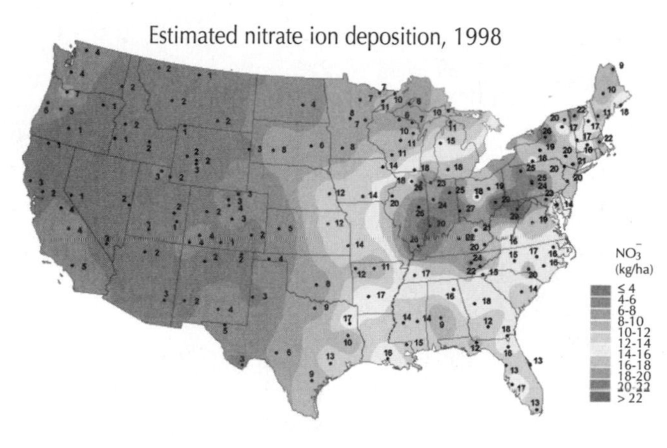
Estimated nitrate ion deposition, 1998
NO_3^-
(kg/ha)
≤ 4
4-6
6-8
8-10
10-12
12-14
14-16
16-18
18-20
20-22
> 22

(e)

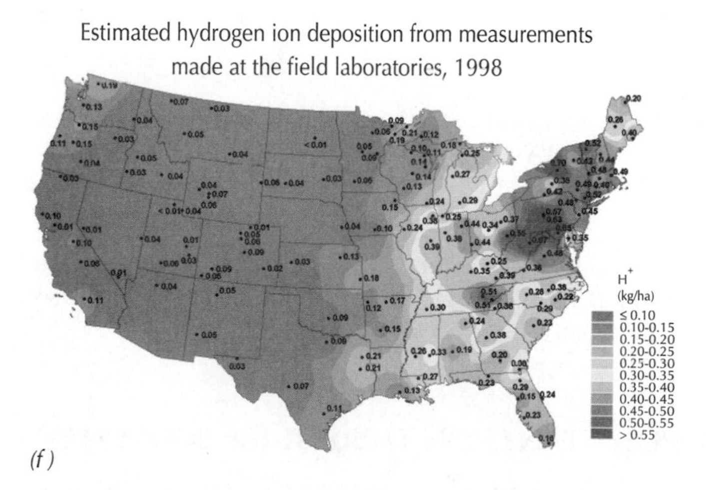
Estimated hydrogen ion deposition from measurements made at the field laboratories, 1998
H^+
(kg/ha)
≤ 0.10
0.10-0.15
0.15-0.20
0.20-0.25
0.25-0.30
0.30-0.35
0.35-0.40
0.40-0.45
0.45-0.50
0.50-0.55
> 0.55

(f)

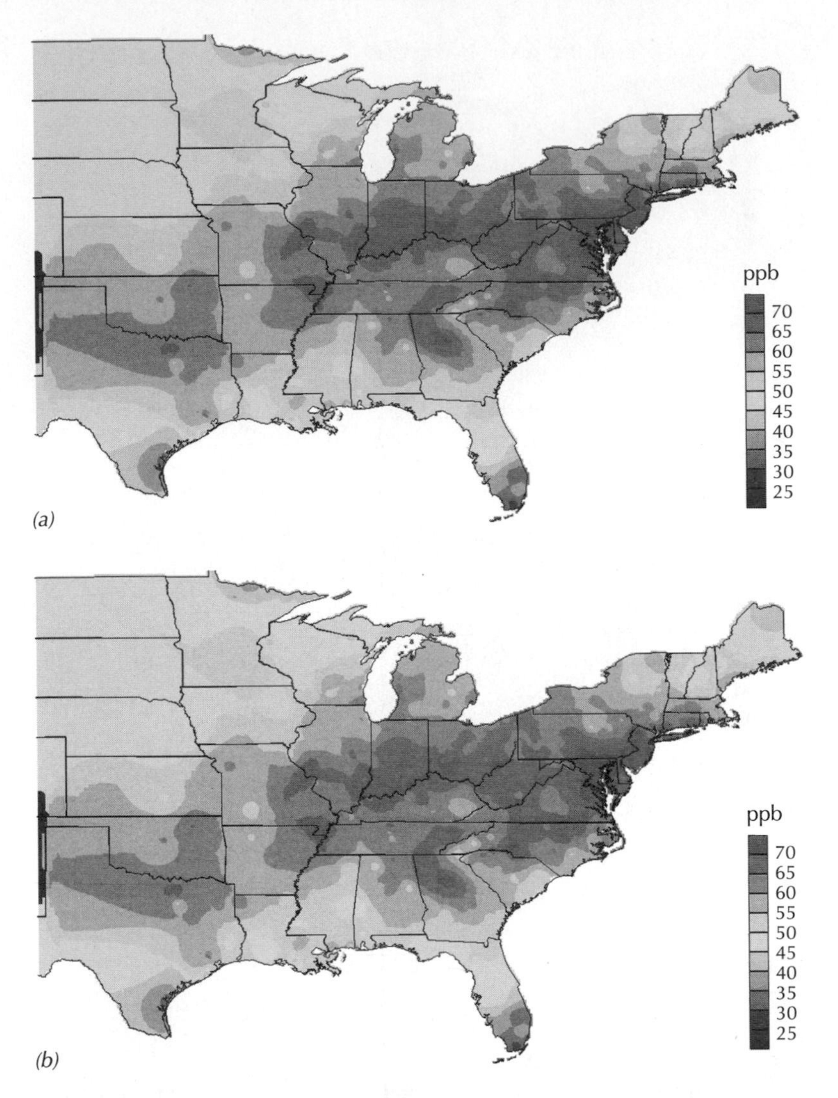

FIGURE 25.6 Distribution of the average daily maximum value for ozone concentration in the eastern United States in (*a*) summer and (*b*) winter. (Heilman et al. 2000) **See plate in color section**

areas. The long-term record of lead emissions from automobiles (Figure 25.8*a*) is found in the pattern of lead concentrations in the wood produced annually by urban trees (Figure 25.8*b*).

EFFECTS OF AIR POLLUTANTS ON TERRESTRIAL ECOSYSTEMS

Some of the most dramatic effects of pollutant deposition have been seen in the immediate vicinity of large smelting plants (Figure 25.9). These plants have released vast quantities of metals and SO_2 that combine to virtually eliminate plant growth in the immediate area. The toxic effects decline with distance from the source (Figure 25.9). Similar scenes of devastation have

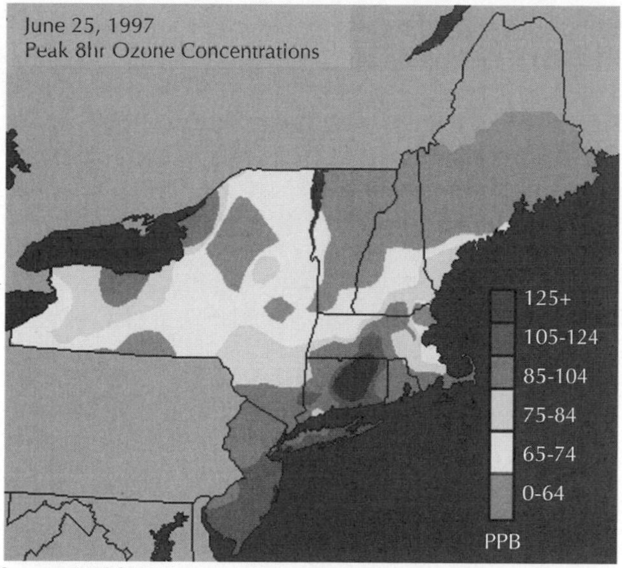

Source: NESCAUM

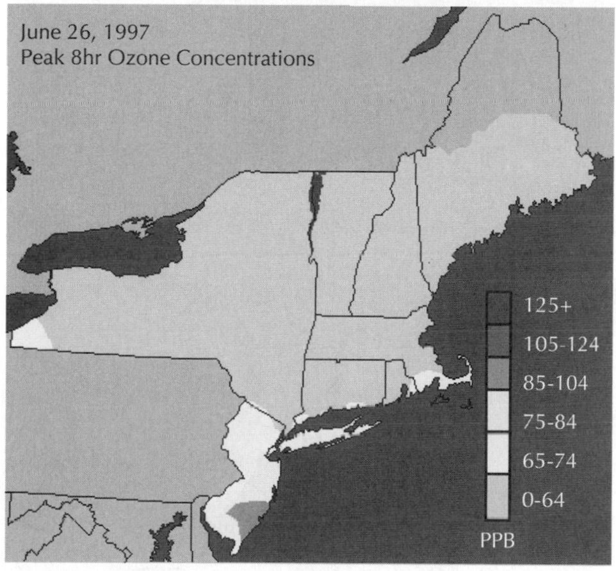

Source: NESCAUM

FIGURE 25.7 Change in maximum daytime ozone concentration on two consecutive days in the northeastern United States. Such large changes generally occur with the passage of a cold front accompanied by decreased temperature and humidity and increased wind speed. (NESCAUM – Northeast States for Coordinated Air Use Management) **See plate in color section**

been recorded in the immediate vicinity of large, inefficient power plants burning tremendous amounts of soft coal, such as the Krusné Hory Mountains on the border between the Czech Republic and Germany. Both of these examples have been turned into environmental success stories, as is described later.

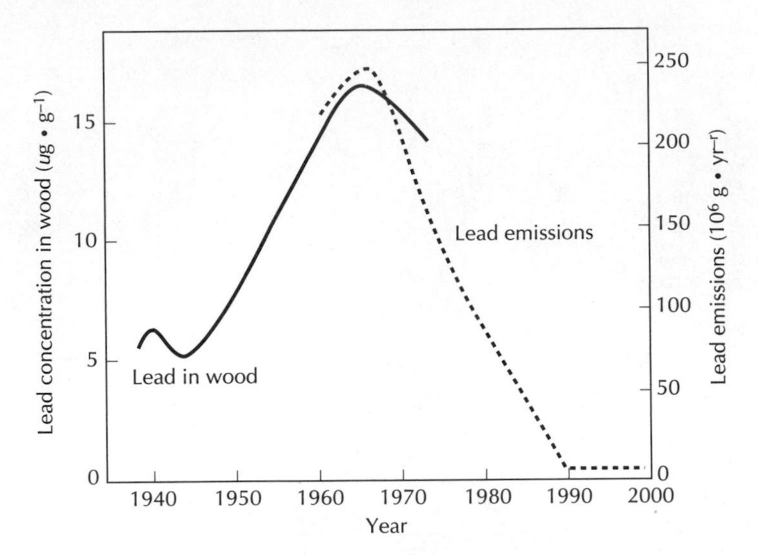

FIGURE 25.8 Effects of the addition and removal of leaded additives to gasoline expressed as concentrations in wood in trees in an urban park in Atlanta. (Ragsdale and Berish 1988. Reprinted with the permission of Kluwer Academic Publishers.)

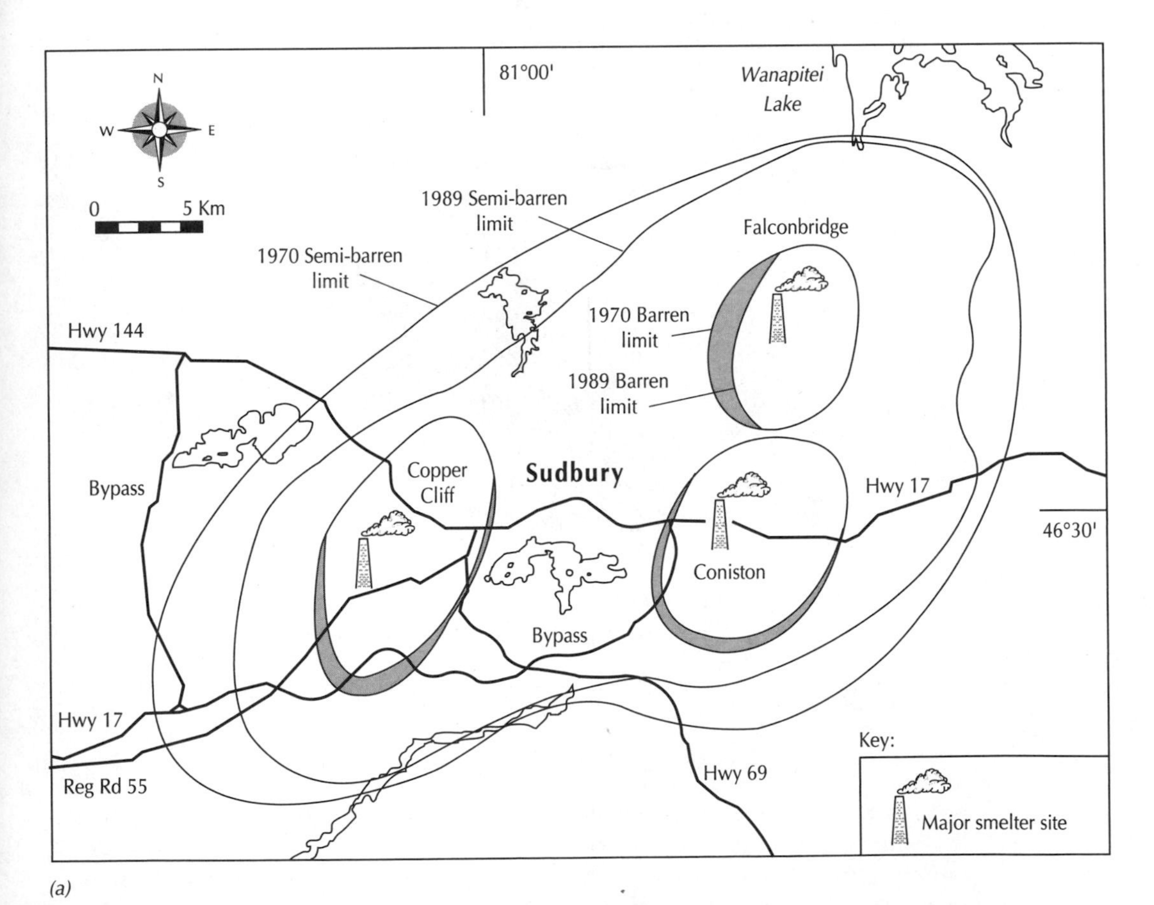

FIGURE 25.9 Sudbury, Ontario, and the impacts of three large smelting plants on vegetation. (*a*) Map of the area showing the extent of completely and partially barren areas resulting from acid and metal deposition.

FIGURE 25.9 *(Continued)* (*b*) Pictures of vegetation in the different zones. (Photos courtesy of Keith Winterhalder. Reprinted with the permission of Kluwer Academic Publishers) **See plate in color section**

For areas not adjacent to large point sources of pollutants, the effects of pollution deposition on the function of terrestrial ecosystems is more complex and subtle, yet the potential for profound disruptions is present. The pollutants listed in Table 25.1 affect nearly every major process that determines resource availability, nutrient cycling, and plant and microbial metabolism. The effects of individual pollutants can be significant. Interactions between pollutants can either offset or enhance individual effects.

EFFECTS OF INDIVIDUAL AIR POLLUTANTS

Oxidant Gases

Of the gaseous forms of air pollution, ozone appears to have the greatest potential for damaging ecosystems outside of urban areas. Its effects are very straightforward. Ozone diffuses into foliage through the stomata. Inside of leaves, the reactive ozone attacks cell membranes, decreasing the efficiency of membrane function. This increases respiration as membrane repair becomes an increased metabolic cost. This, in turn, reduces the rate of net photosynthesis, reducing C accumulation in plants. SO_2 may also have oxidizing effects on plants. In addition, both SO_2 and NO_x can form toxic compounds once in the leaf that further inhibit leaf function.

Effects of the oxidant gases are linear and cumulative in that decreased growth rate in trees, or yield in agricultural crops, can be expressed as a linear function of the cumulative dose (concentration multiplied by exposure time) of the pollutants (Figure 25.10*a*). Rare events of unusually high concentrations do not seem to be crucial. The susceptibility of different species to gaseous pollutants is also a function of the rate at which gases are exchanged between the leaf and the atmosphere. Fast-growing plants have high gas exchange rates (high conductance for rapid CO_2 fixation), so they also have a higher uptake rate for other gases, including pollutants, and a greater reduction in net photosynthesis (Figure 25.10*b*). The predictability of ozone effects on photosynthesis makes the construction of models to predict the effects of different levels of this form of pollution on forest production rela-

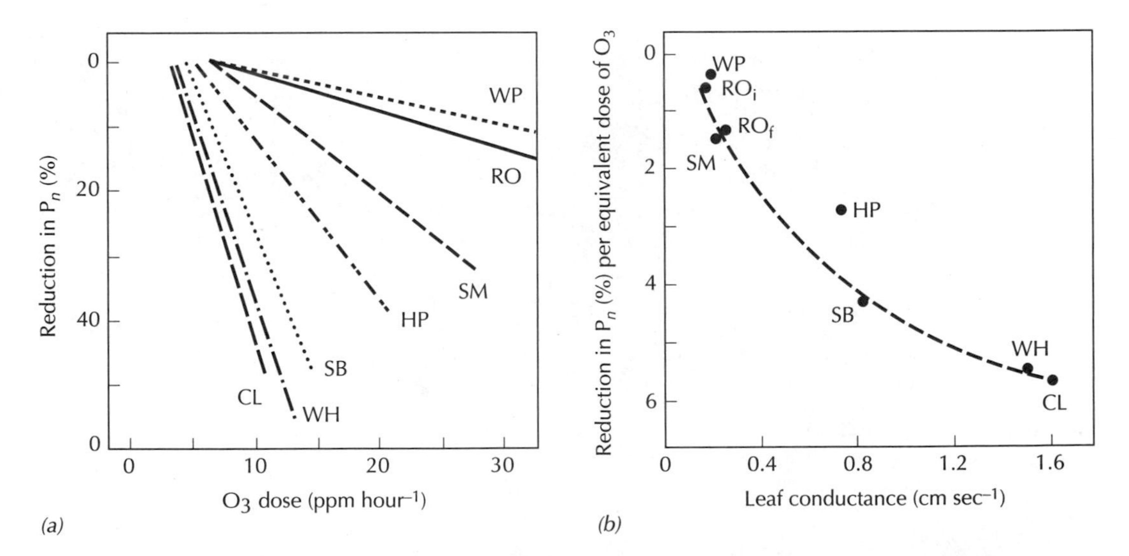

FIGURE 25.10 The effect of ozone, expressed as the total accumulation of concentration times hours of exposure, on net photosynthesis of several species of crops and trees. (*a*) Reduction in rate of net photosynthesis as a function of dose. (*b*) The relative reduction in net photosynthesis as a function of the rate of gas exchange between leaf and atmosphere (conductance). CL = red clover, WH = wheat, SB = soybean, HP = hybrid poplar, SM = sugar maple, RO = northern red oak, WP = white pine. (Reich and Amundson 1985)

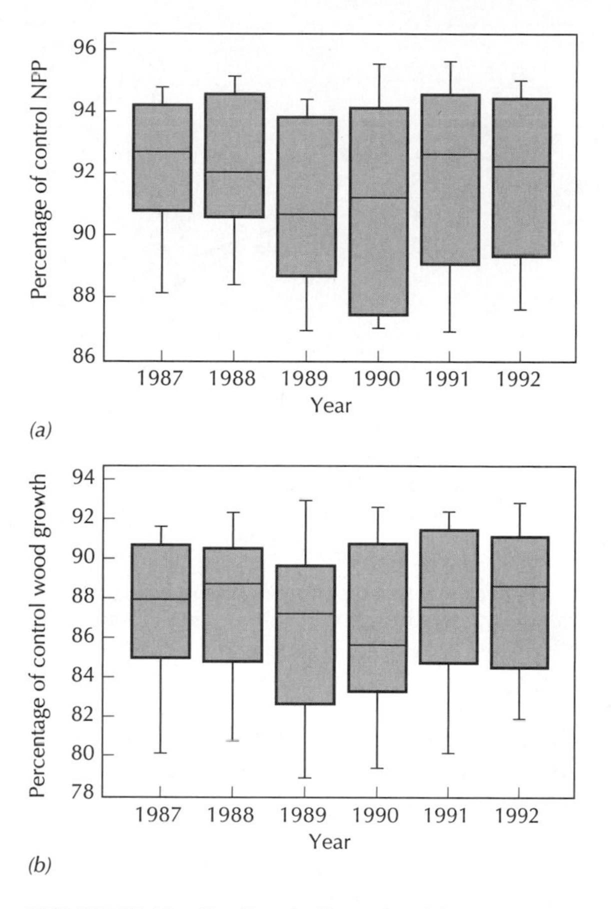

FIGURE 25.11 Predicted effect of ambient ozone concentrations over a 6-year period on total NPP and wood production for forests at 64 sites across the northeastern United States (bar is median value; box includes stands in the 25th–75th percentile; lines show range of data). (Ollinger et al. 1997)

tively straightforward, and predictions of current reductions in forest production have been produced (e.g., Figure 25.11).

Ozone and other gases may be the only air pollutants that can severely affect crop plants because the damage is directly to the plant rather than through effects on soils. The chemistry of agricultural soils in the temperate zone is altered much more by management practices than is ever likely to occur by deposition of pollutants.

Heavy Metals

Heavy metals interfere with many crucial biochemical reactions and so tend to act as general biocides. Again, there are some fairly straightforward relationships between, for example, heavy metal concentration in soil organic matter and the rate at which that organic matter is decomposed (Figure 25.12). Direct effects of metals in artificial soil solutions on seedling root

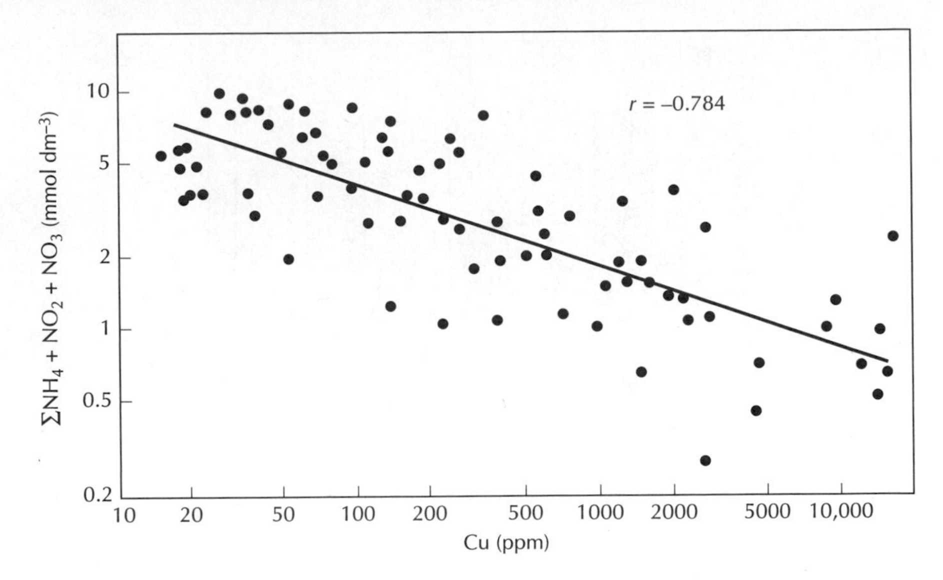

FIGURE 25.12 The effect of increasing copper concentration on rates of nitrogen mineralization in laboratory incubations for soils surrounding a brass mill in Sweden. (Tyler 1975)

growth and nutrient uptake have been shown, but those same effects are difficult to demonstrate in the field. Heavy metals are "fixed" by organic matter and are relatively immobile, especially at higher pH values. This reduces concentrations in soil solutions in organic soil horizons but also leads to continuous accumulations in that organic matter, which could result in reduced decomposition and nutrient mineralization later on.

Nitric and Sulfuric Acids

Both of these acids contribute mobile anions (SO_4^{2-} and NO_3^-) to soils. Biological uptake of NO_3^- and sorption processes for SO_4^{2-} can reduce the concentrations of these anions and reduce their acidifying effects. However, if overall anion leaching below the rooting zone increases due to deposition, cations will be carried along as well. Remember that the nutrient cations Ca, magnesium (Mg), and potassium are leached preferentially (Chapter 9), reducing base saturation and nutrient cation availability and lowering soil pH. In heavily impacted soils where nutrient cation availability has been severely reduced, anion leaching leads to increased concentrations of aluminum and then hydrogen in the soil solution and stream water (Figure 25.13). Elevated aluminum concentrations can cause death in fish by interfering with the operation of gills. Aluminum–Ca ratios of greater than 1.0 in soil solutions have been found to inhibit nutrient uptake and plant growth.

Nitrogen Saturation

In areas of concentrated agricultural and industrial activity, NH_4^+ and NO_3^- deposition combined result in very high rates of total N deposition. The cu-

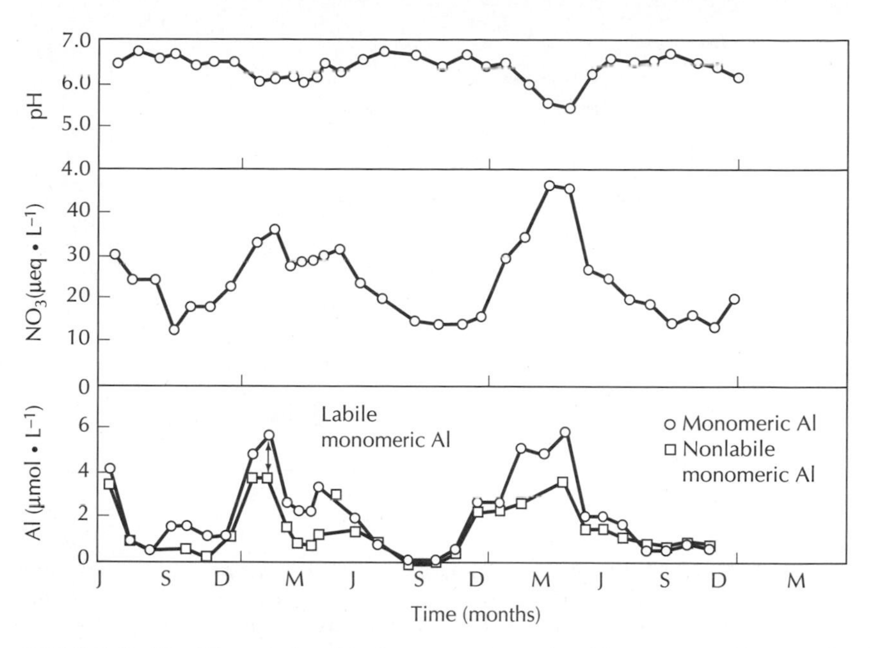

FIGURE 25.13 The relationship between nitrate leaching, water pH, and aluminum content in a small lake in northern New York State. Increased nitrate losses through a soil with low base saturation, particularly during the spring snowmelt season, result in the acidification of stream and lake water and increased mobility and leaching of aluminum. (Driscoll et al. 1987)

mulative effect of high N deposition over many years in these areas is N availability to plants and microbes in excess of their nutritional demands, a condition called **N saturation** (see the discussion of the underlying theory of N saturation in Chapter 18). Excess N availability has been described as a novel form of stress for temperate and boreal forest ecosystems in which N has generally been a limiting nutrient. Plants may not be preadapted to this situation and may not adjust C allocation patterns appropriately. Potential negative effects include pathologically low root and mycorrhizal biomass and loss of frost hardiness in evergreen foliage.

The process by which N deposition leads to N saturation was a topic of concentrated research efforts in the United States and Europe through the 1990s. An initial set of hypotheses predicted that forests would pass through several stages in the transition from N limited to N saturated (Figure 18.4). It was also predicted that most of the changes over time would not be linear; that is, there would be important switch points at which, for example, nitrification and NO_3^- leaching would begin to occur. Extensive research in the conifer forests of Europe clearly identified a critical C:N ratio for forest floors above which nitrification increased linearly. Similar research in the United States found nearly the same result (Figure 25.14). Experiments involving the additions of N to N-limited forests and the extensive surveys in Europe found that forests have a very large and continuing capacity to retain added N (Figure 25.15), although the mechanism by which this occurred was unclear. Forest stands recovering from conversion to agriculture had especially high

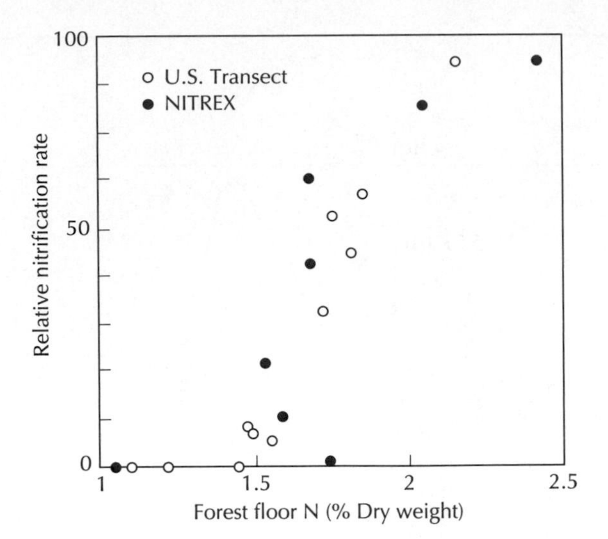

FIGURE 25.14 Relationship between forest floor nitrogen concentration and relative rates of nitrification. U.S. Transect data are spruce–fir forests from 11 areas (total of 161 sites). (McNulty et al. 1991) NITREX data are from eight evergreen conifer stands across Europe. (Tietema and Beier 1995)

capacities for retaining added N and delaying N saturation. On the other hand, several experiments in evergreen conifer forests have shown either reduced forest growth with chronic N additions, or increased growth following the removal of high ambient N deposition loads (see experimental description in Figure 4.9).

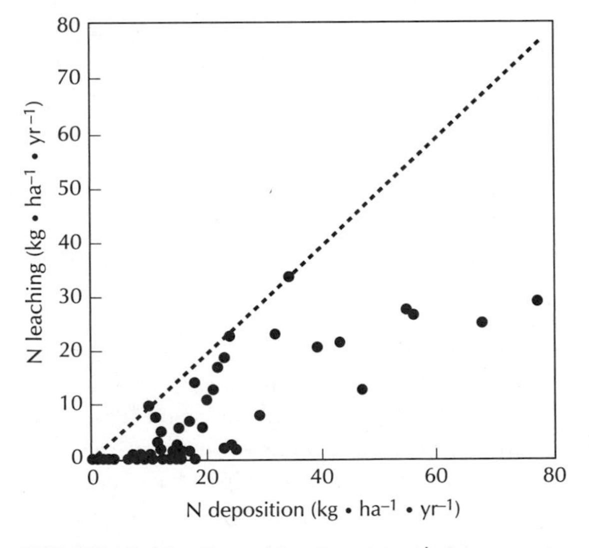

FIGURE 25.15 Deposition inputs and stream water outputs of nitrogen for 65 forest ecosystems in Europe. Note that losses are much less than inputs for most systems (on average, about 75% of nitrogen deposition is retained). (Dise and Wright 1995)

FOREST DECLINE: THE INTERACTIVE EFFECTS OF POLLUTANTS

The direct effects described above and their interactions *might* pose serious threats to terrestrial ecosystems in heavily polluted regions. Are there extensive areas in the landscape where damage can be documented?

Throughout Germany and all of central Europe, the 1970s and 1980s saw an alarming increase in the number of evergreen trees exhibiting symptoms of forest decline. Symptoms included reductions in tree vigor, foliage density, and growth and increases in chlorosis (needle yellowing) and mortality. This region is also one of very concentrated industrial activity, and deposition of several types of air pollutants are quite high. This co-occurrence has led to the general conclusion that air pollution has caused or contributed to forest decline and to a great deal of research to prove or disprove that conclusion.

Similar symptoms of decline have been seen in high elevation spruce–fir forests of the Appalachian Mountains in the eastern United States. In the northeastern states, both ground surveys and remote sensing from satellites have been used to determine the amount of this forest type that, in the 1980s, could be classified into different stages of severity of decline (Table 25.2). In general, the degree of forest decline increased from east to west, increasing in areas that are closer to the major sources of pollution and that experience higher rates of deposition. Again, this suggests that air pollution plays a role in decline.

However, identifying the causes of tree decline is very difficult. Regional die-backs of tree species have occurred frequently in areas or eras when pollution loading has been low. For example, at least two earlier episodes of spruce decline have occurred in the northeastern United States, each trig-

TABLE 25.2 Percentage of Conifer Forests in Different Damage Classes for Mountain Ranges in New York, Vermont, and New Hampshire*
(From Rock et al., 1986)

	% of Total High Elevation Forests in Different Damage Classes		
Site	**Low**	**Medium**	**High**
New York (Adirondack Mountains)			
Whiteface Mountain	8.6	12.7	78.7
High Peaks area	2.3	6.8	90.9
Vermont (Green Mountains)			
Camel's Hump	26.5	20.7	52.8
Mt. Abraham	25.6	25.6	48.9
Broadloaf	37.9	24.6	37.5
New Hampshire (White Mountains)			
Mt. Moosilauke	72.3	16.8	10.9
Lafayette	63.9	19.0	17.0

*Damage estimates for Whiteface Mountain result in part from natural formation of "firwaves" (Spruegel 1984), but the regional trends here have been supported by other studies (e.g., Craig and Friedland 1990).

gered apparently by a period of extreme drought. Reduced growth by spruce in the Northeast began in the 1960s, coincident with another period of drought. Again, this kind of circumstantial evidence cannot prove drought as the cause of the current decline.

Forest decline is a complex process and is often difficult to attribute to a single cause. Rather, decline is often described as the result of a series of stresses. The first is a predisposing stress that weakens the trees and perhaps reduces growth rates without causing visible damage. The second is the triggering stress that causes a sudden loss of vigor and may lead directly to tree death. The third is a secondary or contributing stress, such as insect or pathogen attack on weakened plants, which further reduces vigor or increases mortality. Without being the sole cause for forest decline, air pollution could easily play a role at any of these three levels, with the strength of the air pollution stress being modified by changing climatic or pest infestation conditions.

Combining the complexities of the forest decline phenomenon with the interactions between several sources of air pollution and plant and soil physiology, it is not surprising that definitive causes for a particular decline event cannot always be identified. There are currently several hypotheses as to how air pollution might cause or contribute to forest decline. Most of these relate to interactions between pollutants.

Soil Acidification and Heavy Metals

The longest-standing hypothesis is that acidic precipitation lowers soil pH and reduces cation availability, inhibiting biological function, and reducing forest vigor. A variant of this theory is that low soil pH also increases the solubility of aluminum, which, in turn, interferes with Ca uptake by roots, exacerbating low Ca availability. Initially regarded as a minor threat, soil acidification is increasingly seen as a potential hazard, especially to "cation pump" species, such as sugar maple in the northeastern United States, which occur preferentially in Ca-rich soils and appear to require more of this element than do other species. Reduced soil pH can also bring into solution other heavy metal elements deposited over time.

Nitrogen Saturation, Soil Acidification, and Magnesium Deficiencies

Lower soil pH means reduced base saturation and lower availability of cations, such as potassium and Mg. One of the better documented explanations of forest decline in Europe involves an induced imbalance between N and Mg availability to plants. High N availability fosters increased production of foliage, but low availability of Mg, crucial for chlorophyll formation, results in chlorosis, a yellowing of foliage indicating greatly reduced rates of photosynthesis. Experimental liming of a declining forest in Germany increased growth and vigor of trees, supporting this hypothesis. Similarly, research on red spruce in the northeastern United States has determined that subcritical concentrations of Ca in membranes impairs response to stress at the cellular level, predisposing this species to several different forms of damage (Figure 25.16).

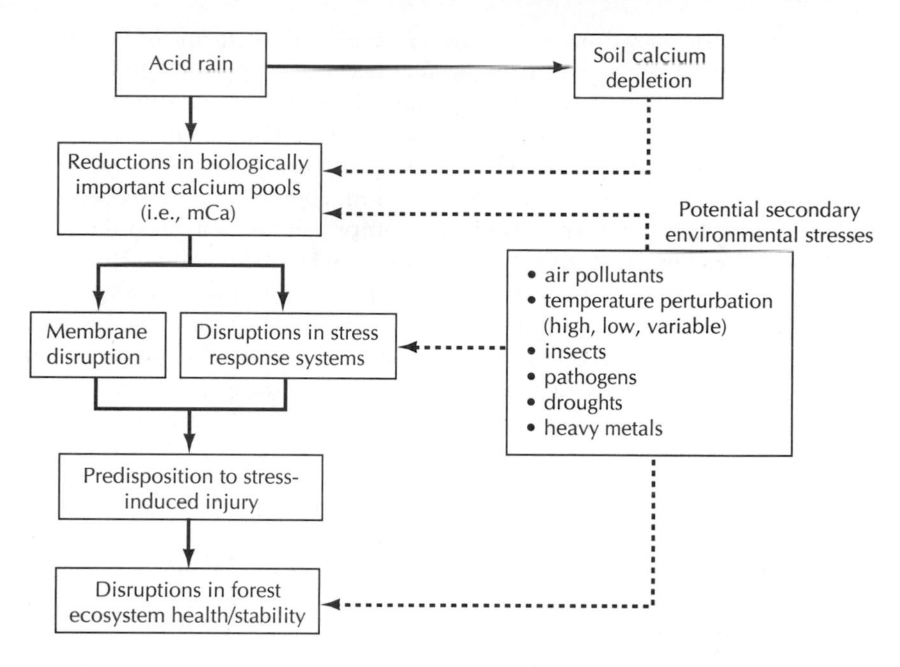

FIGURE 25.16 The effects of acid rain on calcium availability in soils, content of metabolically active calcium in cell membranes, and predisposition to forest decline in red spruce in the northeastern United States. (DeHayes et al. 1999)

Ozone and Nitrogen

Ozone reduces net photosynthesis and so reduces plant growth and vigor. By reducing the amount of mobile C available to plants, elevated ozone levels may intensify internal C–N imbalances resulting from excess N availability. This interaction would also intensify the effects of the N–Mg imbalance described earlier.

Ozone and Acidification

By damaging membrane function in leaves, ozone increases the "leakiness" of cells. Cation nutrients can migrate away from cells to the leaf surface and, there, be removed in cation exchange–type reactions by the H^+ in rainfall. This could increase nutrient leaching from plants and reduce their ability to retain cations in the system against increased anion leaching rates.

Heavy Metals and Nitrogen

The combined effects of ozone and excess N in reducing root biomass and mycorrhizae may be accentuated by the presence of heavy metals in the soil solution, which could again be increased by soil acidification. In contrast, heavy metals in soil organic matter might reduce N mineralization rates and actually reduce excess N availability.

At least in the northeastern United States, several factors combine to make the high elevation spruce–fir forests, those in which forest decline has occurred on a large scale, especially susceptible to air pollution damage. First, they tend to occur on very thin, poorly buffered soils, with low SO_4^{2-} sorption capacity. Second, the cold climate and short growing season limit total annual net photosynthesis and so reduce the potential for N uptake and allocation. Third, the mountaintop locations in which these forests grow tend to experience both long periods of immersion in cloud layers and high wind speeds due to compression of air movement over ridge tops. These interact with the fourth factor, stand structure (high leaf area indices cause canopies to act as very efficient air filtering systems) and increased precipitation at higher elevations, to cause large increases in total pollutant deposition. Ozone concentrations have also been found to remain high for longer periods of time at the elevation zone in which spruce forests grow. All of these factors combine to increase pollutant inputs and increase forest sensitivity to those pollutants.

DETERMINING "CRITICAL LOADS" OF POLLUTION

As with other complex combinations of stress factors discussed in this book, controlled experimentation on the interactive effects of all of the different types of pollution described here across a range of forest conditions is not possible. In the absence of full experimentation, results from single-factor studies in the field, or seedling–sapling–soil core studies in the laboratory can be used to derive computer models that may then be used to predict the effects of pollution.

Modeling changes in stream water chemistry and total element balances in response to atmospheric deposition has been pursued vigorously in both Europe and the United States. In Europe, an international program carried forward through the 1990s was designed to determine the **critical load** for various forms of deposition—that level above which substantial negative changes in ecosystem state or function would occur. A model of forest soil interactions was constructed to determine changes in soil pH and cation balances for different levels of N and S deposition combined with inherent rates of soil weathering and other soil chemical processes. The result of this effort is a map of critical loads for the continent, which, when combined with maps of current deposition rates (Figure 25.17), yields both an indication of where these loads are exceeded and by how much. Such maps are being used to drive policy decisions regarding pollution abatement across the continent.

The United States has no critical loads program. A large and well-funded acid deposition program in the 1980s (called the National Acid Precipitation Assessment Program) focused almost exclusively on the S component of acid rain. One product of this program was a set of models that were used to predict both previous changes in stream water chemistry due to cumulative acid deposition to that point and future trends in water quality under different scenarios of pollution reduction. These models were used as part of the process leading up to the Clear Air Act Amendment of 1990. Since that time, the United States has focused on implementing the goals of that act.

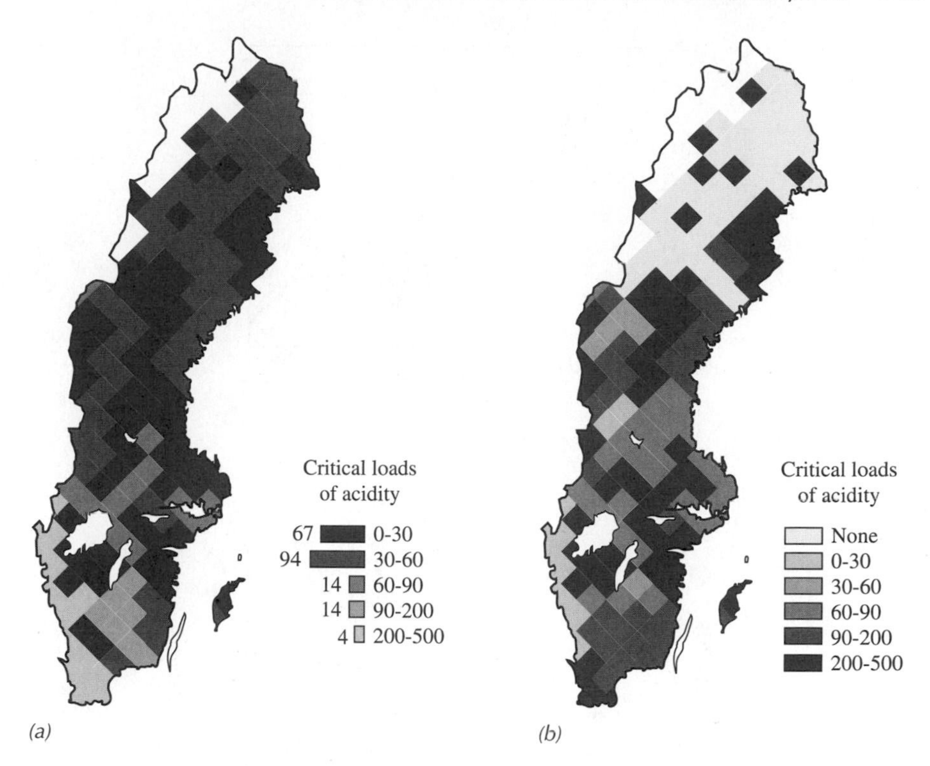

FIGURE 25.17 Calculated critical loads and exceedances for Swedish forests. (*a*) Calculated critical load based on acid neutralizing capacity of different soil types. (*b*) Estimated exceedances—the extent to which current deposition rates exceeds the critical load. (Warfvinge and Sverdrup 1995) **See plate in color section**

ENVIRONMENTAL SUCCESS STORIES: POLLUTION REDUCTIONS IN THE UNITED STATES AND EUROPE

A combination of greater public awareness, increasing scientific understanding, and larger changes in the social fabric of the global community have led to a series of timely and impressive reductions in pollution emissions and ecosystem recoveries in several parts of the world. The removal of lead from gasoline (Figure 25.8) is one example, and there are others.

By 1970, the area immediately surrounding the huge smelters near Sudbury, Ontario, Canada, had been decimated by decades of SO_4^{2-} and heavy metal emissions (see Figure 25.9). Rings of totally lifeless zones around each of three smelters were embedded in a much larger "semibarren" zone, in which a mixture of shrubs and grasses occupied patches, with bare ground in between. The affected area covered more than 200 km^2. Environmental protection regulations, along with the closure of one of the smelters, began to reduce emissions in the 1970s; SO_2 emissions have dropped by nearly 90%; and the acidity of water in a nearby lake has increased in response, although metal concentrations remain high (Figure 25.18). Natural recovery in such contaminated systems is very slow. Boundary shifts in the 20 years between 1970 and 1990 can be seen in Figure 25.9. All recovery in the barren zones has been due to an aggressive program of ecosystem recovery and restoration car-

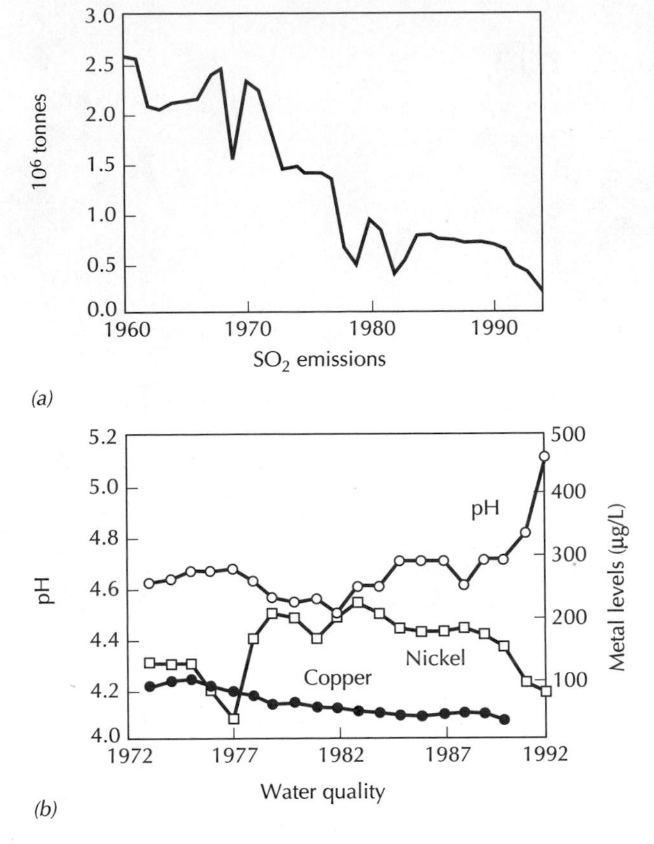

FIGURE 25.18 The beginnings of environmental recovery in the Sudbury, Ontario, area. (*a*) Declines in sulfur dioxide emissions since 1960. (*b*) Response in lake pH and metal concentrations. (Gunn et al. 1995)

ried forward by a consortium of municipal, environmental, and business groups.

Central planning and poor environmental management led to the continued operation for several decades of a very large and extremely inefficient soft coal power plant in the Krusné Hory district of the former Czechoslovakia, adjacent to the German border. The tremendous amounts of S emitted by this plant proved extremely detrimental to both forest and human health in the region. The fall of the Soviet Union and the creation of the Czech Republic brought local control to this plant, some reduction in economic activity, and a significant reduction in levels of pollution. Ambient SO_2 concentrations have been cut in half, as has total S deposition to the forests in the surrounding mountains. In response, SO_4^{2-} concentrations in streams have also declined by nearly one third (Figure 25.19). A combination of liming and reforestation has begun to restore forest cover to the mountains as well.

While efforts to reduce air pollution emissions have been modest in the United States compared to those in Europe, measurable progress at-

(text continues on p. 508)

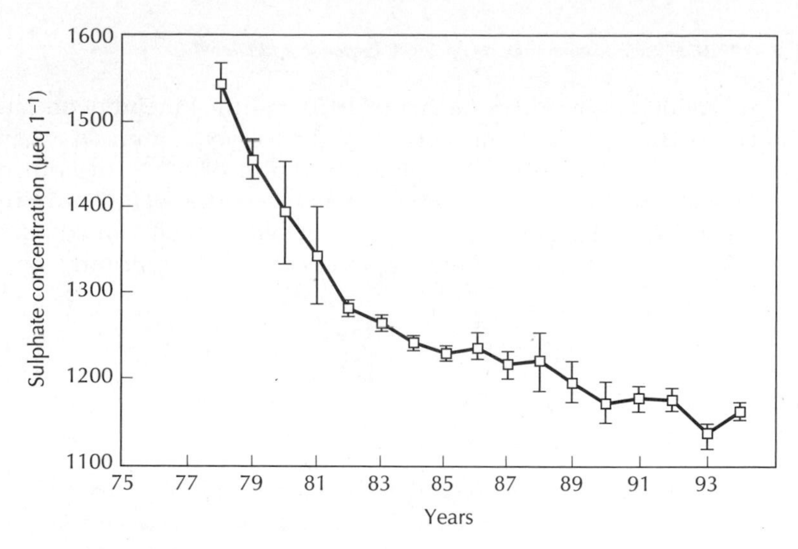

FIGURE 25.19 Recovery of stream water chemistry in the Krusné Hory region of the Czech Republic following reductions in sulfur emissions from a large power plant complex. (Cerny 1995)

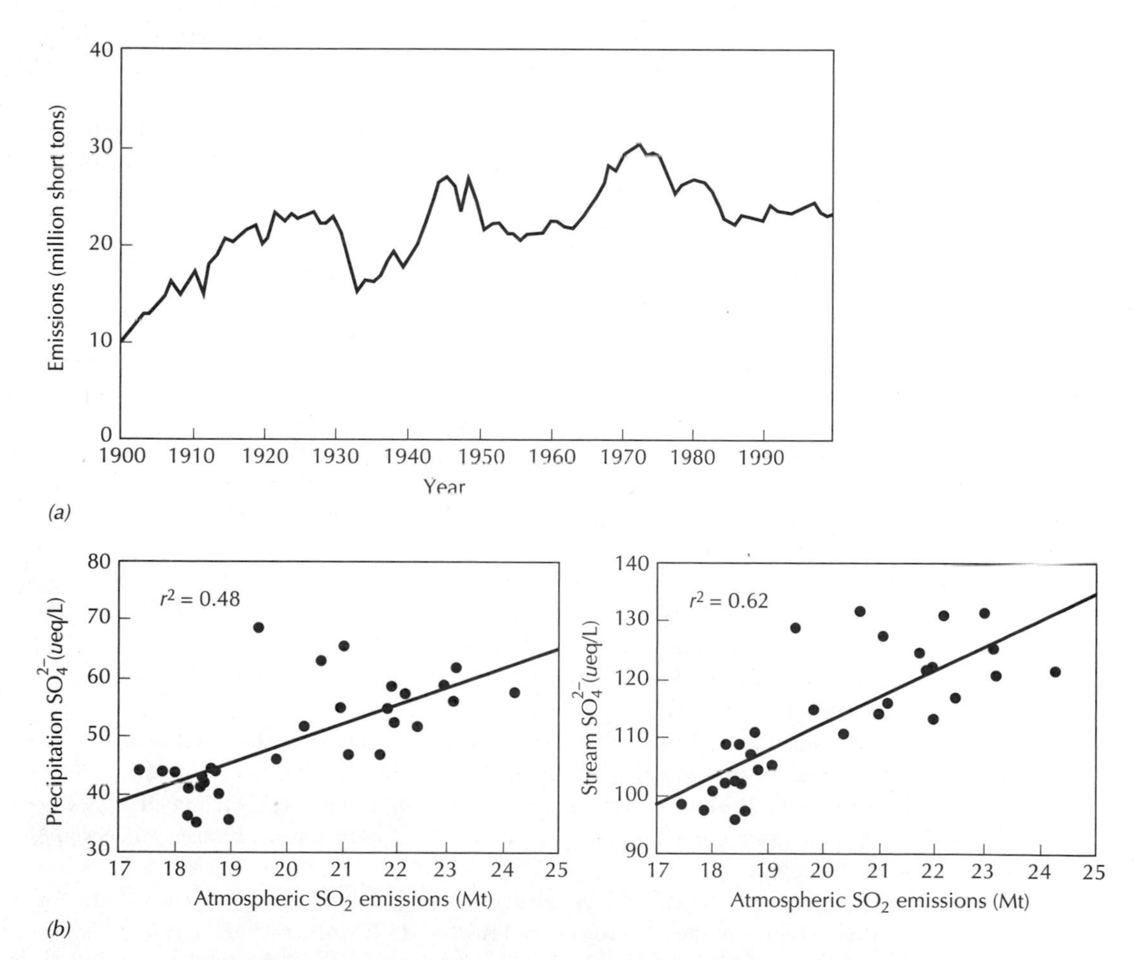

FIGURE 25.20 Sulfur emissions, precipitation, and stream chemistry. (*a*) Trend of total sulfur emissions over time for the United States. (*b*) The relationship between sulfur emissions and precipitation and stream water chemistry in the reference watershed at Hubbard Brook, New Hampshire. (Driscoll et al. 1998)

tributable to the Clean Air Act of 1970 and the amendments passed in 1990 can be seen. Total S emissions have decreased by about one third since the early 1970s (Figure 25.20*a*). Concentration of SO_4^{2-} in both precipitation and stream water in the reference watershed at the Hubbard Brook research site in New Hampshire are directly related to total United States emissions (Figure 25.20*b*), and both have decreased significantly over the same period.

REFERENCES CITED

Cerny, J. 1995. Recovery of acidified catchments in the extremely polluted Krusné Hory Mountains, Czech Republic. *Water, Air and Soil Pollution* 85:589–594.

DeHayes, D. H., P. G. Schaberg, G. J. Hawley, and G. R. Strimbeck. 1999. Acid rain impacts on calcium nutrition and forest health. *BioScience* 49:789–799.

Dise, N. B., and R. F. Wright. 1995. Nitrogen leaching in European forests in relation to nitrogen deposition. *Forest Ecology and Management* 71:153–162.

Driscoll, C. T., G. E. Likens, and M. R. Church. 1998. Recovery of surface waters in the northeastern U.S. from decreases in atmospheric deposition of sulfur. *Water Air and Soil Pollution* 105:319–329.

Driscoll, C. T., C. P. Yatsko, and F. J. Unangst. Longitudinal and temporal trends in the water chemistry of the North Branch of Moose River. *Biogeochemistry* 3:37–62.

Galoway, J. N., G. E. Likens, and M. E. Hawley. 1984. Acid precipitation: natural versus anthropogenic components. *Science* 226:829–831.

Gunn, J., W. Keller, J. Negusanti, R. Potvin, P. Beckett, and K. Winterhalder. 1995. Ecosystem recovery after emission reductions: Sudbury, Canada. *Water, Air and Soil Pollution* 85:1783–1788.

Heilman, W. E., J. Hom, and B. E. Potter. 2000. Climate and atmospheric deposition patterns and trends. In Mickler, R. A., R. A. Birdsey, and J. Hom (eds.). *Responses of Northern Forests to Environmental Change.* Springer, New York.

McNulty, S. G., J. D. Aber, and R. D. Boone. 1991. Spatial changes in forest floor and foliar chemistry of spruce–fir forests across New England. *Biogeochemistry* 14:13–29.

Ollinger, S. V., J. D. Aber, and P. B. Reich. 1997. Simulating ozone effects on forest productivity: Interactions among leaf-, canopy- and stand-level processes. *Ecological Applications* 7:1237–1251.

Ragsdale, H. L., and C. W. Berish. 1988. The decline of lead in tree rings of *Carya* spp. in urban Atlanta, GA, USA. *Biogeochemistry* 6:21–30.

Reich, P. B. and R. G. Amundson. 1985. Ambient levels of ozone reduce net photosynthesis in tree and crop species. *Science* 230:566–570.

Rock, B. N., J. E. Vogelmann, D. L. Williams, A. F. Vogelmann, and T. Hoshizaki. 1986. Remote detection of forest damage. *BioScience* 36:439–445.

Spruegel, D. G. 1984. Density, biomass, productivity and nutrient cycling changes during stand development in wave-regenerated balsam fir forests. *Ecological Monographs* 54:165–186.

Tietema, A., and C. Beier. 1995. A correlative evaluation of nitrogen cycling in the forest ecosystems of the EC projects NITREX and EXMAN. *Forest Ecology and Management* 71:143–152.

Tyler, G. 1975. Heavy metal pollution and mineralisation of nitrogen in forest soils. *Nature* 255:701 702.

Warfvinge, P., and H. Sverdrup. 1995. *Critical Loads of Acidity to Swedish Forest Soils: Methods, Data and Results.* Reports in Ecology and Environmental Engineering Number 5. Lund University, Lund, Sweden. 104 pp.

26

THE GLOBAL CARBON CYCLE AND CLIMATE CHANGE

INTRODUCTION

In Chapter 1, we introduced Lindemann's concept of energy as a "universal currency" in ecosystem studies. You know now that carbon (C) is the element most closely associated with the transfer of energy through ecosystems, and that the availability of carbon dioxide (CO_2) in the atmosphere is a key controller of the rate of photosynthesis. We can add two more important attributes of CO_2: (1) it is a "greenhouse" gas that traps long-wave infrared radiation and reduces the loss of energy from the atmosphere to outer space and (2) it is the primary product formed by the combustion of fossil fuels. Given that the modern global economy is driven by the combustion of fossil fuels and that global warming is a focus of environmental concern, it is not surprising that C is also a key element in the interaction between economic production and environmental quality.

One of the key components of sound policy for the management of atmospheric CO_2 is a quantitative understanding of the fluxes of C between the atmosphere and the terrestrial biosphere. In response to the need for information relevant to the policy issues, there has been a dramatic increase in research on factors controlling the C balance of terrestrial ecosystems.

This chapter begins with some background information on changes in atmospheric CO_2 and implications for climate change. We then present two facets of the interaction between terrestrial ecosystems and the changing global C cycle: (1) physiological interactions between increased CO_2 and ecosystem function and (2) the role of land use in altering the C balance of ecosystems. We then present several different methods for estimating the current role of terrestrial ecosystems in the global C cycle.

ATMOSPHERIC CARBON DIOXIDE AND CLIMATE

The continuous, long-term record of atmospheric CO_2 concentrations made at the Mauna Loa Observatory on the island of Hawaii has become one of the key indicators of human impact on the global environment. The site was chosen because of its isolation from human influences; samples collected at the summit are assumed to represent the average or well-mixed condition of the lower atmosphere (the troposphere) in the northern hemisphere.

Measurements of CO_2 at Mauna Loa (Figure 26.1) show two clear patterns. The first is an annual cycle in concentration, with declines in spring and summer and increases in fall and winter, tied to the seasonality of photosynthesis and decomposition in the northern hemisphere. In this sense, the measurements at Mauna Loa are charting the metabolism of a hemisphere. Imposed on top of this annual cycle is a continuous, long-term increase in CO_2 concentration, indicating an imbalance between sources (processes that emit CO_2 to the atmosphere) and sinks (those that remove CO_2). Why has this increase occurred?

Figure 26.2 is one current view of the C metabolism of the Earth. Before the industrial age, the complimentary processes of photosynthesis and respiration (far left in Figure 26.2) would have been nearly balanced in any 1 year, with losses due to disturbance in some systems offset by gains in aggrading systems elsewhere. In effect, the global terrestrial biosphere was in a state of equilibrium. Changes in land use (arrows directly to the left of NPP and respiration) were also minimal in any 1 year. The high solubility of CO_2 in seawater acted to keep atmospheric CO_2 levels both low and relatively stable. While very long-term changes in CO_2 concentrations in the atmosphere have occurred over geological time, the rapid shifts seen in the past few decades did not.

Human activity has modified this balance. Perhaps the most obvious cause is the increase in combustion of fossil fuels (Figure 26.2, *far right*). However, humans are also altering the Earth's ecosystems by increasing the intensity of forest harvesting and converting forests to agriculture and back again. While different scenarios exist for future concentrations of CO_2, there is a general consensus that this crucial gas will be twice as abundant as it is in the atmosphere currently sometime in the twenty-first century (Figure 26.3).

Increasing CO_2 in the atmosphere may cause significant, long-term changes in climate. Carbon dioxide is transparent to short-wave (including

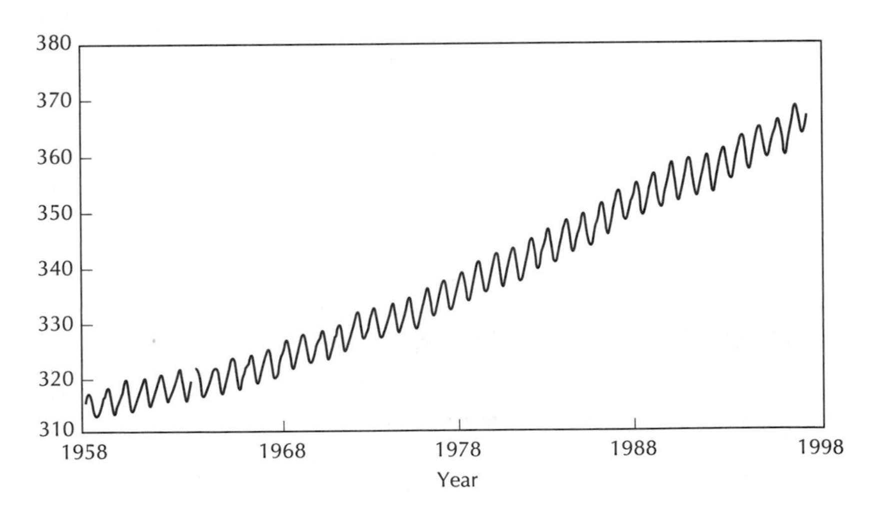

FIGURE 26.1 Trends in the CO_2 concentration in the Earth's atmosphere. (Keeling and Whorf 1999, http://cdiac.esd.ornl.gov/ftp/maunaloa-co2/maunaloa.co2)

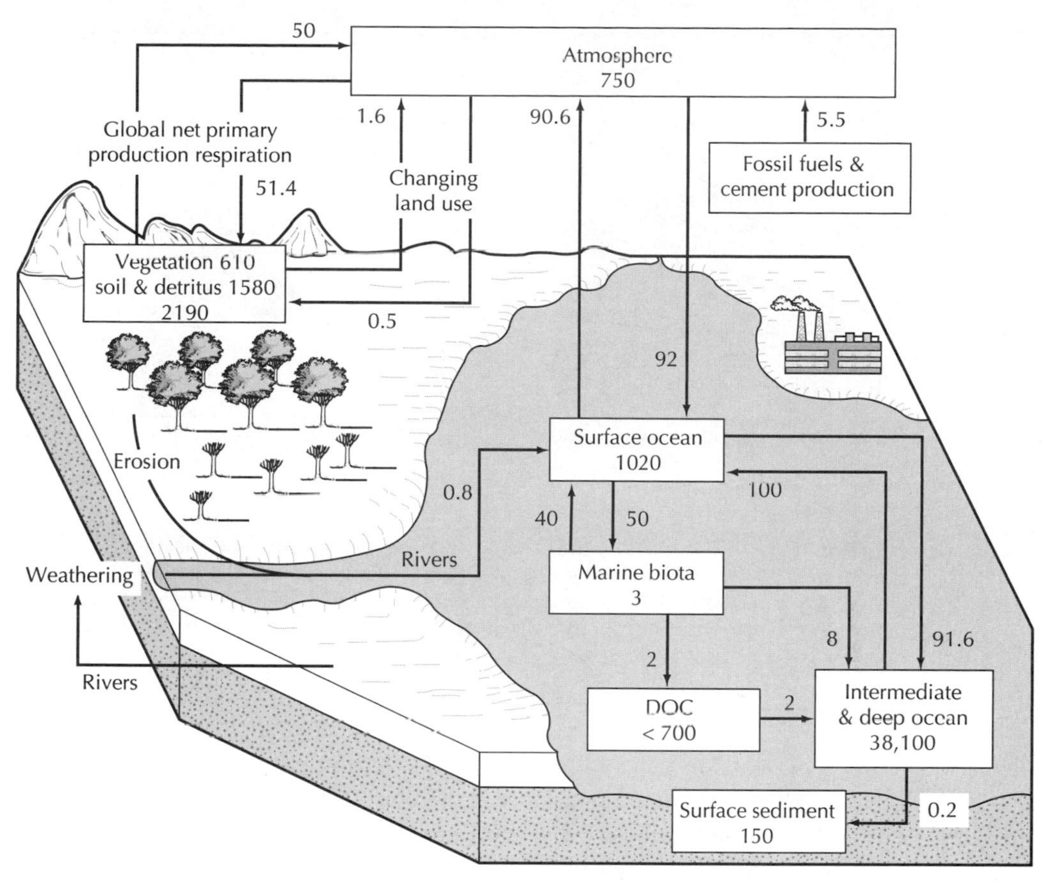

FIGURE 26.2 One view of the current global annual CO_2 budget (all values in petagram [Pg] or 10^{15} g C · year^{-1}). The net gain of 1.4 Pg in net primary production over respiration results from CO_2 and nitrogen fertilization. Clearing of forests releases 1.6 Pg · year^{-1}, while regrowth of abandoned lands takes up 0.5 Pg · year^{-1}. DOC =dissolved organic carbon. (Sarmiento and Wofsy 1999)

visible) radiation but is an efficient absorber of long-wave radiation. Thus, it allows incoming solar radiation, which is mostly short-wave, to penetrate and heat the Earth's surface while trapping long-wave, thermal radiation (Chapter 6) before it can be lost to space. This has been called the **greenhouse effect** because the resulting warming of the atmosphere is similar to the effect of glass in warming a greenhouse on a sunny day (Figure 26.4). The basic physics of greenhouse gas warming has been known since the late 1800s, and early estimates of the cumulative effect of human use of fossil fuels on mean global climate proposed nearly 100 years ago are not too different from those proposed today.

> Any doubling of the percentage of carbon dioxide in the air would raise the temperature of the Earth's surface by 4 degrees. . . . By the influence of the increasing percentage of carbonic acid in the atmosphere, we may hope to enjoy ages with more equable and better climates, es-

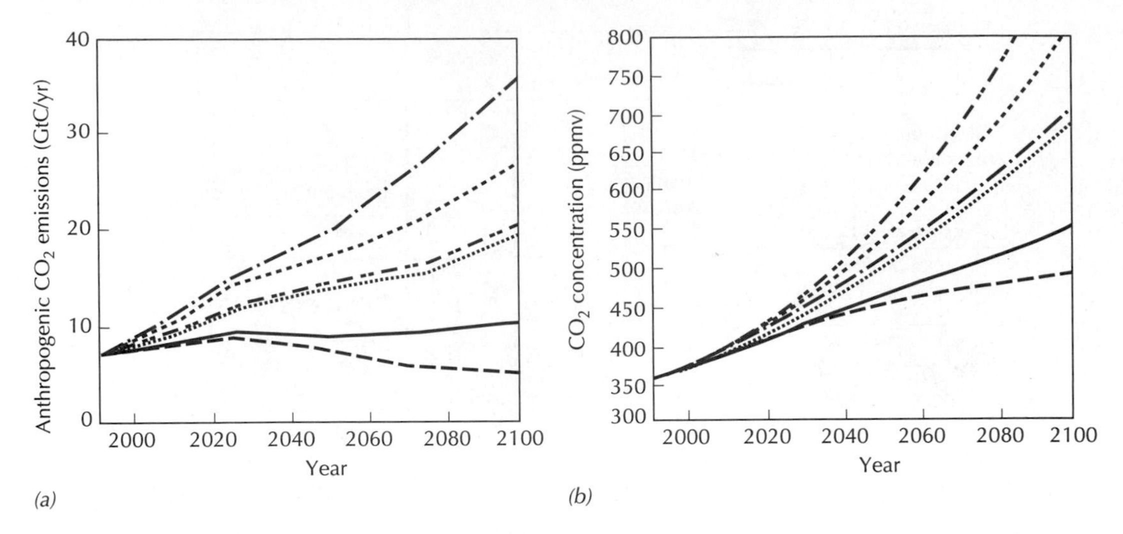

FIGURE 26.3 The future of atmospheric CO_2 concentrations. (*a*) The range of possible future CO_2 emission rates under different assumptions of fossil fuel availability, global economic activity, and global conservation efforts. GtC = gigatons of carbon. (*b*) Expected changes in atmospheric CO_2 in response to the different emission scenarios. ppmv = parts per volume. (Houghton et al. 1996)

> pecially as regards the colder regions of the Earth, ages when the Earth will bring forth much more abundant crops than at present for the benefit of rapidly propagating mankind. (Svante Arrhenius, 1906 and 1908, as cited in Fleming 1998)

However, whether this warming is viewed as good or bad has changed markedly. In periods of relative climatic stability, warming due to CO_2 has been described as a defense against the return of the glacial invasions and a new ice age. In periods such as the current one in which global temperatures are increasing at a rapid rate, serious concerns about globally elevated temperatures have been expressed.

Climate and climate change are not solely atmospheric phenomena. The energy balance of the atmosphere is determined in large measure by interactions with oceans and ocean currents, as well as by the condition of vegetation on land, especially the amount of foliage displayed and the potential to evaporate water. As we discussed in Chapters 6 and 7, conversion of water from liquid to vapor requires a considerable amount of energy. The ratio of AET to PET (Chapter 2) is a measure of not only the degree of water stress on plants but also the fraction of solar energy that goes into the evaporation of water rather than heating the land surface and lower atmosphere. The presence or absence of vegetation, water, ice, and snow also drastically alter the albedo of the land surface (i.e., the fraction of incoming short-wave radiation that is reflected immediately back out to space as short-wave radiation without contributing to warming of the surface or the atmosphere). The climate system, in turn, affects conditions both in the oceans and on land. A definitive model of climate change thus needs to include interactions and feedbacks among the atmosphere, oceans, and land surface.

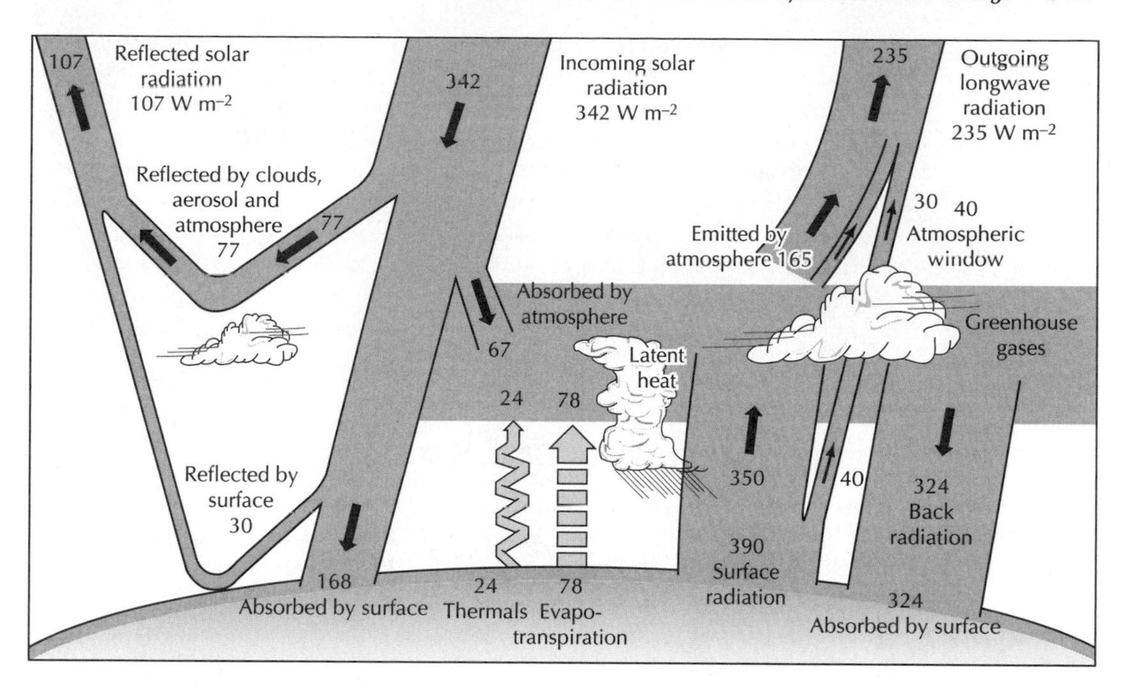

FIGURE 26.4 The energy balance of the Earth's surface and atmosphere. Of the 342 Watts (W) · m^{-2} of incoming solar radiation, 107 W is reflected back into space by the surface and atmosphere, 168 W is absorbed by the surface, and 67 W is absorbed by the atmosphere. The biggest flux between the atmosphere and surface is through long-wave infrared radiation emitted by the surface and absorbed and reradiated downward by gases in the atmosphere. Increased content of CO_2 and other "greenhouse gases" increases the absorption of long-wave radiation and the downward flux of radiant energy to the surface, increasing temperatures. (Houghton et al. 1966)

Models predicting the effects of altered atmospheric chemistry on climate have progressed from one- and two-dimensional versions, calculating changes either for the atmosphere as a whole or for a single latitudinal gradient, to three-dimensional depictions of the atmosphere. At this writing (2000), models that incorporate linkages between the atmosphere and ocean, and so include the time lags inherent in warming the huge amount of water the oceans contain, are beginning to appear. These models predict transient changes in climate over the next century. Linkages to land surface conditions remain less well defined. In particular, recent weather modeling at the regional scale has shown that the local distribution of land uses (agriculture, forest, urban, water) can cause significant changes in precipitation and temperature that are not captured in the coarser-scale global models. Global models can also not yet predict large-scale cyclical phenomena, such as the El Niño/La Niña cycles of warming and cooling in the Pacific Ocean, which drive drought–flood cycles in many parts of the world or the likelihood of severe weather events, such as hurricanes.

With all of these caveats, existing models can be used to predict changes in the mean fields of temperature and precipitation at the global scale. This is a rapidly advancing field of science, and models and predictions change frequently. Validation of these model predictions (see Chapter 5) cannot be made. We cannot run the experiment, or, said in a different way, we are running the experiment at the global scale now but will not know the outcome for 100 years. The value of these models is in helping us envision what the climate might look like at the end of the century if we follow different policy options. Two of the transient models in use at this time (2000) produce quite different predictions over the continental United States using "business-as-usual" scenarios for fossil fuel use (Figure 26.5). Essentially, the Hadley model predicts a moderate increase in temperature and precipitation. The Canadian model predicts larger increases in temperature and lower values for precipitation. The implications of these changes for agricultural productivity, energy demand, and sea level rise are currently under intense study at the national and international levels.

PHYSIOLOGICAL EFFECTS OF CARBON DIOXIDE AND CLIMATE

In Chapter 6, we showed that photosynthesis in plants is CO_2 limited and that higher levels of CO_2 in the atmosphere lead both to faster rates of C fixation and higher water use efficiency, helping to relieve water stress (Figures 6.1 and 6.9). Increased primary production and increased C storage in ecosystems, buffering further atmospheric increases in CO_2, would seem a logical result. However, several arguments against this have been raised.

First, increased NPP must precede greater C storage, and NPP can be limited by the availability of nutrients. For example, several experiments in which plants were exposed to twice the current CO_2 concentration (700 ppm versus 350 ppm) without additional nitrogen (N) show that one response to

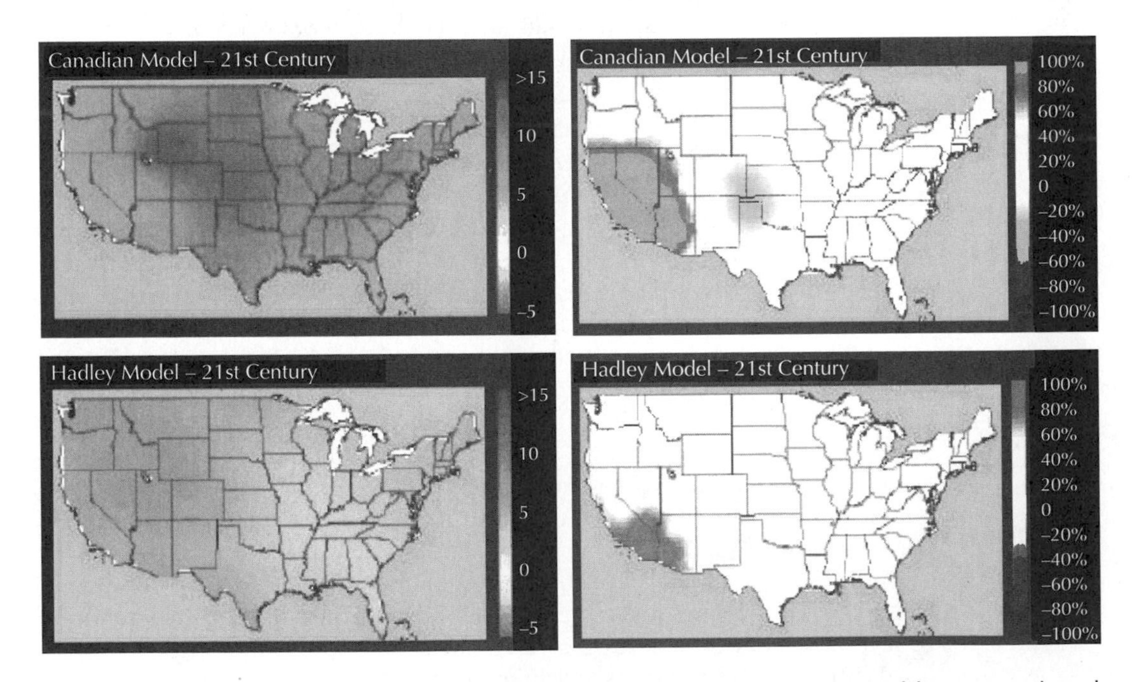

FIGURE 26.5 Changes in mean annual temperature (left top and bottom) and total annual precipitation (right top and bottom) as predicted by two different models (National Assessment Synthesis Team 2000). Temperature changes are in degrees celcius. Precipitation changes are in percentage of current values. **See plate in color section**

increased C fixation is the production of foliage with a lower N content (Figure 26.6). Remembering the relationship between foliar N concentration and photosynthetic rates (Figure 6.3), this dilution of leaf N could offset increases due to CO_2. When shed as litter, this foliage may also contain a reduced N content, resulting in slower decomposition rates, higher N immobilization, and an overall reduction in N availability and cycling. In effect, the positive feedbacks described in Figure 18.10 may exaggerate N limitations in native ecosystems, reducing N availability and limiting increases in C storage.

A second effect is called "down-regulation" and involves physiological adjustments in photosynthetic capacity to match the ability of plants to use the C fixed. In the absence of the nutrients required to build leaf and nonleaf biomass, the demand for C for tissue construction declines and excess carbohydrate can lead to "end-product inhibition," or a reduction in photosynthesis due to the accumulation of starch. In response, maximum photosynthetic rates may be reduced.

One early study in the Alaskan tundra showed only very short-term increases in photosynthesis and negligible changes in net ecosystem C storage. This was due in large part to down-regulation of photosynthetic capacity (Figure 26.7*a*). While foliage grown in either 340- or 680-ppm CO_2 still shows a response to CO_2 concentration, the curve for the plants grown in elevated CO_2 is shifted to the right, such that foliage grown at 680 ppm shows the same rate of C gain in the 680 ppm environment as does the foliage grown at 340 ppm and measured at 340 ppm (Figure 26.7*b*). In mesocosm experiments (using small intact "pieces" of the tundra soil–plant system grown in a controlled environment), net C gain was increased only slightly and for only a few weeks by increased CO_2 alone. Only with the addition of N fertilizer did substantial and prolonged increases in the C gain occur (Figure 26.7*a*).

A major experiment in a field-grown forest ecosystem (Figure 3.15) is

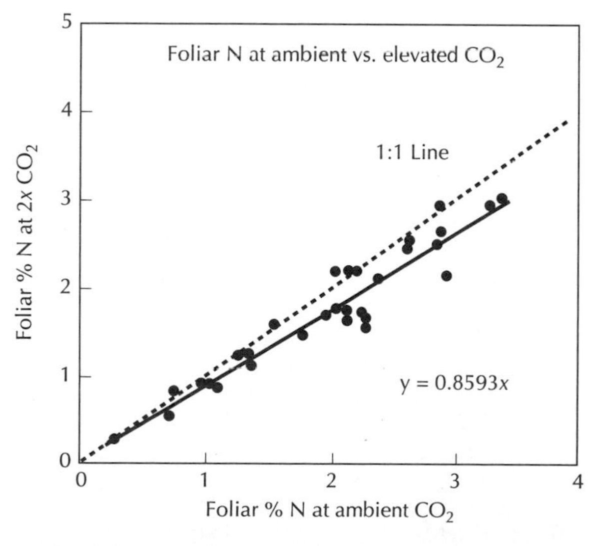

FIGURE 26.6 Data from several studies on the effect of a doubled CO_2 concentration on foliar nitrogen levels. On average, foliar nitrogen concentration was 14% lower in plants growing in the 2 × CO_2 environment. (Ollinger et al. 2000)

currently in progress. Initial results show a significant increase in tree growth and potential C storage. Whether down-regulation will occur in this system as well is still to be determined.

Increased CO_2 concentrations may also limit C accumulation indirectly through effects on climate. While the two model results in Figure 26.5 differ in the changes in temperature and precipitation expected, both predict in-

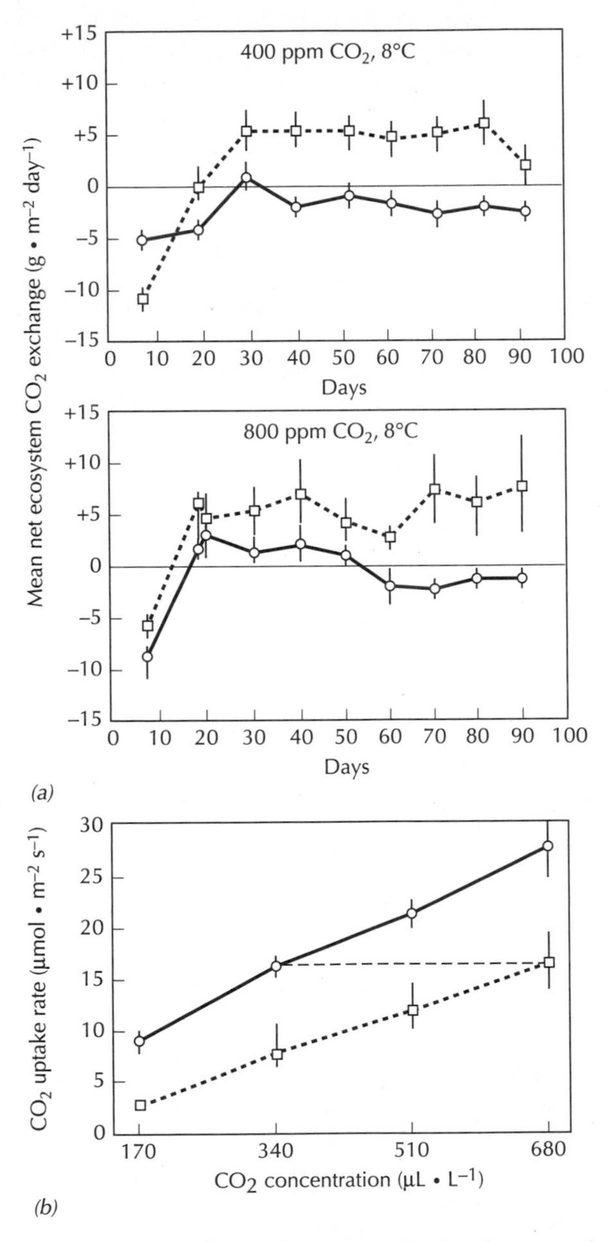

FIGURE 26.7 Effects of negative feedbacks on carbon accumulation in a tundra ecosystem with elevated CO_2. (*a*) Net carbon accumulation at 400 and 800 ppm CO_2 with (*solid lines*) and without (*dashed lines*) nitrogen additions. (*b*) Down-regulation of photosynthetic capacity. (Shaver et al. 1992)

creased temperatures. Higher temperatures will affect both photosynthesis and respiration by plants. Looking back to Chapter 6, we see that photosynthesis shows a bell-shaped response to temperature, while respiration increases exponentially (Figure 6.5). Whether this means an increase or decrease in net photosynthesis depends on how the optimum temperature for photosynthesis compares with the current mean temperatures. Also under intensive study is the speed with which plants respond to altered temperatures by changing basal respiration rates or the optimum temperature for photosynthesis.

Soil respiration also responds to temperature. Increases in temperature generally increase C fluxes from soils. However, another result of this higher respiration rate is increased mineralization of N from organic matter. So, while increased temperatures could lead to increased C losses from soils, additional N availability could also lead to faster plant growth (especially in a CO_2-enriched atmosphere), leading to higher NPP and greater C storage.

Finally, temperature changes affect the length of the growing season in temperate, boreal, and tundra ecosystems. Again, both photosynthesis and respiration are affected, as well as the amount of water required for evapotranspiration. The net effect in terms of NPP and C gain or loss could rest on whether there is enough water to support the higher total photosynthesis that might be possible over a longer growing season and whether plants or decomposers are more susceptible to drought.

The extent to which these contradictory responses occur in native ecosystems remains an important unknown. The kind of whole-ecosystem CO_2 fertilization experiment required to answer the question is very difficult to perform in the field, especially because it must be maintained for several years to allow meaningful responses of ecosystem processes to occur.

HISTORICAL CHANGES IN LAND USE AND CARBON STORAGE

In addition to the physiological effects of C and N fertilization, the C balance of individual units of the landscape is determined by their stage of development and recent disturbance history. As we saw in Chapter 3 (Table 3.2), net CO_2 balance need not be related to total rate of C gain through photosynthesis. Rather, that net balance is the difference between C gain and respiration by plants and decomposers. A highly productive ecosystem, in terms of NPP, may be a smaller sink for C than a low-productivity system if, for example, the productive system has been recently disturbed and contains a large amount of rapidly decomposing litter. That highly productive system may even be a net source of C.

Even agricultural systems that are continuously disturbed in the same way, that is, by plowing and harvest, will eventually reach a new equilibrium or constant C content. This new equilibrium may not be realized for a century or more and will likely be at a very different level of C content (see again Figure 13.14), but once the new equilibrium is reached, the net C balance returns to zero.

At the regional level, net changes in C storage will occur only if the relative distribution of land uses is changing or if changes in C content due to current practices have not been completed. As a simple example (Figure 26.8), imagine a region consisting of three large blocks of forest that are harvested at 90-year intervals, but with the harvests staggered at 30-year intervals. Each block will continually pass through the full spectrum of C content, but as one area is losing C through harvest and decay, another is gaining C by regrowth. The total C storage for the region—the sum of the three curves for the subregions—is relatively stable over time, although reduced from the maximum storage that would occur if all the forests were allowed to regrow to maturity (or to whatever C content the natural disturbance regime would allow). Actually, the net impact of these managed forests on the global C cycle would be determined in large measure by how the harvested products are used.

The same concept applies to regions that contain a mixture of agricultural and native landscapes. As long as the percentage of the region in each land-use category does not change, the total C storage for the region will not show large changes. Conversely, large changes in C storage will occur in regions where dramatic changes in the distribution of land use are occurring.

One straightforward method to estimate the C balance of land ecosys-

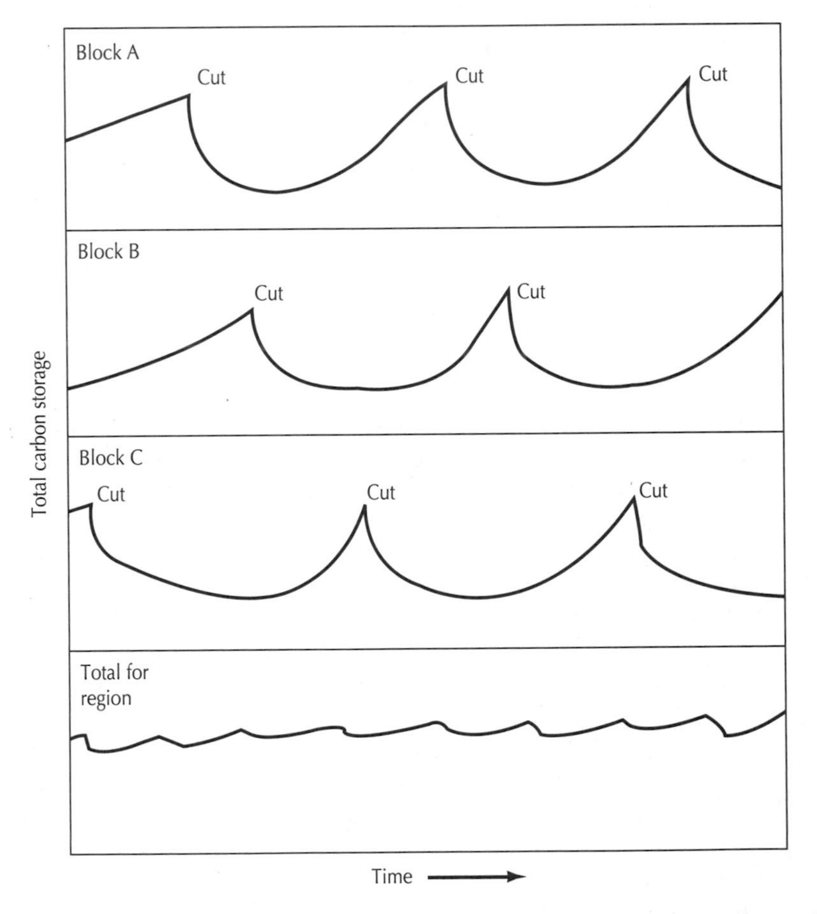

FIGURE 26.8 The effect of continuous forest cutting, with regrowth, on the net carbon balance of a region. Continuous cycles of cutting and regrowth cause individual units of the landscape to be alternately net sources and sinks for carbon. The combined effect of these is very little net change in total carbon storage for the whole region.

tems has combined the estimation of temporal changes in C storage following different types of disturbance in different ecosystems, with annual estimates of land use in these same types and regions throughout the world.

First, the landmass of the world was divided into nine regions (Figure 26.9*a*). Within each zone, several vegetation–land-use types were defined (e.g., tropical rain forest, grasslands, cropland, etc., and not all types occurred in all regions). For each region and land-use type, a set of characteristic curves for changes in C storage in soils and in vegetation over time were defined. For forests, this could include commercial cutting with immediate regrowth or with conversion to agriculture either temporarily or permanently (Figure 26.9*b*). Grasslands and shrublands would be defined similarly, with smaller dynamics for the vegetation component of the budget. While simple in concept, the construction of a single C "trajectory" that captures the tremendous diversity of conditions within each of the continental biome units is a difficult task.

With the characteristic curves established, the next step was to determine how much of the land area in each region and vegetation class was being cleared, cut, converted, or allowed to regrow each year, and to sum the changes in each landscape unit following the characteristic curves. The balance of sources and sinks for all of the regions and vegetation types is the C balance of the world's terrestrial ecosystems. Using historical records of land use, predictions of changes in C fluxes from terrestrial ecosystems over the past century and longer could be made.

Historical trends in land settlement and the spread of agriculture are clearly revealed in the model projections of C balances at the continental scale (Figure 26.10). In North America in the late 1800s, mature forests were still being harvested, and both forests and grasslands were being converted to agricultural land, mainly in the West and Midwest. As a result, large amounts of C were released to the atmosphere, and North America was the largest source of C to the atmosphere through land-use change. In comparison, the long-settled European continent was only a minor source of C at this time and remains so today. The tremendous increases in population levels in Asia, South America, and tropical Africa in the twentieth century drove the expansion of agricultural activity, including the conversion of forests to cropland and pasture, leading to large estimates of current C release.

An earlier estimate of the temporal trend in total C release from terrestrial ecosystems as estimated by this approach can be compared with those from fossil fuel combustion (Figure 26.11). One surprising result is that input of C to the atmosphere from fossil fuels has been greater than the input from human use of terrestrial ecosystems only since the 1950s. As much as 30% of the total increase in CO_2 content of the atmosphere is the result of human use of the landscape rather than fossil fuel consumption.

A COMPARISON OF METHODS FOR ESTIMATING CARBON BALANCES: THE UNITED STATES AS A CASE STUDY

The Kyoto process has put the spotlight on the total CO_2 balances of different countries. If the world community is to control CO_2 fluxes, then the role of each country in contributing to the total flux to the atmosphere is a cru-

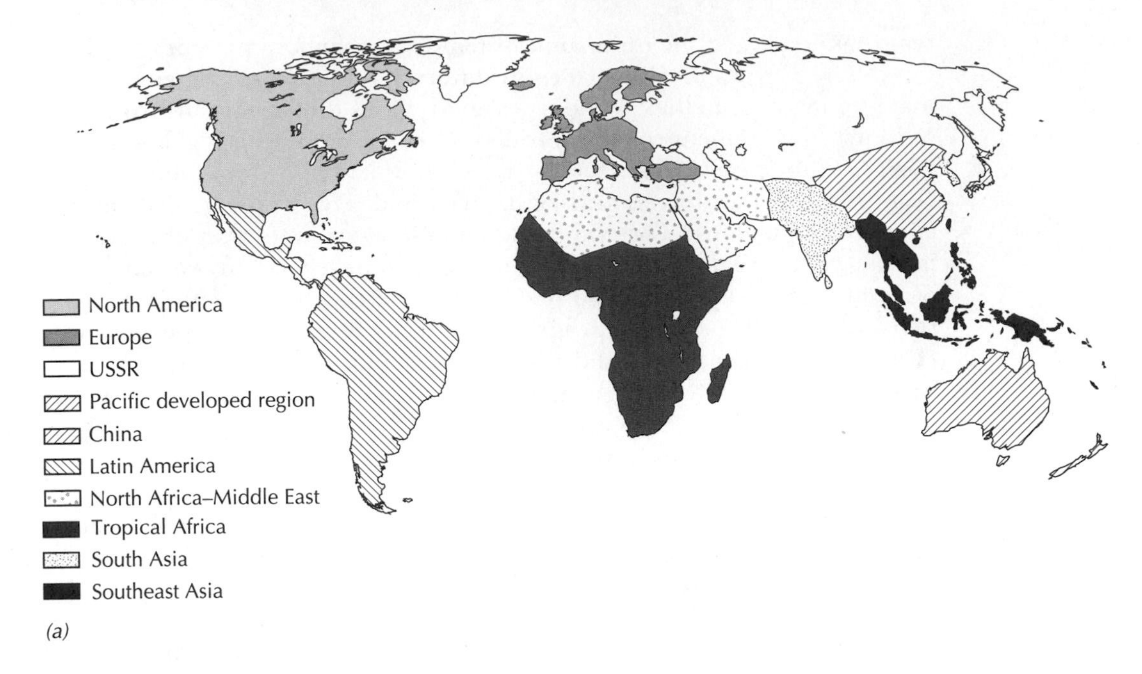

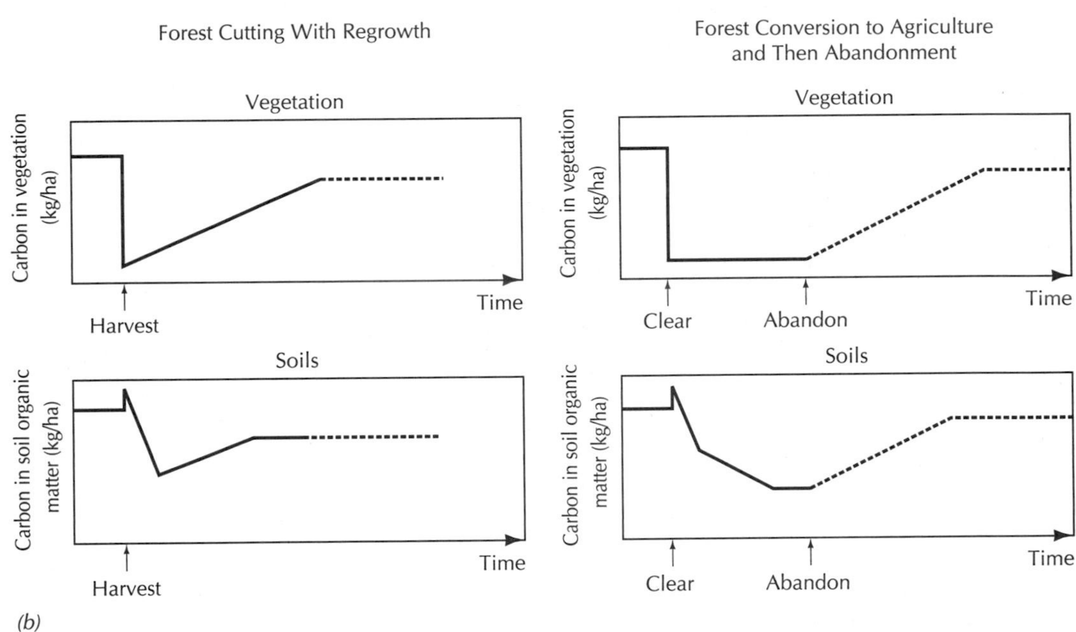

FIGURE 26.9 One approach to the calculation of the carbon balance over the Earth's terrestrial ecosystems. (*a*) The land areas are divided into regions, with several vegetation types within each region. (*b*) For each region and vegetation type combination, a series of curves describing changes in total carbon storage as a function of land use are used in conjunction with records of land-use history to calculate the total balance for the region. (Houghton et al. 1983)

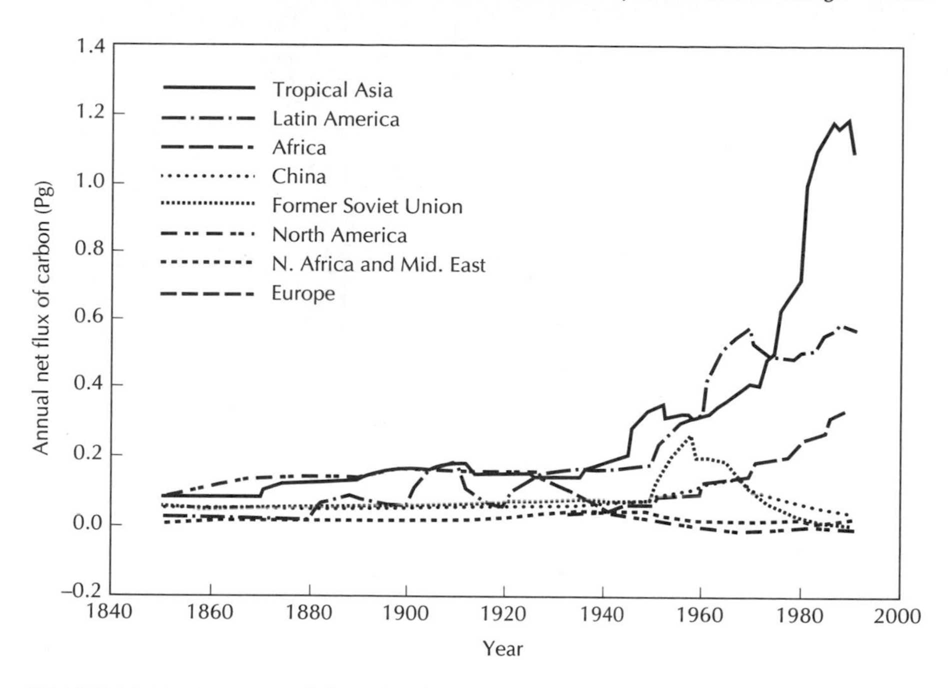

FIGURE 26.10 Net annual flux of carbon to the atmosphere from different continents for 1850–1990. (Houghton 1999)

cial piece of information. It is still unclear how viable the Kyoto process will be. The United States in particular, as the largest single source of CO_2 from fossil fuel combustion, has been slow to recognize Kyoto and cooperate with the global community.

Because of the importance of the terrestrial C balance and the number of disciplines involved in the study of the global C cycle, several different approaches have been taken in estimating the C balance of land systems in the United States, often with conflicting results. We will present results from several different approaches: physiological models, models of land-use change, forest inventory, direct measurements of CO_2 exchange, and estimates based on atmospheric chemistry at the continental scale. Many of the efforts to estimate C balances over native ecosystems have focused on forests because the large pools of C stored in forests are dynamic in response to disturbance and management. All results reported here are for the continental United States.

Physiological Models

In the United States, an integrated data and modeling program has been tasked with predicting these effects. Three models of the biogeochemical responses of ecosystems to climate and CO_2 have been combined with three models predicting shifts in vegetation type over the lower 48 states, and are driven by three different model estimates of mean climate at twice the current CO_2 concentrations in the atmosphere. Results were used to construct maps of predicted changes in C storage over the next century (see subsequent discussion). The same three biogeochemical models were run using actual cli-

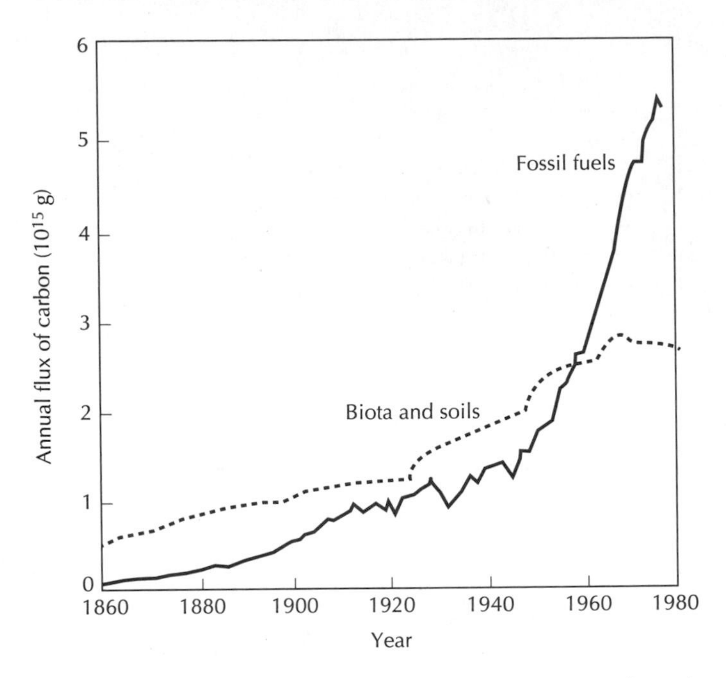

FIGURE 26.11 The annual flux of carbon to the atmosphere through fossil fuel combustion and land-use changes in terrestrial ecosystems from 1860 to 1980. (Houghton et al. 1983)

mate data and current vegetation maps to estimate the effects of climate and CO_2 alone on changes in C storage. None of the feedbacks to nutrient cycling are present in these models, and a steady-state, or undisturbed, vegetation condition was used, such that land-use effects also do not alter the predictions.

Results (Figure 26.12) show gains in all regions of the country for all models due solely to transient climatic effects and increased in atmospheric CO_2. Overall, the total estimated increase in C storage was 80×10^{12} g C $\cdot$ year^{-1}.

A Land-Use Model

The land-use model presented earlier (Figure 26.9) can also be run for the United States alone. Figure 26.13 summarizes historical changes in the net C balance for the continental United States and separates fluxes into several categories. Through the 1800s, conversion of forest and prairie to cropland was the major process driving large transfers of C to the atmosphere. Harvests of wood for fuel peaked at about the same time. As the amount of cropland stabilized and some was allowed to revert to native vegetation types, the C balance from this usage returned to near zero. Forests harvested in the 1800s show C gains in the 1900s as cutover forests regrow. This analysis also describes the suppression of forest fires as an important cause of increased C storage in forests, with a diminishing effect in recent decades. Currently, this analysis assigns a net sink value of 37×10^{12} g C $\cdot$ year^{-1} to the land ecosystems of the United States. Note that this does not include the effects of either climate or CO_2 enrichment included in the physiological models.

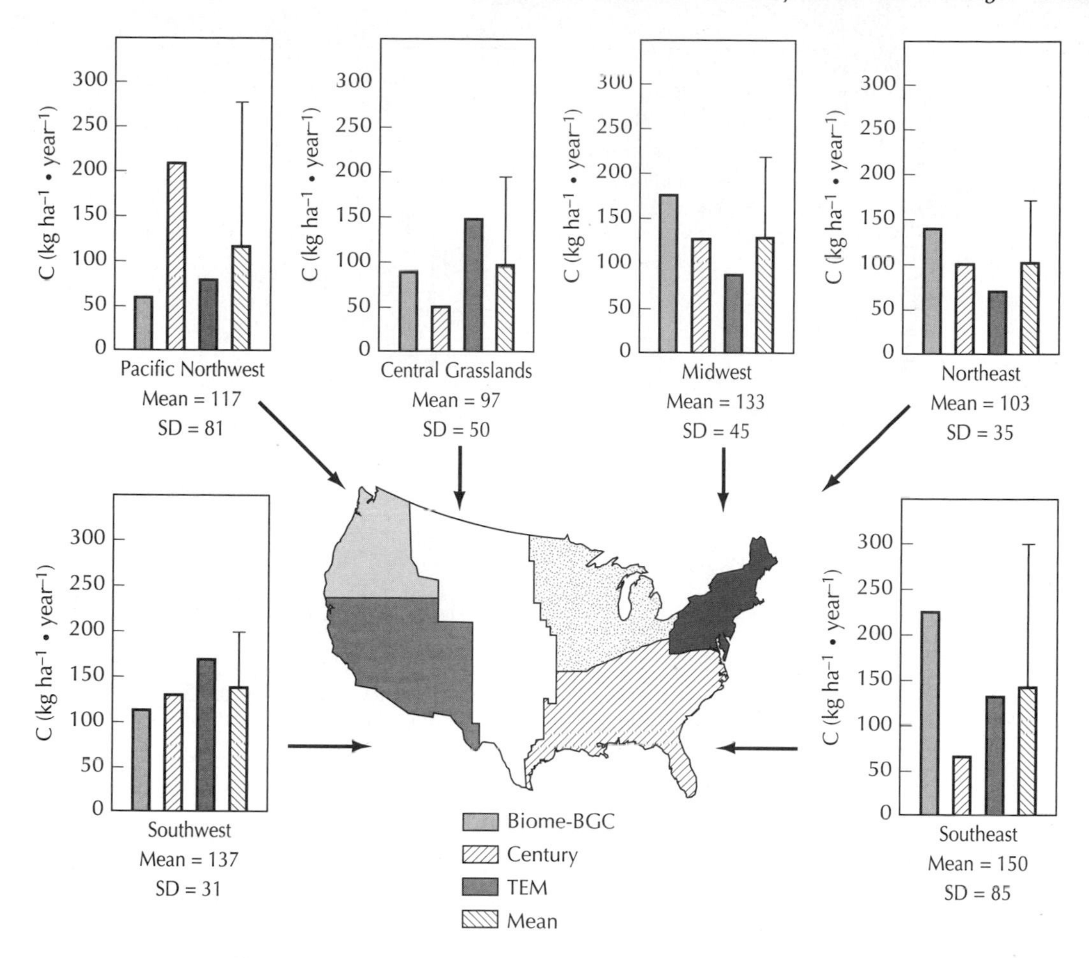

FIGURE 26.12 Estimated changes in total carbon storage by native ecosystems in different regions of the United States between 1980 and 1993 based on predictions by three different biogeochemical models of ecosystem function. Increases are due solely to increased CO_2 concentrations and interannual variability in climate. (Schimel et al. 2000)

Forest Inventory

The value of the forest resource for the wood products industry has led to the establishment of continuous forest inventory programs in many countries. In the United States, the Forest Inventory and Analysis Program has gathered tree- and plot-level data for thousands of individual plots across the country. While traditionally reported at the county level and in terms of growing stock and removals, this same database has recently become extremely valuable in the calculation of total C storage in forests. It represents the summation of thousands of direct measurements. However, because only tree data are collected in most locations, estimates of changes in C storage in litter and soil are estimated from regional or biome means and as a function of stand age or tree biomass.

Two different estimates derived from the Forest Inventory and Analysis

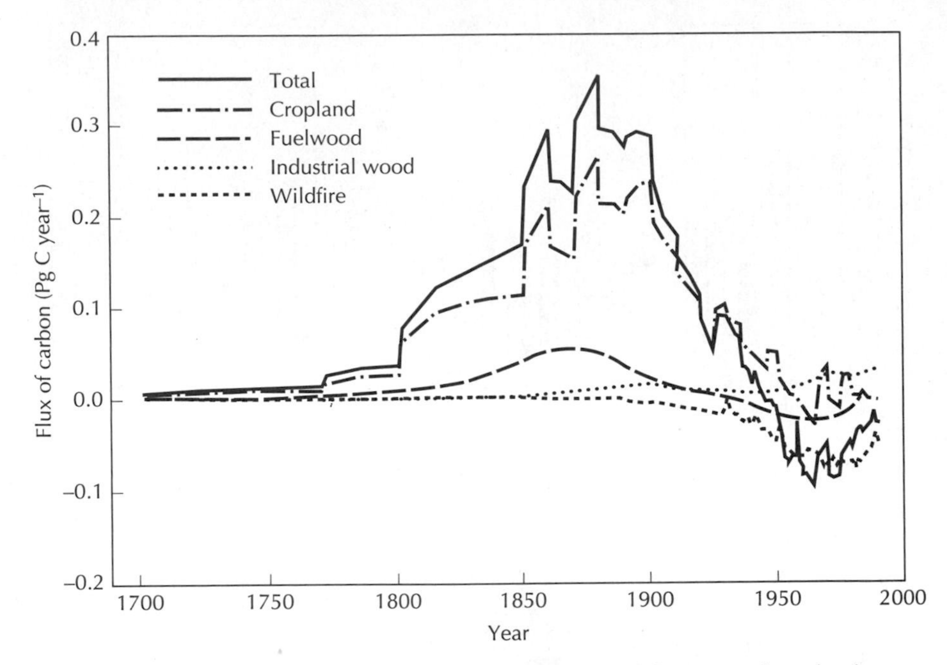

FIGURE 26.13 Fluxes of carbon between ecosystems and the atmosphere for the continental United States over time. Note that positive values are fluxes to the atmosphere. (Houghton et al. 1999)

database differ significantly (Table 26.1). The first (211×10^{12} g C $\cdot$ year^{-1}) used actual measurements of total stand biomass at different measurement intervals and assigned values for litter and soils using equations related to tree biomass for each forest type and biomass. The second estimate (79×10^{12} g C $\cdot$ year^{-1}) used current inventory biomass values and standard forest yield tables (which predict forest growth based on forest type, site quality, and current growth and stocking) to predict current rates of accumulation in trees and to estimate accumulations in litter and soils. This estimate would include the effects of land-use changes and whatever physiological effects of climate and CO_2 have occurred.

Ecosystem-Level Exchanges with the Atmosphere

In Chapter 3, we introduced the eddy covariance technique for measuring gross and net C fluxes over ecosystems. This method is revolutionizing the direct measurement of C metabolism at the land–atmosphere interface. Currently, there are very few locations with multiple-year records obtained by this technique, but large networks are planned in both the Americas and Europe (Chapter 3). The advantage of this technique is that it measures the net effect of both tree growth and respiration and decay in a single measurement. It also can be used to develop estimates in a single year where inventory methods requires 10 years or more of accurate assessment of changes in C storage in trees. Estimating changes in soil C from change in tree biomass, a central feature of the inventory methods, is always problematic.

TABLE 26.1 Estimated Rates of Net Carbon Storage over the Lower 48 United States Produced by Different Methods

Method	Factors Considered	Net C Storage (10^{12} g C · year^{-1})	Study
Physiological models	CO_2, climate	80	Schimel et al. 2000
Land-use model	Land use	37	Houghton et al. 1999
Forest inventory	All factors, forests only	211	Birdsey and Heath 1995
Forest inventory	All factors, forests only	79	Turner et al. 1995
Eddy covariance	All factors, single forest	500	Munger et al. 2001
Atmospheric dynamics	All factors, by difference	1700	Fan et al. 1998

We can use one long-term study of C balance over a forest ecosystem in Massachusetts to derive one estimate of rates of C gain in forests. This aggrading forest in various stages of recovery from both forest harvesting and agriculture shows substantial interannual variability in total C balance but is a net sink in each year, with a mean value of 2.1×10^6 g C · ha^{-1} · year^{-1}. If this was a typical value for all forests in the United States, and with current estimates of a total of about 246×10^6 ha of forested land in the continental United States, then the total accumulation would be about 500×10^{12} g C · year^{-1}. This method does include the physiological effects of climate and CO_2 but is biased in terms of land-use effects because the measurement is made in an aggrading forest with no similar measurement in either a mature or a recently disturbed forest. Similar measurements in an old-growth boreal forest in Canada yielded a near-zero C balance over a 4-year period (Chapter 3). Results from the larger network of eddy covariance towers will help to refine this estimate.

Continental-Scale Dynamics in Carbon Dioxide Concentration

What if the same principles used in eddy covariance could be applied at the continental scale—that by measuring gradients in concentration of CO_2 across continents and predicting atmospheric fluxes, we could estimate the net C flux over the intervening landmass? This approach has been attempted using a very nonrandomly distributed set of CO_2 measurements made over the surface of the Earth (Figure 26.14). In what is known as an **inverse modeling approach,** estimates of the rate of atmospheric movement and mixing across continents is used in conjunction with known sources of CO_2 from combustion and measured atmospheric CO_2 concentration gradients to estimate sources and sinks of CO_2 from all other processes by difference. It is assumed that the C balance of terrestrial ecosystems is the main source or sink in this difference term. This method has been applied for three regions of the world, including North America, for which the atmospheric sampling density is greatest.

This method predicts a very large sink for CO_2 in the northern hemisphere mainly south of 51°N, reducing atmospheric CO_2 concentrations by as much as 2 ppm (out of about 370 ppm) over central North America (Figure

26.14). From this small difference between measured CO_2 concentrations and those predicted by a fairly complex model of atmospheric dynamics, a net sink of 1700×10^{12} g C $\cdot$ year^{-1} is estimated for North America, nearly all of it in the United States.

Comparing The Estimates: Implications for Policy

These different methods for estimating the total C balance over the terrestrial ecosystems of the United States produce a wide range of values (Table 26.1). If the Kyoto process moves forward, the range of estimates has important policy implications. For example, the atmospheric dynamics inverse modeling approach suggests that the C sink in native ecosystems in North America south of 51°N nearly offsets the entire fossil fuel combustion source for the United States and could be used to suggest that the United States is "CO_2-neutral" despite the high rates of fossil fuel use. Direct measurements by eddy covariance provide a smaller estimate, and this value is based on aggrading forests only. Values for C storage derived from forest inventory databases, which include land use, climate, and CO_2 effects, but use rough statistical methods for estimating soil and litter C, are lower still. Models that consider either land-use or physiological effects alone yield the lowest C sink estimates. The high-profile nature of this area of research guarantees continued refinement in each of these methods in the near future.

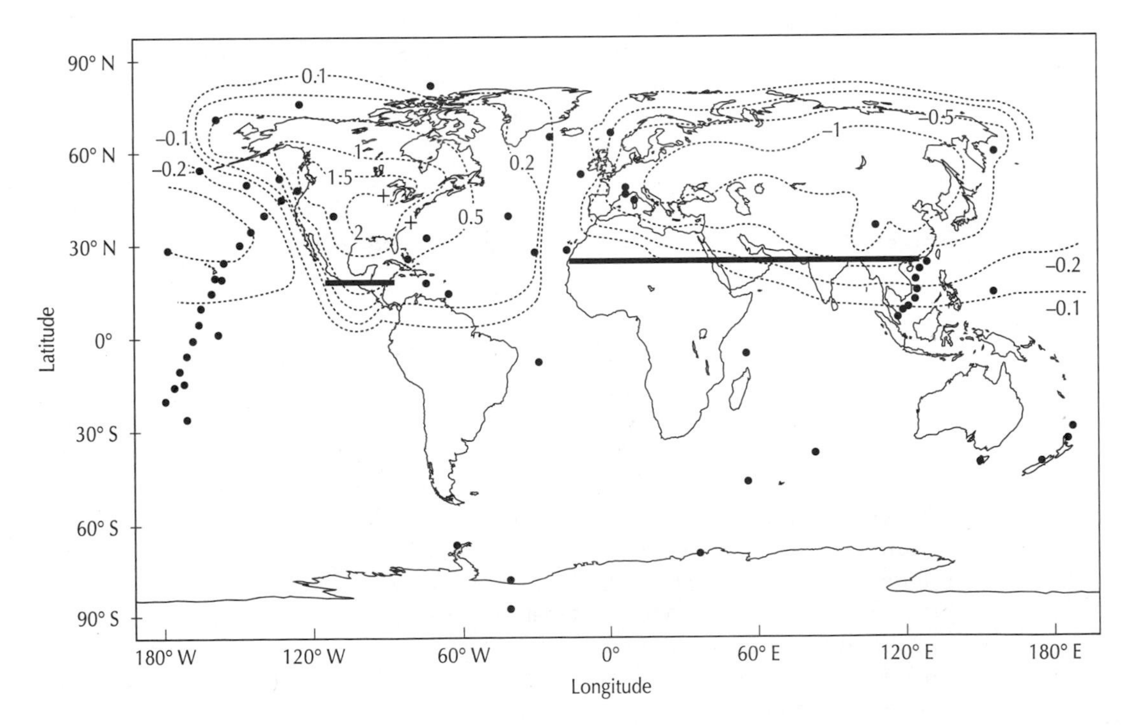

FIGURE 26.14 Location of sampling stations (*dots*) for atmospheric CO_2, used to estimate draw down in atmospheric CO_2 concentration by terrestrial ecosystems (*dashed lines*) by an atmospheric dynamics model and CO_2 emissions data. (Fan et al. 1998)

PREDICTING NET PRIMARY PRODUCTION AND CARBON BALANCES IN THE FUTURE

In contrast to estimating current C balances, there are no direct measurements that allow us to predict how future changes in environmental conditions will affect ecosystem C balances. The full factorial field experiments dealing with changes in CO_2, temperature, precipitation, and nutrient additions are prohibitively expensive to run in even a single location, much less across several ecosystem and land-use classes. In the absence of direct experimentation, models are the only tools available for predicting these interactive effects.

The same three physiological models that were used earlier to estimate current increases in C storage due to climate and CO_2 have also been used to predict changes in NPP and C storage into the future. Differences among the predictions for the different model combinations suggest that considerable uncertainty still exists (Table 26.2). All model combinations predict increased NPP (average of +21%, but with a range of −1% to +40 %). In terms of feedback to the global C cycle, the models predict a wide range in changes in total C storage, varying from −39% to +32%, with a mean value near zero (+2%). This apparently neutral-to-positive picture for the entire nation masks some very large changes NPP predicted for different parts of the country, which are matched by potentially large changes in vegetation distribution (Chapter 24).

Predicting the effects of climate, CO_2, and land use on ecosystems is complex, but other important environmental factors are also changing as rapidly as these. While regional in extent, the fertilizing effects of N deposition can also alter C storage. A simple model of this interaction presents inputs of N deposition globally (Figure 26.15*a*), allocates them among different compartments within ecosystems, and uses C–N ratios in each compartment to estimate increases in C storage. By mapping N deposition and ecosystem types globally, an estimation of the current increase in C storage as a function of

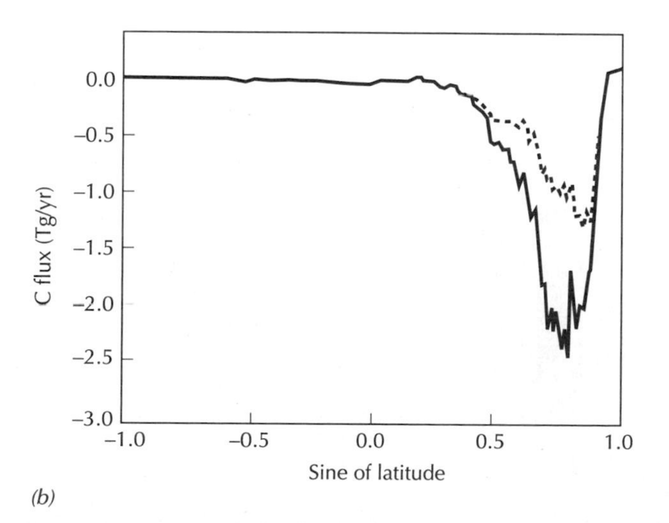

FIGURE 26.15 Estimated global distribution of nitrogen deposition. (Townsend et al. 1996)

TABLE 26.2 Range of Results for Model-Based Predictions of Changes in NPP and Total Carbon Storage Using Three Biogeochemical Models and Run with Three Different Climate-Change Scenarios and Three Models Predicting Vegetation Distribution (total of 9 runs per model). (VEMAP members 1995)

Biogeochemical Model	Change in Annual NPP (%)	Change in Total Carbon Storage (%)
BIOME-BGC	−0.7 to +21.7	−39.4 to −8.3
CENTURY	+14.7 to +26.09	−1.5 to +20.4
TEM	+27.0 to +32.4	0.0 to +32.2

latitude can be derived (Figure 26.15*b*). The large estimated sink in the northern hemisphere is in keeping with estimates based on atmospheric dynamics (Table 26.2), but the magnitude of the estimated sink is smaller. This model does not include the effects of CO_2 and climate.

We also saw in Chapter 25 that ozone can be an important pollutant in industrialized regions and can affect C balances by reducing photosynthesis (see Figures 25.10 and 25.11). One model has attempted to combine the effects of ozone with those of CO_2, N fertilization, and land use and predicts that ozone will partially counteract those fertilizing effects (Figure 26.16). The relative responses to CO_2 and N additions are modified significantly by land-use history.

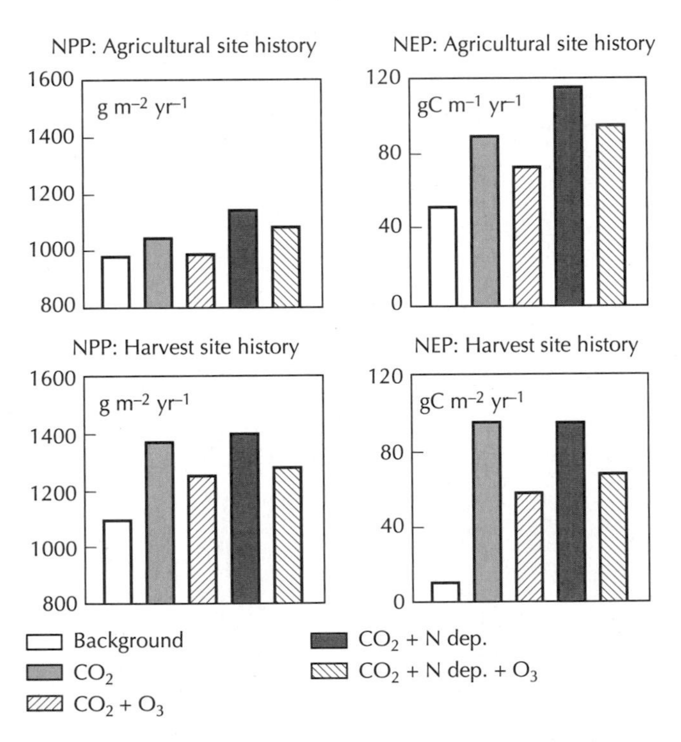

FIGURE 26.16 Predicted effects of CO_2, ozone, nitrogen deposition, and land-use history on net primary production (NPP) and total carbon storage (NEE) in a temperate forest in the northeastern United States (Ollinger et al. 2000)

To date, there have been no model predictions presented that include all of the major, interacting components of environmental change (temperature, precipitation, CO_2, N deposition, ozone, and land use), in a single modeling context.

REFERENCES CITED

Birdsey, R. A., and L. S. Health. 1995. Carbon changes in U. S. Forests. In: Joyce, L. A. (ed.), *Productivity of America's Forests and Climate Change.* USDA Forest Service, General Technical Report RM-271. USDAFS, Ft. Collins, CO.

Fan, S., M. Gloor, S. Pacala, J. Sarmiento, T. Takahashi, and P. Tans. 1998. A large terrestrial carbon sink in North America implied by atmospheric and oceanic carbon dioxide data and models. *Science* 282:442–446.

Fleming, J. R. 1998. *Historical Perspectives on Climate Change.* Oxford University Press, New York.

Houghton, J. T., L. G. Meira, B. A. Callander, N. Harris, A. Kattenberg, and K. Maskell (eds.). 1996. *Climate Change 1995.* Cambridge University Press.

Houghton, R. A. 1999. The annual net flux of carbon in the atmosphere from changes in land use, 1850–1990. *Tellus* 51B:298–313.

Houghton, R. A., J. L. Hackler, and K. T. Lawrence. 1999. The U. S. carbon budget: Contributions from land-use change. *Science* 285: 574–578.

Houghton, R. A., J. E. Hobbie, J. M. Melillo, B. Moore, B. J. Peterson, G. R. Shaver, and G. M. Woodwell. 1983. Changes in the carbon content of terrestrial biota and soils between 1860 and 1980: A net release of CO_2 to the atmosphere. *Ecological Monographs* 53:235–262.

Munger, C., W. Barford, and S. C. Wofsy. Atmospheric exchanges. In Foster, D., and J. D. Aber (eds.), *Forest Landscape Dynamics in New England* (submitted).

National Assessment Synthesis Team. 2000. *Climate Change Impacts of the United States.* U. S. Global Change Research Program, Washington, DC.

Ollinger, S. V., J. D. Aber, P. B. Reich, and R. Freuder. 2001. Tropospheric ozone and land use history affect forest carbon uptake in response to CO_2 and N deposition. *Ecosystems* (submitted).

Sarmiento, J. L., and S. C. Wofsy. 1999. *A U. S. Carbon Cycle Plan.* U. S. Global Change Research Program, Washington, DC.

Schimel, D., J. Melillo, H. Tian, A. D. Maguire, D. Kicklighter, T. Kittel, N. Rosembloom, S. Running, P. Thornton, D. Ojima, W. Parton, R. Kelly, M. Sykes, R. Neilson, and B. Rizzo. 2000. Contribution of increasing CO_2 and climate to carbon storage by ecosystems in the United States. *Science* 287: 2004–2006.

Shaver, G. R., W. D. Billings, F. S. Chapin III, A. E. Giblin, K. J. Nadelhoffer, W. C. Oechel, and E. B. Rastetter. 1992. Global change and the carbon balance of Arctic ecosystems. *BioScience* 42:433–441.

Townsend, A. R., B. H. Braswell, E. A. Holland, and J. E. Penner. 1996. Spatial and temporal patterns in potential terrestrial carbon storage resulting from deposition of fossil fuel derived nitrogen. *Ecological Applications* 6:806–814.

Turner, D. P., G. J. Keorper, M. E. Harmon, and J. J. Lee. 1995. A carbon budget for forests of the coterminus United States. *Ecological Applications* 5:421–436.

VEMAP members. 1995. Vegetation/ecosystem modeling and analysis project: Comparing biogeography and biogeochemistry models in a continental-scale study of terrestrial ecosystem responses to climate change and CO_2 doubling. *Global Biogeochemical Cycles* 9:407–437.

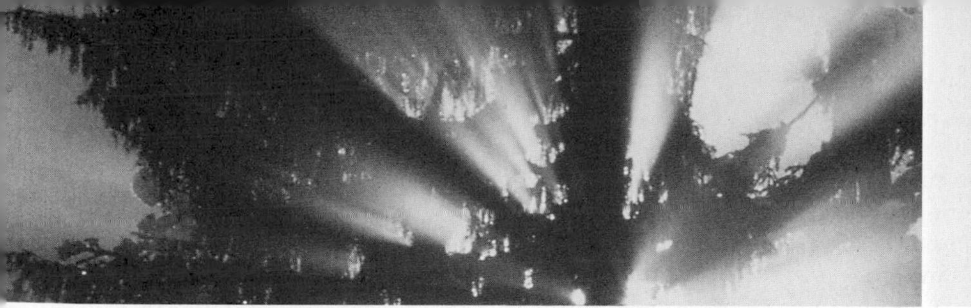

EPILOG

Terrestrial Ecosystems in a Rapidly Changing World

HUMAN IMPACT ON THE GLOBAL ECOSYSTEM

Humans are altering the global environment, driving changes in crucial characteristics at rates largely unprecedented in the history of the Earth. Figure E.1 captures two crucial dimensions of global change, displaying relative rates of increase in both human population and the CO_2 concentration of the atmosphere. CO_2 concentrations have varied widely over geological time, but current rates of change have not been matched. Different animal groups have dominated the Earth's ecosystems and then vanished (dinosaurs are the most spectacular), but no species has been as widely distributed, has increased in number so rapidly, or has been so dominant a force in shaping all of the Earth's ecosystems as humans are today.

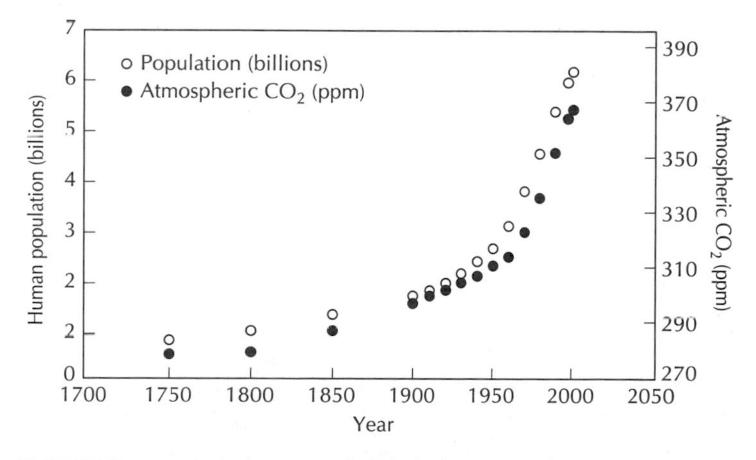

FIGURE E.1 Two indicators of global change. Both human population levels and the concentration of CO_2 in the atmosphere have shown exponential growth at rates that are unprecedented in the history of the Earth. (Population data from the United Nations http://www.popin.org/pop1998/4.htm. CO_2 data from Neftel et al. http://cdiac.esd.ornl.gov/trends/co2/siple.htm and Keeling and Whorf http://cdiac.esd.ornl.gov/trends/co2/sio-mlo.htm)

We can quantify the intensity of the human impact on global ecosystems by estimating what fraction of crucial resources and processes have been **domesticated,** or diverted to human use. In Chapter 1, we described Lindemann's innovation in the comparative study of ecosystems using energy flows as the common currency of ecosystem studies. Table E.1 applies this concept to global NPP and suggests that, by conversion to crops and pastures, harvesting of forests, desertification, and urbanization, humans now control the fate of nearly 40% of the global NPP.

Figure E.2 extends this approach to other resources. Human impact is calculated either as the fraction of the resource utilized (e.g., land conversion and water use), as the fractional change in chemical concentrations and cycles (atmosphere CO_2 and global nitrogen cycling), or in species composition and abundance (plant invasion, bird extinctions, and marine fisheries). This is a diverse set of metrics, all of which support the same conclusion. Humans now appropriate 20–60% of the life-sustaining resources of the Earth.

We have been fortunate that, by the efforts of dedicated scientists, patterns of ecosystem development, responses to disturbance, and the whole diversity of pattern and process in largely natural systems has been recorded, analyzed, and preserved. Fewer of those "undisturbed" sites will exist in the future. What has been recorded will act as a permanent template—a basis for comparison against which we can judge the performance of our managed systems.

Conservation and preservation of ecosystems continues by national and international organizations, national parks, and private consortiums. These wild lands will serve as a continuing source of refreshment and renewal to those who seek them out and to others who draw comfort just from their continued existence. However, in a world in which climate change is likely to shift biome boundaries by hundreds of kilometers, these reserves are not likely to remain unchanged and unchanging "museum pieces" of the natural world. Species invasions and global changes in atmospheric chemistry and climate will drive changes in ecosystem structure and function. In the future, we are

TABLE E.1 An Inclusive Estimation of the Fraction of Total Global Net Primary Production Appropriated for Human Use by Different Activities
(Adapted from Vitousek et al. 1986)

Activity	NPP Affected (10^{15} g)
Cultivation: direct effects and reduced NPP	24.0
Grazing	11.6
Forest harvest and conversion	15.0
Desertification	4.5
Human occupation	3.0
Total	58.1
Total global NPP	149.8
Percentage of total NPP appropriated	38.8%

NPP = net primary production.

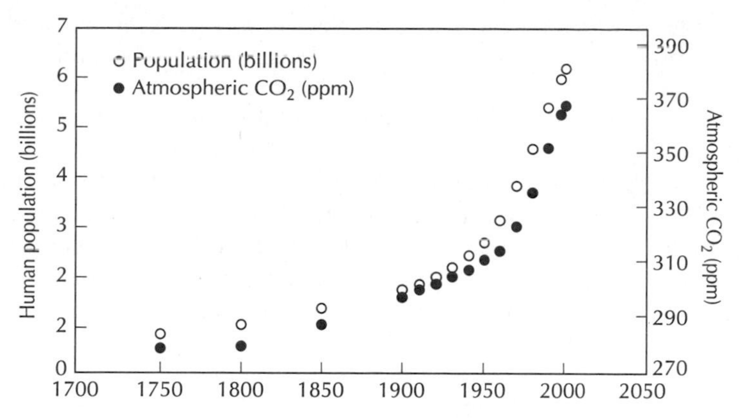

FIGURE E.2 Human appropriation of global resources expressed as the percentage of land surface altered, percentage of atmospheric CO_2 of human origin, percentage of accessible fresh water used, percentage of global N fixation, percentage of invasive plant species in Canada, percentage of bird species driven to extinction in the past 2000 years, and the percentage of major marine fisheries fully exploited. (Vitousek et al. 1997)

likely to be studying the responses of ecosystems to dynamic shifts in the environment—the outcome of our ongoing experiment in environmental modification at the global scale.

THE RELEVANCE OF TERRESTRIAL ECOSYSTEM STUDIES

In the scientific exploration of terrestrial ecosystems presented in this book, we have focused mostly on naturally evolved processes and the patterns in ecosystem development that follow from them. Only in the last eight chapters, and particularly in the last four, have we dealt explicitly with human manipulations in the landscape and the atmosphere. Does this book have any continuing relevance, then, to the future of its intended subject the land ecosystems of the Earth?

We would say yes. While the processes discussed in this book evolved in the absence of human impact, they now determine the response of terrestrial ecosystems to human use. In a world in which the "natural" or equilibrium conditions no longer exist, an understanding of the processes that determined those patterns, and how these processes will respond to a changing environment, will be crucial to predicting and managing ecosystems in the changed world.

As we have seen, not all of those responses are necessarily negative. Increasing CO_2 and N deposition can both fertilize ecosystems and increase production. Is this a good thing? If this results in a change of species composition in "natural" communities, is it still a good thing? The choices will become more complex, and the metrics more difficult to define. Whatever policy decisions need to be made, a clear understanding of how ecosystems will respond to different scenarios (see Chapter 26) is crucial.

An understanding of process is also important in efforts to reclaim and restore the most severely degraded ecosystems. An understanding of the mechanisms by which the Sudbury smelters decimated the surrounding ecosystems was important in the prescription of remedies and restoration procedures. The realization that frequent fires were essential to the successful restoration of tall-grass prairies (Chapter 17) demonstrates the value of both the restoration process as a method for understanding systems and the value of that understanding in repairing the Earth.

An appreciation for ecosystem processes, in addition to technical understanding, may prove a crucial part of the effort to bring human activity into harmony with ecological limitations. Efforts at ecosystem **gardenification,** or the combination of agricultural production and the provision of other environmental services with the conservation of biodiversity call for both knowledge and concern.

> Tropical wildlands and their biodiversity will survive in perpetuity only through their integration into human society. One protocol for integration is to explicitly recognize conserved tropical wildlands and wildland gardens. Gardens . . . are an integral part of *Homo sapiens.* Garden terminology, acceptance, perception, administration and usefulness are deeply imbedded in our cultural codes. . . . How does one facilitate nondamaging or minimally damaging development of wildland biodiversity and its ecosystems? By the generation of wildland garden goods and services. . . . We need to ask less what is the opportunity cost of a wildland and ask more what is the opportunity cost of the urban and agroscape alternatives. (Janzen 1999)

Another adaptation of the ecosystem approach can be found in environments as far removed as possible from high-diversity tropical forests. Major new initiatives are attempting to apply ecosystem analysis in urban environments. A new paradigm has been proposed in which human actions are recognized as the central feature controlling ecosystem structure (Figure E.3).

Both of these last two approaches recognize the role of humans as a "keystone species" in shaping ecosystem structure and directing ecosystem processes. For either of these to succeed, we need to increase our understanding of the basic processes controlling the fluxes of energy, water, nutrients, and pollutants, and how they respond under these very different conditions.

THE STABILITY OF TERRESTRIAL ECOSYSTEMS

The responses of terrestrial ecosystems to both natural change and human use play a large role in the biogeochemical cycles and climate systems at the global level. In a changing world, this question arises continuously: Are the Earth's ecosystems, and the Earth itself as an ecosystem, stable or sustainable in the face of human use?

One point of view on this question has been very eloquently put forth in a popular book entitled *Gaia,* after the Greek term for "Mother Earth":

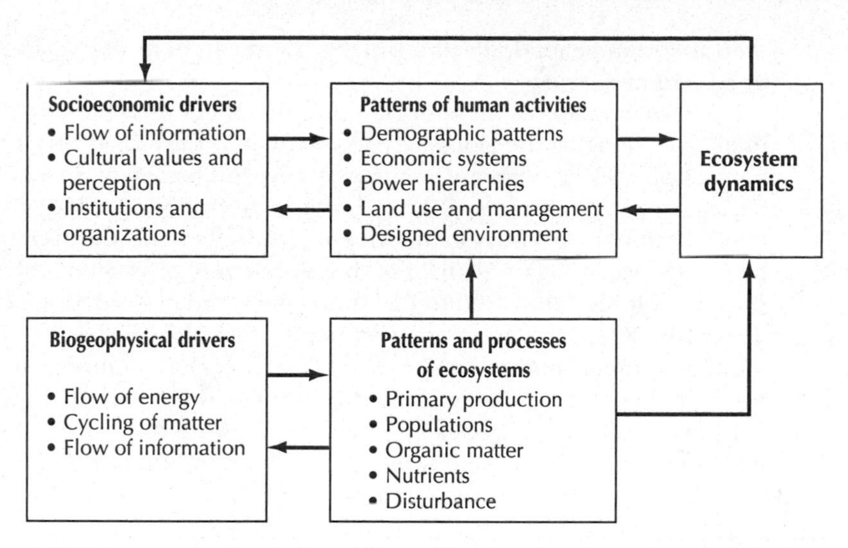

FIGURE E.3 A new paradigm for ecosystem studies. An outline of the approach to the study of urban ecosystems under the National Science Foundation's two new urban sites within the Long-Term Ecological Research Program. (Grimm et al. 2000)

> The chemical composition of the atmosphere bears no relation to the expectations of steady-state chemical equilibrium. The presence of methane, nitrous oxide and even nitrogen in our present oxidizing atmosphere represents violation of the rules of chemistry to be measured in tens of orders of magnitude. Disequilibria on this scale suggest that the atmosphere is not merely a biological product, but more probably a biological construction: not living, but like a cat's fur, a bird's feathers, or the paper of a wasp's nest, an extension of a living system designed to maintain a chosen environment. Thus the atmospheric concentration of gases such as oxygen and ammonia is found to be kept at the optimum value from which even small departures could have disastrous consequences for life. . . . We have since defined *Gaia* as a complex living entity involving the earth's biosphere, atmosphere, oceans and soil; the totality constituting a feedback or cybernetic system which seeks an optimal physical and chemical environment for life on this planet. . . . Gaia has remained a hypothesis but, like other useful hypotheses, she has already proved her theoretical value, if not her existence, by giving rise to experimental questions and answers which were profitable exercises in themselves. (Lovelock 1982)

In this book and later volumes, the author hypothesizes the existence of very strong negative feedbacks in the Earth's biogeochemical systems that have maintained a livable biosphere for over 3 billion years in the face of large changes in total energy input from the sun and dramatic changes in global chemistry. Less comforting, perhaps, is that this permanence is attributed to life in general, life as a process, rather than to the persistence of any one form or species.

There may also be an important mismatch between the rate at which humans are changing the global environment, and the rate at which Gaian feedbacks operate. For example, humans have increased atmospheric CO_2 by 30% since the beginning of the industrial revolution. The long-term sink for this additional carbon is in the deep oceans, but the transfer from atmosphere to ocean lags well behind human-driven production of CO_2. If we stopped adding "new" CO_2 to the atmosphere today, it would take many centuries for CO_2 concentrations to return to pre-industrial levels.

The Gaian hypothesis, in its scientific rather than philosophical form, has been an important paradigm for the study of the Earth. If these feedbacks exist, what are the processes involved, and how do they work? Just how fragile are these feedbacks, and will they continue to function given the current rates of change in land cover, atmospheric chemistry, and other characteristics?

The efforts to predict the future state of terrestrial ecosystems as presented in Part 4 can be viewed as our best-educated, scientific look into the world a century hence. There may be major surprises along the way that make all of these predictions inaccurate. Still, we need to take this look into the future and try to prepare for it with as much understanding of ecosystem dynamics as possible. The experiment continues, and the outcome is in doubt.

Still, we have not yet changed the Earth as much as we might. The development of nuclear energy has "freed" us from the limitations of the purely solar-driven world described in these pages but has the potential to make the world uninhabitable for humans, if not for all forms of life. The direct effects of nuclear weapons are abundantly clear to all, and the end of the cold war has reduced, but not eliminated, world tensions about their imminent use. However, simple model calculations of the secondary effects of a nuclear war on the Earth's energy budget suggest that a "nuclear winter" could result from the tremendous amount of dust, smoke, and ash produced by the explosions. These calculations have been followed by suggestions that similar shifts in climate and major disruptions in ecological communities and even human societies have occurred in the past, and could occur in the future, in response to major volcanic eruptions or meteor impacts.

On a more subtle but no less important level, humans are already altering the Earth's environment. We are living an experiment at the global scale. Whether the Gaian hypothesis holds or does not hold is still an unresolved question. It will be up to the global community to determine how far to let the experiment proceed before intervening to contain environmental change. Hopefully those decisions will be made on the basis of increased understanding of our role in the natural function of the planet, including a better understanding of the function of terrestrial ecosystems.

REFERENCES CITED

Grimm, N. B., J. M. Grove, S. T. A. Pickett, and C. L. Redman. 2000. Integrated approaches to long-term studies of urban ecological systems. *BioScience* 50:571–584.

Janzen, D. 1999. Gardenification of tropical conserved wildlands: Multitasking, multicropping, and multiusers. *Proceedings of the National Academy of Sciences USA* 96:5987–5994.

Lovelock, J. 1982. *Gaia: A New Look at Life on Earth.* Oxford University Press, Oxford.

Vitousek, P. M., P. R. Ehrlich, A. H. Ehrlich, and P. A. Matson. 1986. Human appropriation of the products of photosynthesis. *BioScience* 36:368–373.

Vitousek, P. M., H. A. Mooney, J. Lubchenko, and J. M. Melillo. 1997. Human domination of Earth's ecosystems. *Science* 277:494–499.

Illustration Credits

American Association for the Advancement of Science
3.16 © 1999; 7.9 © 1965; 15.11 © 1985; 16.11 © 1995; 16.14 © 1994; 17.4 © 1973; 18.1 © 1969; 20.8 © 1976; 21.6 © 1975; 23.3 © 1997; 23.15 © 1998; 24.14 © 1996; 25.2 © 1984; 25.10 © 1985; 26.12 © 2000; 26.13 © 1999; E.2 © 1997

Academic Press
1.2 © 1962; 2.8b © 1998; 7.4 © 1998; 11.16 © 1997; 13.11 © 1986; 15.9 © 1983; 15.10 © 1983; 16.5 © 1969; 16.7 © 1969; 16.8 © 1969; 17.2 © 1973

American Geophysical Union
2.3 © 1996; 2.8a © 1996; 4.4 © 1969; 13.16 © 1994

American Institutes of Biological Sciences
11.8 © 1991; 11.16 © 1997; 14.5 © 2000; 16.1 © 1998; 16.4 © 1998; 16.15 © 1988; 17.12 © 1999; 17.13 © 1989; 18.3 © 1975; 18.4 © 1989; 18.6 © 1975; 18.7 © 1982; 19.2 © 1983; 19.3 © 1983; 19.6 © 1983; 19.7 © 1983; 21.18 © 1976; 23.14 © 1984; 24.17 © 2001; 25.16 © 1999; 26.7 © 1992; E.3 © 2000

American Society for Environmental History
13.7a © 1997; 13.16 © 1998

American Society of Agronomy
8.6 © 1997; 8.9 © 1988; 14.7 © 1982; 14.8 © 1982; 17.6a © 1984; 23.3 © 1996

American Zoologist
18.2 © 1975

American Society of Plant Physiologists
6.8 © 1996

Annual Reviews
8.1 © 1973; 18.8 © 1980

Blackwell Scientific
5.2 © 1994; 5.3 © 1995; 6.7 © 1981; 12.15 © 1995; 24.1 © 1996

British Ecological Society
20.3 © 1965

Cambridge University Press
6.3 © 1986

CRC Press
9.9 © 1999; 9.12 © 1999; 10.9 © 1985; 23.9 © 1999

Elsevier Science Publishers
3.5 © 1974; 4.9a © 1995; 4.9b © 1998; 5.9 © 1987; 8.14 © 1997; 12.10 © 1978; 13.6 © 2000; 13.13 © 1998; 13.14 © 1973; 13.15 © 1988; 18.10 © 1993; 25.14 © 1995

Ecological Society of America
1.1 © 1942; 2.5 © 1995; 2.7 © 1974; 7.7 © 1976; 8.10 © 1979; 8.12 © 1982; 11.5b © 1991; 11.7 © 1997; 11.12 © 1989; 13.8a © 1982; 14.4 © 1984; 16.3 © 1991; 16.6 © 1955; 16.13 © 1999; 17.5 © 1993; 17.9 © 1991; 17.11 © 1979; 18.5 © 1995; 21.11 © 1974; 21.12 © 1980; 21.14 © 1998; 21.16 © 1998; 21.17 © 1999; 22.2 © 1995; 22.4 © 1995; 22.7 © 1995; 23.10 © 1987; 24.5b © 1970; 24.10 © 1991; 24.13 © 1995; 25.11 © 1997; 26.3 © 1983; 26.9 © 1983; 26.11 © 1983; 26.15 © 1996

HarperCollins Publishers
24.4 © 1972

Harvard University Press
24.2 © 1992; 24.3 © 1975; 24.6 © 1992; 24.7 © 1992

Heron Publishing
3.7 © 1996; 11.5a © 1998

Holt, Rhinehart & Winston
6.15 © 1972; 7.8 © 1972

Ian Michaels Publishers
12.14 © 1994

International Atomic Energy Commission
12.11 © 1977

Iowa State University Press
9.7a © 1980

John Wiley and Sons
9.4 © 1985; 12.11 © 1972

Kluwer Academic Publishers
6.16 © 1976; 13.7b © 1998; 23.8b © 1982; 25.8 © 1988; 25.9 © 1995; 25.13 © 1998 25.18 © 1995; 25.19 © 1995; 25.20 © 1998

MacMillan Publishers
2.6 © 1968

MacMillan Magazine Publishers
10.4 © 1998; 13.5 © 1995; 19.11 © 2000; 19.12 © 2000; 20.6 © 1973; 22.3 © 1999; 22.5 © 1999; 22.6 © 1999; 22.8 © 1999; 22.9 © 1999; 22.10 © 1999; 23.19 © 1992; 24.9 © 1996; 24.12 © 1994; 25.12 © 1975

Munksgaard Scientific Journals
26.10 © 1999

National Academy of Sciences
6.4 © 1997; 23.1 © 1999; 23.2 © 1999

National Research Council of Canada
11.10 © 1981; 13.2 © 1982; 13.3 © 1988; 13.8b © 1990; 13.9 © 1995; 13.10 © 1998; 13.12 © 1996; 14.4 © 1983; 19.1 © 1983; 19.4 © 1983; 19.5 © 1983; 19.8 © 1983; 19.9 © 1983; 19.10 © 1983; 21.4 © 1989; 21.13 © 1997

New Phytologist
6.2 © 1967

New Zealand Journal of Botany
6.6 © 1969; 11.3 © 1969

Oikos
15.6 © 1999; 16.10 © 1983

Oliver and Boyd
9.5 © 1967

Oxford University Press
17.6b © 1998

Prentice-Hall
2.1a © 1975; 7.3 © 1969; 15.3 © 1982

Princeton University Press
6.12 © 1971; 6.13 © 1971; 24.5a © 1972

Ronald Press Co.
6.8 © 1968

Saunders College Publishing
16.9 © 1971; 2.1b ©1989

Society of American Foresters
15.12 © 1983; 23.6 © 1977

Springer-Verlag, Inc.
3.8 © 1995; 3.10 © 1995; 3.11 © 1995; 3.13 © 1970; 3.14 © 1970; 4.3 © 1977; 4.5 © 1977; 4.6 © 1977; 4.8 © 1979; 8.3 © 1995; 10.2 © 1998; 10.7 © 1985; 10.8 © 1998; 11.4 © 1995; 11.13 © 1985; 11.14 © 2000; 16.2 © 1981; 17.2 © 1979; 18.9 © 2000; 19.13 © 1999; 20.16 © 1990; 21.2 © 1977; 21.3 © 1979; 21.5 © 1979; 21.9 © 1979; 21.10 © 1979; 23.13 © 1998; 24.8 © 1993; 24.10 © 1993; 24.11 © 1993; 25.6 © 2000

Taylor and Francis AS
13.1 © 1986; 13.4 © 1988

University of Chicago Press
11.15 © 1982; 20.1 © 1979; 20.2 © 1977; 20.9 © 1979; 20.11 © 1979; 20.12 © 1979; 20.13 © 1979; 20.14 © 1979; 20.15 © 1979

University of Georgia
23.8a © 1983

University of Hawaii Press
22.1 © 1983

University of Lund
25.17 © 1995

University of Notre Dame
8.13 © 1982

W.H. Freeman and Company
23.7 © 1979; 23.18 © 1979

Wadsworth Publishing
7.2 © 1978

Westview Press
9.13 © 1984; 9.14 © 1976

Index

Note: Page numbers in *italics* indicate figures, page numbers followed by "t" indicate tables.

R